Environmental Science

A Study of Interrelationships

FOURTEENTH EDITION

ELDON D. ENGER
Delta College

BRADLEY F. SMITH
Western Washington University

Reinforced Binding

What does it mean?

Since high schools frequently adopt for several years, it is important that a textbook can withstand the wear and tear of usage by multiple students. To ensure durability, McGraw-Hill has elected to manufacture this textbook with a reinforced binding.

Mc Graw Hill Education

ENVIRONMENTAL SCIENCE: A STUDY OF INTERRELATIONSHIPS, FOURTEENTH EDITION

Published by McGraw-Hill Education, 2 Penn Plaza, New York, NY 10121. Copyright © 2016 by McGraw-Hill Education. All rights reserved. Printed in the United States of America. Previous editions © 2013, 2010, and 2008. No part of this publication may be reproduced or distributed in any form or by any means, or stored in a database or retrieval system, without the prior written consent of McGraw-Hill Education, including, but not limited to, in any network or other electronic storage or transmission, or broadcast for distance learning.

Some ancillaries, including electronic and print components, may not be available to customers outside the United States.

This book is printed on acid-free paper.

2 3 4 5 6 7 8 9 0 DOW 1 0 9 8 7 6 5

ISBN 978-0-07-673262-3
MHID 0-07-673262-2

Senior Vice President, Products & Markets: *Kurt L. Strand*
Vice President, General Manager, Products & Markets: *Marty Lange*
Vice President, Content Design & Delivery: *Kimberly Meriwether David*
Managing Director: *Thomas Timp*
Brand Manager: *Michelle Vogler*
Director of Development: *Rose Koos*
Marketing Manager: *Matthew Garcia*
Product Developer: *Robin Reed*
Director, Content Design & Delivery: *Linda Avenarius*
Program Manager: *Lora Neyens*
Content Project Managers: *Laura Bies & Tammy Juran*
Buyer: *Sandy Ludovissy*
Design: *Tara McDermott*
Content Licensing Specialist: *Lori Hancock*
Cover Image: *©Stephen Emerson / Alamy*
Compositor: *Laserwords Private Limited*
Typeface: *10/12 Times Roman LT STD*
Printer: *R. R. Donnelley*

All credits appearing on page or at the end of the book are considered to be an extension of the copyright page.

Library of Congress Cataloging-in-Publication Data

Environmental science: a study of interrelationships / Eldon D. Enger, Delta College, Bradley F. Smith, Huxley College of the Environment, Western Washington University.—Fourteenth edition.
  pages cm
  Includes index.
  Audience: Age: 18+
  ISBN 978-0-07-673262-3 (acid-free paper)—MHID 0-07-673262-2 (acid-free paper)
  1. Environmental sciences—Textbooks. I. Enger, Eldon D. II. Smith, Bradley F.

GE105.E54   2015
304.2—dc23

2014023634

The Internet addresses listed in the text were accurate at the time of publication. The inclusion of a website does not indicate an endorsement by the authors or McGraw-Hill Education, and McGraw-Hill Education does not guarantee the accuracy of the information presented at these sites.

www.mheonline.com

To Judy, my wife and friend,
for sharing life's adventures

ELDON ENGER

For Josh Fox, whose kind and steady nature
makes me proud to have you as a member of our
family and excited at the prospect of watching and
being part of your family as it grows

BRAD SMITH

Eldon D. Enger is an emeritus professor of biology at Delta College, a community college near Saginaw, Michigan. He received his B.A. and M.S. degrees from the University of Michigan. Professor Enger has over 30 years of teaching experience, during which he has taught biology, zoology, environmental science, and several other courses. He has been very active in curriculum and course development. A major curriculum contribution was the development of an environmental technician curriculum and the courses that support it. He was also involved in the development of learning community courses in stream ecology, winter ecology, and plant identification. Each of these courses involved students in weekend-long experiences in the outdoors that paired environmental education with physical activity—stream ecology and canoeing, winter ecology and cross-country skiing, and plant identification with backpacking.

Professor Enger is an advocate for variety in teaching methodology. He feels that if students are provided with varied experiences, they are more likely to learn. In addition to the standard textbook assignments, lectures, and laboratory activities, his classes included writing assignments, student presentation of lecture material, debates by students on controversial issues, field experiences, individual student projects, and discussions of local examples and relevant current events. Textbooks are very valuable for presenting content, especially if they contain accurate, informative drawings and visual examples. Lectures are best used to help students see themes and make connections, and laboratory activities provide important hands-on activities.

Professor Enger received the Bergstein Award for Teaching Excellence and the Scholarly Achievement Award from Delta College and was selected as a Fulbright Exchange Teacher twice—to Australia and Scotland. He has participated as a volunteer in several Earthwatch Research Programs. These include: studying the behavior of a bird known as the long-tailed manakin in Costa Rica, participating in a study to assess the possibility of reintroducing endangered marsupials from off-shore islands to mainland Australia, and helping with efforts to protect the nesting beaches of the leatherback turtle in Costa Rica, and assisting with on-going research on the sustainable use of fish, wildlife, and forest resources in the Amazon Basin in Peru. He also participated in a People to People program, which involved an exchange of ideas between U.S. and South African environmental professionals.

He has traveled extensively, which has allowed him first-hand experience with coral reefs, ocean coasts, savannas, mangrove swamps, tundra, prairies, tropical rainforests, cloud forests, deserts, temperate rainforests, coniferous forests, deciduous forests, and many other special ecosystems. These experiences have provided opportunities to observe the causes and consequences of many environmental problems from a broad social and scientific perspective.

He volunteers at a local nature center, land conservancy, and Habitat for Humanity affiliate. Since 2005, he and his wife have spent a month each year with other volunteers from their church repairing houses damaged by tornados, floods, and hurricanes throughout the United States.

Professor Enger and his wife Judy have two married sons and three grandchildren. He enjoys a variety of outdoor pursuits such as cross-country skiing, snowshoeing, hiking, kayaking, hunting, fishing, camping, and gardening. Other interests include reading a wide variety of periodicals, beekeeping, singing in a church choir, picking wild berries, and preserving garden produce.

Bradley F. Smith is the Dean Emeritus of Western Washington University in Bellingham, Washington, having served as Dean from 1994 to 2012. Prior to assuming the position as Dean in 1994, he served as the first Director of the Office of Environmental Education for the U.S. Environmental Protection Agency in Washington, D.C., from 1991 to 1994. Dean Smith also served as the Acting President of the National Environmental Education and Training Foundation in Washington, D.C., and as a Special Assistant to the EPA Administrator.

Before moving to Washington, D.C., Dean Smith was a professor of political science and environmental studies for 15 years, and the executive director of an environmental education center and nature refuge for five years.

Dean Smith has considerable international experience. He was a Fulbright Exchange Teacher to England and worked as a research associate for Environment Canada in New Brunswick. He is a frequent speaker on environmental issues worldwide and serves on the International Scholars Program for the U.S. Information Agency. He also served as a U.S. representative on the Tri-Lateral Commission on environmental education with Canada and Mexico. He was awarded a NATO Fellowship to study the environmental problems associated with the closure of former Soviet military bases in Eastern Europe. He is a Fellow of the Royal Institute of Environmental Science in the U.K.

He also served on the Steering Committee of the Commission for Education and Communication for the International Union for the Conservation of Nature (IUCN) from 2004 to 2013.

Dean Smith is a trustee of the National Environmental Education Foundation, a member of the North Pacific Research Board, and is Vice-Chair of the Washington State Fish and Wildlife Commission. He also serves on the board of Washington Sea Grant. Previously, he served as the chair of the Washington Sustainability Council, as President of the Council of Environmental Deans and Directors, and as a member of the National Advisory Council for Environmental Policy and Technology for the EPA. He also served on President Clinton's Council for Sustainable Development (Education Task Force).

Dean Smith holds B.A. and M.A. degrees in Political Science/International Relations and Public Administration and a Ph.D. from the School of Natural Resources and the Environment at the University of Michigan.

Dean Smith and his wife, Daria, live along the shores of Puget Sound in Bellingham, Washington, and spend part of the summer at their summer home on the shores of Lake Huron in the Upper Peninsula of Michigan. He has two grown children and is an avid outdoor enthusiast.

Brief Contents

Contents

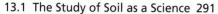

The Role of Environmental Science In Society

We live in a time of great change and challenge. Our species has profoundly altered the Earth. Our use of fossil fuels to provide energy is altering climate, our use of Earth's soil resources to feed ourselves results in extinctions, overexploitation of fish populations has resulted in the population declines of many marine species, and freshwater resources are becoming scarce. At the same time we see significant improvement in other indicators. Energy-efficient and alternative energy technologies are becoming mainstream, population growth is beginning to slow, air and water pollution problems are being addressed in many parts of the world, and issues of biodiversity loss, climate change, and human health are beginning to be addressed on a worldwide basis.

However, there are still major challenges and there are additional opportunities to lighten our impact on Earth. Understanding the fundamental principles that describe how the Earth's systems work is necessary knowledge for everyone, not just scientists who study these systems. It is particularly important for political, industrial, and business leaders because the political, technical, and economic decisions they make affect the Earth.

Why "A Study of Interrelationships"?

Environmental science is an interdisciplinary field. Because environmental problems occur as a result of the interaction between humans and the natural world, we must include both scientific and social aspects when we seek solutions to environmental problems. Therefore, the central theme of this book is interrelatedness. It is important to have a historical perspective, to appreciate economic and political realities, to recognize the role of different social experiences and ethical backgrounds, and to integrate these with the science that describes the natural world and how we affect it. *Environmental Science: A Study of Interrelationships* incorporates all of these sources of information when discussing any environmental issue.

Environmental science is also a global science. While some environmental problems may be local in nature—pollution of a river, cutting down a forest, or changing the flow of a river for irrigation—other problems are truly global—climate change, overfishing of the oceans, or loss of biodiversity. In addition, individual local events often add together to cause a worldwide problem—the actions of farmers in China or Africa can result in

dust storms that affect the entire world, or the individual consumption of energy from fossil fuels increases carbon dioxide concentrations in the Earth's atmosphere. Therefore, another aspect of the interrelationships theme of this text is to purposely include features that highlight problems, issues, and solutions involving a variety of cultures.

This text has been translated and published in Spanish, Chinese, and Korean. Therefore, students in Santiago, Shanghai, Seoul, or Seattle are learning the "hows and whys" involved in thinking and acting sustainably. At the end of the day we all share the same air, water, and one not-so-big planet. It's important for all of us to make it last.

What Makes This Text Unique?

We present a balanced view of issues, diligently avoiding personal biases and fashionable philosophies.

It is not the purpose of this textbook to tell readers what to think. Rather, our goal is to provide access to information and the conceptual framework needed to understand complex issues so that readers can comprehend the nature of environmental problems and formulate their own views. Two features of the text encourage readers to think about issues and formulate their own thoughts:

- The **Issues & Analysis** feature at the end of each chapter presents real-world, current issues and provides questions that prompt students to think about the complex social, political, and scientific interactions involved.

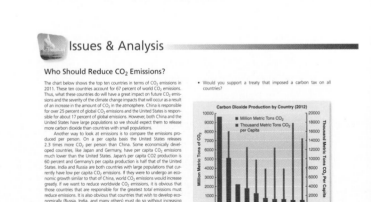

Issues & Analysis

Who Should Reduce CO_2 Emissions?

The chart below shows the top ten countries in terms of CO_2 emissions in 2011. These ten countries account for 67 percent of world CO_2 emissions. Thus, what these countries do will have a great impact on future CO_2 emissions and the severity of the climate change impacts that will occur as a result of an increase in the amount of CO_2 in the atmosphere. China is responsible for over 25 percent of global CO_2 emissions and the United States is responsible for about 17 percent of global emissions. However, both China and the United States have large populations so we should expect them to release more carbon dioxide than countries with small populations.

Another way to look at emissions is to compare the emissions produced per person. On a per capita basis the United States releases 2.3 times more CO_2 per person than China. Some economically developed countries, like Japan and Germany, have per capita CO_2 emissions much lower than the United States. Japan's per capita CO_2 production is 60 percent and Germany's per capita production is half that of the United States. India and Russia are both countries with large populations that currently have low per capita CO_2 emissions. If they were to undergo an economic growth similar to that of China, world CO_2 emissions would increase greatly. If we want to reduce worldwide CO_2 emissions, it is obvious that those countries that are responsible for the greatest total emissions must reduce emissions. It is also obvious that countries that wish to develop economically (Russia, India, and many others) must do so without increasing carbon dioxide emissions.

- What actions could the United States and China take to reduce their carbon dioxide emissions?
- What actions could the international community take to encourage economically emerging nations to develop economically without increasing carbon dioxide emissions?

- Would you support a treaty that imposed a carbon tax on all countries?

Carbon Dioxide Production by Country (2012)
- Million Metric Tons CO_2
- Thousand Metric Tons CO_2 per Capita

China, United States, India, Russia, Japan, Germany, South Korea, Iran, Saudi Arabia, Canada

Source: Data from Global Carbon Atlas

- The **What's Your Take?** feature found in each chapter asks students to take a stand on a particular issue and develop arguments to support their position, helping students develop and enhance their critical thinking skills.

What's Your Take?

Climate change will increase the incidence of flooding in several ways. Rising sea levels will threaten low-lying coastal areas. Intense storms will cause coastal storm surges that can affect areas not normally considered to be in a flood plain. Heavy rain storm events will cause streams and rivers to rise and flood areas adjacent to water courses. Most insurance policies do not cover flood damage. The federal government has made flood insurance available for those in areas at risk of floods; however, many people do not purchase the flood insurance because it is expensive. When flooding occurs victims seek help from government sources and charitable organizations. One way to reduce the cost of flooding is to prevent people from building in areas that are likely to flood. This could be done by federal or state laws or local ordinances. Draw up a law or ordinance that would reduce flood damage to homes and businesses. List at least three criteria that would be used to prevent building in flood-prone areas and justify your selection of each criterion.

We recognize that environmental problems are global in nature.

Three features of the text support this concern:
- Throughout the text, the authors have made a point to use **examples** from around the world as well as those from North America.
- Many of the boxed readings—**Focus On; Going Green; Science, Politics, & Policy;** and **Issues & Analysis**—are selected to provide a global flavor to the basic discussion in the text.

Focus On

Biomass Fuels and the Developing World

Although most of the world uses fossil fuels as energy sources, much of the developing world relies on biomass as its source of energy. The biomass can be wood, grass, agricultural waste, or dung. According to the United Nations, 2 billion people (30 percent of the world's population) use biomass as fuel for cooking and heating dwellings. In developing countries, nearly 40 percent of energy used comes from biomass. In some regions, however, the percentage is much higher. For example, in sub-Saharan Africa, fuelwood provides about 80 percent of energy consumed. Worldwide, about 60 percent of wood removed from the world's forests is used for fuel.

This dependence on biomass has several major impacts:
- Often women and children must walk long distances and spend long hours collecting firewood and transporting it to their homes.
- Because the fuel is burned in open fires or inefficient stoves, smoke contaminates homes and affects the health of the people. The World Health Organization estimates that in the developing world, 40 percent of acute respiratory infections are associated with poor indoor air quality related to burning biomass. A majority of those who become ill are women and children because the children are in homes with their mothers who spend time cooking food for their families.
- Often the fuel is harvested unsustainably. Thus, the need for an inexpensive source of energy is a cause of deforestation. Furthermore, deforested areas are prone to soil erosion.
- When dung or agricultural waste is used for fuel, it cannot be used as an additive to improve the fertility or organic content of the soil. Thus, the use of these materials for fuel negatively affects agricultural productivity.

Nepali woman carrying brushwood.

- The presence of easily accessible **Foldout World Maps** at the back of the text allows students to quickly locate a country or region geographically.

We recognize that many environmental issues involve complex social, economic, and cultural aspects.

- The first three chapters focus on the underlying social, economic, health, and ethical aspects involved in understanding how people view environmental issues.
- The **Science, Politics, & Policy feature** shows how the scientific understanding of environmental problems is filtered through the lens of social and political goals to determine policy.

Science, Politics, & Policy

Disposal of Waste from Nuclear Power Plants

The disposal of spent fuel rods from nuclear power plants has been a continuing issue since the first commercial power reactor went on line in 1957. In the more than 50 years since then, it has been clear that there are only two methods of dealing with the nuclear waste from spent nuclear fuel: reprocessing the fuel to reduce the amount of waste or storing the waste at a safe site. U.S. policy has been to store rather than reprocess nuclear waste. Federal law requires the U.S. government to provide a solution to the storage of spent nuclear fuel. All nuclear power plants in the United States have been operating with the assumption that eventually their waste would be stored in a secure federal facility. The National Academy of Sciences recommended underground storage as the best way to deal with waste from nuclear power plants.

The history of U.S. efforts to establish a repository for high-level radioactive waste is long and complicated. The following provides a brief chronology of major steps in the process:

- 1982—The U.S. Congress passed legislation that gave the responsibility for finding, building, and operating a nuclear waste site to the Department of Energy with completion by 1998.
- 1987—Initially several sites were considered and Yucca Mountain was selected to receive further study.
- 1994–1997—A five-mile-long, U-shaped tunnel (the Exploratory Studies Facility) was constructed to study the suitability of the Yucca Mountain site.
- 2002—President George W. Bush signed a joint resolution of Congress designating Yucca Mountain as the site for the nuclear repository.
- 2008—The Department of Energy filed a license application with the Nuclear Regulatory Commission to construct a repository for spent nuclear fuel and high-level radioactive waste at Yucca Mountain. The citizens and political leaders of Nevada opposed the designation.

- 2010—President Obama withdrew funding for Yucca Mountain and the Department of Energy withdrew its request to the Nuclear Regulatory Commission to operate the facility. President Obama also established the Blue Ribbon Commission on America's Nuclear Future.
- 2012—The report of the Blue Ribbon Commission on America's Nuclear Future 2012 included the following statement:
  Recommendation #1: The United States should undertake an integrated nuclear waste management program that leads to the timely development of one or more permanent deep geological facilities for the safe disposal of spent fuel and high-level nuclear waste.
- 2013—A U.S. Court of Appeals ruling stated that the designation of Yucca Mountain as the nation's nuclear repository is still in effect and the Nuclear Regulatory Commission and the President cannot ignore the law and proceed with plans to close Yucca Mountain.
- The Future—The future is uncertain, but it is clear that no permanent solution for storing spent nuclear fuel is likely for decades.

Yucca Mountain

- Critical Thinking questions appear at the end of each chapter and require students to evaluate information, recognize bias, characterize the assumptions behind arguments, and organize information.

We recognize that it is important to focus on the positive.

Environmental science often seems to focus on the negative, since one of the outcomes of any analysis of an environmental situation is to highlight problems and point out where change is needed. We often overlook the many positive actions of individuals and organizations. Therefore, each chapter has two features that call attention to the positive:
- **Going Green** boxes describe actions that are having a positive environmental impact. Some of these actions are taken by governments, some are by corporations, and some are individual efforts.

Going Green

Increasing Populations of Red-Cockaded Woodpeckers

The red-cockaded woodpecker (*Picoides borealis*) is listed as an endangered species. This medium-sized bird (about the size of a cardinal) is a cooperative colony nester—the dominant male and female raise young with the support of nonbreeding members of the colony. They are only found in the southeastern United States—southern Virginia to eastern Texas—where native southern yellow pine forests occur. Several pine species, including slash pine, shortleaf pine, loblolly pine, and longleaf pine, are typical of this region. The original forests were fire-adapted in that mature trees were able to withstand moderate ground fires. This resulted in a rather open forest type. The woodpeckers typically construct their nesting cavities in older, diseased longleaf pine trees.

The trees these birds use for nesting are also commercially important. Thus, the amount of suitable breeding habitat has been severely reduced as older trees are harvested and natural stands of pines have been replaced with plantations, where large tracts are planted to a single species and the trees are harvested before they reach old age. Since much of the suitable habitat is privately owned, protecting populations of red-cockaded woodpeckers requires the cooperation of private landowners, conservation organizations, state and federal governments, and commercial forest products companies.

In 1998, International Paper entered into an agreement with the U.S Fish and Wildlife Service, which is responsible for monitoring the status of endangered species, to increase the amount of suitable nesting habitat on its lands. International Paper agreed to set aside particular parcels of forest to maintain colonies of red-cockaded woodpeckers. One of those parcels was the Southlands Experimental Forest near Bainbridge, Georgia. When the agreement was signed in 1998, there were three male red-cockaded woodpeckers at the site. By 2008, there were over 50 individuals. The increase is attributable to protection and improvement of the birds' habitat and transfer of birds to the area from other locations. Today there are about 15,000 red-cockaded woodpeckers throughout its range. In 2006, the company decided to sell nearly all of its land holdings in the United States. Many environmentally sensitive lands were sold to conservation organizations such as The Nature Conservancy and the Conservation Fund, as well as state governments. The Southlands Experimental Forest was sold to the state of Georgia with some funding assistance from the Conservation Fund. This land transfer protects the population gains made by this population of red-cockaded woodpeckers.

Red-cockaded woodpecker habitat

Red-cockaded woodpecker

- **Acting Green** is an end-of-chapter feature that asks students to consider making personal changes that are relatively simple and will have a positive environmental impact.

New to This Edition

The fourteenth edition of *Environmental Science: A Study of Interrelationships* is the result of extensive analysis of the text and the evaluation of input from environmental science instructors who conscientiously reviewed chapters during the revision. We have used the constructive comments provided by these professionals in our continuing efforts to enhance the strengths of the text. The following is a list of global changes we have made, along with a description of significantly revised chapters.

New Chapter Opening Feature This feature presents an example of a current issue that is germane to the chapter content. The issues involved in the example are easy to visualize and serve as an introduction to the topics covered in the chapter. In many cases, the material in the opening feature is specifically addressed or expanded upon in the chapter.

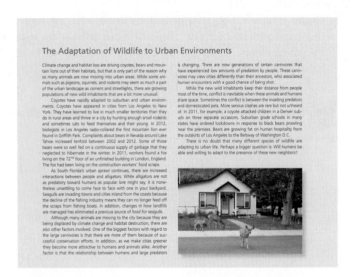

The Adaptation of Wildlife to Urban Environments

Current Content As with previous editions the authors have incorporated the most recent information available at the time of publication.

Revised Art Program More than 100 new photos have been added or substituted throughout the text to depict real-life situations. Over 60 illustrations, graphs, and charts are new or revised to present detailed information in a form that is easier to comprehend than if that same material were presented in text form.

Several Significantly Revised Chapters Every chapter has a new chapter opening feature. In addition, many chapters have other significant changes, including:

Chapter 1 Environmental Interrelationships The section on Emerging Global Issues has three new sections: Population Growth, Maintaining Functional Ecosystems, and Food Security. The section on Environment and Health was completely revised. Sections on air pollution, water pollution, malaria, and accidental deaths were added, since they are major environmental health issues in the developing world. The section on Emerging Diseases was rewritten.

Chapter 2 Environmental Ethics The section on Environmental Ethics was substantially rewritten and there were significant additions to the section on Environmental Justice.

Chapter 3 Risk, Economics, and Environmental Concerns The chapter was completely rewritten around the central theme that risk and cost are intimately intertwined. Environmental risk factors and human health are used throughout the chapter to show how risks and costs are related. The sections on Perception of Risk, Ecosystem Services, Environmental Costs, Cost-Benefit Analysis, and Economics and Sustainable Development were substantially revised. The boxed readings Going Green: Green Collar Jobs and Science, Politics, & Policy: The Developing Green Economy were rewritten and the Issues & Analysis: The Economics and Risks of Mercury Contamination was updated to include recent changes in regulations of emissions from power plants.

Chapter 4 Interrelated Scientific Principles: Matter, Energy, and Environment There is a new Science, Politics, & Policy: The Return of Salmon to the Elwha River and a new Issues & Analysis: The End of the Incandescent Light Bulb.

Chapter 5 Interactions: Environments and Organisms There is a new Issues & Analysis on Wildlife and Climate Change. There are updates to the Going Green: Phosphorus-free Lawn Fertilizer and Science, Politics, & Policy: Emotion and Wolf Management.

Chapter 7 Populations: Characteristics and Issues The content was updated with the most recent data from the Population Reference Bureau and there is a new Science, Politics, & Policy: Funding the Unmet Need for Family Planning. The topic of invasive species is also discussed

Chapters 8, 9, and 10 all deal with aspects of energy. These chapters have been updated with the most current data available. Significant new concepts include the impact of newly industrialized countries on energy demands and evaluating energy alternatives through an accounting of energy return on investment. There are also expanded discussions of hydraulic fracturing, unconventional sources of oil and gas, and the renewable fuel mandate.

Chapter 16 Air Quality Issues Chapter 16 has been significantly changed, since the section on climate change was moved to its own chapter, Chapter 17 Climate Change: A Twenty-first Century Issue. The remaining content was reorganized to create a more logical progression of topics. The section on Control of Air Pollution was moved to follow discussions of Photochemical Smog, Acid Precipitation, and Ozone Depletion. There is a new section, 16.8 Air Pollution in the Developing World, that points out that air pollution is still a major problem in much of the developing world. There is a new Going Green: Going Solvent Free and a new Science, Politics, & Policy: A History of Mercury Regulations. Data on the amounts of air pollutants in the U.S. were updated to best available data.

Chapter 17 Climate Change: A Twenty-first Century Issue

Chapter 17 is a new chapter. Material about climate change was consolidated into a separate chapter as requested by reviewers. In addition, there are several new sections including:

17.1 Earth Is a Greenhouse Planet describes the role of atmospheric gases in making the Earth habitable.

17.2 Geologic Evidence for Global Warming and Climate Change discusses evidence for past climate changes and their relevance to understanding current changes.

17.3 Growth in Knowledge of Climate Change lists the many kinds of research that contributed to our understanding of climate change.

17.5 The Current State of Knowledge about Climate Change incorporates information from the most recent report of IPCC Working Group I—*Climate Change 2013: The Physical Science Basis.*

17.6 Consequences of Climate Change describes the many disruptions to the hydrologic cycle that occur with climate change.

The section on International Agreements has an updated section on the meeting in Doha, Qatar in 2012 in which participating countries were unable to come to agreement on how to limit greenhouse gas emissions.

There is a new Focus On: Doubters, Deniers, Skeptics, and Ignorers that describes reasons why people question the science related to climate change and the techniques people use to refute climate change science.

Other new or significantly updated content occurs throughout the text and includes: information on wolf hunting, conflicting regulations concerning sea lions and salmon, the role of sanitation and safe drinking water to world health, the adaptation of wildlife species to urban environments, the impact of invasive species, concerns about overfishing of marine fisheries, water ownership rights, the growth of megacities, the use of plants to remediate polluted sites, and the economic and political value of biodiversity.

Acknowledgments

The creation of a textbook requires a dedicated team of professionals who provide guidance, criticism, and encouragement. It is also important to have open communication and dialogue to deal with the many issues that arise during the development and production of a text. Therefore, we would like to thank Brand Manager Michelle Vogler; Product Developer Robin Reed; Project Manager Laura Bies; Buyer Sandy Ludovissy; Content Licensing Specialist Lori Hancock; and Designer Tara McDermott for their suggestions and kindnesses. We would like to thank the following individuals who wrote and/or reviewed learning goal-oriented content for LearnSmart.

Sylvester Allred, *Northern Arizona University*
Ray Beiersdorfer, *Youngstown State University*
Anne H. Bower, *Philadelphia University*
Michelle Cawthorn, *Georgia Southern University*
Kathleen Dahl, *University of Kansas*
Dani DuCharme, *Waubonsee Community College*
Tristan Kloss, *University of Wisconsin—Milwaukee*
Arthur C. Lee, *Roane State Community College*
Trent McDowell, *University of North Carolina at Chapel Hill*
Jessica Miles, *Florida Atlantic University*
Brian F. Mooney, *Johnson & Wales University at Charlotte*
Noelle J. Relles, *State University of New York at Cortland*
Gigi Richard, *Colorado Mesa University*
Elise Uphoff
Amy J. Wagner, *California State University at Sacramento*

Finally, we'd like to thank our many colleagues who have reviewed all, or part, of *Environmental Science: A Study of Interrelationships.* Their valuable input has continued to shape this text and help it meet the needs of instructors around the world.

Fourteenth Edition Reviewers

Gwenn Andahazy, *Lawrence Township Public Schools*
Ray Beiersdorfer, *Youngstown State University*
Anne Bower, *Philadelphia University*
Cynthia Carlson, *New England College*
Kip Curtis, *Eckerd College*
Christopher Farrell, *St. Johns River State College*
Brandon Gillette, *Johnson County Community College*
Mandy Hockenbrock, *Wor-Wic Community College*
Kelley Hodges, *Gulf Coast State College*
Susan Hutchins, *Itasca Community College*
David Knotts, *Lindenwood University*
Kathy McCann Evans, *Reading Area Community College*
Brian Mooney, *Johnson & Wales University—Charlotte*
Katherine Winsett, *University of Southern Indiana*

Eldon D. Enger
Bradley F. Smith

Digital Resources

McGraw-Hill offers various tools and technology products to support *Environmental Science,* fourteenth edition.

McGraw-Hill's ConnectPlus (connect.mheducation.com) is a web-based assignment and assessment platform that gives students the means to better connect with their coursework, with their teacher, and with the important concepts that they will need to know for success now and in the future. The following resources are available in ConnectPlus:

- Auto-graded assessments
- LearnSmart, an adaptive diagnostic and learning tool
- SmartBook, an adaptive reading experience
- Powerful reporting against learning outcomes and level of difficulty
- The full textbook as an integrated, dynamic eBook, which teachers can also assign
- An image bank including all of the textbook images
- Base Map and Google Earth exercises
- An extensive Case Studies Library

With ConnectPlus, teachers can deliver assignments, quizzes, and tests online. Teachers can edit existing questions and author entirely new problems; track individual student performance—by question, by assignment, or in relation to the class overall—with detailed grade reports.

By choosing ConnectPlus, teachers are providing their students with a powerful tool for improving academic performance and truly mastering course material. ConnectPlus allows students to practice important skills at their own pace and on their own schedule. Importantly, students' assessment results and teachers' feedback are all saved online, so students can continually review their progress and plot their course to success.

No two students are alike. Why should their learning paths be? LearnSmart uses revolutionary adaptive technology to build a learning experience unique to each student's individual needs. It starts by identifying the topics a student knows and does not know. As the student progresses, LearnSmart adapts and adjusts the content based on his or her individual strengths, weaknesses, and confidence, ensuring that every minute spent studying with LearnSmart is the most efficient and productive study time possible.

LearnSmart also takes into account that everyone will forget a certain amount of material. LearnSmart pinpoints areas that a student is most likely to forget and encourages periodic review to ensure that the knowledge is truly learned and retained. In this way, LearnSmart goes beyond simply getting students to memorize material—it helps them truly retain the material in their long-term memory. Want proof? Students who use LearnSmart are 35 percent more likely to complete their class, are 13 percent more likely to pass their class, and improve their performance by a full letter grade.

SMARTBOOK™

McGraw-Hill SmartBook™ is the first and only adaptive reading experience available for the higher education market. Powered by an intelligent diagnostic and adaptive engine, SmartBook facilitates the reading process by identifying what content a student knows and doesn't know through adaptive assessments. As the student reads, the reading material constantly adapts to ensure the student is focused on the content he or she needs the most to close any knowledge gaps.

create™

With McGraw-Hill, you can create and tailor the course you want to teach. With McGraw-Hill Create, create.mheducation.com, you can easily rearrange chapters, combine material from other content sources, and quickly upload content you have written, like your course syllabus or teaching notes, or arrange your book to fit your teaching style.

ConnectED eBook

This digital version of *Environmental Science: A Study of Interrelationships* offers powerful and instant search capability, and helps students manage notes, highlights and bookmarks all in one place. This downloadable eBook can be viewed on an iPad.

Additional Resources

***Field and Laboratory Exercises in Environmental Science,* Eighth Edition by Enger, Smith, and Lionberger, 978-0-07-759982-9**

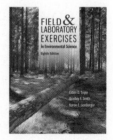

This manual is the ideal resource for the AP Environmental Science course. The major objectives are to provide students with hands-on experiences that are relevant, easy to understand, applicable to the student's life, and presented in an interesting, informative format. Ranging from field and lab experiments to social and personal assessments of the environmental impact of human activities, this manual presents for everyone, regardless of the budget or facilities of each class. These labs are grouped by categories that can be used in conjunction with any environmental science textbook.

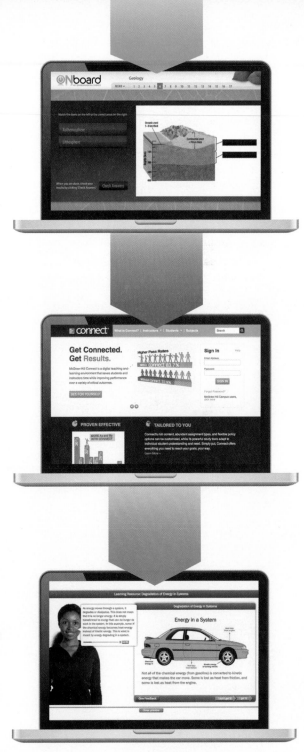

Environmental Interrelationships

Environmental science is the study of interrelationships between humans and the natural world. This farmer in Uganda has cleared a portion of the original forest to create this small farm, which supplies food and income for the family.

OBJECTIVES

After reading this chapter, you should be able to:

- Recognize that the field of environmental science includes social, political, and economic aspects in addition to science.

- Describe examples that illustrate the interrelated nature of environmental science.

- Understand why most social and political decisions are made with respect to political jurisdictions but environmental problems do not necessarily coincide with these human-made boundaries.

- Understand the concept of sustainability.
- Recognize that human population growth contributes to environmental problems.
- Recognize that people rely on the services provided by ecosystems.
- Understand that food security is an issue for many people in the less-developed world.
- Recognize that there are governance issues that make it difficult to solve environmental problems.

- Recognize that the quality of the environment has an important impact on human health.

- Understand that personal security incorporates economic, political, cultural, social, and environmental aspects.

- Describe environmental impacts of globalization.

- Recognize the central role energy use has on environmental problems.

The Important Role of Wolves in Yellowstone

Early explorers of the lands west of the Mississippi River told of a place with fantastic geysers, mud pots, and other thermal features. They also told of abundant wildlife and rivers filled with fish. After several official government expeditions confirmed these tales, Yellowstone National Park was established as the world's first national park in 1872. As more people settled in the west and ranches and farms were established, there was pressure from farmers and ranchers as well as hunters to reduce the number of predator species on public lands in the west. It was also a generally held idea that predators reduced the numbers of elk, deer, and other species preferred by hunters. Thus the U.S. Congress in 1914 provided funding to eliminate wolves and other predators on public lands including national parks. By 1926 wolves had been eliminated from Yellowstone. The lack of wolves led to a cascade of unintended consequences:

- Since hunting of species other than predators was prohibited in the park, the population of elk increased. In addition, coyotes, which are normally killed by wolves, increased greatly. By 1935, park managers felt that overgrazing by the large population of elk was beginning to destroy the park's habitat. Therefore, a program of harvesting elk, bison, and pronghorns was instituted to protect the habitat. This program was discontinued in the 1960s as better knowledge of the habitat indicated that it was not overgrazed.
- Coyotes greatly reduced the number of small mammal species such as mice, squirrels, and rabbits.
- The number of pronghorn antelope also decreased because coyotes killed newborn pronghorns.
- Populations of cottonwood and willows along streams declined substantially due to browsing by elk.

Eventually, as park managers and biologists began to understand the profound changes caused by the elimination of wolves, the decision was made to reintroduced wolves to Yellowstone National Park. The initial introduction of 31 wolves in 1995 and 1996 has resulted in a current population of about 100 wolves. Several changes to the Yellowstone ecosystem can be directly attributed to the alterations brought about by the return of wolves:

- Wolves kill and eat elk. This has resulted in a significant reduction in the size of the elk herd from about 19,000 prior to wolf reintroduction to less than 4,000 now.
- The presence of wolves also has modified the behavior of elk. Because they must be more vigilant and move about more because of the predatory behavior of wolves, elk spend less time feeding on willow, cottonwood, and aspen. Both the change in behavior and the reduced size of the elk herd have allowed the regeneration of stands of cottonwood and willow along rivers. This has in turn resulted in increased numbers of beavers that use these streamside trees for food. The dams built by beavers tend to slow the flow of water and increase the recharge of groundwater. Furthermore, the stands of willow along the banks of streams cool the water and improve fish habitat. The stands of willow also provide needed habitat for some songbirds.
- Wolves directly compete with coyotes and kill them if they have the opportunity. Thus, since the reintroduction of wolves the coyote population has fallen significantly. There is evidence that the populations of the prey of coyotes—voles, mice, and other rodents—have increased. The increased availability of this food source has resulted in an increase in the number of foxes, hawks, and owls.

Thus, it is fair to say that the reintroduction of the wolf has changed how water flows through the landscape and has led to increased populations of many organisms—willow, cottonwood, beaver, songbirds, foxes, certain rodents, hawks, and owls; and to the decline in the population of other organisms—coyote and elk. Truly this is a story that illustrates the point made by the early naturalist John Muir (1838–1914)—*Tug on anything at all and you'll find it connected to everything else in the universe.*

Wolves reintroduced   **Elk decline**   **Willows increase**   **Beavers increase**

1.1 The Nature of Environmental Science

Environmental science is an interdisciplinary field that includes both scientific and social aspects of human impact on the world. The word *environment* is usually understood to mean the surrounding conditions that affect organisms. In a broader definition, **environment** is everything that affects an organism during its lifetime. In turn, all organisms including people affect many components in their environment. **Science** is an approach to studying the natural world that involves formulating hypotheses and then testing them to see if the hypotheses are supported or refuted. However, because humans are organized into complex societies, environmental science also must deal with politics, social organization, economics, ethics, and philosophy. Thus, environmental science is a mixture of traditional science, individual and societal values, economic factors, and political realities that are important to solving environmental problems. (See figure 1.1.)

Although environmental science as a field of study is evolving, it is rooted in the early history of civilization. Many ancient cultures expressed a reverence for the plants, animals, and geographic features that provided them with food, water, and transportation. These features are still appreciated by many modern people. Although the following quote from Henry David Thoreau (1817–62) is over a century old, it is consistent with current environmental philosophy:

> I wish to speak a word for Nature, for absolute freedom and wildness, as contrasted with a freedom and culture merely civil . . . to regard man as an inhabitant, or a part and parcel of Nature, rather than a member of society.

The current interest in the state of the environment began with philosophers like Thoreau and scientists like Rachel Carson and received emphasis from the organization of the first Earth Day on April 22, 1970. Subsequent Earth Days reaffirmed this commitment. As a result of this continuing interest in the state of the world and how people both affect it and are affected by it, environmental science is now a standard course or program at many colleges. It is also included in the curriculum of high schools. Most of the concepts covered by environmental science courses had previously been taught in ecology, conservation, biology, or geography courses. Environmental science incorporates the scientific aspects of these courses with input from the social sciences, such as economics, sociology, and political science, creating a new interdisciplinary field.

FIGURE 1.1 Environmental Science The field of environmental science involves an understanding of scientific principles, economic influences, and political action. Environmental decisions often involve compromise. A decision that may be supportable from a scientific or economic point of view may not be supportable from a political point of view without modification. Often political decisions relating to the environment may not be supported by economic analysis.

Interrelatedness Is a Core Concept

A central factor that makes the study of environmental science so interesting/frustrating/challenging is the high degree of interrelatedness among seemingly unrelated factors. The opening story about the relationship between wolves and elk in Yellowstone National Park illustrates the theme of interrelatedness very well. The absence of wolves led to an increase in elk and coyotes but to a decrease in beaver, streamside stands of willow and cottonwood, and habitat for some birds. The return of wolves resulted in a decrease in elk numbers and changes in elk behavior that allowed the vegetation to rebound and for beaver to increase in numbers. However, this interrelatedness theme does not just relate to the animal and plant actors in this drama. There is an important human-dominated drama as well that involves philosophical, economic, and political actors.

For example, although many biologists and environmentalists argued that it was important to restore the wolf to its former habitat for biological reasons, others looked at the issue in terms of ethics. They felt that humans had an ethical obligation to restore wolves to their former habitat. While park managers could easily see the problems created by a lack of wolves and a huge elk population, they could not simply make the decision to bring back the wolf. A long history of controlling animals that could prey on

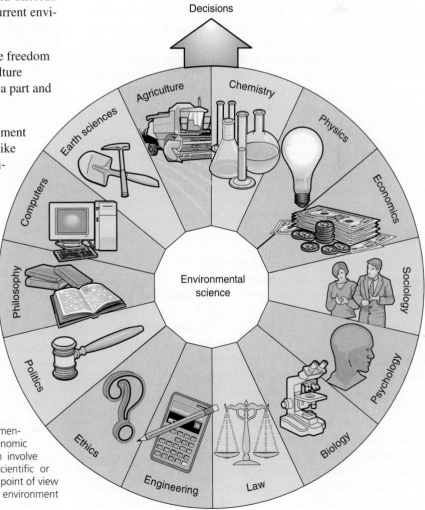

livestock had to be overcome. Ranchers strongly opposed the reintroduction of wolves and saw this as an economic issue. If wolves left the park and killed their livestock, they would lose money. The farm lobby in Congress is very strong and fought long and hard to prevent the reintroduction. After a lengthy period of hearings and many compromises—including a fund to pay ranchers for cattle killed by wolves—the U.S. Fish and Wildlife Service was authorized to proceed with the reintroductions. Thus, the interconnectedness theme associated with the reintroduction of wolves to Yellowstone also applies to social, economic, and political realms of human activity.

An Ecosystem Approach

The idea of interrelatedness is at the core of the ecosystem concept. An **ecosystem** is a region in which the organisms and the physical environment form an interacting unit. Within an ecosystem there is a complex network of interrelationships. For example, weather affects plants, plants use minerals in the soil and are food for animals, animals spread plant seeds, plants secure the soil, and plants evaporate water, which affects weather.

Some ecosystems have easily recognized boundaries. Examples are lakes, islands, floodplains, watersheds separated by mountains, and many others. Large ecosystems always include smaller ones. A large watershed, for example, may include a number of lakes, rivers, streams, and a variety of terrestrial ecosystems. A forest ecosystem may cover hundreds of square kilometers and include swampy areas, openings, and streams as subsystems within it. Often the boundaries between ecosystems are indistinct, as in the transition from grassland to desert. Grassland gradually becomes desert, depending on the historical pattern of rainfall in an area. Thus, defining an ecosystem boundary is often a matter of practical convenience.

However, an ecosystem approach is important to dealing with environmental problems. The task of an environmental scientist is to recognize and understand the natural interactions that take place and to integrate these with the uses humans must make of the natural world.

Political And Economic Issues

Most social and political decisions are made with respect to political jurisdictions, but environmental issues do not necessarily coincide with these artificial political boundaries. For example, Yellowstone National Park is located in the northwest corner of Wyoming. (See figure 1.2.) Therefore, the citizens of the bordering states—Montana and Idaho—as well as the citizens of Wyoming were involved in arguing for or against the reintroduction of wolves to Yellowstone. Citizens recognized that once wolves returned to the park they would migrate to areas surrounding the park. Similarly, air pollution may involve several local units of government, several states or provinces, and even different nations. Air pollution generated in China affects air quality in western coastal states in the United States and in British Columbia, Canada. On a more local level, the air pollution generated in Juarez, Mexico, causes problems in the neighboring city of El Paso, Texas. But the issue is more than air quality and human

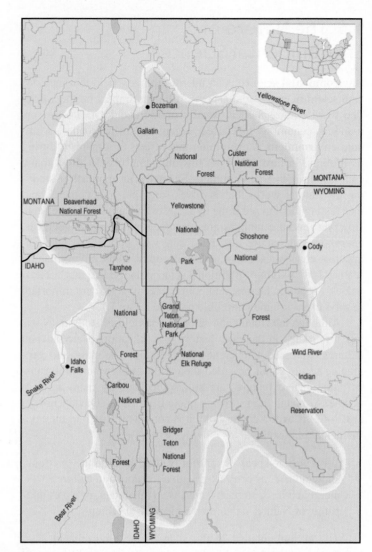

FIGURE 1.2 Environmental Issues often Involve Several Political Jurisdictions The location of Yellowstone National Park in the northwest corner of Wyoming means that citizens of Wyoming and the adjoining states of Idaho and Montana are affected by decisions about the park. In addition, there are several national forests, refuges, and an Indian reservation located near the park. These entities may have goals that differ from those of Yellowstone National Park.
Source: *National Park Service.*

health. Lower wage rates and less strict environmental laws have influenced some U.S. industries to move to Mexico for economic advantages. Mexico and many other developing nations are struggling to improve their environmental image and need the money generated by foreign investment to improve the conditions and the environment in which their people live.

1.2 Emerging Global Issues

Imagine a world in which environmental change threatens people's health, physical security, material needs, and social cohesion. This is a world beset by increasingly intense and frequent storms and by rising sea levels. Some people experience extensive flooding,

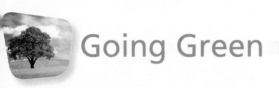

Going Green

Individual Decisions Matter

> **Note to Reader**
>
> *Because environmental science is involved in highlighting problems, the many improvements and positive changes are often overlooked. To call attention to these bits of good news, this book will describe actions that have had a positive environmental impact. Each chapter will have a "Going Green" feature that highlights a particular green initiative. In addition, at the end of each chapter there is an "Acting Green" feature which suggests changes that you can make that collectively can help lead to a sustainable society.*

There is a growing awareness that sustainability needs to be a core value if future generations are to inherit an Earth worth having. Those who support green initiatives are motivated in many different ways. Some are motivated by ethical or moral beliefs that they should "live lightly on the land." Some are motivated by the economic realities of rising energy costs or the costs associated with correcting environmental mistakes. Some simply want to be seen as having green values.

Regardless of their motivation, people around the world are making green decisions. Organizers of conferences and concerts are buying carbon credits to offset the impact of their events. Companies have discovered that consumers seek green products. Governments have passed laws that encourage their citizens to live more sustainably. Ultimately, however, green initiatives depend on individuals making everyday decisions. How many pairs of shoes do I really need? Do I really need the latest electronic gadget? Should I buy products that are produced locally? In the final analysis, most daily decisions have an environmental impact and you have a role to play.

Ten Things You Can Do To Protect Your Environment

1. Reduce your driving (walk, bike, take public transit, carpool). Choose a more efficient car.

2. Do not leave your TV, computers, DVD players on standby. They are using electricity on standby.

3. Recycle everything you can: newspapers, cans, glass bottles, motor oil, etc. Recycling one aluminum can saves enough energy to run a TV set for 3 hours or to light one 100-watt bulb for 20 hours. In 2011, 65 percent of aluminum cans were recycled.

4. Do not leave water running needlessly. Install a water-efficient showerhead and run only full loads in the washing machine or dishwasher. It takes energy to heat the water used by these devices.

5. Do not dispose of gasoline, oil, or weed killers and other lawn and garden pesticides down the drain, into surface water, onto the ground, or in the trash. Check with your local household hazardous waste collection agency for safe disposal of these types of products.

6. Eat a locally produced diet. Grow your own food or support local farmers, natural food stores, and food co-ops.

7. Take unwanted, reusable items to a charitable organization or thrift shop. They don't go to a landfill and someone else is able to use them.

8. Buy in bulk when you can and avoid excess packaging. Even recyclable packaging requires energy and resources to create. Also look for refillable containers.

9. Read labels on pesticides, cleaners, paints, and other products. Choose those with fewer hazardous contents.

10. Become an informed and active citizen. Vote; participate in public forums; get involved in local, state, national or international environmental concerns!

This list is only a start. Go to the website earth911.com. How many additional activities/actions can you add to this list?

---

while others endure intense droughts. Species extinction occurs at rates never before witnessed. Safe water is increasingly limited, hindering economic activity. Land degradation endangers the lives of millions of people.

This is the world today. Yet, as the World Commission on Environment and Development (Brundtland Commission) concluded in its 1987 report, *Our Common Future*, "humanity has the ability to make development sustainable." An important contribution of the report was a concise definition of **sustainability** as "development that meets the needs of the present without compromising the ability of future generations to meet their own needs." Thus the Brundtland Commission addressed the links between development and environment, and challenged policymakers to consider the interrelationships among environmental, economic, and social issues when it comes to solving global problems. Emerging global challenges they identified included continued population growth, maintaining functional ecosystems, food security, environmental governance, health, security, globalization, and energy.

Population Growth

It is fair to say that a core cause of the current environmental crisis is the sheer number of people. If there were fewer people, the pressure on environmental resources and services would be much less. However, the causes of human population growth are not just biological. People have the ability and tools to make decisions about how many children they will have but for a variety of cultural and economic reasons they often have large families. Consequently, the human population continues to grow, with most of the increase in population occurring in poor countries. (See figure 1.3.) This growth puts pressure on resources and leads to the degrading of the environment and often locks people in a cycle of poverty.

Focus On

Campus Sustainability Initiative

The Association for the Advancement of Sustainability in Higher Education (AASHE) was founded in 2006 as a membership organization of colleges and universities in the United States and Canada. There are currently over 800 member colleges and universities. AASHE's mission is to promote sustainability in all aspects of higher education. Its definition of sustainability includes human and ecological health, social justice, secure livelihoods, and a better world for all generations. A core concept of AASHE is that higher education must be a leader in preparing students and employees to understand the importance of sustainability and to work toward achieving it. Furthermore, campuses should showcase sustainability in their operations and curriculum.

To accomplish its goals, AASHE sponsors conferences and workshops to educate members. It also provides networking opportunities and an e-bulletin to facilitate the exchange of information about sustainable practices on campuses.

AASHE has developed a rating system that allows educational institutions to assess their progress toward achieving sustainability. The Sustainability Tracking, Assessment, and Rating System (STARS) focuses on three major categories of activity: education and research, operations, and administration and finance.

Is your college a member? Go to the AASHE website and check its membership list.

FIGURE 1.3 Population Growth Most of the growth in human population is occurring in the less-developed world. The population growth rate for most of the economically developed world is stable or falling.

Poverty is often linked to poor health because of malnutrition and lack of access to affordable health care.

The economically developed countries generally have stable or falling populations and many encourage immigration from the less-developed parts of the world to provide the labor needed for their economies.

Maintaining Functional Ecosystems

As people seek to provide food and other resources for their families, they necessarily affect natural ecosystems. Much of the Earth's surface has been converted to agricultural use for raising crops or grazing cattle. When this conversion occurs natural ecosystems are destroyed or degraded. We are beginning to recognize that biodiversity and functional ecosystems have economic value and their loss can have profound economic consequences.

When a species of organism goes extinct, its loss has a ripple effect throughout its ecosystem. As was described in the chapter opening, the local extinction of wolves resulted in changes in the populations of plants and other animals. There is consensus among scientists that the current rate of extinction is similar to that which occurred in the mass extinctions of the geologic past.

A related concern is that the loss of ecosystems results in a loss or reduction in the services they provide. These **ecosystem services** include *provisioning services,* such as food, minerals, renewable energy, and water; *regulating services,* such as waste decomposition, pollination, purification of water and air, and pest and disease control; *cultural services,* such as spiritual, recreational, and cultural benefits; and *supporting services,* such as nutrient cycling, photosynthesis, and soil formation. (See figure 1.4.) Environmental changes that alter these services affect human security. Although all people rely on ecosystem services, the world's poorest people are especially dependent on environmental goods and services for

FIGURE 1.4 Ecosystem Services Pollination is an important ecosystem service.

their livelihoods, which makes them particularly sensitive and vulnerable to environmental changes.

Food Security

The world is divided into those who have abundant food, those who have adequate food, and those who often lack food. The poor of the world are often subsistence farmers who rely on the food they grow to feed their families. (See figure 1.5.) Environmental disasters such as droughts, floods, or outbreaks of disease in their animals or crops often result in a lack of food and malnutrition.

The amount of food produced in the world is currently able to feed all people adequately. When people face a food shortage, food can be shipped from those that have a surplus to those that need food. However, this is not as simple as it sounds. The poor cannot pay for the food or the cost of shipping it to them. Humanitarian organizations or governments that provide food must fund these emergency programs. A related problem is that the people who need food often must migrate to areas where food is being distributed, which leads to the establishment of refugee camps or increased squatter populations in and around cities.

Therefore, major efforts are being made to provide farmers with better farming methods, improved seeds, and with crops that provide food but do less damage to the land.

Environmental Governance

Despite a greater understanding of the ties between environment and development, real progress toward sustainable development has been slow. Many governments continue to create policies concerned with environmental, economic, and social matters as separate issues. As a result, strategies for economic development often ignore the need to maintain the ecosystem on which long-term development depends. A good example of this disconnect is the continued building of housing on coastlines and floodplains that are subject to flooding. The extent of the damaged caused by hurricane Katrina in 2005 and hurricane Sandy in 2012 was at least partly due to the failure of some government agencies to see

FIGURE 1.5 Food Security In much of the less-developed world, small-scale farms provide the food that people need. If environmental disasters affect crops, people go hungry.

the link between destruction of coastal wetlands and the increased vulnerability of coastal communities to storms.

The issue of declining salmon stocks in the Pacific Northwest of the United States and British Columbia, Canada illustrates another aspect of the problem of governance. (See figure 1.6.) There is typically political and economic friction associated with a resource that crosses political boundaries. From the U.S. perspective alone there are five federal cabinet-level departments, two federal agencies, five federal laws, and numerous tribal treaties that affect decisions about the use of this resource. Furthermore, commercial fishers from several states and provinces are economically affected by any decisions made concerning the harvesting of these fish. They are all politically active and try to influence the laws and rulings of state, provincial, and national governments. It is also safe to say that good science is not always the motivator for the laws and policies.

Environment and Health

The health of countless people around the world is affected by human-induced changes in the environment. According to the World Health Organization (WHO), almost one-quarter of all diseases are caused by environmental exposure. WHO estimates that 13 million deaths worldwide could be prevented every year by environmental improvement.

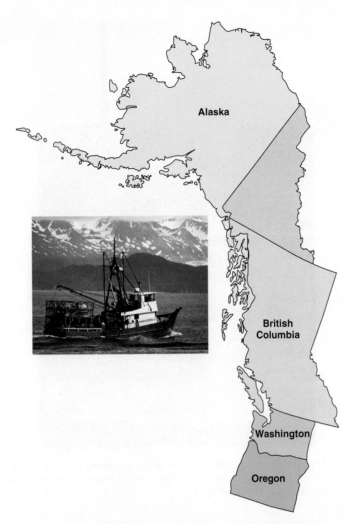

FIGURE 1.6 The Regional Nature of Environmental Problems The regulation of salmon fishing in the Northwest involves several states and the Canadian province of British Columbia. These political entities regulate fishing seasons and the kind of gear that can be used.

In the poorer countries of the world about one third of deaths have environmental causes. Environmental risks including air and water pollution, hazards in the workplace, traffic-related injuries, ultraviolet radiation, noise, and climate and ecosystem change all need to be addressed to generate better global health. The following examples show how environmental conditions and human health are linked.

Air pollution is a serious problem in much of the developing world. In many urban areas the general public is exposed to poor air quality that results from unregulated industrial sources and vehicles with poor pollution control devices. In addition, many people are exposed to high levels of air pollution in their workplaces, and people are exposed in their homes because burning of wood for cooking and heating releases wood smoke. Cigarette smoking exposes the user, and those who live and work with smokers are exposed secondhand. Common diseases related to air pollution are pneumonia, emphysema, and bronchitis, which are responsible for about 6 million deaths per year.

Water pollution results from industrial and municipal releases of pollutants into waterways. Many people in the developing world do not have access to a safe drinking water source or sanitary facilities. When untreated human wastes contaminate water, disease organisms are easily spread from person to person. Diarrhea that results from contaminated drinking water causes dehydration and malnutrition and leads to nearly 2 million deaths per year. The majority of deaths occur in children.

Malaria is caused by a protozoan parasite carried by mosquitos. Programs to protect people from being bitten by infected mosquitos have eliminated the disease in much of the developed world. However, it still results in over half a million deaths per year primarily in children in Sub-Saharan Africa. Once a person is infected with malaria, they continue to have episodes of the disease and when bitten by mosquitos can cause mosquitos to become infected and carry the disease to other persons. Breaking the cycle of disease involves altering the environment so that mosquitos have fewer breeding places and preventing people from being bitten by mosquitos.

Accidents in the home and workplace and those that result from traffic cause about 900,000 deaths per year. Over half the accidental deaths result from traffic accidents. In much of the less-developed world, road conditions are poor, vehicles are poorly maintained, and the mix of pedestrians, bicycles, animals pulling carts, and motor vehicles on roads and streets results in many accidents.

Cancer and coronary heart disease cause about 4 million deaths per year. They are common throughout the world but are most prevalent in developed countries. The environmental causes of these diseases are varied but include exposure to cigarette smoke, exposure to ultraviolet light, and the kinds and amounts of foods eaten. Obesity is a contributing cause.

Emerging diseases result from new organisms or those that become a problem because of environmental changes. Since 1980, more than 35 infectious diseases have emerged or taken on new importance. Often these diseases result from interactions between animals and humans that result in the transfer of animal diseases to humans. The AIDS virus and several flu viruses are examples. In other cases, human changes to the environment lead to changes in organisms that make them more deadly. For example, the wide use of antibiotics has caused the evolution of antibiotic-resistant bacteria such as tuberculosis and methicillin-resistant *Staphylococcus aureus* (MRSA).

Figure 1.7 illustrates several links between environment and health.

Environment and Security

A person's security incorporates economic, political, cultural, social, and environmental aspects. It means having stable and reliable access to resources and the ability to be secure from natural and human disasters. Environmental resources are a critical part

(a) Obesity

(b) Water Pollution

(c) Air Pollution

(d) Traffic Congestion

FIGURE 1.7 Environment and Health Figure 1.7a depicts obesity, which is a major health concern in a growing number of developed countries. Figure 1.7b shows women washing clothes in polluted water in an urban slum in India. People will also use the same water downstream to wash vegetables, increasing the risk of spreading disease. Figure 1.7c shows air pollution in Shanghai, China. Air pollution is a major health issue in China. Figure 1.7d shows traffic congestion in Chengdu, China. This kind of traffic situation often results in accidents that lead to injuries and deaths.

of the livelihoods of millions of people, and when these resources are threatened through environmental change, people's security is also threatened making conflict and social instability common. (See figure 1.8.) For example, disputes over water quantity and quality are ongoing in many parts of the world, and when land ownership is concentrated in the hands of a few rich people, revolution and redistribution of land often result.

Human migration and urbanization have complex relationships with environmental change. Natural disasters such as floods and droughts and human disasters such as war cause many people to migrate to new areas. The local increase in population caused by new immigrants puts increased demands on the local environment to supply resources and provide adequate ecosystem services. Thus, local ecosystems are typically degraded.

Urbanization in particular can cause significant pressure on the environment. Rapid urban growth often overwhelms the ability of cities to provide adequate services to their inhabitants and industrial and human wastes pollute the local environment. On the

(a)

(b)

(c)

FIGURE 1.8 Environment and Security Figure 1.8a shows a young person obtaining water from a well in Bangladesh even though the surrounding area is inundated by flood water. Figure 1.8b shows the No Name Creek forest fire in Colorado in 2002. Figure 1.8c indicates that the beach is unsafe for swimming. All of these situations place persons at risk and reduce their sense of security.

other hand, cities can provide economies of scale, opportunities for sustainable transport, and efficient energy options, which can relieve pressure on the environment.

Environment and Globalization

Globalization is the worldwide exchange of ideas, goods, and cultural views. In particular it allows business transactions to take place across much of the world. The environment and globalization are strongly linked. The globalization of trade has led to the spread of exotic species. For example, the zebra mussel has spread through North America during the last 25 years, resulting in significant ecological and economic impacts. (See figure 1.9.) Its introduction corresponds with a dramatic increase in wheat shipments between the United States, Canada, and the former Soviet Union.

In a globalized world, important decisions related to environmental protection may have more to do with corporate management and outcomes than with political or scientific factors.

Unfortunately, this was perhaps one of the realities of the 2010 Gulf of Mexico oil tragedy. Countries may be reluctant to enforce strict environmental laws, fearing that companies will relocate.

However, it is often forgotten that the environment itself can have an impact on globalization. Resources fuel global economic growth and trade. Solutions to environmental crises, such as climate change, require coordinated global action and greater globalization of governance.

Examples of international activities to address concerns about the Earth's natural systems and how humans are affecting them include the following:

The Earth Summit

The first worldwide meeting of heads of state that was directed to a concern for the environment took place at the Earth Summit, formally known as the United Nations Conference on Environment and Development (UNCED) in Rio de Janeiro in 1992. One of the key outcomes of the conference was a series of

(a)   (b)

FIGURE 1.9 Globalization and Invasive Species Zebra mussels were carried to the Detroit area by European ships that dumped their ballast water. Zebra mussels quickly spread throughout the Great Lakes regions and the Mississippi River drainage. Figure 1.9a shows the striped pattern on the shells for which zebra mussels are named. Figure 1.9b shows a current meter encrusted with zebra mussels.

 # What's Your Take?

The governors of the Great Lakes states have signed an agreement that prohibits the export of water from the Great Lakes. They argue that the water is a valuable resource that is needed by the citizens of their states and that export would deprive the states' citizens of the resource. Regions of the country that are water poor argue that the water in the Great Lakes is a resource that should be shared by all citizens of the country. Develop an argument that supports or refutes the governors' stated policy.

policy statements on sustainable development that were identified as Agenda 21.

More than 178 governments at the 1992 conference adopted three documents related to sustainable development: Agenda 21, the Rio Declaration on Environment and Development, and the Statement of Principles for the Sustainable Management of Forests. The United Nations Commission on Sustainable Development was created in 1993 to monitor and report on implementation of the agreements. Follow-up conferences were held in 1997, 2002, and 2012 to assess progress.

Climate Change

In 1997, representatives from 125 nations met in Kyoto, Japan, for the Third Conference of the United Nations Framework Convention on Climate Change. This conference, commonly referred to as the Kyoto Conference on Climate Change, resulted in commitments from the participating nations to reduce their overall emissions of six greenhouse gases (linked to global warming) by at least 5 percent below 1990 levels and to do so between the years 2008 and 2012. The Kyoto Protocol, as the agreement was called, was viewed by many as one of the most important steps to date in environmental protection and international diplomacy. It is clear that most countries did not meet their goals, because there has been a steady increase in the amount of the most important greenhouse gas, carbon dioxide, in the atmosphere. Additional climate change conferences have not produced a plan that is binding on all countries. Countries have been encouraged to reduce their greenhouse gas emissions but most countries have not made significant progress.

Energy and the Environment

The world is facing twin threats: inadequate and insecure supplies of energy at affordable prices, and environmental damage due to overconsumption of energy. Global demand for energy keeps growing, placing an ever increasing burden on natural resources and on the environment. (See figure 1.10.) During the 10-year

(a)

(c)

(d)

FIGURE 1.10 Energy Use and the Environment Global demand for energy keeps growing, placing an ever-increasing burden on natural resources and on the environment. Figures 1.10a and 1.10b show consumption of energy to provide transportation and light homes. Figure 1.10c is a drilling platform in the Gulf of Mexico that supplies oil as a source of energy. Figure 1.10d shows solar panels on a German home. Germany has pledged to significantly reduce its dependence on fossil fuels and has subsidized the installation of solar panels.

(b)

period from 2003 to 2012, world energy consumption increased by 25 percent. Most of this increase came from countries with developing economies in Asia and South America. China, India, South Korea, Brazil, Argentina, and Mexico are among the countries with large increases in energy use.

Global increases in carbon dioxide emissions are primarily due to fossil fuel use. Traditional biomass (firewood and dung) remains an important energy source in developing countries, where 2.1 billion people rely on it for heating and cooking.

The need to curb growth in energy demand, increase fuel supply diversity, and mitigate climate destabilizing emissions is more urgent than ever. However, expansion of alternative energy sources, such as biofuels, must also be carefully planned. Brazil expects to double the production of ethanol, a "modern" biofuel, in the next two decades. In order to produce enough crops to reach production targets, the cultivated area is increasing rapidly, potentially jeopardizing entire ecosystems.

National Security Policy and Climate Change

Conflict brought on by droughts, famine, and unwelcome migration is as old as history itself. However, a growing number of military analysts think that climate change will exacerbate these problems worldwide and are encouraging countries to prepare to maintain order even as shrinking resources make their citizens desperate. Analysis by the Intergovernmental Panel on Climate Change (IPCC) predicts that climate change will manifest itself in a number of ways. Increases in sea level, melting of glaciers, changes in weather patterns, and altered agricultural productivity will put people at odds with one another as they face a potential scarcity of foods and potable water.

An example of this potential conflict is a high-tech fence that India installed in 2008 along its border with Bangladesh to help deter an influx of illegal immigrants displaced by rising sea levels flooding their low-lying homeland. It is argued that as warming temperatures deplete water supplies and alter land use, already vulnerable communities in Asia and Africa could descend into conflicts and even wars as more people clamor for increasingly scarce resources.

In 2007, a study was undertaken in the United States by the CNA Corporation in Washington, D.C. The study, entitled **National Security and the Threat of Climate Change,** brought together a dozen retired admirals and generals to study how climate change could affect the United States' security over the next 30 to 40 years. The recommendations of the report are:

- The national security consequences of climate change should be fully integrated into national security and national defense strategies.
- The United States should commit to a stronger national and international role to help stabilize climate change at levels that will avoid significant disruption to global security and stability.
- The United States should commit to global partnerships that help less-developed nations build the capacity and resiliency to better manage climate impacts.
- The Department of Defense should enhance its operational capability by accelerating the adoption of improved business processes and innovative technologies that result in improved U.S. combat power through energy efficiency.
- The Department of Defense should conduct an assessment of the impact on U.S. military installations worldwide of rising sea levels, extreme weather events, and other projected climate changes over the next 30 years.

It appears that military planners take the threat of climate change seriously. In the national and international security environment, it is becoming increasingly clear that climate change threatens to add new hostile and stressing factors. On the simplest level, it has the potential to create sustained natural and humanitarian disasters far beyond those we see today. The consequences will likely foster political instability where societal demands exceed the capacity of governments and policymakers to cope.

However, it is up to political leaders in Congress and to the president to evaluate these recommendations and propose legislation to act on these recommendations.

Soldiers of the Border Security Force of India (BSFI) on strict vigil near the International Barbed-Wire Border Fence on India-Bangladesh International Border at Kedar village of Dhubri district of Northeastern Indian State, Assam.

Government Regulation and Personal Property

There are many ways in which government intrudes into your personal lives. Many kinds of environmental regulations require people to modify their behavior. However, one of the most controversial situations occurs when government infringes on personal property rights. The Endangered Species Act requires that people do no harm to threatened and endangered species. They may not be hunted or harvested and often special areas are established to assure their protection.

Many people have found after they have purchased a piece of land that it has endangered species as inhabitants. They are then faced with a situation in which they cannot use the land as they intended, and the land loses much of its value to them. Some argue that they should be allowed to use the land for their original purpose because they did not know it was habitat for an endangered species. Others argue that they have been deprived of a valuable good by the federal government and that the government should compensate them for their loss.

On the other hand, the people charged with enforcing the regulations say that they are simply following the laws of the land and that the landowner must obey.

The threatened California gnatcatcher is a small grey bird that inhabits coastal sagebrush habitats. Developers and those that put forth efforts to preserve the California gnatcatcher reached a compromise. Some land that was originally planned for development has been set aside as habitat for the gnatcatcher.

- Do you think landowners should be compensated for their loss by the federal government?
- Should purchasers of land do better research about the land they want to buy?

Preserved California Gnatcatcher Habitat

California Gnatcatcher

Summary

Environmental science involves the interrelationships among science, economics, ethics, and politics in arriving at solutions to environmental problems. Artificial political boundaries create difficulties in managing environmental problems because most environmental units, or ecosystems, do not coincide with political boundaries. The concept of sustainability requires that we consider future generations when we make decisions about how to use resources. Furthermore, as the population of the world has grown and the exchange of people and goods between countries has increased, many environmental problems have become global in nature. This presents problems with how governments cooperate to assure that their citizens can be healthy, secure, and adequately fed, without compromising the environment that supports them. Energy use continues to be a major issue related to sustainability.

Acting Green

1. Look for locally grown produce in the supermarket—less energy is used to transport locally grown products.
2. Join a local environmental organization.
3. Volunteer for your local Earth Day event in April.
4. Visit a natural area, nature center, or park typical of your region and learn to identify five plants.
5. Go to the website of the League of Conservation Voters, click on the Scorecard tab, and find out the "environmental score" of your senators and representative.

Review Questions

1. Give examples of political, economic, and biological aspects of the reintroduction of wolves to Yellowstone National Park.
2. Describe what is meant by an ecosystem approach to environmental problem solving.
3. Define sustainability.
4. Give examples of several kinds of services provided by natural ecosystems.
5. Give examples that show how social and political factors influence how environmental decisions are made.
6. Explain why the following statement is true. There is enough food in the world to feed everyone but some people do not have enough food.
7. Describe how solving environmental problems is made more difficult by problems with governance.
8. Give examples of environmental conditions that lead to poor health and death for exposed people.
9. Why are wars fought over oil, water, or land?
10. Give examples of the environmental consequences of globalization.
11. What is the major environmental problem related to the use of fossil fuels as an energy source?
12. In what parts of the world is energy use increasing?

Critical Thinking Questions

1. Imagine you are a U.S. congressional representative from a western state and a new wilderness area is being proposed for your district. Who might contact you to influence your decision? What course of action would you take? Why?

2. How do you weigh in on the issue of jobs or the environment? What limits do you set on economic growth? Environmental protection?

3. Imagine you are an environmentalist in your area who is interested in local environmental issues. What kinds of issues might these be?

4. In a discussion with a representative of China, you state that China contributes the largest amounts of greenhouse gases of any country in the world. The representative responds by stating that the developed world including the United States has benefited from 200 years of fossil fuel use and industrialization that has led to a high standard of living. Why shouldn't the citizens of his country experience the same standard of living? How would you respond? Is the situation today different from that 100 years ago?

5. You are the superintendent of Yellowstone National Park and want to move to an ecosystem approach to managing the park. How might an ecosystem approach change the current park? How would you present your ideas to surrounding landowners?

6. Look at the issue of climate change from several different disciplinary perspectives—economics, climatology, sociology, political science, agronomy. What might be some questions that each discipline could contribute to our understanding of global warming?

To access additional resources for this chapter, please visit ConnectPlus® at connect.mheducation.com. There you will find interactive exercises, including Google Earth™, additional Case Studies, and SmartBook™, an adaptive eBook that integrates our LearnSmart® adaptive learning technology.

Environmental Ethics

CHAPTER OUTLINE

Human use alters the natural world. Often we must ask what kind of use is ethical? Is clearing a tropical forest for agriculture ethically supportable?

OBJECTIVES

After reading this chapter, you should be able to:

- Understand the role of ethics in society.
- Recognize the importance of a personal ethical commitment.
- List three conflicting attitudes toward nature.
- Explain the connection between material wealth and resource exploitation.
- Describe the factors associated with environmental justice.

- Explain how corporate behavior connects to the state of the environment.
- Describe how environmental leaders in industry are promoting more sustainable practices.
- Describe the triple bottom line.
- Explain the concept "greenwashing."
- Describe the influence that corporations wield because of their size.

- Explain the relationship between economic growth and environmental degradation.
- Explain some of the relationships between affluence, poverty, and environmental degradation.
- Explain the importance of individual ethical commitments toward environment.
- Explain why global action on the environment is necessary.

Of Sea Lions and Salmon—An Environmental and Ethical Dilemma

Animal rights groups continue to fight in court plans to trap and kill California sea lions that feed on protected fish species near the Bonneville Dam on the Columbia river. Oregon, Washington, and Idaho first asked the National Marine Fisheries Service (NMFS) in 2006 for permission to "take" California sea lions that eat salmon, steelhead, and sturgeon just below the dam, which acts as a bottleneck to fish migrating upstream in the Columbia River. The request was part of an effort to protect threatened or endangered populations of these fish.

But California sea lions are also protected. The Marine Mammals Protection Act bans the taking of all marine mammals, unless individual animals "are having significant impact on the decline or recovery" of threatened or endangered species. Another species of sea lion, the Steller sea lion, also feeds at the dam but is off limits because it is protected by the Marine Mammals Protection Act and the Endangered Species Act.

In 2008 the states were permitted to kill up to 85 California sea lions per year, but the U.S. Ninth Circuit Court vacated that authorization in 2010. The federal appeals court said NMFS failed to explain why sea lions posed a greater threat to endangered fish than commercial fisheries and hydroelectric power plants. In 2012 NMFS granted another round of authorizations, this time to kill or permanently capture up to 92 California sea lions per year. The authorizations are good until June 2016, and allow the states to trap and euthanize individual animals or to shoot them under certain conditions. The Humane Society filed a new lawsuit under the Endangered Species Act and the Marine Mammals Protection Act. They claim NMFS simply recycled its analysis to support the kill authorizations without addressing the problems raised by the Ninth Circuit Court. The battle of sea lions v. salmon continues.

California Sea lion taking a salmon near the Bonneville Dam on the Columbia River.

2.1 The Call for a New Ethic

The most beautiful object I have ever seen in a photograph in all my life is the planet Earth seen from the distance of the moon, hanging in space, obviously alive. Although it seems at first glance to be made up of innumerable separate species of living things, on closer examination every one of its things, working parts, including us, is interdependently connected to all the other working parts. It is, to put it one way, the only truly closed ecosystem any of us know about.

—Lewis Thomas

When we try to pick out anything by itself, we find it hitched to everything else in the Universe.

—John Muir

One of the marvels of recent technology is that we can see the Earth from the perspective of space, a blue sphere unique among all the planets in our solar system. (See figure 2.1.) Looking at ourselves from space, it becomes obvious that a lot of what we do on our home planet connects us to something or somebody else. This means, as Harvard University ecologist William Clark points out, that only as a global species, "pooling our knowledge, coordinating our actions, and sharing what the planet has to offer—do we have any prospect for managing the planet's transformation along pathways of sustainable development."

FIGURE 2.1 The Earth as Seen from Space Political, geographic, and national differences among humans do not seem so important from this perspective. In reality, we all share the same planet.

Some people see little value in an undeveloped river and feel it is unreasonable to leave it flowing in a natural state. It could be argued that rivers throughout the world ought to be controlled to provide power, irrigation, and navigation for the benefit of humans. It could also be argued that to not use these resources would be wasteful.

In the U.S. Pacific Northwest, there is a conflict over the value of old-growth forests. Economic interests want to use the forests for timber production and feel that to not do so would cause economic hardship. They argue that trees are simply a resource to be used in any way deemed necessary for human economic benefit. An opposing view is that all the living things that make up the forest have a kind of value beyond their economic utility. Removing the trees would destroy something ethically significant that took hundreds of years to develop and may be almost impossible to replace.

Interactions between people and their environment are as old as human civilization. The problem of managing those interactions, however, has been transformed today by unprecedented increases in the rate, scale, and complexity of the interactions. At one time, pollution was viewed as a local, temporary event. Today pollution may involve several countries and may affect multiple generations. The debates over chemical and radioactive waste disposal are examples of the increasingly international nature of pollution. Many European countries are concerned about the transportation of radioactive and toxic wastes across their borders. What were once straightforward confrontations between ecological preservation and economic growth have today become complex balancing acts containing multiple economic, political, and ethical dimensions.

The character of the environmental changes that today's technology makes possible and the increased public awareness of the importance of the natural environment mean that we have entered a new age of environmental challenges. Across the world, people are beginning to understand that part of what is needed to meet these challenges is the development of a new and more robust environmental ethic.

2.2 Environmental Ethics

Ethics is one branch of philosophy. Ethics seeks to define what is right and what is wrong. For example, most cultures are ethically committed to the idea that it is wrong to needlessly take life. Many cultures ground this belief on the existence of a right to life. It is considered unethical to deprive humans of this right to their life. Ethics can help us to understand *what* actions are wrong and *why* they are wrong. Many of the issues discussed throughout this book (energy, population, environmental risk, biodiversity, land-use planning, air quality, etc.) have ethical dimensions.

Ethics are a broad way of thinking about what constitutes a good life and how to live one. They address questions of right and wrong, making good decisions, and the character or attributes necessary to live a good life. Applied ethics address these issues with a special emphasis on how they can be lived out in a practical manner. Environmental ethics apply ethical thinking to the natural world and the relationship between humans and the earth. Environmental ethics are a key feature of environmental studies, but they have application in many other fields as human society deals in a more meaningful way with pollution, resource degradation, the threat of extinction, and global climate disruption.

In the most general sense, environmental ethics invites us to consider three key propositions:

1. The Earth and its creatures have moral status, in other words, are worthy of our ethical concern.

2. The Earth and its creatures have intrinsic value, meaning that they have moral value merely because they exist, not only because they meet human needs.

3. Based on the concept of an ecosystem, human beings should consider "wholes" that include other forms of life and the environment.

Environmental ethics has required us to consider far more carefully the actual extent of the range of stakeholders in any ethical decision. These may include the immediate people involved, but the stakeholders may also include the people of future generations who may be affected by environmental decisions made today. The stakeholders may also include people who live far away from where decisions are made but may be impacted by those environmental decisions. And stakeholders may include the natural world itself. This concept makes us consider the moral status or the intrinsic value of each stakeholder—whether the stakeholder is a human being or the animals, plants, and ecosystems of the natural world itself. The concept of "the moral status of nature" is an important feature that distinguishes environmental ethics from social ethics.

When discussing stakeholders the importance of including stakeholders of the future is critical. The future is a category especially pertinent to environmental ethics. In many ethical decisions, the effects of our actions are immediate and apparent. In many environmental ethics decisions, however, the effects of our actions may be cumulative, long-lasting and, at least in the near term, hidden. For instance, the runoff from a new residential subdivision will eventually flow into a nearby river. Because of the impermeable streets and sidewalks, rainwater will be carried to the river rapidly and carry with it fertilizer and herbicides from lawns, oil and other fluids from roadways, and other discards of people living there. At first, the damaging effects may be slight. But, over time, these effects may accumulate until the character of the river is fundamentally and destructively altered.

Environmental ethics also makes us think differently in terms of place. That is, environmental ethics forces us to consider decisions in light of such living realities as the biosphere and ecosystems. Thus, when we trace the possible effects of a particular action, we must pay close attention to how the initial effects near at hand may well create a chain reaction of critical effects.

Conflicting Ethical Positions

Even when people have strong personal ethical commitments, they might find that some of their commitments conflict. For example, a mayor might have an ethical commitment to preserving the land around a city but at the same time have an ethical commitment to bringing in the jobs associated with the construction of a new factory on the outskirts of town. There are often difficult balances to be struck between multiple ethical values.

As you can see, ethics can be very complicated. Ethical issues dealing with the environment are especially complex because sometimes it appears that what is good for people conflicts with what is good for the environment. Saving the forest might mean the loss of some logging jobs. While recognizing that there are some real conflicts involved, it is also important to see that it is not necessarily the case that when the environment wins people lose. In a surprising number of cases it turns out that what is good for the environment is also good for people. For example, even when forest protection reduces logging jobs, a healthier forest might lead to new jobs in recreation, fisheries, and tourism. Searching for genuine "win-win" situations has become a priority in environmental decision making.

The Greening of Religion

For many years, environmental issues were considered to be the concern of scientists, lawyers, and policymakers. Now the ethical dimensions of the environmental crisis are becoming more evident. What is our moral responsibility toward future generations? How can we ensure equitable development that does not destroy the environment? Can religious and cultural perspectives be considered in creating viable solutions to environmental challenges?

Until recently, religious communities have been so absorbed in internal sectarian affairs that they were unaware of the magnitude of the environmental concerns facing the world. Certainly, the natural world figures prominently in the world's major religions: God's creation of material reality in Judaism, Christianity, and Islam; the manifestation of the divine in the karmic processes underlying the recycling of matter in Hinduism and Jainism; the interdependence of life in Buddhism; and the Tao (the Way) that courses through nature in Confucianism and Taoism.

Today, many religious leaders recognize that religions, as enduring shapers of culture and values, can make major contributions to the rethinking of our current environmental impasse. Religions have developed ethics for homicide, suicide, and genocide; now they are challenged to respond to biocide and ecocide. Moreover, the environment presents itself as one of the most compelling concerns for robust interreligious dialogue. The common ground is the Earth itself, along with a shared sense among the world's religions of the interdependence of all life.

Much of the credit for increases in such "faith-based" environmentalism can go the National Religious Partnership for the Environment (NRPE), which was founded in 1993 to "weave the mission of care for God's creation across all areas of organized religion." NRPE has forged relationships with a diverse group of religious organizations, including the U.S. Catholic Conference, the National Council of Churches of Christ, the Coalition on the Environment and Jewish Life, and the Evangelical Environmental Network.

These organizations work with NRPE to develop environmental programs that mesh with their own varied spiritual teachings.

For instance, some 135,000 congregations—counting Catholic parishes, synagogues, Protestant and Eastern Orthodox churches, and evangelical congregations—have been provided with resource kits on environmental issues, including sermons for clergy, lesson plans for Sunday school teachers, and even conservation tips for church and synagogue building managers.

Even Evangelical Christians, known for their conservative stand on most issues, are becoming green. The Colorado-based National Association of Evangelicals is urging its 30 million members to pursue a "biblically balanced agenda" to protect the environment alongside fighting poverty. Indeed it was Evangelical minister Reverend Jim Ball who in 2004 started the "What Would Jesus Drive?" campaign, promoting hybrid cars.

Three Philosophical Approaches to Environmental Ethics

Given the complexity of the issues, environmental philosophers have developed a number of theoretical approaches to help us see more clearly our ethical responsibilities concerning the environment. In these environmentally conscious times, most people agree that we need to be environmentally responsible. Toxic waste contaminates groundwater, oil spills destroy shorelines, and fossil fuels produce carbon dioxide, thus adding to global warming. The goal of environmental ethics, then, is not simply to convince us that we should be concerned about the environment—many already are. Instead, environmental ethics focuses on the moral foundation of environmental responsibility and how far this responsibility extends. There are three primary theories of moral responsibility regarding the environment. Although each can support environmental responsibility, their approaches are different. (See figure 2.2.)

Anthropocentrism

The first of these theories is **anthropocentrism** or human-centered ethics. Anthropocentrism is the view that all environmental responsibility is derived from human interests alone. The assumption here is that only human beings are morally significant and have direct moral standing. Since the environment is crucial to human well-being and human survival, we have an indirect duty toward the environment, that is, a duty derived from human interests. We must ensure that the Earth remains environmentally hospitable for supporting human life and even that it remains a pleasant place for humans to live. Nevertheless, according to this view, the value of the environment lies in its *instrumental worth* for humans. Nature is fundamentally an instrument for human manipulation. Some anthropocentrists have argued that our environmental duties are derived both from the immediate benefit that people receive from the environment and from the benefit that future generations of people will receive. But critics have maintained that since future generations of people do not yet exist, then, strictly speaking, they cannot have rights any more than a dead person can have rights. Nevertheless, both parties to this dispute acknowledge that environmental concern derives solely from human interests.

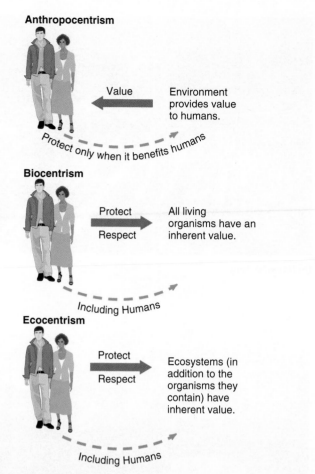

Anthropocentrism

Value ← Environment provides value to humans.

Protect only when it benefits humans

Biocentrism

Protect → All living organisms have an
Respect inherent value.

Including Humans

Ecocentrism

Protect → Ecosystems (in
Respect addition to the organisms they contain) have inherent value.

Including Humans

FIGURE 2.2 Philosophical Approaches Of the three major approaches only anthropocentrism refers all value back to human needs and interests.

Biocentrism

A second theory of moral responsibility to the environment is **biocentrism** or life-centered environmental ethics. According to the broadest version of the biocentric theory, all forms of life have an inherent right to exist. A number of biocentrists recognize a hierarchy of values among species. Some, for example, believe that we have a greater responsibility to protect animal species than plant species and a greater responsibility to protect mammals than invertebrates. Another group of biocentrists, known as "biocentric egalitarians," take the view that all living organisms have an exactly equal right to exist. Since the act of survival inevitably involves some killing (for food and shelter) it is hard to know where biocentric egalitarians can draw the lines and still be ethically consistent.

Ecocentrism

The third approach to environmental responsibility, called **ecocentrism,** maintains that the environment deserves direct moral consideration and not consideration that is merely derived from human or animal interests. In ecocentrism it is suggested that the environment itself, not just the living organisms that inhabit it, has moral worth. Some ecocentrists talk in terms of the systemic value that a

particular ecosystem possesses as the matrix that makes biological life possible. Others go beyond particular ecosystems and suggest that the biological system on Earth as a whole has an integrity to it that gives it moral standing. Another version goes even further and ascribes personhood to the planet, suggesting that Mother Earth or "Gaia" should have the same right to life as any mother.

One of the earliest and most well-known spokespersons for ecocentrism was the ecologist Aldo Leopold. Leopold is most famous for his 1949 book *A Sand County Almanac,* which was written in response to the relentless destruction of the landscape that Leopold had witnessed during his life. *A Sand County Almanac* redefined the relationship between humankind and the Earth. Leopold devoted an entire chapter of his book to "The Land Ethic."

> All ethics so far evolved rest upon a single premise: that the individual is a member of a community of interdependent parts. . . . The land ethic simply enlarges the boundaries of the community to include soils, waters, plants, and animals, or collectively the land . . . a land ethic changes the role of *Homo sapiens* from conqueror of the land-community to plain member and citizen of it. It implies respect for his fellow-members, and also respect for the community as such.
>
> It is inconceivable to me that an ethical relation to land can exist without love, respect, and admiration for land, and a high regard for its value. By value, I of course mean something far broader than mere economic value; I mean value in the philosophical sense.

What Leopold put forth in "The Land Ethic" was viewed by many as a radical shift in how humans perceive themselves in relation to the environment. Originally we saw ourselves as conquerors of the land. Now, according to Leopold, we need to see ourselves as members of a community that also includes the land and the water.

Leopold went on to claim that "a thing is right when it tends to preserve the integrity, stability, and beauty of the biotic community. It is wrong when it tends otherwise. . . . We abuse land because we regard it as a commodity belonging to us. When we see land as a community to which we belong, we may begin to use it with love and respect."

Other Philosophical Approaches

As traditional political and national boundaries fade or shift in importance, new variations of environmental philosophy are fast emerging. Many of these variations are founded on an awareness that humanity is part of nature and that nature's component parts are interdependent. Beyond the three ethical positions discussed previously, other areas of thought recently developed by philosophers to address the environmental crisis include:

Ecofeminism—the view that there are important theoretical, historical, and empirical connections between how society treats women and how it treats the environment.

Social ecology—the view that social hierarchies are directly connected to behaviors that lead to environmental destruction. Social ecologists are strong supporters of the environmental justice movement (see pp. 25–27).

Deep ecology—the generally ecocentric view that a new spiritual sense of oneness with the Earth is the essential starting point for a more healthy relationship with the environment. Deep ecology also includes a biocentric egalitarian world view. Many deep ecologists are environmental activists.

Environmental pragmatism—an approach that focuses on policy rather than ethics. Environmental pragmatists think that a human-centered ethic with a long-range perspective will come to many of the same conclusions about environmental policy as an ecocentric ethic. Consequently they find the emphasis on ethical theories unhelpful.

Environmental aesthetics—the study of how to appreciate beauty in the natural world. Some environmental aesthetics advocates think that the most effective philosophical ground for protecting the natural environment is to think in terms of protecting natural beauty.

Animal rights/welfare—this position asserts that humans have a strong moral obligation to nonhuman animals. Strictly speaking, this is not an environmental position because the commitment is to individual animals and not to ecosystems or ecological health. Animal rights advocates are particularly concerned about the treatment of farm animals and animals used in medical research.

2.3 Environmental Attitudes

Even for those who have studied environmental ethics carefully, it is never easy to act in accordance with one particular ethic in everything. Ethical commitments pull in different directions at different times. Because of these difficulties, it is sometimes easier to talk in terms of general *attitudes* or *approaches* to the environment rather than in terms of particular *ethics*. The three most common approaches are (a) the *development* approach, (b) the *preservation* approach, and (c) the *conservation* approach. (See figure 2.3.)

Development

The **development approach** tends to be the most anthropocentric of the three. It assumes that the human race is and should be master of nature and that the Earth and its resources exist solely for our benefit and pleasure. This approach is reinforced by the capitalist work ethic, which historically dictated that humans should create value for themselves by putting their labor into both land and materials in order to convert them into marketable products. In its unrestrained form, the development approach suggests that improvements in the human condition require converting ever more of nature over to human use. The approach thinks highly of human creativity and ingenuity and holds that continual economic growth is itself a moral ideal for society. In the development approach, the environment has value only insofar as human beings economically utilize it. This mindset has very often accompanied the process of industrialization and modernization in a country.

(a) Development

(b) Preservation

(c) Conservation

FIGURE 2.3 Environmental Attitudes Development, preservation, and conservation are different attitudes toward nature. These attitudes reflect a person's ethical commitments.

Preservation

The **preservationist approach** tends to be the most ecocentric of the three common attitudes toward the environment. Rather than seek to convert all of nature over to human uses, preservationists want to see large portions of nature preserved intact. Nature, they argue, has intrinsic value or inherent worth apart from human uses. Preservationists have various ways of articulating their position.

During the nineteenth century, preservationists often gave openly religious reasons for protecting the natural world. John Muir condemned the "temple destroyers, devotees of ravaging commercialism" who, "instead of lifting their eyes to the God of the mountains, lift them to the Almighty dollar." This was not a call for better cost-benefit analysis: Muir described nature not as a commodity but as a companion. Nature is sacred, Muir held, whether or not resources are scarce.

Philosophers such as Emerson and Thoreau thought of nature as full of divinity. Walt Whitman celebrated a leaf of grass as no less than the "journeywork of the stars." "After you have exhausted what there is in business, politics, conviviality, love, and so on and found that none of these finally satisfy, or permanently wear—what remains? Nature remains," Whitman wrote. These philosophers thought of nature as a refuge from economic activity, not as a resource for it.

While many preservationists adopt an ecocentric ethic, some also include anthropocentric principles in their arguments. These preservationists wish to keep large parts of nature intact for aesthetic or recreational reasons. They believe that nature is beautiful and restorative and should be preserved to ensure that wild places exist for future humans to hike, camp, fish, or just enjoy some solitude.

Conservation

The third environmental approach is the **conservationist approach.** Conservationism tends to strike a balance between unrestrained development and preservationism. Conservationism is anthropocentric in the sense that it is interested in promoting human well-being. But conservationists tend to consider a wider range of long-term human goods in their decisions about environmental management. Conservationist Gifford Pinchot, for example, arguing with preservationist John Muir at the start of the twentieth century about how to manage American forests, thought that the forests should primarily be managed to serve human needs and interests. But Pinchot could see that timber harvest rates should be kept low enough for the forest to have time to regenerate itself. He also realized that water quality for nearby humans was best preserved by keeping large patches of forest intact.

Many hunters are conservationists. Hunters throughout North America and Northern Europe have protected large amounts of habitat for waterfowl, deer, and elk. Hunters have both biocentric and anthropocentric elements to their thinking. Ethically sensitive hunters often see the value of nonhuman animal species and put ethical constraints on the way they hunt them. Even though a hunter tends to think that the human interest in harvesting the meat ultimately overrides the animal's interest in staying alive, he or she often believes that the animal has a place on the landscape and that the world is a better place if it contains healthy populations of wild animals.

Sustainable Development

Sustainable development, a term first used in 1987 in a UN-sponsored document called the Brundtland Report, is often defined as "meeting the needs of current generations without compromising the ability of future generations to meet theirs." Like conservationism, sustainable development is a middle ground that seeks to promote appropriate development in order to alleviate poverty while still preserving the ecological health of the landscape. Sustainable development does not focus solely on environmental issues. The United Nations 2005 World Summit Document refers to the "interdependent and mutually reinforcing pillars" of sustainable development as economic development, social development, and environmental protection. (See figure 2.4.) Indigenous peoples have argued that there are four pillars of sustainable development—the fourth being cultural.

Green development is generally differentiated from sustainable development in that green development prioritizes what its proponents consider to be environmental sustainability over economic and cultural considerations. Proponents of sustainable development argue that it provides a context in which to improve overall sustainability where green development is unattainable. For example, a cutting-edge wastewater treatment plant with extremely high maintenance costs may not be sustainable in regions of the world with fewer financial resources. An environmentally ideal plant that is shut down due to bankruptcy is obviously less sustainable than one that is maintainable by the community, even if it is somewhat less effective from an environmental standpoint.

Still other researchers view environmental and social challenges as opportunities for development action. This is particularly true in the concept of sustainable enterprise that frames these global needs as opportunities for private enterprise to provide innovative and entrepreneurial solutions. This view is now being taught at many business schools.

The United Nations Conference on Environment and Development in Rio de Janeiro, Brazil, in 1992 produced a document entitled Agenda 21, which set out a roadmap for sustainable development. A follow-up conference in South Africa in 2002 drew up a Plan of Implementation for sustainable development. Many observers of the 2002 conference questioned why there had been such a lack of international progress in alleviating poverty and

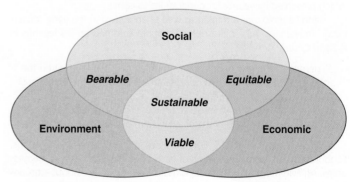

FIGURE 2.4 Scheme of Sustainable Development Sustainable development occurs when social, environmental, and economic concerns are all met.

Focus On

Early Philosophers of Nature

The philosophy behind the environmental movement had its roots in the nineteenth century. Among many notable conservationist philosophers, several stand out: Ralph Waldo Emerson, Henry David Thoreau, John Muir, Aldo Leopold, and Rachel Carson.

In an essay entitled "Nature" published in 1836, Emerson wrote that "behind nature, throughout nature, spirit is present." Emerson was an early critic of rampant economic development, and he sought to correct what he considered to be the social and spiritual errors of his time. In his *Journals,* published in 1840, Emerson stated that "a question which well deserves examination now is the Dangers of Commerce. This invasion of Nature by Trade with its Money, its Credit, its Steam, its Railroads, threatens to upset the balance of Man and Nature."

Henry David Thoreau was a writer and a naturalist who held beliefs similar to Emerson's. Thoreau's bias fell on the side of "truth in nature and wilderness over the deceits of urban civilization." The countryside around Concord, Massachusetts, fascinated and exhilarated him as much as the commercialism of the city depressed him. It was near Concord that Thoreau wrote his classic, *Walden,* which describes the two years he lived in a small cabin on the edge of Walden pond in order to have direct contact with nature's "essential facts of Life." In his later writings and journals, Thoreau summarized his feelings toward nature with prophetic vision:

> But most men, it seems to me, do not care for Nature and would sell their share in all her beauty, as long as they may live, for a stated sum— many for a glass of rum. Thank God, man cannot as yet fly, and lay waste the sky as well as the earth! We are safe on that side for the present. It is for the very reason that some do not care for these things that we need to continue to protect all from the vandalism of a few. (1861)

John Muir combined the intellectual ponderings of a philosopher with the hard-core, pragmatic characteristics of a leader. Muir believed that "wilderness mirrors divinity, nourishes humanity, and vivifies the spirit." Muir tried to convince people to leave the cities for a while to enjoy the wilderness. However, he felt that the wilderness was threatened. In the 1876 article entitled, "God's First Temples: How Shall We Preserve Our Forests?" published in the *Sacramento Record Union,* Muir argued that only government control could save California's finest sequoia groves from the "ravages of fools." In the early 1890s, Muir organized the Sierra Club to "explore, enjoy, and render accessible the mountain regions of the Pacific Coast" and to enlist the support of the government in preserving these areas. His actions in the West convinced the federal government to restrict development in the Yosemite Valley, which preserved its beauty for generations to come.

Aldo Leopold was a thinker as well as an activist in the early conservation movement. As a philosopher, Leopold summed up his feelings in *A Sand County Almanac:*

> Wilderness is the raw material out of which man has hammered the artifact called civilization. No living man will see again the long grass prairie, where a sea of prairie flowers lapped at the stirrups of the pioneer. No living man will see again the virgin pineries of the Lake States, or the flatwoods of the coastal plain, or the giant hardwoods.

While serving in the U.S. Forest Service in New Mexico in the 1920s, Leopold worked for the protection of parts of the forest as early wilderness areas. An avid hunter himself, Leopold learned a great deal about attitudes toward the control of predators as part of his Forest Service job. When he left the Forest Service he pioneered the field of game management. As a professor of game management at the University of Wisconsin, he argued that regulated hunting should be used to maintain a proper balance of wildlife on that landscape. His *Sand County Almanac,* published shortly after his death in 1949, laid down the principles of his land ethic.

While most people talk about what's wrong with the way things are, few actually go ahead and change it. Rachel Carson ranks among those few. A distinguished naturalist and best-selling nature writer, Rachel Carson published in the *New Yorker* in 1960 a series of articles that generated widespread discussion about pesticides. In 1962, she published *Silent Spring,* which dramatized the potential dangers of pesticides to food, wildlife, and humans and eventually led to changes in pesticide use in the United States. *Silent Spring* was more than a study of the effects of synthetic pesticides; it was an indictment of the late 1950s. Humans, Carson argued, should not seek to dominate nature through chemistry in the name of progress.

Although some technical details of her book have been shown to be in error by later research, her basic thesis that pesticides can contaminate and cause widespread damage to the ecosystem has been established. Unfortunately, Carson's early death from cancer came before her book was recognized as one of the most important events in the history of environmental awareness and action in the twentieth century.

Ralph Waldo Emerson

Henry David Thoreau

John Muir

Aldo Leopold

Rachel Carson

protecting the environment. Part of the problem is that people differ in their opinions on how to strike the right balance between the *development* and *preservation* aspects of sustainable development.

2.4 Environmental Justice

Environmental justice is the social justice expression of environmental ethics. The environmental justice movement emerged to challenge the unfair distribution of toxic, hazardous, and dangerous waste facilities, which were disproportionately located in low-income communities of color. This movement is a distinct expression of environmentalism, for it works to improve the protection of human communities and is, in general, less attentive to wild nature. It is environmental protection where people live, work, and play. Over the past two decades it has expanded its scope from community-oriented, anti-toxics activism to address global-scale inequalities in economic development and environmental degradation.

The idea of environmental justice draws heavily from civil rights, public health, and community organizing efforts. As a result, this movement devotes itself to challenging the unfair distribution of environmental risks and the benefits of natural resources, and promotes efforts to prevent pollution from impacting low-income communities. It follows traditional environmentalism's efforts to protect nature by making the poor and marginalized the object of special concern. Its power lies in its appeal to a fundamental ethic of fairness. Members of this movement argue that it is unjust for politically marginalized, low-income communities of color to suffer such a heavy burden.

In 1982, African-American residents of Warren County, North Carolina, protested the use of a landfill in their community as a dumpsite for PCB-contaminated waste. (See figure 2.5.) Some observers asserted that the proposal constituted a form of environmental racism that took advantage of the poorest and least politically influential people in the state. Residents of Warren County took direct action by lying down on the road in front of the trucks headed to the landfill. Hundreds were arrested including Walter Fauntroy, a U.S. Congressional delegate for Washington, D.C., who had traveled to North Carolina in support of the protests.

Following the protests, a number of studies quickly confirmed that toxic waste facilities were indeed disproportionately located in minority neighborhoods. Not only that, the studies also revealed that enforcement of laws was slower and fines levied against violators were lower in areas where residents were made up of poor minorities. By the early 1990s, environmental justice (EJ) was recognized throughout the environmental movement as being a critical component of environmental protection.

In 1998, the Environmental Protection Agency (EPA) in the United States characterized environmental justice as a simple matter of fair treatment. The EPA outlined the federal government's commitment to the principle that "no group of people, including racial, ethnic, or socioeconomic groups, should bear a disproportionate share of the negative environmental consequences resulting from industrial, municipal, and commercial

(a)

(b)

FIGURE 2.5 Environmental Justice (a) The birthplace of the Environmental Justice Movement—Warren County, North Carolina. (b) Residents of Warren County protest the dumping of hazardous PCBs in their local landfill.

operations or the execution of federal, state, local, and tribal programs and policies." The EPA's website on environmental justice (http://www.epa.gov/compliance/environmentaljustice) states that environmental justice also involves "the meaningful involvement of all people regardless of race, color, national origin, or income with respect to the development, implementation, and enforcement of environmental laws, regulations, and policies." Environmental justice is therefore closely related to civil rights. According to the EPA's definition, deliberate discrimination need not be involved. Any action that affects one social group disproportionately is in violation of EPA rules. The difficulty arises in defining what to measure and what should be the standard of comparison.

As a first step in evaluating whether a group is unduly disadvantaged, a policymaker must consider who is affected. Most ethnic data relate to census tracts, zip codes, city boundaries, and counties. If the facility is to be located in a wealthy county but near the county border close to a poor community, how does the policymaker draw the line? Should prevailing winds be considered? Many industrial sites are located where land is cheap; people of low income may choose to live in those areas to minimize living expenses. How are these decisions weighed?

Another difficulty arises in determining whether and how particular groups will be disadvantaged. Landfills, chemical plants, and other industrial works bring benefits to some, although they may harm others. They create jobs, change land values, and generate revenues that are spent in the community. How do officials compare the benefits with the losses? How should potential health risks from a facility be compared with the overall health benefits that jobs and higher incomes bring?

Despite the complications, there are clearly many ways in which governments can act to ensure fairness in how environmental costs and benefits affect their citizens. Governments have established laws and directives to eliminate discrimination in housing, education, and employment, but until the rise of the environmental justice movement they had done very little to address discrimination in environmental practices. One factor that makes environmental justice issues especially troubling is the inverse relationship between who generates the problem and who bears the burden. Studies show that the affluent members of society generate most of the waste, while the impoverished members tend to bear most of the burden of this waste.

Environmental justice encompasses a wide range of issues. In addition to the question of where to place hazardous and polluting facilities, environmental justice questions arise in relation to transportation, safe housing, lead poisoning, water quality, access to recreation, exposure to noise pollution, the viability of subsistence fisheries, access to environmental information, hazardous waste cleanup, and exposure to natural disasters such as Hurricane Katrina (see p. 35). In the United States, environmental justice issues also arise in relation to pollution on Native American Indian reservations. Further questions of environmental justice result from international trade policies that tend to congregate polluting factories in particular countries and their border regions.

The environmental movement in America began as the concern of middle-class and affluent white people. Environmental justice has broadened the demographic of the movement and has raised the profile of many important environmental issues that were simply being ignored. Minorities, indigenous people, and people of color have forced a dialogue about race, class, discrimination, and equity in relation to the environment. They have established beyond doubt that patterns of environmental destruction have important social dimensions.

The movement for environmental justice has been the strongest when community-based organizations have partnered with university researchers. Local groups have more complete knowledge of neighborhood environmental issues, but academics have contributed by bringing their scientific, analytical, and legal expertise to bear on local problems. In collaboration these different kinds of groups bring their own information and can advance more powerful arguments about discriminatory environmental actions and the need for a more equitable approach.

Environmental Justice Highlights

1982

National attention focused on a series of protests in the low-income, minority community of Warren County, North Carolina, over a landfill filled with PCB-contaminated soil from 14 other counties throughout the state. Over 500 people were arrested. The protest prompted the U.S. General Accounting Office (GAO) to initiate a study of hazardous waste landfills in eight southern states. The GAO study concluded that three out of every four landfills were located in minority communities.

1987

The United Church of Christ Commission on Racial Justice published a report showing that race was the most significant factor in the siting of toxic waste facilities throughout the nation. More than 60 percent of African-American and Hispanic people lived in a neighborhood near a hazardous waste site. A similar study by the *National Law Journal* found that polluters in minority communities paid 54 percent lower fines and the EPA took 20 percent longer to place toxic sites on the national priority action list.

1992

EPA created the Office of Environmental Justice to examine the issue of environmental justice in all agency policies and programs. EPA reported that low-income and minority communities were more likely to be exposed to lead, contaminated fish, air pollution, hazardous waste, and agricultural pesticides.

1994

President Clinton signed Executive Order 12898, a directive requiring all federal agencies to begin taking environmental justice into account. The order specified that "Each Federal agency shall make achieving environmental justice part of its mission by identifying and addressing, as appropriate, disproportionately high and adverse human health or environmental effects of its programs, policies, and activities on minority populations and low-income populations."

2003

The U.S. Commission on Civil Rights, an independent group charged with monitoring federal civil rights enforcement, issued to Congress its report titled "Not in My Backyard," which found that several federal agencies (EPA, DOT, HUD, and DOI) have failed to fully implement the 1994 Environmental Justice Executive Order.

2004

The American Bar Association Special Committee on Environmental Justice published *Environmental Justice For All: A Fifty State Survey Of Legislation, Policies, and Initiatives* (2004). The report identifies the statutes, policies, initiatives, or other commitments that states have undertaken to give force of law and/or tangible meaning to the goal of environmental justice.

A March 2004 Office of Inspector General (OIG) report, *EPA Needs to Consistently Implement the Intent of the Executive Order on Environmental Justice,* summed up the treatment of environmental justice under the Bush administration. After a decade, EPA "has not developed a clear vision or a comprehensive strategic plan, and has not established values, goals, expectations, and performance measurements" for integrating environmental justice into its day-to-day operations.

2005

A July 2005 U.S. Government Accountability Office report, *Environmental Justice: EPA Should Devote More Attention to Environmental Justice When Developing Clean Air Rules,* criticized EPA for its handling of environmental justice issues when drafting clean air rules.

Hurricane Katrina hit New Orleans. In the wake of the storm, authorities received reports of 575 oil and toxic chemical spills. Of these, ten major oil spills resulted in a total volume approaching 8 million gallons. The hurricane also generated more than 100 million cubic yards of debris—enough to cover 1,000 football fields with a six-story-high mountain of trash. The massive amounts of debris resulted in hastily permitted urban landfills being established near residential areas of people of color.

2007

The United Church of Christ released *Toxic Wastes and Race at Twenty,* an update of the landmark study in 1987.

Using new methods, this report found "that racial disparities in the distribution of hazardous wastes are greater than previously reported. In fact, these methods show that people of color make up the majority of those living in host neighborhoods within 3 kilometers (1.8 miles) of the nation's hazardous waste facilities."

2009

EPA launched The State Environmental Justice Cooperative Agreement program awarding Alaska, California, Illinois, Pennsylvania, and South Carolina a total of $800,000.

EPA sponsored *Strengthening Environmental Justice Research and Decision Making: A Symposium on the Science of*

Disproportionate Environmental Health Impacts. The symposium was designed to lay the groundwork for developing a systematic and scientifically defensible approach for incorporating environmental justice concerns into EPA's decision-making process.

2010

EPA sponsored the *Conference on Environmental Justice, Air Quality, Goods Movement, and Green Jobs in New Orleans.*

2013

The U.S. Department of Health and Human Services (HHS) released its Environmental Justice Implementation Progress Report. The report identifies the role and contributions of the HHS to environmental justice in the areas of stakeholder engagement; policy development and dissemination; education and training; research and data collection; analysis; and services.

2.5 Societal Environmental Ethics

The environmental ethic expressed by a society is a product of the decisions and choices made by a diverse range of social actors that includes individuals, businesses, and national leaders. Western developed societies have long acted as if the Earth has unlimited reserves of natural resources, an unlimited ability to assimilate wastes, and a limitless ability to accommodate unchecked growth.

The economies of developed nations have been based on a rationale that favors continual growth. Unfortunately, this growth has not always been carefully planned or even desired. This "growth mania" has resulted in the unsustainable use of nonrenewable resources for comfortable homes, well-equipped hospitals, convenient transportation, fast-food outlets, VCRs, home computers, and battery-operated toys, among other things. In economic terms, such "growth" measures out as "productivity." But the question arises, "What is enough?" Poor societies have too little, but rich societies never say, "Halt! We have enough." The Indian philosopher and statesman Mahatma Gandhi said, "The Earth provides enough to satisfy every person's need, but not every person's greed."

Until the last quarter of the twentieth century, **economic growth** and **resource exploitation** were by far the dominant orientations toward the natural environment in industrialized societies. Developing countries were encouraged to follow similar anthropocentric paths. Since the rise of the modern environmental movement in the last 40 years, things have started to change. Some of the most dramatic changes have occurred in corporate business practices.

2.6 Corporate Environmental Ethics

The enormous effects of business on the state of the environment highlight the important need for corporate environmental ethics.

Corporations are legal entities designed to operate at a profit. They possess certain rights and privileges, such as the right to own

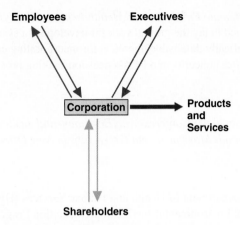

FIGURE 2.6 Corporate Obligations Corporate obligations do not involve the environment unless the shareholders, employees, or executives demand it.

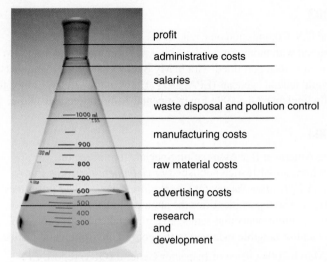

Chemical industry

FIGURE 2.7 Corporate Decision Making Corporations must make a profit. When they look at pollution control and waste disposal they view its cost like any other cost: and reductions in cost increase profits.

property and the limited liability of their shareholders. Though in some legal senses corporations operate as "artificial persons," a corporation's primary purpose is neither to benefit the public nor to protect the environment but to generate a financial return for its shareholders. This does not, however, mean that a corporation has no ethical obligations to the public or to the environment. Shareholders can demand that their directors run the corporation ethically. (See figure 2.6.)

In business, incorporation allows for the organization and concentration of wealth and power far surpassing that of individuals or partnerships. Some of the most important decisions affecting our environment are made not by governments or the public but by executives who wield massive corporate power.

Waste and Pollution

The daily tasks of industry, such as procuring raw materials, manufacturing and marketing products, and disposing of wastes, cause large amounts of pollution. This is not because any industry or company has adopted pollution as a corporate policy. It is simply a fact that industries consume energy and resources to make their products and that they must sell those products profitably to exist.

When raw materials are processed, some waste is usually inevitable. It is often hard to completely control all the by-products of a manufacturing process. Some of the waste material may simply be useless. For example, the food-service industry uses energy to prepare meals. Much of the energy is lost as waste heat. Smoke and odors are released into the atmosphere and spoiled food items must be discarded. Heat, smoke, and food wastes appear to be part of the cost of doing business in the food industry. The cost of controlling this waste can be very important in determining a company's profit margin.

The cheaper it is to produce an item, the greater the possible profit. Ethics are clearly involved when a corporation cuts corners in production quality or waste disposal to maximize profit without regard for public or environmental well-being. Often it is cheaper in the short run to dump wastes into a river than to install a wastewater treatment facility, and it is cheaper to release wastes into the air than it is to trap them in filters. Actions such as these are known as

external costs, since the public or the environment, rather than the corporation, pays these costs. Many people consider such pollution unethical and immoral, but on a corporate balance sheet it can look like just another of the factors that determine **profitability.** (See figure 2.7.) Because stockholders expect a return on an investment, corporations are often drawn toward making decisions based on short-term profitability rather than long-term benefit to the environment or society.

Is There a Corporate Environmental Ethic?

Corporations have certainly made more frequent references to environmental issues over the past several years. Is such concern only rhetoric and social marketing (also called "greenwashing"), or is it the beginning of a new corporate environmental ethic?

Greenwashing is a form of corporate misrepresentation whereby a company presents a green public image and publicizes green initiatives while privately engaging in environmentally damaging practices. Companies are trying to take advantage of the growing public concern and awareness about environmental issues by creating an environmentally responsible image. Greenwashing can help companies win over investors, create competitive advantage in the marketplace, and convince critics that the company is well intentioned. Examples of greenwashing could include: European McDonald's changing the color of their logos from yellow and red to yellow and green; food products that have packaging that evokes an environmentally friendly image even though there has been no attempt made at lowering the environmental impact of its production; and BP, the world's second-largest oil company rebranding its slogan to "beyond petroleum" with a green and yellow sunburst design for their logo.

Although some corporations only want to appear green, others have taken a more ethical approach. Corporations face real choices between using environmentally friendly or harmful production processes. As the idea of an environmental ethic has become more firmly established within society, corporations are being increasingly pressured to adopt more environmentally and socially responsible practices. Real improvements have been made. For example, Ray Anderson of Interface Incorporated led the way in greening the carpet industry by reducing the amount of waste produced by his company by 75 percent. The International Organization for Standardization (www.iso.org) has developed a program it calls ISO 14000 to encourage industries to adopt the most environmentally sensitive production practices.

Reaction by the business community to the 1989 *Exxon Valdez* oil spill in Alaska is a good example of the mixed responses of corporations to the ethical challenges. (See figure 2.8.) The Oil Pollution Act (OPA) was passed in 1990 to reduce the environmental impact of future oil spills and has resulted in a 94 percent reduction in spills. One of the new regulations in OPA was that all large tankers must have double hulls or be phased out of service by 2010. To get around the law, however, many oil carriers shifted their oil transport operations to lightly regulated oil barges. This reduction in oil spill safety led to several barge oil spills including an oil barge that hit a tanker in July 2008 on the Mississippi River near New Orleans. About 1.6 million liters of oil were released.

A group of environmentalists, investors, and companies formed the Coalition for Environmentally Responsible Economics (CERES) group in 1989 and created a set of 10 environmental standards by which their business practices could be measured called the CERES Principles (they were first called the Valdez Principles). CERES companies pledge to voluntarily go beyond the requirements set by the law to strive for environmental excellence through business practices that: (1) protect the biosphere, (2) sustainably use natural resources, (3) reduce and dispose of waste safely, (4) conserve energy, (5) minimize environmental

risks through safe technologies, (6) reduce the use, manufacture, and sale of products and services that cause environmental damage, (7) restore environmental damage, (8) inform the public of any health, safety, or environmental conditions, (9) consider environmental policy in management decisions, and (10) report the results of an annual environmental audit to the public. Today, over 80 companies have publicly endorsed the CERES Principles, including many Fortune 500 firms. In addition, CERES coordinates an investor network, and worldwide, CERES firms have won environmental awards from many organizations in recognition of their approach.

In 1997, the Global Reporting Initiative (GRI) was established. Convened by CERES in partnership with the United Nations Environment Programme, the GRI encourages the active participation of corporations, nongovernmental organizations, accountancy organizations, business associations, and other stakeholders from around the world. The mission of the GRI is to develop globally applicable guidelines for reporting on economic, environmental, and social performance, initially for corporations and eventually for any business, governmental, or nongovernmental organization.

The GRI's *Sustainability Reporting Guidelines* were released in draft form in 1999. The GRI guidelines represent the first global framework for comprehensive sustainability reporting, encompassing the "triple bottom line" of economic, environmental, and social issues.

Improved disclosure of sustainability information is an essential ingredient in the mix of approaches needed to meet the governance challenges in the globalizing economy. Today, at least 2,000 companies around the world voluntarily report information on their economic, environmental, and social policies, practices,

(a)

(b)

FIGURE 2.8 Oil Spill Response The 1989 oil spill in Prince William Sound, Alaska, led to the development of the CERES Principles. (a) The otter is a victim of the spill. (b) Cleaning oil from the Exxon Valdez spill on a beach in Prince William Sound.

and performance. Unfortunately, this information can sometimes be inconsistent and incomplete. Measurement and reporting practices vary widely according to industry, location, and regulatory requirements. The GRI's *Sustainability Reporting Guidelines* are designed to address some of these challenges. Currently, 392 organizations in 33 countries follow GRI's guidelines.

Green Business Concepts

There will be little political enthusiasm for preserving the environment if preservation results in national economic collapse. Neither does it make sense to maintain industrial productivity at the cost of breathable air, drinkable water, wildlife species, parks, and wildernesses. Environmental advocates should consider the corporation's need to make a profit when they demand that businesses take more account of the environmental consequences of their actions. Corporations should recognize that profits must come neither at the cost of the health of current and future generations nor at the cost of species extinctions. **Natural capitalism** is the idea that businesses can both expand their profits and take good care of the environment. Natural capitalism works. The 3M Company is estimated to have saved US $500 million over the last 20 years through its Pollution Prevention Pays (3P) program. Innovations in **ecological** and **environmental economics** promote accounting techniques that make visible all of the social and environmental costs of doing business so that these costs can no longer be externalized by corporate decision makers.

In the mid-1990s, a concept called **industrial ecology** emerged that links industrial production and environmental quality. One of the most important elements of industrial ecology is that it models industrial production on biological production. Industrial ecology forces industry to account for where waste is going. Dictionaries define *waste* as useless or worthless material. In nature, however, nothing is discarded. All materials ultimately get re-used. Industrial ecology makes it clear that discarding or wasting materials taken from the Earth at great cost is a short-sighted view. Materials and products that are no longer in use could be termed *residues* rather than *wastes*. Residues are materials that our economy has not yet learned to use efficiently. In this view, a pollutant is a resource out of place. In industrial ecology, good environmental practices are good economics. It forces us to view pollution and waste in a new way.

Another green business concept is the **triple bottom line.** The triple bottom line has been referred to as the ethical criteria for business success. The traditional measure of success for business has always been profit. The bottom line has typically been a purely financial one. The triple bottom line concept, which is growing in popularity, is about gauging corporate success on three fronts: financial, social, and environmental (also sometimes called people, planet, and profit). In other words, executives are concerned not only about money but also about the impact their business actions have on people and the planet. Proponents of the triple bottom line say that only by making these values part of the core of business operations can companies survive in our changing world. Some businesses have voluntarily adopted a triple bottom line as part of their articles of incorporation or bylaws, and some have advocated for state laws creating a "sustainable corporation" that would grant triple bottom line businesses benefits such as tax breaks.

The environmental movement has effectively influenced public opinion and started to move the business community toward greater environmental responsibility. More complex and stringent environmental and public safety demands will increasingly influence corporate decisions throughout this century. Business itself will expand its horizons and find new ways to make profits while promoting environmental benefits and minimizing environmental costs. However, as population increases and as consumption levels rise around the globe, the environmental burden on the planet will inevitably continue to rise. Changes in business practices alone will not be enough.

2.7 Individual Environmental Ethics

Ethical changes in society and business must start with individuals. We have to recognize that our individual actions have a bearing on environmental quality and that each of us bears some personal responsibility for the quality of the environment in which we live. In other words, environmental ethics must express themselves not only in new national laws and in better business practices but also in significant changes in the ways in which we all live.

Various public opinion polls conducted over the past decade have indicated that Americans think environmental problems can often be given a quick technological fix. The Roper polling organization found that the public believes that "cars, not drivers, pollute, so business should invent pollution-free autos. Coal utilities, not electricity consumers, pollute, so less environmentally dangerous generation methods should be found." It appears that many individuals want the environment cleaned up, but they do not want to make major lifestyle changes to make that happen.

While new technologies will certainly play a major role in the future in lessening the environmental impact of our lifestyles, individual behavioral choices today can also make a significant difference to the health of ecological systems. Environmental ethics must therefore take hold not only at the level of government and business but also at the level of personal choices about consumption.

2.8 The Ethics of Consumption

In 1994, when delegates from around the world gathered in Cairo for the International Conference on Population and Development, representatives from developing countries protested that a baby born in the United States will consume during its lifetime at least 20 times as much of the world's resources as an African or Indian baby. The problem for the world's environment, they argued, is over consumption in the Northern Hemisphere, not just overpopulation in the Southern Hemisphere, China, or India. Do we in the Northern Hemisphere consume too much?

North Americans, only 5 percent of the world's population, consume one-fourth of the world's oil. They use more water and own more cars than anybody else. They waste more food than most people in sub-Saharan Africa eat. It has been estimated that

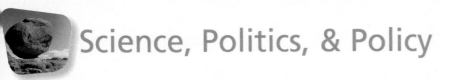
Should Environmental Scientists Be Advocates for Environmental Policy?

Should environmental scientists be advocates for environmental policy? This question has been at the heart of policy debates since the environment became a major policy agenda item. A growing number of environmental scientists are now answering "yes." They argue that scientists, by virtue of being citizens first and scientists second, have a responsibility to advocate to the best of their abilities and in a justified and transparent manner.

Much of what has been written about advocacy looks at its appropriateness without adequately assessing its nature. For example, most of the arguments, whether they are for or against advocacy, often look at only one side of the debate. The question of scientific credibility is often raised. Those who oppose scientists acting as advocates often say that advocacy undermines a scientist's credibility. The counter argument could say that as long as a scientist's work is transparently honest, the scientific community is obligated to support it. Scientific credibility, however, is not the same thing as effectiveness. One may have scientific credibility and be effective or ineffective at advocacy.

There is one area about which many scientists do agree: that as citizens first and scientists second, scientists have a responsibility to use their scientific data and insights to guide policy decisions. The ethicist and the scientist call it an ideal blend of philosophical ethics and scientific commitment to data collection and analysis. A growing number of scientists are calling for more active participation of their profession in matters of policy. They argue that broad participation of scientists in policy issues will undoubtedly result in disagreement among good scientists and will complicate the policy-making process. It is further argued, however, that

the ultimate goal should not be simplicity in the process but rather the betterment of society. Many scientists state that the risk of not participating in policy debates on environmental issues is greater than the risk of participating.

What are your thoughts? Should environmental scientists be policy advocates? Does involvement in policy issues compromise the role of a scientist or enhance it?

---

if the rest of the world consumed at the rate at which people in the United States consume, we would need five more Planet Earths to supply the resources.

Food

Two centuries ago, Thomas Malthus declared that worldwide famine was inevitable as human population growth outpaced food production. In 1972, a group of scholars known as the Club of Rome predicted much the same thing for the waning years of the twentieth century. It did not happen because—so far, at least—human ingenuity has outpaced population growth.

Fertilizers, pesticides, and high-yield crops have more than doubled world food production in the past 40 years. The reason 850 million people go hungry today and 6 million children under the age of 5 die each year from hunger-related causes is not that there is not enough food in the world but that social, economic, and political conditions make it impossible for those who need the food to get it. This tragedy is made more troubling by the fact that in 2000 the world reached the historic landmark of there being the same number of overweight people as those that were malnourished.

Norman Borlaug, who won the Nobel Peace Prize in 1970 for his role in developing high-yield crops, predicts that genetic engineering and other new technologies will keep food production ahead of population increases over the next half century.

New technologies, however, are not free from controversy. The Mexican government recently confirmed that genetically modified corn has escaped into native corn populations, and the European Union ended a five-year ban on U.S. imports by requiring labeling of all foods containing more than 0.9 percent materials from genetically modified organisms.

Adding to the uncertainty about future food production are factors that include decreasing soil fertility caused by repeated chemical applications, desertification and erosion caused by poor farming techniques, and the loss of available cropland as a result of urbanization. The increasing evidence of rapid global climate change also makes extrapolations from past harvests increasingly unreliable indicators of the future. With global population set to peak at around 9 billion people in the middle of the twenty-first century, it remains unclear whether there will be enough food to go around. Even if it turns out that enough food can be produced for the world in the twenty-first century, whether everybody will get a fair share is much less certain.

Energy

If everybody on Earth consumed as much oil as the average American, the world's known reserves would be gone in about 40 years. Even at current rates of consumption, known reserves will not last through the current century. Technological optimists,

Going Green

Do We Consume Too Much?

The desire to consume is nothing new. It has been around for millennia. People need to consume to survive. However, consumption has evolved as people have found new ways to help make their lives simpler and/or to use their resources more efficiently. We consume a variety of resources and products today as we move beyond meeting basic needs to include luxury items and technological innovations to improve efficiency. Such consumption beyond minimal and basic needs is not necessarily a bad thing in and of itself—throughout history we have always sought to find ways to make our lives a bit easier to live. However, increasingly there are important issues around consumerism that need to be understood. For example:

- How are the products and resources we consume actually produced?
- What are the impacts of that process of production on the environment, on society, and on individuals?
- What are the impacts of certain forms of consumption on the environment, on society, and on individuals?

- What is a necessity and what is a luxury?
- Businesses and advertising are major engines in promoting the consumption of products so that they may survive. How much of what we consume is influenced by their needs versus our needs?
- How do material values influence our relationships with other people?
- How does the way we value material posessions influence our relationships with other people?

We can likely think of numerous other questions as well. We can, additionally, see that consumerism and consumption are at the core of many, if not most, societies. The impacts of consumerism—positive and negative—are very significant in all aspects of our lives, as well as on our planet.

---

however, tell us not to worry. New technologies, they say, will avert a global energy crisis.

Already oil companies have developed cheaper and more efficient ways to find oil and extract it from the ground, possibly extending the supply into the twenty-second century. In many regions of the world, natural gas is replacing oil as the primary source of domestic and industrial power. New coal gasification technologies also hold promise for cleaner and extended fossil fuel power. However, it is impossible to ignore the fact that there is a finite amount of fossil fuel on the planet. These fuels cannot be the world's primary power source forever. Even before fossil fuels run out, concerns about global warming may compel the world to stop burning them. Furthermore, since fossil fuels are found in abundance only in particular parts of the world, geopolitical events can suddenly cause fuel prices to spike in ways that can be disastrous for national economies. Natural disasters can also destroy infrastructure and add to the uncertainty surrounding fossil fuel supply.

The more foresighted energy companies are already looking ahead by investing in the technologies that will replace fossil fuels. In some countries, the winds of political change have brought nuclear power back onto the table. In others, solar, wind, wave, and biomass technologies are already meeting increasing proportions of national energy needs. With accelerating global demand, it remains unclear whether there will be enough clean energy supply to meet the world's needs in the years ahead.

Water

The world of the future may not need oil, but without water, humanity could not last more than a few days. Right now, humans use about half the planet's accessible supply of renewable freshwater—the supply regenerated each year and available

for human use. A simple doubling of agricultural production with no efficiency improvements would push that fraction to about 85 percent. Unlike fossil fuels, which could eventually be replaced by other energy sources, there is no substitute for water.

Given its fundamental role for all human survival and the antiquity of our cultural reflection on its importance, one might have expected humans to have developed a broad consensus of thought or a measure of cumulative wisdom about water usage in the ecosystem. But what we might call a *water ethic*—a set of common understandings, shared values, and widely accepted norms governing how humans ought to behave with reference to water—does not appear to be widely thought of in contemporary human affairs. Water itself is far from uniformly appreciated. Some cultures extol its value as priceless, while others behave as if it were worthless.

We live on what has been called the "water planet," yet over 99 percent of Earth's water is either saline or frozen. Humankind depends upon the remaining 1 percent for its survival. Competition for that 1 percent has already become intense in many parts of the world, and even those who live in water-abundant regions are becoming conscious of water as a precious asset. Beyond valuing water sources, however, we are only just beginning to become aware of the downstream impact that our water habits are having upon whole communities of life-forms that inhabit lake, river, estuary, and marine environments, some of which may prove vital for our own survival.

Regarding the world's major rivers, the world atlas no longer tells the truth. Today, dozens of the greatest rivers are dry long before they reach the sea. They include the Nile in Egypt, the Yellow River in China, the Indus in Pakistan, the Rio Grande and the Colorado in the United States, the Murray in Australia, and the Jordan, which is emptied before it can even reach the country that bears its name.

Recent droughts in parts of the United States have focused interest on allocation of water resources for basic needs. In the ongoing California drought, while many cities have been passing emergency water rationing and increasing basic costs to consumers, other cities have been permitting potable water to irrigate golf courses and allowing new swimming pools to be built. Who makes the decisions on how water is allocated? How can a state as diverse as California make ethical decisions about water that treat everyone equally and fairly?

In addition to its role as the substance and medium for all life-forms, water needs to be respected as a geologic force. In the face of recent extreme weather events, cultures around the world are realizing that as humans we did not create, nor can we control, the hydrologic cycle. This can be difficult for technological cultures accustomed to the illusion of dominance over nature. The multiple roles that water plays in the evolving climate system are being investigated with an eye to the future. Any shift in climate is likely to involve new patterns of global water distribution, and these new circumstances will in turn require humankind to devise new habits, policies, and ethical norms to govern water usage. Perhaps now is the time that we begin to discuss a new global water ethic.

More than any other resource, water may limit consumerism during the next century. "In the next century," World Bank Vice President Ismail Serageldin predicted a few years ago, "wars will be fought over water."

Wild Nature

Every day in the United States, somewhere from 1,000 to 2,000 hectares of farmland and natural areas are permanently lost to development. As more and more people around the world achieve modern standards of living, the land area converted to houses, shopping malls, roads, and industrial parks will continue to increase. The planet will labor under rising levels of resource extraction and pollution. Tropical rainforests will be cut and wild lands will become entombed under pavement. Mighty rivers like the Yangtze and Nile, already dammed and diverted, will become even more canal-like. As the new century progresses, more and more of us will live urbanized lives. The few pockets of wild nature that remain will be biologically isolated from each other by development. We will increasingly live in a world of our own making.

2.9 Personal Choices

Threats to supplies of food, energy, water, and wild nature certainly require action on national and international levels. However, in each of these cases ethically responsible action can also begin with the individual. Individuals committed to a strong environmental ethic can make many lifestyle changes to significantly reduce their personal impact on the planet. Food choice is one place to start. Eating food that is produced locally, is low on the food chain, and is grown with a minimum of chemical fertilizers and pesticides not only reduces the environmental impact of food production; it might also lead to better health. Heart disease and certain cancers are increasingly being linked to diets high in animal fats. Buying durable consumer products and reusing or repairing products that still have useful life in them reduces the amount of raw materials that have to be extracted from the ground to meet your needs. Conserving energy at home and on the road can help lessen the amount of fossil fuels you use to support your lifestyle. Living in town rather than out in the suburbs, lobbying for the protection of wild areas, and voting for officials who take environmental issues seriously are all ways that your own environmental ethic can directly contribute to a reduced environmental impact. Consumer behavior is a vote for things you believe in. Lifestyle choices are an expression of your ethical commitments.

The concept of an **ecological footprint** has been developed to help individuals measure their environmental impact on the Earth. One's ecological footprint is defined as "the area of Earth's productive land and water required to supply the resources that an individual demands, as well as to absorb the wastes that the individual produces." Websites exist that allow you to estimate your ecological footprint and to compare it to the footprint of others by answering a few questions about your lifestyle. Running through one of these exercises is a good way to gain a sense of personal responsibility for your own environmental impact. To learn more about ecological footprints visit:

http://www.earthday.org/Footprint/info.asp
http://www.rprogress.org/newprojects/ecolFoot

Finally, think about the words of anthropologist Margaret Mead, who vividly drew attention to the importance of individual action when she stated, "Never doubt that a small group of thoughtful, committed citizens can change the world. Indeed, it's the only thing that ever has."

2.10 Global Environmental Ethics

As human stresses on the environment increase, the stability of the planet's ecological systems becomes more and more uncertain. Small environmental changes can create large-scale and unpredictable disruptions. Increased atmospheric carbon dioxide and methane, whether caused by humans or not, are leading to changes in surface temperatures that will result in major ecological effects. Feedback loops add to the urgency. For example, just a small reduction in seasonal snow and ice coverage in Arctic regions due to global warming can greatly increase the amount of solar energy the Earth absorbs. This additional energy itself raises atmospheric temperature, leading to a further reduction in snow coverage. Once established, this feedback loop continually reinforces itself. Some models predict that ocean currents, nutrient flows, and hydrologic cycles could make radical shifts from historic patterns in a matter of months. Such disruptions would cause catastrophic environmental change by shifting agricultural regions, threatening species with extinction, decimating crop harvests, and pushing tropical diseases into areas where they are currently unknown. Glaciers will continue to melt and ocean waters will rise, flooding heavily populated low-lying places like Bangladesh, the Netherlands, and

Approximate relative ecological footprint of one person
in the developing world.

Approximate relative ecological footprint of one person
in the developed world.

FIGURE 2.9 Lifestyle and Environmental Impact Significant differences in lifestyles and their environmental impact exist between the rich and poor nations of the world. The ecological footprint of those in the industrialized world can be 20 to 40 times higher than the footprint of those in the developing world. What would be the environmental impact on the Earth if the citizens of China and India and other less-developed countries enjoyed the standard of living in North America? Can we deny them that opportunity?

even parts of Florida and the U.S. Gulf Coast. Millions of people would be displaced by famine, flood, and drought.

As environmental justice advocates point out, these changes will hit the poor and those least able to respond to them first. However, the changes predicted are of such magnitude that even the very wealthy countries will suffer environmental consequences that they cannot hope to avoid. These scenarios are not distant, future worries. They are here now. The year 2012 was one of the top two hottest years on record. The record massive storms of 2012 in the United States may be just a small indication of what lies in store.

Many of these problems require global solutions. In 1990, Noel Brown, the director of the United Nations North American Environmental Programme, stated:

Environmental Disasters and Poverty

On August 29, 2005, Hurricane Katrina came ashore on the U.S. Gulf Coast between Mobile, Alabama, and New Orleans, Louisiana. Katrina was an enormous hurricane, just one of the 26 named storms that hit the Americas in the worst Atlantic hurricane season in history. A few hours after the hurricane made landfall, the combination of the storm surge and the torrential rain falling inland overwhelmed the levees that were supposed to protect New Orleans. Up to 80 percent of the city flooded. Close to 1,000 people died in Louisiana alone, with most of those deaths occurring in New Orleans. Mandatory evacuation orders were issued for New Orleans' 500,000 residents in the days that followed the storm.

The devastation caused by Hurricane Katrina starkly illustrated the way in which environmental destruction can cause particular hardship for the poor. Twenty-eight percent of New Orleans residents lived below the poverty line. Information about the hurricane and about evacuation options was harder for the city's poorer residents to access. Compounding the problem, a large number of these poorer households did not own cars. As environmental justice advocate Robert Bullard pointed out, without a car, a driver's license, or a credit card, even a timely evacuation order can be extremely hard to obey. As Hurricane Katrina battered the city, poor and minority residents were generally left behind, forced to flee to crowded and unsanitary temporary shelters like the New Orleans Superdome.

Outside of the New Orleans city center, some of the areas hardest hit by the hurricane were home to many chemical and petrochemical plants. The location of those plants on the Mississippi River close to African-American communities is a legacy of the area's social and economic history that raises its own environmental justice concerns. In the wake of Hurricane Katrina, damage to the industrial infrastructure threatened homes with floodwaters that were laced with a toxic stew of chemicals. Those people left behind were often forced to wade through the contaminated waters to safety. Their saturated homes required demolition.

The images of hardship, suffering, and death in Louisiana that were broadcast worldwide after the disaster illustrated how, even in wealthy countries, the burden of environmental disasters falls particularly hard on the poor, the sick, and the elderly. By 2011 the effects of Katrina were still being felt. While there has been considerable rebuilding of communities there is still much to be done especially in the poorer sections of New Orleans. The lasting effects of Katrina also have been seen in physical health problems. Beyond the short-term concerns over contaminated water, some of the problems that continue to show up in Gulf Coast hospitals include certain skin infections and respiratory problems. In addition, elevated concentrations of lead, arsenic, and other toxic chemicals are present throughout New Orleans, particularly in the poorer areas of the city. It is suggested that widespread cleanup efforts and demolition stirred up airborne toxins known to cause adverse health effects.

- Do you feel there is a direct link between environmental disasters and poverty?
- If you feel there is, can you suggest ways to minimize that effect?

Hurricane Katrina heading toward New Orleans on August 28, 2005.

Flooding in New Orleans immediately after Hurricane Katrina.

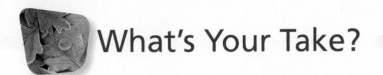

What's Your Take?

The grizzly bear (*Ursus arctos horribilis*) has been receiving federal protection under the U.S. Endangered Species Act for over 30 years. The federal government has now proposed removing that protection on the basis of increasing numbers of bears in the Greater Yellowstone Ecosystem and elsewhere. There is considerable disagreement among conservation biologists about how many bears are needed for the species to be "recovered." When a species is delisted, management is handed over to individual states. If it proposes an acceptable plan, a state may introduce a management plan that includes the hunting of the previously listed species. What kind of ethic underlies the Endangered Species Act? What kind of ethic underlies a management plan that includes hunting? Develop an ethical argument for or against the delisting of the grizzly bear.

Suddenly and rather uniquely the world appears to be saying the same thing. We are approaching what I have termed a consensual moment in history, where suddenly from most quarters we get a sense that the world community is now agreeing that the environment has become a matter of global priority and action.

This new sense of urgency and common cause about the environment is leading to unprecedented cooperation in some areas. Despite their political differences, Arab, Israeli, Russian, and American environmental professionals have been working together for several years. Ecological degradation in any nation almost inevitably impinges on the quality of life in others. For years, acid rain has been a major irritant in relations between the United States and Canada. Drought in Africa and deforestation in Haiti have resulted in waves of refugees. From the Nile to the Rio Grande, conflicts flare over water rights. The growing megacities of the Third World are time bombs of civil unrest.

Much of the current environmental crisis is rooted in and exacerbated by the widening gap between rich and poor nations. Industrialized countries contain only 20 percent of the world's population, yet they control 80 percent of the world's goods and create most of its pollution. (See figure 2.9.) The developing countries are hardest hit by overpopulation, malnutrition, and disease. As these nations struggle to catch up with the developed world and improve the quality of life for their people, a vicious circle begins: Their efforts at rapid industrialization poison their cities, while their attempts to boost agricultural production often result in the destruction of their forests and the depletion of their soils, which lead to greater poverty.

Perhaps one of the most important questions for the future is, "Will the nations of the world be able to set aside their political differences to work toward a global environmental course of action?" The United Nations Conference on Human Environment held in Stockholm, Sweden, in 1972 was a step in the right direction. Out of that international conference was born the UN Environment Programme, a separate department of the United Nations that deals with environmental issues. A second world environmental conference was held in 1992 in Brazil and a third in 2002 in South Africa and again in Brazil in 2012. Each followed up the Stockholm conference with many new international initiatives. A major world conference on climate change was held in Kyoto, Japan, in 1997.

Through organizations and conferences such as these, nations can work together to solve common environmental problems.

Even those who doubt the accuracy of the gloomiest predictions about our environmental future can hardly deny that we live on a changing planet. Uncertainty about the ecological baseline that we will be dealing with 50—or even 5—years from now means that it would be wise to think hard about the environmental consequences of our actions. As John Muir and Lewis Thomas expressed in their remarks quoted at the beginning of this chapter, humans should expect that many of their actions on planet Earth will in time affect something or somebody else. Anytime our actions may harm another, we face serious ethical questions. Environmental ethics suggests that we may have an obligation beyond simply minimizing the harm that we cause to our families, our neighbors, our fellow human citizens, and future generations

FIGURE 2.10 Wangari Maathai. Dr. Maathai was awarded the 2004 Nobel Peace Prize. She began the Green Belt Movement in Africa empowering women in grassroots organizing around tree planting.

of people that will live on Planet Earth. It suggests that we may also have an obligation to minimize the harm we cause to the ecological systems and the biodiversity of the Earth itself. The ecological systems, many now believe, deserve moral consideration for what they are in themselves, quite apart from their undeniable importance to human beings. Recognizing that our treatment of the natural environment is an ethical issue is a good start on the challenges that lie ahead.

Nobel Peace Prize winner Wangari Maathai (see figure 2.10) captured these sentiments about ethics in the speech that she gave shortly after receiving her prize:

Today we are faced with a challenge that calls for a shift in our thinking, so that humanity stops threatening its life-support system. We are called to assist the Earth to heal her wounds and in the process heal our own—indeed, to embrace the whole creation in all its diversity, beauty, and wonder. This will happen if we see the need to revive our sense of belonging to a larger family of life, with which we have shared our evolutionary process. In the course of history, there comes a time when humanity is called to shift to a new level of consciousness, to reach a higher moral ground. A time when we have to shed our fear and give hope to each other. That time is now.

Summary

People of different cultures view their place in the world from different perspectives. Among the things that shape their views are religious understandings, economic pressures, geographic location, and fundamental knowledge of nature. Because of this diversity of backgrounds, different cultures put different values on the natural world and the individual organisms that compose it. Environmental ethics investigates the justifications for these different positions.

Three common attitudes toward nature are the development approach, which assumes that nature is for people to use for their own purposes; the preservationist approach, which assumes that nature has value in itself and should be preserved intact; and the conservationist approach, which recognizes that we must use nature to meet human needs but encourages us to do so in a sustainable manner. The conservationist approach is generally known today as "sustainable development."

Ethical obligations to the environment are usually closely connected to ethical obligations toward people, particularly poor people and minority groups. Environmental justice is about ensuring that no group is made to bear a disproportionate burden of environmental harm. Environmental justice is also about ensuring that governments develop and enforce environmental regulations fairly across different segments of society. The environmental justice movement has forced environmentalists to recognize that you cannot think about protecting nature without also thinking about people.

Recognition that there is an ethical obligation to protect the environment can be made by corporations, by individuals, by nations, and by international bodies. Corporate environmental ethics are complicated by the existence of a corporate obligation to its shareholders to make a profit. Corporations often wield tremendous economic power that can be used to influence public opinion and political will. Many corporations are now being driven to include environmental ethics in their business practices by their shareholders. Natural capitalism and industrial ecology are ideas that promote ways of doing profitable business while also protecting the environment.

Corporations are composed of individuals. An increasing sensitivity of individual citizens to environmental concerns can change the political and economic climate for the whole of society. Individuals must demonstrate strong commitments to environmental ethics in their personal choices and behaviors. The concept of an ecological footprint has been developed to help individuals gauge their personal environmental impact.

Global commitments to the protection of the environment are enormously important. Accelerating international trade and communication technologies means that the world is getting smaller while the potential impact of humanity on the planet is getting larger and more uncertain. Tens of millions of people are added to the world's population each year while economic development increases the environmental impact of those already here. Opportunities for global cooperation and agreement are of critical importance in facing these real and increasing challenges. Environmental ethics has a role to play in shaping human attitudes toward the environment from the smallest personal choice to the largest international treaty.

Acting Green

1. Calculate your ecological footprint.
2. Participate in sustainability activities in your community or university.
3. Give up the use of your car for part of each week. Walk, bicycle, or take public transportation.
4. Read *Silent Spring, A Sand County Almanac,* and *Walden.*
5. Work with a local business in helping it apply green business concepts.
6. "Adopt" an elderly neighbor. Help them become more sustainable in their yard, home, and food buying decisions.
7. Take a long walk in a natural area. Leave all phones, radios, etc. at home. Listen to the silence.

Review Questions

1. Why does the environmental crisis demand a new ethic?
2. Describe three types of environmental ethics developed by philosophers.
3. Describe three common attitudes toward the environment found in modern society.
4. Why is environmental justice part of the environmental movement?
5. What are the conflicts between corporate behavior and environmental ethics?
6. How can individuals direct business toward better environmental practices?
7. How can individuals implement environmental ethics in their own lives?
8. Where does global environmental ethics fit in the broad scheme of environmental protection?
9. Is the triple bottom line a realistic concept in the corporate world?

Critical Thinking Questions

1. Give three different ethical justifications for protecting a forest using an anthropocentric, a biocentric, and an ecocentric viewpoint.
2. Which approach to the environment—development, preservation, or conservation—do you think you adopt in your own life? Do you think it appropriate for everybody in the world to share the same attitude you hold?
3. What ethical obligations do you personally feel toward wolves and whales? What ethical obligations do you feel toward future generations of people? What ethical obligations do you feel toward future generations of wolves and whales?
4. Until recently, it was generally believed that growth and development were unquestionably good. Now, early in the twenty-first century, some are beginning to question that belief. Are those questions appropriate ones to address to a citizen of the developing world? Would you ever make the argument that development has gone far enough?
5. Imagine you are a business executive who wants to pursue an environmental policy for your company that limits pollution and uses fewer raw materials but would cost more. What might be the discussion at your next board of directors meeting? How would you make your case to your directors and your shareholders?
6. In 1997 Ojibwa Indians in northern Wisconsin sat on the railroad tracks to block a shipment of sulfuric acid from crossing their reservation on its way to a controversial copper mine in Michigan. Try to put yourself in their position. What values, beliefs, and perspectives might have contributed to their actions? Was it the right thing to do? How would you try and mediate a heated conversation between an Ojibwa protestor and a Michigan copper miner?
7. Wangari Maathai led a protest movement against the stripping of African forests. Are environmental issues important enough for citizens to become activists and perhaps break the law? How could an environmental activist respond to the pro-development position that it is more important to feed people and lift them out of poverty than it is to save a few trees? Are there any ethical principles that you think an environmental activist and a pro-development advocate share?
8. Reread Focus On Early Philosophers of Nature. Is the environmental crisis in certain respects a problem in ethics? Does philosophy have a role to play in helping to solve the problem? Should scientists, business leaders, and politicians study environmental ethics? What role is environmental ethics going to play in your own life from this point on?

To access additional resources for this chapter, please visit ConnectPlus® at connect.mheducation.com. There you will find interactive exercises, including Google Earth™, additional Case Studies, and SmartBook™, an adaptive eBook that integrates our LearnSmart® adaptive learning technology.

chapter 3

Risk, Economics, and Environmental Concerns

Many environmental questions are framed by the interrelated concepts of risk and cost. The cost of a helmet and life vest greatly reduces the risk of injury or death when kayaking in rough water.

OBJECTIVES

After reading this chapter, you should be able to:

- Recognize that the probability, consequences, and economic cost of a risk are important in evaluating a risk.
- Describe the tools used to assess risk.
- Understand the difference between risk assessment and risk management.
- Describe examples of risk management activities.
- Understand factors that cause people to have a distorted view of risks.
- Define natural resources and describe the difference between renewable and nonrenewable natural resources.

- Describe why changes in supply or demand affect the price.
- Describe ecosystem services and explain why it is difficult to place an economic value on them.
- Give examples of environmental costs.
- Explain the concepts of deferred costs, external costs, and opportunity costs.
- Describe how pollution costs and pollution-prevention costs differ.
- Understand how and why cost-benefit analysis is used.
- Describe ways in which economic systems and ecological systems differ.

- Explain why resources that are owned and used in common by many people are typically overexploited.
- Give examples of subsidies and market-based approaches that have been used to improve environmental conditions.
- Explain how life cycle analysis and extended product responsibility are used to determine the total cost of producing a good or service.
- Understand the concept of sustainable development.
- Understand market approaches to solving environmental problems.

39

Drinking Water, Sanitation, and Disease

According to the World Health Organization there are about 1.7 billion cases of diarrheal disease each year resulting in about 760,000 deaths of children under the age of five. Diarrhea leads to dehydration, which is the primary cause of death. The major cause of diarrhea is infections of the digestive tract caused by the ingestion of disease organisms in food or water. About 780 million people (11 percent of the population) lack access to safe drinking water and about 2.5 billion people (35 percent of the population) do not have access to toilets or latrines. This means that they defecate on the street, in fields, or near their homes. These two factors—lack of access to safe drinking water and to sanitary facilities—lead to the easy spread of disease from one person to another.

The diseases caused by lack of safe drinking water and sanitation have several economic impacts. The treatment of disease, even though it may be minimal, costs families and governments money. In addition, sick adults are not able to work or may be less productive in their work and thus provide less income for their families. Another economic aspect related to safe drinking water is the time that people expend in getting water to their homes. If people do not have access to safe drinking water in their homes, someone from the household—usually women or children—must walk to a safe water source and carry water back to their homes. To satisfy the basic needs of washing, cooking, and drinking, each person needs about 20 liters of water per day (about 5.2 gallons per person per day). Although this does not seem like much water, consider the following. Twenty liters of water weighs 20 kilograms (44 pounds). A family of four would require 80 kilograms (176 pounds) of water each day. Because of the work and time involved, in general, if people must walk more than 30 minutes round trip to get water, they are not likely to get enough water to satisfy all their daily needs. The time spent by women and children providing families with water detracts from the time they could spend doing more productive activities such as attending school, raising crops, or earning money from other activities.

The simplest way to remove this disease burden is to provide safe drinking water and access to toilets or latrines to people who do not have them. So why is this not done? One of the primary barriers to accomplishing this goal is economic. About 20 percent of the people in the world live on $1.25 per day or less, which makes it impossible for individuals to provide their own safe drinking water supplies and sanitary facilities. Thus, some level of the national or local government or some outside agency must provide the funding. Governments in countries with many poor people are themselves often short on funds and find it difficult to shift money from other programs to initiate programs that provide water and sanitary facilities.

Children in India obtaining water from an unsafe source.

3.1 Making Decisions

Whether we recognize it or not, most life decisions involve a consideration of two factors: risk and cost. We commonly ask such questions as "How likely is it that someone will be hurt physically, emotionally, or economically?" or "What will it cost me if I choose to take this action?" Typically, the two factors of risk and cost are interrelated and must be considered at the same time. For example, risky decisions that lead to physical harm are often reduced to economic terms when medical care costs or legal fees must be paid. Similarly, risky business decisions result in a loss of money when the business fails.

Most decisions related to environmental issues involve characterizing the risks incurred by people and other living things and evaluating the cost of taking action to eliminate the risk as well as the cost of doing nothing. For example, suppose an air pollution regulation is proposed to reduce the public's exposure to a chemical that is thought to cause disease in a small percentage of the exposed public. In such a case, industry will be sure to point out that it will cost a considerable amount of money to put these controls in place and will reduce their profitability. Citizens also may point out that their tax money will have to support a larger governmental bureaucracy to ensure that the regulations are followed. On the other side, advocates will point out that the regulations will lead to reduced risk of illness and reduced health care costs for people who live in areas affected by the pollutant. Since understanding risk and economics is important to evaluating options to address environmental issues, we will explore the nature of risk and economics in the following sections.

3.2 Characterizing Risk

Risk is the probability that a condition or action will lead to an injury, damage, or loss. When we consider any activity or situation that poses a risk, we generally think about three factors: the probability of a bad outcome, the consequences of a bad outcome, and the cost of dealing with a bad outcome.

- **Probability** is a mathematical statement about how likely it is that something will happen. Probability is often stated in terms like, "The probability of developing a particular illness is 1 in 10,000," or "The likelihood of winning the lottery is 1 in 5 million." It is important to make a distinction between *probability* and *possibility*. When we say something is *possible*, we are just saying that it could occur. It is a very inexact term. *Probability* specifically defines in mathematical terms how likely it is that a *possible* event will occur.
- The *consequences* of a bad outcome resulting from accepting a risk may be minor or catastrophic. For example, ammonia is a common household product. Exposure to ammonia will result in 100 percent of people reacting with watery eyes and other symptoms. The probability of an exposure and the probability of an adverse effect are high; however, the consequences are not severe, and there are no lasting effects after the person recovers. Therefore, we are willing to use ammonia in our homes and accept the high probability of an annoying exposure. By contrast, if a large dam were to fail, it would cause extensive property damage and the deaths of thousands of people downstream. Because the consequences of a failure are great, we insist on very high engineering standards so that the probability of a failure is extremely low.
- The *economic cost* of dealing with bad outcomes is one of the consequences of accepting a risk. If people become ill or are

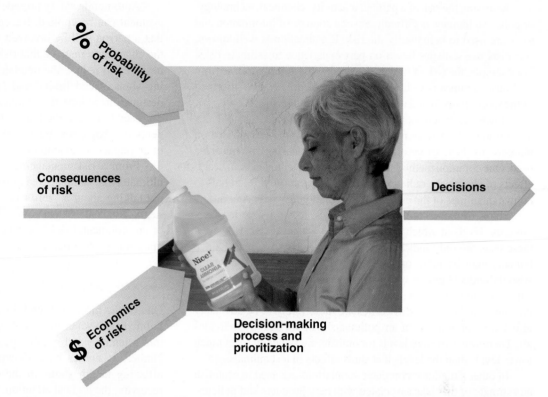

FIGURE 3.1 Assessing Risk We make many decisions on a daily basis that involve assessing the probability, consequences, and cost of accepting a risk.

injured, health care costs are likely to be associated with the acceptance of the risk. If a dam fails and a flood occurs downstream, there will be loss of life and property, which ultimately is converted to an economic cost. The assessment and management of risk involve an understanding of the probability, consequences, and economic costs of decisions. (See figure 3.1.)

Risk Assessment

Environmental **risk assessment** is the use of facts and assumptions to estimate the probability of harm to people or the environment from particular environmental factors or conditions. The World Health Organization has identified major environmental factors that lead to injury and death. (See table 3.1)

Table 3.1 Major Causes of Death Due to Environmental Factors

Cause of Death	Primary Environmental Factors	Deaths per Year (2011)
Lower Respiratory Infections (Bronchitis, pneumonia, and similar diseases)	Indoor and outdoor air pollution	3.2 million
Chronic Obstructive Pulmonary Diseasey (Emphysema, chronic bronchitis, and similar diseases)	Dusts and fumes in workplaces and homes Smoking	3 million
Road Traffic Injuries	Poor urban design or poor design of transport systems Inadequate laws to address speed, drunk-driving, helmets, seat-belts and child restraints	2.3 million
Diarrheal Disease	Unsafe drinking water and lack of sanitation	1.9 million
Malaria	Exposure to disease-carrying mosquitos	589,000

Source: Data from World Health Organization *Global Health Estimates 2013*

Assessing the risk of a particular activity, chemical, technology, or policy to humans is difficult. Several sources of information and tools are used to help clarify the risk. If a situation is well known, scientists use statistics based on past experience to estimate risks. For example, the risk of miners developing black lung disease from coal dust in mines is well established, and people can be informed of the risks involved and of actions that can reduce the risk.

However, when new technologies, chemicals, or policies are introduced, there is no established history of their effects on humans. In these cases other methods of assessing risk must be used. One common method for modeling the potential human health effects of chemicals, radiation, or drugs is to expose test animals to known quantities of the chemicals or radiation. By exposing different groups of animals to different dosages, it is possible to determine the levels at which negative effects are observed. However, these tests are only indicators of potential harm to humans, since humans may be more or less sensitive than laboratory animals. Also individual humans vary in their sensitivity, so what may present no risk to one person may be a high risk to others. For example, persons with breathing difficulties are more likely to be adversely affected by high levels of air pollutants than are healthy individuals. Therefore, exposure levels for humans are typically set at much lower levels than the levels that show effects in test animals.

In other situations, computer simulations are used to establish an estimate of the risks associated with new products and policies. For example, in an attempt to understand the risks associated with global climate change, complex computer models of climate have been used to assess the potential harmful effects of current climate trends. These computer models can also be used to assess the potential impact of policy changes designed to lessen the impact of climate change.

Although assessing the risk of environmental factors to humans is extremely important, environmental scientists are also interested in the effects of human-initiated changes on organisms and ecosystems. If human activities cause the extinction of species, there is a negative environmental impact, although direct human impact may not be obvious. Similarly, unwise policy decisions may lead to the unsustainable harvest of forest products, fish, wildlife, or other resources, thus depleting the resource for future generations.

Risk Management

Risk management is a decision-making process that involves using the results of risk assessment, weighing possible responses to the risk, and selecting appropriate actions to minimize or eliminate the risk. It is included as part of all good environmental management systems within business and industry. A risk management plan includes:

1. Evaluating the scientific information regarding various kinds of risks;
2. Deciding how much risk is acceptable;
3. Deciding which risks should be given the highest priority;
4. Deciding where the greatest benefit would be realized by spending limited funds;
5. Deciding how the plan will be enforced and monitored.

Automobile safety provides a good example of how risk management can be applied. Recognizing that the probability is high that a person will be involved in an automobile accident leads to the changes to manage that risk. Some management activities are designed to reduce the number of accidents. Traffic lights, warning signs, speed limits, and laws against drunk driving are all designed to reduce the number of accidents. Other activities are designed to reduce the trauma to people who are involved in accidents. Air bags, seat belts, and car designs that absorb the energy of an impact are examples.

From a risk management standpoint, the deciding question ultimately is: What degree of risk is acceptable? In general, we are not talking about a "zero risk" standard but rather the concept of **negligible risk:** At what point is there really no significant health or environmental risk? At what point is there an adequate safety margin to protect public health and the environment?

Perception of Risk

One of the most profound dilemmas facing decision makers and public health scientists is how to address the discrepancy between the scientific and public perceptions of environmental risks. Numerous studies have shown that environmental hazards truly affecting health status in the country are not necessarily those receiving the highest attention. (See figure 3.2.)

For example, indoor air pollution, in its various forms, receives relatively little attention compared with outdoor sources and yet probably accounts for as much, if not more, poor health. Hazardous waste dumps, on the other hand, which are difficult to associate with any measurable ill health, attract much attention and resources. The same chemicals that are found in hazardous waste sites, when present in common consumer products, such as household cleaners, pesticides, and gasoline, account for much more exposure and ill health and yet raise comparatively little concern from the public.

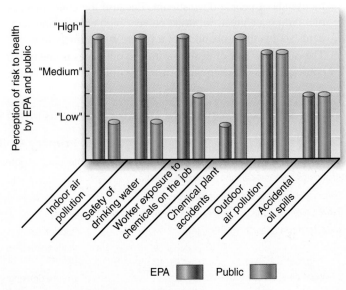

FIGURE 3.2 Perception of Risk Professional regulators and the public do not always agree on what risks are.

Table 3.2 Factors That Affect How People Perceive Risk

More acceptable	Less acceptable	Examples
Risk voluntarily accepted	Risk imposed by others	Risk of exceeding speed limits vs. Risk of being in an accident caused by drunk driver
Risk under personal control	Risk controlled by others	Risk of air pollution caused by campfire vs. Risk of air pollution caused by industry
Risk has clear benefit	Risk has little or no benefit	Risk of doing a dangerous job for high pay vs. Risk of climbing to the top of a tree to get an apple
Risk is fairly distributed	Risk is unfairly distributed	Risk of getting the flu vs. Risk of exposure to lead paint in older homes
Risk generated by a trusted source	Risk generated by an untrusted source	Risk of dam failure stated by an engineering firm vs. Risk of future property value falling stated by developer
Risk is part of nature	Risk is human created	Risk of exposure to sunlight (tanning) vs. Risk of exposure to X-rays
Risk is familiar	Risk is unfamiliar	Risk of consuming alcohol vs. Risk of exposure to nuclear radiation
Risk to adults	Risk to children	Risk of adults smoking vs. Risk of injury to children on a playground

Many factors affect the way a person perceives a risk. Table 3.2 shows how various factors affect how the public is likely to perceive the significance of a risk.

As an example of how the public's perception may be irrational we will look at the reaction of people to the presence of asbestos in schools. When it was discovered that many schools and other buildings had asbestos as a part of some of their building materials, there was a public outcry to immediately rid the buildings of this "dangerous, carcinogenic material" to protect children. The perception of many was that children were at risk and simply being near asbestos in a building that had asbestos-containing material would result in lung cancer and death. However, this was not the case. It is true that asbestos is dangerous and can cause asbestosis and lung and other cancers. However, to become ill with these or other diseases, you must first be exposed. To be exposed, you must inhale or swallow asbestos fibers. Furthermore, for a person to develop illness, they typically must be regularly exposed for many years. In most cases, the best way to minimize the risk of exposure to asbestos is to leave it in place and encapsulate it with a coating such as paint.

Asbestos in buildings does not become a problem unless it is disturbed during renovation or demolition and then protective actions are taken. Spraying with water to prevent dust, which can contain asbestos, from being distributed and careful clean-up during renovation or demolition is typically all that is needed. If you look at the list in table 3.2 you find that there are many aspects of the risk of disease from exposure to asbestos that fall in the less acceptable column and the perception of the risk of asbestos in buildings was much greater than the actual risk.

Whatever the issue, it is hard to ignore the will of the people, particularly when sentiments are firmly held and not easily changed. A fundamental issue surfaces concerning the proper role of government and other organizations in a democracy when it comes to matters of risk. Should the government focus available resources and technology where they can have the greatest tangible impact on human and ecological well-being, or should it focus them on problems about which the public is most upset? What is the proper balance? For example, would adequate prenatal health care for all pregnant women have a greater effect on the health of children than removing asbestos from all school buildings?

Throughout this discussion of risk assessment and management, it was implied that there is a cost to eliminating risk. In many cases, as risk is eliminated, the cost of the product or service increases. Business and industry constantly lobby to prevent or modify the passage of laws and regulations that reduce risk but raise their cost of doing business. Many environmental risks, such as loss of biodiversity, conversion of forest to farmland, or the impact of climate change, are difficult to evaluate from a purely economic point of view, but economics is one of the tools used to help analyze and clarify environmental problems.

3.3 Environmental Economics

Economics is the study of how people choose to use resources to produce goods and services and how goods and services are distributed to the public. In other words, economics is an allocation process that determines the purposes to which resources are put. In many respects, environmental problems are primarily economic problems. While this may be an overstatement, it is not possible to view environmental issues outside the normal economic process that is central to our way of life. Companies that mine coal, drill for oil, harvest lumber, burn fuels, and transport goods, as well as many other business activities have a negative effect on the environment. However, they only engage in these activities because they are able to sell their products to the public. Government regulations are often imposed to moderate the environmental damage done. Regulations typically increase the cost to business and industry. To appreciate the interplay between environmental issues and economics, it is important to have an understanding of some basic economic concepts.

Resources

Economists look at **resources** as the available supply of something that can be used. Classically, there are three kinds of resources: labor, capital, and land. Labor is commonly referred to as a human resource. Capital is anything that enables the efficient production of goods and services (technology and knowledge are examples). Land can be thought of as the natural resources of the planet. **Natural resources** are structures and processes that humans can use for their own purposes but cannot create. The agricultural productivity of the soil, rivers, minerals, forests, wildlife, and weather (wind, sunlight, rainfall) are all examples of natural resources. The landscape is also a natural resource, as we see in countries with a combination of mountainous terrain and high rainfall that can be used to generate hydroelectric power or in those that have beautiful scenery or biotic resources that foster tourism.

Natural resources are usually categorized as either renewable or nonrenewable. **Renewable resources** can be formed or regenerated by natural processes. Soil, vegetation, animals, air, and water are renewable primarily because they naturally undergo processes that repair, regenerate, or cleanse them when their quality or quantity is reduced. Just because a resource is renewable, however, does not mean that it is inexhaustible. Overuse of renewable resources can result in their irreversible degradation. **Nonrenewable resources** are not replaced by natural processes, or the rate of replacement is so slow as to be ineffective. For example, iron ore, fossil fuels, and mountainous landscapes are nonrenewable on human timescales. Therefore, when nonrenewable resources are used up, they are gone, and a substitute must be found or we must do without.

Supply and Demand

An economic good or service can be defined as anything that is scarce. Scarcity exists whenever the demand for anything exceeds its supply. We live in a world of general scarcity, where resources are limited relative to the desires of humans to consume them. The mechanism by which resources are allocated involves the establishment of a price for a good or service. The price describes how we value goods and services and is set by the relationship between the supply of a good or service and society's demand for it.

The **supply** is the amount of a good or service people are willing to *sell* at a given price. **Demand** is the amount of a good or service that consumers are willing and able to *buy* at a given price. The **price** of a good or service is its monetary value. One of the important mechanisms that determines the price is the relationship between the supply and demand, which is often illustrated with a **supply/demand curve.** For any good or service, there is a constantly shifting relationship among supply, demand, and price. The price of a product or service reflects the strength of the demand for and the availability of the commodity. When demand exceeds supply, the price rises. (See figure 3.3.) The increase in price results in a chain of economic events. Price increases cause people to seek alternatives or to decide not to use a product or service, which results in a lower quantity demanded.

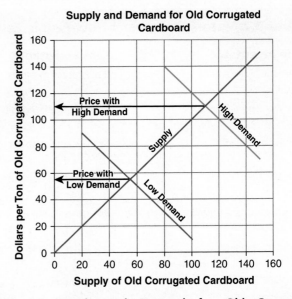

FIGURE 3.3 Supply and Demand for Old Corrugated Cardboard The price of a commodity is determined by the interplay between the supply and demand. Demand for old corrugated cardboard is high when many goods are being shipped or much old corrugated cardboard is being exported. When demand is high, businesses are willing to pay more for old corrugated cardboard. When demand falls, so does the price.

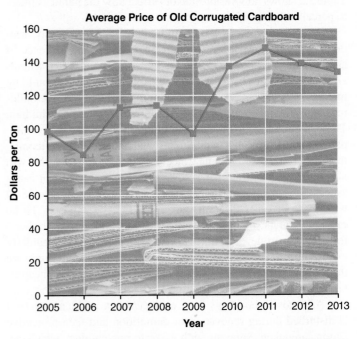

FIGURE 3.4 Recent Prices for Old Corrugated Cardboard The economic downturn of the early 2000s caused lower demand for old corrugated cardboard, and the price manufacturers were willing to pay fell. In more recent years, as the economy improved, the price paid for old corrugated cardboard has risen.

For example, the price of old corrugated cardboard, fluctuates significantly based on supply and demand. (See figure 3.4.) The supply of old corrugated cardboard is well understood, since about 90 percent of cardboard packaging is recycled. The supply is primarily determined by the amount of packaging needed to ship products. When the economy is strong and people are buying things, the supply of old corrugated cardboard rises. A strong economy also stimulates demand because when people buy things, their purchases are typically shipped in corrugated cardboard containers. A greater demand leads to an increase in the price cardboard manufacturers are willing to pay for old corrugated cardboard. Conversely, when people are not buying things, less packaging is needed, demand falls, and the price falls as well. A second factor that determines demand is the strength of the export market. When other countries are buying old corrugated cardboard, less is available for the domestic market and the price increases. Finally, when the price of old corrugated cardboard gets too high, cardboard producers can begin to use pulpwood to make corrugated cardboard.

The price paid for food items is a good example of how supply and demand determine price. Farmers use many kinds of inputs to raise crops. These include tractors and other machinery, the purchase or rental of land, specially developed plant seeds, fertilizer, pest-control agents, and fuel for machines. Thus, when the price of any of these inputs changes, the economics of raising a crop changes, which causes farmers to evaluate the kinds and quantities of crops they raise. (See figure 3.5.)

FIGURE 3.5 Factors That Affect Food Prices Many factors affect the prices consumers pay for food. The land, machinery, fuel, and chemicals required by farmers are important. Weather can also influence crop production and affect price.

In recent years the United States government has created an artificial demand for corn by mandating that a certain amount of ethanol fuel be produced from corn. This increased demand resulted in higher prices being paid by cattle farmers, who bought corn to feed to their cattle, which resulted in an increase in the price of meat at the butcher counter. In addition to purely economic forces, farmers are greatly affected by weather. Drought, frost, cold spring temperatures, hail, and many other weather conditions can lead to poor crop yields, whereas warm summers with ample rainfall result in high yields. Low crop production results in a lower supply and farmers receive a higher price for their crops and food prices at the supermarket rise. Conversely, high yields result in oversupply and the prices farmers receive for their crops fall.

Assigning Value to Natural Resources and Ecosystem Services

Nature provides both natural resources and ecosystem services. As stated earlier, natural resources are structures and processes that humans can use for their own purposes but cannot create. **Ecosystem services** are beneficial effects of functioning ecosystems for people and society. Both are valuable but it is easier to assign an economic value to natural resources.

Natural Resources

We assign value to natural resources based on our perception of their relative scarcity. We are willing to pay for goods or services we value highly and are unwilling to pay for things we think there are plenty of. For example, we will readily pay for a warm, safe place to live but would be offended if someone suggested that we pay for the air we breathe.

If a natural resource has always been rare, it is expensive. Pearls and precious metals are expensive because they have always been rare. If the supply of a resource is very large and the demand for it is low, the resource may be thought of as free. Sunlight, oceans, and air are often not even thought of as natural resources because their supply is so large. However, modern technologies have allowed us to exploit natural resources to a much greater extent than our ancestors were able to achieve and resources that were once considered limitless are now rare. For example, in the past land and its covering of soil were considered limitless natural resources, but as the population grew and the demand for food, lodging, and transportation increased, we began to realize that land is a finite, nonrenewable resource. The economic value of land is highest in metropolitan areas, where open land is unavailable. Unplanned, unwise, or inappropriate use can result in severe damage to the land and its soil. (See figure 3.6.)

Even renewable resources can be overexploited. If the overexploitation is severe and prolonged, the resource itself may be destroyed. For example, overharvesting of fish, wildlife, or forests can change the natural ecosystem so much that it cannot recover, and a resource that should have been renewable becomes a depleted nonrenewable resource.

FIGURE 3.6 Mismanagement of a Renewable Resource Although soil is a renewable resource, poor farming practices can cause soil loss and reduce the value of the land. In this photograph, water erosion has carried soil away and formed a gully, reducing the value of the land.

Ecosystem Services

Functioning ecosystems are a tangible source of economic wealth. Ecosystem services include such things as pollination by insects, regrowth of forests, cleansing of water and air, decomposition of wastes, hydropower, and recreational opportunities. Because ecosystems services are not traded in conventional markets, economists lack information about their economic value—we don't explicitly pay a price for a "glorious view" or the value of decay organisms in our compost bin. Of course, just because something doesn't have a price doesn't mean it is not valuable; the challenge, then, is to determine ways to get people to reveal the value they place on an ecosystem service. There are several ways to carry this out.

- First, we can get people to state their preferences by asking them questions designed to elicit value. Would you prefer to live near a city park? Do you enjoy seeing songbirds in your backyard?
- Second, we can look to people's behavior and infer the value they place on ecosystem services from that behavior. Houses in scenic settings sell for more than houses without scenery, for example. When people spend time and money traveling to places where they can observe nature, they are assigning value to those recreational opportunities.
- Another way to place a value on ecosystem services is to determine the maximum amount of other goods and services individuals are willing to give up to preserve or use an ecosystem service. For example, the public can be asked if they are willing to give up a section of a river with a trout fishery for the electricity that will be generated by the building of a dam.

In 2010, a United Nations report stated that ecosystems such as freshwater, coral reefs, and forests accounted for between 47 percent and 89 percent of what the UN calls the "Gross Domestic Product (GDP)" of the rural poor. A recent estimate of the value of the world's coral reefs was about $30 billion per year. If coral reef ecosystems collapse because of climate change and other factors, millions of people will lose their source of livelihood.

The economic value of pollination services by honeybees and other animals is estimated to be about $200 billion per year worldwide. This service has been brought into sharp focus by worldwide problems in the survival of domesticated honeybees.

Environmental Costs

Air pollution, water pollution, plant and animal extinctions, depletion of a resource, and loss of scenic quality are all examples of the **environmental costs** of resource exploitation. Often environmental costs are difficult to measure, since they are not easily converted to monetary values. This is especially true with the loss of ecosystem services just discussed. Although economists have not traditionally discussed environmental costs, they have described two common aspects of environmental costs.

Deferred costs are those that are ignored, not recognized, or whose effects accumulate over time, but that eventually must be paid. For example, when dams were built on the Colorado River to provide electric power and irrigation water, planners did not anticipate that the changes in the flow of the river would reduce habitat for endangered bird species, lead to the loss of native fish species because the water is colder, and result in increased salinity in lower regions of the river. Soil erosion is another example of a deferred cost. The damage done by practices that increase soil erosion may not be felt immediately, but eventually as the amount of damage accumulates and soil fertility declines, the cost becomes obvious to future generations.

External costs are those that are borne by someone other than the individuals who produce or consume a good or service. For example, when a logging operation removes so many trees from a hillside that runoff from the hillside destroys streams and causes mudslides, the logging operation has transferred a cost to the public. Another example is the thousands of hazardous waste sites produced by industries that are no longer in business. The cleanup of these abandoned hazardous waste sites became the responsibility of government and the taxpayers. The entities that created the sites avoided paying for their cleanup. Similarly, when a new shopping complex is built, many additional external costs are paid for by the public and the municipality. Additional roads, police and fire protection, sewer and water services, runoff from parking lots, and pressure to convert the remaining adjacent land to shopping are all external costs typically borne by the taxpayer.

Opportunity costs are those that occur when a decision precludes other potential uses for a resource. For example, if houses are built in a forested region, the land's possible use as a natural area for hiking or hunting is lost. Similarly, when strip mining for coal takes place in farmland, the local landscape is severely altered so that its use as farmland is no longer possible. (See figure 3.7.) Probably most environmental costs include external, deferred, and opportunity cost aspects.

A good example of a problem that includes all of these kinds of costs is the damage caused by the use of high-sulfur coal as an inexpensive way to produce electricity. The sulfur compounds released into the atmosphere resulted in acid rain that caused a decline in the growth of forests and damage to buildings and other structures. The damage accumulated over time, so the cost of acid rain was a deferred cost. The cost of the damage was paid for by the public as fewer scenic vistas, by forest products industries with fewer trees to harvest, and by property owners as repair costs for buildings and other structures, so it was an external cost not paid for by the electric utilities directly. The production of coal requires large amounts of water to wash the coal to remove unwanted material. The water quality is degraded, so it is unsuitable for drinking, irrigation, or recreation, which represents an opportunity cost.

FIGURE 3.7 Opportunity Cost The decision to mine this farmland for coal prevents the use of the land for farming in the future.

Water pollution. This sign indicates it is unsafe to swim in this area because of high levels of bacteria.

Smoke. Smoke contains small particles that can cause lung problems.

Odors. Feedlots create an odor problem that many people find offensive.

Visual pollution. Unsightly surroundings are annoying but not hazardous to your health.

Health hazard

Smog. Smog that develops when air pollution is trapped is a serious health hazard.

Solvents. Solvents evaporate and cause localized pollution.

Thermal pollution. Cooling towers release heat into the atmosphere and can cause localized fog.

Junkyard. This is unsightly but constitutes only a minor safety hazard.

Annoyance

FIGURE 3.8 Examples of Pollution There are many kinds of pollution. Some are major health concerns, whereas others merely annoy.

As people recognize the significance of environmental costs, these costs are being converted to economic costs as stricter controls on pollution and environmental degradation are enforced. It takes money to clean up polluted water and air or to reclaim land that has been degraded, and the people who cause the damage should not be allowed to defer the cost or escape paying for the necessary cleanup or remediation.

The Economics of Pollution

Pollution is any addition of matter or energy that degrades the environment for humans and other organisms. When we think about pollution, however, we usually mean something that people produce in large enough quantities that it interferes with our health or well-being. Two primary factors that affect the amount of damage done by pollution are the size of the population and the development of technology that "invents" new forms of pollution.

It is not always easy to agree on what constitutes pollution. To some, the smell of a little wood smoke in the air is pleasant; others do not like the odor. A business may consider advertising signs valuable and necessary; others consider them to be visual pollution. Finally, it is important to recognize that it is impossible to eliminate all the negative effects produced by humans and our economic processes. The difficult question is to determine the levels of pollution that are acceptable. (See figure 3.8.)

As people began to appreciate the cost of pollution in terms of diminished living conditions and disease, individuals and their governments began to demand that pollution be controlled. The cost of controlling pollution can be divided into two kinds of activities. **Pollution costs** are private or public expenditures to correct pollution damage once pollution has occurred. Pollution costs include: the cost for health care for those whose health was impaired because of pollution, expenditures to correct past pollution such as the clean-up of polluted industrial sites, and the loss of the use of public resources, such as clear air, clean water, or quiet surroundings, because of pollution. As was discussed in the previous section these are *external costs* and in many cases are also *deferred costs*.

Pollution-prevention costs are those incurred either in the private sector or by government to prevent, either entirely or partially, the pollution that would otherwise result from some production or consumption activity. The cost incurred by local government to treat its sewage before releasing it into a river is a pollution-prevention cost; so is the cost incurred by an electric utility to prevent air pollution by installing new equipment on their exhaust stacks or to switch to less polluting sources of fuel.

The philosophy of **pollution prevention** is that pollution should be prevented or reduced at the source whenever feasible. An important aspect of this philosophy is that pollution should not be an external cost paid for by the public, but that the potential pollution cost should be internalized and reflected in the price of the

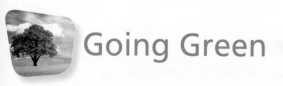

Going Green

Green-Collar Jobs

A general definition of a green-collar job is one that is in some way related to environmental improvement. They range from those that involve the installation of energy saving devices and new energy technology to corporate environmental management. The Bureau of Labor Statistics defines green-collar jobs as those that are involved in renewable energy, energy efficiency, pollution reduction, natural resource conservation, and environmental compliance, education, and public awareness. They reported that in 2011 there were about 4.3 million jobs that fit their definition of green. Two areas that have seen considerable growth in the past and are expected to continue to grow in the future are solar and wind energy jobs.

Part of the growth in green-collar jobs will come from government initiatives. In 2007, the U.S. House of Representatives passed the Green

Jobs Act, which provides $125 million annually to train people for green vocational fields that offer living wages and upward mobility for low-income communities.

In the private sector, many corporations are establishing offices of sustainability and have added Chief Sustainability Officer to their list of top-level executives. So the variety of jobs and salaries are quite varied. To explore some of the green job possibilities, use your browser to connect to one of the following job boards.

- Ecojobs
- Green jobs network
- jobs.greenbiz.com

---

good or service being provided. Although some business and industrial organizations continually lobby for less stringent environmental regulations and oppose new regulations, many companies have shown that preventing pollution can actually cut business costs and thus increase profits. Pollution prevention, then, does make cents!

For example, several years ago, 3M's European chemical plant in Belgium switched from a polluting solvent to a safer but more expensive water-based substance to make the adhesive for its Scotch™ Brand Magic™ Tape. The switch was not made to satisfy any environmental law in Belgium or the European Union. 3M managers were complying with company policy to adopt the strictest pollution-control regulations that any of its subsidiaries is subject to—even in countries that have no pollution laws at all. Part of the policy is founded on corporate public relations, a response to growing customer demand for "green" products and environmentally responsible companies. But as many North American multinationals with similar global environmental policies are discovering, cleaning up waste, whether voluntarily or as required by law, can cut costs dramatically. In 1975, 3M established its "Pollution Prevention Pays" (P3) program. This corporate philosophy has cut the company's air, water, and waste pollution significantly. Recent company data show that greenhouse gases have been reduced 72 percent from 1990, energy use has been reduced by 42 percent since 2000, water use has been reduced by 35 percent since 2005, waste has been reduced by 44 percent since 2000, and volatile organic compound releases to the air have been reduced by 98 percent since 1990. Less waste has meant less spending to comply with pollution control laws. But, in many cases, 3M actually has made money selling wastes it formerly hauled away. And because of recycling prompted by the 3P program, it has saved money by not having to buy as many raw materials.

Xerox Corporation has focused on recycling materials in its global environment efforts. It provides buyers of its copiers with free United Parcel Service pickup of used copier cartridges, which contain metal-alloy parts that otherwise would wind up in landfills.

The cartridges and other parts are now cleaned and used to make new ones. Other recycled Xerox copier parts include power supplies, motors, paper transport systems, printed wiring boards, and metal rollers. In 2012, about 1,700 metric tons of parts were reused. Remanufacturing, recycling, and waste reduction results in significant savings because fewer raw materials need to be purchased.

Cost-Benefit Analysis

Because natural resources and finances are limited and there is competition for their use, it is essential that a process be used to help make decisions about the wise use of natural resources and finances. **Cost-benefit analysis** is a formal process for calculating the costs and benefits of a project or course of action, to decide if benefits are greater than costs. It has long been the case in many developed countries that major projects, especially those undertaken by the government, require some form of cost-benefit analysis with respect to environmental impacts and regulations. In the United States, for example, such requirements were established by the National Environmental Policy Act of 1969, which mandates environmental impact statements for major government-supported projects. Increasingly, similar analyses are required for projects supported by national and international lending institutions such as the World Bank.

An Example—Providing Clean Water and Improved Sanitation

In 2004 the World Health Organization published a cost-benefit analysis of providing clean water sources and "improved" sanitary facilities to the people who do not have them. The study assessed the cost-benefit of five different levels of intervention. The medium level was quite modest and assessed the impact of providing clean drinking water and at least access to a pit latrine within a short walking distance of each home. The authors used a variety

Table 3.3 Cost-Benefit Analysis of Gaining Access to Clean Water and Sanitary Facilities

(A clean water source and an "improved" sanitary facility both available near the home)

Major Impact on Population	Comment	Economic benefits of providing clean water and sanitary facilities near each home	Cost of providing clean water and sanitary facilities near each home
Population affected	3 billion people impacted		US $22.6 billion per year (US $5.30 per person per year)
Less expenditure by health care system on treatment of diarrheal disease	900 million cases of diarrhea avoided	US $11.6 billion/year saved	
Fewer treatment costs borne by family of patients	Travel to clinic, purchase medication, etc.	US $0.565 billion/year saved	
Fewer worker days lost due to their illness or to care for sick relatives	5.6 billion additional working days/year Additional income calculated at minimum wage for the country (An average of US $4.66/day)	US $1.2 billion/year	
Early deaths avoided	Early deaths result in reduction in total lifetime earnings	US $5.6 billion	
Additional days babies and infants are healthy	2.4 billion more healthy days Better lifetime health	US $11.2 billion/year	
More time for children to attend school	443 million days/year Better achievement in school	US $2.1 billion/year	
Less time needed to collect water and access sanitary facilities	595 billion hours per year. These hours could be used to earn wages (calculated at minimum wage for country)	US $229 billion/year	
Other benefits	Higher home values More leisure time		
Total		US $262 billion	US $22.6 billion

Cost-Benefit Ratio = 11.6 (For this analysis there is a return of US $11.60 for each US $1.00 invested.)

of sources of information and made several assumptions about the dollar value of changes that would result from improving water and sanitary services. Table 3.3 provides a synopsis of the kinds of assessments made by the authors of the study. For the medium level of service, they determined that for every dollar spent to provide services, there would be about 11 dollars in benefits.

Concerns About The Use of Cost-Benefit Analysis

There is clearly value in using cost-benefit analysis in the decision-making process. However, there are several kinds of concerns about how cost-benefit analysis is used and difficulties in assigning economic value to many kinds of benefits. Some people argue that if the only measure of value is economic, many simple noneconomic values such as beauty, quiet, recreational opportunities, or cleanliness will be overlooked or ignored. Furthermore, different cultures and even different people within a culture may have very different opinions of the value of a noneconomic resource. (See figure 3.9.)

In cases of Third World development projects, these already difficult environmental issues are made more difficult by cultural and socioeconomic differences. A less-developed country, for

example, may place a lower value on clean air and safe working conditions if a project will provide jobs that are greatly needed.

One particularly compelling critique of cost-benefit analysis is that those doing the study must decide which preferences count—that is, which preferences are the most important for cost-benefit analysis. In theory, cost-benefit analysis should count all benefits and costs, regardless of who benefits or bears the costs. In practice, however, this is not always done. For example, if a cost is spread thinly over a large population, it may not be recognized as a cost at all. The cost of air pollution in many parts of the world could fall into such a category. Debates over how to count benefits and costs for future generations, inanimate objects such as rivers, and nonhumans, such as endangered species, are also common.

3.4 Comparing Economic and Ecological Systems

One of the problems associated with matching economic processes with environmental resources is the great differences in the way economic systems and ecological systems function. For most

Which is more valuable: a hiking trail or a trail for motor bikes?

Which is more valuable: a shoreline for hunting or a shoreline for commerce?

Which is more valuable: a commercial street or a quiet residential street?

FIGURE 3.9 Assigning Economic Values to Resource Use The way we use resources is based on the perceived value of the resource. Not all people see the same value for a resource, and values are not always easy to measure.

natural scientists, current crises such as biodiversity loss, climate change, and many other environmental problems are symptoms of an imbalance between the socioeconomic system and the natural world. While it is true that humans have always changed the natural world, it is also clear that this imprint is much greater now than anything experienced in the past. One reason for the profound effect of human activity on the natural world is the fact that there are so many of us that rely on natural resources and ecological services for our well-being.

How Economic and Ecological Systems Differ

As mentioned earlier, many natural functions of ecological systems are overlooked because they are not easily converted to economic terms. There are several fundamental ways in which economic and ecological systems differ.

Time frame An obvious difference between economics and ecology is the great difference in the time frame of markets and ecosystems. Many ecosystem processes take place over hundreds, thousands, or even millions of years. The time frame for market decisions is short. It may be as short as minutes for stock trades or as long as a few years for the development and construction of a factory. Where U.S. economic policy is concerned, two- to four-year election cycles are the frame of reference. For investors and dividend earners, performance time frames of three months to one year are the rule.

Location For ecosystems, place is critical. For example, groundwater resources are not transferable from one location

to another. The size and location of groundwater reservoirs are influenced by soil quality, hydrogeological conditions, regional precipitation rates, plants that live in the region, and losses from evaporation, transpiration, and groundwater flow. For economic activities, place is increasingly irrelevant. Topography, location, and function within a bioregion or local ecological features do not enter into economic calculations except as simple functions of transportation costs or comparative economic advantage. Production is transferable, and the preferred location is anywhere production costs are the lowest.

Units of measure Different units are used to measure changes and values in economics and ecology. Ecological systems are measured in physical units such as calories of energy, carbon dioxide absorption, centimeters of rainfall, or parts per million of nitrate contamination. The unifying measure of market economics is money. Change is measured by the increase or decrease in the value of a good or service. Focusing only on the economic value of resources may mask serious changes in environmental quality or function.

Complexity Ecological systems have a much greater degree of complexity than do economic systems. To understand an ecological system one must consider thousands of kinds of organisms interacting in specific ways with one another, as well as the impact of physical features such as air, water, and soil. While economic systems are not simple, the number of ingredients that must be assessed is much less.

The loss of biodiversity is an example that illustrates the conflicting frameworks of economics and ecology. For example, from an economic point of view, the value of land used for beef production is measured according to the economic value of the beef produced. However, the decision to use land to raise cattle typically has a significant effect on the diversity of grass species, the numbers and kinds of microorganisms in the soil, the kinds of animals that share the land with cattle, and groundwater quality. The costs resulting from the simplification of an ecosystem or the changes in processes like carbon dioxide absorption are not measured in economic terms as long as production of beef gives a positive economic return. However, it is possible to develop economic plans that include an understanding of ecological processes. For example, in Zimbabwe and other African nations, some ranchers now earn more money managing native species of wildlife for ecotourism in a biodiverse landscape than they would from raising cattle in a landscape with reduced biodiversity. The wildlife species have been a part of the local landscape for thousands of years and are better adapted to the location than cattle.

Common Property Resource Problems—The Tragedy of the Commons

When everybody shares ownership of a resource, there is a strong tendency to overexploit and misuse that resource. The problems inherent in common ownership of resources were outlined by biologist Garrett Hardin in a classic essay entitled "The Tragedy of the Commons" (1968). The original "commons" were areas of pastureland in England that were provided free by the king to anyone who wished to graze cattle.

There are no problems on the commons as long as the number of animals is small in relation to the size of the pasture. From the point of view of each herder, however, the optimal strategy is to enlarge his or her herd as much as possible because if his animals do not eat the grass, someone else's will. Thus, the size of each herd grows, and the density of stock increases until the commons becomes overgrazed. The result is that everyone eventually loses as the animals die of starvation. The tragedy is that even though the eventual result should be perfectly clear, no one acts to avert disaster.

The ecosphere is one big commons stocked with air, water, and irreplaceable mineral resources—a "people's pasture," to be used in common, but it is a pasture with very real limits. Each nation attempts to extract as much from the commons as possible without regard to other countries. Furthermore, the United States and other industrial nations consume far more than their fair share of the total world resource harvest each year, much of it imported from less-developed nations.

One clear modern example of this problem involves the overharvest of marine organisms. Since no one owns the oceans, many countries feel they have the right to exploit the fisheries resources present. As in the case of grazing cattle, individuals seek to get as many fish as possible before someone else does. Currently, the UN estimates that nearly all of the marine fisheries of the world are being fished at or above capacity.

Finally, common ownership of land resources, such as parks and streets, is the source of other environmental problems. People who litter in public parks do not generally dump trash on their own property. Common ownership of the ocean makes it inexpensive for ships and oil drilling platforms to use the ocean as a dump for their wastes. (See figure 3.10.)

FIGURE 3.10 The Ocean Is a Common Property Resource Since the oceans of the world are a shared resource that nobody owns, there is a tendency to use the resource unwisely. Coastal communities often dump their waste into the ocean and many ships dispose of waste at sea.

The tragedy of the commons also operates on an individual level. Most people are aware of air pollution, but they continue to drive their automobiles. Many families claim to need a second or third car. It is not that these people are antisocial; most would be willing to drive smaller or fewer cars if everyone else did, and they could get along with only one small car if public transport were adequate. But people frequently get "locked into" harmful situations, waiting for others to take the first step, and many unwittingly contribute to tragedies of the commons. After all, what harm can be done by the birth of one more child, the careless disposal of one more beer can, or the installation of one more air conditioner?

3.5 Using Economic Tools to Address Environmental Issues

The traditional way of dealing with environmental issues is to develop regulations that prohibit certain kinds of behavior. This is often called a "command and control" approach. It has been very effective at reducing air and water pollution, protecting endangered species, and requiring that environmental concerns be addressed by environmental impact statements. However, there are also tools that use economic incentives to encourage environmental stewardship.

Subsidies

A **subsidy** is a gift from government to individuals or private enterprise to encourage actions considered important to the public interest. Subsidies may include income tax rebates for purchases of energy efficient appliances, low-interest loans or grants to businesses to encourage them to invest in specific technologies to control pollution, or tax advantages to business and industry to

Mercury contamination of our lakes, streams, oceans, soil, and groundwater is an ongoing issue. Because of surface water contamination, many state and federal agencies have issued health advisories to limit the amounts of certain fish and shellfish that people eat. If you like to eat fish or shellfish, would you limit your consumption based on warnings from state or federal health agencies? Develop an argument for or against consuming fish that potentially contain mercury.

develop new technologies. These gifts, whether loans, favorable tax situations, or direct grants, are all paid for by taxes on the public, so in effect they are an external cost.

When subsidy programs have a clear purpose and are used for short periods to move to new ways of doing business, they can be very useful. In recent years, subsidies that encourage people and corporations to improve their energy efficiency have significantly reduced energy usage. Government payments to farmers that encourage them to permanently take highly erodable land from production reduces erosion and the buildup of sediment in local streams. The better water quality benefits fish, and the return of land to more natural vegetation benefits wildlife.

One common effect of a subsidy is to keep the price of a good or service below its true market price. The actual cost of a subsidized good or service is higher than the subsidized market price because subsidy costs must be added to the market price to arrive at the product's true cost.

Once subsidies become a part of the economic fabric of a country, they are very difficult to eliminate. In 1996, the U.S. Congress passed the Freedom to Farm Act, which eliminated many agricultural subsidies and was hailed as the end of agricultural subsidies. During 1996 and 1997 farm subsidies were about $8 billion annually. However, subsequent legislation led to changes that increased subsidies to nearly $15 billion by 2012. The Agricultural Act of 2014 contains provisions that amount to subsidies of about $20 billion per year to farmers.

The building of roads and bridges is a major part of the U.S. federal budget. Currently about $14 billion is spent per year on the building and improvement of roads. This constitutes a subsidy for automobile transportation. Higher taxes on automobile use to cover the cost of building and repairing highways would encourage the use of more energy-efficient public transport.

Market-Based Instruments

Historically governments have sought to achieve environmental improvements by establishing regulations with penalties for those who do not comply with the regulations. However, increasingly governments are looking to economic tools commonly known as market-based instruments to encourage people to meet environmental goals. Instead of issuing inflexible directives, the use of market-based policies allows people and businesses to choose solutions that make sense to them economically. For example, a price can be established for pollution-causing activities. Companies are then allowed to decide for themselves how best to achieve the required level of environmental protection.

Several kinds of market-based instruments are currently in use:

Information programs provide consumers with information about the environmental consequences of purchasing decisions. Information about the environmental consequences of choices makes clear to consumers that it is in their personal interest to change their decisions or behaviors. Examples include information tags on electric appliances that inform the public about the energy efficiency of the product, the mileage ratings of various automobiles, and labeling on pesticide products that describes safe use and disposal.

Tradable emissions permits give companies the right to emit specified quantities of pollutants. Companies that emit less than the specified amounts can sell their permits to other firms or "bank" them for future use. Thus, businesses responsible for pollution have an incentive to internalize the external cost they were previously imposing on society: If they clean up their pollution sources, they can realize a profit by selling their permit to pollute. Once a business recognizes the possibility of selling its permit, it sees that reducing pollution can have an economic benefit. The establishment of tradable sulfur dioxide pollution permits for coal-fired power plants has resulted in huge reductions in the amount of sulfur dioxide released.

Emissions fees and taxes make environmentally damaging activity or products more expensive, which provides an economic incentive to reduce environmental harm. Businesses and individuals reduce their level of pollution wherever it is cheaper to reduce the pollution than to pay the charges.

Deposit-refund programs place a surcharge on the price of a product that is refunded when the used product is returned for reuse or recycling. Deposit-refund schemes have been widely used to encourage recycling of bottles and cans, which reduces trash and the cost of its clean-up.

Performance bonds are fees that are collected to ensure that proper care is taken to protect environmental resources. Some nations—including Indonesia, Malaysia, and Costa Rica—use performance bonds to ensure that reforestation takes place after timber harvesting. The United States also has used this approach to ensure that strip-mined lands are reclaimed. The money collected as a performance bond is not fully released until all performance standards have been met.

The Developing Green Economy

In the United States environmental policy tends to be fragmented because the development of a coherent policy is highly politicized. For example, climate scientists emphasize that climate change is real. However, many in Congress deny that climate change is a problem and interfere with actions to deal with it. The fossil-fuel industry continues to receive government subsidies, but subsidies for renewable energy are considered wasteful. The Environmental Protection Agency has successfully introduced programs that improve air quality and reduce exposure to toxic chemicals and recently received a Supreme Court ruling that allows the EPA to regulate carbon dioxide as an air pollutant. President Obama used an executive order to establish higher fuel efficiency standards for cars

and trucks. Despite this piecemeal approach to environmental policy, there is a great deal of evidence that the science and technology related to environmental issues is being accepted by the business community and the public.

Driving the green economy is the development of sustainability policies and goals by corporations and a general recognition by business and industry that at some point there will be a price on carbon emissions. There is also a growing interest by the public in environmental improvements and energy efficiency.

Economic sectors such as information technology, packaging, and green building all have momentum regardless of the lack of federal policy. For example, Cisco Systems, Inc., designs, manufactures, sells, and services networking equipment. In 2008, the CEO of Cisco Systems stated the company was committed to reducing its greenhouse gas emissions by 25 percent by 2013. It met that goal in 2012. In 2013, Cisco Systems announced additional goals to reduce greenhouse gases by 40 percent, energy use by 15 percent, and to purchase at least 25 percent of its electricity from renewable sources. Cisco's commitment extends beyond the company's own footprint. In 2009, they undertook a supply chain innovation program to pressure Cisco suppliers to adhere to similar environmental standards.

Many other business sectors have not been as successful at fitting into the green economy. While many companies have clear environmental goals, about half of the top 600 U.S. publicly traded companies do not publish a sustainability statement. Energy companies, companies that produce raw materials, and the food processing companies generally have had a poor environmental record. Although it may be more difficult for these kinds of companies to become "green," the lack of sustainability goals means the environment is not a priority.

Life Cycle Analysis and Extended Product Responsibility

Life cycle analysis is the process of assessing the environmental effects associated with the production, use, reuse, and disposal of a product over its entire useful life. Life cycle analysis can help us understand the full cost of new products and their associated technologies.

The various stages in the product chain include raw material acquisition, manufacturing processes, transportation, use by the consumer, and ultimately disposal of the used product. When this approach is used, it is possible to identify changes in product design and process technology that would reduce the ultimate environmental impact of the production, use, and disposal of the product. All factors along the product chain share responsibility for the life cycle environmental impacts of the product, from the upstream impacts inherent in the selection of materials and impacts from the manufacturing process itself to downstream impacts from the use and disposal of the product. (See figure 3.11.)

A logical extension of life cycle analysis is extended product responsibility. **Extended product responsibility** is the concept that the producer of a product is responsible for all the negative effects involved in its production, including the ultimate disposal of the product when its useful life is over. The logic behind extended product responsibility is that if manufacturers pay for the post-consumer impacts of products, they will design them differently to reduce waste.

Many people identify the German packaging ordinance as one of the first instances of extended product responsibility. Under the German packaging ordinance, consumers, retailers, and packaging manufacturers all share this responsibility, with the financial burden of waste management falling on the retailers and packaging manufacturers. This is thought to be one of the first instances of the concept of "take back," or taking the product back for disposal to the place that made it. This made companies look very hard at how they would manufacture something. The thought was: "This thing is coming back to me someday; how will I recycle/reuse or dispose of it?"

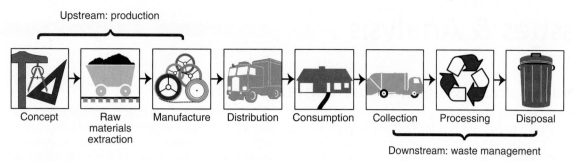

Upstream: production

Concept · Raw materials extraction · Manufacture · Distribution · Consumption · Collection · Processing · Disposal

Downstream: waste management

FIGURE 3.11 The Life Cycle of a Typical Product When life cycle analysis is undertaken, it is important to identify all the steps in the process—from obtaining raw materials, through the manufacturing process, to the final disposal of the item.

Although no national legislation in the United States mandates extended product responsibility, there are several instances in which manufacturers of specific products have implemented it. When several states passed legislation that required manufacturers of nickel-cadmium batteries to take back the worn-out batteries, manufacturers instituted a national take back program. The chemical industry has instituted a program known as Responsible Care®. The concept originated in Canada and has since spread to 46 countries. The primary goals of Responsible Care® are to improve chemical processes and ensure the safe production, transport, use, and disposal of the products of the industry.

3.6 Economics and Sustainable Development

A commonly used definition of the term *sustainable development* is one that originated with the 1987 report *Our Common Future* by the World Commission on Environment and Development (known as the Bruntland Commission). It states that "**sustainable development** is development that meets the needs of the present without compromising the ability of future generations to meet their own needs." This definition includes the concept of economic development, which involves activities that improve the standard of living of people and the economic health of a community or country. Usually economic development includes some degree of natural resource exploitation. However, it also includes improved education, health care, roads, communication, and many other factors. Sustainability is an environmental concept that recognizes that there are limits to the exploitation of natural resources but that if the degree of exploitation is managed properly the resource can be used for generations.

Gaylord Nelson, the founder of the first Earth Day, listed five characteristics that define sustainability:

1. *Renewability:* A community must use renewable resources, such as water, topsoil, and energy sources no faster than they can replace themselves. The rate of consumption of renewable resources cannot exceed the rate of regeneration.

2. *Substitution:* Whenever possible a community should use renewable resources instead of nonrenewable resources. This can be difficult because of barriers to substitution. To be sustainable a community has to make the transition before the nonrenewable resources become prohibitively scarce.

3. *Interdependence:* A sustainable community recognizes that it is a part of a larger system and that it cannot be sustainable unless the larger system is also sustainable. A sustainable community does not import resources in a way that impoverishes other communities, nor does it export its wastes in a way that pollutes other communities.

4. *Adaptability:* A sustainable community can absorb shocks and adapt to take advantage of new opportunities. This requires a diversified economy, educated citizens, and a spirit of solidarity. A sustainable community invests in and uses research and development.

5. *Institutional commitment:* A sustainable community adopts laws and political processes that mandate sustainability. Its economic system supports sustainable production and consumption. Its educational systems teach people to value and practice sustainable behavior.

Today we see mixed results with respect to the goal of sustainability. Many corporations have sustainability objectives that include reduction in pollution, reduction in waste, improved energy efficiency, and many other changes to their business. In most cases, these are sincere efforts that involve the investment of significant amounts of money in new technology and processes that reduce their environmental impact. Likewise, the 28 member countries of the European Union have made many policy changes which commit them to a 20 percent reduction of greenhouse gas emissions from 1990 levels. By 2011 greenhouse gas emissions had fallen by about 18 percent. By contrast, both China and the United States, which together account for about 45 percent of greenhouse gas emissions, have had increases during the period 1990 to 2011. Yet both countries continue to subsidize their fossil fuel industries that are responsible for the production of a majority of greenhouse gases. Deforestation also presents a confusing picture. The amount of forest land in the United States and many European countries has increased while forest land in tropical Africa, Asia, and Central and South America is declining as forests are converted to farmland. Clearly some countries have embraced the concept of sustainability while others have not. In addition, others have picked aspects of sustainability that are politically or economically easy to meet and ignored the more difficult sustainability questions.

Issues & Analysis

The Economics and Risks of Mercury Contamination

Mercury is a chemical element that is used in many industrial processes and products. Since it is a liquid metal, it is used in many kinds of electrical applications. Common uses of mercury include fluorescent lightbulbs, mercury vapor lights, electrical tilt switches, and certain kinds of small batteries. It also is alloyed with silver and other metals to produce fillings in teeth. Elemental mercury by itself is poisonous, but so few people have direct access to elemental mercury that it constitutes a minor risk. Certain bacteria, however, are able to convert elemental mercury to methylmercury, which can be easily taken up by living things and stored in the body.

Some of the mercury available for conversion to methylmercury is from natural rock. Thus, some parts of the world naturally have high mercury levels compared to others. A common human-generated source of mercury is the combustion of coal in electric power plants and similar facilities. The mercury is released into the air and distributed over the landscape downwind from the power plant.

Contamination with methylmercury is a particular problem in aquatic food chains where fish at higher trophic levels accumulate high quantities of methylmercury in their tissues. Fish are an important source of omega-3 fatty acids that are associated with reductions in heart disease. In many places where methylmercury is present, people are advised to severely restrict their consumption of fish. If fish are declared unfit for human consumption because of methylmercury levels, there is a major economic impact on the people who catch and process fish and consumers' food choices are limited.

In 2000, under the Clinton administration, the EPA announced that it would require reductions in the amount of mercury released from coal-fired power plants. This would require expensive additions to power plants to capture the mercury so that it was not released into the atmosphere. The power industry protested that changes would be too expensive, and the program was never initiated. In December 2003, under the George W. Bush administration, the EPA announced a mercury emission trading proposal to deal with the problem. Under this proposal, coal-fired power plants would be issued permits to release specific amounts of mercury. If they did not release much mercury, they could sell their permit to other power plants that were not able to reduce the amount of mercury they released. Proponents point out that a similar emissions trading program for sulfur dioxide resulted in major reductions in sulfur dioxide released. Critics of the mercury emissions program say that the program is a concession to the power industry. The program was never instituted.

In December 2011 the EPA published a rule that set a limit to the amount of mercury power plants could release. Power plants were given four years in which to comply with the ruling. As a result many proposed coal-fired power plants have been shelved and some older (more polluting) plants are being shut down.

- Why were power plants successfully able to stop previous efforts to force them to control mercury emissions?
- Should all pollution be prevented regardless of the cost, or should risk assessment and economic analysis be a part of the decision-making process?

Summary

Risk is the probability that a condition or action will lead to an injury, damage, or loss. Risk assessment is the use of facts and assumptions to estimate the probability of harm to human health or the environment that may result from exposures to pollutants or toxic agents. While it is difficult to calculate risks, risk assessment is used in risk management, which analyzes the risk factors in decision making. Perception of risk by people is altered by many factors, which causes them to have an inaccurate measure of the real risk.

To a large degree, environmental problems can be viewed as economic problems that revolve around decisions about how to use resources. A major problem is how to give an economic value to ecosystem services. Many environmental costs are deferred (paid at a later date) or external (paid by someone other than the entity that causes the problem). Pollution is a good example of both a deferred and an external cost. Cost-benefit analysis is used to determine whether a policy generates more social benefits than social costs. Criticism of cost-benefit analysis is based on the question of whether everything has an economic value. It has been argued that if economic thinking dominates society, then noneconomic values, such as beauty, can survive only if a monetary value is assigned

to them. There is a strong tendency to overexploit and misuse resources that are shared by all. This concept was developed by Garrett Hardin in his essay "The Tragedy of the Commons."

Economic policies and concepts, such as supply and demand and subsidies, play important roles in environmental decision making. The balance between the amount of a good or service available for purchase and the demand for that commodity determines the price. Subsidies are gifts from government to encourage desired behaviors. Recently, several kinds of market-based approaches have been developed to deal with the economic costs of environmental problems. These approaches include information programs, tradable emissions, emissions fees, deposit refund programs, and performance bond programs. The goal of all these mechanisms is to introduce a profit motive for institutions and individuals to use resources wisely.

A newer school of economic thought is referred to as sustainable development. Sustainable development has been defined as actions that address the needs of the present without compromising the ability of future generations to meet their own needs. Sustainable development requires choices based on values.

Acting Green

1. If you smoke, quit! If you do not smoke, help a friend who does smoke to quit.
2. Research the "green marketing" claims for several products that you use.
3. Work on sustainability projects in your college.
4. Develop a cost-benefit analysis for a local issue.
5. Buy products that come in a container that is reusable or that requires a deposit.

Review Questions

1. What factors are used to characterize a risk?
2. What tools are used to help characterize the risk of environmental factors such as toxins?
3. List three environmental factors that have major worldwide health impacts.
4. Give examples of risk management activities.
5. Define natural resources and give examples of renewable and nonrenewable resources.
6. Describe how the supply of a good or service and the demand for it interact to determine the price.
7. Define and give examples of ecosystem services.
8. Give environmental examples of deferred costs, external costs, and opportunity costs.
9. List four ways in which environmental systems differ from economic systems.
10. What is incorporated in a cost-benefit analysis?
11. What are some of the concerns about the use of cost-benefit analysis in environmental decision making?
12. Differentiate between pollution costs and pollution-prevention costs.
13. Define the problem of common property resource ownership. Provide some examples.
14. Give examples of subsidies, market-based instruments, and life cycle analysis.
15. What kinds of risks are most willingly accepted by people?

Critical Thinking Questions

1. If you were a regulatory official, what kind of information would you require to make a decision about whether a certain chemical was "safe" or not? What level of risk would you deem acceptable for society? For yourself and your family?
2. Why do you suppose some carcinogenic agents, such as those in cigarettes, are so difficult to regulate?
3. Imagine you were assessing the risk of a new chemical plant being built along the Mississippi River in Louisiana. Identify some of the risks that you would want to assess. What kinds of data would you need in order to assess whether or not the risk was acceptable? Do you think that some risks are harder to quantify than others? Why?
4. Granting polluting industries or countries the right to buy and sell emissions permits is a controversial idea. Some argue that the market is the best way to limit pollution. Others argue that trade in permits allows polluting industries to continue to pollute and concentrates that pollution. What do you think?
5. Imagine you are an independent economist who is conducting a cost-benefit analysis of a hydroelectric project. What might be the costs of this project? The benefits? How would you quantify the costs of the project? The benefits? What kinds of costs and benefits might be hard to quantify or might be too tangential to the project to figure into the official estimates?
6. Do you think environmentalists should or should not stretch traditional cost-benefit analysis to include how development affects the environment? What are the benefits to this? The risks?
7. Looking at your own life, what kinds of risks do you take? What kinds are you unwilling to take? What criteria do you use to make a decision about acceptable and unacceptable risk?
8. Is current worldwide growth and development sustainable? If there were less growth, what would be the effect on developing countries? How could we achieve a just distribution of resources and still limit growth?
9. Should our policies reflect an interest in preserving resources for future generations? If so, what level of resources should be preserved? What would you be willing to do without in order to save for the future?
10. If you owned a small business in the United States and were looking to expand your business within your home state, would you consider purchasing a contaminated piece of property in a downtown or urban area? Why or why not? Would you be concerned about the environmental liabilities associated with it? Would you be worried about the health and safety of your workers if you located there? Why? What might be your concerns, or how could you find out? Would you conduct an environmental assessment prior to applying for financing the purchase and construction? Knowing what you know now, what type of risk tolerance do you have concerning your finances, your workers' health and safety, and the environment?

|ENVIRONMENTAL SCIENCE

To access additional resources for this chapter, please visit ConnectPlus® at connect.mheducation.com. There you will find interactive exercises, including Google Earth™, additional Case Studies, and SmartBook™, an adaptive eBook that integrates our LearnSmart® adaptive learning technology.

Interrelated Scientific Principles: Matter, Energy, and Environment

Many of the processes that occur in the natural world involve interactions between matter and energy. The photo above shows topsoil being washed away by the earthquake and resulting tsunami wave that hit Japan in 2011.

OBJECTIVES

After reading this chapter, you should be able to:

- Understand that science is usually reliable because information is gathered in a manner that requires impartial evaluation and continuous revision.
- Understand that matter is made up of atoms that have a specific subatomic structure of protons, neutrons, and electrons.
- Recognize that each element is made of atoms that have a specific number of protons and electrons and that isotopes of the same element may differ in the number of neutrons present.
- Recognize that atoms may be combined and held together by chemical bonds to produce molecules.
- Understand that rearranging chemical bonds results in chemical reactions and that these reactions are associated with energy changes.
- Recognize that matter may be solid, liquid, or gas, depending on the amount of kinetic energy contained by the molecules.
- Realize that energy can be neither created nor destroyed, but when energy is converted from one form to another, some energy is converted into a less useful form.
- Understand that energy can be of different qualities.

Wood Stoves and Air Pollution: A Cause and Effect Relationship

As prices for oil and gas push more people toward alternative fuels, concern over the health effects of burning wood in your home are being raised. Scientists have long known that wood smoke contains carbon monoxide and cancer causing chemicals. Smoke produced from woodstoves and fireplaces contains over 100 different chemical compounds, many of which are harmful and potentially carcinogenic. Wood smoke pollutants include fine particulates, nitrogen oxides, sulfur oxides, carbon monoxide, volatile organic compounds, dioxins, and furans. Breathing air containing wood smoke can cause a number of serious respiratory and cardiovascular health problems. Those at greatest health risk from wood smoke include infants, children, and pregnant women, the elderly, and those suffering from allergies, asthma, bronchitis, emphysema, pneumonia, or any other heart or lung disease.

Pollution from wood stoves is a particular concern in the winter when cold, stagnant air and temperature inversions limit air movement. Communities located in valleys are more strongly affected. As wood burning increases on cold, clear, calm nights, smoke is unable to rise and disperse. Pollutants are trapped and concentrated near the ground, and the small size of the particles allows them to seep into houses through closed doors and windows. In many communities 30 to 80 percent of the wintertime particle pollution is attributed to wood burning in the home.

The Environmental Protection Agency (EPA) estimates that wood stoves are responsible for 5 percent of the smallest, deadliest particles emitted in the United States. That may not seem much; however, many big industrial sources of particles are cleaning up their emissions, and thus residential wood smoke becomes a very important source of particle pollution. The return to wood burning stoves could be especially bad for the 200-plus counties in the United States where levels of particle pollution are higher than federal safety limits.

The EPA set standards for wood stoves in 1990. Stoves cannot be sold to consumers in the U.S. unless they meet certain emission standards for particulate matter and carry the EPA Emission Certification label. Certified stoves reduce smoke emissions by as much as 90 percent, compared with conventional stoves, and are much more efficient. EPA-certified stoves often include design features that promote secondary combustion aimed at burning off dangerous chemicals and toxic substances before they leave the firebox. The problem, however, is that three-quarters of the wood stoves in the U.S. were installed before 1990.

Examples of community and state laws pertaining to wood smoke include:

Denver, Colorado, has mandatory burning bans on "red" advisory days during the high-pollution season.

San Joaquin County, California, requires wood stoves to be replaced with an EPA certified wood stove when a home is sold. It also limits the number of wood stoves or fireplaces that can be installed in new residential units.

The state of Idaho offers taxpayers who buy new wood stoves a tax deduction to replace old, uncertified wood stoves.

The state of Washington established wood stove emission performance standards that are more restrictive than the federal rule. The state also assesses a flat fee on the sale of every wood burning device to fund the education of citizens about wood smoke and health and air impacts.

The EPA has an education campaign regarding using older wood stoves.

4.1 The Nature of Science

Since environmental science involves the analysis of data, it is useful to understand how scientists gather and evaluate information. It is also important to understand some chemical and physical principles as a background for evaluating environmental issues. An understanding of these scientific principles will also help you to appreciate the ecological concepts in the chapters that follow.

The word *science* creates a variety of images in the mind. Some people feel that it is a powerful word and are threatened by it. Others are baffled by scientific topics and have developed an unrealistic belief that scientists are brilliant individuals who can solve any problem. For example, there are those who believe that the conservation of fossil fuels is unnecessary because scientists will soon "find" a replacement energy source. Similarly, many are convinced that if government really wanted to, it would allocate sufficient funds to allow scientists to find a cure for AIDS. Such images do not accurately portray what science is really like.

Science is a process used to solve problems or develop an understanding of nature that involves testing possible answers. Science is distinguished from other fields of study by how knowledge is acquired rather than by what is studied. The process has become known as the *scientific method*. The **scientific method** is a way of gaining information (facts) about the world by forming possible solutions to questions, followed by rigorous testing to determine if the proposed solutions are valid.

Basic Assumptions in Science

When using the scientific method, scientists make several fundamental assumptions. They presume that:

1. There are specific causes for events observed in the natural world;
2. The causes can be identified;
3. There are general rules or patterns that can be used to describe what happens in nature;
4. An event that occurs repeatedly probably has the same cause each time;
5. What one person perceives can be perceived by others; and
6. The same fundamental rules of nature apply regardless of where and when they occur.

For example, we have all observed lightning associated with thunderstorms. According to the assumptions just stated, we should expect that there is an explanation that would account for all cases of lightning regardless of where or when they occur and that all people could make the same observations. We know from scientific observations and experiments that lightning is caused by a difference in electrical charge, that the behavior of lightning follows general rules that are the same as those seen with static electricity, and that all lightning that has been measured has the same cause wherever and whenever it occurred.

Cause-and-Effect Relationships

Scientists distinguish between situations that are merely correlated (happen together) and those that are correlated and show **cause-and-effect relationships.** Many events are correlated, but not all correlations show cause-and-effect. When an event occurs as a direct result of a previous event, a cause-and-effect relationship exists. For example, lightning and thunder are correlated—thunder follows lightning—but they also have a cause-and-effect relationship—lightning causes thunder.

The relationship between autumn and trees dropping their leaves is more difficult to sort out. Because autumn brings colder temperatures, many people assume that the cold temperature causes leaves to turn color and fall. Cold temperatures are correlated with falling leaves. However there is no cause-and-effect relationship. The cause of the change in trees is actually the shortening of days that occurs in the autumn. Experiments have shown that artificially shortening the length of days in a greenhouse will cause trees to drop their leaves with no change in temperature. Knowing that a cause-and-effect relationship exists enables us to predict what will happen should that same set of circumstances occur in the future.

Elements of the Scientific Method

The scientific method requires a systematic search for information and a continual checking and rechecking to see if previous ideas are still supported by new information. If the new evidence is not supportive, scientists discard or change their original ideas. Scientific ideas undergo constant reevaluation, criticism, and modification. The scientific method involves several important identifiable components, including:

- careful observation,
- asking questions about observed events,
- the construction and testing of hypotheses,
- an openness to new information and ideas, and
- a willingness to submit one's ideas to the scrutiny of others.

Underlying all of these activities is constant attention to accuracy and freedom from bias.

Observation

Scientific inquiry often begins with an observation that an event has occurred. An **observation** occurs when we use our senses (smell, sight, hearing, taste, touch) or an extension of our senses (microscope, tape recorder, X-ray machine, thermometer) to record an event. Observation is more than a casual awareness. You may hear a sound or see an image without really observing it. Do you know what music was being played in the shopping mall? You certainly heard it, but if you are unable to tell someone else what it was, you didn't "observe" it. If you had prepared yourself to observe the music being played, you would be able to identify it. When scientists talk about their observations, they are referring to careful, thoughtful recognition of an event—not just casual notice. Scientists train themselves to improve their observational skills, since careful observation is important in all parts of the scientific method. (See figure 4.1.)

FIGURE 4.1 Observation Careful observation is an important part of the scientific method. This scientist is taking water samples to check for signs of pollution entering a stream.

Because many of the instruments used in scientific investigations are complicated, we might get the feeling that science is incredibly complex, when in reality these sophisticated tools are being used simply to answer questions that are relatively easy to understand. For example, a microscope has several knobs to turn and a specially designed light source. It requires considerable skill to use properly, but it is essentially a fancy magnifying glass that allows small objects to be seen more clearly. The microscope has enabled scientists to answer some relatively fundamental questions such as: Are there living things in pond water? and Are living things made up of smaller subunits? Similarly, chemical tests allow us to determine the amounts of specific materials dissolved in water, and a pH meter allows us to determine how acidic or basic a solution is. Both are simple activities, but if we are not familiar with the procedures, we might consider the processes hard to understand.

Questioning and Exploring

Observations often lead one to ask questions about the observations. Why did this event happen? Will it happen again in the same circumstances? Is it related to something else? Some questions may be simple speculation, but others may inspire you to further investigation. The formation of the questions is not as simple as it might seem because the way the questions are asked will determine how you go about answering them. A question that is too broad or too complex may be impossible to answer; therefore, a great deal of effort is put into asking the question in the right way. In some situations, this can be the most time-consuming part of the scientific method; asking the right question is critical to how you look for answers. For example, you observe that robins eat the berries of many plants but avoid others. You could ask the following questions:

1. Do the robins dislike the flavor of some berries?
2. Will robins eat more of one kind of berry if given a choice between two kinds of berries?

The second question is obviously easier to answer.

Once a decision has been made about what question to ask, scientists *explore other sources of knowledge* to gain more information. Perhaps the question already has been answered by someone else or several possible answers already have been rejected. Knowing what others have already done saves one time. This process usually involves reading appropriate science publications, exploring information on the Internet, or contacting fellow scientists interested in the same field of study. Even if the particular question has not already been answered, scientific literature and other scientists can provide insights that may lead to a solution. After exploring the appropriate literature, a decision is made about whether to continue to explore the question. If the scientist is still intrigued by the question, a formal hypothesis is constructed, and the process of inquiry continues at a different level.

Constructing Hypotheses

A **hypothesis** is a statement that provides a possible answer to a question or an explanation for an observation that can be tested. A good hypothesis must be logical, account for all the relevant information currently available, allow one to predict future events relating to the question being asked, and be testable. Furthermore, if one has the choice of several competing hypotheses, one should use the simplest hypothesis with the fewest assumptions. Just as deciding which questions to ask is often difficult, the formation of a hypothesis requires much critical thought and mental exploration. If the hypothesis is not logical or does not account for all the observed facts in the situation, it must be rejected. If a hypothesis is not testable it is mere speculation.

Testing Hypotheses

Keep in mind that a hypothesis is based on observations and information gained from other knowledgeable sources. It predicts how an event will occur under specific circumstances. Scientists test the predictive ability of a hypothesis to see if the hypothesis is supported or is disproved. If you disprove the hypothesis, it is rejected, and a new hypothesis must be constructed. However, if you cannot disprove a hypothesis, it increases your confidence in the hypothesis, but it does not prove it to be true in all cases and for all time. Science always allows for the questioning of ideas and the substitution of new ones that more completely describe what is known at a particular point in time. It could be that an alternative hypothesis you haven't thought of explains the situation or that you have not made the appropriate observations to indicate that your hypothesis is wrong.

The test of a hypothesis can take several forms. It may simply involve the collection of pertinent information that already exists from a variety of sources. For example, if you visited a cemetery and observed from reading the tombstones that an unusually large number of people of different ages died in the same year, you could hypothesize that there was an epidemic of disease or a natural disaster that caused the deaths. Consulting historical newspaper accounts would be a good way to test this hypothesis.

In other cases, a hypothesis may be tested by simply making additional observations. For example, if you hypothesized that a certain species of bird used cavities in trees as places to build nests, you could observe many birds of the species and record the kinds of nests they built and where they built them.

Another common method for testing a hypothesis involves devising an experiment. An **experiment** is a re-creation of an event or occurrence in a way that enables a scientist to support or disprove a hypothesis. This can be difficult because a particular event may involve a great many separate happenings called **variables.** The best experimental design is a **controlled experiment** in which two groups differ in only one way. For example, tumors of the skin and liver occur in the fish that live in certain rivers *(observation).* This raises the question: What causes the tumors? Many people feel that the tumors are caused by toxic chemicals that have been released into the rivers by industrial plants *(hypothesis).* However, it is possible that the tumors are caused by a virus, by exposure to natural substances in the water, or are the result of genes present in the fish. The following experiment could be conducted to test the hypothesis that industrial contaminants cause the tumors: Fish could be collected from the river and placed in one of two groups. One group (the control group) would be raised in a container through which the normal river water passes. The second group (experimental group) would be raised in an identical container through which water from the industrial facility passes. There would need to be large numbers of fish in both groups. This kind of experiment is called a controlled experiment. If the fish in the experimental group develop a significantly larger number of tumors than the control group, something in the water from the plant is the probable cause of the tumors. This is particularly true if the chemicals present in the water are already known to cause tumors. After the data have been evaluated, the results of the experiment would be published.

The results of a well-designed experiment should be able to support or disprove a hypothesis. However, this does not always occur. Sometimes the results of an experiment are inconclusive. This means that a new experiment must be conducted or that more information must be collected. Often, it is necessary to have large amounts of information before a decision can be made about the validity of a hypothesis. The public often fails to appreciate why it is necessary to perform experiments on so many subjects or to repeat experiments again and again.

The concept of **reproducibility** is important to the scientific method. Because it is often not easy for scientists to eliminate unconscious bias, independent investigators must be able to reproduce the experiment to see if they get the same results. To do this, they must have a complete and accurate written document to work from. That means the scientists must publish the methods and results of their experiment. This process of publishing one's work for others to examine and criticize is one of the most important steps in the process of scientific discovery. The results of experiments are only considered reliable if they are supported by many experiments and by different investigators.

The Development of Theories and Laws

When broad consensus exists about an area of science, it is known as a theory or law. A **theory** is a widely accepted, plausible generalization about fundamental concepts in science that explains *why* things happen. An example of a scientific theory is the **kinetic molecular theory,** which states that all matter is made up of tiny, moving particles. As you can see, this is a very broad statement,

and it is the result of years of observation, questioning, experimentation, and data analysis. Because we are so confident that the theory explains the nature of matter, we use this concept to explain why materials disperse in water or air, why materials change from solids to liquids, and why different chemicals can interact during chemical reactions.

Theories and hypotheses are different. A hypothesis provides a possible explanation for a specific question; a theory is a broad concept that shapes how scientists look at the world and how they frame their hypotheses. Because they are broad, unifying statements, there are few theories. However, just because a theory exists does not mean that testing stops. As scientists continue to gain new information, they may find exceptions to a theory or even in rare cases disprove a theory.

It is important to recognize that the word *theory* is often used in a much less restrictive sense. Often it is used in ordinary conversation to describe a vague idea or a hunch. This is not a theory in the scientific sense. So when you see or hear the word *theory,* you must look at the context to see if the speaker or writer is referring to a theory in the scientific sense.

A **scientific law** is a uniform or constant fact of nature that describes *what* happens in nature. An example of a scientific law is the **law of conservation of mass,** which states that matter is not gained or lost during a chemical reaction. While laws describe what happens and theories describe why things happen, laws and theories are similar in a way. They have both been examined repeatedly and are regarded as excellent predictors of how nature behaves.

Communication

At several points in the discussion of the scientific method the significance of communication has arisen. Communication is a central characteristic of the scientific method. Science is conducted openly, under the critical eyes of others who are interested in the same questions. An important part of the communication process involves the publication of articles in scientific journals about one's research, thoughts, and opinions. This communication can occur at any point during the process of scientific discovery.

People may ask questions about unusual observations. They may publish preliminary results of incomplete experiments. They may publish reports that summarize large bodies of material. And they often publish strongly held opinions that may not always be supportable with current data. This provides other scientists with an opportunity to criticize, make suggestions, or agree. (See figure 4.2.) Scientists also talk to one another at conferences and by phone, e-mail, and the Internet. The result is that science is subjected to examination by many minds as it is discovered, discussed, and refined.

4.2 Limitations of Science

Science is a powerful tool for developing an understanding of the natural world, but it cannot analyze international politics, decide if family-planning programs should be instituted, or evaluate the significance of a beautiful landscape. These tasks are beyond the scope

FIGURE 4.2 Communication One important way in which scientists communicate is through publication in scientific journals.

It is important to recognize that scientific knowledge can be used by different people to support opinions that may not be valid. For example, the following statements are all factual.

1. Many of the kinds of chemicals used in modern agriculture are toxic to humans and other animals.

2. Small amounts of agricultural chemicals have been detected in some agricultural products.

3. Low levels of some toxic materials have been strongly linked to a variety of human illnesses.

This does not mean that *all* foods grown with the use of *any* chemicals are less nutritious or are dangerous to health or that "organically grown" foods are necessarily more nutritious or healthful because they have been grown without agricultural chemicals. The idea that something that is artificial is necessarily bad and something natural is necessarily good is an oversimplification. After all, many plants such as tobacco, poison ivy, and rhubarb leaves naturally contain toxic materials. Furthermore, the use of chemical fertilizers has contributed to the health and well-being of the human population because increased crop yields due to fertilizer use account for about one-third of the food grown in the world, thus reducing malnutrition. However, it is appropriate to question if the use of agricultural chemicals is always necessary or if trace amounts of specific agricultural chemicals in food are dangerous.

The scientific method is not, however, an inflexible series of steps that must be followed in a specific order. Figure 4.3 shows how these steps may be linked.

An example of applying the scientific method would be acid rain. In the 1970s people in North America began to *observe* that in the northeastern United States and adjacent Canada, certain bodies of water and certain forested areas were changing. Fish were disappearing and trees were dying. They began to ask *questions* about why this was occurring. At the same time people began to make *observations* about the pH of rain and snow. They found that in many places in the Northeast the pH of rain was very low—it was acid. Normal rain has a pH of about 5.7, and measurements of rain indicated that the pH was often much more acid than normal. People began to ask if there was a *cause-and-effect relationship* between the acid precipitation and the decline of forests and certain lakes.

When the pH of the water in many northeastern lakes was measured it was determined that they were acid and that there was probably a *cause-and-effect relationship* between the acid condition of the water in the lakes and the decline of fish. (Although forest decline was a more difficult problem to understand, research eventually established a *cause-and-effect relationship* between forest decline and acid rain.) The next *question* that was asked was, what was the cause of the acidification of the lakes? Since many of the lakes were downwind of industrial centers, people asked the *question* Is the low pH of the rain the result of industrial activities? In particular, what activities might cause the acid rain?

Experimental evidence linking industrial emissions to acid rain came from several kinds of studies.

1. Measurements of the emissions from industrial smokestacks revealed that there were a variety of compounds in the emissions that were acidic in nature. In particular, sulfur

of scientific investigation. This does not mean that scientists cannot comment on such issues. They often do. But they should not be regarded as more knowledgeable on these issues just because they are scientists. Scientists may know more about the scientific aspects of these issues, but they struggle with the same moral and ethical questions that face all people, and their judgments on these matters can be just as biased as anyone else's. Consequently, major differences of opinion often exist among lawmakers, regulatory agencies, special interest groups, and members of scientific organizations about the significance or value of specific scientific information.

It is important to differentiate between the scientific data collected and the opinions scientists have about what the data mean. Scientists form and state opinions that may not always be supported by fact, just as other people do. Equally reputable scientists commonly state opinions that are in direct contradiction. This is especially true in environmental science, where predictions about the future must be based on inadequate or fragmentary data. The issue of climate change (covered in chapter 17) is an example of this.

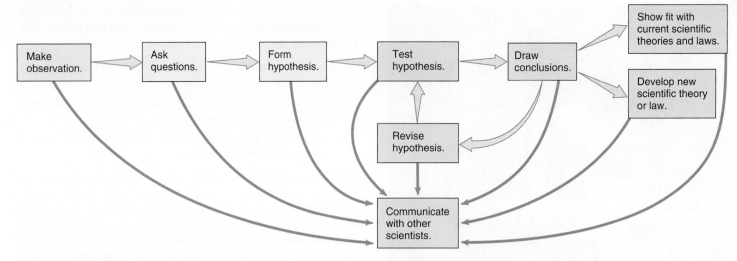

FIGURE 4.3 Elements of the Scientific Method The scientific method consists of several kinds of activities. Observation of a natural phenomenon is usually the first step. Observation often leads people to ask questions about the observation they have made or to try to determine why the event occurred. This questioning is typically followed by the construction of a hypothesis that attempts to explain why the phenomenon occurred. The hypothesis is then tested to see if it is supported. Often this involves experimentation. If the hypothesis is not substantiated, it is modified and tested in its new form. It is important at all times that others in the scientific community be informed by publishing observations of unusual events, their probable cause, and the results of experiments that test hypotheses. Occasionally, this method of inquiry leads to the development of theories that tie together many bits of information into broad statements that state why things happen in nature and serve to guide future thinking about a specific area of science. Scientific laws are similar broad statements that describe how things happen in nature.

compounds from the burning of high-sulfur coal provided compounds that formed acids.

2. The rain in areas that were not downwind of industrial settings had a more normal pH.

3. Reducing emissions resulted in less acid rain.

Since this is a complex problem and involves many scientists working on particular aspects of the problem, there was a great deal of *communication* among scientists as these studies were occurring.

Ultimately, most scientists *concluded* that the cause of the acidification of lakes and the decline of forests was acid rain and

Testing for Acid Rain

the cause of the acid rain was industrial emissions. During the 1990s the U.S. Congress passed legislation that led to the reduction of acid emissions from industrial sources. Efforts to reduce emissions of acid-forming substances continue today. However, as sulfur emissions have been reduced the emphasis today is on nitrogen compounds that result from the burning of fuels by industry, automobiles, and homes.

4.3 Pseudoscience

Pseudoscience (*pseudo* = false) is a deceptive practice that uses the appearance or language of science to convince, confuse, or mislead people into thinking that something has scientific validity when it does not. When pseudoscientific claims are closely examined, it is found that they are not supported by unbiased tests.

Often facts are selected to support a particular point of view. For example, certain kinds of ionizing radiation are known to cause cancer. Electrical devices and power lines emit electromagnetic radiation. These facts have caused many people to see a link between exposure to electromagnetic radiation and health, when there is none. They maintain that living near electric power lines causes cancer or other health effects. However, careful scientific studies, which compared people who lived near power lines with matched groups of people who did not live near power lines, show that those who live near power lines have no more health problems than those who do not. Often it is interesting to look at the motivation of those making pseudoscientific claims. Are they selling something? Are they seeking money for presumed injuries?

Going Green

Evaluating Green Claims

Businesses have discovered that there is money to be made by marketing their products as ecologically friendly, Earth friendly, or green. But how do you know when a product is truly green? Most of us do not have the expertise to evaluate highly technical statements about products, so it is useful to have some general guidelines that help separate the charlatans from those that have a sincere interest in marketing a green product. Be a skeptical consumer.

1. Look for statements that are specific. Statements like "contains all natural ingredients," or "biodegradable," or "nontoxic" are not useful unless there is other evidence provided to support the claim. A "natural ingredients" statement should be backed up with a list of the actual ingredients and their sources. A claim of being biodegradable is not a very useful term, since most things will biodegrade if given enough time and the proper microorganisms. However, if there are statements about how quickly materials break down or that they break down in septic systems or sewage treatment plants, you can have more confidence. A claim of being nontoxic is meaningless—nearly everything is toxic in high enough concentrations, even table salt, ethanol, and vinegar.

2. Look for evidence that companies have a general environmental commitment. Do they market in recycled containers? Do they sell concentrated products that can be diluted at home, thus saving energy? Do they sell with a minimum of packaging?

3. Look for statements about the use of recycled materials in the product. Saying a product or container is recyclable is not very useful if there is no easy way to recycle it. Does the business have a program for recycling its products? If a claim is made that the product is made from recycled materials, evaluate the claim. What percent is post-consumer waste? Many products that claim to be made from recycled materials are made from materials salvaged during the manufacturing process.

4. Look for evidence that the claims are highlighting one positive element while ignoring other important issues. A statement that a product contains no chlorine is not useful if the other ingredients are as bad or worse.

5. Look for statements that take credit for things they are legally required to do. "Meets all FDA requirements" and "contains no CFCs" are meaningless statements because it would be illegal to sell the product if it didn't meet FDA requirements and it is illegal to sell products containing CFCs.

4.4 The Structure of Matter

Now that we have an appreciation for the methods of science, it is time to explore some basic information and theories about the structure and function of various kinds of matter. **Matter** is anything that takes up space and has mass. Air, water, trees, cement, and gold are all examples of matter. As stated earlier, the kinetic molecular theory is a central theory that describes the structure and activity of matter. This theory states that all matter is made up of tiny objects that are in constant motion. Although different kinds of matter have different properties, they all are similar in one fundamental way. They are all made up of one or more kinds of smaller subunits called *atoms.*

Atomic Structure

Atoms are fundamental subunits of matter. There are 92 kinds of atoms found in nature. Each kind forms a specific type of matter known as an **element.** Gold (Au), oxygen (O), and mercury (Hg) are examples of elements. All atoms have a central region known as a **nucleus,** which is composed of two kinds of relatively heavy particles: positively charged particles called **protons** and uncharged particles called **neutrons.** Surrounding the nucleus of the atom is a cloud of relatively lightweight, fast-moving, negatively charged particles called **electrons.**

The atoms of each element differ from one another in the number of protons, neutrons, and electrons present. For example,

a typical mercury atom contains 80 protons and 80 electrons; gold has 79 of each, and oxygen only eight of each. (See figure 4.4.) (Appendix 1, page 471, contains a periodic table of the elements.) All atoms of an element always have the same number of protons and electrons, but the number of neutrons may vary from one atom to the next.

Atoms of the same element that differ from one another in the number of neutrons they contain are called **isotopes.** For example,

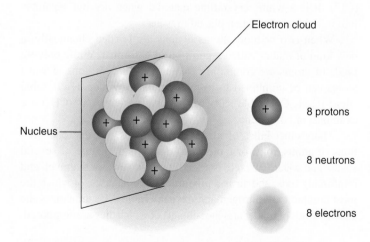

FIGURE 4.4 Diagrammatic Oxygen Atom Most oxygen atoms are composed of a nucleus containing eight positively charged protons and eight neutrons without charges. Eight negatively charged electrons move in a cloud around the nucleus.

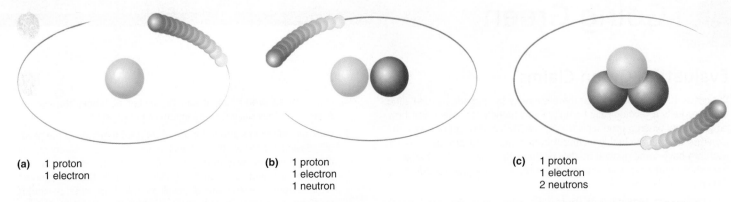

(a) 1 proton
1 electron

(b) 1 proton
1 electron
1 neutron

(c) 1 proton
1 electron
2 neutrons

FIGURE 4.5 Isotopes of Hydrogen (a) The most common form of hydrogen is the isotope that has one proton and no neutrons in its nucleus. (b) The isotope deuterium has one proton and one neutron. (c) The isotope tritium has two neutrons and one proton. Each of these isotopes of hydrogen also has one electron. Most scientists use the term *hydrogen* in a generic sense—that is, the term is not specific but might refer to any or all of these isotopes.

there are three isotopes of the element hydrogen. All hydrogen atoms have one proton and one electron, but one isotope of hydrogen has no neutrons, one has one neutron, and one has two neutrons. These isotopes behave the same chemically but have different masses since they contain different numbers of neutrons. (See figure 4.5.)

The Molecular Nature of Matter

The kinetic molecular theory states that all matter is made of tiny particles that are in constant motion. However, there are several kinds of these tiny particles. In some instances, atoms act as individual particles. In other instances, atoms bond to one another chemically to form stable units called **molecules.** In still other cases, atoms or molecules may gain or lose electrons and thus become electrically charged particles called **ions.** Atoms or molecules that lose electrons are positively charged because they have more protons (+) than electrons (−). Those that gain electrons have more electrons (−) than protons (+) and are negatively charged.

Oppositely charged ions are attracted to one another and may form stable units similar to molecules; however, they typically split into their individual ions when dissolved. For example, table salt (NaCl) is composed of sodium ions (Na^+) and chloride ions (Cl^-). It is a white, crystalline material when dry but separates into individual ions when placed in water.

When two or more atoms or ions are bonded chemically, a new kind of matter called a **compound** is formed. While only 92 kinds of atoms are commonly found, there are millions of ways atoms can be combined to form compounds. Water (H_2O), table sugar ($C_6H_{12}O_6$), table salt (NaCl), and methane gas (CH_4) are examples of compounds.

Many other kinds of matter are **mixtures,** variable combinations of atoms, ions, or molecules. Honey is a mixture of several sugars and water; concrete is a mixture of cement, sand, gravel, and reinforcing rods; and air is a mixture of several gases, of which the most common are nitrogen and oxygen. Table 4.1 summarizes the various kinds of matter and the subunits of which they are composed.

A Word About Water

The temperature of the Earth is such that water can exist in all three phases—solid, liquid, and gas. If we look at Earth from space, it is a blue planet with patches of white. The blue color is caused by the three-quarters of the surface of the Earth covered with water in the oceans. The wisps of white are caused by water droplets in clouds. Water in all its forms determines the weather and climate of regions of the Earth. Earth's surface is shaped by the flow of water and ice. Life on Earth is based on water. Most kinds of organisms live in water and the most common molecule found in living things is water.

Water molecules are polar molecules with positive and negative ends. Since unlike charges attract each other, water molecules tend to stick together. This contributes to several of water's unusual properties. Because water molecules are polar, it takes a great deal of energy to cause water molecules to separate from one another—to go from the liquid form to the water vapor. Thus, the evaporation of water has the ability to cool the surroundings. The polar nature of water molecules also contributes to its ability to dissolve most substances. It is often described as the universal solvent. Almost everything dissolves, to some extent, in water. Many important solutions are formed as a result of water.

Acids, Bases, and pH

Acids and bases are two classes of compounds that are of special interest. Their characteristics are determined by the nature of their chemical bonds. When acids are dissolved in water, hydrogen ions (H^+) are set free. A *hydrogen ion* is positive because it has lost its electron and now has only the positive charge of its proton. Therefore, a hydrogen ion is a proton. An **acid** is any compound that releases hydrogen ions (protons) in a solution. Some familiar examples of common acids are sulfuric acid (H_2SO_4) in automobile batteries and acetic acid (HCH_2COOH) in vinegar.

A **base** is the opposite of an acid in that it accepts hydrogen ions in solution. Many common bases release **hydroxide ions** (OH^-). This ion is composed of an oxygen atom and a hydrogen atom bonded together but with an additional electron. The hydroxide ion is negatively charged. It is a base because it is able to accept hydrogen ions in solution to form water ($H^+ + OH^- \rightarrow H_2O$). A very strong base often used in oven cleaners is sodium hydroxide (NaOH). Sometimes people refer to a base as an alkali and solutions that are basic often are called alkaline solutions.

Table 4.1 Relationships Between the Kinds of Subunits Found in Matter

Category of Matter	Subunits	Characteristics
Subatomic particles	protons	Positively charged
		Located in nucleus of the atom
	neutrons	Have no charge
		Located in nucleus of the atom
	electrons	Negatively charged
		Located outside the nucleus of the atom
Elements	atoms	Atoms of an element are composed of specific arrangements of protons, neutrons, and electrons. Atoms of different elements differ in the number of protons, neutrons, and electrons present.
Compounds	molecules or ions	Compounds are composed of two or more atoms or ions chemically bonded together.
		Different compounds contain specific atoms or ions in specific proportions.
Mixtures	atoms, molecules, or ions	The molecular particles in mixtures are not chemically bonded to each other.
		The number of each kind of molecular particle present is variable.

The concentration of an acid or base solution is given by a number called its **pH.** The pH scale is a measure of hydrogen ion concentration. However, the pH scale is different from what you might expect. First, it is an inverse scale, which means that the lower the pH, the greater the number of hydrogen ions present. Second, the scale is logarithmic, which means that a difference between two consecutive pHs is really a difference by a factor of 10. For example, a pH of 7 indicates that the solution is neutral and has an equal number of H^+ ions and OH^- ions, but a pH of 6 means that the solution has 10 times more hydrogen ions than it would at a pH of 7.

As the number of hydrogen ions in the solution increases, the pH gets smaller. A number higher than 7 indicates that the solution has more OH^- than H^+. As the number of hydroxide ions increases, the pH gets larger. The higher the pH, the more concentrated the hydroxide ions. (See figure 4.6.)

Inorganic and Organic Matter

Inorganic and organic matter are usually distinguished from one another by one fact: Organic matter consists of molecules that contain carbon atoms that are usually bonded to form chains or rings. Consequently, organic molecules can be very large. Many different kinds of organic compounds exist. Inorganic compounds generally consist of small molecules and combinations of ions, and relatively few kinds exist. All living things contain molecules of organic compounds. They must either be able to manufacture organic compounds from inorganic compounds or to modify organic compounds they obtain from eating organic material. Typically, chemical bonds in organic molecules contain a large amount of chemical energy that can be released when the bonds

are broken and new inorganic compounds are produced. Salt, water, metals, sand, and oxygen are examples of inorganic matter. Sugars, proteins, and fats are examples of organic compounds that are produced and used by living things. Natural gas, oil, and coal are all examples of organic substances that were originally produced by living things but have been modified by geologic processes.

Chemical Reactions

When atoms or ions combine to form compounds, they are held together by chemical bonds, as in the water molecule in figure 4.7. **Chemical bonds** are attractive forces between atoms resulting from the interaction of their electrons. Each chemical bond contains a certain amount of energy. When chemical bonds are broken or formed, a chemical reaction occurs. During chemical reactions, the amount of energy within the chemical bonds changes. If the chemical bonds in the new compounds have less chemical energy than the previous compounds, some of the energy is released, often as heat and light. These kinds of reactions are called **exothermic reactions.** In other cases, the newly formed chemical bonds contain more energy than was present in the compounds from which they were formed. Such reactions are called **endothermic reactions.** For such a reaction to occur, the additional energy must come from an external source.

A common example of an exothermic reaction is the burning of natural gas. The primary ingredient in natural gas is the compound methane. When methane and oxygen are mixed together and a small amount of energy is used to start the reaction, the chemical bonds in the methane and oxygen (reactants) are rearranged to form two different compounds, carbon dioxide and water (products). In this kind of reaction, the products have less energy than the reactants. The leftover energy is released as light and heat. (See figure 4.8.) In every reaction, the amount of energy in the reactants and in the products can be compared and the differences accounted for by energy loss to, or gain from, the surroundings.

Even energy-yielding reactions usually need an input of energy to get the reaction started. This initial input of energy is called **activation energy.** In certain cases, the amount of activation energy required to start the reaction can be reduced by the use of a catalyst. A **catalyst** is a substance that alters the rate of a reaction, but the catalyst itself is not consumed or altered in the process. Catalysts are used in catalytic converters, which are attached to automobile exhaust systems. The purpose of the catalytic converter is to bring about more complete burning of the fuel, thus resulting in less air pollution. Most of the materials that are

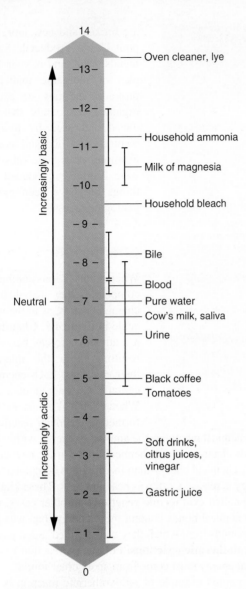

FIGURE 4.6 **The pH Scale** The concentration of hydrogen ions (H⁺) is greatest when the pH is lowest. At a pH of 7.0, the concentrations of H⁺ and OH⁻ are equal. We usually say as the pH gets smaller, the solution becomes more acidic. As the pH gets larger, the solution becomes more basic or alkaline.

not completely burned by the engine require high temperatures to react further; with the presence of catalysts, these reactions can occur at lower temperatures.

Endothermic reactions require an input of energy in order to occur. For example, nitrogen gas and oxygen gas can be combined to form nitrous oxide ($O_2 + N_2 + heat \rightarrow 2NO$). Another important endothermic reaction is the process of photosynthesis, which is discussed in the following section.

Chemical Reactions in Living Things

Living things are constructed of cells that are themselves made up of both inorganic and organic matter in very specific arrangements. The chemical reactions that occur in living things are regulated by protein molecules called **enzymes** that reduce the activation energy needed to start the reactions. Enzymes are important since the high temperatures required to start these reactions without enzymes would destroy living organisms. Many enzymes are arranged in such a way that they cooperate in controlling a chain of reactions, as in photosynthesis and respiration.

Photosynthesis is the process plants use to convert inorganic material into organic matter, with the assistance of light energy. Light energy enables the smaller inorganic molecules (water and carbon dioxide) to be converted into organic sugar molecules. In the process, molecular oxygen is released.

$$6CO_2 + 6H_2O \xrightarrow[\text{in chloroplasts}]{\text{light energy}} C_6H_{12}O_6 + 6O_2$$

$$\text{Carbon dioxide} + \text{Water} \longrightarrow \text{Sugar} + \text{Oxygen}$$

Molecules of the green pigment chlorophyll are found in cellular structures called chloroplasts. Chlorophyll is responsible for trapping the sunlight energy needed in the process of photosynthesis. Therefore, photosynthesis takes place in the green portions of the plant, usually the leaves. (See figure 4.9.) The organic molecules produced as a result of photosynthesis can be used as an energy source by the plants and by organisms that eat the plants.

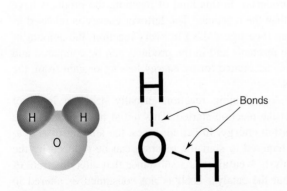

FIGURE 4.7 **Water Molecule** A water molecule consists of an atom of oxygen bonded to two atoms of hydrogen. The molecule is not symmetrical. The hydrogen atoms are not on opposite sides of the oxygen atom.

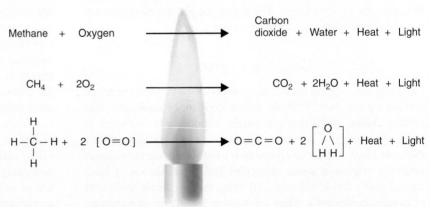

FIGURE 4.8 **A Chemical Reaction** When methane is burned, chemical bonds are rearranged, and the excess chemical bond energy is released as light and heat. The same atoms are present in the reactants as in the products, but they are bonded in different ways, resulting in molecules of new substances.

Science, Politics, & Policy

Return of the Salmon

In 2012 the Elwha River in the Olympic mountains of Washington State flowed freely for the first time in over 100 years. The river was historically one of the most productive salmon streams for its size in the Pacific Northwest. Four hundred thousand salmon once swam its length each year but, in the century since the dam's construction, that number had fallen to a few thousand. Within months of the dam's removal, nature rushed back: over 2,000 salmon returned in 2013. The prospect of a river teeming with salmon weighing over 45 kilograms each may no longer remain a memory of local Native American tribes.

The Elwha Dam removal project stands as one of the first large dams ever removed. The intent of removing the dams is to fully restore the river ecosystem and its native migratory fish species. In doing so, the Elwha Dam project revived the debate of how to balance the conflicting demands of humans for both clean energy and healthy ecosystems. Previously, that debate has been weighted decisively in favor of dam projects.

Construction of the Elwha Dam began in 1910 for the sole purpose of generating the first electricity in the region. The electricity powered lumber mills and fueled economic development, resulting in the construction of a second dam, the Glines Canyon Dam, farther upstream in 1927. The lower dam did not have fish passage and the salmon runs declined from 400,000 per year to about 3,000 fish. A fishery that could be worth over $10 million was lost. The near disappearance of salmon in the watershed also had a cascading effect on the terrestrial ecosystem, where some 22 species of resident wildlife were affected, and over 90 species of migratory birds.

Dams—for so long seen as symbols of development and progress—were increasingly being criticized for their social and environmental impact. The issue came into focus internationally following protests over the forced evictions of hundreds of thousands of people in India and China due to dam construction.

In September of 2011 work began on removing both dams. Removing the dam was only the first step of the real goal of restoring salmon fisheries. During the decade prior to removal, scientists surveyed fish populations in the river to inventory populations of native and migrating fish species. Fisheries biologists also captured Elwha River fish stock for transport to hatcheries and nearby streams for rearing in order to preserve genetic diversity.

Fish stocks will recover following complete deconstruction of the dams, stabilization of sediment transport, and the recovery of the ecosystem food chain that provides food for juvenile salmon that will grow in the Elwha River before migrating to the ocean. However, even in the short time following removal of the Elwha Dam, some wild salmon found the new habitat and spawned. The largest run of Chinook salmon in decades returned to the Elwha River in the fall of 2013. The salmon were found entering into stretches of the river and its tributaries that were formerly blocked by the dam.

The removal of dams on the Elwha River offers a unique opportunity to evaluate the effects of large dam removal and subsequent recovery of a formerly productive aquatic ecosystem that supported large

Prior to Elwah Dam removal.

During Elwah River Dam removal.

populations of salmon and a related complex ecosystem. Although international dam removal of this magnitude is unique, it could become more common as those in the United States and other nations manage an aging system of dams. A comprehensive plan designed to evaluate the effects of dam removal on existing fish populations, food webs and habitats, sediment flow, and many other factors, however, is essential before removing dams.

Sunlight energy

Carbon dioxide (CO₂) enters through leaf surface

Carbon dioxide + Water → Chlorophyll → Sugar + Oxygen

Oxygen (O₂) exits through leaf surface

Water (H₂O) is absorbed by the roots and enters leaf through stem

Sugar (C₆H₁₂O₆) is stored in roots

FIGURE 4.9 Photosynthesis This reaction is an example of one that requires an input of energy (sunlight) to combine low-energy molecules (CO₂ and H₂O) to form sugar (C₆H₁₂O₆) with a greater amount of chemical bond energy. Molecular oxygen (O₂) is also produced.

Respiration involves the use of atmospheric oxygen to break down large, organic molecules (sugars, fats, and proteins) into smaller, inorganic molecules (carbon dioxide and water). This process releases energy the organisms can use.

$$C_6H_{12}O_6 + 6O_2 \longrightarrow 6CO_2 + 6H_2O + Energy$$

$$Sugar + Oxygen \longrightarrow Carbondioxide + Water + Energy$$

(See figure 4.10.) All organisms, including plants, must carry on some form of respiration, since all organisms need a source of energy to maintain life.

Chemistry and the Environment

Chemistry is extremely important in discussions of environmental problems. The chemicals of fertilizer and pesticides are extremely important in food production, but they also present environmental problems. Photochemical smog is a problem created by chemical reactions that take place between pollutants and other components of the atmosphere. Persistent organic chemicals are those that are not broken down by microorganisms and remain in the environment for long periods. Organic molecules in water reduce the amount of oxygen available to animals that need it to stay alive. These and

other examples of environmental chemistry will be discussed in later chapters on agriculture, air pollution, and water pollution.

4.5 Energy Principles

The "Chemical Reactions in Living Things" section started out with a description of matter, yet it used the concept of energy to describe chemical bonds, chemical reactions, and molecular motion. That is because energy and matter are inseparable. It is difficult to describe one without the other. **Energy** is the ability to do work. Work is done when an object is moved over a distance. This occurs even at the molecular level.

Kinds of Energy

There are several kinds of energy. Heat, light, electricity, and chemical energy are common forms. The energy contained by moving objects is called **kinetic energy.** The moving molecules in air have kinetic energy, as does water running downhill or a dog chasing a ball. In contrast, **potential energy** is the energy matter has because of its position. The water behind a dam has potential energy by virtue of its elevated position. (See figure 4.11.) An electron moved to a position farther from the nucleus has

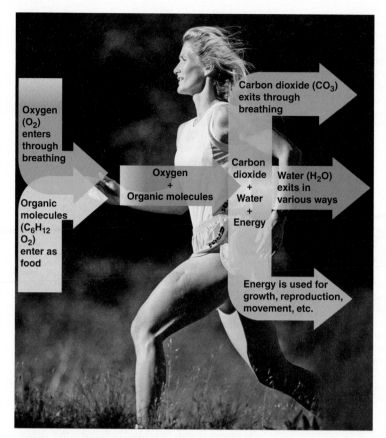

FIGURE 4.10 Respiration Respiration involves the release of energy from organic molecules when they react with oxygen. In addition to providing energy in a usable form, respiration produces carbon dioxide and water.

increased potential energy due to the increased distance between the electron and the nucleus.

States of Matter

Depending on the amount of energy present, matter can occur in three common states: solid, liquid, or gas. The physical nature of matter changes when a change occurs in the amount of kinetic energy its molecular particles contain, but the chemical nature of matter and the kinds of chemical reactions it will undergo remain the same. For example, water vapor, liquid water, and ice all have the same chemical composition but differ in the arrangement and activity of their molecules. The amount of kinetic energy molecules have determines how rapidly they move. (See figure 4.12.) In solids, the molecular particles have comparatively little energy, and they vibrate in place very close to one another. In liquids, the particles have more energy, are farther apart from one another, and will roll, tumble, and flow over each other. In gases, the molecular particles move very rapidly and are very far apart. All that is necessary to change the physical nature of a substance is an energy change. Heat energy must be added or removed.

When two forms of matter have different temperatures, heat energy will flow from the one with the higher temperature to the one with the lower temperature. The temperature of the cooler matter increases while that of the warmer matter decreases. You

FIGURE 4.11 Kinetic and Potential Energy Kinetic and potential energy are interconvertible. The potential energy possessed by the water behind a dam is converted to kinetic energy as the water flows to a lower level.

experience this whenever you touch a cold or hot object. This is referred to as a **sensible heat** transfer. When heat energy is used to change the state of matter from solid to liquid at its melting point or liquid to gas at its boiling point, heat is transferred, but the temperature of the matter does not change. This is called a **latent heat** transfer. You have experienced this effect when water evaporates from your skin. Your body supplies the heat necessary to convert liquid water to water vapor. While the temperature of the water does not change, the physical state of the water does. The heat that was transferred to the water caused it to evaporate, and your body cooled. When substances change from gas to liquid at the boiling point or liquid to solid at the freezing point, there is a corresponding release of heat energy without a change in temperature.

First and Second Laws of Thermodynamics

Energy can exist in several different forms, and it is possible to convert one form of energy into another. However, the total amount of energy remains constant.

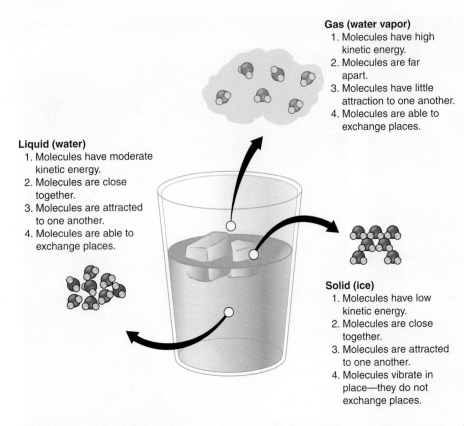

Gas (water vapor)
1. Molecules have high kinetic energy.
2. Molecules are far apart.
3. Molecules have little attraction to one another.
4. Molecules are able to exchange places.

Liquid (water)
1. Molecules have moderate kinetic energy.
2. Molecules are close together.
3. Molecules are attracted to one another.
4. Molecules are able to exchange places.

Solid (ice)
1. Molecules have low kinetic energy.
2. Molecules are close together.
3. Molecules are attracted to one another.
4. Molecules vibrate in place—they do not exchange places.

FIGURE 4.12 States of Matter Matter exists in one of three states, depending on the amount of kinetic energy the molecules have. The higher the amount of energy, the greater the distance between molecules and the greater their degree of freedom of movement.

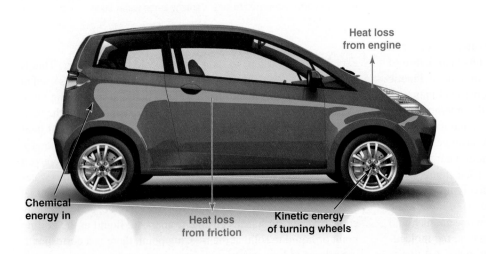

Heat loss from engine

Chemical energy in

Heat loss from friction

Kinetic energy of turning wheels

FIGURE 4.13 Second Law of Thermodynamics Whenever energy is converted from one form to another, some of the useful energy is lost, usually in the form of heat. The burning of gasoline produces heat, which is lost to the atmosphere. As the kinetic energy of moving engine parts is transferred to the wheels, friction generates some additional heat. All of these steps produce low-quality heat in accordance with the second law of thermodynamics.

The **first law of thermodynamics** states that energy can neither be created nor destroyed; it can only be changed from one form into another. From a human perspective, some forms of energy are more useful than others. We tend to make extensive use of electrical energy for a variety of purposes, but there is very little electrical energy present in nature. Therefore, we convert other forms of energy into electrical energy.

The **second law of thermodynamics** states that whenever energy is converted from one form to another, some of the useful energy is lost. The energy that cannot be used to do useful work is called **entropy.** Therefore, another way to state the second law of thermodynamics is to say that when energy is converted from one form to another, entropy increases. An alternative way to look at the idea of entropy is to say that entropy is a measure of disorder and that the amount of disorder (entropy) typically increases when energy conversions take place. Obviously, it is possible to generate greater order in a system (living things are a good example of highly ordered things), but when things become more highly ordered, the disorder of the surroundings must increase. For example, all living things release heat to their surroundings.

It is important to understand that when energy is converted from one form to another, there is no loss to *total* energy, but there is a loss of *useful* energy. For example, gasoline, which contains chemical energy, can be burned in an automobile engine to provide the kinetic energy to turn the wheels. The heat from the burning gasoline is used to move the pistons, which turns the crankshaft and eventually causes the wheels to turn. At each step in the process, some heat energy is lost from the system. During combustion, heat is lost from the engine. Friction results in further loss of energy as heat. Therefore, the amount of useful energy (kinetic energy of the wheels turning) is much less than the total amount of chemical energy present in the gasoline that was burned. (See figure 4.13.)

Within the universe, energy is being converted from one form to another continuously. Stars are converting nuclear energy into heat and light. Animals are converting the chemical potential energy found in food into kinetic energy that allows them to move. Plants are converting sunlight energy into the chemical bond energy of sugar molecules. In each of these cases, some energy is produced that is not able to do useful work. This is generally in the form of heat lost to the surroundings.

4.6 Environmental Implications of Energy Flow

If we wish to understand many kinds of environmental problems, we must first understand the nature of energy and energy flow.

Entropy Increases

The heat produced when energy conversions occur is dissipated throughout the universe. This is a common experience. All machines and living things that manipulate energy release heat. It is also true that organized matter tends to become more disordered unless an external source of energy is available to maintain the ordered arrangement. Houses fall into ruin, automobiles rust, and appliances wear out unless work is done to maintain them. In reality, all of these phenomena involve the loss of heat. The organisms that decompose the wood in our houses release heat. The chemical reaction that causes rust releases heat. Friction, caused by the movement of parts of a machine against each other, generates heat and causes the parts to wear.

Ultimately, orderly arrangements of matter, such as clothing, automobiles, or living organisms, become disordered. There is an increase in entropy. Eventually, nonliving objects wear out and living things die and decompose. This process of becoming more disordered coincides with the constant flow of energy toward a dilute form of heat. This dissipated, low-quality heat has little value to us, since we are unable to use it.

Energy Quality

It is important to understand that some forms of energy are more useful to us than others. Some forms, such as electrical energy, are of high quality because they can be easily used to perform a variety of useful actions. Other forms, such as the heat in the water of the ocean, are of low quality because we are not able to use them for useful purposes. Although the total *quantity* of heat energy in the ocean is much greater than the total amount of electrical energy in the world, little useful work can be done with the heat energy in the ocean because it is of *low quality*. Therefore, it is not as valuable as other forms of energy that can be used to do work for us.

The reason the heat of the ocean is of little value is related to the small temperature difference between two sources of heat. When two objects differ in temperature, heat will flow from the warmer to the cooler object. The greater the temperature difference, the more useful the work that can be done. For example, fossil fuel power plants burn fuel to heat water and convert it to steam. High-temperature steam enters the turbine, while cold cooling water condenses the steam as it leaves the turbine. This steep temperature gradient also provides a steep pressure gradient as heat energy flows from the steam to the cold water, which causes a turbine to turn, which generates electricity. Because the average temperature of the ocean is not high, and it is difficult to find another object that has a greatly lower temperature than the ocean, it is difficult to use the huge heat content of the ocean to do useful work for us.

It is important to understand that energy that is of low quality from our point of view may still have significance to the world in which we live. For example, the distribution of heat energy in the ocean tends to moderate the temperature of coastal climates, contributes to weather patterns, and causes ocean currents that are extremely important in many ways. It is also important to recognize that we can sometimes figure new ways to convert low-quality energy to high-quality energy. For example, it is possible to use the waste heat from power plants to heat cities if the power plants are located in the cities. Scientists have recently made major improvements in wind turbines and photovoltaic cells that allow us to economically convert low-quality light or wind to high-quality electricity.

Biological Systems and Thermodynamics

From the point of view of ecological systems, organisms such as plants do photosynthesis and are able to convert low-quality light energy to high-quality chemical energy in the organic molecules they produce. Eventually, they will use this stored energy for their needs, or it will be used by some other organism that has eaten the plant. In accordance with the second law of thermodynamics, all organisms, including humans, are in the process of converting high-quality energy into low-quality energy. Waste heat is produced when the chemical-bond energy in food is converted into the energy needed to move, grow, or respond. The process of releasing chemical-bond energy from food by organisms is known as cellular respiration. From an energy point of view, it is comparable to the process of **combustion,** which is the burning of fuel to obtain heat, light, or some other form of useful energy. The efficiency of cellular respiration is relatively high. About 40 percent of the energy contained in food is released in a useful form. The rest is dissipated as low-quality heat.

Pollution and Thermodynamics

An unfortunate consequence of energy conversion is pollution. The heat lost from most energy conversions is a pollutant. The wear of the brakes used to stop cars results in pollution. The emissions from power plants pollute. All of these are examples of the effect of the second law of thermodynamics. If each person on Earth used less energy, there would be less waste heat and other forms of pollution that result from energy conversion. The amount of energy in the universe is limited. Only a small portion of that energy is of high quality. The use of high-quality energy decreases the amount of useful energy available, as more low-quality heat is generated. All life and all activities are subject to these important physical principles described by the first and second laws of thermodynamics.

The End of the Incandescent Light Bulb

Many governments around the world, including the United States, have passed legislation to replace incandescent light bulbs for lighting with more energy-efficient lighting alternatives. Brazil and Venezuela were the first countries to phase-out the bulbs in 2005, followed by the European Union in 2009 and several more countries including the United States and Canada in 2014 and China in 2016.

Objections to replacement of incandescent lights include the higher cost of the new bulbs and the different quality of light produced by the new more efficient bulbs. Some also argue that the free market is preferable to government regulation while others state that only government intervention will improve energy efficiency. There are also concerns about mercury contamination with compact fluorescent lamps (CFLs).

Replacements for incandescent bulbs include CFLs and light-emitting diode (LED) lights.

A compact fluorescent lamp uses a fluorescent lamp tube that will fit into the fixture for an incandescent bulb and contains a compact electronic ballast in the base of the lamp. CFL's use 65 to 75 percent less energy than incandescent bulbs and can last eight to fifteen times longer. Newer phosphor formulations have improved the color of the light with "soft white" CFLs similar to standard incandescent lamps.

Light-emitting diode lamps are used for both general and special-purpose lighting. They use at least 75 percent less energy than an equivalent incandescent bulb and last up to 20 years. Unlike CFLs, LEDs contain

Kind of Light	Light Output per Watt of Electricity Used
100-watt incandescent light bulb	About 17 lumens/watt
White LED	About 50 lumens/watt
Compact fluorescent bulb	About 50 lumens/watt
Standard 40-watt fluorescent tube	About 60 lumens/watt

no mercury, they turn on instantly at any temperature, their very long lifetime is not affected by cycling on and off, they have no glass to break, they don't emit UV rays that fade, and they can shine directionally without reflectors. The disadvantages include diminished color shades, supporting electronic circuitry failure, and with traffic lights in snowy areas they fail to melt the snow that may accumulate due to the low heat output.

- Should the government set standards for light bulbs? Why or why not?
- Have you switched your light bulbs at home?

Left: incandescent light bulb; middle: LED bulb; right: CFL bulb

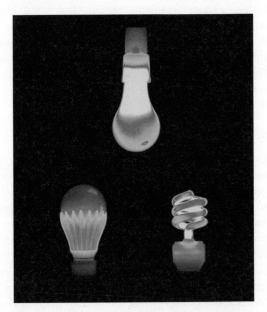

Thermal images of the three light bulbs.

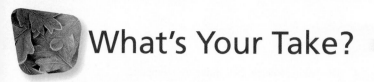

What's Your Take?

Diesel Engine Trade-offs

In the United States nearly all automobiles have gasoline engines. Only about 1 percent have diesel engines. In Europe more than 50 percent of new automobiles sold have diesel engines. Automobile diesel engines are much more efficient than gasoline engines. They have an efficiency of 35 to 42 percent compared to gasoline engines with an efficiency of 25 to 30 percent. This means that they go much farther on a liter of fuel than do equivalent gasoline engines. Furthermore diesel engines last longer than gasoline engines.

Since gasoline engines are less efficient than diesel engines, gasoline engines produce 20 to 40 percent more carbon dioxide than diesel engines for the same distance driven. Increased carbon dioxide is known to cause a warming of the Earth's atmosphere.

Current diesel engines produce more particulate matter and nitrogen oxides than do gasoline engines. However, current air quality guidelines for particulate matter and nitrogen oxides in Europe are more stringent than those in the United States. The World Health Organization has estimated that thousands of people die each year because of particulate air pollution. Changes are being made in fuels, engine design, and pollution control devices that reduce the amount of particulate matter and nitrogen oxides produced.

- Should U.S. automakers switch to diesel engines as the Europeans have?
- Is global climate change an important reason to use diesel engines?
- Should problems with particulate emissions prevent the development of diesel engines for passenger vehicles?

Summary

Science is a method of gathering and organizing information. It involves observation, asking questions, exploring alternative sources of information, hypothesis formation, the testing of hypotheses, and publication of the results for others to evaluate. A hypothesis is a logical prediction about how things work that must account for all the known information and be testable. The process of science attempts to be careful, unbiased, and reliable in the way information is collected and evaluated. This often involves conducting experiments to test the validity of a hypothesis. If a hypothesis is continually supported by the addition of new facts, it may be incorporated into a theory. A theory is a broadly written, widely accepted generalization that ties together large bodies of information and explains why things happen. Similarly, a law is a broad statement that describes what happens in nature.

The fundamental unit of matter is the atom, which is made up of protons and neutrons in the nucleus surrounded by a cloud of moving electrons. The number of protons for any one type of atom is constant, but the number of neutrons in different atoms of the same type of atom may vary. The number of electrons is equal to the number of protons. Protons have a positive charge, neutrons lack a charge, and electrons have a negative charge.

Molecules are units made of a combination of two or more atoms bonded to one another. Chemical bonds are physical attractions between atoms resulting from the interaction of their electrons. When chemical bonds are broken or formed, a chemical reaction occurs, and the amount of energy within the chemical bonds is changed. Chemical reactions require activation energy to get the reaction started.

Matter that is composed of only one kind of atom is known as an element. Matter that is composed of small units containing different kinds of atoms bonded in specific ratios is known as a compound. An atom or molecule that has gained or lost electrons so that it has an electric charge is known as an ion.

Matter can occur in three states: solid, liquid, and gas. These three differ in the amount of energy the molecular units contain and the distance between the units. Kinetic energy is the energy contained by moving objects. Potential energy is the energy an object has because of its position.

The first law of thermodynamics states that the amount of energy in the universe is constant, that energy can neither be created nor destroyed. The second law of thermodynamics states that when energy is converted from one form to another, some of the useful energy is lost (entropy increases). Some forms of energy are more useful than others. The quality of the energy determines how much useful work can be accomplished by expending the energy. Low-temperature heat sources are of poor quality, since they cannot be used to do useful work.

Acting Green

1. Replace household chemicals with those that are more environmentally friendly.
2. Find out when the next household hazardous waste cleanup is in your community and get rid of unwanted hazardous materials.
3. Read the labels on your household cleaning products.
4. Use baking soda as a substitute for toothpaste.
5. Reduce your consumption of electricity by 10 percent.
6. Take a long walk in a park or natural area and list five observations and develop a hypothesis for each.

Review Questions

1. How do scientific disciplines differ from nonscientific disciplines?
2. What is a hypothesis? Why is it an important part of the way scientists think?
3. Why are events that happen only once difficult to analyze from a scientific point of view?
4. What is the scientific method, and what processes does it involve?
5. How are the second law of thermodynamics and pollution related?
6. Diagram an atom of oxygen and label its parts.
7. What happens to atoms during a chemical reaction?
8. State the first and second laws of thermodynamics.
9. How do solids, liquids, and gases differ from one another at the molecular level?
10. List five kinds of energy.
11. Are all kinds of energy equal in their capacity to bring about changes? Why or why not?

Critical Thinking Questions

1. You observe that a high percentage of frogs, which are especially sensitive to environmental poisons, in small ponds in your agricultural region have birth defects. Suspecting agricultural chemicals present in runoff to be the culprit, state the hypothesis in your own words. Next, devise an experiment that might help you support or reject your hypothesis.
2. Given the experiment you proposed in Critical Thinking Question 1, imagine some results that would support that hypothesis. Now imagine you are a different scientist, one who is very skeptical of the initial hypothesis. How convincing do you find these data? What other possible explanations (hypotheses) might there be to explain the results? Devise a different experiment to test this new hypothesis.
3. Increasingly, environmental issues such as global climate change are moving to the forefront of world concern. What role should science play in public policy decisions? How should we decide between competing scientific explanations about an environmental concern such as global climate change? What might be some of the criteria for deciding what is "good science" and what is "bad science"?
4. How important are the first and second laws of thermodynamics to explaining environmental issues? Using the concepts in these laws of thermodynamics, try to explain a particular environmental issue. How does an understanding of thermodynamics change your conceptual framework regarding this issue?
5. The text points out that incandescent lightbulbs are only 5–10 percent efficient at using energy to accomplish their task, while new, initially more expensive, compact fluorescent lighting uses significantly less electricity to provide the same quantity of light. Examine the contextual framework of those who advocate for new lighting methods and the contextual framework of those who continue to design and build using the old methods. What are the major differences in perspective? What could you suggest be done to help bring these different perspectives closer together?
6. Some scientists argue that living organisms constantly battle against the principles of the second law of thermodynamics using the principles of the first law of thermodynamics. What might they mean by this? Do you think this is accurate? What might be some of the implications of this for living organisms?

To access additional resources for this chapter, please visit ConnectPlus® at connect.mheducation.com. There you will find interactive exercises, including Google Earth™, additional Case Studies, and SmartBook™, an adaptive eBook that integrates our LearnSmart® adaptive learning technology.

Interactions: Environments and Organisms

Human activities change ecosystems and alter the way atoms are cycled through an ecosystem. Clear-cutting of forests removes the carbon tied up in the vegetation and also increases the rate of erosion causing the depletion of soil nutrients.

OBJECTIVES

After reading this chapter, you should be able to:

- Identify and list abiotic and biotic factors in an ecosystem.
- Explain the significance of limiting factors.
- Distinguish between *habitat* and *niche*.
- Distinguish between a *population* and a *species*.
- Describe how the process of natural selection operates to shape the ecological niche of an organism.
- Describe the processes that lead to speciation, extinction, and coevolution.
- Describe predator-prey, parasite-host, competitive, mutualistic, and commensalistic relationships.

- Differentiate between a community and an ecosystem.
- Define the roles of producer, consumer, and decomposer.
- Distinguish among the following kinds of consumers: herbivore, carnivore, omnivore, and parasite.
- Describe why some organisms are considered to be keystone species in their ecosystems.
- Describe energy flow through an ecosystem.
- Relate the concepts of food webs and food chains to trophic levels.

- Describe the role of producers, consumers, and decomposers in the cycling of carbon atoms through ecosystems.
- Identify kinds of carbon sinks that tie up carbon for long periods.
- Describe how human activities have altered the carbon cycle.
- Describe the role of nitrogen-fixing bacteria, decomposer organisms, nitrifying bacteria, and denitrifying bacteria in the cycling of nitrogen through ecosystems.
- Describe how human activities have altered the nitrogen cycle.
- Describe how phosphorus is cycled through ecosystems.

The Adaptation of Wildlife to Urban Environments

Climate change and habitat loss are driving coyotes, bears and mountain lions out of their habitats, but that is only part of the reason why so many animals are now moving into urban areas. While some animals such as pigeons, squirrels, and rodents may seem as much a part of the urban landscape as cement and streetlights, there are growing populations of new wild inhabitants that are a bit more unusual.

Coyotes have rapidly adapted to suburban and urban environments. Coyotes have appeared in cities from Los Angeles to New York. They have learned to live in much smaller territories than they do in rural areas and thrive in a city by hunting enough small rodents and sometimes cats to feed themselves and their young. In 2012, biologists in Los Angeles radio-collared the first mountain lion ever found in Griffith Park. Complaints about bears in Nevada around Lake Tahoe increased tenfold between 2002 and 2012. Some of those bears were so well fed on a continuous supply of garbage that they neglected to hibernate in the winter. In 2011, workers found a fox living on the 72nd floor of an unfinished building in London, England. The fox had been living on the construction workers' food scraps.

As South Florida's urban sprawl continues, there are increased interactions between people and alligators. While alligators are not as predatory toward humans as popular lore might say, it is nonetheless unsettling to come face to face with one in your backyard. Seagulls are invading towns and cities inland from the coasts because the decline of the fishing industry means they can no longer feed off the scraps from fishing boats. In addition, changes in how landfills are managed has eliminated a previous source of food for seagulls.

Although many animals are moving to the city because they are being displaced by climate change and habitat destruction, there are also other factors involved. One of the biggest factors with regard to the large carnivores is that there are more of them because of successful conservation efforts. In addition, as we make cities greener they become more attractive to humans and animals alike. Another factor is that the relationship between humans and large predators is changing. There are now generations of certain carnivores that have experienced low amounts of predation by people. These carnivores may view cities differently than their ancestors, who associated human encounters with a good chance of being shot.

While the new wild inhabitants keep their distance from people most of the time, conflict is inevitable when these animals and humans share space. Sometimes the conflict is between the invading predators and domesticated pets. More serious clashes are rare but not unheard of. In 2011, for example, a coyote attacked children in a Denver suburb on three separate occasions. Suburban grade schools in many states have ordered lockdowns in response to black bears prowling near the premises. Bears are growing fat on human hospitality from the outskirts of Los Angeles to the Beltway of Washington D.C.

There is no doubt that many different species of wildlife are adapting to urban life. Perhaps a bigger question is: Will humans be able and willing to adapt to the presence of these new neighbors?

5.1 Ecological Concepts

The science of **ecology** is the study of the ways organisms interact with each other and with their nonliving surroundings. This is a broad field of study that deals with the ways in which organisms are adapted to their surroundings, how they make use of these surroundings, and how an area is altered by the presence and activities of organisms. These interactions involve the flow of energy and matter among organisms. If the flow of energy and matter ceases, the organisms die.

All organisms are dependent on other organisms in some way. One organism may eat another and use it for energy and raw materials. One organism may temporarily use another without harming it. One organism may provide a service for another, such as when animals distribute plant seeds or bacteria break down dead organic matter for reuse. The study of ecology can be divided into many specialties and be looked at from several levels of organization. (See figure 5.1.) Before we can explore the field of ecology in greater depth, we must become familiar with some of the standard vocabulary of this field.

Environment

Everything that affects an organism during its lifetime is collectively known as its **environment.** Environment is a very broad concept. For example, during its lifetime, an animal such as a raccoon is likely to interact with millions of other organisms (bacteria, food organisms, parasites, mates, predators), drink copious amounts of water, breathe huge quantities of air, and respond to daily changes in temperature and humidity. This list only begins

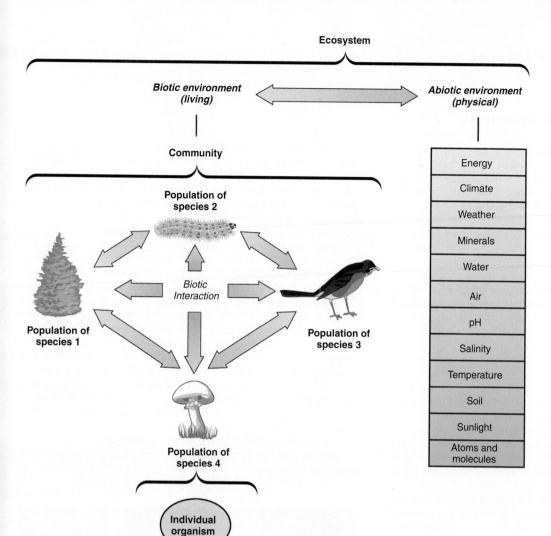

FIGURE 5.1 Levels of Organization in Ecology Ecology is the science that deals with the study of interactions between organisms and their environment. This study can take place at several different levels, from the broad ecosystem level through community interactions to population studies and the study of individual organisms. Ecology also involves study of the physical environment, which includes such things as weather, climate, sunlight, and the atoms and molecules that organisms require to construct their bodies.

to describe the various components that make up the raccoon's environment. Because of this complexity, it is useful to subdivide the concept of environment into abiotic (nonliving) and biotic (living) factors.

Abiotic Factors

Abiotic factors are nonliving things that influence an organism. They can be organized into several broad categories: energy, nonliving matter, and the physical characteristics of the place an organism lives.

Energy is required by all organisms to maintain themselves. The ultimate source of energy for almost all organisms is the sun; in the case of plants, the sun directly supplies the energy necessary for them to maintain themselves. Animals obtain their energy by eating plants or other animals that eat plants. Ultimately, the amount of living material that can exist in an area is determined by the amount of energy plants, algae, and bacteria can trap.

Nonliving matter in the form of atoms such as carbon, nitrogen, and phosphorus and molecules such as water provides the structural framework of organisms. Organisms constantly obtain these materials from their environment. The atoms become part of an organism's body structure for a short time, and eventually all of them are returned to the environment through respiration, excretion, or death and decay.

Physical characteristics of the space organisms inhabit vary greatly. Each space has a particular structure and location that is also an important abiotic aspect of an organism's environment. Some organisms live in the ocean; others live on land at sea level; still others live on mountaintops or fly through the air. Some spaces are homogeneous and flat; others are a jumble of rocks of different sizes. Some are close to the equator; others are near the poles.

The weather and climate (average weather patterns over a number of years) of an area present several kinds of abiotic factors. These include the amount of solar radiation, proximity to the equator, prevailing wind patterns, and closeness to water. The intensity and duration of sunlight in an area cause daily and seasonal changes in temperature. Differences in temperature generate wind. Solar radiation is also responsible for the evaporation into the atmosphere of water that subsequently falls as precipitation. Depending on the climate, precipitation may be of several forms:

Science, Politics, & Policy

Emotion and Wolf Management

For centuries, wolf populations have been under attack by humans. It is an ancient dispute over territory and food, and its battleground spreads from the upper Great Lakes to the Pacific. The effort to reduce or eliminate the species was justified as a way to protect humans, their livestock, or game species such as elk and deer from wolf predation. The possibility of attack on humans was in reality nonexistent; however, attacks on livestock and game species did occur and wildlife managers supported the elimination of wolves as a way to improve big game hunting. Wolves were eventually eliminated in most states.

In 1967, the gray wolf was listed as endangered under the Endangered Species Preservation Act, which was reauthorized as the Endangered Species Act of 1973. In 1978, the gray wolf was relisted as endangered throughout the lower 48 states, with the exception of Minnesota, where it was listed as threatened. In the 1990s, actions were taken to reintroduce the wolf into areas where it had been eradicated.

No reintroduction was more controversial than that in the greater Yellowstone ecosystem where many scientists agreed that elk populations—a favorite wolf prey—had reached harmful levels. By 1996, the U.S. Fish and Wildlife Service had released 31 gray wolves from Canada into the Yellowstone area and an additional 35 in central Idaho. The number of wolves grew rapidly given the large elk herd to prey upon. By 2013 the wolf population in the park was around 100 and there were a total of 1,900 wolves in the states of Wyoming, Idaho, Oregon, and Montana. Wolves are also dispersing into the state of Washington from Oregon, Idaho, and British Columbia. Washington now has nearly 100 wolves.

Wildlife enthusiasts and many in the scientific community are pleased. In Yellowstone alone, tens of thousands come to watch wolves each year, adding an estimated $35 million to the area's economy. Scientists are documenting ecological changes to the Yellowstone ecosystem as the wolves reduce elk populations and vegetation rebounds. These changes have returned the ecosystem to a more stable and biologically diverse condition.

On the other hand, some local residents say they no longer feel as safe taking their families into the woods. Hunters also strongly complain. To them wolves are four-legged killing machines that ravage game populations. Ranchers are also strongly opposed to the wolves' return. During 2012 there were 424 cattle and sheep deaths from wolves throughout the west—less than one percent of livestock deaths in the region—and 274 wolves were killed for attacking livestock.

In recent years, as the wolf population has increased, the federal government proposed removing wolves from the endangered species list and allowing states to establish wolf management programs. In 2009, in western states with approved wolf management plans, wolves were removed from the endangered species list and states implemented their wolf management plans. Most management plans involved regulated hunting of wolves. However, in August 2010, a federal court judge ruled that the wolf must stay on the endangered species list. In April 2011, Congress approved a budget bill that included a provision that removed wolves from the endangered species list in western states. The bill was approved and signed by President Obama.

Many states now allow limited hunting of wolves. Montana and Idaho started to allow hunting in 2011 and Minnesota and Wisconsin in 2013. In 2013, Michigan developed a wolf management plan that would have allowed hunters to remove 43 wolves from a population of about 700. That same year groups opposed to any hunting of wolves collected enough signatures to put wolf management in Michigan up for a public vote. Some argue that this action would deny professional game managers the use of hunting as a management tool. They argue that you cannot manage the state's largest predator through 30-second sound bites and TV commercials. They make the point that wildlife management should be done by scientists and not decided at the ballot box. Others defend the ballot approach as representing the voice of the citizens of the state. The 2013 wolf hunt in Michigan resulted in 23 wolves being harvested out of a quota of 43.

Should the limited hunting of wolves be allowed?

How do you feel about issues such as the hunting of wolves being decided by public vote?

rain, snow, hail, or fog. Furthermore, there may be seasonal precipitation patterns.

The kind of soil present is determined by prevailing weather patterns, local topography, and the geologic history of the region. These factors interact to produce a variety of soils that range from those that are coarse, sandy, dry, and infertile to fertile soils composed of fine particles that hold moisture.

Biotic Factors

The **biotic factors** of an organism's environment include all forms of life with which it interacts. Some broad categories are: plants that carry on photosynthesis; animals that eat other organisms; bacteria and fungi that cause decay; bacteria, viruses, and other parasitic organisms that cause disease; and other individuals of the same species.

Limiting Factors

Although organisms interact with their surroundings in many ways, certain factors may be critical to a particular species' success. A shortage or absence of a specific factor restricts the success of the species; thus, it is known as a **limiting factor.** Limiting factors may be either abiotic or biotic and can be quite different from one species to another. Many plants are limited by scarcity of water, light, or specific soil nutrients such as nitrogen or phosphorus. Monarch butterflies are limited by the number of available milkweed plants, since their developing caterpillars use this plant as their only food source. (See figure 5.2.)

Climatic factors such as temperature range, humidity, periods of drought, or length of winter are often limiting factors. For example, many species of snakes and lizards are limited to the warmer parts of the world because they have difficulty maintaining their body temperature in cold climates and cannot survive long periods of cold. If we look at the number of species of snakes and lizards, we see that the number of species declines as one moves from warmer to colder climates.

While this is a general trend, there is much variation among species. The **range of tolerance** of a species is the degree to which it is able to withstand environmental variation. Some species have a broad range of tolerance, whereas others have a narrow range of tolerance. The common garter snake (*Thamnophis sirtalis*) consists of several subspecies and is found throughout the United States and Southern Canada and has scattered populations in Mexico. Thus, it is able to tolerate a wide variety of climates. Conversely, the green anole (*Anolis carolinensis*) has a much narrower range of tolerance and is found only in the southeastern part of the United States where temperatures are mild. (See figure 5.3.)

Habitat and Niche

It is impossible to understand an organism apart from its environment. The environment influences organisms, and organisms affect the environment. To focus attention on specific elements of this interaction, ecologists have developed two concepts that need to be clearly understood: habitat and niche.

Habitat—Place

The **habitat** of an organism is the space that the organism inhabits, the place where it lives (its address). We tend to characterize an organism's habitat by highlighting some prominent physical or biological feature of its environment such as soil type, availability of water, climatic conditions, or predominant plant species that exist in the area. For example, mosses are small plants that must be covered by a thin film of water in order to reproduce. In addition, many kinds dry out and die if they are exposed to sunlight, wind, and drought. Therefore, the typical habitat of moss is likely to be cool, moist, and shady. (See figure 5.4.) Likewise, a rapidly flowing, cool, well-oxygenated stream with many bottom-dwelling insects is good trout habitat, while open prairie with lots of grass is preferred by bison, prairie dogs, and many kinds of hawks and falcons. The particular biological requirements of an organism determine the kind of habitat in which it is likely to be found.

Niche—Role

The **niche** of an organism is the functional role it has in its surroundings (its profession). A description of an organism's niche includes all the ways it affects the organisms with which it interacts as well as how it modifies its physical surroundings. In addition, the description of a niche includes all of the things that happen to the organism. For example, beavers frequently flood areas by building dams of mud and sticks across streams. (See figure 5.5.) The flooding has several effects. It provides beavers with a larger area of deep water, which they need for protection; it provides a pond habitat for many other species of animals such as ducks and fish; and it kills trees that cannot live with their roots under water. The animals that are attracted to the pond and the beavers often become food for predators. After the beavers have eaten all the suitable food, such as aspen, they abandon the pond, migrate to other areas along the stream, and begin the whole process over again.

In this recitation of beaver characteristics, we have listed several effects that the animal has on its local environment. It changes the physical environment by flooding, it kills trees, it enhances the environment for other animals, and it is a food source for predators. This is only a superficial glimpse of the many aspects of the beaver's interaction with its environment. A complete catalog of all aspects of its niche would make up a separate book.

Adult monarch on milkweed

Caterpillar feeding on milkweed

FIGURE 5.2 Milkweed Is a Limiting Factor for Monarch Butterflies Monarch butterflies lay their eggs on various kinds of milkweed plants. The larvae eat the leaves of the milkweed plant. Thus, monarchs are limited by the number of milkweeds in the area.

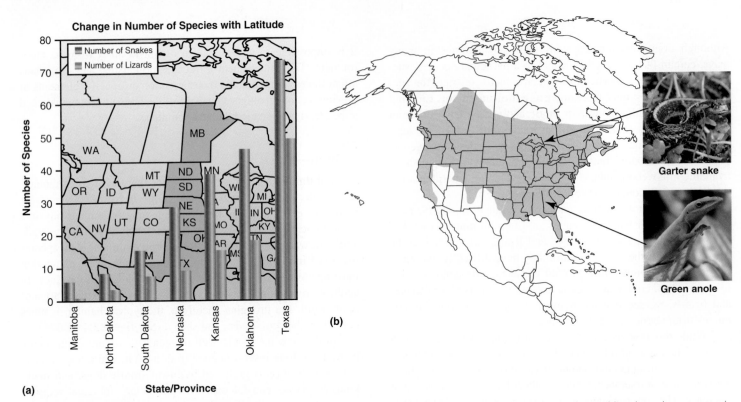

FIGURE 5.3 Temperature Is a Limiting Factor Cold temperature is a limiting factor for many kinds of reptiles. Snakes and lizards are less common in cold regions than in warm regions. (a) The graph shows the number of species of snakes and lizards in regions of central North America. Note that the number of species declines as one proceeds from south to north. (b) Some species, like the common garter snake (*Thamnophis sirtalis*), have a broad range of tolerance and are not limited by cold temperature. It is found throughout the United States and several Canadian provinces. However, the green anole (*Anolis carolinensis*) has a very narrow range of tolerance and is found only in the warm, humid southeastern states.

FIGURE 5.4 Moss Habitat The habitat of mosses is typically cool, moist, and shady, since many mosses die if they are subjected to drying. In addition, mosses must have a thin layer of water present in order to reproduce sexually.

Another familiar organism is the dandelion. (See figure 5.6.) Its niche includes the fact that it is an opportunistic plant that rapidly becomes established in sunny, disturbed sites. It can produce thousands of parachutelike seeds that are easily carried by the wind over long distances. (You have probably helped this process by blowing on the fluffy, white collections of seeds of a mature dandelion fruit.) Furthermore, it often produces several sets of flowers per year. Since there are so many seeds and they are so easily distributed, the plant can easily establish itself in any sunny, disturbed site, including lawns. Since it is a plant, one major aspect of its niche is the ability to carry on photosynthesis and grow. It uses water and nutrients from the soil to produce new plant parts. Since dandelions need direct sunlight to grow successfully, mowing lawns helps provide just the right conditions for dandelions because the vegetation is never allowed to get so tall that dandelions are shaded. Many kinds of animals, including some humans, use the plant for food. The young leaves may be eaten in a salad, and the blossoms can be used to make dandelion wine. Bees visit the flowers regularly to obtain nectar and pollen.

Beavers eat woody plants

Beaver dam

Beaver gnawing

Beaver ponds provide nesting habitat

Beaver ponds provide habitat

FIGURE 5.5 The Ecological Niche of a Beaver The ecological niche of an organism is a complex set of interactions between an organism and its surroundings, which includes all of the ways an organism influences its surroundings as well as all of the ways the organism is affected by its environment. A beaver's ecological niche includes building dams and flooding forested areas, killing trees, providing habitat for waterbirds and other animals, serving as food for predators, and many other effects.

FIGURE 5.6 The Niche of a Dandelion A dandelion is a familiar plant that commonly invades disturbed sites because it produces many seeds that are blown easily to new areas. It serves as food for various herbivores, supplies nectar to bees, and can regrow quickly from its root if its leaves are removed.

5.2 The Role of Natural Selection and Evolution

Since organisms generally are well adapted to their surroundings and fill a particular niche, it is important that we develop an understanding of the processes that lead to this high degree of adaptation. Furthermore, since the mechanisms that result in adaptation occur within a species, we need to understand the nature of a species.

Genes, Populations, and Species

We can look at an organism from several points of view. We can consider an individual, groups of individuals of the same kind, or groups that are distinct from other groups. This leads us to discuss three interrelated concepts.

Genes are distinct pieces of DNA that determine the characteristics an individual displays. There are genes for structures such as leaf shape or feather color, behaviors such as cricket chirps or migratory activity, and physiological processes such as photosynthesis or muscular contractions. Each individual has a particular set of genes.

A **population** is considered to be all the organisms of the same kind found within a specific geographic region. (See figure 5.7.) The individuals of a population will have very similar sets of genes, although there will be some individual variation. Because some genetic difference exists among individuals in a population, a population contains more kinds of genes than any individual within the population. Reproduction also takes place among individuals in a population so that genes are passed from one generation to the next. The concept of a species is an extension of this thinking about genes, groups, and reproduction.

FIGURE 5.7 A Penguin Population This population of Emperor penguins is from South Georgia Island near Antarctica.

A **species** is a population of all the organisms potentially capable of reproducing naturally among themselves and having offspring that also reproduce. Therefore, the concept of a species is a population concept. *An individual organism is not a species but a member of a species.* It is also a genetic concept, since individuals that are of different species are not capable of exchanging genes through reproduction. (See figure 5.8.)

SPECIES

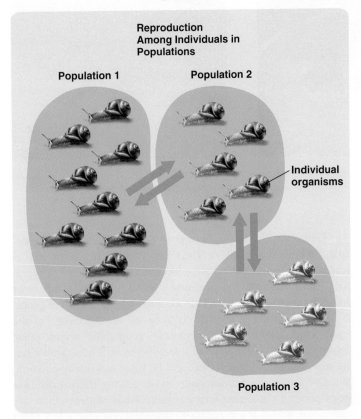

FIGURE 5.8 The Species Concept A species is all the organisms of a particular kind capable of interbreeding and producing fertile offspring. Often a species consists of many distinct populations that show genetic differences from one another.

Some species are easy to recognize. We easily recognize humans as a distinct species. Most people recognize a dandelion when they see it and do not confuse it with other kinds of plants that have yellow flowers. Other species are not as easy to recognize. Most of us cannot tell one species of mosquito from another or identify different species of grasses. Because of this, we tend to lump organisms into large categories and do not recognize the many subtle niche differences that exist among the similar-appearing species. However, species of mosquitoes are quite distinct from one another genetically and occupy different niches. Only certain species of mosquitoes carry and transmit the human disease malaria. Other species transmit the dog heartworm parasite. Each mosquito species is active during certain portions of the day or night. And each species requires specific conditions to reproduce.

Natural Selection

As we have seen, each species of organism is specifically adapted to a particular habitat in which it has a very specific role (niche). But how is it that each species of plant, animal, fungus, or bacterium fits into its environment in such a precise way? Since most of the structural, physiological, and behavioral characteristics organisms display are determined by the genes they possess, these characteristics are passed from one generation to the next when individuals reproduce. The process that leads to this close fit between the characteristics organisms display and the demands of their environment is known as natural selection.

Charles Darwin is generally credited with developing the concept of natural selection. (See figure 5.9.) Although he did not understand the modern gene concept, he understood that

FIGURE 5.9 Charles Darwin Charles Darwin observed that each organism fit into its surroundings. He developed the theory of natural selection to explain how adaptation came about.

characteristics were passed from parent to offspring. He also observed the highly adaptive nature of the relationship between organisms and their environment and developed the concept of natural selection to explain how this adaptation came about.

Natural selection is the process that determines which individuals within a species will reproduce and pass their genes to the next generation. The changes we see in the genes and characteristics displayed in successive generations of a population of organisms over time is known as **evolution.** Thus, natural selection is the mechanism that causes evolution to occur. Several conditions and steps are involved in the process of natural selection.

1. *Individuals within a species show genetically determined variation; some of the variations are useful and others are not.* For example, individual animals that are part of the same species show variation in color, size, or susceptibility to disease. (See figure 5.10.) Some combinations of genes are more valuable to the success of the individual than others.

2. *Organisms within a species typically produce many more offspring than are needed to replace the parents when they die. Most of the offspring die.* One blueberry bush may produce hundreds of berries with several seeds in each berry, or a pair of rabbits may have three to four litters of offspring each summer, with several young in each litter. Few of the seeds or baby rabbits become reproducing adults. (See figure 5.11.)

3. *The excess number of individuals results in a shortage of specific resources.* Individuals within a species must compete with each other for food, space, mates, or other requirements that are in limited supply. If you plant 100 bean seeds in a pot, many of them will begin to grow. Since each needs water and minerals from the soil, these resources will become scarce. Wood ducks nest in hollow trees. Often these nesting spots are in short supply and some wood ducks do not find suitable places to lay their eggs.

4. *Because of variation among individuals, some have a greater chance of obtaining needed resources and, therefore, have a greater likelihood of surviving and reproducing than others.*

FIGURE 5.10 **Genetic Variation** Since genes are responsible for many facial characteristics, the differences in facial characteristics shown in this photograph are a good visual illustration of genetic variation. This population of students also differs in genes for height, skin color, susceptibility to disease, and many other characteristics.

FIGURE 5.11 **Excessive Reproduction** One blueberry bush produces thousands of seeds (offspring) each year. Only a few will actually germinate and only a few of the seedlings will become mature plants.

Individuals that have genes that allow them to obtain needed resources and avoid threats to their survival will be more likely to survive and reproduce. Even if less well-adapted individuals survive, they may mature more slowly and not be able to reproduce as many times as the more well-adapted members of the species.

The degree to which organisms are adapted to their environment influences their reproductive success and is referred to as fitness. It is important to recognize that fitness does not necessarily mean the condition of being strong or vigorous. In this context, it means how well the organism fits in with all the aspects of its surroundings so that it successfully passes its genes to the next generation. For example, in a lodgepole pine forest, many lodgepole pine seedlings become established following a fire. Some grow more rapidly and obtain more sunlight and nutrients. Those that grow more rapidly are more likely to survive. These are also likely to reproduce for longer periods and pass more genes to future generations than those that die or grow more slowly.

5. *As time passes and each generation is subjected to the same process of natural selection, the percentage of individuals showing favorable variations will increase and those having unfavorable variations will decrease.* Those that reproduce more successfully pass on to the next generation the genes for the characteristics that made them successful in their environment, and the genes that made them successful become more common in future generations. Thus, each species of organism is continually refined to be adapted to the environment in which it exists.

One modern example of genetic change resulting from natural selection involves the development of pest populations that are resistant to the pesticides previously used to control them. Figure 5.12 shows a graph of the number of species of weeds that have populations that are resistant to commonly used herbicides. When an herbicide is first used against weed pests, it kills most of them. However, in many cases, some individual weed plants within the species happen to have genes that allow them to resist the effects of the herbicide. These individuals are better adapted to survive in the presence of the herbicide and have a higher likelihood of surviving. When they reproduce, they pass on to their offspring the same genes that contributed to their survival. After several generations of such selection, a majority of the individuals in the species will contain genes that allow them to resist the herbicide, and the herbicide is no longer effective against the weed.

Evolutionary Patterns

When we look at the effects of natural selection over time, we can see considerable change in the characteristics displayed by species and also in the kinds of species present. Some changes take thousands or millions of years to occur. Others, such as resistance to pesticides, can occur in a few years. (See figure 5.12.) Natural selection involves the processes that bring about change in species, and the end result of the natural selection process observable in organisms is called evolution.

Scientists have continuously shown that this theory of natural selection can explain the development of most aspects of the structure, function, and behavior of organisms. It is the central idea that helps explain how species adapt to their surroundings. When we discuss environmental problems, it is helpful to understand that species change. Furthermore, it is important to understand that as the environment is changed, either naturally or by human action, some species will adapt to the new conditions while others will not.

There are many examples that demonstrate the validity of the process of natural selection and the evolutionary changes that result from natural selection. The many species of insects, weeds, and bacteria that have become resistant to the insecticides, herbicides, and antibiotics that formerly were effective against them are common examples.

When we look at the evolutionary history of organisms in the fossil record over long time periods, it becomes obvious that new species have come into being and that others have gone extinct. The same processes continue today.

Speciation

Speciation is the production of new species from previously existing species. It is thought to occur as a result of a species dividing into two isolated subpopulations. If the two subpopulations contain some genetic differences and their environments are somewhat different, natural selection will work on the two groups differently and they will begin to diverge from each other. Eventually, the differences may become so great that the two subpopulations are no longer able to interbreed. At this point, they are two different species. This process has resulted in millions of different species. Figure 5.13 shows a collection of a few of the over 400,000 species of beetles.

Among plants, another common mechanism known as *polyploidy* results in new species. **Polyploidy** is a condition in which the number of sets of chromosomes in the cells of the plant is

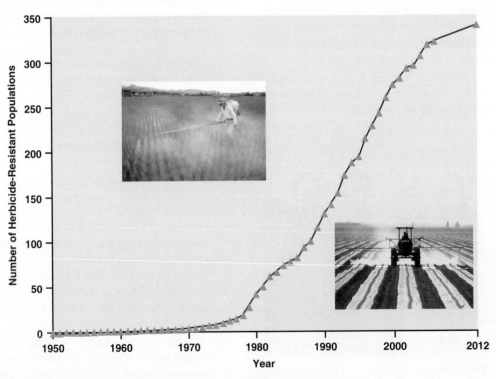

FIGURE 5.12 Change Resulting from Natural Selection Populations of weed plants that have been subjected repeatedly to herbicides often develop resistant populations. Those individual weed plants that were able to resist the effects of the herbicide lived to reproduce and pass on their genes for resistance to their offspring; thus, resistant populations of weeds developed. This graph shows the number of weeds with resistance to commonly used herbicides.
Source: Data from Ian Heap, "The International Survey of Herbicide Resistant Weeds," [cited 31 December 2010], Internet.

FIGURE 5.13 Speciation This photo shows a few of the many different species of beetles.

Others, such as humans, horses, and many kinds of plants, adapted to the new conditions and so survive to the present.

It is also possible to have the local extinction of specific populations of a species. Most species consist of many different populations that may differ from one another in significant ways. Often some of these populations have small local populations that can easily be driven to extinction. While these local extinctions are not the same as the extinction of an entire species, local extinctions often result in the loss of specific gene combinations. Many of the organisms listed on the endangered species list are really local populations of a more widely distributed species.

Natural selection is constantly at work shaping organisms to fit a changing environment. It is clear that humans have had a significant impact on the extinction of many kinds of species. Wherever humans have modified the environment for their purposes (farming, forestry, cities, hunting, and introducing exotic organisms), species are typically displaced from the area. If large areas are modified, entire species may be displaced. Ultimately, humans are also subject to evolution and the possibility of extinction as well.

Although extinction is a common event, there have been several instances of past mass extinctions in which major portions of the living world went extinct in a relatively short time. The background rate of extinction is estimated to be about 10 species per year. However, the current rate is much higher than that, leading many people to suggest that we may be in the middle of a mass

increased. Most organisms are diploid; that is, they have two sets of chromosomes. They got a set from each parent, one set in the egg and one set in the sperm. Polyploid organisms may have several sets of chromosomes. The details of how polyploidy comes about are not important for this discussion. It is sufficient to recognize that many species of plants appear to have extra sets of chromosomes, and they are not able to reproduce with closely related species that have a different number of sets of chromosomes.

Extinction

Extinction is the loss of an entire species and is a common feature of the evolution of organisms. The environment in which organisms exist does not remain constant over long time periods. Those species that lack the genetic resources to cope with a changing environment go extinct. Of the estimated 500 million species of organisms that are believed to have ever existed on Earth since life began, perhaps only 5 million to 10 million are currently in existence. This represents an extinction rate of 98 to 99 percent. Obviously, these numbers are estimates, but the fact remains that extinction has been the fate of most species of organisms. In fact, studies of recent fossils and other geologic features show that only thousands of years ago, huge glaciers covered much of Europe and the northern parts of North America. Humans coexisted with many species of large mammals such as giant bison, mammoths, saber-toothed tigers, and giant cave bears. However, their environment changed as the climate became warmer and the glaciers receded. These environmental changes along with continued human predation caused the extinction of many of the large mammals including mammoths, giant bison, saber-toothed tigers, and giant cave bears that could not adapt to a changing environment. (See figure 5.14.)

FIGURE 5.14 Extinction This figure shows a scene that would have been typical at the end of the last ice age. The mammoths, giant bison, and armadillo-like glyptodont as well as many other large mammals are extinct.

extinction caused by human influences on the Earth. See chapter 11 on biodiversity for a more complete discussion.

Coevolution

Coevolution is the concept that two or more species of organisms can reciprocally influence the evolutionary direction of the other. In other words, organisms affect the evolution of other organisms. Since all organisms are influenced by other organisms, this is a common pattern.

For example, grazing animals and the grasses they consume have coevolved. Grasses that are eaten by grazing animals grow from the base of the plant near the ground rather than from the tips of the branches as many plants do. Furthermore, grasses have hard materials in their cell walls that make it difficult for animals to crush the cell walls and digest them. Grazing animals have different kinds of adaptations that overcome these deterrents. Many grazers have teeth that are very long or grow continuously to compensate for the wear associated with grinding hard cell walls. Others, such as cattle, have complicated digestive tracts that allow microorganisms to do most of the work of digestion.

Similarly, the red color and production of nectar by many kinds of flowers is attractive to hummingbirds, which pollinate the flowers at the same time as they consume nectar from the flower. In addition to the red color that is common for many flowers that are pollinated by hummingbirds, many such flowers are long and tubular. The long bill of the hummingbird is a matching adaptation to the shape of the flower. (See figure 5.15.)

The "Kinds of Organism Interactions" section will explore in more detail the ways that organisms interact and the results of long periods of coevolution.

5.3 Kinds of Organism Interactions

Ecologists look at organisms and how they interact with their surroundings. Perhaps the most important interactions occur between organisms. Ecologists have identified several general types of organism-to-organism interactions that are common in all ecosystems. When we closely examine how organisms interact, we see that each organism has specific characteristics that make it well suited to its role. An understanding of the concept of natural selection allows us to see how interactions between different species of organisms can result in species that are finely tuned to a specific role.

As you read this section, notice how each species has special characteristics that equip it for its specific role (niche). Because these interactions involve two kinds of organisms interacting, we should expect to see examples of coevolution. If the interaction between two species is the result of a long period of interaction, we should expect to see that each species has characteristics that specifically adapt it to be successful in its role.

Predation

One common kind of interaction called **predation** occurs when one organism, known as a **predator,** kills and eats another, known as the **prey.** (See figure 5.16.) The predator benefits from killing

FIGURE 5.15 Coevolution The red tubular flowers and the nectar they produce are attractive to hummingbirds. The long bill of the hummingbird allows it to reach the nectar at the base of the flower. The two kinds of organisms have coevolved.

FIGURE 5.16 Predator-Prey Relationship Cheetahs are predators of impalas. Cheetahs can run extremely fast to catch the equally fast impalas. Impalas have excellent eyesight, which helps them see predators. The chameleon hunts by remaining immobile and ambushing prey. The long, sticky tongue is its primary means of capturing unsuspecting insects as they fly past.

FIGURE 5.17 Intraspecific Competition Whenever a needed resource is in limited supply, organisms compete for it. Intraspecific competition for sunlight among these pine trees has resulted in the tall, straight trunks. Those trees that did not grow fast enough died.

and eating the prey and the prey is harmed. Some examples of predator-prey relationships are lions and zebras, robins and earthworms, wolves and moose, and toads and beetles. Even a few plants show predatory behavior. The Venus flytrap has specially modified leaves that can quickly fold together and trap insects that are then digested.

To succeed, predators employ several strategies. Some strong and speedy predators (cheetahs, lions, sharks) chase and overpower their prey; other species lie in wait and quickly strike prey that happen to come near them (many lizards and hawks); and some (spiders) use snares to help them catch prey. At the same time, prey species have many characteristics that help them avoid predation. Many have keen senses that allow them to detect predators, others are camouflaged so they are not conspicuous, and many can avoid detection by remaining motionless when predators are in the area.

An adaptation common to many prey species is a high reproductive rate compared to predators. For example, field mice may have 10 to 20 offspring per year, while hawks typically have two to three. Because of this high reproductive rate, prey species can endure a high mortality rate and still maintain a viable population. Certainly,

the *individual* organism that is killed and eaten is harmed, but the prey *species* is not, since the prey individuals that die are likely to be the old, the slow, the sick, and the less well-adapted members of the population. The healthier, quicker, and better-adapted individuals are more likely to survive. When these survivors reproduce, their offspring are more likely to have characteristics that help them survive; they are better adapted to their environment.

At the same time, a similar process is taking place in the predator population. Since poorly adapted individuals are less likely to capture prey, they are less likely to survive and reproduce. The predator and prey species are both participants in the natural selection process. This dynamic relationship between predator and prey species is a complex one that continues to intrigue ecologists.

Competition

A second type of interaction between species is **competition,** in which two organisms strive to obtain the same limited resource. Ecologists distinguish two different kinds of competition. **Intraspecific competition** is competition between members of the same species. **Interspecific competition** is competition between members of different species. In either case both organisms involved in competition are harmed to some extent. However, this does not mean that there is no winner or loser. If two organisms are harmed to different extents, the one that is harmed less is the winner in the competition, and the one that is harmed more is the loser.

Examples of Intraspecific Competition

Lodgepole pine trees release their seeds following a fire. Thus, a large number of lodgepole pine seedlings begin growing close to one another and compete for water, minerals, and sunlight. None of the trees grows as rapidly as it could because its access to resources is restricted by the presence of the other trees. Eventually, because of differences in genetics or specific location, some of the pines will grow faster and will get a greater share of the resources. The taller trees will get more sunlight and the shorter trees will receive less. In time, some of the smaller trees die. They lost the competition. (See figure 5.17.)

Similarly, when two robins are competing for the same worm, only one gets it. Both organisms were harmed because they had to expend energy in fighting for the worm, but one got some food and was harmed less than the one that fought and got nothing. Other examples of intraspecific competition include corn plants in a field competing for water and nutrients, male elk competing with one another for the right to mate with the females, and wood ducks competing for the holes in dead trees to use for nesting sites.

Examples of Interspecific Competition

Many species of predators (hawks, owls, foxes, coyotes) may use the same prey species (mice, rabbits) as a food source. If the supply of food is inadequate, intense competition for food will occur and certain predator species may be more successful than others. (See figure 5.18.)

In grasslands, the same kind of competition for limited resources occurs. Rapidly growing, taller grasses get more of the

FIGURE 5.18 Interspecific Competition The lion and vultures are involved in interspecific competition for the carcass of the zebra. At this point, the lion is winning in the competition.

water, minerals, and sunlight, while shorter species are less successful. Often the shorter species are found to be more abundant when the taller species are removed by grazers, fire, or other activities.

Competition and Natural Selection

Competition among members of the same species is a major force in shaping the evolution of a species. When resources are limited, less well-adapted individuals are more likely to die and less likely to reproduce. Because the most successful organisms are likely to have larger numbers of offspring, each succeeding generation will contain more of the genetic characteristics that are favorable for survival of the species in that particular environment. Since individuals of the same species have similar needs, competition among them is usually very intense. A slight advantage on the part of one individual may mean the difference between survival and death.

As with intraspecific competition, one of the effects of interspecific competition is that the species that has the larger number of successful individuals emerges from the interaction better adapted to its environment than the species with which it competes.

The **competitive exclusion principle** is the concept that no two species can occupy the same ecological niche in the same place at the same time. The more similar two species are, the more intense will be the competition between them. If one of the two competing species is better adapted to live in the area than the other, the less-fit species must evolve into a slightly different niche, migrate to a different geographic area, or become extinct.

When the niche requirements of two similar species are examined closely, we usually find significant differences between the niches of the two species. The difference in niche requirements reduces the intensity of the competition between the two species. For example, many small forest birds eat insects. However, they may obtain them in different ways; a flycatcher sits on a branch and makes short flights to snatch insects from the air, a woodpecker excavates openings to obtain insects in rotting wood, and many warblers flit about in the foliage capturing insects.

Even among these categories there are specialists. Different species of warblers look in different parts of trees for their insect food. Because of this niche specialization, they do not directly compete with one another. (See figure 5.19).

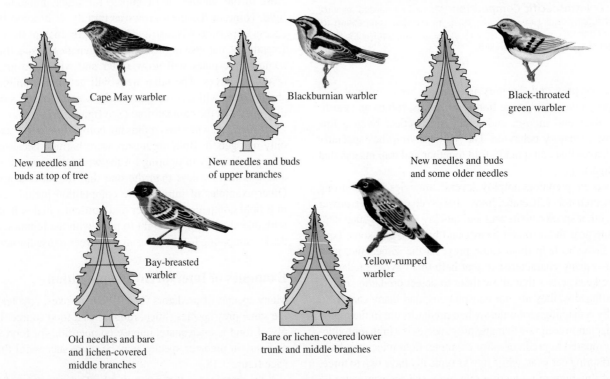

Cape May warbler

New needles and buds at top of tree

Blackburnian warbler

New needles and buds of upper branches

Black-throated green warbler

New needles and buds and some older needles

Bay-breasted warbler

Old needles and bare and lichen-covered middle branches

Yellow-rumped warbler

Bare or lichen-covered lower trunk and middle branches

FIGURE 5.19 Niche Specialization Although all of these warblers have similar feeding habits, the intensity of competition is reduced because they search for insects on different parts of the tree.

Symbiotic Relationships

Symbiosis is a close, long-lasting, physical relationship between two different species. In other words, the two species are usually in physical contact and at least one of them derives some sort of benefit from this contact. There are three different categories of symbiotic relationships: parasitism, commensalism, and mutualism.

Parasitism

Parasitism is a relationship in which one organism, known as the **parasite,** lives in or on another organism, known as the **host,** from which it derives nourishment. Generally, the parasite is much smaller than the host. Although the host is harmed by the interaction, it is generally not killed immediately by the parasite, and some host individuals may live a long time and be relatively little affected by their parasites. Some parasites are much more destructive than others, however.

Newly established parasite-host relationships are likely to be more destructive than those that have a long evolutionary history. With a long-standing interaction between the parasite and the host, the two species generally evolve in such a way that they can accommodate one another. It is not in the parasite's best interest to kill its host. If it does, it must find another. Likewise, the host evolves defenses against the parasite, often reducing the harm done by the parasite to a level the host can tolerate.

Many parasites have complex life histories that involve two or more host species for different stages in the parasite's life cycle. Many worm parasites have their adult, reproductive stage in a carnivore (the definitive host), but they have an immature stage that reproduces asexually in another animal (the intermediate host) that the carnivore uses as food. Thus, a common dog tapeworm is found in its immature form in certain internal organs of rabbits.

Other parasite life cycles involve animals that carry the parasite from one host to another. These carriers are known as **vectors.** For example, many blood-feeding insects and mites can transmit parasites from one animal to another when they obtain blood meals from successive hosts. Malaria, Lyme disease, and sleeping sickness are transmitted by these kinds of vectors.

Parasites that live on the surface of their hosts are known as **ectoparasites.** Fleas, lice, ticks, mites, and some molds and mildews are examples of ectoparasites. Many other parasites, such as tapeworms, malaria parasites, many kinds of bacteria, and some fungi, are called **endoparasites** because they live inside the bodies of their hosts. A tapeworm lives in the intestines of its host, where it is able to resist being digested and makes use of the nutrients in the intestine. If a host has only one or two tapeworms, it can live for some time with little discomfort, supporting itself and its parasites. If the number of parasites is large, the host may die.

Even plants can be parasites. Mistletoe is a flowering plant that is parasitic on trees. It establishes itself on the surface of a tree when a bird transfers the seed to the tree. It then grows down into the water-conducting tissues of the tree and uses the water and minerals it obtains from these tissues to support its own growth.

Some flowering plants, such as beech drops and Indian pipe, lack chlorophyll and are not able to do photosynthesis. They derive their nourishment by obtaining nutrients from the roots of trees or soil fungi and grow above ground for a short period when they flower. Indian pipe is interesting in that it is parasitic on the fungi that assist tree roots in absorbing water. The root fungi receive nourishment from the tree and the Indian pipe obtains nourishment from the fungi. So the Indian pipe is an indirect parasite on trees.

Parasitism is a very common life strategy. If we were to categorize all the organisms in the world, we would find many more parasitic species than nonparasitic species. Each organism, including you, has many others that use it as a host. (See figure 5.20.)

Commensalism

Commensalism is a relationship between organisms in which one organism benefits while the other is not affected. It is possible to visualize a parasitic relationship evolving into a commensal one. Since parasites generally evolve to do as little harm to their host as possible and the host is combating the negative effects of the parasite, they might eventually evolve to the point where the host is not harmed at all.

(a) A tapeworm is an internal parasite **(b) A black-legged tick is an external parasite** **(c) Indian pipe is a plant that is a parasite**

FIGURE 5.20 Parasitism Parasites benefit from the relationship because they obtain nourishment from the host. (a) Tapeworms are internal parasites in the guts of their host, where they absorb food. (b) The tick is an external parasite that sucks body fluids from its host. (c) Indian pipe (*Monotropa uniflora*) is a flowering plant that lacks chlorophyll and is parasitic on fungi that have a mutualistic relationship with tree roots. The host in any of these three situations may not be killed directly by the relationship, but it is often weakened, becoming more vulnerable to predators and diseases.

Many examples of commensal relationships exist. Many orchids, ferns, mosses, and vines use trees as a surface upon which to grow. Those plants that live on the surface of other plants are called epiphytes. The tree is not harmed or helped, but the epiphyte needs a surface upon which to establish itself and also benefits by being close to the top of the tree, where it can get more sunlight and rain.

In the ocean, many sharks have a smaller fish known as a remora attached to them. Remoras have a sucker on the top of their heads that they can use to attach to the shark. In this way, they can hitch a ride as the shark swims along. When the shark feeds, the remora frees itself and obtains small bits of food that the shark misses. Then, the remora reattaches. The shark does not appear to be positively or negatively affected by remoras. (See figure 5.21.)

Many commensal relationships are rather opportunistic and may not involve long-term physical contact. For example, many birds rely on trees of many different species for places to build their nests but do not use the same tree year after year. Similarly, in the spring, bumblebees typically build nests in underground mouse nests that are no longer in use.

Mutualism

Mutualism is another kind of symbiotic relationship and is actually beneficial to both species involved. In many mutualistic relationships, the relationship is obligatory; the species cannot live without each other. In others, the species can exist separately but are more successful when they are involved in a mutualistic relationship. Some species of *Acacia,* a thorny tree, provide food in the form of sugar solutions in little structures on their stems. Certain species of ants feed on the solutions and live in the tree, which they will protect from other animals by attacking any animal that begins to feed on the tree. Both organisms benefit; the ants receive food and a place to live, and the tree is protected from animals that would use it as food.

One soil nutrient that is usually a limiting factor for plant growth is nitrogen. Many kinds of plants, such as legumes (beans, clover, and acacia trees) and alder trees, have bacteria that live in their roots in little nodules. The roots form these nodules when they are infected with certain kinds of bacteria. The bacteria do not cause disease but provide the plants with nitrogen-containing molecules that the plants can use for growth. The nitrogen-fixing bacteria benefit from the living site and nutrients that the plants provide, and the plants benefit from the nitrogen they receive. (See figure 5.22.)

Similarly, many kinds of fungi form an association with the roots of plants. The root-fungus associations are called **mycorrhizae.** The fungus obtains organic molecules from the roots of the plant, and the branched nature of the fungus assists the plant in obtaining nutrients such as phosphates and nitrates. In many cases, it is clear that the relationship is obligatory.

Some Relationships are Difficult to Categorize

Sometimes it is not easy to categorize the relationships that organisms have with each other. For example, it is not always easy to say whether a relationship is a predator-prey relationship or a host-parasite relationship. How would you classify a mosquito or a tick? Both of these animals require blood meals to live and reproduce. They don't kill and eat their prey. Neither do they live in or on a host for a long period of time. This question points out the difficulty encountered when we try to place all kinds of organism interactions into a few categories. However, we can eliminate this problem if we call them temporary parasites or blood predators.

Another relationship that doesn't fit well is the relationship that certain birds such as cowbirds and European cuckoos have with other birds. Cowbirds and European cuckoos do not build nests but lay their eggs in the nests of other species of birds, who are left to care for a foster nestling at the expense of their own nestlings, who generally die. This situation is usually called nest parasitism or brood parasitism. (See figure 5.23.)

What about grazing animals? Are they predators or parasites on the plants that they eat? Sometimes they kill the plant they eat,

(a) A remora hitching a ride on a shark

(b) This bird's nest fern is an epiphyte that benefits from living on the surface of trees.

FIGURE 5.21 Commensalism (a) Remoras hitch a ride on sharks and feed on the scraps of food lost by the sharks. This is a benefit to the remoras. The sharks do not appear to be affected by the presence of remoras. (b) The bird's nest fern does not harm the tree but benefits from using the tree surface as a place to grow.

while at other times they simply remove part of the plant and the rest continues to grow. In either case, the plant has been harmed by the interaction and the grazer has benefited.

(a) Root nodules containing nitrogen-fixing bacteria

(b) Oxpeckers removing parasites

FIGURE 5.22 Mutualism (a) The growths on the roots of this plant contain beneficial bacteria that make nitrogen available to the plant. The relationship is also beneficial to the bacteria, since the bacteria obtain necessary raw materials from the plant. It is a mutually beneficial relationship. (b)The oxpecker (bird) and the impala have a mutualistic relationship. The oxpecker obtains food by removing parasites from the surface of the impala and the impala benefits from a reduced parasite load.

FIGURE 5.23 Nest (Brood) Parasitism This cowbird chick in the nest is being fed by its host parent, a yellow warbler. The cowbird chick and its cowbird parents both benefit but the host is harmed because it is not raising any of its own young.

There are also mutualistic relationships that do not require permanent contact between the participants in the relationship. Bees and the flowering plants they pollinate both benefit from their interactions. The bees obtain pollen and nectar for food and the plants are pollinated. But the active part of the relationship involves only a part of the life of any plant, and the bees are not restricted to any one species of plant for the food. They must actually switch to different flowers at different times of the year.

5.4 Community and Ecosystem Interactions

Thus far, we have discussed specific ways in which individual organisms interact with one another and with their physical surroundings. However, often it is useful to look at ecological relationships from a broader perspective. Two concepts that focus on relationships that involve many different kinds of interactions are community and ecosystem.

A **community** is an assemblage of all the interacting populations of different species of organisms in an area. Some species play minor roles, while others play major roles, but all are part of the community. For example, the grasses of the prairie have a major role, since they carry on photosynthesis and provide food and shelter for the animals that live in the area. Grasshoppers, prairie dogs, and bison are important consumers of grass. Meadowlarks consume many kinds of insects, and though they are a conspicuous and colorful part of the prairie scene, they have a relatively minor role and little to do with maintaining a prairie community. Bacteria and fungi in the soil break down the bodies of dead plants and animals and provide nutrients to plants. Communities consist of interacting populations of different species, but these species interact with their physical world as well.

An **ecosystem** is a defined space in which interactions take place between a community, with all its complex interrelationships, and the physical environment. The physical world has a major impact on what kinds of plants and animals can live in an area. We do not expect to see a banana tree in the Arctic or a walrus in the Mississippi River. Banana trees are adapted to warm, moist, tropical areas, and walruses require cold ocean waters. Some ecosystems, such as grasslands and certain kinds of forests, are shaped by periodic fires. The kind of soil and the amount of moisture also influence the kinds of organisms found in an area.

While it is easy to see that the physical environment places limitations on the kinds of organisms that can live in an area, it is also important to recognize that organisms impact their physical surroundings. Trees break the force of the wind, grazing animals form paths, and earthworms create holes that aerate the soil. While the concepts of community and ecosystem are closely related, an ecosystem is a broader concept because it involves physical as well as biological processes.

Every system has parts that are related to one another in specific ways. A bicycle has wheels, a frame, handlebars, brakes, pedals, and a seat. These parts must be organized in a certain way or the system known as a bicycle will not function. Similarly, ecosystems have parts that must be organized in specific ways or the systems will not operate. To more fully develop the concept of ecosystem, we will look at ecosystems from three points of view: the major roles played by organisms, the way energy is utilized within ecosystems, and the way atoms are cycled within ecosystems.

Major Roles of Organisms in Ecosystems

Ecologists have traditionally divided organisms' roles in ecosystems into three broad categories: producers, consumers, and decomposers.

Producers

Producers are organisms that are able to use sources of energy to make complex, organic molecules from the simple inorganic substances in their environment. In nearly all ecosystems, energy is supplied by the sun, and organisms such as plants, algae, and certain bacteria use light energy to carry on photosynthesis. In terrestrial ecosystems, plants are the major producers. In aquatic ecosystems, many kinds of single-celled organisms (particularly algae and bacteria) are the predominant producers. These tiny, floating organisms are collectively known as **phytoplankton.** Since producers are the only organisms in an ecosystem that can trap energy and make new organic material from inorganic material, all other organisms rely on producers as a source of food, either directly or indirectly.

Consumers

Consumers are organisms that require organic matter as a source of food. They consume organic matter to provide themselves with energy and the organic molecules necessary to build their own bodies. An important part of their role is the process of respiration in which they break down organic matter to obtain energy while they release inorganic matter.

However, consumers can be further subdivided into categories based on the kinds of things they eat and the way they obtain food.

Primary consumers, also known as **herbivores,** are animals that eat producers (plants or phytoplankton) as a source of food. Herbivores, such as leaf-eating insects, seed-eating birds, and animals that filter phytoplankton from the water are usually quite numerous in ecosystems, where they serve as food for the next organisms in the chain.

Secondary consumers or **carnivores** are animals that eat other animals. Secondary consumers can be further subdivided into categories based on what kind of prey they capture and eat. Some carnivores, such as ladybird beetles, primarily eat herbivores, such as aphids; others, such as eagles, primarily eat fish that are themselves carnivores. While these are interesting conceptual distinctions, most carnivores will eat any animal they can capture and kill.

In addition, many animals, called **omnivores,** include both plants and animals in their diet. Even animals that are considered to be carnivores (foxes, bears) regularly include large amounts of plant material in their diets. Conversely, animals often thought of as herbivores (mice, squirrels, seed-eating birds) regularly consume animals as a source of food.

Parasites are also consumers that have a special way of obtaining their food.

Decomposers

Decomposers are organisms that use nonliving organic matter as a source of energy and raw materials to build their bodies. Whenever an organism sheds a part of itself, excretes waste products, or dies, it provides a source of food for decomposers. Since decomposers carry on respiration, they are extremely important in recycling matter by converting organic matter to inorganic material. Many small animals, fungi, and bacteria fill this niche. (See table 5.1.)

Table 5.1 Roles in an Ecosystem

Category	Major Role or Action	Examples
Producer	Converts simple inorganic molecules into organic molecules by the process of photosynthesis	Trees, flowers, grasses, ferns, mosses, algae
Consumer	Uses organic matter as a source of food	Animals, fungi, bacteria
Herbivore	Eats plants directly	Grasshopper, elk, human vegetarian
Carnivore	Kills and eats animals	Wolf, pike, dragonfly
Omnivore	Eats both plants and animals	Rats, raccoons, most humans
Scavenger	Eats meat but often gets it from animals that died by accident or illness, or were killed by other animals	Coyote, vulture, blowflies
Parasite	Lives in or on another living organism and gets food from it	Tapeworm, many bacteria, some insects
Decomposer	Returns organic material to inorganic material; completes recycling of atoms	Fungi, bacteria, some insects and worms

Keystone Species

Ecosystems typically consist of many different species interacting with each other and their physical surroundings. However, some species have more central roles than others. In recognition of this idea, ecologists have developed the concept of keystone species.

A **keystone species** is one that has a critical role to play in the maintenance of specific ecosystems. In prairie ecosystems, grazing animals are extremely important in maintaining the mix of species typical of a grassland. Without the many influences of the grazers, the nature of the prairie changes. A study of the American tallgrass prairie indicated that when bison are present, they increase the biodiversity of the site. (See figure 5.24.) Bison typically eat grasses and, therefore, allow smaller plant species that would normally be shaded by tall grasses to be successful. In ungrazed plots, the tall grasses become the dominant vegetation and biodiversity decreases. Bison also dig depressions in the soil, called wallows, to provide themselves with dust or mud with which they can coat themselves. These wallows retain many species of plants that typically live in disturbed areas. Bison urine has also been shown to be an important source of nitrogen for the plants.

The activities of bison even affect the extent and impact of fire, another important feature of grassland ecosystems. Since bison prefer to feed on recently burned sites and revisit these sites several times throughout the year, they tend to create a patchwork of grazed and ungrazed areas. The grazed areas are less likely to be able to sustain a fire, and fires likely will be more prevalent in ungrazed patches.

The concept of keystone species has also been studied in marine ecosystems. The relationship among sea urchins, sea otters, and kelp forests suggests that sea otters are a keystone species. Sea otters eat sea urchins, which eat kelp. A reduction in the number of otters results in an increase in the number of sea urchins. Increased numbers of sea urchins lead to heavy grazing of the kelp by sea urchins. When the amount of kelp is severely reduced, fish and many other animals that live within the kelp beds lose their habitat and biodiversity is significantly reduced.

The concept of keystone species is useful to ecologists and resource managers because it helps them to realize that all species cannot be treated equally. Some species have pivotal roles, and their elimination or severe reduction can significantly alter ecosystems. In some cases, the loss of a keystone species can result in the permanent modification of an ecosystem into something considerably different from the original mix of species.

Energy Flow Through Ecosystems

An ecosystem is a recognizable, relatively stable unit composed of organisms and their physical surroundings. This does not mean that an ecosystem is unchanging. The organisms within it are growing, reproducing, dying, and decaying. In addition, an ecosystem must have a continuous input of energy to retain its stability. The only significant source of energy for most ecosystems is sunlight. Producers are the only organisms that are capable of trapping solar energy through the process of photosynthesis and making it available to the ecosystem. The energy is stored in the form of chemical bonds in large organic molecules such as carbohydrates (sugars, starches), fats, and proteins. The energy stored in the molecules of producers is transferred to other organisms when the producers are eaten.

Trophic Levels

Each step in the flow of energy through an ecosystem is known as a **trophic level.** Producers (plants, algae, phytoplankton) constitute the first trophic level, and herbivores constitute the second trophic level. Carnivores that eat herbivores are the third trophic level, and carnivores that eat other carnivores are the fourth trophic level. Omnivores, parasites, and scavengers occupy different trophic levels, depending on what they happen to be eating at the time. If we eat a piece of steak, we are at the third trophic level; if we eat celery, we are at the second trophic level. (See figure 5.25.)

Energy Relationships

The second law of thermodynamics states that whenever energy is converted from one form to another, some of the energy is converted to a nonuseful form (typically, low-quality heat). Thus, there is always less useful energy following an energy conversion. When energy passes from one trophic level to the next, there is less useful energy left with each successive trophic level. This loss of low-quality heat is dissipated to the surroundings and warms the air, water, or soil. In addition to this loss of heat, organisms must expend energy to maintain their own life processes. It takes energy to grow, chew food, defend nests, walk to waterholes, or produce and raise offspring. Therefore, the amount of energy contained in higher trophic levels is considerably less than that at lower levels. Approximately 90 percent of the useful energy is lost with each transfer to the next highest trophic level. So in any ecosystem, the amount of energy contained in the herbivore trophic level is only about 10 percent of the energy contained in the producer trophic level. The amount of energy at the third trophic level is approximately 1 percent of that found in the first trophic level. (See figure 5.26.)

Because it is difficult to actually measure the amount of energy contained in each trophic level, ecologists often use other measures to approximate the relationship between the amounts of energy at

FIGURE 5.24 A Keystone Species The bison is a keystone species in prairie ecosystems. They affect the kinds of plants that live in an area, provide nutrients in the form of urine and manure, and produce wallows that provide special habitats for certain plants and animals.

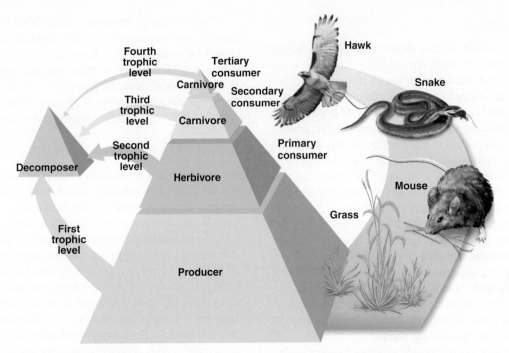

FIGURE 5.25 Trophic Levels in an Ecosystem Organisms within ecosystems can be placed in several broad categories, known as trophic levels, based on how they obtain the energy they need for their survival.

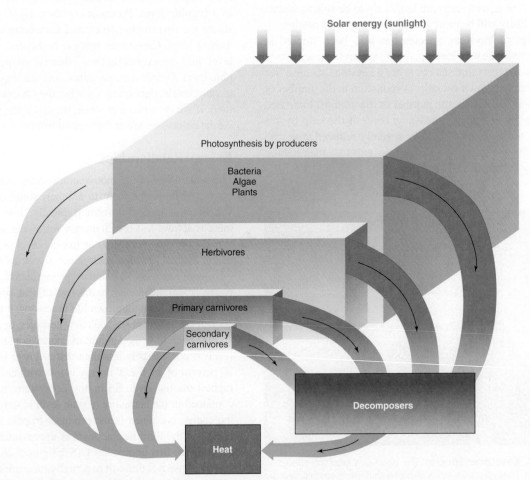

FIGURE 5.26 Energy Flow Through Ecosystems As energy flows through an ecosystem, it passes through several levels known as trophic levels. Each trophic level contains a certain amount of energy. Each time energy flows to another trophic level, approximately 90 percent of the useful energy is lost, usually as heat to the surroundings. Therefore, in most ecosystems, higher trophic levels contain less energy and fewer organisms.

each level. One of these is the biomass. The **biomass** is the weight of living material in a trophic level. It is often possible in a simple ecosystem to collect and weigh all the producers, herbivores, and carnivores. The weights often show the same 90 percent loss from one trophic level to the next as happens with the amount of energy.

Food Chains and Food Webs

A **food chain** is a series of organisms occupying different trophic levels through which energy passes as a result of one organism consuming another. For example, cattails, reeds, and some grasses grow well in very moist soil, perhaps near a pond. The plants' leaves capture sunlight and convert carbon dioxide and water into sugars and other organic molecules. The leaves serve as a food source for insects, such as caterpillars, grasshoppers, and leaf beetles, that have chewing mouthparts and a digestive system adapted to plant food. Some of these insects are eaten by spiders, which fall from the trees into the pond below, where they are consumed by a frog. As the frog swims from one lily pad to another, a large bass consumes the frog. A human may use an artificial frog as a lure to entice the bass from its hiding place. A fish dinner is the final step in this chain of events that began with the leaves of plants beside the pond. (See figure 5.27.) This food chain has six trophic levels. Each organism occupies a specific niche and has special abilities that fit it for its niche, and each organism in the food chain is involved in converting energy and matter from one form to another.

Some food chains rely on a constant supply of small pieces of dead organic material coming from situations where photosynthesis is taking place. The small bits of nonliving organic material are called **detritus.** Detritus food chains are found in a variety of situations. The bottoms of the deep lakes and oceans are too dark for photosynthesis. The animals and decomposers that live there rely on a steady rain of small bits of organic matter from the upper layers of the water where photosynthesis does take place. Similarly, in most streams, leaves and other organic debris serve as the major source of organic material and energy.

In another example, the soil on a forest floor receives leaves, which fuel a detritus food chain. A mixture of insects, crustaceans, worms, bacteria, and fungi cooperates in the breakdown of the large pieces of organic matter, while at the same time feeding on one another. When a leaf dies and falls to the forest floor, it is colonized by bacteria and fungi, which begins the breakdown process. An earthworm will also feed on the leaf and at the same time consume the bacteria and fungi. If that earthworm is eaten by a bird, it becomes part of a larger food chain that includes material from both a detritus food chain and a photosynthesis-driven food chain. When several food chains overlap and intersect, they make up a **food web.** (See figure 5.28.) This diagram is typical of the kinds of interactions that take place in a community. Each organism is likely to be a food source for several other kinds of organisms. Even the simplest food webs are complex.

Nutrient Cycles in Ecosystems— Biogeochemical Cycles

All matter is made up of atoms that are cycled between the living and nonliving portions of an ecosystem. The activities involved in the cycling of atoms include biological, geological, and

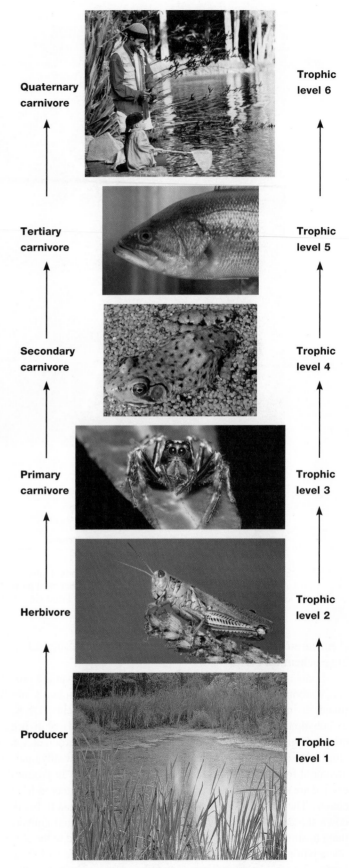

FIGURE 5.27 Trophic Levels in a Food Chain As one organism feeds on another organism, energy flows through the series. This is called a food chain.

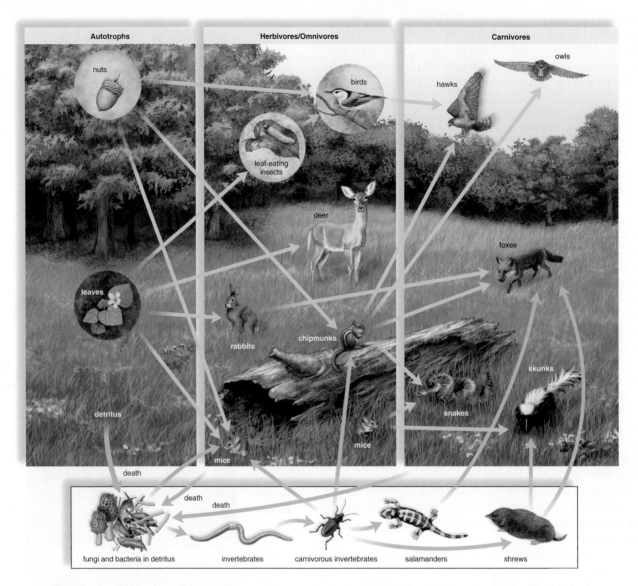

Autotrophs — nuts, birds, leaf-eating insects, leaves, detritus, rabbits, chipmunks, mice, death

Herbivores/Omnivores — deer, mice

Carnivores — owls, hawks, foxes, skunks, snakes

fungi and bacteria in detritus — invertebrates — carnivorous invertebrates — salamanders — shrews

FIGURE 5.28 Food Web A food web describes the many kinds of interactions among the organisms that feed on one another. Plants are the base of most food webs, with a variety of animals feeding on leaves, fruits, and other plant parts. These herbivores are in turn eaten by carnivores. Dead organisms or parts of organisms (leaves, logs, hair, feathers) and waste products (urine and feces) are acted upon by a variety of decomposer organisms, which in turn are eaten by animals.

chemical processes. Therefore, these nutrient cycles are often called **biogeochemical cycles.**

Some atoms are more common in living things than are others. Carbon, nitrogen, oxygen, hydrogen, and phosphorus are found in important organic molecules such as proteins, DNA, carbohydrates, and fats, which are found in all kinds of living things. Organic molecules contain large numbers of carbon atoms attached to one another. These organic molecules are initially manufactured from inorganic molecules by photosynthesis in producers and are transferred from one living organism to another in food chains. The processes of respiration and decay ultimately break down the complex organic molecules of organisms and convert them to simpler, inorganic constituents that are returned to the abiotic environment. In this section, we will look at the flow of three kinds of atoms, carbon, nitrogen, and phosphorus, within communities and between the biotic and abiotic portions of an ecosystem.

Carbon Cycle

All living things are composed of organic molecules that contain atoms of the element carbon. The **carbon cycle** includes the processes and pathways involved in capturing inorganic carbon-containing molecules, converting them into organic molecules that are used by organisms, and the ultimate release of inorganic carbon molecules back to the abiotic environment. (See figure 5.29.)

The same carbon atoms are used over and over again. In fact, you are not exactly the same person today that you were yesterday. Some of your carbon atoms are different. Furthermore, those carbon atoms have been involved in many other kinds of living things over the past several billion years. Some of them were temporary residents in dinosaurs, extinct trees, or insects, but at this instant, they are part of you.

1. The Role of Producers Producers use the energy of sunlight to convert inorganic carbon compounds into organic compounds

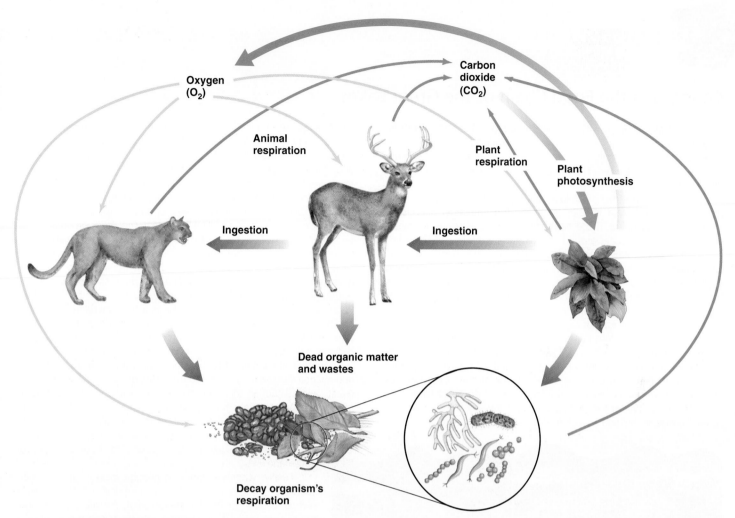

FIGURE 5.29 Carbon Cycle Carbon atoms are cycled through ecosystems. Plants can incorporate carbon atoms from carbon dioxide into organic molecules when they carry on photosynthesis. The carbon-containing organic molecules are passed to animals when they eat plants or other animals. Organic wastes or dead organisms are consumed by decay organisms. All organisms, plants, animals, and decomposers return carbon atoms to the atmosphere when they carry on respiration. Oxygen atoms are being cycled at the same time that carbon atoms are being cycled.

through the process of photosynthesis. In this process, sunlight energy is converted to chemical-bond energy in organic molecules, such as sugar. Their source of inorganic carbon is carbon dioxide (CO_2), which is present in small quantities as a gas in the atmosphere and dissolved in water. During photosynthesis, carbon dioxide is combined with hydrogen from water molecules (H_2O) to form organic molecules like carbohydrates. In terrestrial plants, photosynthesis typically takes place in the leaves. The carbon dioxide is absorbed into the leaves from the air and the water is absorbed from the soil by the roots and transported to the leaves. Aquatic producer organisms, such as algae and bacteria, absorb carbon dioxide and water molecules from the water in which they live. (Actually about 50 percent of photosynthetic activity takes place in the oceans.)

An important by-product of the process of photosynthesis is oxygen. Oxygen molecules (O_2) are released into the atmosphere or water during the process of photosynthesis because water molecules are split to provide hydrogen atoms necessary to manufacture carbohydrate molecules. The remaining oxygen is released as a waste product of photosynthesis. It is important to recognize that plants

and other producer organisms use the organic molecules they produce for growth and to provide energy for other necessary processes.

2. The Role of Consumers Herbivores use the complex organic molecules in the bodies of producers as food. When an herbivore eats plants or algae, it breaks down the complex organic molecules into simpler organic molecular building blocks, which can be reassembled into the specific organic molecules that are part of the herbivore's chemical structure. The carbon atom, which was once part of an organic molecule in a producer, is now part of an organic molecule in an herbivore. Consumers also use organic molecules as a source of energy. Through the process of respiration, oxygen from the atmosphere is used to break down large organic molecules into carbon dioxide and water, with the release of energy. Much of the chemical-bond energy released by respiration is lost as heat, but the remainder is used by the herbivore for movement, growth, and other activities.

In similar fashion, when an herbivore is eaten by a carnivore, some of the carbon-containing molecules of the herbivore become incorporated into the body of the carnivore. The remaining organic

Changes in the Food Chain of the Great Lakes

Many kinds of human activity aid the distribution of species from one place to another. Today there are over 100 exotic species in the Great Lakes. Some, such as smelt, brown trout, and several species of salmon, were purposely introduced. However, most exotic species entered accidentally as a result of human activity.

Prior to the construction of locks and canals, the Great Lakes were effectively isolated from invasion of exotic fish and other species by Niagara Falls. Beginning in the early 1800s, the construction of canals allowed small ships to get around the falls. This also allowed some fish species such as the sea lamprey and alewife to enter the Great Lakes. The completion of the St. Lawrence Seaway in 1959 allowed ocean-going ships to enter the Great Lakes. Because of the practice of using water as ballast, ocean-going vessels are a particularly effective means of introducing species. They pump water into their holds to provide ballast when they do not have a full load of cargo. (Ballast adds weight to empty ships to make their travel safer.) Ballast water is pumped out when cargo is added. Since these vessels may add water as ballast in Europe and empty it in the Great Lakes, it is highly likely that organisms will be transported to the Great Lakes from European waters. Some of these exotic species have caused profound changes in the food chain of the Great Lakes.

The introduction of the zebra and quagga mussels is correlated with several changes in the food web of the Great Lakes. Both mussels reproduce rapidly and attach themselves to any hard surface, including other mussels. They are very efficient filter feeders that remove organic matter and small organisms from the water. Measurements of the abundance of diatoms and other tiny algae show that they have declined greatly—up to 90% in some areas where zebra or quagga mussels are common. There has been a corresponding increase in the clarity of the water. In many places, people can see objects two times deeper than they could in the past.

Diporeia is a bottom-dwelling crustacean that feeds on organic matter. Populations of *Diporeia* have declined by 70% in many places in the Great Lakes. Many feel that this decline is the result of a reduction in their food sources, which are being removed from the water by zebra and quagga mussels. Since *Diporeia* is a major food organism for many kinds of bottom-feeding fish, there has been a ripple effect through the food chain. Recently, whitefish that rely on *Diporeia* as a food source have shown a decline in body condition. Other bottom-feeding fish that eat *Diporeia* serve as a food source for larger predator fish and there have been recent declines in the populations and health of some of these predator fish.

Another phenomenon that is correlated with the increase in zebra and quagga mussels is an increased frequency of toxic algal blooms in the Great Lakes. Although there are no clear answers to why this is occurring, two suggested links have been tied to mussels. The clarity of the water may be encouraging the growth of the toxic algae or the mussels may be selectively rejecting toxic algae as food, while consuming the nontoxic algae. Thus, the toxic algae have a competitive advantage.

Finally, wherever zebra or quagga mussels are common, species of native mussels and clams have declined. There may be several reasons for this correlation. First, the zebra and quagga mussels are in direct competition with the native species of mussels and clams. Zebra and quagga mussels are very efficient at removing food from the water and may be out-competing the native species for food. Secondly, since zebra and quagga mussels attach to any hard surface, they attach to native clams, essentially burying them.

A new threat to the Great Lakes involves the potential for exotic Asian carp (bighead and silver carp) to enter through a canal system that connects Lake Michigan at Chicago to the Mississippi River. These and other species of carp were introduced into commercial fish ponds in the southern United States. However, they soon escaped and entered the Mississippi River and have migrated upstream and could easily enter Lake Michigan. Both bighead and silver carp are filter feeders that consume up to 40% of their body weight in plankton per day. They could have a further impact on the base of the Great Lakes food web, which has already been greatly modified by zebra and quagga mussels.

Phytoplankton

Sport fish (trout, salmon, walleye, whitefish)

Zebra mussels
Quagga mussels

Forage fish (alewife, smelt, sculpin)

Diporeia

Actual Size 7-8 mm

---

molecules are broken down in the process of respiration to obtain energy, and carbon dioxide and water are released.

3. The Role of Decomposers The organic molecules contained in animal waste products and dead organisms are acted upon by decomposers that use these organic materials as a source of food. The decay process of decomposers involves respiration and releases carbon dioxide and water. Thus, the carbon atoms in the organic molecules of dead organisms and waste products are returned to the atmosphere as carbon dioxide to be used again.

4. Carbon Sinks Processes or situations that remove atoms from active, short-term nutrient cycles are known as *sinks*. There are several kinds of sinks for carbon atoms. Some plants, particularly

Whole Ecosystem Experiments

Many environmental issues are difficult to resolve because, although there are hypotheses about what is causing a problem, the validity of the hypotheses has not been tested by experiments. Therefore, when governments seek to set policy, there are always those who argue that there is little hard evidence that the problem is real or that the cause of the problem has not been identified. Several examples include: What causes eutrophication of lakes? What causes acidification of lakes and rivers? What are the causes of global climate change? What is the likelihood that emissions from coal-fired power plants are causing increased mercury in fish? The most powerful tests of hypotheses related to these problems are experiments that take place on a large scale in natural settings. Several such experiments have been crucial in identifying causes of environmental problems and led to policy changes that have alleviated environmental problems.

Beginning in 1966, the Canadian government established the Experimental Lakes Area in western Ontario. Many lakes were designated for experiments that would help answer questions about environmental issues. One experiment tested the hypothesis that phosphorus was responsible for eutrophication (excessive growth of algae and plants) of lakes. Laboratory studies had suggested that carbon, nitrogen, or phosphorus could be responsible. To help answer which of these three nutrients was the cause of eutrophication, a dumbbell-shaped lake was divided in two at its narrow "waist" by placing a plastic curtain across the lake. One portion had carbon, nitrogen, and phosphorus added to it and the other portion had only carbon and nitrogen added. The results were clear. The portion of the lake with the added phosphorus had an abundant growth of algae and turned green. The other portion of the lake with carbon and nitrogen but no phosphorus did not (see photo). As a result of this experiment, governments were justified in requiring detergent manufacturers to remove phosphorous compounds from their products and requiring sewage treatment plants to eliminate phosphorus from their effluent.

Other experiments on whole lakes have investigated:

- The effects of acid deposition on food webs in lakes—predator fish starve as their prey disappear.

- The effects of flooding of land by dams—there is an increase in the mercury content of fish and carbon dioxide and methane are released into the atmosphere.

- The effects of removal of aquatic vegetation—northern pike populations declined.

After each experiment, the Canadian government requires that the lake be returned to its pristine condition.

long-lived trees, can tie up carbon in their bodies for centuries. Organic matter in soil and sediments is the remains of once living organisms that have been removed from the active, short-term carbon cycle. Thus, situations where decomposition is slow (tundra, northern forests, grasslands, swamps, marine sediments) constitute a sink for carbon that can tie up carbon for hundreds to thousands of years.

Fossil fuels (coal, petroleum, and natural gas) were formed from the remains of organisms and are a longer-term sink that involves hundreds of millions of years. The carbon atoms in fossil fuels at one time were part of the active carbon cycle but were removed from the active cycle when the organisms accumulated without decomposing. The organisms that formed petroleum and natural gas are thought to be the remains of marine organisms that got covered by sediments. Coal was formed from the remains of plants that were buried by sediments. Once the organisms were buried, their decomposition slowed, and heat from the Earth and pressure from the sediments helped to transform the remains of living things into fossil fuels.

Oceans are a major carbon sink. Carbon dioxide is highly soluble in water. Many kinds of carbonate sedimentary rock are formed from the precipitation of carbonates from solution in oceans. In addition, many marine organisms form skeletons or shells of calcium carbonate. These materials accumulate on the ocean floor as sediments that over time can be converted to limestone. Limestone typically contains large numbers of fossils. The huge amount of carbonate rock is an indication that there must have been higher amounts of carbon dioxide in the Earth's atmosphere in the past.

5. *Human Alteration of the Carbon Cycle* Humans are involved in modifying the carbon cycle in several ways.

Burning fossil fuels releases large amounts of carbon dioxide into the atmosphere. One consequence of this action is that the amount of carbon dioxide in the atmosphere has been increasing steadily since humans began to use fossil fuels extensively. It has become clear that increasing carbon dioxide is causing changes in the climate of the world, and many nations are seeking to reduce energy use to minimize the increase in carbon dioxide in the atmosphere. This topic will be discussed in more detail in chapter 16, which deals with air pollution.

Altering ecosystems has also contributed to a change in the carbon cycle. Converting forests or grasslands to agriculture replaces long-lived plants with those that live for only one year. In addition, the removal of the crop from the field and frequent tilling of the soil tends to reduce the amount of organic matter in the soil. Thus, agricultural ecosystems do not store carbon as effectively as natural ecosystems. Often the clearing of forests to provide farmland involves the burning of the trees, which contributes carbon dioxide to the atmosphere. Figure 5.30 illustrates the importance of carbon sinks and the effects human activities have on the carbon cycle.

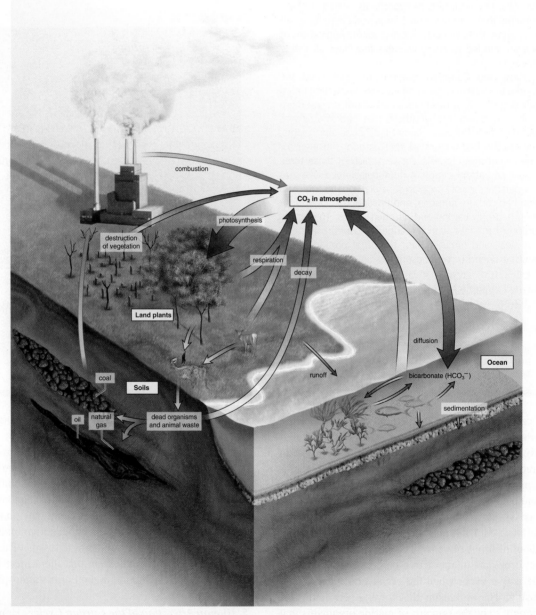

FIGURE 5.30 Carbon Sinks and Human Effects Carbon sinks are situations that tie up carbon for extended periods. Fossil fuels, long-lived vegetation, soil organic matter, carbon dioxide dissolved in the oceans, and ocean sediments are examples of sinks. Human activities have affected the carbon cycle in several ways. The burning of fossil fuels, destruction of forests, and farming activities that reduce soil organic matter all increase the flow of carbon dioxide to the atmosphere. Increased carbon dioxide in the atmosphere increases the amount of carbon dioxide dissolved in the oceans.

Nitrogen Cycle

The **nitrogen cycle** involves the cycling of nitrogen atoms between the abiotic and biotic components and among the organisms in an ecosystem. Seventy-eight percent of the gas in the air we breathe is made up of molecules of nitrogen gas (N_2). However, the two nitrogen atoms are bound very tightly to each other, and very few organisms are able to use nitrogen in this form. Since plants and other producers are at the base of nearly all food chains, they must make new nitrogen-containing molecules, such as proteins and DNA. Plants and other producers are unable to use the nitrogen in the atmosphere and must get it in the form of nitrate (NO_3^-) or ammonia (NH_3).

1. The Role of Nitrogen-Fixing Bacteria

Because atmospheric nitrogen is not usable by plants, nitrogen-containing compounds are often in short supply and the availability of nitrogen is often a factor that limits the growth of plants. The primary way in which plants obtain nitrogen compounds they can use is with the help of bacteria that live in the soil.

Nitrogen-fixing bacteria are able to convert the nitrogen gas (N_2) that enters the soil into ammonia that plants can use. Certain kinds of these bacteria live freely in the soil and are called **free-living nitrogen-fixing bacteria.** Others, known as **symbiotic nitrogen-fixing bacteria,** have a mutualistic relationship with certain plants and live in nodules in the roots of plants known as legumes (peas, beans, and clover) and certain trees such as alders. Some grasses and evergreen trees appear to have a similar relationship with certain root fungi that seem to improve the nitrogen-fixing capacity of the plant.

2. The Role of Producers and Consumers

Once plants and other producers have nitrogen available in a form they can use, they can construct proteins, DNA, and other important nitrogen-containing organic molecules. When herbivores eat plants, the plant protein molecules are broken down to smaller building blocks called amino acids. These amino acids are then reassembled to form proteins typical for the herbivore. This same process is repeated throughout the food chain. During metabolic processes involving the modifying or breakdown of proteins, consumers generate nitrogen-containing waste products.

3. The Role of Decomposers and Other Bacteria

Bacteria and other types of decay organisms are involved in the nitrogen cycle also. Dead organisms and their waste products contain molecules, such as proteins, urea, and uric acid, that contain nitrogen. Decomposers break down these nitrogen-containing organic molecules, releasing ammonia, which can be used directly by many kinds of plants. Several other kinds of soil bacteria called **nitrifying bacteria** are able to convert ammonia to nitrate. Plants can use nitrate as a source of nitrogen for synthesis of nitrogen-containing organic molecules.

Finally, under conditions where oxygen is absent, bacteria known as **denitrifying bacteria** are able to convert nitrate or other nitrogen-containing compounds to nitrogen gas (N_2), which is ultimately released into the atmosphere. These nitrogen atoms can reenter the cycle with the aid of nitrogen-fixing bacteria. (See figure 5.31.)

4. Unique Features of the Nitrogen Cycle

Although a cyclic pattern is present in both the carbon cycle and the nitrogen cycle, the nitrogen cycle shows two significant differences. First, most of the difficult chemical conversions are made by bacteria and other microorganisms. Without the activities of bacteria, little nitrogen would be available and the world would be a very different place. Second, although nitrogen enters organisms by way of nitrogen-fixing bacteria and returns to the atmosphere through the actions of denitrifying bacteria, there is a secondary loop in the cycle that recycles nitrogen compounds directly from dead organisms and wastes directly back to producers.

5. Nitrogen Sinks

The primary sink for nitrogen is nitrogen gas in the atmosphere. However, there are a couple of minor nitrogen sinks.

Since living things have nitrogen as a part of protein, nitrogen that was once part of the active nitrogen cycle was removed when organic material accumulated. In ecosystems in which large amounts of nonliving organic matter accumulates (swamps, humus in forests, and marine sediments), nitrogen can be tied up for relatively long time periods. In addition, some nitrogen may be tied up in sedimentary rock and in some cases is released with weathering. Nitrogen compounds are very soluble in water so when sedimentary rock is exposed to water, these materials are dissolved and reenter the active nitrogen cycle.

6. Human Alteration of the Nitrogen Cycle

In naturally occurring soil, nitrogen is often a limiting factor for plant growth. To increase yields, farmers provide extra sources of nitrogen in several ways. Inorganic fertilizers are a primary method of increasing the nitrogen available. These fertilizers may contain ammonia, nitrate, or both.

Since the manufacture of nitrogen fertilizer requires a large amount of energy and uses natural gas as a raw material, fertilizer is expensive. Therefore, farmers use alternative methods to supply nitrogen and reduce their cost of production. Several different techniques are effective. Farmers can alternate nitrogen-yielding crops such as soybeans with nitrogen-demanding crops such as corn. Since soybeans are legumes that have symbiotic nitrogen-fixing bacteria in their roots, if soybeans are planted one year, the excess nitrogen left in the soil can be used by the corn plants grown the next year. Some farmers even plant alternating strips of soybeans and corn in the same field. A slightly different technique involves growing a nitrogen-fixing crop for a short period of time and then plowing the crop into the soil and letting the organic matter decompose. The ammonia released by decomposition serves as fertilizer to the crop that follows. This is often referred to as green manure. Farmers can also add nitrogen to the soil by spreading manure from animal production operations or dairy farms on the field and relying on the soil bacteria to decompose the organic matter and release the nitrogen for plant use.

Agricultural runoff has altered many aquatic ecosystems. Agricultural runoff consists of a mixture of compounds that serve as nutrients. The two primary nutrients are nitrogen and phosphorus, both of which stimulate the growth of aquatic plants, algae, and phytoplankton.

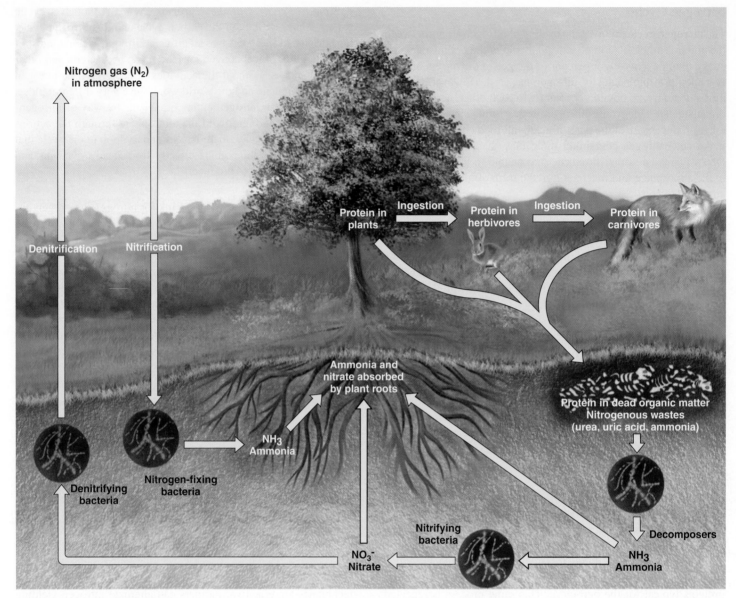

FIGURE 5.31 Nitrogen Cycle Nitrogen atoms are cycled in ecosystems. Atmospheric nitrogen is converted by nitrogen-fixing bacteria to a form that plants can use to make protein and other compounds. Nitrogen is passed from one organism to another when proteins are eaten. Dead organisms and waste products are acted on by decay organisms to form ammonia, which may be reused by plants or converted to other nitrogen compounds by other kinds of bacteria. Denitrifying bacteria are able to convert nitrate and other nitrogen-containing compounds into atmospheric nitrogen.

The nutrients in fertilizers are intended to become incorporated into the bodies of the plants we raise for food. However, if too much fertilizer is applied or if it is applied at the wrong time, much of the fertilizer is carried into aquatic ecosystems. In addition, raising large numbers of animals for food in concentrated settings results in huge amounts of animal wastes that contain nitrogen and phosphorus compounds. These wastes often enter local water sources.

The presence of large amounts of these nutrients, in either freshwater or salt water, results in increased rates of growth of bacteria, algae, and aquatic plants. Increases in the number of these organisms can have many different effects. Many algae are toxic, and when their numbers increase significantly fish are killed and incidents of human poisoning occur. An increase in the number of plants and algae in aquatic ecosystems also can lead to low oxygen concentrations in the water. When these organisms die, decomposers use oxygen from the water as they break down the dead organic matter. This lowers the oxygen concentration, and many organisms die.

The burning of fossil fuels has also altered the nitrogen cycle. When fossil fuels are burned, the oxygen and nitrogen in the air are heated to high temperatures and a variety of nitrogen-containing compounds are produced. (See chapter 16 for a discussion of the smog-producing effects of these nitrogen compounds.) These compounds are used by plants as nutrients for growth. Many people suggest that these sources of nitrogen, along with that provided by fertilizers, have doubled the amount of nitrogen available today as compared to preindustrial times.

Phosphorus Cycle

Phosphorus is another element common in the structure of living things. It is present in many important biological molecules

Going Green

Phosphorus-Free Lawn Fertilizer

Many lakes and streams experience extreme growth of algae and aquatic plants during the warm summer months. The presence of large amounts of vegetation in water interferes with boating, fishing, swimming, and other recreational activities. Some cyanobacteria (microscopic blue-green bacteria) actually produce toxic compounds that can sicken and kill. Human exposure to such toxins usually results in mild symptoms—skin rash, hay fever-like symptoms, or gastrointestinal upset. However, poisonings leading to deaths of pets, livestock, and wildlife are common when the animals drink water from lakes or streams that contain toxic cyanobacteria.

Excessive growth of aquatic plants and algae is stimulated by the warm water and abundant supplies of nutrients—particularly phosphorus. This has led many communities that have close ties to ocean, lakes, or streams to consider regulations on the use of phosphorus in lawn fertilizer. In 2002, Minnesota became the first state to regulate the use of phosphorus in lawn fertilizer. The law is predicated on the fact that soil tests in the state show that nearly all soils have adequate phosphorus and additional phosphorus in fertilizer is not needed. Excess phosphorus is not taken up by plants and much of it runs off into streams and lakes where it stimulates the growth of plants and algae. The regulations prohibit or severely restrict the use of phosphorus in lawn fertilizers. Exceptions are made for newly established lawns, which can use phosphorus-containing fertilizer during the first year of growth or places where soil tests show low phosphorus. The lawn-care industry responded immediately by producing phosphorus-free lawn fertilizer.

While many states now have regulations limiting fertilizers made with phosphates, phosphate fertilizers are still widely used. There are alternatives. The following are several lawn and garden tips that could be used:

- Use compost. Composited fruits and vegetables provide natural phosphorus for your garden.
- Adding organic matter helps the soil release natural phosphorus to your plants, making it and other nutrients more absorbable.
- When you need additional phosphorus, use an organic source such as bone meal, soy meal, or manure.
- If you do buy fertilizer make sure it is phosphate-free. You can tell by looking at the number on the bag. The phosphate number is the middle number and it should read "0."

What's Your Take?

In the United States more than 90 million hectares—an area eight times the size of New Jersey—are planted with lawns, and we are adding 1,000 square kilometers of turfgrass every year. Maintaining all that lawn is a huge undertaking. Annually, the average U.S. homeowner spends the equivalent of at least a full workweek pushing or driving a mower. Lawns soak up massive amounts of water and pesticides, and they use more than 10 times the amount of pesticides applied for farming purposes. The toxic runoff percolates into groundwater, threatening wildlife and human health.

In the United States, about $20 billion is spent each year on the care of lawns and gardens. A significant amount of this cost involves the use of fertilizer to stimulate the growth of grass. Fertilizer runoff from lawns is a significant water pollution problem. Although most lawns will grow without fertilizer, the quality of the lawn is often not as nice. It is often possible to substitute groundcovers or natural landscaping for lawns.

Should people maintain grass lawns? Choose one side of the question and develop arguments that support your position.

such as DNA and in the membrane structure of cells. In addition, the bones and teeth of animals contain significant quantities of phosphorus.

The phosphorus cycle differs from the carbon and nitrogen cycles in one important respect. Phosphorus is not present in the atmosphere as a gas. The ultimate source of phosphorus atoms is rock. In nature, new phosphorus compounds are released by the erosion of rock and become dissolved in water. Plants use the dissolved phosphorus compounds to construct the molecules they need. Animals obtain the phosphorus they need when they consume plants or other animals. When an organism dies or excretes waste products, decomposer organisms recycle the phosphorus compounds back into the soil. While phosphorus moves rapidly through organisms in food chains, phosphorus ions are not very soluble in water and tend to precipitate in the oceans to form sediments that eventually become rock on the ocean floor. Once this has occurred, it takes the process of geologic uplift followed by erosion to make phosphorus ions available to terrestrial

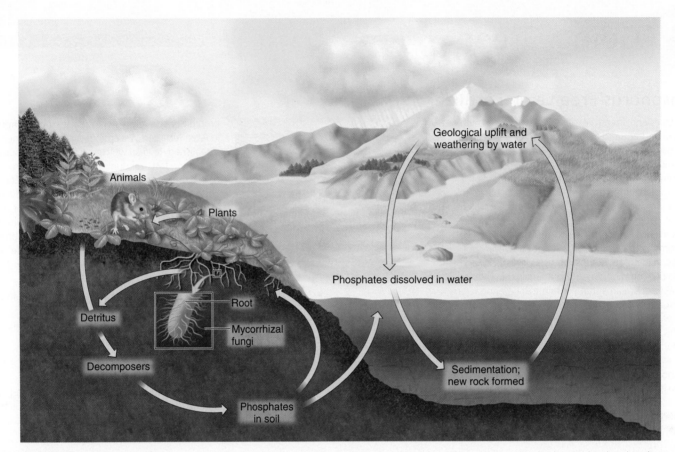

FIGURE 5.32 Phosphorus Cycle The ultimate source of phosphorus in the phosphorus cycle is rock that, when dissolved, provides the phosphate used by plants and animals. Phosphorus is important in molecules such as DNA and molecules in cell membranes. Plants obtain phosphorus from the soil through their roots and incorporate it into their bodies. Animals eat plants and obtain phosphorus. Dead organisms and waste materials are broken down by decomposers that release phosphorus. The long-term cycle involves precipitation of phosphorus-containing compounds in oceans in sediments that form rock. The uplift of mountains elevates these deposits and makes phosphorus available through the process of weathering and erosion.

ecosystems. Thus, sediments in the ocean constitute a sink for phosphrous.

Waste products of animals often have significant amounts of phosphorus. (See figure 5.32.) In places where large numbers of seabirds or bats congregate for hundreds of years, the thickness of their droppings (called guano) can be a significant source of phosphorus for fertilizer. (See Issues & Analysis, page 107.) In many soils and in aquatic ecosystems, phosphorus is a limiting factor and must be provided to crop plants to get maximum yields. Since phosphorus compounds come from rock, mining of phosphate-containing rock is necessary to provide the phosphate for fertilizer.

Agricultural runoff is a significant source of phosphrous in aquatic ecosystems that stimulate the growth of aquatic producers and cause a variety of environmental problems. (See discussion of agricultural runoff in the nitrogen cycle section of this chapter.)

Summary

Everything that affects an organism during its lifetime is collectively known as its environment. The environment of an organism can be divided into biotic (living) and abiotic (nonliving) components.

The space an organism occupies is known as its habitat, and the role it plays in its environment is known as its niche. The niche of a species is the result of natural selection directing the adaptation of the species to a specific set of environmental conditions.

Organisms interact with one another in a variety of ways. Predators kill and eat prey. Organisms that have the same needs compete with one another and do mutual harm, but one is usually harmed less and survives. Symbiotic relationships are those in which organisms live in physical contact with one another. Parasites live in or on another organism and derive benefit from the relationship, harming the host in the process. Commensal organisms derive benefit from another organism but do not harm the host. Mutualistic organisms both derive benefit from their relationship.

A community is the biotic portion of an ecosystem that is a set of interacting populations of organisms. Those organisms and their abiotic environment constitute an ecosystem. In an ecosystem, energy is trapped by producers and flows from

Issues & Analysis

Wildlife and Climate Change

Climate is one of the primary factors that determines the distribution of wild plants and animals around the world. There is evidence from the past of how species respond when the climate changes. About 18,000 years ago at the peak of the last glacial period, the vegetation over most of the northern half of the United States was primarily boreal forest and tundra. Today, those vegetation types and their associated animal species are found in central and northern Canada. As the world warmed following glaciation, species moved to higher latitudes, or upslope in mountainous areas, following a climate to which they adapted.

While we have a sense of how species will respond, there are some major differences between the former and current climate changes that threaten many species. The current rate of change is many times faster than what occurred coming out of the ice age. That warming took roughly 8,000 years, while we will likely see the same magnitude of global temperature increase in less than 300 years this time. Many species with limited ability to move, such as plants and nonflying invertebrates, will simply not be able to keep up as the climate to which they are adapted moves on.

In addition, the natural landscape is now more fragmented by human development such as cropland, highways, and cities. This development forms a barrier to the movement of many species and will inhibit their ability to respond to climate change.

The changing climate also is affecting the timing of annual events in the life cycle of species. Studies have documented recent shifts in the timing of migration, insect emergence, flowering and leaf out, all driven by the earlier arrival of spring. Not all species are responding in the same way, and this can lead to the uncoupling of ecological relationships among species. For example, insect emergence is occurring up to three weeks earlier in the Arctic, while migratory songbirds are not leaving their wintering grounds earlier because they rely on day length, not temperature, as a cue to begin migration. Thus, they may be arriving in the Arctic as a key food resource for nesting is declining, resulting in less successful reproduction.

Scientists predict that one-fourth of Earth's species will be headed for extinction by 2050 if the warming trend continues at its current rate. Many species are already feeling the heat:

- In 1999, the death of the last Golden Toad in Central America marked the first documented species extinction driven by climate change.
- Pacific salmon, usually restricted by cold water temperatures, have been found far outside their normal ranges.
- Coral reefs cover less than 1 percent of the ocean floor, but provide food and shelter for over one-third of all marine fish. Coral reefs have undergone global mass-bleaching events whenever sea temperatures have exceeded summer averages by more than 1 degree centigrade for several weeks. Since 1979, mass-bleaching events have become increasingly more frequent.
- Bowhead whales feed on zooplankton. Without sea ice their principle food source may be threatened.

- Due to melting ice in the Artic, polar bears may be gone from the planet in as little as 100 years.

Climate change is a global challenge with many questions associated with it.

- What do you think are the likely specific ecological effects of climate change on wildlife and fish?
- Is this something we should be concerned with? Why or why not?
- Is there an ethical or moral aspect to caring for our fellow creatures?

Golden Toad

Bowhead Whale

producers through various trophic levels of consumers (herbivores, carnivores, omnivores, and decomposers). About 90 percent of the energy is lost as it passes from one trophic level to the next. This means that the amount of biomass at higher trophic levels is usually much less than that at lower trophic levels. The sequence of organisms through which energy flows is known as a food chain. Several interconnecting food chains constitute a food web.

The flow of atoms through an ecosystem involves all the organisms in the community. The carbon, nitrogen, and phosphorus cycles are examples of how these materials are cycled in ecosystems.

Acting Green

1. Purchase detergents that do not contain phosphorus (phosphate).
2. Compost food waste—do not use the garbage disposal.
3. Reduce the amount of carbon dioxide you produce by reducing energy consumption.
4. Choose local, native plants for landscaping.
5. Reduce or eliminate the amount of yard planted to grass.
6. Provide habitat for wildlife with plantings that provide food or shelter.
7. Do not use fertilizer or pesticides on your lawn.
8. Leave a legacy. Plant 10 trees in your community.
9. Grow some of your own food. A garden can begin with one tomato plant! Or better yet organize a community garden!

Review Questions

1. List three abiotic and three biotic factors of your environment.
2. Describe a primary limiting factor for reptiles.
3. How is an organism's niche different from its habitat?
4. Describe, in detail, the niche of a human.
5. How are the concepts of population and species similar?
6. Describe how genetic differences, number of offspring, and death are related to the concept of natural selection.
7. How is natural selection related to the concept of niche?
8. What is speciation and why does it occur?
9. Why does extinction occur?
10. Give an example of coevolution.
11. List five predators and their prey organisms.
12. Describe the difference between interspecific and intraspecific competition.
13. What do parasitism, mutualism, and commensalism have in common? How are they different?
14. How do the concepts of ecosystem and community differ?
15. What roles do producers, consumers, and decomposers fulfill in an ecosystem?
16. Give examples of organisms that are herbivores, carnivores, and omnivores.
17. What distinguishes a keystone species from other species in an ecosystem?
18. How is the concept of trophic levels related to energy flow in an ecosystem?
19. Describe a food chain and a food web.
20. Describe how each of the following is involved in the carbon cycle: carbon dioxide, producer, organic compounds, consumer, respiration, and decomposer.
21. List three changes to the carbon cycle caused by human activity.
22. Describe how each of the following is involved in the nitrogen cycle: atmospheric nitrogen, nitrogen-fixing bacteria, nitrifying bacteria, denitrifying bacteria, producer, protein, consumer, and decomposer.
23. List three ways humans have altered the nitrogen cycle.
24. Describe how each of the following is involved in the phosphorus cycle: phosphorus in rock, producer, consumer, animal waste, respiration, and decomposer.

Critical Thinking Questions

1. Many people in the world have very little protein in their diet. They are often able to grow crops to feed themselves but do not raise cattle or other sources of meat. Describe why these people are not likely to use some of the crops they raise to feed to cattle.
2. Some people predict that the available sources of phosphorus from mines will be exhausted in the next 50 years. Describe what changes are likely to occur in ecosystems if phosphorus is not available.
3. Polar bears hunt seals from ice and have been placed on the endangered species list due to warming temperatures. Why has the habitat of the polar bear changed?

 | ENVIRONMENTAL SCIENCE

To access additional resources for this chapter, please visit ConnectPlus® at connect.mheducation.com. There you will find interactive exercises, including Google Earth™, additional Case Studies, and SmartBook™, an adaptive eBook that integrates our LearnSmart® adaptive learning technology.

Kinds of Ecosystems and Communities

Ecosystems are functional units that involve interactions between organisms and their physical environment. In forest ecosystems, the trees are the dominant producer organisms that trap sunlight and modify the physical environment for other organisms.

OBJECTIVES

After reading this chapter, you should be able to:

- Recognize the difference between primary and secondary succession.
- Describe the process of succession from pioneer to climax community in both terrestrial and aquatic situations.
- Describe how the concepts of succession and climax community developed and how they are used today.

- Identify the major environmental factors that determine the kind of community that develops in an area.
- State major abiotic and biotic characteristics of the various terrestrial biomes.
- State abiotic differences between marine and freshwater ecosystems.

- State how pelagic and benthic ecosystems differ.
- Describe the role of phytoplankton and zooplankton in aquatic ecosystems.
- State biotic differences between marine and freshwater ecosystems.

Overfishing of Marine Ecosystems—A Global Disaster

Factory fishing ships can capture and process large amounts of fish.

Fish can be captured faster than they can reproduce.

According to marine ecologists, overfishing is the greatest threat to ocean ecosystems today. Overfishing occurs because fish are captured faster than they can reproduce. Advanced fishing technology and an increased demand for fish have led to overfishing, causing several marine species to become extinct or endangered as a result. In the long-term, overfishing can have a devastating impact on ocean communities as it destabilizes the food chain and destroys the natural habitats of many aquatic species.

In the past, fishing was more sustainable because fishermen could not access every location and because they had a limited capacity for processing and storing fish aboard their vessels. Today, however, small fishing boats have been replaced by massive factory ships that can capture and process large amounts of fish. These ships use sonar instruments and global positioning systems (GPS) to rapidly locate large schools of fish. Once fish are located highly efficient methods are used to capture fish. Depending on the species of fish sought, fishing lines with thousands of hooks can be deployed or huge nets 50 meters wide with a capacity to pull the weight of a medium-sized plane can be dragged behind a boat to capture fish. The captured fish can then be transferred to factory ships that have plants for processing and packing fish. Because these ships have all the equipment to freeze and can fish, they only need to return to their base once they are full.

Today industrial fishing has expanded considerably and fishermen can now explore new shores and deeper waters to keep up with the increased demand for seafood. The United Nations Food and Agriculture Organization (FAO) stated that over 70 percent of the world's fisheries are "fully exploited," "over exploited," or "significantly depleted." The annual total global catch of fish is 125 million metric tons and growing.

Fishing gear is often nonselective in the fish it targets. Fish that are too big to get through the mesh of a net are captured. Therefore, overfishing does not only threaten the species of fish that is targeted for food, but also many nontarget species. For example, for every ton of shrimp caught, three tons of other fish are killed and thrown away. This is known as by-catch. Many of the fish caught this way include endangered and over exploited species, 95 percent of which are eventually thrown away.

Many modern fishing methods are also irreversibly destructive. For example, bottom trawling, a technique that uses wide nets with heavy metal rollers, that can crush everything in the path of the gear, destroy fragile corals, smash rock formations, and kill fish and animals as by-catch. These practices can wreak havoc on delicate marine ecosystems.

Given that fishing is a food source for millions of people, attempting to solve the problem of overfishing is not easy, especially for developing countries. Sustainable fishing will be a necessary goal of counterbalancing depletion in fisheries and restabilizing coastal ecosystems.

We can reverse most of the destruction. In some situations it might only take a decade, in other situations it might take many centuries. In the end we can have productive and healthy oceans again as is shown in many examples around the world. To do so, every long-term successful and sustainable fishery, nearshore or high-seas, needs to be managed according to some basic rules:

- **Safe catch limits.** A constantly reassessed, scientifically determined limit on the total number of fish caught and landed by a fishery. Politics and short-term economic incentives should have no role in this.

- **Controls on by-catch.** The use of techniques or management rules to prevent the unintentional killing and disposal of fish, crustaceans, and other oceanic life not part of the target catch.

- **Protection of pristine and important habitats.** The key parts in ecosystems need full protection from destructive fisheries; those parts include the spawning and nursing grounds of fish, delicate sea floor, unique unexplored habitats, and corals, among others.

- **Monitoring and enforcement.** A monitoring system to make sure fishermen do not land more than they are allowed to or fish in closed areas. Enforcement activities that impose significant fines are needed to make it uneconomical to cheat.

Are these goals realistic in today's world? Is it possible to remove politics from commercial fishing? Do you know what kind of seafood you eat? Is it caught in a sustainable way? If you do not know, look at the *Guide to Good Fish Guides* (www.fishonline.org) for some tips.

6.1 Succession

When we look at ecosystems around the world, we can easily recognize certain types, like forests, coral reefs, deserts, or swamps. Although these ecosystems appear to be unchanging, they are not. On a daily basis, plants grow and die, animals feed on plants and on one another, and decomposers recycle the chemical elements that make up the biotic portion of any ecosystem. Organisms even influence abiotic factors. Trees modify the local climate by providing shade and evaporating water, and the organic matter from decaying plants helps to hold moisture in the soil.

We can also recognize that some ecosystems are not stable but change significantly within a human lifetime. For example, abandoned fields become forests and shallow ponds become swamps or marshes.

When we observe ecosystems over long time periods, it is possible to see trends in the way the structure of a community changes. Generally, this series of changes eventually results in a relatively long-lasting, stable combination of species that is self-perpetuating. These observations led ecologists to develop the concepts of succession and climax communities.

Succession is the concept that communities proceed through a series of recognizable, predictable changes in structure over time. Succession occurs because the activities of organisms cause changes in their surroundings that make the environment less suitable for themselves and better suited for other kinds of organisms. When new species become established, they compete with the original inhabitants. In some cases, the original species may be replaced completely. In other cases, early species may not be replaced but may become less numerous as invading species take a dominant role. Slowly, over time, it is recognized that a significantly different community has become established. Typically, succession eventually leads to what is called a climax community.

A **climax community** is a relatively stable, long-lasting community that is the result of succession. In the traditional view of succession, the kind of climax community that developed was primarily determined by climate. However, today we recognize that in addition to weather and climate, soil characteristics, availability of water, locally available seed sources, frequency of disturbance, invasions of organisms from outside the area, and many other factors affect the course of succession and the kind of climax community that develops in a place. (See figure 6.1.)

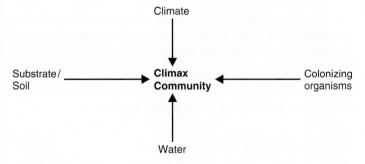

FIGURE 6.1 Factors That Determine the Kind of Climax Community There are many biotic and abiotic factors that interact to determine the kind of climax community that will develop in any place.

Ecologists have traditionally recognized two kind of succession: primary and secondary succession.

Primary succession is a successional progression that begins with a total lack of organisms and bare mineral surfaces or water. Such conditions can occur when volcanic activity causes lava flows or glaciers scrape away the organisms and soil. Similarly, a lowering of sea level exposes new surfaces for colonization by terrestrial organisms. Primary succession takes an extremely long time because often there is no soil and few readily available nutrients for plants to use for growth.

Secondary succession is a successional progression that begins with the destruction or disturbance of an existing ecosystem. It is much more commonly observed and generally proceeds more rapidly than primary succession. This is true because, although fire, flood, windstorm, or human activity can destroy or disturb a community of organisms, there is at least some soil and often there are seeds or roots from which plants can begin growing almost immediately. In addition, animals can migrate to an area from nearby undisturbed ecosystems.

Primary Succession

Primary succession can begin on a bare rock surface, in pure sand, or in standing water. Since succession on rock and sand is somewhat different from that which occurs with watery situations, we deal with them separately. We discuss terrestrial succession first.

Terrestrial Primary Succession

Several factors determine the rate of succession and the kind of climax community that will develop in an area. The kind of substrate (rock, sand, clay) will greatly affect the kind of soil that will develop. The kinds of spores, seeds, or other reproductive structures will determine the plant species available to colonize the area. The climate will determine the species that will be able to live in an area and how rapidly they will grow. The rate of growth will determine how quickly organic matter will accumulate in the soil. The kind of substrate, climate, and amount of organic matter will influence the amount of water available for plant growth. Finally, the kinds of plants will determine the kinds of animals able to live in the area. Let's look at a specific example of how these factors are interrelated in an example of primary succession from bare mineral surfaces.

1. Pioneer Stages Bare rock or sand is a very inhospitable place for organisms to live. The temperature changes drastically, there is no soil, there is little moisture, the organisms are exposed to the damaging effect of the wind, few nutrients are available, and few places are available for organisms to attach themselves or hide. However, windblown spores or other tiny reproductive units of a few kinds of organisms can become established and survive in even this inhospitable environment. This collection of organisms is known as the **pioneer community** because it is the first to colonize bare rock. (See figure 6.2.)

Lichens are common organisms in many pioneer communities. Lichens are actually mutualistic relationships between two

FIGURE 6.2 Pioneer Organisms The lichens growing on this rock are able to accumulate bits of debris, carry on photosynthesis, and aid in breaking down the rock. All of these activities contribute to the formation of a thin layer of soil, which is necessary for plant growth in the early stages of succession.

kinds of organisms: algae or bacteria that carry on photosynthesis and fungi that attach to the rock surface and retain water. The growth and development of lichens is often a slow process. It may take lichens 100 years to grow as large as a dinner plate. Lichens are the producers in this simple ecosystem, and many tiny consumer organisms may be found associated with lichens. Some feed on the lichen and many use it as a place of shelter, since even a drizzle is like a torrential rain for a microscopic animal. Since lichens are firmly attached to rock surfaces, they also tend to accumulate bits of airborne debris and store small amounts of water that would otherwise blow away or run off the rock surface. Acids produced by the lichen tend to cause the breakdown of the rock substrate into smaller particles. This fragmentation of rock, aided by physical and chemical weathering processes, along with the trapping of debris and the contribution of organic matter by the death of lichens and other organisms, ultimately leads to the accumulation of a very thin layer of soil.

2. Intermediate Stages   The thin layer of soil produced by the pioneer community is the key to the next stage in the successional process. The soil layer can retain some water and support some fungi, certain small worms, insects, bacteria, protozoa, and perhaps a few tiny annual plants that live for only one year but produce flowers and seeds that fall to the soil and germinate the following growing season. Many of these initial organisms or their reproductive

structures are very tiny and will arrive as a result of wind and rain. As these organisms grow, reproduce, and die, they contribute additional organic material for the soil-building process, and the soil layer increases in thickness and is better able to retain water.

This stage, which is dominated by annual plants, eliminates the lichen community because the plants are taller and shade the lichens, depriving them of sunlight. This annual plant stage is itself replaced by a community of small perennial grasses and herbs. The perennial grasses and herbs are often replaced by larger perennial woody shrubs, which are often replaced by larger trees that require lots of sunlight, which often are replaced in turn by trees that can tolerate shade. Sun-loving (shade-intolerant) trees are replaced by shade-tolerant trees because the seedlings of shade-intolerant trees cannot grow in the shade of their parents, while seedlings of shade-tolerant trees can.

3. Climax Community   Eventually, a relatively stable, long-lasting, complex, and interrelated climax community of plants, animals, fungi, and bacteria is produced. The climax community in this example consists of a forest in which the trees are shade tolerant. Each step in this process from pioneer community to climax community is called a **successional stage,** or **seral stage,** and the entire sequence of stages—from pioneer community to climax community—is called a **sere.** (See figure 6.3.)

Although in this example we have described a successional process that began with a lichen pioneer community and ended with a climax forest, it is important to recognize that the process of succession can stop at any point along this continuum. In certain extreme climates, lichen communities may last for hundreds of years and must be considered climax communities. Others reach a grass-herb stage and proceed no further.

Regardless of the kind of climax community that develops in an area, it is possible to make some general statements about how successional and climax communities differ.

1. Climax communities maintain their mix of species for a long time, while successional communities show changes in the kinds of species present.
2. Climax communities tend to have many specialized ecological niches, while successional communities tend to have more generalized niches.
3. Climax communities tend to have many more kinds of organisms and kinds of interactions among organisms than do successional communities.
4. Climax communities tend to recycle nutrients and maintain a relatively constant biomass, while successional communities tend to accumulate large amounts of new material.

The general trend in succession is toward increasing complexity and more efficient use of matter and energy compared to the successional communities that preceded them.

Aquatic Primary Succession

The principal concepts of land succession can be applied to aquatic ecosystems. Except for the oceans, most aquatic ecosystems are considered temporary. Certainly, some are going to be

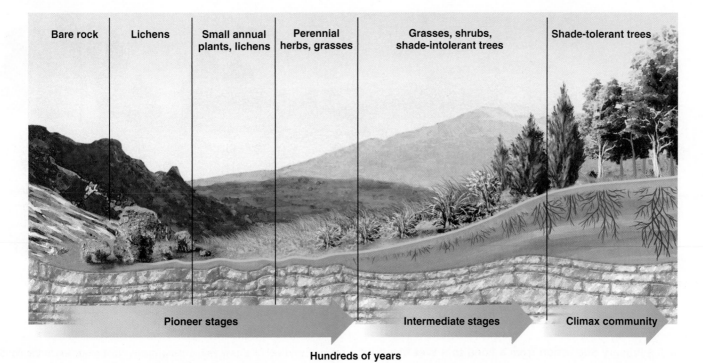

| Bare rock | Lichens | Small annual plants, lichens | Perennial herbs, grasses | Grasses, shrubs, shade-intolerant trees | Shade-tolerant trees |

Pioneer stages → **Intermediate stages** → **Climax community**

Hundreds of years

FIGURE 6.3 Primary Succession on Land The formation of soil is a major step in primary succession. Until soil is formed, the area is unable to support large amounts of vegetation, which modify the harsh environment. Once soil formation begins, the site proceeds through an orderly series of stages toward a climax community.

around for thousands of years, but eventually they will disappear and be replaced by terrestrial ecosystems as a result of normal successional processes. All aquatic ecosystems receive a continuous input of soil particles and organic matter from surrounding land, which results in the gradual filling in of shallow bodies of water such as ponds and lakes.

1. Early Stages—Aquatic Vegetation In deep portions of lakes and ponds, only floating plants and algae can exist. However, as sediment accumulates, the water depth becomes less, and it becomes possible for certain species of submerged plants to establish their roots in the sediments of the bottom of shallow bodies of water. They carry on photosynthesis, which results in a further accumulation of organic matter. These plants also tend to trap sediments that flow into the pond or lake from streams or rivers, resulting in a further decrease in water depth. Eventually, as the water becomes shallower, emergent plants become established. They have leaves that float on the surface of the water or project into the air. The network of roots and stems below the surface of the water results in the accumulation of more material, and the water depth decreases as material accumulates on the bottom. As the process continues, a wet soil is formed and grasses and other plants that can live in wet soil become established. This successional stage is often called a wet meadow.

2. Later Stages—Transition to Terrestrial Communities The activities of plants in a wet meadow tend to draw moisture from the soil, and, as more organic matter is added to the top layer of the soil, it becomes somewhat drier. Once this occurs, the stage is

set for a typical terrestrial successional series of changes, eventually resulting in a climax community typical for the climate of the area. (See figure 6.4.)

3. Observing Aquatic Succession Since the shallower portions of most lakes and ponds are at the shore, it is often possible to see the various stages in aquatic succession from the shore. In the central, deeper portions of the lake, there are only floating plants and algae. As we approach the shore, we first find submerged plants such as *Elodea* and algal mats, then emergent vegetation such as water lilies and cattails, then grasses and sedges that can tolerate wet soil, and on the shore, the beginnings of a typical terrestrial succession resulting in the climax community typical for the area.

In many northern ponds and lakes, sphagnum moss forms thick, floating mats. These mats may allow certain plants that can tolerate wet soil conditions to become established. The roots of the plants bind the mat together and establish a floating bog, which may contain small trees and shrubs as well as many other smaller, flowering plants. (See figure 6.5.) Someone walking on such a mat would recognize that the entire system was floating when they noticed the trees sway or when they stepped through a weak zone in the mat and sank to their hips in water. Eventually, these bogs will become increasingly dry and the normal climax vegetation for the area will succeed the more temporary bog stage.

Secondary Succession

The same processes and activities drive both primary and secondary succession. The major difference is that secondary succession

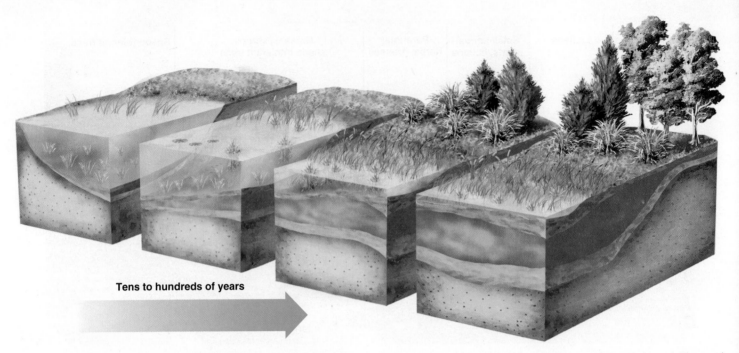

Tens to hundreds of years

FIGURE 6.4 Primary Succession from a Pond to a Wet Meadow A shallow pond will fill slowly with organic matter from producers in the pond. Eventually, a wet soil will form and grasses will become established. In many areas, this will be succeeded by a climax forest.

FIGURE 6.5 Floating Bog In many northern regions, sphagnum moss forms a floating mat that can be colonized by plants that tolerate wet soils. A network of roots ties the mat together to form a floating community.

occurs when an existing community is destroyed but much of the soil and some of the organisms remain. A forest fire, a flood, or the conversion of a natural ecosystem to agriculture may be the cause. Since much of the soil remains and many of the nutrients necessary for plant growth are still available, the process of succession can advance more rapidly than primary succession. In addition, because some plants and other organisms may survive the disturbance and continue to grow and others will survive as roots or seeds, they can quickly reestablish themselves in the area. Furthermore, undamaged communities adjacent to the disturbed area can serve as sources of seeds and animals that migrate into

the disturbed area. Thus, the new climax community is likely to resemble the one that was destroyed. Figure 6.6 shows old field succession—the typical secondary succession found on abandoned farmland in the southeastern United States.

Similarly, when beavers flood an area, the existing terrestrial community is replaced by an aquatic ecosystem. As the area behind the dam fills in with sediment and organic matter, it goes through a series of changes that may include floating plants, submerged plants, emergent plants, and wet meadow stages, but it eventually returns to the typical climax community for the area.

Many kinds of communities exist only as successional stages and are continually reestablished following disturbances. A variety of woodlands along rivers exist only where floods remove vegetation, allowing specific species to become established on the disturbed floodplain. Some kinds of forest and shrub communities exist only if fire occasionally destroys the mature forest. Windstorms such as hurricanes are also important in causing openings in forests that allow the establishment of certain kinds of plant communities.

Modern Concepts of Succession and Climax

Our understanding of the concepts of succession and climax communities has been modified considerably since these concepts were first developed. Some historical perspective will help to clarify how ecologists have altered their concept of successional change. When European explorers traveled across the North American continent, they saw huge expanses of land dominated by specific types of communities: hardwood forests in the east, evergreen forests in the north, grasslands in central North America, and deserts in the southwest. These regional communities came to be considered the steady state or normal situation for those parts

Mature oak/hickory forest destroyed	Farmland abandoned	Annual plants	Grasses and biennial herbs	Perennial herbs and shrubs begin to replace grasses and biennials	Pines begin to replace shrubs	Young oak and hickory trees begin to grow	Pines die and are replaced by mature oak and hickory trees	Mature oak/hickory forest
		1–2 years	3–4 years	4–15 years	5–15 years	10–30 years	50–75 years	

FIGURE 6.6 **Secondary Succession on Land** A plowed field in the southeastern United States shows a parade of changes over time, involving plant and animal associations. The general pattern is for annual weeds to be replaced by grasses and other perennial herbs, which are replaced by shrubs, which are replaced by trees. As the plant species change, so do the animal species.

of the world. When ecologists began to explore the way in which ecosystems developed over time, they began to think of these ecosystems as the end point or climax of a long journey, beginning with the formation of soil and its colonization by a variety of plants and other organisms.

As settlers removed the original forests or grasslands and converted the land to farming, the original "climax" community was destroyed. Eventually, as poor farming practices destroyed the soil, many farms were abandoned and the land was allowed to return to its "original" condition. This secondary succession often resulted in forests that resembled those that had been destroyed. However, in most cases, these successional forests contained fewer species and in some cases were entirely different kinds of communities from the originals. These new stable communities were also called climax communities, but they were not the same as the original climax communities.

In addition, the introduction of species from Europe and other parts of the world changed the mix of organisms that might colonize an area. Many grasses and herbs that were introduced either on purpose or accidentally have become well established. Today, some communities are dominated by these introduced species. Even diseases have altered the nature of climax communities. Chestnut blight and Dutch elm disease have removed tree species that were at one time dominant species in certain plant communities.

Ecologists began to recognize that there was no fixed, predetermined community for each part of the world, and they began to modify the way they looked at the concept of climax communities. The concept today is a more plastic one. It is still used to talk about a stable stage following a period of change, but ecologists no longer feel that land will eventually return to a "preordained" climax condition. They have also recognized in recent years that the type of climax community that develops depends on many factors other than simply climate. One of these is the availability of seeds to colonize new areas. Some seeds may lie dormant in the

soil for a decade or more, while others may be carried to an area by wind, water, or animals. Two areas with very similar climate and soil characteristics may develop very different successional and "climax" communities because of the seeds that were present in the area when the lands were released from agriculture.

Furthermore, we need to recognize that the only thing that differentiates a "climax" community from a successional one is the timescale over which change occurs. "Climax" communities do not change as rapidly as successional ones. However, all communities are eventually replaced, as were the swamps that produced coal deposits, the preglacial forests of Europe and North America, and the pine forests of the northeastern United States.

Many human activities alter the nature of the successional process. Agricultural practices obviously modify the original community to allow for the raising of crops. However, several other management practices have also significantly altered communities. Regular logging returns a forest to an earlier stage of succession. The suppression of fire in many forests has also changed the mix of organisms present. When fire is suppressed, those plants that are killed by regular fires become more common and those that are able to resist fire become less common. Changing the amount of water present will also change the kind of community. Draining an area makes it less suitable for the original inhabitants and more suitable for those that live in drier settings. Similarly, irrigation and flooding increase the amount of water present and change the kinds of organisms that can live in an area.

So what should we do with these concepts of succession and climax communities? Although the climax concept embraces a false notion that there is a specific end point to succession, it is still important to recognize that there is an identifiable, predictable pattern of change during succession and that later stages in succession are more stable and longer lasting than early stages. Whether we call a specific community of organisms a climax community is not really important.

6.2 Biomes Are Determined by Climate

Biomes are terrestrial climax communities with wide geographic distribution. (See figure 6.7.) The primary environmental factor that determines the kind of biome present is climate. Although the concept of biomes is useful for discussing general patterns and processes, it is important to recognize that when different communities within a particular biome are examined, they will show differences in the exact species present. For example, although the general characteristics of the grasslands of Africa, Australia, and North America are similar, the species of plants and animals are different.

Precipitation and Temperature

Patterns of precipitation and temperature are two primary nonbiological factors that have major impacts on the kind of climax community that develops in any part of the world. Several aspects of precipitation are important: the total amount of precipitation per year, the form in which it arrives (rain, snow, sleet), and its seasonal distribution. Precipitation may be evenly spaced throughout the year, or it may be concentrated at particular times so that there are wet and dry seasons.

The temperature patterns are also important and can vary considerably in different parts of the world. Tropical areas have warm, relatively unchanging temperatures throughout the year. Areas near the poles have long winters with extremely cold temperatures and relatively short, cool summers. Other areas are more evenly divided between cold and warm periods of the year. (See figure 6.8.)

Although temperature and precipitation are of primary importance, several other factors may influence the kind of climax community present. Periodic fires are important in maintaining some grassland and shrub climax communities because the fires prevent the establishment of larger, woody species. Some parts of the world have frequent, strong winds that prevent the establishment of trees and cause rapid drying of the soil. The type of soil present is also very important. Sandy soils tend to dry out quickly and may not allow the establishment of more water-demanding species such as trees, while extremely wet soils may allow only certain species of trees to grow. Obviously, the kinds of organisms currently living in the area are also important, since their offspring will be the ones available to colonize a new area.

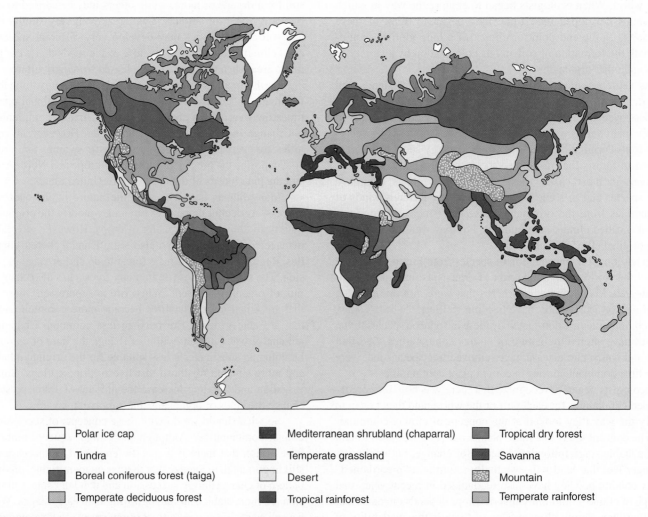

Polar ice cap

Tundra

Boreal coniferous forest (taiga)

Temperate deciduous forest

Mediterranean shrubland (chaparral)

Temperate grassland

Desert

Tropical rainforest

Tropical dry forest

Savanna

Mountain

Temperate rainforest

FIGURE 6.7 Biomes of the World Although most biomes are named for a major type of vegetation, each also includes a specialized group of animals adapted to the plants and the biome's climatic conditions.

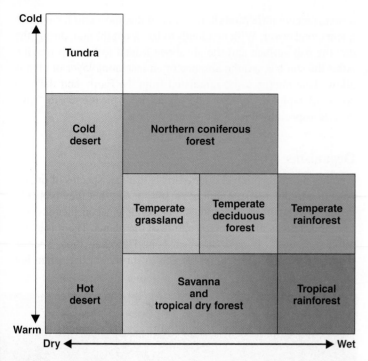

FIGURE 6.8 **Influence of Precipitation and Temperature on Vegetation** Temperature and moisture are two major factors that influence the kind of vegetation that can occur in an area. Areas with low moisture and low temperatures produce tundra; areas with high moisture and freezing temperatures during part of the year produce deciduous or coniferous forests; dry areas produce deserts; moderate amounts of rainfall or seasonal rainfall support temperate grasslands or savannas; and areas with high rainfall and warm temperatures support tropical rainforests.

The Effect of Elevation on Climate and Vegetation

The distribution of terrestrial ecosystems is primarily related to precipitation and temperature. The temperature is warmest near the equator and becomes cooler toward the poles. Similarly, as the height above sea level increases, the average temperature decreases. This means that even at the equator, it is possible to have cold temperatures on the peaks of tall mountains. As one proceeds from sea level to the tops of mountains, it is possible to pass through a series of biomes that is similar to what would be encountered as one traveled from the equator to the North Pole. (See figure 6.9.)

6.3 Major Biomes of the World

In the next sections, we will look at the major biomes of the world and highlight the abiotic and biotic features typical of each biome.

Desert

Deserts are found throughout the world where rainfall is low.

Climate

A lack of water is the primary factor that determines that an area will be a desert. **Deserts** are areas that generally average less than 25 centimeters (10 inches) of precipitation per year. (See

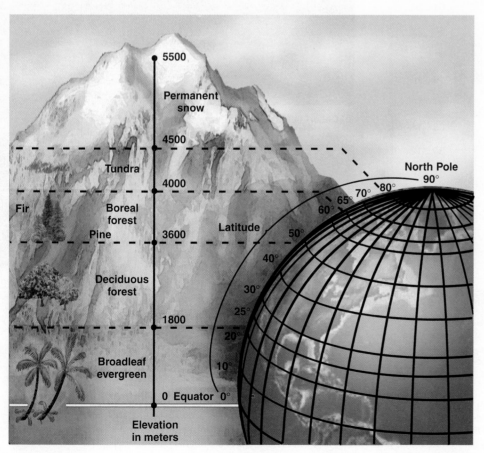

FIGURE 6.9 **Relationship Among Height above Sea Level, Latitude, and Vegetation** As one travels up a mountain, the climate changes. The higher the elevation, the cooler the climate. Even in the tropics, tall mountains can have snow on the top. Thus, it is possible to experience the same change in vegetation by traveling up a mountain as one would experience traveling from the equator to the North Pole.

figure 6.10.) When and how precipitation arrives is quite variable in different deserts. Some deserts receive most of the moisture as snow or rain in the winter months, while in others rain comes in the form of thundershowers at infrequent intervals. If rain comes as heavy thundershowers, much of the water does not sink into the ground but runs off into gullies. Also, since the rate of evaporation is high, plant growth and flowering usually coincide with the periods when moisture is available. Deserts are also likely to be windy.

We often think of deserts as hot, dry wastelands devoid of life. However, many deserts are quite cool during a major part of the year. Certainly, the Sahara Desert and the deserts of the southwestern United States and Mexico are hot during much of the year, but the desert areas of the northwestern United States and the Gobi Desert in Central Asia can be extremely cold during winter months and have relatively cool summers. Furthermore, the temperature can vary greatly during a 24-hour period. Since

deserts receive little rainfall, it is logical that most will have infrequent cloud cover. With no clouds to block out the sun, during the day the soil surface and the air above it tend to heat up rapidly. After the sun has set, the absence of an insulating layer of clouds allows heat energy to be reradiated from the Earth, and the area cools off rapidly. Cool to cold nights are typical even in "hot" deserts, especially during the winter months.

Organisms

Another misconception about deserts is that few species of organisms live in the desert. There are many species, but they typically have low numbers of individuals. For example, a conspicuous feature of deserts is the dispersed nature of the plants. There is a significant amount of space between them. However, those species that are present are specially adapted to survive in dry, often hot

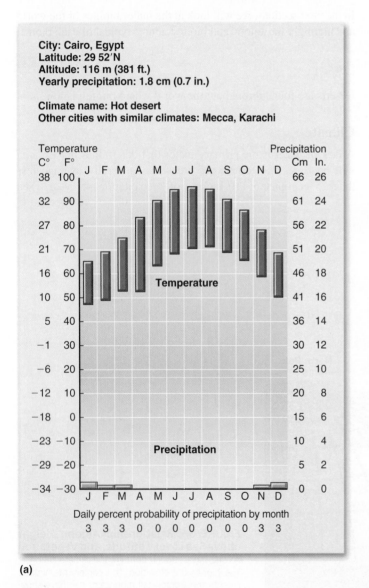

(a)

(b) Desert landscape

(c) Coyote

(d) Collared lizard

FIGURE 6.10 Desert (a) Climagraph for Cairo, Egypt. (b) The desert receives less than 25 centimeters (10 inches) of precipitation per year, yet it teems with life. Cactus, sagebrush, lichens, snakes, small mammals, birds, and insects inhabit the desert. (c) Coyotes are common in North American deserts. (d) Collared lizards are common reptiles in many deserts of the United States. Because daytime temperatures are often high, most animals are active only at night, when the air temperature drops significantly. Cool deserts also exist in many parts of the world, where rainfall is low but temperatures are not high.

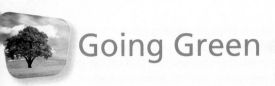

Going Green

Conservation Easements

There are over 1,600 private organizations in the United States that are involved in conservation of land. Some are small, single-purpose organizations that protect a small parcel of land with special conservation value. On the other hand, The Nature Conservancy is an international organization that has protected millions of acres.

People often develop an attachment to their land and wish to see it preserved even after they have died. People may have a long family history of using the land for farming or ranching and want to see that use continue. Others may recognize that their land has special conservation value because of its geology, scenic value, or biodiversity and wish to see it protected for the public good. Others may simply have a purely emotional reason for wanting to preserve their land. One of the tools used by land conservation organizations is a legal tool known as a *conservation easement.*

A conservation easement is a legally binding agreement placed on a piece of privately held land that limits the future use or development even when the land is passed to heirs or sold. For example, a conservation easement may prohibit the subdividing of a piece of land or restrict buildings to a specific portion of the property, or an easement may specify that the public must have access to view significant biological or geological features. Alternatively, the easement may restrict access to protect endangered species or archeological sites.

Regardless of their motivation, when people enter into a conservation easement they give up something. In some cases, people donate a conservation easement and receive no financial benefit. In other cases, they may sell a conservation easement to an organization that agrees to provide stewardship of the property into the future. In nearly all cases the

placement of a conservation easement on property diminishes its economic value, since its future use is restricted. Yet, thousands of people have entered into such arrangements.

As of 2013, in the United States over 20 million acres of land had been protected by conservation easements. You can see the conservation easements in your state by visiting the National Conservation Easement Database (NCED) website www.conservationeasement.us/. The NCED is the national database of conservation easement information, compiling records from land trusts and public agencies throughout the United States.

---

environments. For example, water evaporates from the surfaces of leaves. As an adaptation to this condition, many desert plants have very small leaves that allow them to conserve water. Some even lose their leaves entirely during the driest part of the year. Some, such as cactus, have the ability to store water in their spongy bodies or their roots for use during drier periods. Other plants have parts or seeds that lie dormant until the rains come. Then they germinate, grow rapidly, reproduce, and die, or become dormant until the next rains. Even the perennial plants are tied to the infrequent rains. During these times, the plants are most likely to produce flowers and reproduce. Many desert plants are spiny. The spines discourage large animals from eating the leaves and young twigs.

Similarly, animals do not have large, dense populations. The desert has many kinds of animals. However, they are often overlooked because their populations are low, numerous species are of small size, and many are inactive during the hot part of the day. They also aren't seen in large, conspicuous groups. Many insects, lizards, snakes, small mammals, grazing mammals, carnivorous mammals, and birds are common in desert areas. All of the animals that live in deserts are able to survive with a minimal amount of water. Some receive nearly all of their water from the moisture in the food they eat. They generally have an outer skin or cuticle that resists water loss, so they lose little water by evaporation.

They also have physiological adaptations, such as extremely efficient kidneys, that allow them to retain water. They often limit their activities to the cooler part of the day (the evening), and small mammals may spend considerable amounts of time in underground burrows during the day, which allows them to avoid extreme temperatures and to conserve water.

Human Impact

Throughout history, deserts have been regions where humans have had little impact. The harshness of the climate does not allow for agriculture. Therefore, hunter-gatherer societies were the most common ones associated with deserts. Some deserts support nomadic herding in which herders move their livestock to find patches of vegetation for grazing. Modern technology allows for the transport of water to the desert. This has resulted in the development of cities in some desert areas and some limited agriculture as a result of irrigation.

Temperate Grassland

Temperate grasslands, also known as **prairies** or **steppes,** are widely distributed over temperate parts of the world.

Climate

As with deserts, the major factor that contributes to the establishment of a temperate grassland is the amount of available moisture. Grasslands generally receive between 25 and 75 centimeters (10 to 30 inches) of precipitation per year. These areas are windy with hot summers and cold-to-mild winters. In many grasslands, fire is an important force in preventing the invasion of trees and releasing nutrients from dead plants to the soil.

Organisms

Grasses make up 60 to 90 percent of the vegetation. Many other kinds of flowering plants are interspersed with the grasses. (See figure 6.11.) Typically, the grasses and other plants are very close together, and their roots form a network that binds the soil together. Trees, which generally require greater amounts of water, are rare in these areas except along watercourses.

The primary consumers are animals that eat the grasses, such as large herds of migratory, grazing mammals such as bison, wildebeests, wild horses, and various kinds of sheep, cattle, and goats. While the grazers are important as consumers of the grasses, they also supply fertilizer from their dung and discourage invasion by woody species of plants because they eat the young shoots.

In addition to grazing mammals, many kinds of insects, including grasshoppers and other herbivorous insects, dung beetles (which feed on the dung of grazing animals), and several kinds of flies are common. Some of these flies bite to obtain blood. Others lay their eggs in the dung of large mammals. Some feed on dead animals and lay their eggs in carcasses. Small herbivorous mammals, such as mice and ground squirrels, are also common.

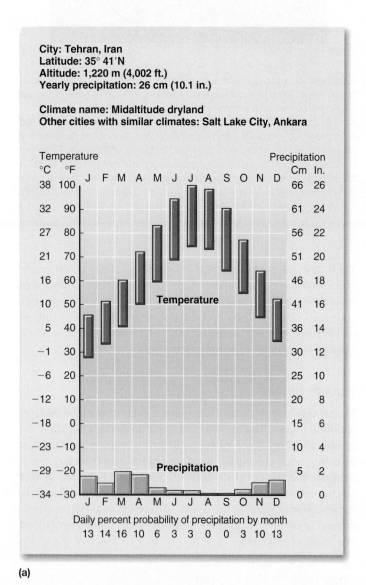

(a)

(b) Prairie landscape

(c) Pronghorn

(d) Grasshopper

(e) Meadowlark

FIGURE 6.11 Temperate Grassland (a) Climagraph for Tehran, Iran. (b) Grasses are better able to withstand low water levels than are trees. Therefore, in areas that have moderate rainfall, grasses are the dominant plants. (c & d) Pronghorns and grasshoppers are common herbivores and (e) meadowlarks are common consumers of insects in North American grasslands.

Birds are often associated with grazing mammals. They eat the insects stirred up by the mammals or feed on the insects that bite grazers. Other birds feed on seeds and other plant parts. Reptiles (snakes and lizards) and other carnivores such as coyotes, foxes, and hawks feed on small mammals, birds, and insects.

Human Impact

Most of the moist grasslands of the world have been converted to agriculture, since the rich, deep soil that developed as a result of the activities of centuries of soil building is useful for growing cultivated grasses such as corn (maize) and wheat. The drier grasslands have been converted to the raising of domesticated grazers such as cattle, sheep, and goats. Therefore, little undisturbed grassland is left, and those fragments that remain need to be preserved as refuges for the grassland species that once occupied huge portions of the globe.

Savanna

Savannas are found in tropical parts of Africa, South America, and Australia and are characterized by extensive grasslands spotted with occasional trees or patches of trees. (See figure 6.12.)

Climate

Although savannas receive 50 to 150 centimeters (20 to 60 inches) of rain per year, the rain is not distributed evenly throughout

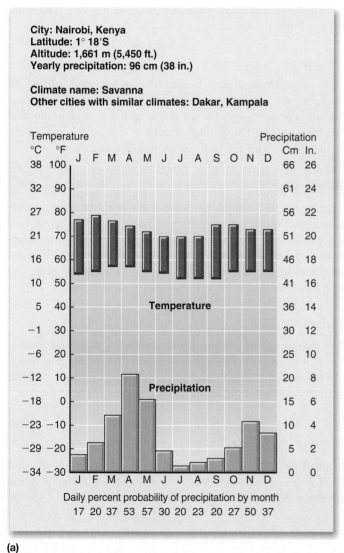

(a)

Savanna landscape

(c) Lion

(d) Secretary birds

(e) Thomson's gazelles

FIGURE 6.12 Savanna (a) Climagraph for Nairobi, Kenya. (b) Savannas develop in tropical areas that have seasonal rainfall. They typically have grasses as the dominant vegetation with drought- and fire-resistant trees scattered through the area. Grazing animals such as elephants and gazelles (b & e) are common herbivores and lions and secretary birds (c & d) are common carnivores in African savannas.

Kinds of Ecosystems and Communities   121

the year. Typically, a period of heavy rainfall is followed by a prolonged drought. This results in a very seasonally structured ecosystem.

Organisms

The plants and animals time their reproductive activities to coincide with the rainy period, when limiting factors are less severe. The predominant plants are grasses, but many drought-resistant, flat-topped, thorny trees are common. As with grasslands, fire is a common feature of the savanna, and the trees present are resistant to fire damage. Many of these trees are particularly important because they are legumes that are involved in nitrogen-fixation. They also provide shade and nesting sites for animals. As with grasslands, the predominant mammals are the grazers. Wallabies in Australia, wildebeests, zebras, elephants, and various species of antelope in Africa, and capybaras (rodents) in South America are examples. In Africa, the large herds of grazing animals provide food for many different kinds of large carnivores (lions, hyenas, leopards). Many kinds of rodents, birds, insects, and reptiles are associated with this biome. Among the insects, mound-building termites are particularly common.

Human Impact

Savannas have been heavily impacted by agriculture. Farming is possible in the more moist regions, and the drier regions are used for the raising of livestock. Because of the long periods of drought, the raising of crops is often difficult without irrigation. Some areas support nomadic herding. In Africa there are extensive areas set aside as parks and natural areas and ecotourism is an important source of income. However, there is a constant struggle between the people who want to use the land for agriculture or grazing and those who want to preserve it in a more natural state.

Mediterranean Shrublands (Chaparral)

The **Mediterranean shrublands** are located near oceans and are dominated by shrubby plants.

Climate

Mediterranean shrublands have a climate with wet, cool winters and hot, dry summers. Rainfall is 40 to 100 centimeters (15 to 40 inches) per year. As the name implies, this biome is typical of the Mediterranean coast and is also found in coastal southern California, the southern tip of Africa, a portion of the west coast of Chile, and southern Australia.

Organisms

The vegetation is dominated by woody shrubs that are adapted to withstand the hot, dry summer. (See figure 6.13.) Often the plants are dormant during the summer. Fire is a common feature of this biome, and the shrubs are adapted to withstand occasional fires. The kinds of animals vary widely in the different regions of the world with this biome. Many kinds of insects, reptiles, birds, and

mammals are found in these areas. In the chaparral of California, rattlesnakes, spiders, coyotes, lizards, and rodents are typical inhabitants.

Human Impact

Very little undisturbed Mediterranean shrubland still exists. The combination of moderate climate and closeness to the ocean has resulted in all Mediterranean shrublands being heavily altered by human activity. Agriculture is common, often with the aid of irrigation, and many major cities are located in this biome.

Tropical Dry Forest

The **tropical dry forest** is another biome that is heavily influenced by seasonal rainfall. Tropical dry forests are found in parts of Central and South America, Australia, Africa, and Asia (particularly India and Myanmar).

Climate

Many of the tropical dry forests have a monsoon climate in which several months of heavy rainfall are followed by extensive dry periods ranging from a few to as many as eight months. (See figure 6.14.) The rainfall may be as low as 50 centimeters (20 inches) or as high as 200 centimeters (80 inches).

Organisms

Since the rainfall is highly seasonal, many of the plants have special adaptations for enduring drought. In many of the regions that have extensive dry periods, many of the trees drop their leaves during the dry period. Many of the species of animals found here are also found in more moist tropical forests of the region. However, there are fewer kinds in dry forests than in rainforests.

Human Impacts

Many of these forests occur in areas of very high human population. Therefore, the harvesting of wood for fuel and building materials has heavily affected these forests. In addition, many of these forests have been converted to farming or the grazing of animals.

Tropical Rainforest

Tropical rainforests are located near the equator in Central and South America, Africa, Southeast Asia, and some islands in the Caribbean Sea and Pacific Ocean. (See figure 6.15.)

Climate

The temperature is normally warm and relatively constant. There is no frost, and it rains nearly every day. Most areas receive in excess of 200 centimeters (80 inches) of rain per year. Some receive 500 centimeters (200 inches) or more. Because of the warm temperatures and abundant rainfall, most plants grow very rapidly; however, soils are usually poor in nutrients because water

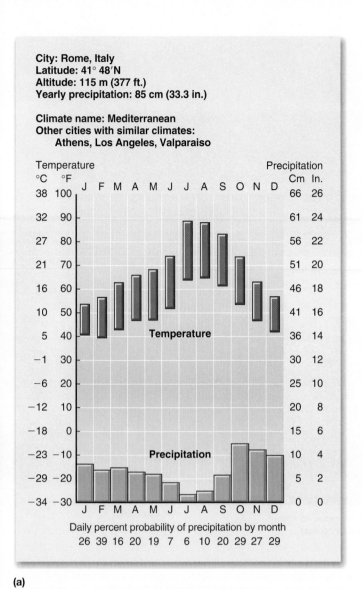

City: Rome, Italy
Latitude: 41° 48′N
Altitude: 115 m (377 ft.)
Yearly precipitation: 85 cm (33.3 in.)

Climate name: Mediterranean
Other cities with similar climates:
    Athens, Los Angeles, Valparaiso

Daily percent probability of precipitation by month
26 39 16 20 19 7 6 10 20 29 27 29

(a)

(b) Chaparral landscape

(c) California quail

(d) Black-tailed jack rabbit

FIGURE 6.13 Mediterranean Shrubland (a) Climagraph for Rome, Italy. (b) Mediterranean shrublands are characterized by a period of winter rains and a dry, hot summer. The dominant plants are drought-resistant, woody shrubs. (c & d) Common animals in the Mediterranean shrubland (chaparral) of California are the California quail and black-tailed jack rabbit.

tends to carry away any nutrients not immediately taken up by plants. Many of the trees have extensive root networks, associated with fungi (mycorrhizae), near the surface of the soil that allow them to capture nutrients from decaying vegetation before the nutrients can be carried away.

Organisms

Tropical rainforests have a greater diversity of species than any other biome. More species are found in the tropical rainforests of the world than in the rest of the world combined. A small area of a few square kilometers is likely to have hundreds of species of trees. Furthermore, it is typical to have distances of a kilometer or more between two individuals of the same species. Balsa, teak-wood, and many other ornamental woods are from tropical trees.

Each of those trees is home to a set of animals and plants that use it as food, shelter, or support. The canopy, which forms a solid wall of leaves between the sun and the forest floor, consists of two or three levels. A few trees, called emergent trees, protrude above the canopy. Below the canopy is a layer of understory tree species.

Since most of the sunlight is captured by the trees, only shade-tolerant plants live beneath the trees' canopy. In addition, the understory has many vines that attach themselves to the tall trees and grow toward the sun. When the vines reach the canopy, they can compete effectively with their supporting tree for available sunlight. In addition to supporting various vines, each tree serves as a surface for the growth of ferns, mosses, and orchids.

Recently, biologists discovered a whole new community of organisms that live in the canopy of these forests. Rainfall is a source of new nutrients, since atmospheric particles and gases dissolve as the rain falls. The canopy contains many kinds of epiphytic plants (plants that live on the surfaces of other plants) that trap many of these nutrients in the canopy before they can reach the soil.

Kinds of Ecosystems and Communities   123

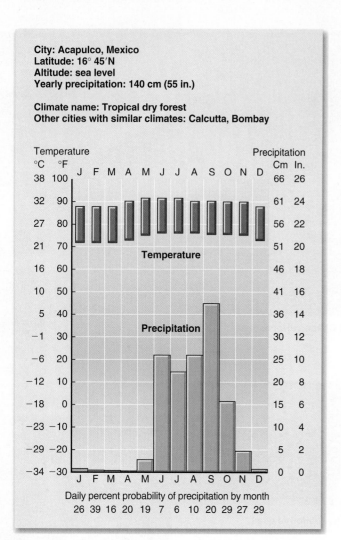

City: Acapulco, Mexico
Latitude: 16° 45'N
Altitude: sea level
Yearly precipitation: 140 cm (55 in.)

Climate name: Tropical dry forest
Other cities with similar climates: Calcutta, Bombay

Temperature

Temperature

Precipitation

Daily percent probability of precipitation by month
26 39 16 20 19 7 6 10 20 29 27 29

(a)

(b) Tropical dry forest landscape

(c) White-faced coati in Costa Rica

(d) Rhinoceros in Indian tropical dry forest

FIGURE 6.14 Tropical Dry Forest (a) Climagraph for Acapulco, Mexico. (b) Tropical dry forests typically have a period of several months with no rain. In places where the drought is long, many of the larger trees lose their leaves. The coati (c) is a common animal in the tropical dry forests of the Americas. The endangered one-horned rhinoceros (d) is an inhabitant of the tropical dry forest of Asia.

Associated with this variety of plants is an equally large variety of animals. Insects, such as ants, termites, moths, butterflies, and beetles, are particularly abundant. Birds also are extremely common, as are many climbing mammals, lizards, and tree frogs. The insects are food for many of these species. Since flowers and fruits are available throughout the year, there are many kinds of nectar and fruit-feeding birds and mammals. Their activities are important in pollination and spreading seeds throughout the forest. Because of the low light levels and the difficulty of maintaining visual contact with one another, many of the animals communicate by making noise.

Human Impact

Tropical rainforests are under intense pressure from logging and agriculture. Many of the countries where tropical rainforests are present are poor and seek to obtain jobs and money by exploiting this resource. Generally, agriculture has not been successful because most of the nutrients in a tropical rainforest are tied up in the biomass, not in the soil, and the high rainfall quickly carries away nutrients. However, poor people will still try to raise food by burning the forest and raising crops for a year or two. Many other areas have been cleared for cattle ranching. Forestry can be a sustainable activity, but in many cases, it is not. The forests are being cut down with no effort to protect them for long-term productivity.

Temperate Deciduous Forest

Temperate deciduous forests have a winter–summer change of seasons and have trees that lose their leaves during the winter and replace them the following spring. This kind of forest is typical of the eastern half of the United States, parts of south central and

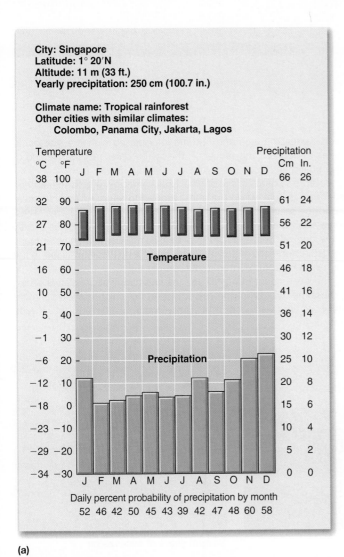

City: Singapore
Latitude: 1° 20′N
Altitude: 11 m (33 ft.)
Yearly precipitation: 250 cm (100.7 in.)

Climate name: Tropical rainforest
Other cities with similar climates:
    Colombo, Panama City, Jakarta, Lagos

Temperature
°C   °F   J F M A M J J A S O N D   Precipitation
                                            Cm In.
38 100   66 26
32   90
27   80
21   70
16   60   **Temperature**   46 18
10   50   41 16
 5   40   36 14
-1   30   30 12
-6   20   **Precipitation**   25 10
-12 10   20   8
-18   0   15   6
-23 -10   10   4
-29 -20   5   2
-34 -30   J F M A M J J A S O N D   0   0

Daily percent probability of precipitation by month
52 46 42 50 45 43 39 42 47 48 60 58

(a)

(b) Tropical rainforest landscape

(c) Blue morpho butterfly

(d) Squirrel monkey

FIGURE 6.15 Tropical Rainforest (a) Climagraph for Singapore. (b-d) Tropical rainforests develop in areas with high rainfall and warm temperatures. They have an extremely diverse mixture of plants and animals such as birds, butterflies, and monkeys.

southeastern Canada, southern Africa, and many areas of Europe and Asia.

Climate

These areas generally receive 75 to 100 centimeters (30 to 60 inches) of relatively evenly distributed precipitation per year. The winters are relatively mild, and plants are actively growing for about half the year.

Organisms

In the temperate deciduous forest biome, each area of the world has certain species of trees that are the major producers for the biome. (See figure 6.16.) In contrast to tropical rainforests, where individuals of a tree species are scattered throughout the forest,

temperate deciduous forests generally have many fewer species, and many forests may consist of two or three dominant tree species. In deciduous forests of North America and Europe, common species are maples, aspen, birch, beech, oaks, and hickories. These tall trees shade the forest floor, where many small flowering plants bloom in the spring. These spring wildflowers store food in underground structures. In the spring, before the leaves come out on the trees, the wildflowers can capture sunlight and reproduce before they are shaded. Many smaller shrubs also are found in the understory of these forests.

These forests are home to a great variety of insects, many of which use the leaves and wood of trees as food. Beetles, moth larvae, wasps, and ants are examples. The birds that live in these forests are primarily migrants that arrive in the spring of the year, raise their young during the summer, and leave in the fall. Many

Grassland Succession

Because there are many kinds of grasslands, it is difficult to generalize about how succession takes place in these areas. Most grasslands in North America have been heavily influenced by agriculture and the grazing of domesticated animals. The grasslands reestablished in these areas may be quite different from the original ecosystem. However, there appear to be several stages typically involved in grassland succession.

After land is abandoned from cultivation, a short period of one to three years elapses in which the field is dominated by annual broadleaf weeds. In this respect, grassland succession is like deciduous forest succession. The next stage varies in length (10 or more years) and is dominated by annual grasses. Usually, in these early stages, the soil is in poor condition, lacking organic matter and nutrients. After several years, the soil fertility increases as organic material accumulates from the death and decay of annual grasses. This leads to the next stage in development, perennial grasses. Eventually, a mature grassland develops as prairie flowers invade the area and become interspersed with the grasses. In general, throughout this sequence, the soil becomes more fertile and of higher quality.

Because so much of the original North American grassland has been used for agriculture, when the land is allowed to return to a prairie, there may not be seeds of all of the original plants native to the area. Thus, the grassland that results from secondary succession may not be exactly like the original; some species may be missing. Consequently, in many managed restorations of prairies, seeds that are no longer available in the local soils are introduced from other sources.

The low amount of rainfall and the fires typical of grasslands generally cause the successional process to stop at this point. However, if more water becomes available or if fire is prevented, woody trees may invade moist sites.

Actively farmed

Recently abandoned

Several years of succession

of these birds rely on the large summer insect population for their food. Others use the fruits and seeds that are produced during the summer months. A few kinds of birds, including woodpeckers, grouse, turkeys, and some finches, are year-round residents. Amphibians (frogs, toads, salamanders) and reptiles (snakes and lizards) prey on insects and other small animals. Several kinds of small and large mammals inhabit these areas. Mice, squirrels, deer, shrews, moles, and opossums are common examples. Major predators on these mammals are foxes, badgers, weasels, coyotes, and birds of prey.

Human Impact

Most of the temperate deciduous forests have been heavily affected by human activity with much of it having been cleared for farming or is subjected to periodic logging. Furthermore, the major population centers of eastern North America and Europe are in areas that were originally temperate deciduous forest.

Temperate Rainforest

Temperate rainforests exist in the coastal areas of northern California, Oregon, Washington, British Columbia, and southern Alaska. New Zealand and the southwest coast of Chile also have temperate rainforests.

Climate

In these coastal areas, the prevailing winds from the west blow over the ocean and bring moisture-laden air to the coast. As the air meets the coastal mountains and is forced to rise, it cools and the moisture falls as rain or snow. Temperate rainforests typically receive at least 130 cm (50 inches) of rain each year. Most areas receive much more than that—often 300 cm (120 inches) or more. Furthermore, rain occurs throughout the year and the cool climate slows evaporation, so things are generally damp. This abundance of water, along with fertile soil and mild temperatures, results in a lush growth of plants.

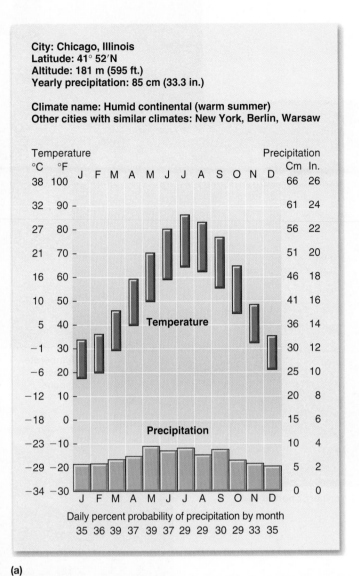

City: Chicago, Illinois
Latitude: 41° 52′N
Altitude: 181 m (595 ft.)
Yearly precipitation: 85 cm (33.3 in.)

Climate name: Humid continental (warm summer)
Other cities with similar climates: New York, Berlin, Warsaw

Daily percent probability of precipitation by month
35 36 39 37 39 37 29 29 30 29 33 35

(a)

(b) Temperate deciduous forest in summer

(c) Temperate deciduous forest in fall

(d) Raccoon

FIGURE 6.16 Temperate Deciduous Forest (a) Climagraph for Chicago, Illinois. (b & c) A temperate deciduous forest develops in areas that have significant amounts of moisture throughout the year but where the temperature falls below freezing for parts of the year. During this time, the trees lose their leaves. (d) Raccoons are common animals in the temperate deciduous forest of North America. This kind of forest once dominated the eastern half of the United States and southeastern Canada.

Organisms

Sitka spruce, Douglas fir, and western hemlock are typical evergreen coniferous trees in the North American temperate rainforest. Undisturbed (old-growth) forests of this region have trees as old as 800 years that are nearly 100 meters (300 feet) tall. Deciduous trees of various kinds (e.g., red alder, big leaf maple, black cottonwood) also exist in open areas where they can get enough light. All the trees are covered with mosses, ferns, and other plants that grow on the surface of the trees. The dominant color is green, because most of the surfaces have a photosynthetic organism growing on them. (See figure 6.17.)

When a tree dies and falls to the ground, it rots in place and serves as a site for the establishment of new trees. This is such a common feature of the forest that the fallen, rotting trees are called *nurse trees*.

A wide variety of animals lives in the temperate rainforest. Insects use the vegetation as food. Many kinds of birds, such as woodpeckers, chickadees, juncos, and warblers, use the insects and fruits as food. Slugs are a common sight on the forest floor. A wide variety of larger animals such as elk, blacktail deer, bears, beavers, and owls are also common. Several species of salmon migrate seasonally up the streams and rivers to spawn.

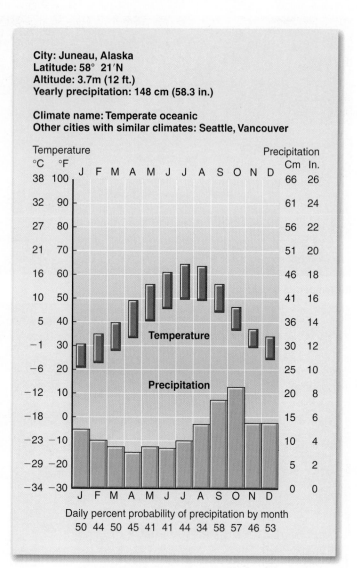

City: Juneau, Alaska
Latitude: 58° 21′N
Altitude: 3.7m (12 ft.)
Yearly precipitation: 148 cm (58.3 in.)

Climate name: Temperate oceanic
Other cities with similar climates: Seattle, Vancouver

Daily percent probability of precipitation by month
50 44 50 45 41 41 44 34 58 57 46 53

(a)

(b) Temperate rainforest landscape

(c) Northern spotted owl

(d) Blacktail deer

FIGURE 6.17 Temperate Rainforest Biome (a) Climagraph for Juneau, Alaska. (b) The temperate rainforest is characterized by high levels of rainfall, which support large evergreen trees and the many mosses and ferns that grow on the surface of the trees. The blacktail deer (d) is common in this biome, which is also the home of the endangered northern spotted owl (c).

Human Impact

Because of the rich resource of trees, at least half of the original temperate rainforest has already been logged. Many of the remaining areas are scheduled to be logged, although some patches have been protected, because they are home to the endangered northern spotted owl and marbled murrelet (a seabird).

Taiga, Northern Coniferous Forest, or Boreal Forest

Throughout the southern half of Canada, parts of northern Europe, and much of Russia, there is an evergreen coniferous forest known as the **taiga, northern coniferous forest,** or **boreal forest.** (See figure 6.18.)

Climate

The climate is one of short, cool summers and long winters with abundant snowfall. The winters are extremely harsh and can last as long as six months. Typically, the soil freezes during the winter. Precipitation ranges between 25 and 100 centimeters (10 to 40 inches) per year. However, the climate is typically humid because there is a great deal of snowmelt in the spring and generally low temperatures reduce evaporation. The landscape is typically dotted with lakes, ponds, and bogs.

Organisms

Conifers such as spruces, firs, and larches are the most common trees in these areas. These trees are specifically adapted to winter

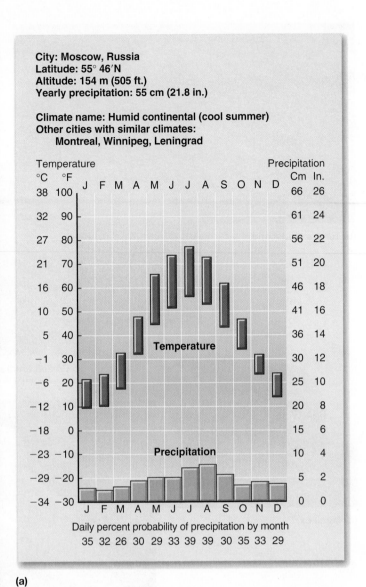

City: Moscow, Russia
Latitude: 55° 46′N
Altitude: 154 m (505 ft.)
Yearly precipitation: 55 cm (21.8 in.)

Climate name: Humid continental (cool summer)
Other cities with similar climates:
    Montreal, Winnipeg, Leningrad

Temperature   Precipitation

Daily percent probability of precipitation by month
35 32 26 30 29 33 39 39 30 35 33 29

(a)

(b) Taiga landscape

(c) Lynx and snow-shoe hare

(d) Taiga in winter

FIGURE 6.18 Taiga, Northern Coniferous Forest, or Boreal Forest (a) Climagraph for Moscow. (b & d) The taiga, northern coniferous forest, or boreal forest occurs in areas with long winters and heavy snowfall. The trees have adapted to these conditions and provide food and shelter for the animals that live there. (c) In North America the snowshoe hare and lynx are common animals.

conditions. Winter is relatively dry as far as the trees are concerned because the moisture falls as snow and stays above the soil until it melts in the spring. The needle-shaped leaves of conifers are adapted to prevent water loss; in addition, the larches lose their needles in the fall. The branches of these trees are flexible, allowing them to bend under a load of snow so that the snow slides off the pyramid-shaped trees without greatly damaging them. As with the temperate deciduous forest, many of the inhabitants of this biome are temporarily active, during the summer. Most birds are migratory and feed on the abundant summer insect population, which is not available during the long, cold winter. A few birds, such as woodpeckers, owls, and grouse, are permanent residents. Typical mammals are deer, caribou, moose, wolves, weasels, mice, snowshoe hares, and squirrels. Because of the cold, few reptiles and amphibians live in this biome.

Human Impact

Human impact is less severe than with many other biomes because population density is generally low in this region. Logging is a common activity, and some herding of reindeer occurs in northern Scandinavia. Many native peoples rely on subsistence hunting for food.

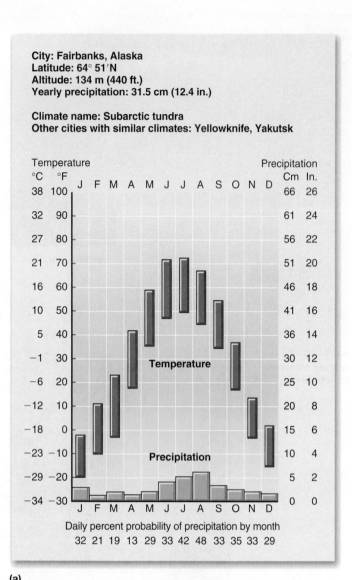

City: Fairbanks, Alaska
Latitude: 64° 51′N
Altitude: 134 m (440 ft.)
Yearly precipitation: 31.5 cm (12.4 in.)

Climate name: Subarctic tundra
Other cities with similar climates: Yellowknife, Yakutsk

Temperature — Precipitation chart (Temperature and Precipitation)

Daily percent probability of precipitation by month
32 21 19 13 29 33 42 48 33 35 33 29

(a)

(b) Tundra landscape

(c) Frozen tundra

(d) Musk ox

(e) Snowy owl

FIGURE 6.19 Tundra (a) Climagraph for Fairbanks, Alaska. (b & c) In the northern latitudes and on the tops of some mountains, the growing season is short and plants grow very slowly. Trees are unable to live in these extremely cold areas, in part because there is a permanently frozen layer of soil beneath the surface, known as the permafrost. Because growth is so slow, damage to the tundra can still be seen generations later. (d & e) Musk ox and snowy owls are common in the tundra.

Tundra

North of the taiga is the **tundra,** an extremely cold region that lacks trees and has a permanently frozen subsurface soil. This frozen soil layer is known as **permafrost.** (See figure 6.19.)

Climate

Because of the permanently frozen soil and extremely cold, windy climate (up to 10 months of winter), no trees can live in the area. Although the amount of precipitation is similar to that in some deserts—less than 25 centimeters (10 inches) per year—the short summer is generally wet because the winter snows melt in the spring and summer temperatures are usually less than 10°C (50°F), which reduces the evaporation rate. Since the permafrost does not let the water sink into the soil, waterlogged soils and many shallow ponds and pools are present.

Organisms

When the top few centimeters (inches) of the soil thaw, many plants (grasses, dwarf birch, dwarf willow) and lichens, such as reindeer moss, grow. The plants are short, usually less than 20 centimeters (8 inches) tall.

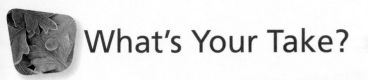

What's Your Take?

Clouds of insects are common during the summer and serve as food for migratory birds. Many waterfowl such as ducks and geese migrate to the tundra in the spring; there, they mate and raise their young during the summer before migrating south in the fall. Permanent resident birds are the ptarmigan and snowy owl. No reptiles or amphibians survive in this extreme climate. A few hardy mammals such as musk oxen, caribou (reindeer), arctic hare, and lemmings can survive by feeding on the grasses and other plants that grow during the short, cool summer. Arctic foxes, wolves, and owls are the primary predators in this region.

Scattered patches of tundralike communities also are found on mountaintops throughout the world. These are known as **alpine tundra.** Although the general appearance of the alpine tundra is similar to true tundra, many of the species of plants and animals are different. Many of the birds and large mammals migrate up to the alpine tundra during the summer and return to lower elevations as the weather turns cold.

Human Impact

Few people live in this region. Local native people often rely on subsistence hunting for food. However, because of the very short growing season, damage to this kind of ecosystem is slow to heal, so the land must be handled with great care.

6.4 Major Aquatic Ecosystems

Terrestrial biomes are determined by the amount and kind of precipitation and by temperatures. Other factors, such as soil type and wind, also play a part. Aquatic ecosystems also are shaped by key environmental factors. Several important factors are the ability of the sun's rays to penetrate the water, the depth of the water, the nature of the bottom substrate, the water temperature, and the amount of dissolved salts.

An important determiner of the nature of aquatic ecosystems is the amount of salt dissolved in the water. Those that have little dissolved salt are called **freshwater ecosystems,** and those that have a high salt content are called **marine ecosystems.**

Marine Ecosystems

Like terrestrial ecosystems, marine ecosystems are quite diverse. Ecologists recognize several categories of marine ecosystems.

Pelagic Marine Ecosystems

In the open ocean, many kinds of organisms float or swim actively. Crustaceans, fish, and whales swim actively as they pursue food. Organisms that are not attached to the bottom are called **pelagic organisms,** and the ecosystem they are a part of is called a **pelagic ecosystem.**

In a pelagic ecosystem, plankton is an important category of organisms. The term **plankton** is used to describe aquatic organisms that are so small and weakly swimming that they are simply carried by currents. Some planktonic organisms, known as **phytoplankton,** carry on photosynthesis and are the base of the food web in a pelagic ecosystem. In the open ocean, a majority of these organisms are small, microscopic, floating algae and bacteria. Since they carry on photosynthesis, they are found in the upper layer of the ocean, where the sun's rays penetrate, known as the **euphotic zone.** The thickness of the euphotic zone varies with the degree of clarity of the water. In clear water it can be up to 150 meters (500 feet) in depth.

Zooplankton are small, floating or weakly swimming protozoa and animals of many kinds that feed on the phytoplankton. Zooplankton are often located at a greater depth in the ocean than the phytoplankton but migrate upward at night and feed on the large population of phytoplankton. The zooplankton are in turn eaten by larger animals such as fish and larger shrimp, which are eaten by larger fish such as salmon, tuna, sharks, and mackerel. (See figure 6.20.)

A major factor that influences the nature of a marine community is the kind and amount of material dissolved in the water. Of particular importance is the amount of dissolved, inorganic nutrients available to the organisms that carry on photosynthesis. Phosphorus, nitrogen, and carbon are all required for the construction of new living material. In water, these are often in short supply. Therefore, the most productive aquatic ecosystems are those in which these essential nutrients are most common. These areas include places in oceans where currents bring up nutrients that

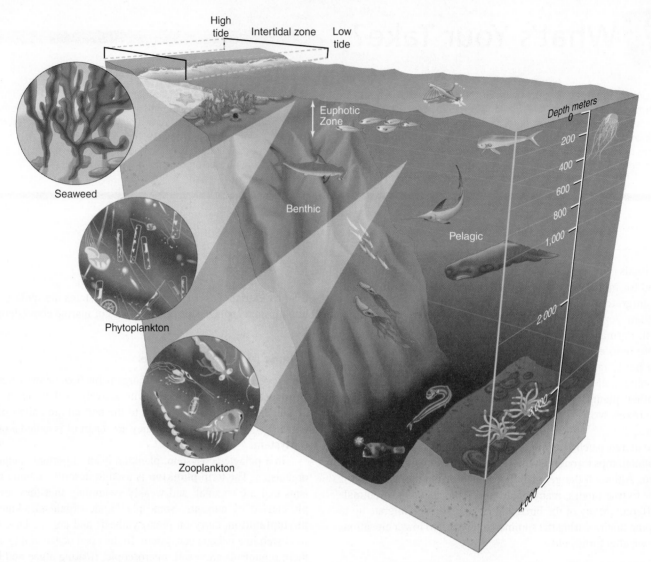

FIGURE 6.20 Marine Ecosystems All of the photosynthetic activity of the ocean occurs in the shallow water called the euphotic zone, either in attached algae near the shore or in minute phytoplankton in the upper levels of the open ocean. Consumers are either free-swimming pelagic organisms or benthic organisms that live on the bottom. Small protozoa and animals that feed on phytoplankton are known as zooplankton.

have settled to the bottom and places where rivers deposit their load of suspended and dissolved materials.

Benthic Marine Ecosystems

Organisms that live on the ocean bottom, whether attached or not, are known as **benthic** organisms, and the ecosystem of which they are a part is called a **benthic ecosystem.** Some fish, clams, oysters, various crustaceans, sponges, sea anemones, and many other kinds of organisms live on the bottom. In shallow water, sunlight can penetrate to the bottom, and a variety of attached photosynthetic organisms frequently called seaweeds are common. Since they are attached and some, such as kelp, can grow to very large size, many other bottom-dwelling organisms, such as sea urchins, worms, and fish, are associated with them.

The substrate is very important in determining the kind of benthic community that develops. Sand tends to shift and move, making it difficult for large plants or algae to become established,

although some clams, burrowing worms, and small crustaceans find sand to be a suitable habitat. Clams filter water and obtain plankton and detritus or burrow through the sand, feeding on other inhabitants. Mud may provide suitable habitats for some kinds of rooted plants, such as mangrove trees or sea grasses. Although mud usually contains little oxygen, it still may be inhabited by a variety of burrowing organisms that feed by filtering the water above them or that feed on other animals in the mud. Rocky surfaces in the ocean provide a good substrate for many kinds of large algae. Associated with this profuse growth of algae is a large variety of animals. (See figure 6.21.)

Temperature also has an impact on the kind of benthic community established. Some communities, such as coral reefs or mangrove swamps, are found only in areas where the water is warm.

Coral reef ecosystems are produced by coral animals that build cup-shaped external skeletons around themselves. Corals protrude from their skeletons to capture food and expose themselves to the sun. Exposure to sunlight is important because corals

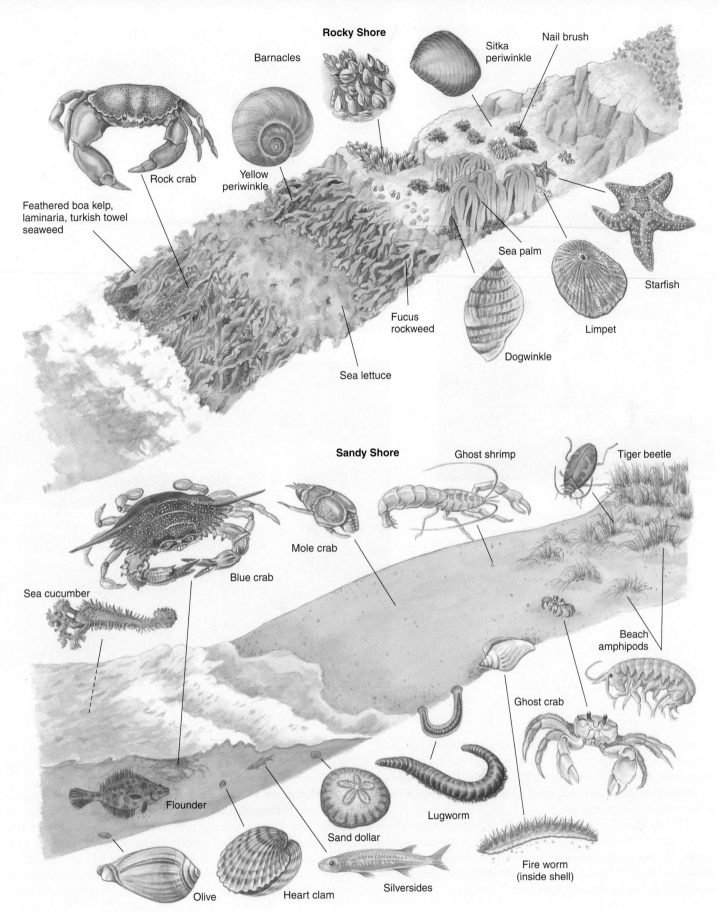

Rocky Shore

Barnacles

Sitka periwinkle

Nail brush

Rock crab

Yellow periwinkle

Feathered boa kelp, laminaria, turkish towel seaweed

Sea palm

Starfish

Fucus rockweed

Dogwinkle

Limpet

Sea lettuce

Sandy Shore

Ghost shrimp

Tiger beetle

Mole crab

Blue crab

Sea cucumber

Beach amphipods

Ghost crab

Flounder

Lugworm

Sand dollar

Olive

Heart clam

Silversides

Fire worm (inside shell)

FIGURE 6.21 Types of Shores The kind of substrate determines the kinds of organisms that can live near the shore. Rocks provide areas for attachment that sands do not, since sands are constantly shifting. Muds usually have little oxygen in them; therefore, the organisms that live there must be adapted to those kinds of conditions.

Coral reef organisms

FIGURE 6.22 Coral Reef Corals are small sea animals that secrete external skeletons. They have a mutualistic relationship with certain algae, which allows both kinds of organisms to be very successful. The skeletal material serves as a substrate upon which many other kinds of organisms live. Coral organisms are found in warmer waters of the ocean. The map shows the distribution of corals is limited to the warmer waters near the equator.

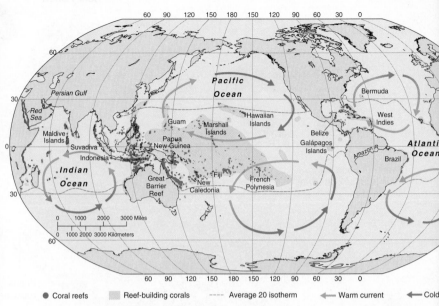

● Coral reefs   ▨ Reef-building corals   ---- Average 20 isotherm   ← Warm current   ← Cold

Great barrier reef

contain single-celled algae within their bodies. These algae carry on photosynthesis and provide both themselves and the coral animals with the nutrients necessary for growth. This mutualistic relationship between algae and coral is the basis for a very productive community of organisms.

The skeletons of the corals provide a surface upon which many other kinds of animals live. Some of these animals feed on corals directly, while others feed on small plankton and bits of algae that establish themselves among the coral organisms. Many kinds of fish, crustaceans, sponges, clams, and snails are members of coral reef ecosystems.

Because they require warm water, coral ecosystems are found only near the equator. Coral ecosystems also require shallow, clear water, since the algae must have ample sunlight to carry on photosynthesis. Coral reefs are considered to be among the most productive ecosystems on Earth. (See figure 6.22.)

Mangrove swamp ecosystems are tropical forest ecosystems that occupy shallow water near the shore and the adjacent land. The dominant organisms are special kinds of trees that can tolerate the high salt content of the ocean because the trees can excrete salt from their leaves. In areas where the water is shallow and wave action is not too great, the trees can become established. They have seeds that actually begin to germinate on the tree. When the germinated seed falls from the tree it floats in the water. When the seeds become trapped in mud, they take root.

FIGURE 6.23 Mangrove Swamp Mangroves are tropical trees that are able to live in very wet, salty muds found along the ocean shore. Since they can trap additional sediment, they tend to extend farther seaward as they reproduce.

The trees also have extensively developed roots that extend above the water, where they can obtain oxygen and prop up the plant. The tree roots trap sediment and provide places for oysters, crabs, jellyfish, sponges, and fish to live. The trapping of sediment and the continual extension of mangroves into shallow areas result in the development of a terrestrial ecosystem in what was once shallow ocean. Mangroves are found in south Florida, the Caribbean, Southeast Asia, Africa, and other parts of the world where tropical mudflats occur. (See figure 6.23.)

An **abyssal ecosystem** is a benthic ecosystem that occurs at great depths in the ocean. In such deep regions of the ocean there is no light to support photosynthesis. Therefore, the animals must rely on a continuous rain of organic matter from the euphotic zone. Essentially, all of the organisms in this environment are scavengers that feed on whatever drifts their way. Many of the animals are small and generate light that they use for finding or attracting food.

Estuaries

An **estuary** is a special category of aquatic ecosystem that consists of shallow, partially enclosed areas where freshwater enters the ocean. The saltiness of the water in the estuary changes with tides and the flow of water from rivers. The organisms that live here are specially adapted to this set of physical conditions, and the number of species is less than in the ocean or in freshwater.

Estuaries are particularly productive ecosystems because of the large amounts of nutrients introduced into the basin from the rivers that run into them. This is further enhanced by the fact that the shallow water allows light to penetrate to most of the water in the basin. Phytoplankton and attached algae and plants are able to use the sunlight and the nutrients for rapid growth. This photosynthetic activity supports many kinds of organisms in the estuary. Estuaries are especially important as nursery sites for fish and crustaceans such as flounder and shrimp.

The adults enter these productive, sheltered areas to reproduce and then return to the ocean. The young spend their early life in the estuary and eventually leave as they get larger and are more able to survive in the ocean. Estuaries also trap sediment. This activity tends to prevent many kinds of pollutants from reaching the ocean and also results in the gradual filling in of the estuary, which may eventually become a salt marsh and then part of a terrestrial ecosystem.

Human Impact on Marine Ecosystems

Since the oceans cover about 70 percent of the Earth's surface, it is hard to imagine that humans can have a major impact on them. However, we use the oceans in a wide variety of ways. The oceans provide a major source of protein in the form of fish, shrimp, and other animals. However, overfishing has destroyed many of the traditional fishing industries of the world. Fish farming results in the addition of nutrients and has caused diseases to spread from farmed species to wild fish. Estuaries are important fishing areas but are affected by the flow of fertilizer, animal waste, and pesticides down the rivers that drain farmland and enter estuaries. The use of the oceans as transportation results in oil pollution and trash regularly floating onto the shore. Coral reefs are altered by fishing and siltation from rivers. Mangrove swamps are converted to areas for the raising of fish. It is clear that humans have a great impact on marine ecosystems.

Freshwater Ecosystems

Freshwater ecosystems differ from marine ecosystems in several ways. The amount of salt present is much less, the temperature of the water can change greatly, the water is in the process of moving to the ocean, oxygen can often be in short supply, and the organisms that inhabit freshwater systems are different. In particular, flowering plants and insects have major roles to play in freshwater ecosystems.

Freshwater ecosystems can be divided into two categories: those in which the water is relatively stationary, such as lakes, ponds, and reservoirs, and those in which the water is running downhill, such as streams and rivers.

Lakes and Ponds

Large lakes have many of the same characteristics as the ocean. Therefore, many of the same terms are used to characterize these ecosystems as were used to discuss marine ecosystems. If the lake is deep, there is a *euphotic zone* at the top, with many kinds of *phytoplankton,* and *zooplankton* that feed on the phytoplankton. Small fish feed on the zooplankton and are in turn eaten by larger fish. The species of organisms found in freshwater lakes are different from those found in the ocean, but the roles played are similar.

Along the shore and in the shallower parts of lakes, many kinds of flowering plants are rooted in the bottom. Some have leaves that float on the surface or protrude above the water and are called **emergent plants.** Cattails, bulrushes, arrowhead plants, and water lilies are examples. Rooted plants that stay submerged below the surface of the water are called **submerged plants.** *Elodea* and *Chara* are examples.

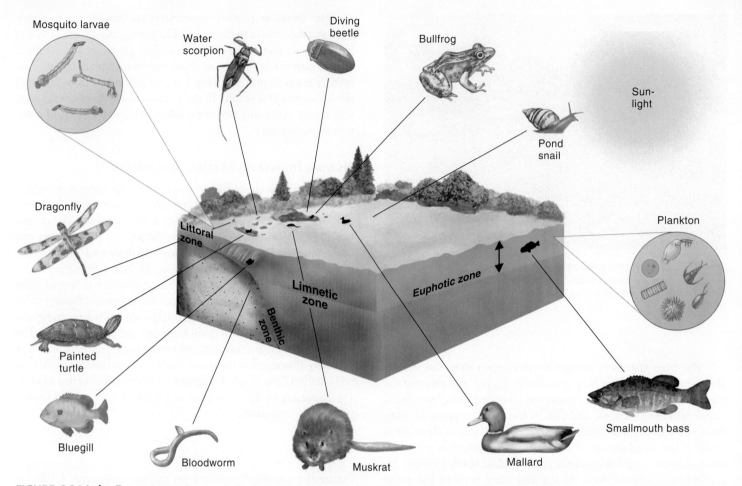

Mosquito larvae

Water scorpion

Diving beetle

Bullfrog

Sun-light

Pond snail

Dragonfly

Plankton

Littoral zone

Limnetic zone

Euphotic zone

Benthic zone

Painted turtle

Bluegill

Bloodworm

Muskrat

Mallard

Smallmouth bass

FIGURE 6.24 Lake Ecosystem Lakes are similar in structure to oceans except that the species are different because most marine organisms cannot live in freshwater. Insects are common organisms in freshwater lakes, as are many kinds of fish, zooplankton, and phytoplankton.

Many kinds of freshwater algae also grow in the shallow water, where they may appear as mats on the bottom or attached to vegetation and other objects in the water. Associated with the plants and algae are a large number of different kinds of animals. Fish, crayfish, clams, and many kinds of aquatic insects are common inhabitants of this mixture of plants and algae. This region, with rooted vegetation, is known as the **littoral zone,** and the portion of the lake that does not have rooted vegetation is called the **limnetic zone.** (See figure 6.24.)

The productivity of the lake is determined by several factors. Temperature is important, since cold temperatures tend to reduce the amount of photosynthesis. Water depth is important because shallow lakes will have light penetrating to the lake bottom, and therefore, photosynthesis can occur throughout the entire water column. Shallow lakes also tend to be warmer as a result of the warming effects of the sun's rays. A third factor that influences the productivity of lakes is the amount of nutrients present. This is primarily determined by the rivers and streams that carry nutrients to the lake. River systems that run through areas that donate many nutrients will carry the nutrients to the lakes. Farming and construction expose soil and release nutrients, as do other human activities such as depositing sewage into streams and lakes. Deep,

clear, cold, nutrient-poor lakes are low in productivity and are called **oligotrophic lakes.** Shallow, murky, warm, nutrient-rich lakes are called **eutrophic lakes.**

The dissolved oxygen content of the water is important, since the quantity of oxygen determines the kinds of organisms that can inhabit the lake. When organic molecules enter water, they are broken down by bacteria and fungi. These decomposer organisms use oxygen from the water as they perform respiration. The amount of oxygen used by decomposers to break down a specific amount of organic matter is called the **biochemical oxygen demand (BOD).**

Organic materials enter aquatic ecosystems in several ways. The organisms that live in the water produce the metabolic wastes. When organisms that live in or near water die or shed parts, their organic matter is contributed to the water. The amount of nutrients entering the water is also important, since the algae and plants whose growth is stimulated will eventually die and their decomposition will reduce oxygen concentration. Many bodies of water experience a reduced oxygen level during the winter, when producers die. The amount and kinds of organic matter determine, in part, how much oxygen is left to be used by other organisms, such as fish, crustaceans, and snails. Many lakes may experience

Focus On

Varzea Forests—Where the Amazon River and Land Meet

The Amazon River and its many tributaries constitute the largest drainage basin (about 40 percent of South America) and the highest volume of flow of any river system in the world—about 20 percent of all river flow in the world. The water is supplied by abundant rainfall—many areas receive over 300 cm (100 in.) of rain per year—in the basin and snowmelt from the Andes. Because the snowmelt, and to a certain extent the rainfall, is seasonal, the Amazon and its tributaries are characterized by seasonal flooding.

Much of the river basin is very flat. The city of Iquitos is about 3600 kilometers (2,200 miles) from the ocean but the river at that point is only 100 meters (300 feet) above sea level. When the river floods, extensive areas along the river are flooded under several meters of water due to the flat terrain. The area flooded extends several kilometers from the river. This creates a seasonal wetland forest known as the *varzea*. The land farther from the river that does not flood is known as the *terra firme*.

This seasonally flooded area accounts for about 4 percent of the total area of the Amazon rainforest. The vegetation of the varzea is different from that of the terra firme because the trees and other vegetation must be able to withstand extensive periods of flooding.

The animals of the river and the varzea are greatly affected by the flooding. Animals of the river move into the forest with the flood and use forest resources as food. Varzea forest areas are critical to the freshwater fisheries of the Amazon Basin, since many fish actually change their diet and become fruit eaters when they are able to enter the flooded forest. In the dryer portions of the year when the river recedes, they return to the main river channel and are carnivores. In addition to using the forest for food, the fish also distribute the seeds of fruits in their feces. Other river animals such as the caimans and the giant river otter also move into the forest with the flood.

The terrestrial animals of the forest face a different problem. As the river rises, they are forced to retreat to higher ground and often become trapped on islands. This results in intense competition for food. Monkeys and birds are less troubled by the flooding. Many of them rely on fruits of trees as their primary food source, which is available even during the flood. The monkeys can simply travel from tree to tree and the birds can fly over the water.

The periodic flooding of the area deposits silt, which provides a fertile soil. Therefore, the varzea is affected by human activity as farmers use the dry season to raise crops. Often the crops are a mixture of normal forest plants along with crops like bananas, rice, and root crops. Because of the flooding, people who live along the river build their houses on high ground and often on stilts. The rivers are also the primary highways of the region and small boats are the most common form of transportation.

Boats are primary form of transportation. Here bananas are being loaded to go to market.

The river nearly reaches to the top of this bank during floods that occur every year.

Varzea forest

periods when oxygen is low, resulting in the death of fish and other organisms.

Streams and Rivers

Streams and rivers are a second category of freshwater ecosystem. Since the water is moving, planktonic organisms are less important than are attached organisms. Most algae grow attached to rocks and other objects on the bottom. This collection of attached algae, animals, and fungi is called the **periphyton.** Since the water is shallow, light can penetrate easily to the bottom (except for large or extremely muddy rivers). Even so, it is difficult for photosynthetic organisms to accumulate the nutrients necessary for growth, and most streams are not very productive. As a matter of fact, the major input of nutrients is from organic matter that falls into the stream from terrestrial sources. These are primarily the

Science, Politics, & Policy

Preventing Asian Carp from Entering the Great Lakes

Several species of carp have been introduced into the waters of the United States. Two species (bighead carp and silver carp) have become the focus of a policy debate involving the federal government, several states, the province of Ontario, and the city of Chicago. Both species filter plankton from the water as they swim and can consume huge amounts of plankton daily. In the 1970s, Asian bighead carp were imported into the United States where their filter-feeding habits were used to remove algae from catfish ponds. But flooding in the 1990s allowed the fish to escape the ponds and enter the Mississippi River. Since then, they have spread through much of the Mississippi River drainage, which includes tributaries that flow from Chicago.

An important ingredient in this policy battle is the presence of the Chicago Sanitary and Ship Canal that connects Lake Michigan to the Mississippi River drainage. Electric barriers were installed to prevent the migration of the fish and still allow boat traffic. But these barriers have not been effective, and Asian carp have been found above the barriers.

Many scientists and policymakers are concerned that the presence of the plankton-eating Asian carp in the Great Lakes could greatly alter the food chain in the Great Lakes and threaten the sport and commercial fishing industries that are worth more than $7 billion a year.

A lawsuit by Michigan and four other states, plus Ontario, Canada, requested the closing of the locks in the canal to block Asian carp from entering Lake Michigan. Commercial interests, including tour boat operators, barge companies, and the recreational boating industry, oppose the closing of the connecting waterways. The argument then develops around how you weigh the economic consequences of closing the locks for the shipping industry against the potentially greater economic impacts of allowing the carp to enter the Great Lakes.

In 2010, the U.S. Supreme Court ruled that it would not order the closure of locks between Chicago area waterways and Lake Michigan. This action resulted in two responses at the federal level of government. A bill entitled the Stop Asian Carp Act 2011 was introduced in the Senate to study the feasibility of closing the connection between the Illinois River drainage and Lake Michigan. It was referred to committee. In addition, the executive branch established the Asian Carp Regional Coordinating Committee (ACRCC) with a budget of $7 million. The ACRCC has conducted surveys to determine if Asian carp have gotten past the electric barriers and have removed Asian carp from areas downstream from the barriers. One of the survey methods was to sample water for Asian carp DNA. One sample that was collected above the barrier was positive for Asian carp DNA.

A report released in 2013 turned up 58 positive DNA hits for bighead or silver carp in the Chicago Area Waterway System. The study also found six DNA hits in western Lake Erie. Some of the Chicago DNA was found in Lake Calumet, where a live bighead carp was caught in 2010.

The Army Corps of Engineers contends that the electric barrier is preventing the carp from getting through, even though their DNA has turned up repeatedly on the other side of the barrier. The federal agencies stated the genetic material could have been transported by bird feces, fish sampling gear, barges, and storm sewers.

That argument is even less plausible for the DNA found in Lake Erie. The DNA that was found there was more than 160 kilometers from waterways infested with Asian carp. So if birds were the source, it seemingly would mean they ate carp, flew a long distance and excreted feces within a few hours of when the researchers collected the water samples.

A study released by the Army Corps of Engineers in 2014 stated that physical separation of the waterways would require 25 years to complete and cost between $15 billion to $18 billion. Much of the cost, $14 billion,

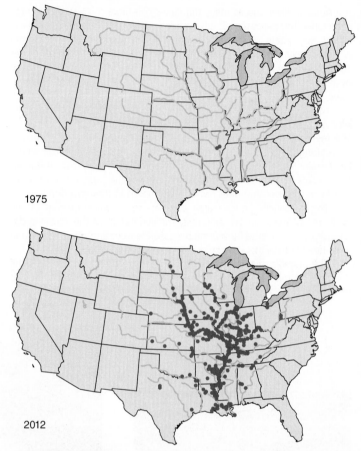

1975

2012

Reported Asian Carp from 1975–2012.

Asian Carp in the Mississippi River.

would be for flood management basins and water runoff tunnels that would have to be built to reduce increased risks of flooding in the Chicago area.

Environmental groups are not satisfied. They point to the presence of Asian carp DNA above the barrier as proof that the fish have the ability to get past the barrier and call for the complete closure of the canal.

Now, put yourself in the position of the individual who makes the final decision regarding the fate of the canal. Do you permanently close it or not? Can you defend your decision to the general public? Is it possible to prevent the spread of invasive species?

Is the Cownose Ray a Pest or a Resource?

Ecosystems involve complex interactions between organisms. Because it is so large and difficult to study, the ocean is one of the least well understood ecosystems. We use many of the organisms of the ocean as sources of food but in many cases do not understand many of the forces that affect the distribution and abundance of the resource. Recent changes in the populations of cownose rays have led to reverberations both up and down the food chain.

The cownose ray is a common predator of shellfish along the east coast of the United States. The rays migrate northward to the coast of the Carolinas and to Chesapeake Bay in the summer, when the waters warm. They arrive in large number and feed on oysters, scallops, and other marine organisms. In recent years, increased populations of cownose rays

have caused great losses in the oyster and scallop fisheries along the east coast, losses that have led fisheries managers to make adjustments and suggest changes that would protect the shellfish industry. North Carolina actually closed its scallop fishery for a period because scallop numbers had become so low in part because of cownose ray predation. The state of Virginia has promoted fishing for cownose rays as a way to decrease the population and protect the shellfish industry of Chesapeake Bay. They even changed the common name for the ray from bullfish to Chesapeake Ray and provide recipes for their preparation.

Various scientists point to the harvesting of sharks as a primary cause of the increased cownose ray populations. In the last 50 years, the harvest of sharks for food has more than tripled and many species of sharks have seen major population declines. Sharks eat rays, and reduced numbers of sharks means that more rays survive and populations grow. Some species of sharks are being overharvested and their capture is regulated or prohibited in U.S. waters. However, many kinds of sharks are highly migratory and are subject to harvesting outside U.S. waters.

However, harvesting of cownose rays is opposed by many because they feel that the ray does not reproduce fast enough to support a fishery. They point out that the rays do not reach sexual maturity until about 7 years of age and that the females produce only one young per year.

Thus, there are several important human players in this drama: local people that make a livelihood from harvesting shellfish, local and international fishers that harvest sharks, fisheries managers who are trying to regulate a resource, and state officials who respond to the needs and desires of their constituents.

- Should large portions of the coastal ocean be regulated to produce shellfish?
- Should shellfish be protected from cownose ray predation?
- Who should decide whether to harvest cownose rays?

leaves from trees and other vegetation, as well as the bodies of living and dead insects.

Within the stream is a community of organisms that are specifically adapted to use the debris as a source of food. Bacteria and fungi colonize the organic matter, and many kinds of insects shred and eat this organic matter along with the fungi and bacteria living on it. The feces (intestinal wastes) of these insects and the tiny particles produced during the eating process become food for other insects that build nets to capture the tiny bits of organic matter that drift their way. These insects are in turn eaten by carnivorous insects and fish.

Organisms in larger rivers and muddy streams, which have less light penetration, rely in large part on the food that drifts their way from the many streams that empty into the river. These larger rivers tend to be warmer and to have slower-moving water. Consequently, the amount of oxygen is usually less, and the species of plants and animals change. Any additional organic matter added to the river system adds to the BOD, further reducing the oxygen in the water. Plants may become established along the

river bank and contribute to the ecosystem by carrying on photosynthesis and providing hiding places for animals.

Just as estuaries are a bridge between freshwater and marine ecosystems, swamps and marshes are a transition between aquatic and terrestrial ecosystems. **Swamps** are wetlands that contain trees that are able to live in places that are either permanently flooded or flooded for a major part of the year. **Marshes** are wetlands that are dominated by grasses and reeds. Many swamps and marshes are successional states that eventually become totally terrestrial communities.

Human Impact on Freshwater Ecosystems

Freshwater resources in lakes and rivers account for about 0.02 percent of the world's water. Most freshwater ecosystems have been heavily affected by human activity. Any activity that takes place on land ultimately affects freshwater because of runoff from the land. Agricultural runoff, sewage, sediment, and trash all find their way into streams and lakes. Chapter 15 covers these issues in greater detail.

Summary

Ecosystems change as one kind of organism replaces another in a process called succession. Ultimately, a relatively stable stage is reached, called the climax community. Succession may begin with bare rock or water, that is referred to as primary succession, or may occur when the original ecosystem is destroyed, in which case it is called secondary succession. The stages that lead to the climax are called successional stages.

Major regional terrestrial climax communities are called biomes. The primary determiners of the kinds of biomes that develop are the amount and yearly distribution of rainfall and the yearly temperature cycle. Major biomes are desert, temperate grassland, savanna, Mediterranean shrublands, tropical dry forest, tropical rainforest, temperate deciduous forest, temperate rainforest, taiga, and tundra. Each has a particular set of organisms that is adapted to the climatic conditions typical for the area. As one proceeds up a mountainside, it is possible to witness the same kind of change in biomes that occurs if one were to travel from the equator to the North Pole.

Aquatic ecosystems can be divided into marine (saltwater) and freshwater ecosystems. In the ocean, some organisms live in open water and are called pelagic organisms. Light penetrates only the upper layer of water; therefore, this region is called the euphotic zone. Tiny photosynthetic organisms that float near the surface are called phytoplankton. They are eaten by small protozoa and animals known as zooplankton, which in turn are eaten by fish and other larger organisms.

The kind of material that makes up the shore determines the mixture of organisms that live there. Rocky shores provide surfaces to which organisms can attach; sandy shores do not. Muddy shores are often poor in oxygen, but marshes and swamps may develop in these areas. Coral reefs are tropical marine ecosystems dominated by coral animals. Mangrove swamps are tropical marine shoreline ecosystems dominated by trees. Estuaries occur where freshwater streams and rivers enter the ocean. They are usually shallow, very productive areas. Many marine organisms use estuaries for reproduction.

Insects are common in freshwater and absent in marine systems. Lakes show a structure similar to that of the ocean, but the species are different. Deep, cold-water lakes with poor productivity are called oligotrophic, while shallow, warm-water, highly productive lakes are called eutrophic. Streams differ from lakes in that most of the organic matter present in them falls into them from the surrounding land. Thus, organisms in streams are highly sensitive to the land uses that occur near the streams.

Acting Green

1. Learn to identify five plants native to your area.
2. Visit a nature center, wildlife refuge, or state nature preserve. List 10 organisms you identified.
3. Participate in a local program to restore a habitat or eliminate invasive species.
4. Participate in Earth Day (April 22) and Arbor Day (in the spring but the date varies by state) activities in your community.
5. Visit the National Wildlife Federation website and learn about its Backyard Wildlife Habitat Program.
6. Visit a disturbed site—vacant lot, road side, abandoned farmland. What evidence do you see that succession is taking place?
7. Participate in a local river or shoreline clean-up program.
8. Make an attempt to spend some time every week in nature. When you do, do so "un-plugged."

Review Questions

1. Describe the process of succession. How does primary succession differ from secondary succession?
2. How does a climax community differ from a successional community?
3. List two abiotic characteristics typical of each of the following biomes: tropical rainforest, desert, tundra, taiga, savanna, Mediterranean shrublands, tropical dry forest, temperate grassland, temperate rainforest, and temperate deciduous forest.
4. List two biotic characteristics typical of each of the following biomes: tropical rainforest, desert, tundra, taiga, savanna, Mediterranean shrublands, tropical dry forest, temperate grassland, temperate rainforest, and temperate deciduous forest.
5. What two primary factors determine the kind of terrestrial biome that will develop in an area?
6. How does height above sea level affect the kind of biome present?
7. What areas of the ocean are the most productive?
8. What is the role of each of the following organisms in a marine ecosystem: phytoplankton, zooplankton, algae, coral animals, and fish?
9. Which of the following organisms functions only in the euphotic zone: seaweed, crabs, phytoplankton, fish?
10. List three differences between freshwater and marine ecosystems.
11. What is an estuary? Why are estuaries important?

Critical Thinking Questions

1. Does the concept of a "climax community" make sense? Why or why not?

2. What do you think about restoring ecosystems that have been degraded by human activity? Should it be done or not? Why? Who should pay for this reconstruction?

3. Identify the biome in which you live. What environmental factors are instrumental in maintaining this biome? What is the current health of your biome? What are the current threats to its health? How might your biome have looked 100, 1,000, 10,000 years ago?

4. Imagine you are a conservation biologist who is being asked by local residents what the likely environmental outcomes of development would be in the tropical rainforest in which they live. What would you tell them? Why do you give them this evaluation? What evidence can you cite for your claims?

5. The text says that about half of the old-growth temperate rainforest in the Pacific Northwest has been logged. What to do with the remaining forest is still a question. Some say it should be logged, and others say it should be preserved. What values, beliefs, and perspectives are held by each side? What is your ethic regarding logging old-growth in this area? What values, beliefs, and perspectives do you hold regarding this issue?

6. Much of the old-growth forest in the United States has been logged, economic gains have been realized, and second-growth forests have become established. This is not the case in the tropical rainforests, although they are being lost at alarming rates. Should developed countries, which have already "cashed in" on their resources, have anything to say about what is happening in developing countries? Why do you think the way you do?

To access additional resources for this chapter, please visit ConnectPlus® at connect.mheducation.com. There you will find interactive exercises, including Google Earth™, additional Case Studies, and SmartBook™, an adaptive eBook that integrates our LearnSmart® adaptive learning technology.

chapter 7

Populations: Characteristics and Issues

Populations are collections of organisms of the same species. This crowd of people at a market in Sao Paulo, Brazil, is a local population with characteristics that differ from other human populations elsewhere in the world and even within Brazil. The human population has altered the planet profoundly.

CHAPTER OUTLINE

OBJECTIVES

After reading this chapter, you should be able to:

- Describe at least five ways that two populations of the same species could differ.
- Understand that birth rate and death rate are both important in determining the population growth rate.
- Define the following characteristics of a population: natality, mortality, sex ratio, age distribution, biotic potential, and spatial distribution.
- Explain the significance of biotic potential to the rate of population growth.
- Describe the lag, exponential growth, deceleration, and stable equilibrium phases of a population growth curve. Explain why each of these stages occurs.
- Describe how limiting factors determine the carrying capacity for a population.

- Distinguish between extrinsic and intrinsic limiting factors.
- Distinguish between density-dependent and density-independent limiting factors.
- List the four categories of limiting factors.
- List four characteristic differences between K-strategists and r-strategists.
- Recognize that populations can fluctuate in cycles.
- Understand the implications of overreproduction.
- Describe the relationship between stage of economic development and rate of population growth.
- Explain the meaning of $I = P \times A \times T$.

- Describe the ecological footprint concept.
- Explain how human population growth is influenced by economic, social, theological, philosophical, and political thinking.
- Explain why the age distribution and the status and role of women affect population growth projections.
- Recognize that most countries of the world have a rapidly growing population.
- Describe the implications of the demographic transition concept.
- Recognize that rapid population growth and poverty are linked.
- Explain why people in the less-developed world do not eat much meat.

Invasive Species

The sizes of the populations of most organisms are stable. In other words, although there are births and deaths and there may be seasonal or short-term fluctuations, on average, the number of organisms remains about the same. The populations of some organisms are declining. These populations are usually small and are often labeled as threatened or endangered. There are also many species of organisms that have rapidly growing populations. These species typically have arrived in an area where the species were not present before. Not all introductions of species are successful, but those that are wildly successful are often called invasive species. Invasive species, whether they are plants, animals, or microorganisms, have certain characteristics.

Invasive species arrive in an area where they were not present previously. Then, because they lack natural enemies in their new habitat and because of their specific reproductive characteristics, they go through a population explosion. Some key characteristics of invasive species include:

- Almost all are non-native (exotic) species that lack natural predators or disease-causing organisms in their new habitat. Therefore, they lack natural checks to population growth.
- Most invasive species are generalists that can live in a variety of habitats.
- They grow and mature rapidly, produce many offspring, and many of the offspring survive.
- They have very effective dispersal mechanisms.
- They can successfully outcompete native species.

Examples of exotic invasive species abound.

Autumn olive (*Elaeagnus umbellata*) is native to Asia and was purposely introduced into the United States as a wildlife food source and to help control erosion. It has currently spread throughout the eastern U.S. and three western states. It can live in a variety of open habitats, grows quickly, and produces large numbers of berries containing seeds. The berries are eaten by birds that disperse the seeds when they defecate. Once established, it is able to suppress native species because native species cannot get established in the shade of the autumn olive. Once established, it is nearly impossible to eliminate.

Emerald ash borer (*Agrilus planipennis*) is an exotic Asian beetle accidentally introduced into North America. It probably arrived in wood products or wooden packing material. It was first discovered near Detroit, Michigan, in 2002. The beetles had probably been in the Detroit area for 10 or more years as a small population prior to 2002. By 2013 it had spread to 22 eastern states and Ontario and Quebec. It has no natural enemies and affects all species of ash trees. The adults emerge in the late spring and mate. The female lays as many as 100 eggs on the bark of ash trees. The eggs hatch and the larvae bore through the bark and feed on the actively growing tissue just under the bark. This eventually kills the tree. Although the adults can fly a distance up to a kilometer, humans that transport logs and firewood are the primary means of dispersal.

Humans (*Homo sapiens*) also fit the definition of an invasive species. Humans began as a small population in Africa and have since spread throughout the world. It is highly adaptable, can live in a variety of habitats, and its offspring have a high success rate. It can travel long distances and can outcompete any organisms in the new habitats it invades.

Autumn olive

Emerald ash borer

Humans

7.1 Population Characteristics

A **population** can be defined as a group of individuals of the same species inhabiting an area. Examples are the dandelions in a yard, the rat population in your city sewer, and the number of students in a biology class. On a larger scale, all the people of the world constitute the world human population. The terms *species* and *population* are interrelated, because a species is a population—the largest possible population of a particular kind of organism. The term *population,* however, is often used to refer to portions of a species by specifying a space and time. Just as individuals within a population are recognizable, different populations of the same species have specific characteristics that distinguish them from one another. Some important ways in which populations differ include genetics, natality (birth rate), mortality (death rate), sex ratio, age distribution, growth rates, density, and spatial distribution.

Genetic Differences

Because each local population is a small portion of its species, and each population is adapted to its local conditions, it is common that populations of the same species show characteristics that differ from those of other populations. For example, the black bear (*Ursus americanus*) is found in forested areas throughout North America. With such a large range, it inhabits a variety of habitats from the tundra of Alaska and Canada to the dry southwestern United States and Mexico. Within that range there are about 16 distinctly different populations. Many of them are partially isolated from other populations on peninsulas or islands. Others are found in mountains or other partially isolated wooded areas. Biologists recognize several genetic differences among these different populations and call them subspecies. One subspecies is the cinnamon bear which is found in the drier regions of southwestern Canada and northwestern United States. It is easily distinguished from the more common subspecies of black bear by its reddish-brown coat. (See figure 7.1.)

Natality—Birth Rate

Natality refers to the number of individuals added to the population through reproduction over a particular time period. There are two ways in which new individual organisms are produced—asexual reproduction and sexual reproduction.

Asexual Reproduction

Bacteria and other tiny organisms reproduce primarily asexually when they divide to form new individuals that are identical to the original parent organism. Plants and many kinds of animals, such as sponges, jellyfish, and several kinds of worms, reproduce asexually by dividing into two parts or by budding off small portions of themselves that become independent individuals. Even some insects and lizards have a special kind of asexual reproduction in which the females lay unfertilized eggs that are genetically identical to the female.

(a) Cinnamon-colored black bear

(b) Common black bear

FIGURE 7.1 Genetic Differences between Black Bear Populations The reddish-brown color of its fur is one of the genetic differences that distinguishes the cinnamon subspecies (population) from other subspecies of black bear.

Sexual Reproduction

However, most species have some stage in their life cycle in which they reproduce sexually to produce genetically different offspring. In plant populations, sexual reproduction results in the production of numerous seeds, but the seeds must land in appropriate soil conditions before they will germinate to produce a new individual. Animal species also typically produce large numbers of offspring as a result of sexual reproduction.

In human populations, natality is usually described in terms of the **birth rate,** the number of individuals born per 1,000 individuals per year. For example, if a population of 2,000 individuals produced 20 offspring during one year, the birth rate would be 10 per thousand per year. The natality for most species is typically quite high. Most species produce many more offspring than are needed to replace the parents.

Mortality—Death Rate

It is important to recognize that the growth of a population is not determined by the birth rate (natality) alone. **Mortality,** the number of deaths in a population over a particular time period, is also important. For most species, mortality rates are very high, particularly among the younger individuals. For example, of all the seeds that plants produce, very few will result in a mature plant that itself will produce offspring. Many seeds are eaten by animals, some do not germinate because they never find proper soil conditions, and those that germinate must compete with other organisms for nutrients and sunlight.

In human population studies, mortality is usually discussed in terms of the **death rate,** the number of people who die per 1,000 individuals per year. Compared to the high mortality of the young of most species, the infant death rate of long-lived animals such as humans is relatively low. In order for the size of a population to grow, the number of individuals added by reproduction must be greater than the number leaving it by dying. (See figure 7.2.)

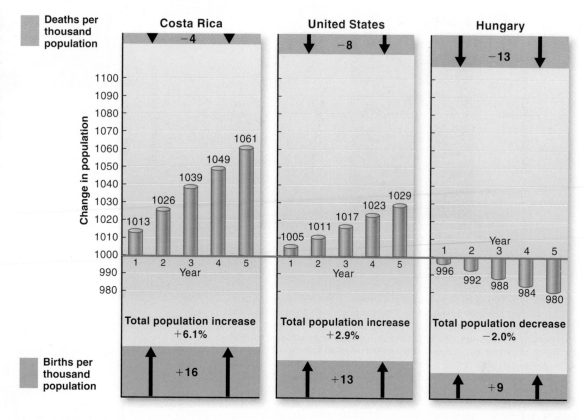

FIGURE 7.2 Effect of Birth Rate and Death Rate on Population Size For a population to grow, the birth rate must exceed the death rate for a period of time. These three human populations illustrate how the combined effects of births and deaths would change population size if birth rates and death rates were maintained for a five-year period.

Source: Data from World Population Data Sheet 2013, Population Reference Bureau, Inc., Washington, D.C.

Another way to view mortality is to view how likely it is that an offspring will survive to a specific age. One way of visualizing this is with a survivorship curve. A **survivorship curve** shows the proportion of individuals likely to survive to each age. While each species is different, three general types of survivorship curves can be recognized: species that have high mortality among their young, species in which mortality is evenly spread over all age groups, and species in which survival is high until old age, when mortality is high. Figure 7.3 gives examples of species that fit these three general categories.

Population Growth Rate

The **population growth rate** is the birth rate minus the death rate. In human population studies, the population growth rate is usually expressed as a percentage of the total population. For example, in the United States, the birth rate is 13 births per thousand individuals in the population. The death rate is 8 per thousand. The difference between the two is 5 per thousand, which is equal to an annual population increase of 0.5 percent (5/1,000).

Sex Ratio

The population growth rate is greatly influenced by the sex ratio of the population. The **sex ratio** refers to the relative numbers of males and females. (Many kinds of organisms, such as earthworms and most plants, have both kinds of sex organs in the same body; sex ratio has no meaning for these species.) The number of females

is very important, since they ultimately determine the number of offspring produced in the population. There are a few species of animals that are truly monogamous. In monogamous species, a male and female pair up, mate, and raise their young together. Unpaired females are less likely to be fertilized and raise young. Even if an unpaired female is fertilized, she will be less successful in raising young. In polygamous species, one male may mate with many females. Therefore, the number of males is less important to the population growth rate than the number of females.

It is typical in most species that the sex ratio is about 1:1 (one female to one male). However, there are populations in which this is not true. In populations of many species of game animals, the males are shot (have a higher mortality) and the females are not. This results in an uneven sex ratio in which the females outnumber the males. In many social insect populations (bees, ants, and wasps), the number of females greatly exceeds the number of males at all times, though most of the females are sterile. In humans, about 106 males are born for every 100 females. However, in the United States, by the time people reach their mid-twenties, a higher death rate for males has equalized the sex ratio. The higher male death rate continues into old age, when women outnumber men.

Age Distribution

The **age distribution** is the number of individuals of each age in the population. Age distribution greatly influences the population

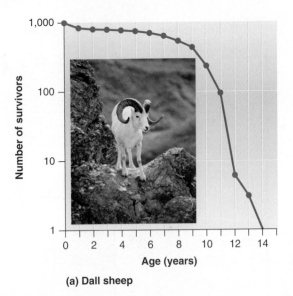

(a) Dall sheep

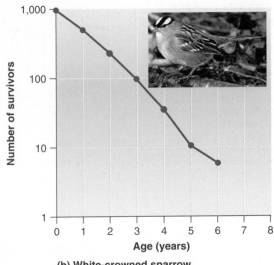

(b) White-crowned sparrow

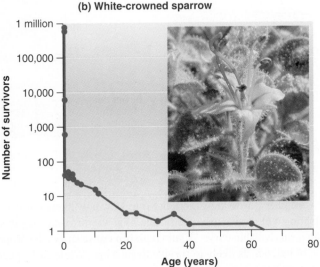

(c) *Cleome droserifolia*

FIGURE 7.3 Types of Survivorship Curves (a) The Dall sheep is a large mammal that produces relatively few young. Most of the young survive, and survival is high until individuals reach old age, when they are more susceptible to predation and disease. (b) The curve shown for the white-crowned sparrow is typical of that for many kinds of birds. After a period of high mortality among the young, the mortality rate is about equal for all ages of adult birds. (c) Many small animals and plants, such as the Mediterranean shrub *Cleome droserifolia,* produce enormous numbers of offspring (seeds). Mortality is very high in the younger individuals. Few seeds find conditions favorable for germination and many seedlings and young plants die. Few individuals reach old age.

growth rate. As you can see in figure 7.4, some are prereproductive juveniles, some are reproducing adults, and some are postreproductive adults. If the population has a large number of prereproductive juveniles, it would be expected to grow in the future as the young become sexually mature. If the majority of a population is made up of reproducing adults, the population should be growing. If the population is made up of old individuals whose reproductive success is low, the population is likely to fall.

Many species, particularly those that have short life spans, have age distributions that change significantly during the course of a year. Species typically produce their young during specific parts of the year. Annual plants (those that live for only one year) produce seeds that germinate in the spring or following a rainy period of the year. Therefore, during one part of the year, most of the individuals are newly germinated seeds and are prereproductive. As time passes, nearly all of those seedlings that survive become reproducing adults and produce seeds. Later in the year, they all die. A similar pattern is seen in many insects that go through their entire life cycle in a year. They emerge from eggs as larvae, transform into adults, mate and lay eggs, and die. Animals that live for several years typically produce their young at a time when food is abundant. In northern climates, this is generally in the spring of the year. In regions where rainfall is sporadic (deserts) or highly seasonal (savannas and some forests), the production of offspring usually occurs following rain. Thus, there is a surge in the number of prereproductive individuals at specific times of the year.

In species that live a long time, it is possible for a population to have an age distribution in which the proportion of individuals in these three categories is relatively constant. Such populations typically have more prereproductive individuals than reproductive individuals and more reproductive individuals than postreproductive individuals.

Human populations exhibit several types of age distribution. (See figure 7.4.) Kenya's population has a large prereproductive and reproductive component. This means that it will continue to increase rapidly for some time. The United States has a very large reproductive component with a declining number of prereproductive individuals. Eventually, if there were no immigration, the U.S. population would begin to decline if current trends in birth rates and death rates continued. Italy has an age distribution with high postreproductive and low prereproductive portions of the population. With low numbers of prereproductive individuals entering their reproductive years, the population of Italy has begun to decline, but part of that decline is offset by immigration.

Population Density and Spatial Distribution

Because of such factors as soil type, quality of habitat, and availability of water, organisms normally are distributed unevenly. Some populations have many individuals clustered into a small

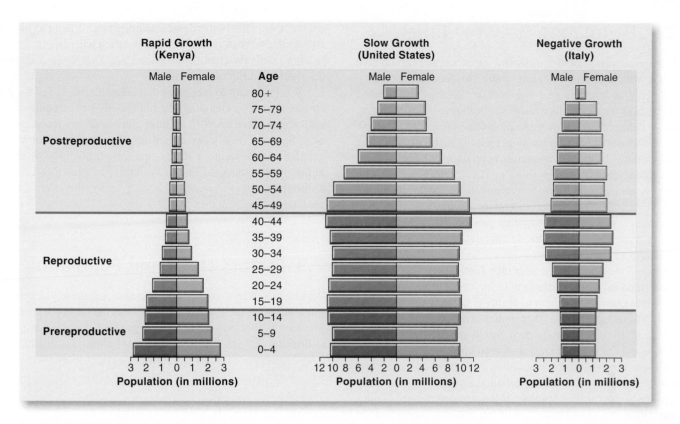

FIGURE 7.4 Age Distribution in Human Populations The relative numbers of individuals in each of the three categories (prereproductive, reproductive, and postreproductive) are good clues to the future growth of a population. Kenya has a large number of young individuals who will become reproducing adults. Therefore, this population is likely to grow rapidly. The United States has a large proportion of reproductive individuals and a moderate number of prereproductive individuals. Therefore, the population is likely to grow slowly. Italy has a declining number of reproductive individuals and a very small number of prereproductive individuals. Therefore, its population has begun to decline.

Source: Data from United States Census Bureau International Data Base.

space, while other populations of the same species may be widely dispersed. **Population density** is the number of organisms per unit area. For example, fruitfly populations are very dense around a source of rotting fruit, while they are rare in other places. Similarly, humans are often clustered into dense concentrations we call cities, with lower densities in rural areas.

When the population density is too great, all individuals within the population are injured because they compete severely with each other for necessary resources. Plants may compete for water, soil nutrients, or sunlight. Animals may compete for food, shelter, or nesting sites. In animal populations, overcrowding might cause some individuals to explore and migrate to new areas. This movement from densely populated locations to new areas is called **dispersal.** It relieves the overcrowded conditions in the home area and, at the same time, increases the population in the places to which they migrate. Often, it is juvenile individuals that relieve overcrowding by leaving. The pressure to migrate from a population **(emigration)** may be a result of seasonal reproduction leading to a rapid increase in population size or environmental changes that intensify competition among members of the same species. For example, as water holes dry up, competition for water increases, and many desert birds migrate to areas where water is still available.

The organisms that leave one population often become members of a different population. This migration into an area **(immigration)** may introduce characteristics that were not in the population originally. When Europeans immigrated to North America, they brought genetic and cultural characteristics that had a tremendous impact on the existing Native American population. Among other things, Europeans brought diseases that were foreign to the Native Americans. These diseases increased the death rate and lowered the birth rate of Native Americans, resulting in a sharp decrease in the size of their populations.

Summary of Factors that Influence Population Growth Rates

Populations have an inherent tendency to increase in size. However, as we have just seen, many factors influence the rate at which a population can grow. At the simplest level, the rate of increase is determined by subtracting the number of individuals leaving the population from the number entering. Individuals leave the population either by death or emigration; they enter the population by birth or immigration. Birth rates and death rates are influenced by several factors, including the number of females in the population and their age. In addition, the density of a population may encourage individuals to leave because of intense competition for a limited supply of resources.

7.2 A Population Growth Curve

Each species has a **biotic potential** or inherent reproductive capacity, which is its biological ability to produce offspring. Reproducing individuals of some species, such as watermelon plants or moths, may produce hundreds or thousands of offspring (seeds or caterpillars) per year, while others, such as geese, typically produce about 5 young per year. (See figure 7.5.) Some large animals, such as bears or elephants, may produce one young every two to three years. Although there are large differences among species, generally, adults produce many more offspring during their lifetimes than are needed to replace themselves when they die. However, among organisms that produce large numbers of offspring, most of the young die, so only a few survive to become reproductive adults themselves.

Because most species have a high biotic potential, there is a natural tendency for populations to increase. If we consider a hypothetical situation in which mortality is not a factor, we could have the following situation. If two mice produced four offspring and they all lived, eventually they would produce offspring of their own, while their parents continued to reproduce as well. Under these conditions, the population will grow exponentially. Exponential growth results in a population increasing by the same percentage each year. For example, if the population were to double each year, we would have 2, 4, 8, 16, 32, etc., individuals in the population. While populations cannot grow exponentially forever, they often have an exponential period of growth.

Population growth often follows a particular pattern, consisting of a lag phase, an exponential growth phase, a deceleration phase, and a stable equilibrium phase. Figure 7.6 shows a typical population growth curve. During the first portion of the curve, known as the **lag phase,** the population grows very slowly because there are few births, since the process of reproduction and growth of offspring takes time. Organisms must mature into adults before they can reproduce. When the offspring have matured and begin to mate and have young, the parents may be producing a second set of offspring. Since more organisms now are reproducing, the population begins to increase at an accelerating rate. This stage is known as the **exponential growth phase (log phase).** The population will continue to grow as long as the birth rate exceeds the death rate. Eventually, however, the population growth rate will begin to slow as the death rate and the birth rate come to equal

one another. This is the **deceleration phase.** When the birth rate and death rate become equal, the population will stop growing and reach a relatively stable population size. This stage is known as the **stable equilibrium phase.**

It is important to recognize that although the size of the population may not be changing, the individuals are changing. As new individuals enter by birth or immigration, others leave by death or emigration. For most organisms, the first indication that a population is entering a stable equilibrium phase is an increase in the death rate. A decline in the birth rate may also contribute to the stabilizing of population size. Usually, this occurs after an increase in the death rate.

7.3 Factors That Limit Population Size

Populations cannot continue to increase indefinitely. Eventually, some factor or set of factors acts to limit the size of a population. The factors that prevent unlimited population growth are known as **limiting factors.** All of the different limiting factors that act on a population are collectively known as **environmental resistance.**

Extrinsic and Intrinsic Limiting Factors

Some factors that control populations come from outside the population and are known as **extrinsic limiting factors.** Predators, loss of a food source, lack of sunlight, or accidents of nature are all extrinsic factors. However, the populations of many kinds of organisms appear to be regulated by factors from within the populations themselves. Such limiting factors are called **intrinsic limiting factors.** For example, a study of rats under crowded living conditions showed that as conditions became more crowded, abnormal social behavior became common. There was a decrease in litter size, fewer litters per year were produced, mothers were more likely to ignore their young, and adults killed many young. Thus changes in the behavior of the members of the rat population itself resulted in lower birth rates and higher death rates that limited population size. Among populations of white-tailed deer, it is well known that reproductive success is reduced when the deer experience a series of severe winters. When times are bad, the female deer are more likely to have single offspring than twins.

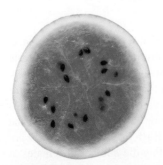
Watermelon offspring (seeds)

Moth offspring (caterpillars)

Geese

FIGURE 7.5 Biotic Potential The ability of a species to reproduce greatly exceeds the number necessary to replace those who die. Here are some examples of the prodigious reproductive abilities of some species.

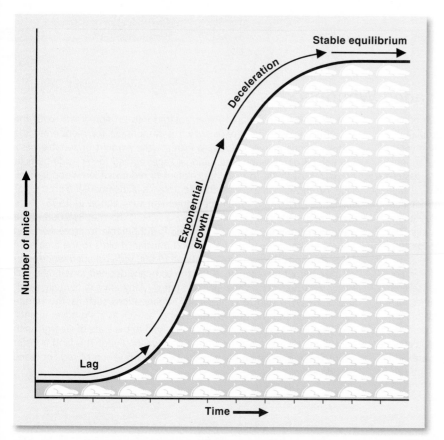

FIGURE 7.6 A Typical Population Growth Curve In this mouse population, there is little growth during the lag phase. During the exponential growth phase, the population rises rapidly as increasing numbers of individuals reach reproductive age. Eventually, the population growth rate begins to slow during the deceleration phase and the population reaches a stable equilibrium phase, during which the birth rate equals the death rate.

Density-Dependent and Density-Independent Limiting Factors

Density-dependent limiting factors are those that become more effective as the density of the population increases. For example, the larger a population becomes, the more likely it is that predators will have a chance to catch some of the individuals. A prolonged period of increasing population allows the size of the predator population to increase as well. Disease epidemics are also more common in large, dense populations because dense populations allow for the easy spread of parasites from one individual to another. The rat example discussed previously is another good example of a density-dependent limiting factor because the amount of abnormal behavior increased as the density of the population increased. In general, whenever there is competition among members of a population, its intensity increases as the population density increases. Large organisms that tend to live a long time and have relatively few young are most likely to be controlled by density-dependent limiting factors.

Density-independent limiting factors are population-controlling influences that are not related to the density of the population. They are usually accidental or occasional extrinsic factors in nature that happen regardless of the density of a population. A sudden rainstorm may drown many small plant seedlings and soil organisms. Many plants and animals are killed by frosts that come late in spring or early in the fall. A small pond may dry up, resulting in the death of many organisms. The organisms most likely to be controlled by density-independent limiting factors are small, short-lived organisms that can reproduce very rapidly.

7.4 Categories of Limiting Factors

For most populations, limiting factors recognized as components of environmental resistance can be placed into four broad categories: (1) the availability of raw materials, (2) the availability of energy, (3) the accumulation of waste products, and (4) interactions among organisms.

Availability of Raw Materials

Raw materials come in many forms. For example, plants need nitrogen and magnesium from the soil as raw materials for the manufacture of chlorophyll. If these minerals are not present in sufficient quantities, the plant population cannot increase. The application of fertilizers is a way of preventing certain raw materials from being a limiting factor in agriculture. When one limiting factor is removed, some new primary limiting factor will emerge. Perhaps it will be the amount of water, the number of insects that feed on the plants, or competition for sunlight. Animals also require certain minerals as raw material, which they obtain in their diets. They may also require objects with which to build nests, to provide places for escape, or to serve as observation sites.

Availability of Energy

Energy sources are important to all organisms. Plants require energy in the form of sunlight for photosynthesis, so the amount of light can be a limiting factor for many plants. When small plants are in the shade of trees, they often do not grow well and have small populations because they do not receive enough sunlight. Animals require energy in the form of the food they eat, and if food is scarce, many die.

Accumulation of Waste Products

The accumulation of waste products is not normally a limiting factor for plants, since they produce few wastes, but it can be for other kinds of organisms. Bacteria, other tiny organisms, and many kinds of aquatic organisms that live in small ecosystems such as puddles, pools, or aquariums may be limited by wastes.

Going Green

Increasing Populations of Red-Cockaded Woodpeckers

The red-cockaded woodpecker (*Picoides borealis*) is listed as an endangered species. This medium-sized bird (about the size of a cardinal) is a cooperative colony nester—the dominant male and female raise young with the support of nonbreeding members of the colony. They are only found in the southeastern United States—southern Virginia to eastern Texas—where native southern yellow pine forests occur. Several pine species, including slash pine, shortleaf pine, loblolly pine, and longleaf pine, are typical of this region. The original forests were fire-adapted in that mature trees were able to withstand moderate ground fires. This resulted in a rather open forest type. The woodpeckers typically construct their nesting cavities in older, diseased longleaf pine trees.

The trees these birds use for nesting are also commercially important. Thus, the amount of suitable breeding habitat has been severely reduced as older trees are harvested and natural stands of pines have been replaced with plantations, where large tracts are planted to a single species and the trees are harvested before they reach old age. Since much of the suitable habitat is privately owned, protecting populations of red-cockaded woodpeckers requires the cooperation of private landowners, conservation organizations, state and federal governments, and commercial forest products companies.

In 1998, International Paper entered into an agreement with the U.S Fish and Wildlife Service, which is responsible for monitoring the status of endangered species, to increase the amount of suitable nesting habitat on its lands. International Paper agreed to set aside particular parcels of forest to maintain colonies of red-cockaded woodpeckers. One of those parcels was the Southlands Experimental Forest near Bainbridge, Georgia. When the agreement was signed in 1998, there were three male red-cockaded woodpeckers at the site. By 2008, there were over 50 individuals. The increase is attributable to protection and improvement of the birds' habitat and transfer of birds to the area from other locations. Today there are about 15,000 red-cockaded woodpeckers throughout its range. In 2006, the company decided to sell nearly all of its land holdings in the United States. Many environmentally sensitive lands were sold to conservation organizations such as The Nature Conservancy and the Conservation Fund, as well as state governments. The Southlands Experimental Forest was sold to the state of Georgia with some funding assistance from the Conservation Fund. This land transfer protects the population gains made by this population of red-cockaded woodpeckers.

Red-cockaded woodpecker habitat

Red-cockaded woodpecker

When a small number of a species of bacterium is placed on a petri plate with nutrient agar (a jellylike material containing food substances), the population growth follows a curve shown in figure 7.7. As expected, it begins with a lag phase, continues through an exponential growth phase, and eventually levels off in a stable equilibrium phase. However, in this small, enclosed space, there is no way to get rid of the toxic waste products, which accumulate, eventually killing the bacteria. This decline in population size is known as the **death phase.**

Interactions Among Organisms

Interactions among organisms are also important in determining population size. For example, white-tailed deer and cottontail rabbits eat the twigs of many species of small trees and shrubs. Repeated browsing by herbivores

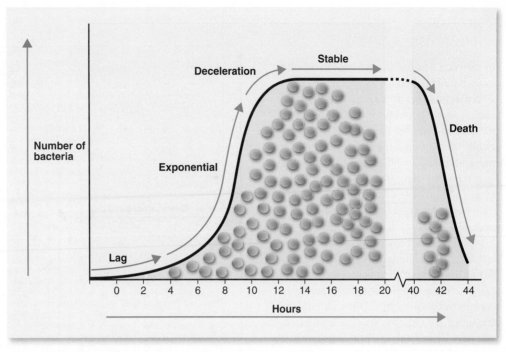

FIGURE 7.7 A Bacterial Growth Curve with a Death Phase The initial change in population size follows a typical population growth curve until waste products become lethal. When a population begins to decline, it enters the death phase.

retards growth and can cause death of trees and shrubs. Thus, damage by herbivores can limit the size of some tree and shrub populations. Many single-celled aquatic organisms produce waste products that build up to toxic levels and result in the death of fish. Parasites and predators weaken or cause the premature death of individuals, thus limiting the size of the population.

Some studies indicate that populations can be controlled by interaction among individuals within the population. A study of laboratory rats shows that crowding causes a breakdown in normal social behavior, which leads to fewer births and increased deaths. The changes observed include abnormal mating behavior, decreased litter size, fewer litters per year, lack of maternal care, and increased aggression in some rats or withdrawal in others. Thus, limiting factors can reduce birth rates as well as increase death rates. Many other kinds of animals have shown similar reductions in breeding success when population densities were high.

7.5 Carrying Capacity

The populations of many organisms are at their maximum size when they reach the stable equilibrium phase. This suggests that the environment sets an upper limit to the size of the population. Ecologists have developed a concept for this observation, called the *carrying capacity.* **Carrying capacity** is the maximum sustainable population for a species in an area. The carrying capacity is determined by a set of limiting factors. (See figure 7.8.)

Carrying capacity is not an inflexible number, however. Often such environmental differences as successional changes, climate variations, disease epidemics, forest fires, or floods can change the carrying capacity of an area for specific species. In aquatic ecosystems one of the major factors that determine the carrying capacity is the amount of nutrients in the water. In areas where nutrients are abundant, the numbers of various kinds of organisms are high. Often nutrient levels fluctuate with changes in current or runoff from the land, and plant and animal populations fluctuate as well. In addition, a change that negatively affects the carrying capacity for one species may increase the carrying capacity for another. For example, the cutting down of a mature forest followed by the growth of young trees increases the carrying capacity for deer and rabbits, which use the new growth for food, but decreases the carrying capacity for squirrels, which need mature, fruit-producing trees as a source of food and old, hollow trees for shelter.

Wildlife management practices often encourage modifications to the environment that will increase the carrying capacity for the designated game species. The goal of wildlife managers is to have the highest sustainable population available for harvest by hunters. Typical habitat modifications include creating water holes, cutting forests to provide young growth, planting food plots, and building artificial nesting sites.

7.6 Reproductive Strategies and Population Fluctuations

So far, we have talked about population growth as if all organisms reach a stable population when they reach the carrying capacity. That is an appropriate way to begin to understand population changes, but the real world is much more complicated.

K-Strategists and r-Strategists

Species can be divided into two broad categories based on their reproductive strategies. **K-strategists** are organisms that typically reach a stable population as the population reaches the carrying capacity. K-strategists usually occupy relatively stable environments and tend to be large organisms that have relatively long lives, produce few offspring, and provide care for their offspring. Their reproductive strategy is to invest a great deal of energy in producing a few offspring that have a good chance of living to reproduce. Deer, lions, and swans are examples of this kind of organism. Humans generally produce single offspring, and even in countries with high infant mortality, 80 percent of the children survive beyond one year of age, and the majority of these will reach adulthood. Generally, populations of K-strategists are controlled by density-dependent limiting factors that become more severe as the size of the population increases. For example, as the size of the hawk population increases, the competition among hawks for available food becomes more severe. The increased competition for food is a density-dependent limiting factor that leads to less food for the young in the nest. Therefore, many of the young die, and the population growth rate slows as the carrying capacity for the area is reached.

The **r-strategists** are typically small organisms that have a short life, produce many offspring, exploit unstable environments, and do not reach a carrying capacity. Examples are bacteria, protozoa, many insects, and some small mammals. The reproductive strategy of r-strategists is to expend large amounts of energy producing many offspring but to provide limited care (often none) for them. Consequently, there is high mortality among the young. For example, one female oyster may produce a million eggs, but of those that become fertilized and grow into larvae, only a few find suitable places to attach themselves and grow into mature oysters. Typically, populations of r-strategists are limited by density-independent limiting factors. These factors can include changing weather conditions that kill large numbers of organisms, habitat loss such as occurs when a pond dries up or fire destroys a forest, or an event such as a deep snow or flood that buries sources of food and leads to the death of entire populations. The population size of r-strategists is likely to fluctuate wildly. They reproduce rapidly, and the size of the population increases until some density-independent factor causes the population to crash; then they begin the cycle all over again. (See figure 7.9.)

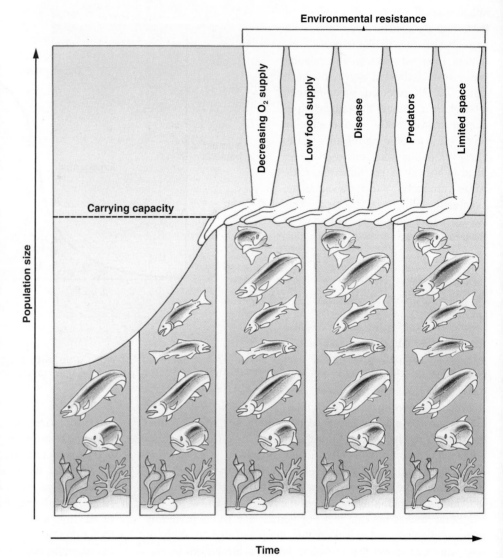

FIGURE 7.8 Carrying Capacity A number of factors in the environment, such as oxygen supply, food supply, diseases, predators, and space, determine the number of organisms that can survive in a given area—the carrying capacity of that area. The environmental factors that limit populations are known collectively as environmental resistance.

The concepts of K- or r-strategists describe idealized situations. (See table 7.1.) (The letters K and r in *K-strategists* and *r-strategists* come from a mathematical equation in which K represents the carrying capacity of the environment and r represents the biotic potential of the species.) In the real world many organisms don't fit clearly into either category. For example, many kinds of mammals provide care for their offspring but have short life spans. On the other hand, many reptiles such as turtles may live for many years, but they produce large numbers of eggs and do not care for them.

Since humans are K-strategists, it may be difficult for us to appreciate that the r-strategy can be viable from an evolutionary point of view. Resources that are present only for a short time can be exploited most effectively if many individuals of one species monopolize the resource, while denying other species access to it. Rapid reproduction can place a species in a position to compete against other species that are not able to increase numbers as rapidly. Obviously, most of the individuals will die, but not before they have left some offspring or

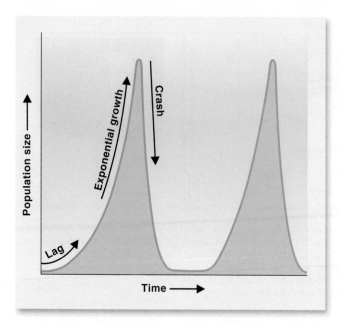

FIGURE 7.9 A Population Growth Curve for Short-Lived Organisms Organisms that are small and only live a short time often have the kind of population growth curve shown here. There is a lag phase followed by an exponential growth phase. However, instead of entering into a stable equilibrium phase, some density-independent limiting factor causes the population to crash. Then the population begins to increase again.

resistant stages that will be capable of exploiting the resource should it become available again.

Even K-strategists, however, have some fluctuations in their population size for a variety of reasons. One reason is that even in relatively stable ecosystems, there will be variations from year to year. Floods, droughts, fires, extreme cold, and similar events may affect the carrying capacity of an area, thus causing fluctuations in population size. Epidemic disease or increased predation may also lead to populations that vary in size from year to year. Many endangered species have reduced populations because their normal environment has been altered either naturally or as a result of human activity. (See chapter 11 on biodiversity issues.)

One idea is that heavy feeding by large populations of herbivores causes the plants to produce increased amounts of chemicals, which taste bad or are toxic. A second thought is that when an herbivore population is large, many different predators shift to eating them, causing the herbivore population to crash. Another idea is that interactions between a prey organism and a specialized predator naturally lead to population cycles. The length of the population cycle depends on the reproductive biology of the prey and their predators.

A study of the population biology of the collared lemming (*Dicrostonyx groenlandicus*) on Greenland illustrates the population interactions between lemmings and four different predators. Lemmings have a very high biotic potential. They produce two to three litters of young per year. Their population is held in check by four different predators. Three of these predators—the snowy owl, arctic fox, and longtailed skua (a bird that resembles a gull)—are generalist predators whose consumption of lemmings is directly related to the size of the lemming population. They constitute a density-dependent limiting factor for the lemming population. When lemming numbers are low, these predators seek other prey. For example, the snowy owl often migrates to other regions when lemming numbers are low in a particular region. The fourth predator is the shorttail weasel (*Mustela ermina*), a specialist predator on lemmings. The weasels are much more dependent on lemmings for food than the other predators. Since the weasels mate once a year, their populations increase at a slower rate than those of the lemmings. As weasel populations increase, however, they eventually become large enough that they drive down the lemming populations. The resulting decrease in lemmings leads to a decline in the number of weasels, which allows greater survival of lemmings, which ultimately leads to another cycle of increased weasel numbers. (See figure 7.10.)

7.7 Human Population Growth

The human population growth curve has a long lag phase followed by a sharply rising exponential growth phase that has only recently shown signs of slowing. (See figure 7.11.) A major reason for the

Population Cycles

In northern regions of the world, many kinds of animals show distinct population cycles—periods of relatively large populations followed by periods of small populations. This is generally thought to be the case because the ecosystems are relatively simple, with few kinds of organisms affecting one another. Many of these cycles are quite regular. Biologists have been studying these cycles since the 1920s, and they have developed several theories about why northern populations cycle.

Table 7.1 A Comparison of Life History Characteristics of Typical K- and r-Strategists

Characteristic	K-Strategist	r-Strategist
Environmental stability	Stable	Unstable
Size of organism	Large	Small
Length of life	Long, most live to reproduce	Short, most die before reproducing
Number of offspring	Small number produced, parental care provided	Large number produced, no parental care
Primary limiting factors	Density-dependent limiting factors	Density-independent limiting factors
Population growth pattern	Exponential growth followed by a stable equilibrium stage at the carrying capacity	Exponential growth followed by a population crash
Examples	Alligators, humans, redwood trees	Protozoa, mosquitoes, annual plants

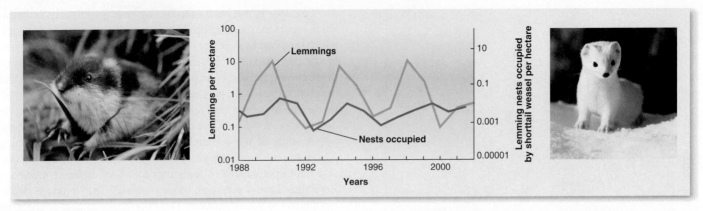

FIGURE 7.10 Population Cycles In many northern regions of the world, population cycles are common. In the case of collared lemmings and shorttail weasels on Greenland, interactions between the two populations result in population cycles of about four years. The graph shows the population of lemmings per hectare and the number of lemming nests occupied by weasels per hectare. A hectare is 10,000 square meters—about 2.47 acres. The researchers used the number of lemming nests occupied by weasels as an indirect measure of the size of the weasel population because of difficulties in measuring the size of the weasel population by other means. Notice that, in general, the population size of the weasels lags behind the population size of the lemmings.

Source: Data from O. Gilg, I. Hamski, and B. Sittler, "Cyclic Dynamics in a Simple Predator-Prey Community," *Transpol'air* Online Magazine, www.transpolair.com.

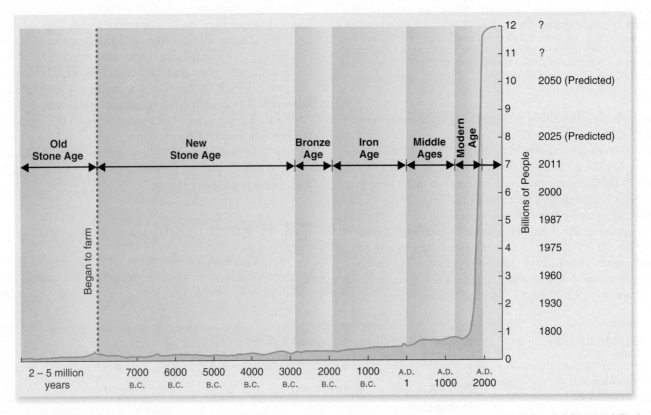

FIGURE 7.11 The Historical Human Population Curve From A.D. 1800 to A.D. 1930, the number of humans doubled (from 1 billion to 2 billion) and then doubled again by 1975 (4 billion) and is projected to double again (8 billion) by the year 2025. How long can this pattern continue before the Earth's ultimate carrying capacity is reached?

Source: Data from Jean Van Det Tak, et al., "Our Population Predicament: A New Look," *Population Bulletin*, vol. 34, no. 5 (December 1979), Population Reference Bureau, Washington, D.C.: and more recent data from the Population Reference Bureau.

continuing increase in the size of the human population is that the human species has lowered its death rate. When various countries reduce environmental resistance by increasing food production or controlling disease, they share this technology throughout the world. Developed countries send health care personnel to all parts of the globe to improve the quality of life for people in less-developed countries. Physicians offer advice on nutrition, and engineers develop wastewater treatment systems. Improved sanitary facilities in India and Indonesia, for example, decreased deaths caused by cholera. These advancements tend to reduce death rates while birth rates remain high. Thus, the size of the human population increases rapidly.

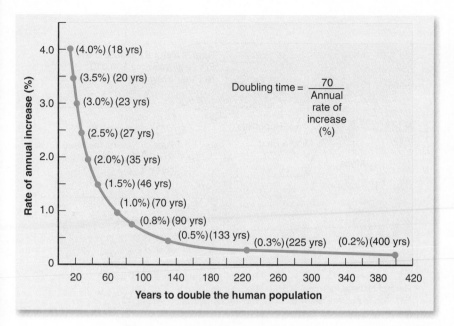

FIGURE 7.12 Doubling Time for the Human Population This graph shows the relationship between the rate of annual increase in percent and doubling time. A population growth rate of 1 percent per year would result in the doubling of the population in about 70 years. A population growth rate of 3 percent per year would result in a population doubling in about 23 years.

Let us examine the human population situation from a different perspective. The world population is currently increasing at an annual rate of 1.2 percent. That may not seem like much, but even at 1.2 percent, the population is growing rapidly. It can be difficult to comprehend the impact of a 1 or 2 percent annual increase. Remember that a growth rate in any population compounds itself, since many of the additional individuals eventually reproduce, thus adding more individuals. One way to look at this growth is to determine how much time is needed to double the population. This is a valuable method because most of us can appreciate what life would be like if the number of people in our locality were doubled, particularly if the doubling were to occur within our lifetime.

Figure 7.12 shows the relationship between the rate of annual increase for the human population and the number of years it would take to double the population if that rate were to continue. The doubling time for the human population is easily calculated by dividing the number 70 by the annual rate of increase. Thus, at a 1 percent rate of annual increase, the population will double in 70 years (70/1). At a 2 percent rate of annual increase, the human population will double in 35 years (70/2). The current worldwide annual increase of about 1.2 percent will double the world human population in about 58 years.

7.8 Human Population Characteristics and Implications

The human population dilemma is very complex. To appreciate it, we must understand current population characteristics and how they are related to social, political, and economic conditions.

Economic Development

The world can be divided into two segments based on the state of economic development of the countries. The **more-developed countries** of the world typically have a per capita income that exceeds US $25,000; they include all of Europe, Canada, the United States, Australia, New Zealand, and Japan, with a combined population of about 1.25 billion people. The remaining countries of the world are referred to as **less-developed countries** and typically have a per capita income of less than US $15,000. The population of these countries totals about 6 billion people, nearly 3 billion of whom live on less than US $2 per day. While these definitions constitute an oversimplification and some countries are exceptions, basically this means that the majority of Asian, Latin American, and African citizens are much less well off economically than those who live in the more-developed countries. Collectively, the more-developed countries of the world have relatively stable populations and are expected to grow by less than 5 percent between 2014 and 2050. The less-developed regions of the world, however, have high population growth rates and are expected to grow by about 40 percent between 2014 and 2050. (See figure 7.13.) If these trends continue, the total population of the less-developed world will increase from the current 6 billion to about 8.4 billion by 2050, when this region will contain over 86 percent of the world's people.

Measuring the Environmental Impact of a Population

Human population growth is tied to economic development and is a contributing factor to nearly all environmental problems. Current population growth has led to

1. famine in areas where food production cannot keep pace with increasing numbers of people;
2. political unrest in areas with great disparities in the availability of resources (jobs, goods, food);
3. environmental degradation (erosion, desertification, strip mining, oil spills, groundwater mining) caused by poor agricultural practices and the destructive effects of exploitation of natural resources;
4. water pollution caused by human and industrial waste;
5. air pollution caused by the human need to use energy for personal and industrial applications; and
6. extinctions caused by people converting natural ecosystems to managed agricultural ecosystems.

Several factors interact to determine the impact of a society on the resources of its country. These include the land and other natural resources available, the size of the population, the amount of resources consumed per person, and the environmental damage caused by using resources. The following equation is often used as a

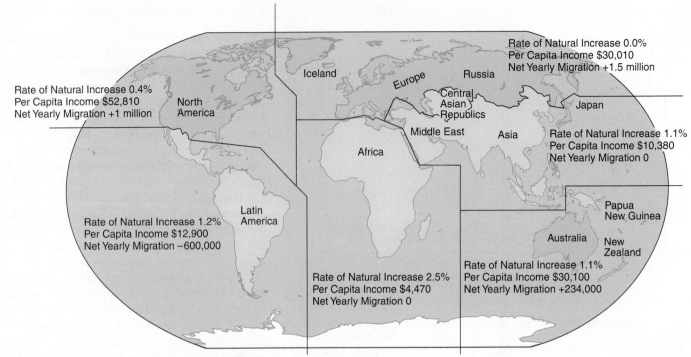

FIGURE 7.13 Population Growth and Economic Development (2014) The population of the world can be divided into the economically more-developed and less-developed nations. The more-developed nations are indicated in green and the less-developed in tan. Currently, about 82.5 percent of the world's population is in the less-developed nations of Latin America, Africa, and Asia. These areas also have the highest rates of population increase. Because of the high birth rates, they are likely to remain less developed and are projected to constitute about 86 percent of the world's population by the year 2050. In general, there is a net migration of people to the developed nations of the world from less-developed nations.
Data from: Population Reference Bureau, 2014 Population Data Sheet.

shorthand way of stating these relationships: $I = P \times A \times T$ (*Impact on the environment = Population* size $\times$ *Affluence* (amount of resources consumed per person) $\times$ *Technology* (effects of methods used to provide items consumed).

Population

As the population of a country increases, it puts a greater demand on its resources. Some countries have abundant natural resources, such as good agricultural land, energy resources, or mineral resources. Others are resource poor. Thus, some countries can sustain high populations while others cannot.

Population density, the number of people per unit of land area, relates the size of the population to the resources available. A million people spread out over the huge area of the Amazon Basin have much less impact on resources than that same million people in a small island country because the impact is distributed over a greater land surface. Countries with abundant resources can sustain higher population densities than resource-poor countries.

Affluence

People in highly developed countries consume huge amounts of resources. Citizens of these countries eat more food, particularly animal protein, which requires larger agricultural inputs than does a vegetarian diet. They have more material possessions and

consume vast amounts of energy compared to people in the less-developed world. (See figure 7.14.)

Technology

The technology used to provide the things people consume and use is an important contributor to environmental impact. Some methods are efficient and have minor impacts; others are very damaging. For example, the use of firewood to heat homes and provide fuel for cooking can lead to deforestation. Similarly, the use of inefficient, polluting, coal-fired power plants has a high environmental impact. More efficient power plants or the use of wind or solar energy to provide energy lowers environmental impact.

The affluence and technology components of this equation are very difficult to tease apart. Often the per-capita energy consumption is used as a measure of the combined effects of affluence and technology.

The Ecological Footprint Concept

The environmental impact of the developed world is often underestimated because the population in these countries is relatively stable and local environmental conditions are good. However, developed countries purchase goods and services from other parts of the world, often degrading environmental conditions

Possessions of a family from Ahraura Village, India

Possessions of a family from Cologne, Germany

FIGURE 7.14 Differences in Affluence These two families have greatly different numbers of possessions. Notice the great difference in the number of persons in the family, the kinds of possessions they have, and the kinds of energy-consuming items present.

in less-developed countries. Thus, the environmental impact of highly developed regions such as North America, Japan, Australia, New Zealand, and Europe is often felt in distant places, while the impact on their own resources may be minimal. For example, deforestation may occur in Indonesia to grow oil palm plants. The palm oil extracted from the plants is used in many food products sold in the United States. So the quantity of palm oil used in products sold in the U.S. contributes to deforestation in Indonesia.

This has led to the development of the concept of the ecological footprint of a society. The **ecological footprint** is a measure of the land area required to provide the resources and absorb the wastes of a population. Most of the more-developed countries of the world have a much larger ecological footprint than is represented by their land area. For example, Japan has a highly developed economy but few resources. Thus, it must import most of the materials it needs. One study calculated that the ecological impact of Japan is nearly five times larger than its locally available resources. The same study estimated that the ecological footprint of the United States is 1.5 times locally available resources.

It is clear that as the world human population continues to increase, it will become more difficult to limit the environmental degradation that accompanies it. Since much of the population growth will occur in the less-developed areas of the world that have weak economies, the money to invest in pollution control, health programs, and sustainable agricultural practices will not be present.

While controlling world population growth would not eliminate all environmental problems, it could reduce the rate at which environmental degradation is occurring. It is also generally believed that the quality of life for many people in the world would improve if their populations grew less rapidly. Why, then, does the human population continue to grow at such a rapid rate?

7.9 Factors That Influence Human Population Growth

Human populations are subject to the same biological factors discussed earlier in this chapter. There is an ultimate carrying capacity for the human population. Eventually, limiting factors will cause human populations to stabilize. However, unlike other kinds of organisms, humans are also influenced by social, political, economic, and ethical factors. We have accumulated knowledge that allows us to predict the future. We can make conscious decisions based on the likely course of events and adjust our lives accordingly. Part of our knowledge is the certainty that as populations continue to increase, death rates and birth rates will become equal. This can happen by allowing the death rate to rise or by choosing to limit the birth rate. Controlling human population would seem to be a simple process. Once people understand that lowering the birth rate is more humane than allowing the death rate to rise, they should make the "correct" decision and control their birth rates; however, it is not quite that simple.

Focus On

Thomas Malthus and His Essay on Population

In 1798, Thomas Robert Malthus, an Englishman, published an essay on human population. In it, he presented an idea that was contrary to popular opinion. His basic thesis was that human population increased in a geometric or exponential manner (2, 4, 8, 16, 32, 64, etc.), while the ability to produce food increased only in an arithmetic manner (1, 2, 3, 4, 5, 6, etc.). The ultimate outcome of these two different rates would be that population would outgrow the ability of the land to produce food. He concluded that wars, famines, plagues, and natural disasters would be the means of controlling the size of the human population. His predictions were hotly debated by the intellectual community of his day. His assumptions and conclusions were attacked as erroneous and against the best interest of society. At the time he wrote the essay, the popular opinion was that human knowledge and "moral constraint" would be able to create a world that would supply all human needs in abundance. One of Malthus's basic postulates was that "commerce between the sexes" (sexual intercourse) would continue unchanged, while other philosophers

Thomas Robert Malthus

of the day believed that sexual behavior would take less procreative forms and human population would be limited. Only within the past 50 years, however, have really effective conception-control mechanisms become widely accepted and used, and they are used primarily in developed countries.

Malthus did not foresee the use of contraception, major changes in agricultural production techniques, or the exporting of excess people to colonies in the Americas, Australia, and other parts of the world. These factors, as well as high death rates, prevented the most devastating of his predictions from coming true. However, in many parts of the world today, people are experiencing the forms of population control (famine, epidemic disease, wars, and natural disasters) predicted by Malthus in 1798. Many people feel that his original predictions were valid—only his timescale was not correct—and that we are seeing his predictions come true today.

Biological Factors

The scientific study of human populations, their characteristics, how these characteristics affect growth, and the consequences of that growth is known as **demography.** Demographers can predict the future growth of a population by looking at several biological indicators. Table 7.2 shows several population characteristics for the ten most populous countries in the world. These countries represent different levels of economic development and different regions of the world.

Birth Rate and Death Rate

If the birth rate exceeds the death rate, the population will grow. In general the greater the differences between birth rate and death rate, the faster the population will grow. Nigeria and Pakistan have birth rates that greatly exceed the death rate and have very rapidly growing populations. At the other extreme Japan, Russia, and many countries of the former Soviet Union have low birth rates and death rates that equal or exceed the birth rate. Their populations are declining. Most countries in Europe have stable or declining populations. In general the regions of Latin America, Asia, and Africa have high birth rates and low death rates and rapidly growing populations.

Total Fertility Rate

The most important determinant of the rate at which human populations grow is related to how many women in the population

are having children and the number of children each woman will have. The **total fertility rate** of a population is the number of children born per woman in her lifetime. A total fertility rate of 2.1 is known as **replacement fertility,** since parents produce 2 children who will replace the parents when they die. Eventually, if the total fertility rate is maintained at 2.1, population growth will stabilize. A rate of 2.1 is used rather than 2.0 because some children do not live very long after birth and therefore will not contribute to the population for very long. When a population is not growing, and the number of births equals the number of deaths, it is said to exhibit **zero population growth.**

For several reasons, however, a total fertility rate of 2.1 will not necessarily immediately result in a stable population with zero growth. First, the death rate may fall as living conditions improve and people live longer. If the death rate falls faster than the birth rate, there will still be an increase in the population even though it is reproducing at the replacement rate. Note in Table 7.2 that those countries with high total fertility rates have high population growth rates.

Age Distribution

The **age distribution,** the number of people of each age in the population, also has a great deal to do with the rate of population growth. If a population has many young people who are raising families or who will be raising families in the near future, the population will continue to increase even if the families limit themselves to two children. Depending on the number of young people

Table 7.2 Population Characteristics of the Ten Most Populous Countries, 2014

Country	Current Population (Millions)	Births per 1,000 Individuals	Deaths per 1,000 Individuals	Total Fertility Rate (Children per Woman per Lifetime)	Rate of Natural Increase (Annual %)	% Population Less Than 15 Years Old	Migration Rate (Persons per Year)	Projected Population Change 2014–30 (%)
World	7,238	20	8	2.5	1.2%	26%		+ 16.7%
Japan	127.1	8	10	1.4	(−0.2%)	13%	+127,300	(−8.3%)
Russia	143.7	13	13	1.7	0	16%	+287,400	0
China	1,364.1	12	7	1.6	0.5%	16%	0	+2.6%
United States	317.7	13	8	1.9	0.4%	19%	+953,100	+11.6%
Brazil	202.8	15	6	1.8	0.9%	24%	0	+10%
Indonesia	251.5	20	6	2.6	1.4%	29%	(−251,500)	+22%
India	1,296.3	22	7	2.4	1.5%	31%	0	+16.5%
Bangladesh	158.5	20	6	2.2	1.5%	29%	(−475,500)	+16.8%
Pakistan	194.0	28	8	3.8	2.0%	38%	(−582,000)	+31.3%
Nigeria	177.5	39	13	6.5	2.5%	44%	0	+47.4%

Source: Population Reference Bureau, 2014 Population Data Sheet.

in a population, it may take 20 years to a century for the population of a country to stabilize so that there is no net growth even if they limit their family size to two children. Note in table 7.2 that those countries with the largest number of young people have the highest projected population growth rates.

Social Factors

It is clear that populations in economically developed countries of the world have low fertility rates and low rates of population growth and that the less-developed countries have high fertility rates and high population growth rates. It also appears obvious that reducing fertility rates would be to everyone's advantage; however, not everyone in the world feels that way. Several factors influence the number of children a couple would like to have. Some of these factors are religious, some are traditional, some are social, and some are economic.

Culture and Traditions

A major social factor that determines family size is the status of women in the culture. In many male-dominated cultures, the traditional role of women is to marry and raise children. Often this role is coupled with strong religious input as well. Typically, little value is placed on educating women, and early marriage is encouraged. In these cultures, women are totally dependent on their husbands or their grown children to provide for them in old age. Because early marriage is encouraged, fertility rates are high, since women are exposed to the probability of pregnancy for more of their fertile years. (See figure 7.15.)

Lack of education reduces options for women in these cultures. Because they lack an education, they have reduced chances for jobs, find it difficult to delay marriage, and typically have

many children. By contrast, in much of the developed world, women are educated, delay marriage, and have fewer children. It has been said that the single most important activity needed to reduce the world population growth rate is to educate women. Whenever the educational level of women increases, fertility rates fall. Figure 7.16 compares total fertility rates and educational levels of women in the 10 most populous countries of the world. The educational level of women is strongly correlated with the total fertility rate and economic well-being of a population.

Even childrearing practices have an influence on population growth rates. In countries where breast-feeding is practiced, several benefits accrue. Breast milk is an excellent source of nutrients for the infant as well as a source of antibodies against some diseases. Furthermore, since many women do not return to a normal reproductive cycle until after they have stopped nursing, during the months a woman is breast-feeding her child, she is less likely to become pregnant again. Since in many cultures breast-feeding may continue for one to two years, it serves to increase the time between successive births. Increased time between births results in a lower mortality among women of childbearing age.

Attitudes Toward Birth Control

As women become better educated and obtain higher-paying jobs, they become financially independent and can afford to marry later and consequently have fewer children. Better-educated women are also more likely to have access to and use birth control. In economically advanced countries, a high proportion (over 70 percent) of women typically use contraception. In many of the less-developed countries, contraceptive use is much lower—about 34 percent in Africa and about 57 percent in Asia if China's rate of 85 percent is excluded.

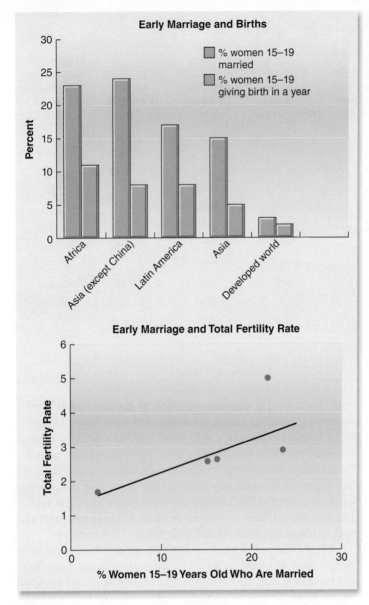

FIGURE 7.15 Early Marriage and Fertility Rate In cultures where early marriage is encouraged, women are exposed to the possibility of pregnancy for a longer part of their lives and total fertility rates are high.

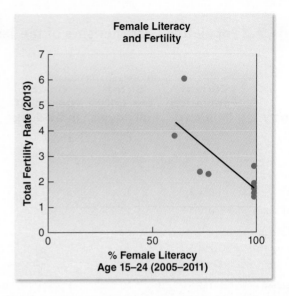

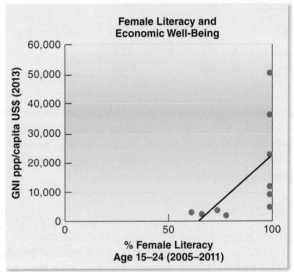

FIGURE 7.16 Female Literacy, Total Fertility Rate, and Economic Well-Being These graphs show how female literacy is related to the total fertility rate and per capita income. The data are from the ten most populous countries of the world. In general the higher the female literacy rate the lower the total fertility rate and the higher the per capita income.

Source: Data from Population Reference Bureau, 2013 Data Sheet, and *World Development Indicators 2013* from The World Bank.

It is important to recognize that access to birth control alone will not solve the population problem. What is most important is the desire of parents to limit the size of their families. In developed countries, the use of birth control is extremely important in regulating the birth rate. This is true regardless of religion and previous historical birth rates. For example, Italy and Spain are both traditionally Catholic countries and have low total fertility rates of 1.4 and 1.3, respectively. The average for the developed countries of the world is 1.6. Obviously, women in these countries make use of birth control to help them regulate the size of their families. By contrast, Mexico, which is also a traditionally Catholic country, has a total fertility rate of 2.2, which is somewhat less than 2.6, which is typical for the less-developed world regardless of religious tradition.

Women in the less-developed world typically have more children than they think is ideal, and the number of children they have is higher than the replacement fertility rate of 2.1 children. Access to birth control will allow them to limit the number of children they have to their desired number and to space their children at more convenient intervals, but they still desire more children than the 2.1 needed for replacement. Why do they desire large families? There are several reasons. In areas where infant mortality is high, it is traditional to have large families, since several of a woman's children may die before they reach adulthood. This is particularly important in the less-developed world, where there is no government program of social security. Parents are more secure in old age if they have several children to contribute to their needs when they become elderly and can no longer work.

Economic Factors

In less-developed countries, the economic benefits of children are extremely important. Even young children can be given jobs that contribute to the family economy. They can protect livestock from predators, gather firewood, cook, or carry water. In the developed world, large numbers of children are an economic drain. They are prevented by law from working, they must be sent to school at great expense, and they consume large amounts of the family income. Many parents in the developed world make an economic decision about having children in the same way they buy a house or car: "We are not having children right away. We are going to wait until we are better off financially."

Political Factors

Two other factors that influence the population growth rate of a country are government policies on population growth and immigration.

Government Population Policy

Many economically developed countries have official policies that state their population growth rates are too low. As their populations age and there are few births, they are concerned about a lack of working-age people in the future and have instituted programs that are meant to encourage people to have children. For example, Hungary, Sweden, and several other European countries provide paid maternity leave for mothers during the early months of a child's life and the guarantee of a job when the mother returns to work. Many countries provide childcare facilities and other services that make it possible for both parents to work. This removes some of the economic barriers that tend to reduce the birth rate. The tax system in many countries, such as the United States, provides an indirect payment for children by allowing a deduction for each child. Canada pays a bonus to couples on the birth of a child.

By contrast, most countries in the developing world publicly state that their population growth rates are too high. To reduce the birth rate, they have programs that provide information on maternal and child health and on birth control. The provision of free or low-cost access to contraceptives is usually a part of their population-control effort as well.

China and India are the two most populous countries in the world, each with over a billion people. China has taken steps to control its population and now has a total fertility rate of 1.6 children per woman while India has a total fertility rate of 2.4. This difference between these two countries is the result of different policy decisions over the last 50 years. The history of China's population policy is an interesting study of how government policy affects reproductive activity among its citizens. When the People's Republic of China was established in 1949, the official policy of the government was to encourage births, because more Chinese would be able to produce more goods and services, and production was the key to economic prosperity. The population grew from 540 million to 614 million between 1949 and 1955. However, economic progress was slow. Consequently, the government changed its policy and began to promote population control.

Since 1955 China has had a series of family-planning programs all aimed at reducing the number of births. Although many believe these programs include human rights violations, the birth rate has fallen steadily, and the current total fertility rate is 1.6, well below the replacement rate. Eighty-five percent of couples use contraception; the most commonly used forms are male and female sterilization and the intrauterine device. Abortion is also an important aspect of this program.

By contrast, during the same 50 years, India has had little success in controlling its population. In the past, the emphasis of government programs was on meeting goals of sterilization and contraceptive use, but this has not been successful. Today, only about 55 percent of couples use contraceptives. In 2000, a new plan was unveiled that had the goal of bringing the total fertility rate from 3.1 children per woman to 2 (replacement rate) by 2010.

This new plan emphasizes improvements in the quality of life of the people. The major thrusts are to reduce infant and maternal death, immunize children against preventable disease, and encourage girls to attend school. It is hoped that improved health will remove the perceived need for large numbers of births. Currently, about 65 percent of the women in India can read and write. The emphasis on improving the educational status of women is related to the experiences of other developing countries. In many other countries, it has been shown that an increase in the education level of women is linked to lower fertility rates. Although the goal of a total fertility rate of 2.0 by 2010 was not met, the total fertility rate did fall to 2.4 by 2014, which is a positive trend.

Immigration

The immigration policies of a country also have a significant impact on the rate at which the population grows. Birth rates are currently so low in several European countries, Japan, and China that these countries will likely have a shortage of working-age citizens in the near future. One way to solve this problem is to encourage immigration from other parts of the world.

The developed countries are under great pressure to accept immigrants. The standard of living in these countries is a tremendous magnet for refugees or people who seek a better life than is possible where they currently live. In the United States, approximately one-third of the population increase experienced each year is the result of immigration. Canada encourages immigrants and has set a goal of accepting between 250,000 and 300,000 new immigrants each year.

7.10 Population Growth Rates and Standard of Living

There appears to be an inverse relationship between the rate at which the population of a country is growing and its standard of living. The **standard of living** is an abstract concept that attempts to quantify the quality of life of people. Standard of living is a difficult concept to quantify, since various cultures have different attitudes and feelings about what is desirable. However, several factors can be included in an analysis of standard of living: economic well-being, health conditions, and the ability to change one's status in the society. Figure 7.17 lists several factors that are important in determining standard of living and compares three countries with very different standards of living (the United States, Argentina, and Kenya). One important economic measure of standard of living is the average purchasing power per person. One index of purchasing power is the **gross national income (GNI).** The GNI is an index that measures the total goods and services generated within a country as well as income earned and sent home by citizens of the country who are living in other countries. Since the prices of goods and services vary from one country to another, a true comparison of purchasing power requires some adjustments. Therefore, a technique used to compare economic well-being across countries is a measure called the GNI PPP (gross national income purchasing power parity). Finally, the GNI PPP can be divided by the number of people in the country to get a per capita (per person) GNI PPP. As you can see from figure 7.17, a wide economic gap exists between economically advanced countries and those that are less developed. Yet the people of less-developed countries aspire to the same standard of living enjoyed by people in the developed world.

Health criteria reflect many aspects of standard of living. Access to such things as health care, safe drinking water, and adequate food are reflected in life expectancy, infant mortality, and growth rates of children. The United States and Argentina have similar life expectancies (over 75 years) and adequate nutrition. Kenya has a low life expectancy (62 years), many undernourished children (20 percent

are underweight), and a high infant mortality rate (47 per 1,000). The United States has a low infant mortality rate (5.4 per 1,000). Argentina has an intermediate infant mortality rate (11.7 per 1,000).

Finally, the educational status of people determines the kinds of jobs that are available and the likelihood of being able to improve one's status. In general, men are more likely to receive an education than women, but the educational status of women has a direct bearing on the number of children they will have and, therefore, on the economic well-being of the family. In the United States the average adult has over 13 years of education. Argentina has about 9 years and Kenya has 7. Obviously, tremendous differences exist in the standard of living among these three countries. What the average U.S. citizen would consider poverty level would be considered a luxurious life for the average person in Kenya.

7.11 Hunger, Food Production, and Environmental Degradation

As the human population increases, the demand for food rises. People must either grow food themselves or purchase it. Most people in the developed world purchase what they need and have more than enough food to eat. Most people in the less-developed world must grow their own food and have very little money to purchase additional food. Typically, these farmers have very little

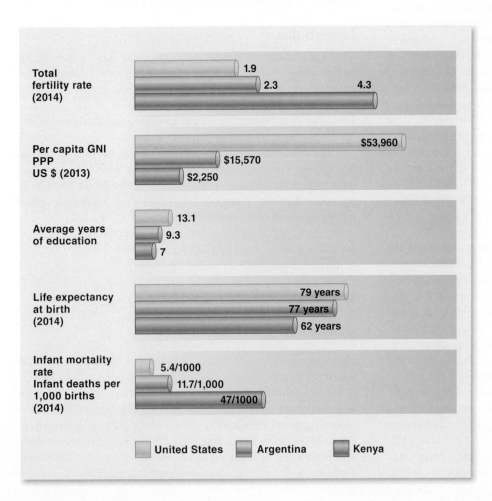

FIGURE 7.17 Standard of Living and Population Growth in Three Countries Standard of living is a measure of how well one lives. It is not possible to get a precise definition, but when we compare the United States, Argentina, and Kenya, it is obvious that there are great differences in how the people in these countries live.

Source: Data from The Population Reference Bureau, 2014. Population Data Sheet; and United Nations Development Program *International Human Development Indicators*.

surplus. If crops fail, people starve. Even in countries with the highest population (China and India), a large proportion of the people live on the land and farm. (Forty-six percent of Chinese and 70 percent of Indians live in rural areas.)

Environmental Impacts of Food Production

The human population can increase only if the populations of other kinds of plants and animals decrease. Each ecosystem has a maximum biomass that can exist within it. There can be shifts within ecosystems to allow an increase in the population of one species, but this always adversely affects certain other populations because they are competing for the same basic resources.

When humans need food, they convert natural ecosystems to artificially maintained agricultural ecosystems. The natural mix of plants and animals is destroyed and replaced with species useful to humans. If these agricultural ecosystems are mismanaged, the region's total productivity may fall below that of the original ecosystem. The desertification in Africa and destruction of tropical rainforests are well-known examples. In countries where food is in short supply and the population is growing, pressure is intense to convert remaining natural ecosystems to agriculture. Typically, these areas are the least desirable for agriculture and will not be productive. However, to a starving population, the short-term gain is all that matters. The long-term health of the environment is sacrificed for the immediate needs of the population.

The Human Energy Pyramid

A consequence of the basic need for food is that people in less-developed countries generally feed at lower trophic levels than do those in the developed world. (See figure 7.18.) Converting the less-concentrated carbohydrates of plants into more nutritionally valuable animal protein and fat is an expensive process. During the process of feeding plants to animals and harvesting animal products, approximately 90 percent of the energy in the original plants is lost. (See chapter 5.) Although many modern agricultural practices in the developed world obtain better efficiencies than this, most of the people in the developing world are not able to use such sophisticated systems. Thus, their conversion rates approach the 90 percent loss characteristic of natural ecosystems. Therefore, in terms of economics and energy, people in less-developed countries must consume the plants themselves rather than feed the plants to animals and then consume the animals. In most cases, if the plants were fed to animals, many people would starve to death. On the other hand, a lack of protein in diets that consist primarily of plants can lead to

malnutrition. It is possible to get adequate protein from a proper mixture of plant foods. In regions where food is in short supply, however, the appropriate mixture of foods is often not available. Thus, many people in the less-developed world suffer from a lack of adequate protein, which stunts their physical and mental development.

In contrast, in most of the developed world, meat and other animal protein sources are important parts of the diet. Many people suffer from overnutrition (they eat too much); they are "malnourished" in a different sense. About 66 percent of North Americans are overweight and 30 percent are obese. The ecological impact of one person eating at the carnivore level is about 10 times that of a person eating at the herbivore level. If people in the developed world were to reduce their animal protein intake, they would significantly reduce their demands on world resources. Almost all of the corn and soybeans grown in the United States are used as animal feed or to produce biofuels. Instead, if these grains were used to feed people, less grain would have to be grown and the impact on farmland would be reduced.

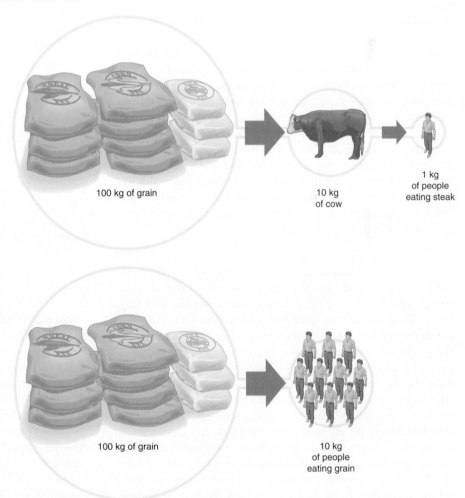

FIGURE 7.18 **Population and Trophic Levels** The larger a population, the more energy it takes to sustain the population. Every time one organism is eaten by another organism, approximately 90 percent of the energy is lost. Therefore, when countries are densely populated, they usually feed at the herbivore trophic level because they cannot afford the 90 percent energy loss that occurs when plants are fed to animals. The same amount of grain can support 10 times more people at the herbivore level than at the carnivore level.

Economics and Politics of Hunger

Some countries, such as the United States, Canada, Australia, New Zealand, and many countries in South America, have an abundance of agricultural land and are net food exporters. Many countries, such as India and China, are able to grow enough food for their people but do not have any left for export. Others, including Japan, several European nations, Russia and numerous other nations of the former Soviet Union, and many countries in Africa, are not able to grow enough to meet their own needs and, therefore, must import food.

A country that is a net food importer is not necessarily destitute. Japan and several European countries are net food importers but have enough economic assets to purchase what they need. Hunger occurs when countries do not produce enough food to feed their people and cannot obtain food through purchase or humanitarian aid.

The current situation with respect to world food production and hunger is very complicated. It involves the resources needed to produce food, such as arable land, labor, and machines; appropriate crop selection; and economic incentives. It also involves the maldistribution of food within countries. This is often an economic problem, since the poorest in most countries have difficulty finding the basic necessities of life, while the rich have an excess of food and other resources. In addition, political activities often determine food availability. War, payment of foreign debt, corruption, and poor management often contribute to hunger and malnutrition.

The areas of greatest need are in sub-Saharan Africa. Africa is the only major region of the world where per capita grain production has decreased over the past few decades. People in these regions are trying to use marginal lands for food production, as forests, scrubland, and grasslands are converted to agriculture. Often, this land is not able to support continued agricultural production. This leads to erosion and desertification.

Solving the Problem

What should be done about countries that are unable to raise enough food for their people and are unable to buy the food they need? This is not an easy question. A simple humanitarian solution to the problem is for the developed countries to supply food. Many religious and humanitarian organizations do an excellent service by taking food to those who need it and save many lives. However, the aim should always be to provide temporary help and insist that the people of the country develop mechanisms for solving their own problem. Often, emergency food programs result in large numbers of people migrating from their rural (agricultural) areas to cities, where they are unable to support themselves. They become dependent on the food aid and stop working to raise their own food, not because they do not want to work but because they need to leave their fields to go to the food distribution centers. Many humanitarian organizations now recognize the futility of trying to feed people with gifts from the developed world. They try to provide food aid in local villages rather than in large cities and support projects that provide incentives for improving the local agricultural economy. The emphasis must be on self-sufficiency.

7.12 The Demographic Transition Concept

Clearly there is a relationship between the standard of living and the population growth rate. Countries with the highest standard of living have the lowest population growth rate, and those with the lowest standard of living have the highest population growth rate. This has led many people to suggest that countries naturally go through a series of stages called **demographic transition.**

The Demographic Transition Model

The demographic transition model is based on the historical, social, and economic development of Europe and North America. In a demographic transition, the following four stages occur (See figure 7.19.):

1. Initially, countries have a stable population with a high birth rate and a high death rate. Death rates often vary because of famine and epidemic disease.

2. Improved economic and social conditions (control of disease and increased food availability) bring about a period of rapid population growth as death rates fall, while birth rates remain high.

3. As countries develop an industrial economy, birth rates begin to drop because people desire smaller families and use contraceptives.

4. Eventually, birth rates and death rates again become balanced. However, the population now has *low* birth rates and *low* death rates.

This is a very comfortable model because it suggests that if a country can develop a modern industrial economy, then social, political, and economic processes will naturally cause its population to stabilize.

Applying the Model

However, the model leads to some serious questions. Can the historical pattern exhibited by Europe and North America be repeated in the less-developed countries of today? Europe, North America, Japan, and Australia passed through this transition period when world population was lower and when energy and natural resources were still abundant. It is doubtful whether these supplies are adequate to allow for the industrialization of the major portion of the world currently classified as less developed.

Furthermore, when the countries of Europe and North America passed through the demographic transition, they had access to large expanses of unexploited lands, either within their boundaries or in their colonies. This provided a safety valve for expanding populations during the early stages of the transition. Without this safety valve, it would have been impossible to deal adequately with the population while simultaneously encouraging economic development. Today, less-developed countries may be unable to accumulate the necessary capital to develop economically, since they do not have uninhabited places to which their people can migrate and an ever-increasing population is a severe economic drain.

A second concern is the time element. With the world population increasing as rapidly as it is, industrialization probably cannot occur fast enough to have a significant impact on population growth. As long as people in less-developed countries are poor, there is a strong incentive to have large numbers of children. Children are a form of social security because they take care of their elderly parents. Only people in developed countries can save money for their old age. They can choose to have children, who are expensive to raise, or to invest money in some other way.

Today, most people feel that this model provides important insight into why some populations stabilize, but that most countries will require assistance in the form of economic development funds, education, and birth control information and technology if they are to be able to make the transition.

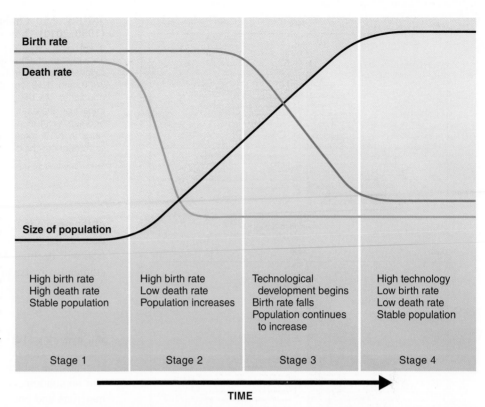

Birth rate

Death rate

Size of population

Stage 1	Stage 2	Stage 3	Stage 4
High birth rate High death rate Stable population	High birth rate Low death rate Population increases	Technological development begins Birth rate falls Population continues to increase	High technology Low birth rate Low death rate Stable population

TIME

FIGURE 7.19 Demographic Transition The demographic transition model suggests that as a country develops technologically, it automatically experiences a drop in the birth rate. This certainly has been the experience of the developed countries of the world. However, the developed countries make up less than 20 percent of the world's population. It is doubtful whether the less-developed countries can achieve the kind of technological advances experienced in the developed world.

7.13 The U.S. Population Picture

In many ways, the U.S. population, which has a total fertility rate that is low (1.9), is similar to those of other developed countries of the world with low birth rates. One might expect the population to be stabilizing under these conditions. However, two factors are operating to cause significant change over the next 50 years. One factor has to do with the age structure of the population, and the other has to do with immigration policy.

The U.S. population includes a postwar baby boom component, which has significantly affected population trends. These baby boomers were born during an approximately 15-year period (1947–61) following World War II, when birth rates were much higher than today, and constitute a bulge in the age distribution profile. (See figure 7.20.) As members of this group have raised families, they have had a significant influence on how the U.S. population has grown. As this population bulge ages and younger people limit their family size, the population will gradually age. By 2030, about 20 percent of the population will be 65 years of age or older compared to about 12 percent in 1980.

Both legal and illegal immigration significantly influence future population growth trends. Even with the current total fertility rate of 1.9 children per woman, the population is still growing by about 0.7 percent per year. About 70 percent is the result of natural increases due to the difference between birth rates and death rates. The remainder is the result of immigration into the United States.

Current immigration policy in the United States is difficult to characterize. Strong measures are being taken to reduce illegal immigration across the southern border. This is in part due to pressures placed on Congress by states that receive large numbers of illegal immigrants. Illegal immigrants add to the education and health care costs that states must fund. At the same time, some segments of the U.S. economy (agriculture, tourism) maintain that they are unable to find workers to do certain kinds of work. Consequently, special guest workers are allowed to enter the country for limited periods to serve the needs of these segments of the economy. There is also a consistent policy of allowing immigration that reunites families of U.S. residents. Obviously, the families that fall into this category are likely to include U.S. citizens who were recent immigrants themselves. Most immigration policy is the result of political decisions and has little to do with concern about the rate at which the U.S. population is growing or other demographic issues.

Projections based on the 2010 census indicate that the population will continue to grow and does not seem to be moving toward zero growth despite the fact that the total fertility rate of 1.9 is slightly less than the replacement rate of 2.1. A primary reason for this situation is that immigration adds about a million people per year, and new immigrants typically are young and have larger numbers of children than nonimmigrants. This is likely to result in an increase from about 318 million people in 2014 to about 395 million by 2050, an increase of about 24 percent.

Differences in family size exist between different segments of the U.S. population. The Hispanic and Asian-American portions of the population tend to have larger families. Furthermore, many

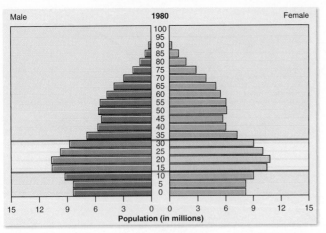

Male — 1980 — Female

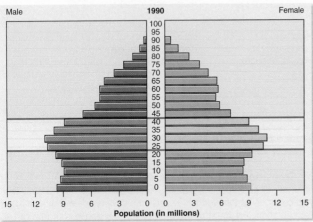

Male — 1990 — Female

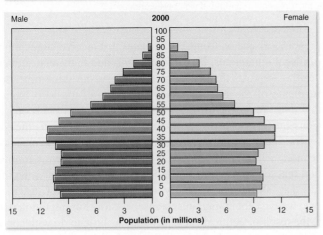

Male — 2000 — Female

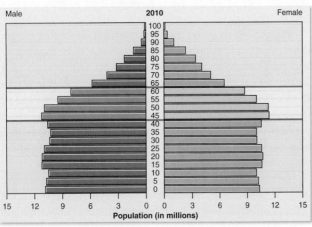

Male — 2010 — Female

FIGURE 7.20 Changing Age Distribution of U.S. Population (1980–2010) These graphs show the number of people in the United States at each age level. Notice that in the year 1980 a bulge begins to form at age 15 and ends at about age 34. These people represent the "baby boom" that followed World War II. As you compare the age distribution for 1980, 1990, 2000, and 2010, you can see that this group of people moves through the population. As this portion of the population has aged, it has had a large impact on the nature of the U.S. population. In the 1970s and 1980s, baby boomers were in school. In the 1990s, they were in their middle working years. In 2010, some of them were beginning to retire, and many more will do so throughout the decade.

Source: Data from the U.S. Department of Commerce, U.S. Census Bureau, International Data Base.

of the people in these groups are recent immigrants, so these portions of the population will grow rapidly, and the Caucasian portion will decline. Thus, the U.S. population will be much more ethnically diverse in the future.

7.14 What Does the Future Hold?

Humans are subject to the same limiting factors as other species. Our population cannot increase beyond our ability to acquire raw materials and energy and safely dispose of our wastes. We also must remember that interactions with other species and with other humans will help determine our carrying capacity. Furthermore, when we think about the human population and carrying capacity, we need to distinguish between the biological carrying capacity, which describes how many people the Earth can support, and a cultural carrying capacity, which describes how many people the Earth can support with a reasonable standard of living. Regardless of which of these two concepts we are thinking about, the same four basic factors are involved. Let us look at these four factors in more detail.

Available Raw Materials

Many of us think of raw materials simply as the amount of food available. However, we have become increasingly dependent on technology, and our lifestyles are directly tied to our use of other kinds of resources, such as irrigation water, genetic research, and antibiotics. Food production is becoming a limiting factor for some segments of the world's human population. Malnutrition is a serious problem in many parts of the world because sufficient food is not available. Currently, about 870 million people (one-eighth of the world's population) suffer from a lack of adequate food.

Available Energy

The second factor, available energy, involves problems similar to those of raw materials. Essentially, all species on Earth are ultimately dependent on sunlight for their energy. Currently, the world's human population depends on fossil fuels to raise food, modify the environment, and move from place to place. When energy prices increase, much of the world's population is placed in jeopardy because incomes are not sufficient to pay the increased

Focus On

North America—Population Comparisons

The three countries that make up North America (Canada, the United States, and Mexico) interact politically, socially, and economically. The characteristics of their populations determine how they interact. Canada and the United States are both wealthy countries with similar age structures and low total fertility rates. The United States has a total fertility rate of 1.9, and Canada has a total fertility rate of 1.6. Both countries have a relatively small number of young people (less than 20 percent of the population is under 15 years of age) and a relatively large number of older people (about 14 to 15 percent are 65 or older). Without immigration, these countries would have stable or falling populations. Therefore, they must rely on immigration to supply additional people to fill the workplace. This is particularly true for jobs with low pay. The United States currently receives about 950,000 immigrants per year. Canada currently receives about 240,000 immigrants per year.

Mexico, on the other hand, has a young, rapidly growing population. About 30 percent of the population is under 15 years of age, and the total fertility rate is 2.2. At that rate, the population will double in less than 50 years. Furthermore, the average Mexican has a purchasing power about 40 percent of that in Canada and about 30 percent of that in the United States. These conditions create a strong incentive for individuals to migrate from Mexico to other parts of the world. The United States is the usual country of entry. The number is hard to evaluate, since many enter the United States illegally or as seasonal workers and eventually plan to return to Mexico. Current estimates are that about 380,000 people enter the United States from Mexico each year.

In response to the large number of illegal immigrants from Mexico, the United States has erected fences and increased surveillance. In 2012, about 360,000 people were apprehended attempting to cross the border from Mexico to the United States. Thus, instead of crossing at normal points of entry, illegal immigrants are likely to cross the border in remote desert areas, which has resulted in numerous deaths due to exposure.

The economic interplay between Mexico and the United States has several components. Working in the United States allows Mexican immigrants to improve their economic status. Furthermore, their presence in the United States has a significant effect on the economy of Mexico, since many immigrants send much of their income to Mexico to support family members. The North American Free Trade Agreement (NAFTA) allows for relatively free exchange of goods and services among Canada, the United States, and Mexico. Consequently, many Canadian, U.S., and European businesses have built assembly plants in Mexico to make use of the abundant inexpensive labor, particularly along the U.S.-Mexican border. Labor leaders in Canada and the United States complain that many high-paying jobs have been moved to Mexico, where labor costs are lower.

Country	Population Size 2014 (Millions)	Birth Rate per 1,000	Death Rate per 1,000	Rate of Natural Increase	Total Fertility Rate (Children per woman per lifetime)	Life Expectancy (Years)	Infant Mortality Rate (Deaths per 1,000 births)	Per Capita GNI PPP 2013 US $	% Under 15 Years of Age	% Over 65 Years of Age	Annual Immigration/ Emigration
Canada	35.5	11	7	0.4	1.6	81	4.8	$42,590	16%	15%	+248,500
United States	317.7	13	8	0.4	1.9	78	5.4	$53,960	19%	14%	+953,100
Mexico	119.7	19	6	1.4	2.2	74	13	$16,110	28%	6%	(−239,400)

Source: Population Reference Bureau, 2014 Population Data Sheet.

costs for energy and other essentials. New, less disruptive methods of harnessing this energy must be developed to support an increasing population. Increases in the efficiency with which energy is used could reduce demands on fossil fuels. In addition, the development of more efficient solar and wind energy conversion systems could reduce the need for fossil fuels.

Waste Disposal

Waste disposal is the third factor determining the carrying capacity for humans. Most pollution is, in reality, the waste product of human activity. Lack of adequate sewage treatment and safe drinking water causes large numbers of deaths each year. Some people are convinced that disregard for the quality of our environment will be a major limiting factor. In any case, it makes good sense to control pollution and to work toward cleaning our environment.

Interaction with Other Organisms

The fourth factor that determines the carrying capacity of a species is interaction with other organisms. We need to become aware that we are not the only species of importance. When we

Funding the Unmet Need for Family Planning

The World Health Organization has stated that *"Meeting the need for family planning is one of the most cost-effective investments to alleviate poverty and improve health."*

One of the most important determiners of the birth rate of a country is the desire of women to have or not have children. The unmet need for family planning is determined by asking women: (1) if they would like to limit the number of children they have or (2) if they would like to increase the time between successive births and (3) if they are using birth control. Using these criteria, worldwide, over 200 million women have an unmet family planning need. In other words they are exposed to the risk of a pregnancy they do not want.

Researchers have surveyed women and found that there are several reasons they do not use birth control. Some women fear the side effects of using birth control. Some women or their partners are opposed to birth control. Some are unaware of birth control methods. Some are unable to afford birth control methods. Thus, there are two aspects to a family planning program: providing accurate information to women and their partners and providing access to contraceptives. Both cost money.

The United Nations Fund for Population Activities (UNFPA) and various non-governmental organizations seeking to improve the reproductive health of women, rely on donations to fund their activities. The United States State Department provides funds for a variety of development projects including those that address maternal and child health.

During the Nixon, Ford, Carter, and the first Reagan administrations (1969–1985), the U.S. provided between 25 and 80 percent of the funding for UNFPA. In 1985, the passage of the Kemp/Kasten amendment forbid the spending of U.S. funds by any agency or country that uses coercion to force people to have abortions or be sterilized. In 1986 the Reagan administration declared that UNFPA was in violation of the Kemp/Kasten amendment and cut off funding for UNFPA. This continued during the George H. W. Bush administration (1989–1992) when Congress allocated $235 million over 7 years but the funding was withheld by the president. Following the election of President Clinton (1993–2000) funding for UNFPA was restored. With the election of George W. Bush (2001–2008) funding for UNFPA was again curtailed by the president.

In 2009, President Obama restored funding to UNFPA. In recent years, some members of Congress have sought to end funding for the family planning activities of UNFPA. However, compromises have allowed funding through 2013. It is obvious that from 1968 through 1985 the majority in Congress and the executive branch supported funding of family planning. However, with the passage of the Kemp/Kasten amendment things changed and support became political, with Republican administrations cutting off funding and Democratic administrations restoring funding. What had begun as bipartisan support for family planning to improve the living conditions of many in the world has become a partisan issue revolving around abortion, even though 77 percent of U.S. women use birth control and abortion is legal in the United States. (See graph.) In addition, polls of the U.S. public show that 75 percent of the U.S. public supports international family planning activities. The ideological debate in Congress continues, and it is clear that the United States has no clear policy regarding funding family planning activities in poor countries.

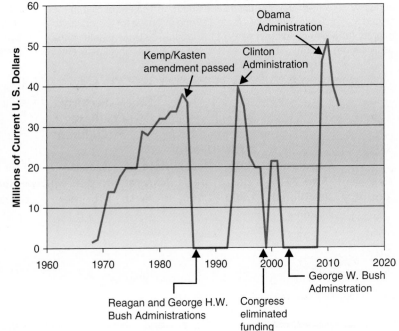

United States Funding of UNFDA

---

convert land to meet our needs, we displace other species from their habitats. Many of these displaced organisms are not able to compete with us successfully and must migrate or become extinct. Unfortunately, as humans expand their domain, the areas available to these displaced organisms become rarer. Parks and natural areas have become tiny refuges for the plants and animals that once occupied vast expanses of land. If these refuges fall to the developer's bulldozer or are converted to agricultural use, many organisms will become extinct. What today seems like an unimportant organism, one that we could easily do without, may someday be seen as an important link to our very survival. It is also important to recognize that many organisms provide services that we enjoy

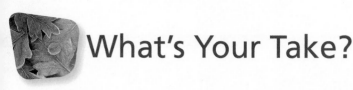

What's Your Take?

Like much of the developed world, the United States has an aging population and would cease to grow without immigration. Immigrants (both legal and illegal) have larger families than nonimmigrants. Immigrants often take low-paying jobs that the rest of the population does not want. Illegal immigrants from Mexico constitute a major problem, and the United States spends over a billion dollars each year to try to control illegal immigration.

Consequently, many people support a guest-worker program in which immigrants could enter the country for specific time periods but must eventually return to their home countries. Choose to support or oppose the concept of a guest-worker program, and develop arguments to support your point of view.

without thinking about them. Forest trees release water and moderate temperature changes, bees and other insects pollinate crops, insect predators eat pests, and decomposers recycle dead organisms. All of these activities illustrate how we rely on other organisms. Eliminating the services of these valuable organisms would be detrimental to our way of life.

Social Factors Influence Human Population

Human survival depends on interaction and cooperation with other humans. Current technology and medical knowledge are available to control human population growth and to improve the health of the people of the world. Why, then, does the population continue to increase, and why do large numbers of people continue to live in poverty, suffer from preventable diseases, and endure malnutrition? Humans are social animals who have freedom of choice and frequently do not do what is considered "best" from an unemotional, uninvolved, biological point of view. People make decisions based on history, social situations, ethical and religious considerations, and personal desires. The biggest obstacles to controlling human population are not biological but are the province of philosophers, theologians, politicians, and sociologists. People in all fields need to understand that the cause of the population problem has both biological and social components if they are to successfully develop strategies for addressing it.

Ultimate Size Limitation

The human population is subject to the same biological constraints as are other species of organisms. We can say with certainty that our population will ultimately reach its carrying capacity and stabilize. There is disagreement about how many people can exist when the carrying capacity is reached. Some people suggest we are already approaching the carrying capacity, while others maintain

that we could more than double the population before the carrying capacity is reached. Furthermore, uncertainty exists about what the primary limiting factors will be and about the quality of life the inhabitants of a more populous world would have. If the human population continues to grow at its current rate of 1.2 percent per year, the population will reach nearly 10 billion by 2050.

As with all K-strategist species, when the population increases, density-dependent limiting factors will become more significant. Some people suggest that a lack of food, a lack of water, or increased waste will ultimately control the size of the human population. Still others suggest that, in the future, social controls will limit population growth. These social controls could be either voluntary or involuntary. In the economically developed portions of the world, families have voluntarily lowered their birth rates to fewer than two children per woman. Most of the poorer countries of the world have higher birth rates. What kinds of measures are needed to encourage them to limit their populations? Will voluntary compliance with stated national goals be enough, or will enforced sterilization and economic penalties become the norm? Others are concerned that countries will launch wars to gain control of limited resources or to simply eliminate people who compete for the use of those resources.

It is also important to consider the age structure of the world population. In most of the world, there are many reproductive and prereproductive individuals. Since most of these individuals are currently reproducing or will reproduce in the near future, even if they reduce their rate of reproduction, there will be a sharp increase in the number of people in the world in the next few years.

No one knows what the ultimate human population size will be or what the most potent limiting factors will be, but most agree that we are approaching the maximum sustainable human population. If the human population continues to increase, eventually the amount of agricultural land available will not be able to satisfy the demand for food.

The Lesser Snow Goose—A Problem Population

Lesser snow geese use breeding grounds in the Arctic and subarctic regions of Canada and wintering grounds on salt marshes of the Gulf of Mexico, California, and Mexico. (See map.) One population of the lesser snow goose has breeding grounds around Hudson Bay and wintering grounds on the Gulf of Mexico. Because these birds migrate south from Hudson Bay through the central United States west of the Mississippi River to the west coast of the Gulf of Mexico, they are called the midcontinent population. Populations of the midcontinent lesser snow goose have grown from about 800,000 birds in 1969 to an estimated 4.6 million in 2013. (Some experts believe the number may be closer to 6 million.) These numbers are so large that the nesting birds are destroying their breeding habitat.

What has caused this drastic population increase? A primary reason appears to be a change in feeding behavior. Originally, during the winter snow geese fed primarily on the roots and tubers of aquatic vegetation in the salt marshes of the Gulf of Mexico. Since this habitat and the food source it provided was limited, there was a natural control on the size of the population. However, many of these wetland areas have been destroyed. Simultaneously, agriculture (particularly rice farming) in the Gulf Coast region provided an alternative food source. As the geese began to use agricultural areas for food, they had an essentially unlimited source of food during the winter months, which led to very high winter survival. Furthermore, concerns about erosion have resulted in farmers using no-till and reduced tillage practices in the grain fields along the route used by migrating geese. The fields provided food for the fall migration that improved the survival of young snow geese. During the spring migration, geese used the same resources to store food as body fat that they took with them to their northern breeding grounds. Other factors that may have played a role are the decline in the number of people who hunt geese and a warming of the climate that leads to greater breeding success on the cold northern breeding grounds.

The large goose population is causing habitat destruction on their Canadian breeding grounds. Since snow geese feed by ripping plants from the ground, and they nest in large colonies, they have destroyed large areas of the coastal tundra vegetation around Hudson Bay. (See photo.) Because there is a short growing season, the areas are slow to regenerate.

What can be done to control the snow goose population and protect their breeding habitat? There are two possibilities: let nature take its course or use population management tools. If nothing is done, the geese will eventually destroy so much of their breeding habitat that they will be unable to breed successfully, and the population will crash. It could take decades for the habitat to recover, however. The alternative is to use population management tools to try to resolve the problem. Several management activities have been instituted. These include allowing hunters to kill more geese, increasing the length of the hunting season, allowing hunts during the spring as well as the fall migration seasons, and allowing more harvest by Canadian native communities. In recent years, the harvest in the United States and Canada has increased, varying between 1 million and 1.5 million geese killed. Preliminary evidence suggests that the increased harvest is starting to have an effect. It will take several more years, however, to determine if the population can be controlled.

- Do you think the killing of geese is justified to protect the coastal habitat of Hudson Bay?

- Do you think it is better to have the population crash as a result of natural forces or to reduce the size of the population by increasing the harvest?

Legend:

- Wintering grounds
- Nesting grounds
- Principal staging areas
- Pacific flyway migration
- Greater snow goose migration
- Midcontinental spring migration
- Midcontinental fall migration
- Western Canadian Arctic population

The map shows the migration routes between the traditional nesting and wintering areas of different snow goose populations.

This photo shows the kind of damage done by geese. The exclosures were established by Dawn R. Bazely and Robert L. Jefferies of the Hudson Bay Project.

Summary

A population is a group of organisms of the same species that inhabits an area. The birth rate (natality) is the number of individuals entering the population by reproduction during a certain period. The death rate (mortality) measures the number of individuals that die in a population during a certain period. Population growth is determined by the combined effects of the birth rate and death rate.

The sex ratio of a population is a way of stating the relative number of males and females. Age distribution and the sex ratio have a profound impact on population growth. Most organisms have a biotic potential much greater than that needed to replace dying organisms.

Interactions among individuals in a population, such as competition, predation, and parasitism, are also important in determining population size. Organisms may migrate into (immigrate) or migrate out of (emigrate) an area as a result of competitive pressure.

A typical population growth curve shows a lag phase followed by an exponential growth phase, a deceleration phase, and a stable equilibrium phase at the carrying capacity. The carrying capacity is determined by many limiting factors that are collectively known as environmental resistance. The four major categories of environmental resistance are available raw materials, available energy, disposal of wastes, and interactions among organisms. Some populations experience a death phase following the stable equilibrium phase.

K-strategists typically are large, long-lived organisms that reach a stable population at the carrying capacity. Their population size is usually controlled by density-dependent limiting factors. Organisms that are r-strategists are generally small, short-lived organisms that reproduce very quickly. Their populations do not generally reach a carrying capacity but crash because of some density-independent limiting factor.

Currently, the world's human population is growing very rapidly. The causes for human population growth are not just biological but also social, political, philosophical, and theological. Many of the problems of the world are caused or made worse by an increasing human population. Most of the growth is occurring in the less-developed areas of the world (Africa, Asia, and Latin America), where people have a low standard of living. The more-developed regions of the world, with their high standard of living, have relatively slow population growth and, in some instances, declining populations.

Demography is the study of human populations and the things that affect them. Demographers study the sex ratio and age distribution within a population to predict future growth. Population growth rates are determined by biological factors such as birth rate, which is determined by the number of women in the population and the age of the women, and death rate. Sociological and economic conditions are also important, since they affect the number of children desired by women, which helps set the population growth rate. In more-developed countries, women usually have access to jobs. Couples marry later, and they make decisions about the number of children they will have based on the economic cost of raising children. In the less-developed world, women marry earlier, and children have economic value as additional workers, as future caregivers for the parents, and as status for either or both parents.

The demographic transition model suggests that as a country becomes industrialized, its population begins to stabilize. However, there is little hope that the Earth can support the entire world's population in the style of the industrialized nations. It is doubtful whether there are enough energy resources and other natural resources to develop the less-developed countries or whether there is enough time to change trends of population growth. Highly developed nations should anticipate increased pressure in the future to share their wealth with less-developed countries.

Acting Green

1. Participate in a local effort to eliminate alien species.
2. Make a donation to an organization devoted to improving living conditions in the developing world.
3. Express your views on population policy to your congressional representatives.
4. Eat at the herbivore trophic level for one week.
5. Trace your family back three generations. Start with your grandparents and list all of their descendants.
6. Determine the number of siblings for each of the following: each of your grandparents, each parent, you.
7. Much of the world's population lives on less than $2.00 a day. For one day, feed yourself with food you can buy for $2.00. It can be done.

Review Questions

1. How is biotic potential related to the rate at which a population will grow?
2. List three characteristics populations might have.
3. Why do some populations grow? What factors help to determine the rate of this growth?
4. Draw and label a population growth curve.
5. Under what conditions might a death phase occur?
6. List four factors that could determine the carrying capacity of an animal species.
7. How do the concepts of birth rate and population growth differ?
8. How does the population growth curve of humans compare with that of bacteria on a petri dish?
9. Give examples of intrinsic, extrinsic, density-dependent, and density-independent limiting factors.
10. How do K-strategists and r-strategists differ?
11. As the human population continues to increase, what might happen to other species?
12. All successful organisms overproduce. What advantage does this provide for the species? What disadvantages may occur?
13. What is demographic transition? What is it based on?
14. Interpret the meaning of $I = P \times A \times T$.
15. Why is your ecological footprint larger than that of a person in Africa?
16. How does the age distribution of a population affect the rate at which a population grows?
17. Why do economic well-being and the status of women influence the number of children born in a country?
18. List ten differences between your standard of living and that of someone in a less-developed country.
19. Why do people who live in overpopulated countries use plants as their main source of food?
20. Which three areas of the world have the highest population growth rate? Which three areas of the world have the lowest standard of living?
21. Describe three reasons why women in the less-developed world might desire more than two children.
22. How are age distribution, total fertility rate, and immigration affecting the way the U.S. population is changing?

Critical Thinking Questions

1. Why do you suppose some organisms display high natality and others display lower natality? For example, why do cottontail rabbits show high natality and wolves relatively low natality? Why wouldn't all organisms display high natality?
2. Consider the differences between K-strategists and r-strategists. What costs are incurred by adopting either strategy? What evolutionary benefits does each strategy enjoy?
3. Do you think it is appropriate for developed countries to persuade less-developed countries to limit their population growth? What would be appropriate and inappropriate interventions, according to your ethics? Now imagine you are a citizen of a less-developed country. What might be your reply to those who live in more-developed countries? Why?
4. Population growth causes many environmental problems. Identify some of these problems. What role do you think technology will play in solving these problems? Are you optimistic or pessimistic about these problems being solved through technology? Why?
5. Do you think that demographic transition will be a viable option for world development? What evidence leads you to your conclusions? What role should the developed countries play in the current demographic transition of developing countries? Why?
6. Imagine a debate between an American and a Sudanese person about human population and the scarcity of resources. What perspectives do you think the American might bring to the debate? What perspectives do you think the Sudanese would bring? What might be their points of common ground? On what might they differ?
7. Many people in developing countries hope to achieve the standard of living of those in the developed world. What might be the effect of this pressure on the environment in developing countries? On the political relationship between developing countries and already developed countries? What ethical perspective do you think should guide this changing relationship?
8. The demographic changes occurring in Mexico have an influence on the United States. What problems does Mexico face regarding its demographics? Should the United States be involved in Mexican population policy?

To access additional resources for this chapter, please visit ConnectPlus® at connect.mheducation.com. There you will find interactive exercises, including Google Earth™, additional Case Studies, and SmartBook™, an adaptive eBook that integrates our LearnSmart® adaptive learning technology.

Energy and Civilization: Patterns of Consumption

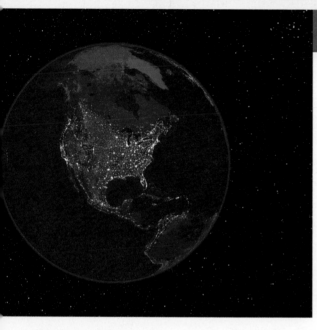

The amount of energy consumed by modern civilizations is illustrated by the lights of North America as seen from space.

OBJECTIVES

After reading this chapter, you should be able to:

- Explain why all organisms require a constant input of energy.
- Describe how per capita energy consumption increased as civilization developed from hunting and gathering to primitive agriculture to advanced cultures.
- Describe how advanced modern civilizations developed as new fuels were used to run machines.
- Correlate the Industrial Revolution with social and economic changes.
- Explain how the automobile changed people's lifestyles.

- Recognize that fossil fuels provide over 80 percent of the energy used in the United States.
- Describe the major categories of energy use in countries.
- Describe the central role electrical production has in the economies of most countries.
- Recognize that electricity can be produced in many different ways and that countries differ in how they generate electricity.
- Explain why energy consumption is growing more rapidly in developing

countries than in the industrialized world.
- Describe the role of OPEC in determining oil prices.
- Describe actions countries take to alter how energy is used.
- Recognize that as demand for energy increases, increased competition for supplies will increase prices and create international tension.

The Impact of Newly Industrialized Countries

Economists identify countries as being on a continuum from least developed to modern industrialized countries like the United States and many countries of Europe. Least developed countries are defined by poverty. Per capita income is less than US $1,000 per year and economic conditions are unstable. Subsistence farming is common, there is little industry, and there are poor education and health services. Developing countries are doing better economically than least developed countries and are considered to be making progress toward a modern economy. This is a very large group of countries with different levels of economic development. An important subcategory consists of countries known as newly industrialized countries and includes: Brazil, India, China, Indonesia, South Africa, Mexico, Philippines, Malaysia, Thailand, and Turkey.

Newly industrialized countries share several characteristics. They have growing economies characterized by low labor costs, large national corporations, and capital investment from foreign countries. They have strong political leadership, and the people have improving education, health services, and civil rights. As a country shifts to an industrialized economy people tend to leave agricultural jobs in the countryside and migrate to the cities to engage in manufacturing jobs with higher pay.

Another consequence of industrialization and improved standard of living is an increase in energy consumption. The graph contrasts "old" industrialized economies—the United States and the European Union—with newly industrialized economies. In the ten-year period from 2002–2012 the United States and the European Union have had falling energy consumption while the newly industrialized countries have rapidly increasing energy demands. China has grown by about 155 percent and now consumes more energy than any other country.

India's energy demand has increased by 80 percent. As China, India, and other newly industrialized countries continue to improve their economies, their demand for energy will increase, both because of industrial use and because of their citizens' improving standard of living will require increased energy use.

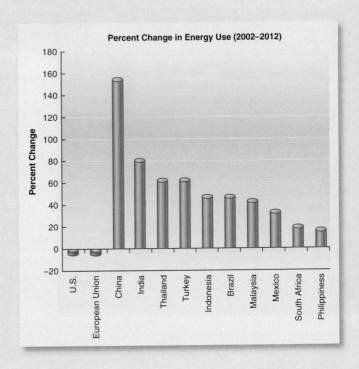

8.1 History of Energy Consumption

Every form of life and all societies require a constant input of energy. If the flow of energy through organisms or societies ceases, they stop functioning and begin to disintegrate. Some organisms and societies are more energy efficient than others. In general, complex industrial civilizations use more energy than simple hunter-gatherer or primitive agricultural cultures. If modern societies are to survive, they must continue to expend energy. However, they may need to change their pattern of energy consumption as current sources become limited.

Biological Energy Sources

Energy is essential to maintain life. In every ecosystem, the sun provides that energy. (See chapter 5.) The first transfer of energy occurs during photosynthesis, when plants convert light energy into the chemical energy in the organic molecules they produce. Herbivorous animals utilize the food energy in the plants. The herbivores, in turn, are a source of energy for carnivores.

Because nearly all of their energy requirements were supplied by food, primitive humans were no different from other animals in their ecosystems. In such hunter-gatherer cultures, nearly all human energy needs were met by using plants and animals as food, tools, and fuel. (See figure 8.1.)

Early in human history, people began to use additional sources of energy to make their lives more comfortable. They domesticated plants and animals to provide a more dependable supply of food. They no longer needed to depend solely on gathering wild plants and hunting wild animals for sustenance. Domesticated animals also furnished a source of energy for transportation, farming, and other tasks. (See figure 8.2.) Wood or other plant material was used for building materials and other cultural uses but also was the primary source of fuel for heating and cooking. Eventually, this biomass energy was used in simple technologies, such as shaping tools and extracting metals.

The development of complex early civilizations, such as the Aztecs, Chinese, Indians, Greeks, Egyptians, and Romans, resulted in the development of cities and led to an increased demand for energy. Although they were culturally advanced, they

FIGURE 8.1 Hunter-Gatherer Society In a hunter-gatherer society, people obtain nearly all of their energy from the collection of wild plants and the hunting of animals.

FIGURE 8.2 Animal Power The domestication of animals was an important stage in the development of human civilization. Domesticated animals provided people with a source of power other than their own muscles.

still primarily used human muscle, animal muscle, and the burning of biomass as sources of energy. These civilizations were usually stratified into a ruling class and various lower classes. The muscles of slaves were an important source of energy for many physically demanding tasks such as farming, building, mining, and transporting goods. Slaves were also employed in more skilled activities such as cooking, managing property, and skilled crafts.

Increased Use of Wood

When dense populations of humans made heavy use of wood for fuel and building materials, they eventually used up the readily available sources and had to import wood or seek alternative forms of fuel. Because of a long history of high population density, India

and some other parts of the world experienced a wood shortage hundreds of years before Europe and North America did. In many of these areas, animal dung replaced wood as a fuel source. It is still used today in some parts of the world.

Western Europe and North America were able to use wood as a fuel for a longer period of time. The forests of Europe supplied sufficient fuel until the thirteenth century. In North America, vast expanses of virgin forests supplied adequate fuel until the late nineteenth century. Fortunately for the developing economies of Europe and North America, when local supplies of wood declined, coal, formed from fossilized plant remains, was available as an alternative energy source. By 1890, coal had replaced wood as the primary energy source in North America.

Fossil Fuels and the Industrial Revolution

Fossil fuels are the modified remains of plants, animals, and microorganisms that lived millions of years ago. (The energy in these fuels is stored sunlight, just as the biomass of wood represents stored sunlight.) Coal was formed during the Carboniferous period, 299 million to 359 million years ago, when the Earth's climate was warm and humid. There were extensive swamps with forests of primitive land plants. These conditions were conducive to the accumulation of large amounts of plant material that was converted into deposits of coal by heat and pressure.

Oil and natural gas formed primarily from tiny marine organisms whose bodies accumulated in large quantities on the seafloor and became compressed over millions of years. Heat and pressure from overlying sediment layers eventually converted the organic matter into oil and gas.

Historically, the first fossil fuel to be used extensively was coal. In the early eighteenth century, regions of the world that had readily available coal deposits were able to switch to this new fuel and participate in a major cultural change known as the **Industrial Revolution.** (See figure 8.3.) The Industrial Revolution began in England and spread to much of Europe and North America. It involved the invention of machines that replaced human and animal labor in manufacturing and transporting goods. Central to this change was the invention of the steam engine, which could convert heat energy into the energy of motion. The steam engine made possible the large-scale mining of coal. Steam engines were used to pump water from mines to prevent mines from flooding. The engines were also used to move coal to the surface. The source of energy for steam engines was either wood or coal; wood was quickly replaced by coal in most cases. Nations without an easily exploited source of coal did not participate in the Industrial Revolution.

Prior to the Industrial Revolution, Europe and North America were predominately rural. Goods were manufactured on a small scale in the home. However, as machines and the coal to power them became increasingly available, the factory system of manufacturing products replaced the small home-based operation. Because expanding factories required a constantly increasing labor supply, people left the farms and congregated in areas surrounding the factories. Villages became towns, and towns became cities. Widespread use of coal in cities resulted in increased air pollution.

FIGURE 8.3 The Industrial Revolution The invention of the steam engine led to the development of many machines during the Industrial Revolution.

In spite of these changes, the Industrial Revolution was viewed as progress. Energy consumption increased, economies grew, and people prospered. Within a span of 200 years, the daily per capita energy consumption of industrialized nations increased eightfold.

The Role of the Automobile

Although the Chinese used some oil and natural gas as early as 1000 B.C., the oil well that Edwin L. Drake, an early oil prospector, drilled in Pennsylvania in 1859 was the beginning of the modern petroleum era. For the first 60 years of production, the principal use of oil was to make kerosene, a fuel burned in lamps to provide lighting. The gasoline produced was discarded as a waste product. During this time, oil was abundant relative to its demand, and thus, it had a low price.

The invention of the internal combustion engine and the automobile dramatically increased the demand for oil products. In 1900, the United States had only 8,000 automobiles. By 1950, it had over 40 million cars, and by 2010, about 250 million. More oil was needed to make automobile fuel and lubricants. During this period, the percentage of energy provided by oil increased from about 2 percent in 1900 to about 40 percent by 1950 and has remained near 40 percent to the present. (See figure 8.4.)

The growth of the automobile industry, first in the United States and then in other industrialized countries, led to roadway construction, which required energy. Thus, the energy costs of driving a car were greater than just the fuel consumed in travel. As roads improved, higher speeds were possible. Bigger and faster cars required more fuel and even better roads. So roads were continually being improved, and better cars were being produced. A cycle of *more chasing more* had begun. In North America and

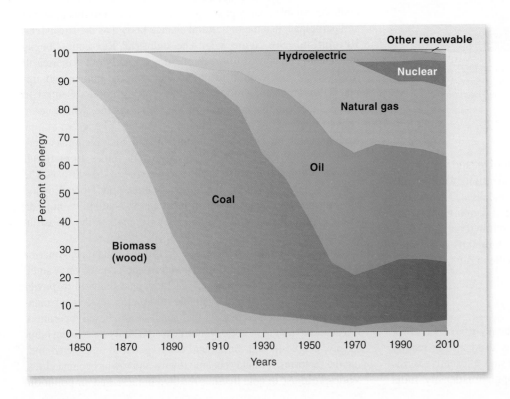

FIGURE 8.4 Changes in Energy Sources This graph shows how the percentage of energy obtained from various sources in the United States has changed over time. Until 1880 wood (biomass) was the most used source of energy. It was slowly replaced by coal which was the primary form of energy used until about 1940. Since 1960, oil and natural gas have provided over 60 percent of the energy, while coal has fallen to about 20 percent and biomass to about 3 percent. Nuclear power has grown to about 9 percent, and hydroelectric power provides about 4 percent. Other renewable sources (geothermal, solar, and wind) collectively account for nearly 2 percent.

Source: Data from *Annual Energy Review, 2009* and online statistics, Energy Information Administration.

FIGURE 8.5 Energy-Demanding Lifestyle (a) Building homes in suburbs some distance from shopping areas and places of employment is directly related to the heavy use of the automobile as a mode of transportation. (b) Highway systems provide the means to get from place to place. (c) Parking lots are a major feature of any shopping mall. Heating and cooling a large, enclosed shopping mall, along with the gasoline consumed in driving to the shopping center, increase the demand for energy.

much of Europe, the convenience of the automobile encouraged two-car families that created a demand for more energy.

More cars meant more jobs in the automobile industry, the steel industry, the glass industry, and hundreds of other industries. Constructing thousands of kilometers of roads created additional jobs. Thus, the automobile industry played a major role in the economic development of the industrialized world. All this wealth gave people more money for cars and other necessities of life. The car, originally a luxury, was now considered a necessity.

The car also altered people's lifestyles. Vacationers could travel greater distances. New resorts and chains of motels, restaurants, and other businesses developed to serve the motoring public, creating thousands of new jobs. Because people could live farther from work, they began to move to the suburbs. (See figure 8.5.)

As people moved to the suburbs, they also changed their buying habits. Labor-saving, energy-consuming devices became essential in the home. The vacuum cleaner, dishwasher, garbage disposal, and automatic garage door opener are only a few of the ways human power has been replaced with electrical power. About 30 percent of the electrical energy produced in North America is used in homes to operate lights and appliances. Other aspects of our lifestyles illustrate our energy dependence. Regardless

of where we live, we expect Central American bananas, Florida oranges, California lettuce, Texas beef, Hawaiian pineapples, Ontario fruit, and Nova Scotia lobsters to be readily available at all seasons. What we often fail to consider is the amount of energy required to process, refrigerate, and transport these items. The car, the modern home, the farm, and the variety of items on our grocery shelves are only a few indications of how our lifestyles are based on a continuing supply of cheap, abundant energy.

Growth in the Use of Natural Gas

Initially, natural gas was a waste product of oil production that was burned at oil wells because it was difficult to store and transport. Before 1940, natural gas accounted for less than 10 percent of energy consumed in the United States. In 1943, in response to World War II, a federally financed pipeline was constructed to transport oil within the United States.

After the war, the federal government sold these pipelines to private corporations. The corporations converted the pipelines to transport natural gas. Thus, a direct link was established between the natural gas fields in the Southwest and the markets in the Midwest and East. By 1970, about 30 percent of energy needs were being met by natural gas. Currently, over 25 percent of the energy consumed in the United States is from natural gas, primarily for home heating and industrial purposes.

8.2 How Energy Is Used

The amount of energy consumed by countries of the world varies widely. This is easily seen when we compare the amount of energy consumed per person in the population. (See figure 8.6.) The highly industrialized countries consume much more energy than less-developed countries. However, even among countries with the same level of development, great differences exist in the amount of energy they use as well as in how they use it. To maintain their style of living, individuals in the United States use about twice as much energy as people in France, Germany, or Japan, about 3.5 times more energy than the people of China, and about 16 times more energy than the people of India.

Differences also exist in the purposes for which people use energy. Industrialized nations use energy about equally for three purposes: (1) residential and commercial uses, (2) industrial uses, and (3) transportation. Less-developed nations with little industry use most of their energy for residential purposes (cooking and heating). Countries that are making the transition from less-developed to industrial economies use large amounts of energy to develop their industrial base. Figure 8.7 shows for the United States the proportion of energy obtained from various sources and the uses to which this energy is put.

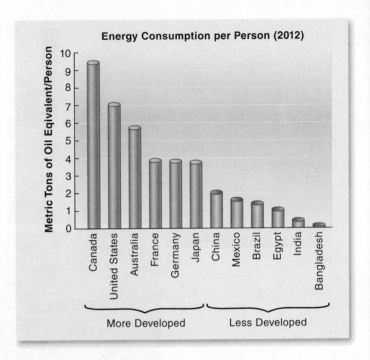

FIGURE 8.6 Per Capita Energy Consumption Countries of the developed world use much more energy per person than do countries of the less-developed world. However, there are great differences in energy use among developed nations. People in the United States use nearly twice the energy per person as people in France, Germany, and Japan.

Source: Data from *BP Statistical Review of World Energy*, June 2013 and *Population Reference Bureau, 2013 World Population Data Sheet.*

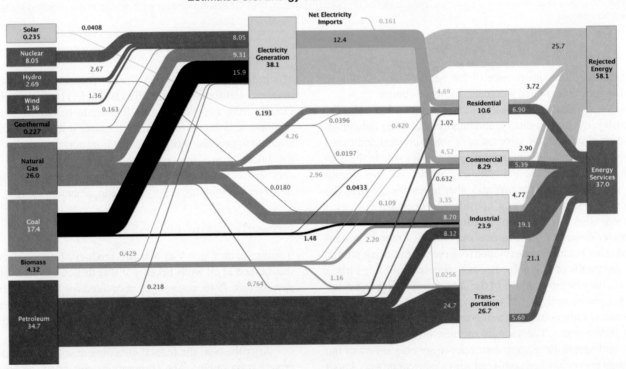

FIGURE 8.7 Energy Sources and Uses in the United States Although each country is different, this energy flow diagram for the United States shows the typical sources and uses of energy. The primary sources are coal, oil, natural gas. and nuclear power. A great deal of energy is used to make electricity that is then used along with other sources of energy for residential, commercial, industrial, and transportation purposes.

Source: Department of Energy, Lawrence Livermore National Laboratory.

Residential and Commercial Energy Use

The amount of energy required for residential and commercial use varies greatly throughout the world. For example, about 16 percent of the energy used in North America is for residential purposes and 12 percent for commercial purposes, while in India, 40 percent of the energy is for residential uses and about 3 percent for commercial purposes. The ways that residential energy is used also vary widely. In the United States, about 40 percent is used for space heating. In Canada, which has a cold climate, about 60 percent of the residential energy is used for heating. However, in many parts of Africa and Asia much of the energy used in the home is for cooking, and much of that energy comes from firewood.

Computer systems and the Internet are a relatively new segment of the economy that consumes energy. Although early estimates suggested that this segment of the economy would consume over 10 percent of the U.S. electrical energy supply, more recent estimates put the energy consumed at about 3 percent of the electrical energy supply.

Industrial Energy Use

The amount of energy countries use for industrial processes varies considerably. Nonindustrial countries use little energy for industry. Their economies are based on subsistence farming and small manufacturing activities in homes. Countries that are developing new industries dedicate a high percentage of their energy use to them. They divert energy to the developing industries at the expense of other sectors of their economy. For example, Brazil and China devote over 40 percent of their energy use to industrial purposes. Highly industrialized countries use a significant amount of their energy in industry, but their energy use is high in other sectors as well. In the United States, industry claims about 20 percent of the energy used.

Transportation Energy Use

As with residential, commercial, and industrial uses, the amount of energy used for transportation varies widely throughout the world. In some of the less-developed nations, transportation uses are very small. Per capita energy use for transportation is larger in developing countries and highest in highly developed countries. (See figure 8.8.) There are also great differences in the percentage of energy devoted to transportation. Bangladesh, India, and China use about 12 percent or less of their energy for transportation, while most industrialized countries use 25 to 40 percent for the same purpose.

Once a country's state of development has been taken into account, the specific combination of bus, rail, waterways, and private automobiles is the main factor in determining a country's energy use for transportation. In Europe, Latin America, and many other parts of the world, rail and bus transport are widely used because they are more efficient than private automobile travel, governments support these transportation methods, or a large part of the populace is unable to afford an automobile. In countries with high population densities, rail and bus transport is particularly

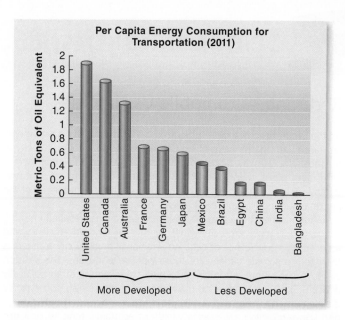

FIGURE 8.8 Per Capita Energy Use for Transportation (2011) Per capita energy use for transportation is highest among developed countries. However, there are great differences among countries. Most European countries and Japan use less than half the energy per person compared to the United States and Canada.

Source: International Energy Agency online data 2013 and *2011 World Population Data Sheet*, Population Reference Bureau.

FIGURE 8.9 Public Transportation People will use public transportation in situations where automobile travel is inconvenient, time consuming, or expensive. Public transportation is more energy efficient than travel by private automobile.

efficient. (See figure 8.9.) In general, automobiles require significantly more energy per passenger kilometer than does bus or rail transport. In addition, most of these countries have a high tax on fuel, which raises the cost to the consumer and encourages the use of public transport.

In North America, the situation is different. Government policy has kept the cost of energy artificially low and supported the automobile industry while removing support for bus and rail transport. Consequently, the automobile plays a dominant role, and public transport is used primarily in metropolitan areas.

Science, Politics, & Policy

Reducing Automobile Use in Cities

As the world becomes more urbanized, national and city governments are faced with the problem of how to move people through the city in the most efficient way. Studies of efficiency show that public transport—particularly buses, trains, and trollies—are the most energy-efficient way to move people within cities. Although people like the convenience of the personal automobile, cars are the least energy-efficient means of transportation in cities. Ironically, the desire to use cars has resulted in traffic congestion in most large cities that has further reduced efficiency and reduced the convenience of cars. An additional issue for city planners is the need to provide parking for those who drive in cities. The decision to move away from automobiles to public transport is often politically difficult. The construction and maintenance of public transport systems is expensive and many are unwilling to pay taxes to support public transport. Governments have used a variety of economic tools to convince people to use public transport, reduce automobile use, and relieve congestion. The restriction on the use of cars is coupled with an increase in the availability of trams, buses, and trains. Several different mechanisms have been successful.

LONDON

Automobiles entering the Congestion Charging Zone in central London have their license plates photographed. The image of the license plate is linked to computers that record when a driver enters the Congestion Charging Zone. Drivers of cars in the central city must pay a fee of £10 a day. Those who do not pay by midnight must pay a fine. There is also an automatic pay option which directly bills your debit or credit card.

STOCKHOLM

Cars entering central Stockholm have their license plate number recorded and are subject to a tax depending on the time of day. The tax can be paid by a variety of methods including automatic deduction from a person's bank account.

HONG KONG

An initial registration tax equal to 40 percent of the value of the car was instituted in 2011. Electronic sensors on cars record highway travel and time of day. Drivers have the fee deducted directly from a prepaid account (commuter hours are the most expensive).

SINGAPORE

Automobiles and motorcycles are equipped with an in-vehicle unit that automatically deducts money from a cash card when the vehicle passes through a sensor and enters the restricted zone of downtown Singapore.

GOTHENBURG, SWEDEN

To encourage pedestrian traffic, the central business district has been divided like a pie into zones, with cars prohibited from moving directly from one zone to another. Autos may move from zone to zone by way of a peripheral ring road. Parking spaces were also reduced and parking fees were increased.

ROME

All traffic except buses, taxis, delivery vehicles, and cars belonging to area residents have been banned between 6:30 A.M. and 6 P.M.

TOKYO

Before closing the sale, the buyer of a vehicle must show evidence that a permanent parking space is available for the car. Many cars are parked in mechanical lifts that allow cars to be parked above one another. Monthly costs for parking range from $300 to $800.

(See figure 8.10.) Private automobiles in North America, with about 4.5 percent of the world's population, consume about 40 percent of the gasoline produced in the world. Air travel is relatively expensive in terms of energy, although it is slightly more efficient than an automobile carrying a single passenger. Passengers, however, are paying for the convenience of rapid travel over long distances.

8.3 Electrical Energy

Electrical energy is such a large proportion of energy consumed in most countries that it deserves special comment. Electricity is both a way that energy is consumed and a way that it is supplied. Almost all electrical energy is produced as a result of burning fossil fuels. In the same way we use natural gas to heat homes, we can use natural gas to produce electricity. Because the transportation of electrical energy is so simple and the uses to which it can be put are so varied, electricity is a major form in which energy is supplied to people of the world.

Electrical energy can be produced in many ways. The primary methods of generating electricity are burning fossil fuels, nuclear power plants, hydroelectric plants, and other renewable methods (geothermal, wind, tidal, solar). The combination of methods used to generate electricity in any country depends on the natural

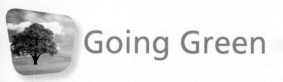

Going Green

Saving Energy at Home

The pie chart shows how energy is used in a typical home in the United States. Looking for ways to reduce energy use in each of these categories can lead to lower energy uses and lower energy bills. The cheapest energy is the energy you don't use. Here are some ways to conserve energy.

Home Heating and Cooling

- Install a programmable thermostat. Lowering the temperature during the night when you are sleeping or during the day when no one is at home can reduce energy consumption by 5–15 percent. The energy savings will pay for the cost of the programmable thermostat in less than a year.

- Lower the temperature to 20°C (68°F) during the heating season and raise the temperature during the summer cooling period. When the heating system or air conditioner is not running, it is not using energy.

Lighting

- Replace any remaining incandescent light bulbs with compact fluorescent light bulbs. Compact fluorescent bulbs are 3 to 4 times more efficient and will pay for themselves in less than a year. Turn off lights in a room if you are going to be gone for more than 15 minutes.

Water Heating

- Lower the thermostat on your water heater to 49°C (120°F). Higher temperatures lead to higher heat loss and greater energy consumption.

- Insulate your water heater.

- Take short showers instead of baths. The less hot water used the less energy used.

- Wash only full loads of dishes and clothes. The fewer times you run the washer the less total hot water used.

Appliances

- Air dry dishes instead of using your dishwasher's drying cycle.

- Look for the ENERGY STAR label on home appliances and products. ENERGY STAR products meet strict efficiency guidelines set by the U.S. Department of Energy and the Environmental Protection Agency.

Electronics

- Turn off your computer and monitor if you're not going to use your PC for more than 2 hours. Make sure your monitors, printers, and other accessories are on a power strip/surge protector and turn off the power strip when they are not in use. In standby mode, they are constantly using energy.

- Plug home electronics, such as TVs and DVD players, into power strips; turn the power strips off when the equipment is not in use (TVs and DVDs in standby mode still use several watts of power).

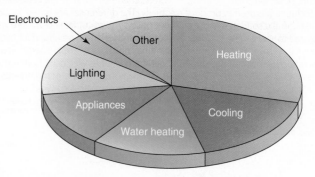

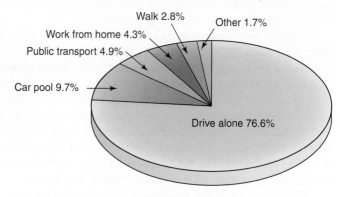

FIGURE 8.10 How Americans Get to Work The vast majority of Americans commute to work in an automobile with one person in it. (*Latest data available from the U.S. Census Bureau.*)

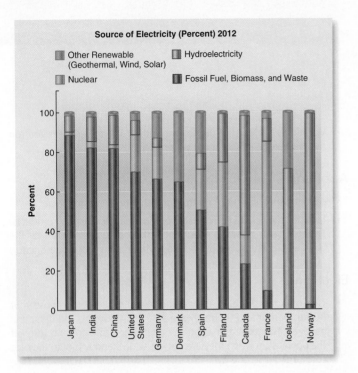

FIGURE 8.11 Sources of Electricity Electricity is generated from fossil fuels, nuclear power, hydropower, and other renewable forms of energy (wind, solar, geothermal, tidal). However, countries differ in the combination of technologies used to produce electricity. The differences are based on available natural resources and government policy. The countries shown here were selected to show these differences. Norway and Iceland have abundant hydroelectricity resources. In addition, Iceland has much geothermal energy. France has made the political decision to generate most of its electricity from nuclear power plants. Denmark, Spain, and Germany have made commitments to produce electricity from renewable technologies—particularly wind.

Note: Japan normally had over 20 percent of its electricity produced by nuclear reactors. However, the earthquake in 2011 caused the shutdown of nuclear reactors and normal production had not returned as of 2013.

Sources: International Energy Agency online data 2013.

resources of the country and government policy. Figure 8.11 shows several countries that were selected to demonstrate how individual countries use different combinations of technologies to produce electricity. Norway gets nearly all of its electricity from hydroelectric plants. Iceland gets its electricity from a combination of hydroelectricity and geothermal energy. France relies heavily on nuclear power. Most countries have a substantial amount of their energy produced from fossil fuels, primarily coal.

As with other forms of energy use, electrical consumption in different regions of the world varies widely. The industrialized countries of the world, with about 20 percent of the world's population, consume about 50 percent of the world's electricity. The per capita use of electricity in North America is about 9½ times greater than average per capita use in the less-developed countries. In Bangladesh, the annual per capita use of electricity is about 140 kilowatt hours, which is enough to light a 100-watt light bulb for less than two months. The per capita consumption of electricity in North America is about 90 times greater than in Bangladesh.

The production and distribution of electricity is a major step in the economic development of a country. In developed nations,

about a quarter of the electricity is used by industry. The remainder is used primarily for residential and commercial purposes. In nations that are developing their industrial base, much of their electricity is used by industry. For example, industries consume about 50 percent of the electricity used in South Korea and China.

8.4 The Economics and Politics of Energy Use

A direct link exists between economic growth and the availability of inexpensive energy. The replacement of human and animal energy with fossil fuels began with the Industrial Revolution and was greatly accelerated by the supply of cheap, easy-to-handle, and highly efficient fuels. Because the use of inexpensive fossil fuels allows each worker to produce more goods and services, productivity increased. The result was unprecedented economic growth in Europe, North America, and the rest of the industrialized world.

Because of this link between energy and productivity, most industrial societies want to ensure a continuous supply of affordable energy. The higher the price of energy, the more expensive goods and services become. To keep costs down, many countries have subsidized their energy industries and maintained energy prices at artificially low levels. International trade in fossil fuels (particularly oil and natural gas) has a major influence on the world economy and politics. The price of fuels has a huge impact on consumption: low prices encourage consumption and high prices discourage consumption.

Fuel Economy and Government Policy

Governments fashion policies that influence how people use energy. Automobile fuel efficiency is one area in which government policy has had significant impact. For example, the price of a liter of gasoline is determined by two major factors: (1) the cost of purchasing and processing crude oil into gasoline and transporting it to the station and (2) various taxes. Most of the differences in gasoline prices among countries are a result of taxes and reflect differences in government policy toward motor vehicle transportation. The cost of taxes to the U.S. consumer is about 11 percent of the retail gasoline price, and in Canada, about 30 percent of the price of gasoline is taxes, while in Japan and many European countries, taxes account for 40 percent to 60 percent of the cost of gasoline. (See figure 8.12.) When we compare the kinds of automobiles driven, we find a direct relationship between the cost of fuel and fuel efficiency. In the United States and Canada, the average fleet fuel economy is about 8.1 liters per 100 kilometers (29 miles per gallon). This compares with a European average of about 5.5 liters per 100 kilometers (43 miles per gallon). The average European car driver pays more than twice as much for fuel as U.S. and Canadian drivers and uses about 30 percent less fuel to drive the same distance as a U.S. driver. Since taxes make up the majority of the price of gasoline in Europe, government tax policy has provided an incentive for people to purchase fuel-efficient automobiles.

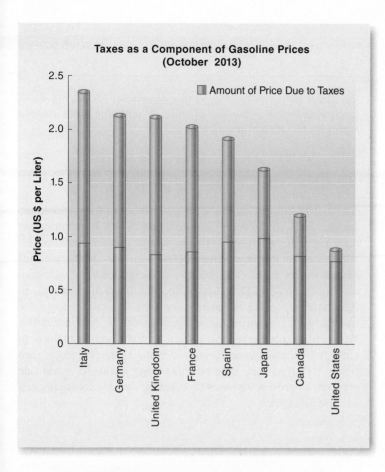

Taxes as a Component of Gasoline Prices (October 2013)

Price (US $ per Liter)

☐ Amount of Price Due to Taxes

Italy, Germany, United Kingdom, France, Spain, Japan, Canada, United States

FIGURE 8.12 Gasoline Taxes and Fuel Efficiency The price paid for fuel is greatly influenced by the amount of tax paid. High fuel prices cause consumers to choose automobiles with greater fuel efficiency.
Source: International Energy Agency online data.

Another objective of governments is to have a mechanism for generating the money needed to build and repair roads. Many European countries raise more money from fuel taxes than they spend on building and repairing roads. The United States, on the other hand, raises approximately 50 percent of the monies needed for roads from fuel taxes, tolls, and other sources of revenue. The relatively low cost of fuel in the United States encourages more travel, which increases road repair costs.

The European Union has also taken steps to comply with the Kyoto Treaty, which mandates reductions in the amount of carbon dioxide released into the atmosphere. To achieve carbon dioxide reduction targets, European automobile manufacturers had voluntarily agreed to reduce their carbon dioxide emissions to 140 grams of carbon dioxide per kilometer by 2008. They did reach 135.7 grams of CO_2 per km (g CO_2/km) in 2011. In 2013 the European Union set a new target of 95 g CO_2/km for 2020. This is equivalent to 59 U.S mpg. Many European countries have instituted taxes based on the amount of carbon dioxide emitted by a car. People who own cars that emit higher amounts pay much higher taxes. By contrast, the United States has not signed the Kyoto Treaty and had falling fuel efficiency throughout the 1990s as more people bought and drove SUVs until about 2005. There have been tiny gains in fuel efficiency in the United States

since 2005. New fuel-efficiency standards announced by President Obama in 2011 call for an increase in fuel economy to 54.5 mpg by 2025.

Electricity Pricing

Since electricity is such an important source of energy for many uses, it is interesting to look at how different countries structure electricity prices. Because of the nature of the electrical industry, most countries regulate the industry and influence the price utilities are able to charge. Furthermore, the cost to industrial users is typically about half that charged to residential customers. In general, higher prices discourage use.

The Importance of OPEC

The Organization of Petroleum Exporting Countries (OPEC) began in September 1960, when the governments of five of the world's leading oil-exporting countries agreed to form a cartel. Three of the original members—Saudi Arabia, Iraq, and Kuwait— were Arab countries, while Venezuela and Iran were not. Today, 13 countries belong to OPEC. These include seven Arab states— Saudi Arabia, Kuwait, Libya, Algeria, Iraq, Qatar, and United Arab Emirates—and six non-Arab members—Iran, Indonesia, Nigeria, Ecuador, Angola, and Venezuela. OPEC nations control about 72 percent of the world's estimated oil reserves of 1,200 billion barrels of oil. Middle Eastern OPEC countries control about 50 percent of this total, which makes OPEC and the Middle East important world influences. Today, OPEC countries control more than 40 percent of the world's oil production and are a major force in determining price. (See table 8.1.)

Increased solidarity among OPEC countries, continuing political instability in the Middle East, and increased demand by countries such as China and India coupled with changes in the value of the dollar and activities by oil speculators, leads to large fluctuations in oil prices. From 2011 through 2013 oil has traded around $100 per barrel.

8.5 Energy Consumption Trends

From a historical point of view, it is possible to plot changes in energy consumption. Economics, politics, public attitudes, and many other factors must be incorporated into an analysis of energy use trends.

Growth in Energy Use

In 2012, world energy consumption was around 12,500 million metric tons of oil equivalent, an increase of more than 20 percent over 10 years. Of this total, conventional fossil fuels—oil, natural gas, and coal—accounted for nearly 90 percent.

Nearly half of world energy is consumed by the 25 countries that are members of the Organization for Economic Cooperation and Development (OECD). These countries (Australia, New Zealand, Japan, Canada, Mexico, the United States, and the

What's Your Take?

The price of gasoline is a hot topic. Although Americans pay less than half as much for gasoline as people who live in other economically developed countries, many feel that U.S prices are too high. Other people feel that increasing the price of gasoline is necessary to stimulate change in the way Americans use energy. Choose to support either the idea of keeping gasoline prices low or the idea of artificially increasing prices. Develop arguments that support your position.

Table 8.1 Major World Oil-Producing Countries

	Country	Million Metric Tons 2013	Percent of Total
OPEC	Saudi Arabia	547.0	13.3
	Russian Federation	526.2	12.8
	United States	394.9	9.6
	China	207.5	5.0
OPEC	Iran	174.9	4.2
	Canada	186.6	4.4
OPEC	United Arab Emirates	154.1	3.7
OPEC	Kuwait	152.5	3.7
OPEC	Iraq	152.4	3.7
	Mexico	143.9	3.5
OPEC	Venezuela	139.7	3.4
OPEC	Nigeria	116.2	2.8
	Brazil	112.2	2.7
	Norway	87.5	2.1
OPEC	Angola	86.9	2.1
OPEC	Qatar	83.3	2.0
	Kazakhstan	81.3	2.0
OPEC	Algeria	73.0	1.8
OPEC	Libya	71.1	1.7
	Columbia	49.9	1.2
	Oman	45.8	1.1
	United Kingdom	45.0	1.1
OPEC	Indonesia	44.6	1.1
	Azerbaijan	43.3	1.1
	India	42.0	1.0
	Egypt	35.4	0.9
	Argentina	31.0	0.8
	Malaysia	29.7	0.7
OPEC	Ecuador	27.1	0.7
	World	4118.9	

Source: *BP Statistical Review of World Energy*, 2013.

countries of Europe) are the developed nations of the world. In the decade (2003-2012), energy consumption in OECD countries has been essentially unchanged. This is in large part due to the worldwide recession that began about 2008. There has also been a shift toward service-based economies, with energy-intensive industries moving to non-OECD countries. In contrast, many countries in Asia have expanding economies and have increased energy use by nearly 70 percent in the ten years between 2003 and 2012. We should expect to see this pattern continue and countries with emerging economies increasingly demanding more energy. (See figure 8.13.) As they demand more energy and the worldwide economy improves, we should see greater demand for energy and an increase in price.

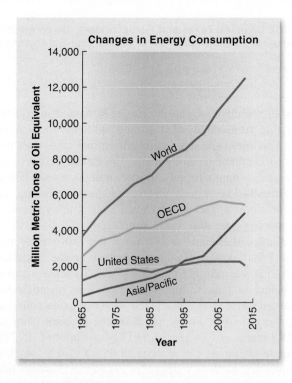

FIGURE 8.13 Changes in World Energy Consumption by Region World energy consumption has increased steadily. Currently the energy consumption of the economically developed nations (OECD) accounts for 45 percent of world consumption. In recent years the fastest growth in energy use has been in the Asia Pacific region while energy use in the developed countries has been declining.

Source: Data from *BP Statistical Review of World Energy*, 2013.

Available Energy Sources

Oil remains the world's major source of energy, accounting for about 33 percent of primary energy demand. Coal accounts for 30 percent and natural gas for 24 percent; the remainder is supplied mainly by nuclear energy and hydropower. These percentages are likely to continue into the near future.

Political and Economic Factors

Political and economic factors have a great deal of influence on energy consumption. The primary factors that determine energy use are worldwide economic conditions, political stability in parts of the world that supply oil, and the price of that oil. Since OPEC and countries of the Middle East control over 40 percent of the world's oil production and over 70 percent of the oil reserves, political stability in this region is very important. (See figure 8.14.)

The energy consumption behavior of most people is motivated by economics rather than by a desire to use energy resources wisely. When the price of energy increases, the price of everything else increases as well and eventually consumption falls. Conversely, when energy prices fall, consumption increases.

During the 1980s, energy costs declined and people in North America and Europe became less concerned about their energy consumption. They used more energy to heat and cool their homes and buildings, bought and used more home appliances, and bought bigger cars. Governments can manipulate energy prices by increasing or decreasing taxes on energy, granting subsidies to energy producers, and using other means.

Over the past several years, world oil prices have been extremely volatile. As recently as 1999, consumers benefited from oil prices that fell to under US $13 per barrel—a result of oversupply caused by lower demand for oil in both southeast Asia, which was suffering from an economic recession, and North America and Western Europe, which had warmer than expected winters. Since then there has been an increase in the price of oil due to increased demand from China, India, and other rapidly industrializing countries, political instability in the Middle East—in part the result of the wars in Iraq and Afghanistan—and increased solidarity among OPEC countries. In 2003, the price of oil was about $32 per barrel. In 2008, it exceeded $147 per barrel—a 360 percent increase in five years before falling at the end of 2008. As economic conditions improve and demand increases, the price is expected to rise.

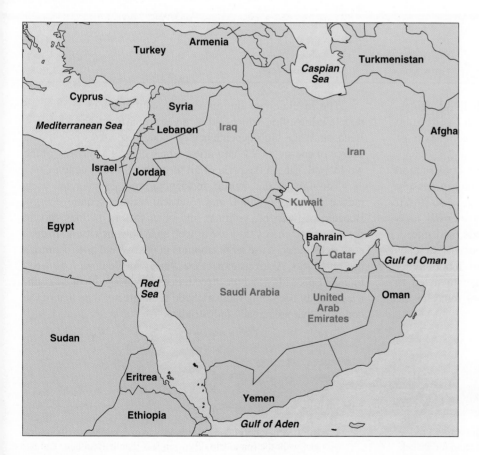

FIGURE 8.14 Persian Gulf OPEC Countries The six OPEC countries of the Persian Gulf (Saudi Arabia, Iraq, Iran, Kuwait, Qatar, and United Arab Emirates) control a major portion of the oil reserves and oil production of the world. Political instability in this region leads to increased oil prices.

Issues & Analysis

Government Action and Energy Policy

Governments use economic tools to encourage desired behaviors. Taxes on goods or services artificially increase the price and encourage a search for alternatives. Subsidies are a gift from government to encourage individuals or corporations to pursue certain kinds of activities. Government regulations specify particular standards or actions and fines or fees are charged to those who do not meet the standards.

With respect to energy policy, there are several ways in which these tools are used:

1. The issue of global warming has generated the idea of a carbon tax. Carbon dioxide is released when fossil fuels are used, so a carbon tax would be a tax on energy. This tax would artificially raise the price of energy and cause people to change their behavior to use less energy or shift to forms of energy that release less carbon dioxide.

2. Taxes on gasoline and diesel fuel can be used to encourage consumers to change their driving habits or the kinds of vehicles they drive.

3. Oil companies receive large subsidies to encourage them to explore for oil. Small subsidies are also provided for the exploration of certain alternative energy sources such as solar energy and wind energy.

4. Tax credits to individuals who make energy-efficiency changes to their homes encourage investment in energy-saving technology.

5. Fuel economy standards for motor vehicles require manufacturers to meet standards by a certain date or they must pay a fine.

All of these activities have their supporters and detractors. How do you feel?

- Do you support subsidies to companies to explore for oil?
- Would you support increased taxes on gasoline?
- Should government impose a carbon tax on all fossil fuels?
- Should automobile manufacturers be forced to make more fuel-efficient vehicles?
- Should government provide tax breaks to individuals who change the way they use energy?

Summary

A constant supply of energy is required by all living things. Energy has a major influence on society. A direct correlation exists between the amount of energy used and the complexity of civilizations.

Wood furnished most of the energy and construction materials for early civilizations. Heavy use of wood in densely populated areas eventually resulted in shortages, so fossil fuels replaced wood as a prime source of energy. Fossil fuels were formed from the remains of plants, animals, and microorganisms that lived millions of years ago. Fossil-fuel consumption in conjunction with the invention of labor-saving machines resulted in the Industrial Revolution, which led to the development of technology-oriented societies today in the developed world.

The invention of the automobile caused major changes in people's lifestyles, which led to greater consumption of energy. Because of the high dependence of modern societies on oil as a source of energy, OPEC countries, which control a majority of the world's oil, can set the price of oil through collective action.

Throughout the world, residential and commercial uses, industries, transportation, and electrical utilities require energy. Because of financial, political, and other factors, nations vary in the amount of energy they use as well as in how they use it. In general, rich countries use large amounts of energy and poor countries use much less. Analysts expect the worldwide demand for energy to increase steadily and the growth in energy usage by those countries that are becoming industrialized to be greater than that of the countries that are already industrialized.

Acting Green

1. List ten activities you engage in each day that require fossil fuel or electrical energy. Choose one that you will abstain from for a week.

2. Keep a record of your transportation activities for one week. Record miles driven, subway fares paid, taxi fares, etc. How many of the trips were really necessary?

3. Write your governmental representative expressing your opinion on energy policy.

4. If a trip to a neighborhood store is less than 1 kilometer (0.6 mile), you can walk the round trip (2 km) in less than half an hour and not use any fossil-fuel energy. If you used a car, how long would it take to make the same round trip? (Time yourself from the time you leave your room until you enter the store and reverse the process on your return.)

Review Questions

1. How did the domestication of animals change energy use in early cultures?
2. In addition to food, what energy requirements does a civilization have?
3. How was the availability of coal important in determining if a country participated in the Industrial Revolution?
4. What factors caused a shift from wood to coal as a source of energy?
5. What major factor caused a shift to the use of oil as a source of energy?
6. Describe two factors that have led to the dominance of automobiles as a form of transportation in the United States.
7. Describe two actions governments take that cause changes in how citizens use energy.
8. What advantages does electrical energy have over other kinds of energy?
9. State two reasons the cost of electricity differs from one country to another.
10. List the three primary categories of energy use in industrialized societies.
11. Why is OPEC important in the world's economy?
12. Give examples of how political and economic events affect energy prices and usage.
13. Based on current trends, what is likely to happen to the availability and price of energy in the next ten years?

Critical Thinking Questions

1. Imagine you are a historian writing about the Industrial Revolution. Imagine that you also have your new knowledge of environmental science and its perspective. What kind of a story would you tell about the development of industry in Europe and the United States? Would it be a story of triumph or tragedy, or some other story? Why?
2. What might be some of the effects of raising gasoline taxes in the United States to the rate that most Europeans pay for gasoline? Why? What do you think about this possibility?
3. Some argue that the price of gasoline in the United States is artificially low because it does not take into account all of the costs of producing and using gasoline. If you were to figure out the "true" cost of gasoline, what kinds of factors would you want to take into account?
4. How has the ubiquitous nature of automobiles changed the United States? Do you feel these changes are, on balance, positive or negative? What should the future look like regarding automobile use in the United States? How can this be accomplished?
5. The Organization of Petroleum Exporting Countries (OPEC) controls over 70 percent of the known oil reserves. What political and economic effects do you think this has? Does this have any effect on energy use?
6. How do you think projected energy consumption will affect world politics and economics, given current concerns about global warming?

To access additional resources for this chapter, please visit ConnectPlus ® at connect.mheducation.com. There you will find interactive exercises, including Google Earth™, additional Case Studies, and SmartBook™, an adaptive eBook that integrates our LearnSmart® adaptive learning technology.

chapter
9
Nonrenewable Energy Sources

Both coal and oil are depicted in this photograph. The huge coal train is passing an oil well. Oil is typically transported by pipelines.

OBJECTIVES

After reading this chapter, you should be able to:

- Identify fossil fuels as the major sources of energy for industrialized nations.
- Differentiate between nonrenewable and renewable sources of energy.
- Differentiate between resources and reserves.
- Describe three factors that cause the amount of reserves to change.
- Identify peat, lignite, bituminous coal, and anthracite coal as steps in the process of coal formation.
- Recognize that natural gas and oil are formed from ancient marine deposits.
- List the regions of the world that have large reserves of coal, oil, or natural gas.
- Explain why surface mining of coal is used in some areas and underground mining in other areas.

- Describe three negative environmental impacts associated with the extraction and use of coal.
- Explain why it is more expensive to find and produce oil today than it was in the past.
- Recognize that secondary recovery methods have been developed to increase the proportion of oil and natural gas obtained from deposits.
- Recognize that transport of natural gas is still a problem in some areas of the world.
- Describe the major environmental issues related to the extraction and use of oil and natural gas.
- Explain why the increased cost of oil and concerns about climate change have influenced attitudes toward nuclear power.

- Describe the process of nuclear fission.
- Describe a nuclear chain reaction.
- Describe how a nuclear reactor produces electricity.
- Describe the basic types of nuclear reactors.
- Explain how a breeder reactor differs from other nuclear reactors.
- List the steps in the nuclear fuel cycle.
- Describe the kinds of radiation produced by nuclear fission.
- Describe how the concepts of time, distance, and shielding are related to protection from radiation.
- Compare the effects of the accidents at Chernobyl, Three Mile Island, and Fukushima Dai-ichi.
- Explain the process of decommissioning a nuclear plant.

Hydraulic Fracturing

Oil and natural gas deposits are not present in pools in the ground but are present as small accumulations in the spaces between sand grains or within cracks in rocky deposits deep underground. Oil and gas can move relatively easily through porous sediments like sand but do not flow easily through nonporous materials like shale. Shale is sedimentary rock. It formed millions of years ago, when clay and silt particles settled from slow moving water and were subsequently converted to rock by pressure and heat. Since these deposits contained aquatic organisms (particularly microorganisms) the organic compounds from these organisms were converted to oil and gas during the process of shale formation. Hydraulic fracturing (also known as *hydrofracking* or simply *fracking*) is an advanced technology for obtaining oil and natural gas from geologic formations like shale that are not porous.

Hydraulic fracturing is not a new technology. It has been used in over a million wells since the 1940s, mostly in vertically drilled oil wells. However, since the 1990s it has become extremely important in producing oil and gas from shale. The technology involves modifications to how a well is drilled and the pumping of materials into the well to improve production. To allow for hydraulic fracturing, a vertical bore hole is drilled down to the layer of shale but then is turned horizontally so that a large area of the shale deposit is intersected by the bore hole. The next step in the process is to pump huge volumes (millions of liters) of water with a small amount of sand and some chemicals under high pressure into the well. The pressure increases the size of existing fractures in the rock and creates new ones. When the water is removed, the sand is left behind in the shale and props open the fractures. This combination of techniques releases gas or oil from the shale and allows it to be pumped to the surface.

All extraction methods for fossil fuels have environmental effects, and hydraulic fracturing is no different.

- Since millions of liters of water are used for each well, management of water resources is an issue. This is particularly true in arid regions that have little surface water and where aquifers are already being fully utilized.
- Since a portion of the water used in hydraulic fracturing is returned to the surface and contains contaminants, it must be treated to remove the contaminants or disposed of in an environmentally acceptable way. Often these fluids are pumped into disposal wells.
- Some oil or gas wells leak, and if the leak is near the ground's surface, aquifers can be contaminated.

- Shale deposits may contain harmful materials that are transported to the surface when drilling fluids are removed and when gas or oil is extracted. These may include radioactive materials, salts, or other chemicals.

Regardless of the environmental implications, hydraulic fracturing has had an enormous economic impact. In 2000 about 1 percent of natural gas produced in the United States was from hydraulic fracturing. By 2011 this had increased to 25 percent because thousands of new wells were drilled. The price of natural gas has decreased, allowing power companies and other industries to switch from coal or oil to natural gas. This has resulted in lower carbon dioxide emissions. The use of hydraulic fracturing in North Dakota and Montana has been used to release oil from shale and has provided a major new source of oil for the United States. Since the production of oil and natural gas by hydraulic fracturing involves the drilling of thousands of wells, thousands of jobs have been created.

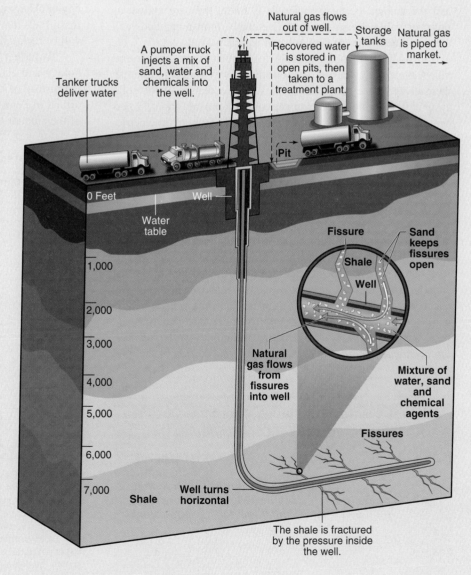

9.1 Major Energy Sources

Chapter 8 outlined the historical development of energy consumption and how advances in civilizations were closely linked to the availability and exploitation of energy. New manufacturing processes relied on dependable sources of energy. Technology accelerated in the twentieth century. Between 1900 and 2010, world energy consumption increased by a factor of about 17 and economic activity increased about 40 times, but population increased only slightly more than four times. (See table 9.1.)

The energy sources most commonly used by industrialized nations are the fossil fuels: oil, coal, and natural gas, which supply about 80 percent of the world's energy. Fossil fuels were formed hundreds of millions of years ago. They result from the accumulation of energy-rich organic molecules produced by organisms as a result of photosynthesis over millions of years. We can think of fossil fuels as concentrated, stored solar energy. The rate of formation of fossil fuels is so slow that no significant amount of fossil fuels will be formed over the course of human history. Since we are using these resources much faster than they can be produced and the amount of these materials is finite, they are known as **nonrenewable energy sources.** Eventually, human demands will exhaust the supplies of coal, oil, and natural gas.

In addition to nonrenewable fossil fuels, there are several renewable energy sources. **Renewable energy sources** replenish themselves or are continuously present as a feature of the solar system. Renewable energy sources currently provide about 12 percent of the energy used worldwide primarily from hydroelectricity and firewood. Renewable energy sources will be discussed in detail in chapter 10.

9.2 Resources and Reserves

When discussing deposits of nonrenewable resources, such as fossil fuels, iron ore, or uranium ore, it is important that we differentiate between deposits that can be extracted and those that cannot. From a technical point of view, a **resource** is a naturally occurring substance of use to humans that can *potentially* be extracted. **Reserves** are known deposits from which materials *can* be extracted profitably with existing technology *under prevailing economic conditions*. Thus, reserves are a portion of the total resource. It is important to recognize that the concept of *reserves* is an economic idea and is only loosely tied to the total quantity of a material present in the world. (See figure 9.1.)

Both terms are used when discussing the amount of fossil-fuel or ore deposits a country has at its disposal. This can cause considerable confusion if the difference between these concepts is not understood. The total amount of a *resource* such as coal or oil changes only by the amount used each year. The amount of a *reserve* changes as technology advances, new deposits are discovered, and economic conditions vary. Furthermore, countries often restate the amount of their reserves for political reasons. Thus, there can be large increases in the amount of *reserves,* while the total amount of the *resource* falls.

When we read about the availability of fossil fuels, we must remember that if the cost of removing and processing a fuel is greater than the fuel's market value, no one is going to produce it. Also, if the amount of energy used to produce, refine, and transport a fuel is greater than the energy produced when it is burned, the fuel will not be produced. A net useful energy yield is necessary to exploit the resource. However, in the future, new technology or changing prices may permit the profitable removal of some fossil fuels that currently are not profitable. If so, those resources will be reclassified as reserves.

To further illustrate the concept of reserves and how technology and economics influence their magnitude, let us look at the history of oil. When the first oil well in North America was drilled in Pennsylvania in 1859, it greatly expanded the estimate of the amount of oil in the Earth. There was a sudden increase in

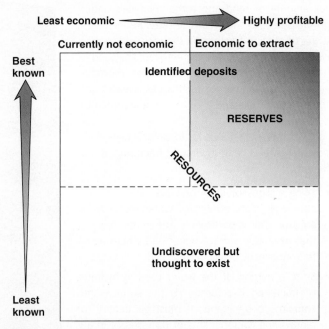

FIGURE 9.1 Resources and Reserves A resource is the quantity of a material thought to exist and includes undiscovered deposits and deposits that currently cannot be profitably used, although it might be feasible to do so in the future if technology or market conditions change. A reserve is the portion of a resource that can be profitably obtained using current technology under current economic conditions. Reserves are shown in the box in the upper right-hand corner in this diagram. The darker the color, the better known the deposit and the more profitable it is to extract.
Source: Adapted from the U.S. Bureau of Mines.

Table 9.1 World Population, Economic Output, and Fossil Fuel Consumption

	Population (Billions)	Gross World Product (Trillion 1990 International $)	Fossil-Fuel Consumption (Billion Metric Tons Coal Equivalent)
1900	1.6	1.1	1
1950	2.5	4.1	3
2010	6.9	44.0	17

the known oil reserves. In the years that followed, new deposits were discovered. Better drilling techniques led to the discovery of deeper oil deposits, and offshore drilling established the location of oil under the ocean floor. At the time of their discovery, these deep deposits and the offshore deposits added to the estimated size of the world's oil resources. But they did not necessarily add to the reserves because it was not always profitable to extract the oil. With advances in drilling and pumping methods and increases in oil prices, it eventually became profitable to obtain oil from many of these deposits. As it became economical to extract them, they were reclassified as reserves. (See figure 9.2.)

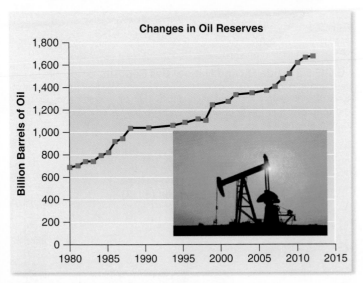

FIGURE 9.2 Changes in Proved Oil The figure shows the changes in proved oil reserves from 1980 to 2012. The major changes in the 1980s were primarily due to more accurate reporting rather than new discoveries.
Source: Data from *BP Statistical Review of World Energy*, 2013.

9.3 Fossil-fuel Formation

Fossil fuels are the remains of once-living organisms that were preserved and altered as a result of geologic forces. Significant differences exist in the formation of coal from that of oil and natural gas.

Coal

Coal was formed from plant material that had been subjected to heat and pressure. Freshwater swamps covered many regions of the Earth 300 million years ago. Conditions in these swamps favored extremely rapid plant growth, resulting in large accumulations of plant material. Because this plant material collected under water, decay was inhibited, and a spongy mass of organic material formed. It is thought that the chemical nature of these ancient plants and the lack of many kinds of decay organisms at that time also contributed to the accumulation of plant material. Today we see deposits of plant materials, known as peat, being formed in bogs.

Due to geologic changes in the Earth, some of these organic deposits were submerged by seas. The plant material that had collected in the swamps was then covered by sediment. The weight of the sediment on top of the deposit compressed it and heat from the Earth caused the evaporation of water and other volatile compounds. Thus, the original plant material was transformed into coal. Depending on the amount of time the organic matter has been subjected to geologic processes, several different grades of coal are produced. (See figure 9.3.) Figure 9.4 describes the distribution of world coal deposits.

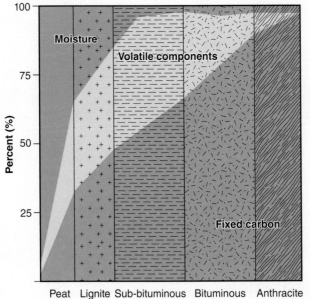

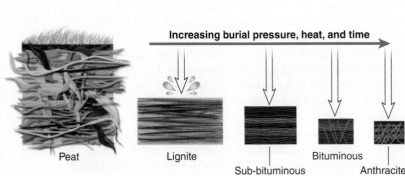

FIGURE 9.3 Different Kinds of Coal Coal is formed from plant material that has been buried and subjected to pressure and heat. Thus, there are several kinds of coal based on the extent of changes that have occurred. Peat is recently dead plant material that has been altered very little and has a high moisture content. Lignite, sub-bituminous, bituminous, and anthracite coal have progressively less water and higher carbon content.

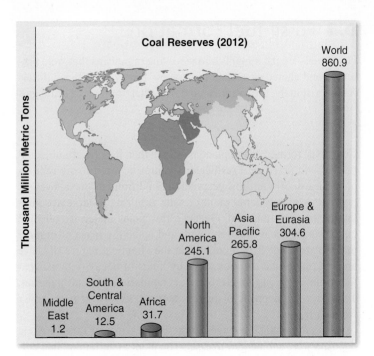

FIGURE 9.4 **Recoverable Coal Reserves of the World 2012** The majority of the coal deposits of the world are in North America, Europe, Eurasia, and the Asia/Pacific regions.
Source: Data from *BP Statistical Review of World Energy,* 2013.

Oil and Natural Gas

Oil and natural gas, like coal, are products from the past. They probably originated from microscopic marine organisms. When these organisms died and accumulated on the ocean bottom and were buried by sediments, their breakdown released oil droplets. Gradually, the muddy sediment formed rock called shale, which contained dispersed oil droplets. Although shale is common and contains a great deal of oil, extraction from shale is difficult because the oil is not concentrated. However, in instances where a layer of porous sandstone formed on top of the oil-containing shale and an impermeable layer of rock formed on top of the sandstone, concentrations of oil often form. Usually, the trapped oil does not exist as a liquid mass but rather as a concentration of oil within sandstone pores, where it accumulates because water and gas pressure force it out of the shale. (See figure 9.5.) These accumulations of oil are more likely to occur if the rock layers were folded by geological forces.

Natural gas, like oil, forms from fossil remains. If the heat generated within the Earth reached high enough temperatures, natural gas could have formed along with or instead of oil. This would have happened as the organic material changed to lighter, more volatile (easily evaporated) hydrocarbons than those found in oil. The most common hydrocarbon in natural gas is the gas methane (CH_4). Water, liquid hydrocarbons, and other gases may be present in natural gas as it is pumped from a well.

The conditions that led to the formation of oil and gas deposits were not evenly distributed throughout the world. Figure 9.6 illustrates the geographic distribution of oil reserves. The Middle East has about 50 percent of the world's oil reserves. Figure 9.7 shows the geographic distribution of natural gas reserves. Eurasia (primarily Russia) and the Middle East have about 70 percent of the world's natural gas reserves.

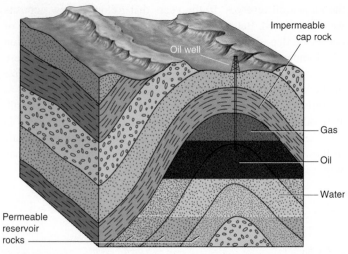

FIGURE 9.5 **Crude Oil and Natural Gas Deposits** Oil and natural gas are found in particular geological conditions where folding or movement of layers of sedimentary rock create conditions that allow water and gas pressure to force oil and gas out of shale and into porous sandstone capped by impermeable rock.

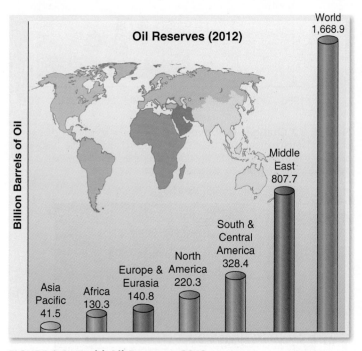

FIGURE 9.6 **World Oil Reserves 2012** The world's supply of oil is not distributed equally. The Middle East controls about 50 percent of the world's oil reserves.
Source: Data from *BP Statistical Review of World Energy,* 2013.

9.4 Issues Related to the Use of Fossil Fuels

The three nonrenewable fossil-fuel resources—coal, oil, and natural gas—supply over 80 percent of the energy consumed worldwide. Each fuel has advantages and disadvantages and requires special techniques for its production and use.

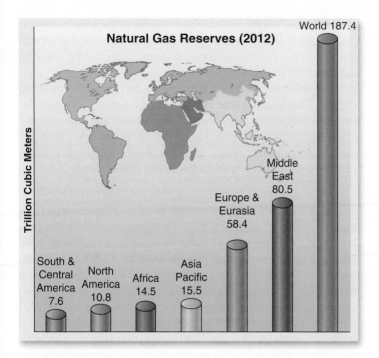

FIGURE 9.7 World Natural Gas Reserves 2012 Natural gas reserves, like oil and coal, are concentrated in certain regions of the world. The Middle East and Europe and Eurasia have over 70 percent of the world's natural gas reserves.

Source: Data from *BP Statistical Review of World Energy, 2013.*

Coal Use

Coal is the world's most abundant fossil fuel, but it supplies less than 30 percent of the energy used in the world. It varies in quality and is generally classified in four categories: lignite, sub-bituminous, bituminous, and anthracite. Lignite (brown) coal has a high moisture content and is crumbly in nature, which makes it the least desirable form. It has a low energy content that makes transportation over long distances uneconomic. Therefore, most lignite is burned in power plants built near the coal mine. Over 60 percent of the lignite used is from Europe. Sub-bituminous coal has a lower moisture content and a higher carbon content (46–60 percent) than lignite and is typically used as fuel for electric power plants. Bituminous (soft) coal has a low moisture content and a high carbon content (60–86 percent). It is primarily used in electrical power generation but is also used in other industrial applications such as cement production and steel making. Bituminous coal is the most widely used because it is the easiest to mine and the most abundant. It supplies about 20 percent of the world's energy requirements. Anthracite (hard) coal is 86–98 percent carbon. It is relatively rare and is used primarily in heating of buildings and for specialty uses.

Extraction Methods

Because coal was formed as a result of plant material being buried under layers of sediment,

it must be mined. There are two methods of extracting coal: surface mining and underground mining. (See figure 9.8.) **Surface mining** (strip mining) involves removing the material located on top of a vein of coal, called **overburden,** to get at the coal beneath. Coal is usually surface mined when the overburden is less than 100 meters (328 feet) thick. This type of mining operation is efficient because it removes most of the coal in a vein and can be profitably used for a seam of coal as thin as half a meter. For these reasons, surface mining results in the best utilization of coal reserves. Advances in the methods of surface mining and the development of better equipment have increased surface-mining activity in the United States from 30 percent of the coal production in 1970 to more than 70 percent today.

If the overburden is thick, surface mining becomes too expensive, and the coal is extracted through **underground mining.** The deeply buried coal seam can be reached in two ways. In flat country where the vein of coal lies buried beneath a thick overburden, the coal is reached by a vertical shaft. In hilly areas where the coal seam often comes to the surface along the side of a hill, the coal is reached from a drift-mine opening.

Health and Safety Issues

Health and safety are important concerns related to coal mining, which is one of the most dangerous jobs in the world. In the United States during the decade of 2000–2010, there was an average of over 60 deaths per year and about 12,000 injuries related to all mining activities—not just coal mining. This is particularly true with underground mining. Explosions of methane gas and accidents kill many miners each year. Many miners suffer from **black lung disease,** a respiratory condition that results from the accumulation of fine coal-dust particles in the miners' lungs. (Over 8 percent of miners who worked 25 or more years in mines tested positive for black lung disease.) The coal particles inhibit the exchange of gases between the lungs and the blood. The health

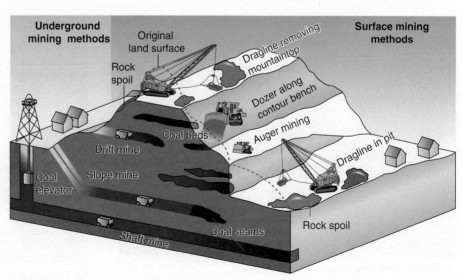

FIGURE 9.8 Coal Mining Methods Surface mining (strip mining) involves the use of huge machines to remove the overlying soil and rock to get at the layer of coal beneath. Underground mining may involve a shaft going down into the Earth or, if the vein of coal comes to the surface on the side of a hill, a drift mine is possible.

Nonrenewable Energy Sources   193

care costs and death benefits related to black lung disease are an indirect cost of coal mining. Since these health care costs are partially paid by the federal government, the total health care cost is not reflected in the price of coal but is paid by taxpayers in the form of federal taxes and higher health insurance premiums.

Transportation Issues

Because coal is bulky, shipping presents a problem. Generally, the coal can be used most economically near where it is produced. Rail shipment is the most economic way of transporting coal from the mine. Rail shipment costs include the expense of constructing and maintaining the tracks, as well as the cost of the energy required to move the long strings of railroad cars. In some areas, the coal is transferred from trains to ships.

Environmental Issues

The mining, transportation, and use of coal as an energy source present several significant environmental problems.

1. Landscape disturbance Surface mining (strip mining) disrupts the landscape, as the topsoil and overburden are moved to access the coal. One of the most controversial surface-mining methods is **mountaintop removal.** In this method, the top of a mountain is removed to get at the coal vein, and the unwanted soil and rock is pushed into the adjacent valley. This causes problems in addition to the disturbance of the landscape. For example, valleys have streams that drain them, and the soil and rock that is pushed into the valley contaminates the water.

It is possible to minimize the effect of landscape disturbance by reclaiming the area after mining operations are completed. (See figure 9.9.) However, reclamation rarely, if ever, returns the land to its previous level of productivity. The cost of reclamation is included in the price of the coal, so mining companies often seek to do as little as possible. Underground mining methods do not disrupt the surface environment as much as surface mining does,

but subsidence (sinking of the land) occurs if the mine collapses. (See figure 9.10.) In addition, large waste heaps are produced around the mine entrance from the debris that must be removed and separated from the coal.

The processing of coal requires that it be washed before being shipped. This results in the production of a liquid waste material known as coal slurry. It is a mixture of coal and other particles and water. Typically this liquid waste is stored at the processing site behind dams in open ponds. Leakage from such ponds pollutes local streams, and failure of the dams has caused floods.

2. Acid mine drainage Since coal is a fossil fuel formed from plant remains, it contains sulfur, which was present in the proteins of the original plants. Sulfur is associated with **acid mine drainage** and air pollution. (See figure 9.11.) Acid mine drainage occurs when the combined action of oxygen, water, and certain bacteria causes the sulfur in coal to form sulfuric acid. Sulfuric acid can seep out of a vein of coal even before the coal is mined. However, the problem becomes worse when the coal is mined and the overburden is disturbed, allowing rains to wash the sulfuric acid into streams. Streams may become so acidic that they can support only certain species of bacteria and algae. Today, many countries regulate the amount of runoff allowed from mines, but underground and surface mines abandoned before these regulations were enacted continue to contaminate rivers and streams.

3. Air pollution Dust is generated by mining and transportation of coal. The large amounts of coal dust released into the atmosphere at the loading and unloading sites can cause local air pollution problems. If a boat or railroad car is used to transport coal, there is the expense of cleaning it before other types of goods can be shipped.

Air pollution from coal burning releases millions of metric tons of material into the atmosphere and is responsible for millions

(a) Unreclaimed strip mine

(b) Reclaimed strip mine

FIGURE 9.9 Surface-Mine Reclamation (a) This photograph shows a large area that has been surface mined with little effort to reclaim the land. By contrast, (b) is an example of effective surface-mining reclamation. The site has been graded and revegetated so that it provides wildlife habitat.

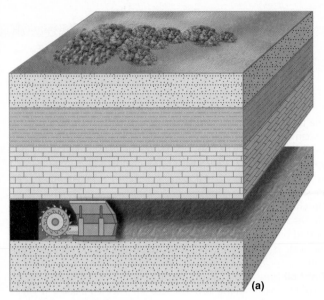 

(a)   (b)

FIGURE 9.10 Subsidence (a) When the coal is removed it leaves a space. It is typical for the underground mines to collapse and cause the land above them to subside. (b) The holes and depressions shown in the photo are the result of subsidence.

of dollars of damage to the environment. The burning of coal for electric generation is the prime source of this type of pollution.

Acid deposition is one of the problems associated with the burning of coal. Acid deposition occurs when coal is burned and sulfur oxides are released into the atmosphere, causing acid-forming particles to accumulate. Each year, over 150 million metric tons of sulfur dioxide are released into the atmosphere worldwide—most comes from coal-fired power plants. This problem is discussed in greater detail in chapter 16.

Mercury is released into the air when coal is burned, since mercury is present in the coal. Mercury is a problem because it tends to accumulate in the bodies of animals and can be in high concentrations in the bodies of predators such as fish. Thus, the federal and state governments have issued advisories on the amount and kinds of fish it is safe to consume. See chapter 16 for additional information on mercury pollution.

Carbon dioxide and climate change are other problems associated with the burning of coal. Increasing amounts of carbon dioxide in the atmosphere are responsible for warming of the planet and the changes in climate that occur as a result. Since burning coal releases more carbon dioxide than other fossil fuels per unit of energy obtained, many industries have switched to other sources of energy.

Oil Use

Worldwide about 33 percent of the energy consumed comes from oil. Oil has several characteristics that make it superior to coal as a source of energy. Its extraction causes less environmental damage than does coal mining. It is a more concentrated source of energy than coal, it burns with less pollution, and it can be moved easily through pipes. Almost half of the oil used in the United States is as gasoline for cars.

FIGURE 9.11 Acid Mine Drainage The red color of the river is a common characteristic of acid mine drainage.

Extraction Methods

Today, geologists use a series of tests to locate underground formations that may contain oil. When a likely area is identified, a test well is drilled to determine if oil is actually present. Since the many easy-to-reach oil fields have already been tapped, drilling now focuses on smaller amounts of oil in less accessible sites, which means that the cost of oil from most recent discoveries is higher than that from the large, easy-to-locate sources of the past. As oil deposits located below land have become more difficult to find, geologists have widened the search to include the ocean

Focus On

Unconventional Sources of Oil and Natural Gas

As the easy-to-reach conventional sources of oil have been more difficult to find and the price of oil has risen, unconventional sources of oil have become economical to develop. These unconventional sources are more difficult to develop because they will not readily flow either because they consist of mixtures of hydrocarbons with a very high viscosity (thick) or because the substrate rock has few channels that allow for their movement. Several of these unconventional sources are: tar sands (oil sands), oil shale, and tight oil.

Tar sands (also referred to as *oil sands*) are a combination of clay, sand, water, and a thick oil called *bitumen*. The bitumen is so thick that it cannot be pumped from the ground but must be mined—usually an open pit or strip mine. The mixture of bitumen and sand must be processed to extract the bitumen, which then must be altered to produce oil that will flow through a pipeline. All of the steps in this process take energy. The net return on energy is about 5 to 1. In other words, for every unit of energy put into the process you get back about five units. Although tar sands are found in many places in the world, the only major commercial production of oil from tar sands is in Alberta, Canada.

The production of oil from tar sands in Alberta became a political issue when a Canadian pipeline company proposed the Keystone XL pipeline that would carry oil from Alberta through Montana, Kansas, and Nebraska to Steele City, Nebraska. Environmental groups, members of Congress, and President Obama were all involved in determining if the pipeline should be built.

Oil shale contains a high viscosity mixture of hydrocarbons that must be heated to extract the oil. Typically the shale is mined and transported to a processing site. Because the cost of producing oil from oil shale is high, few countries have developed oil shale processing plants.

Tight oil (light tight oil) consists of a mixture of hydrocarbons with low viscosity which cannot flow because the rock containing it (shale or tight sandstone) is not porous (tight). Horizontal drilling and hydraulic fracturing are used to open channels that allow the oil to flow. (See chapter opening on **Hydraulic Fracturing**.) The production of tight oil in North Dakota and Montana has become a major source of crude oil for the United States.

FIGURE 9.12 Offshore Drilling Once the drilling platform is secured to the ocean floor, a number of wells can be sunk to obtain the gas or oil.

floor. Building an offshore drilling platform can cost millions of dollars. To reduce the cost, as many as 70 wells may be drilled from a single platform. (See figure 9.12.)

Once a source of oil has been located, the greatest technological problems involve techniques used to extract the oil and transport it to the surface. If the water or gas pressure associated with an oil deposit is great enough, the oil is forced to the surface when a well is drilled. When the natural pressure is not great enough, the oil must be pumped to the surface. These techniques are often referred to as *primary recovery methods* and can extract 5 to 30 percent of the oil depending on geologic characteristics of the source and the viscosity of the oil. In most oil fields, secondary recovery is used to recover more of the oil. *Secondary recovery methods* include pumping water or gas into the well to drive the oil out of the pores in the rock. (See figure 9.13.) These techniques typically result in up to 40 percent of the oil being extracted. As oil prices increase, more expensive and aggressive recovery methods become economical. *Tertiary recovery methods* include pumping steam into the well to lower the viscosity of the oil and allow it to flow more readily. Other techniques include more aggressive pumping of gases or chemicals into wells. All of these methods are expensive and are only used if the price of oil is high and the likelihood of getting significant additional production is great.

Processing Crude Oil

Oil, as it comes from the ground, is not in a form suitable for most uses. It must be refined. The various components of crude oil can

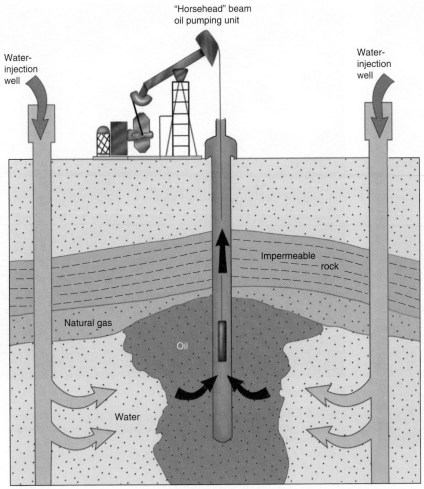

"Horsehead" beam oil pumping unit

Water-injection well

Water-injection well

Impermeable rock

Natural gas

Oil

Water

FIGURE 9.13 Secondary Recovery A common secondary recovery method for obtaining oil is to pump water into the rock that holds the oil. This forces the oil out of the rock and allows it to be pumped to the surface.

be separated and collected by heating the oil in a distillation tower. After distillation, the products may be further refined by "cracking." In this process, heat, pressure, and catalysts are used to produce a higher percentage of volatile chemicals, such as gasoline, from less volatile liquids, such as diesel fuel and furnace oils. In addition, the refining process may be used to produce petrochemicals that serve as raw materials for a variety of synthetic compounds such as plastics, fabrics, paints, and rubber.

Environmental Issues

Although there are disturbances of the land at the drilling site, oil spills are the primary environmental problem associated with the production, transportation, and use of oil. Since oil is a liquid, it is generally moved from place to place through pipes. If these pipes break or leak, the oil contaminates the soil, groundwater, or surface water. This can happen anywhere from the drilling site to where the oil is used.

When oil must be transported across oceans, giant supertankers are used. (See figure 9.14.) Oil spills in the oceans from supertankers have been widely reported by the news media. Because of new regulations, changes in tanker hull design, and greater attention to safety, the number of tanker spills has declined over the last few decades, while the amount of oil being transported has increased. (See figure 9.15.) Although major shipping accidents are spectacular and release large amounts of oil, it is estimated that worldwide only about 10 percent of human-caused oil pollution comes from tanker accidents.

The use of platforms to drill oil wells on the ocean floor has resulted in some huge oil spills. Since 1955, there have been about 50 failures or accidents on ocean drilling rigs that have resulted in the release of oil and natural gas into the ocean. The two largest occurred in the Gulf of Mexico. In 1979, the *Ixtoc 1* oil spill is estimated to have released about 700 metric tons of oil. In 2010, the *Deepwater Horizon* oil rig caught fire and sank. It is estimated that about 10,000 metric tons of oil were released before it was eventually sealed. This is the largest oil spill on record.

The environmental damage caused by oil spills is related to several factors including the amount of oil released, the size of the area affected, the climate and weather, and other factors. Although oil spills have been occurring in the ocean for over 50 years, their effects are still poorly understood. The oil on the ocean surface can coat birds, reptiles, and mammals that must surface to breathe. Since the oil is toxic, fish and other marine life that consume oil can be affected. When the oil reaches the shore it coats surfaces and kills vegetation and marine life along the shore, and also makes the area unsuitable for swimming, boating, fishing, and other human uses.

Although the accidental release of large amounts of oil into the ocean is a sensational event and the local environment may be greatly affected, accidental releases are a small component of the oil released into the oceans of the world. The Ocean Studies Board of the National Research Council published a report that puts ocean oil pollution in perspective. (See figure 9.16.) The vast majority of the oil entering the oceans results from natural seeps and pollution from the use of oil as lubricants and fuel in machinery (cars, trucks, boats, lawn mowers, jet skis, etc.). The residue from these uses washes into streams and is carried to the oceans.

As the most accessible oil drilling sites have been developed, there is pressure to drill in sites with difficult environmental conditions such as on the ocean floor or in the arctic. In addition, sites that were off-limits to drilling have come under pressure. See Science, Politics, & Policy: The Arctic National Wildlife Refuge.

The evaporation of oil products and the incomplete burning of oil fuels contribute to air pollution. These problems are discussed in chapter 16.

(a) Trans-Alaska pipeline

(b) Oil tanker

FIGURE 9.14 **Transportation of Oil** (a) Oil pipelines and (b) oil tankers are the primary methods used to transport oil. When accidents occur, oil leaks contaminate the soil or water.

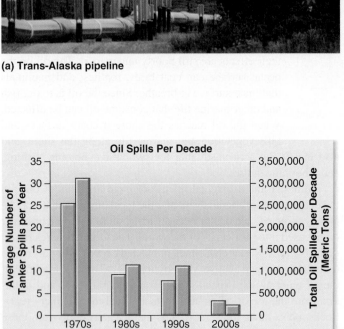

FIGURE 9.15 **Oil Spills From Tankers** The establishment of stronger regulations, better design of supertankers, and attention to safety have greatly reduced spills from oil tankers.

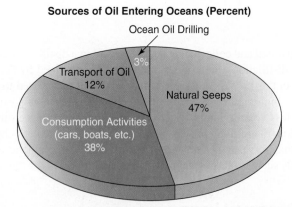

FIGURE 9.16 **Sources of Marine Oil Pollution** Although oil spills can release large amounts of oil in a relatively small area, the majority of oil in the oceans comes from dispersed sources—natural seeps and the consequence of use of oil on land and in shipping.

Natural Gas Use

Natural gas, the third major source of fossil-fuel energy, supplies over 20 percent of the world's energy.

Extraction Methods

The drilling operations to obtain natural gas are similar to those used for oil. In fact, a well may yield both oil and natural gas. As with oil, secondary recovery methods that pump air or water into a well are used to obtain the maximum amount of natural gas from a deposit. See chapter opening, Hydraulic Fracturing, for a discussion of the increased role of hydraulic fracturing in natural gas production. After processing, the gas is piped to the consumer for use.

Transport Methods

Transport of natural gas still presents a problem in some parts of the world. In the Middle East, Mexico, Venezuela, and Nigeria, wells are too far from consumers to make pipelines practical, so much of the natural gas is burned as a waste product at the wells. However, new methods of transporting natural gas and converting it into other products are being explored. At $-162°$ C ($-126°$ F), natural gas becomes a liquid and has only 1/600 of the volume of its gaseous form. Tankers have been designed to transport **liquefied natural gas** from the area of production to an area of demand. In 2012, about 328 billion cubic meters (about 11,500 billion cubic feet) of natural gas were shipped between countries as liquefied natural gas. This is about 10 percent of the natural gas consumed in the world.

A major public concern about liquefied natural gas is the safety at loading and unloading facilities. When new ports are suggested, there is concern about explosions that could result from accidents or the actions of terrorists. Because of these concerns, the loading and unloading facilities are often located several kilometers offshore.

Environmental Issues

Of the three fossil fuels, natural gas is the least disruptive to the environment. A natural gas well does not produce any unsightly waste, although there may be local odor problems. Except for the danger of an explosion or fire, natural gas poses no harm to the environment during transport. Since it is clean burning, it causes almost no air pollution. The products of its combustion are carbon dioxide and water. Although the burning of natural gas produces carbon dioxide, which is a greenhouse gas it produces less carbon dioxide than does coal or oil. The role of carbon dioxide in climate change is discussed in chapter 17.

Although natural gas is used primarily for heat energy, it does have other uses, such as the manufacture of petrochemicals and fertilizer. Methane contains hydrogen atoms that are combined with nitrogen from the air to form ammonia, which can be used as fertilizer.

9.5 Nuclear Power

Although nuclear power does not come from a fossil fuel, it is fueled by uranium that is obtained by mining. Like fossil fuels, uranium is nonrenewable. Worldwide, energy from nuclear power plants is the fifth most important energy source after coal, oil, natural gas, and hydroelectric power.

Forces that Influence the Growth of Nuclear Power

Many interacting forces have influenced the development of nuclear power. Primary among them are environmental and economic issues that become transformed by the political process into governmental policy. Plans to decommission plants, extend the life of plants, or build new plants are political decisions made by national governments in light of the opinions of their citizens. In many parts of the world, there is a strong anti-nuclear sentiment. There are many reasons people take an anti-nuclear stance. Some oppose all things nuclear because they see it as a threat to world peace. Others are concerned about the environmental issues of potential nuclear contamination and the problem of waste disposal.

The acceptance of the threat of climate change has had major implications for nuclear power. Since nuclear power plants do not produce carbon dioxide, many people, including some environmental organizations, have reevaluated the value of nuclear power and see it as a continuing part of the energy equation.

Another strong social force is economics. As the cost of oil and natural gas has increased, electricity generation from these sources has become more costly. This has made nuclear power more attractive. Countries that have few fossil fuel reserves and those with developing economies are most likely to build nuclear power plants.

Democratic governments set policy based on how their citizens vote. Many governments have responded to these forces and declared themselves to be non-nuclear countries. For example, Australia, New Zealand, Denmark, Ireland, Portugal, Luxembourg, Greece, and Austria have declared that they will not pursue nuclear energy as an option. Others have stated an intention to phase out nuclear power as a source of electricity. Even so, these policies can change as economic and environmental realities confront policy.

The Current Status of Nuclear Power

As of October 2013, there were 432 nuclear power reactors in operation. They account for about 11 percent of the electrical energy generated in the world. There were also 70 nuclear power plants under construction in 13 countries. A further 173 were in the planning stages. Most are in Asia, where China, India, Japan, and South Korea are projected to add about 89 new plants over the next ten years. Russia has plans for an additional 28 plants, and electric utilities in the United States have requested approval for 9 new

Science, Politics, & Policy

The Arctic National Wildlife Refuge

The Arctic National Wildlife Refuge (ANWR) has been the center of a continuing struggle around the need for oil. In 1960, 3.6 million hectares (8.9 million acres) were set aside as the Arctic National Wildlife Range. Passage of the Alaskan National Interest Lands Conservation Act in 1980 expanded the range to 8 million hectares (19.8 million acres) and established 3.5 million hectares (8.6 million acres) as wilderness. The act also renamed the area the Arctic National Wildlife Refuge (ANWR). A key provision of the law has made the status of ANWR intensely political. The act requires specific authorization from Congress before oil drilling or other development activities can take place on the coastal plain in the refuge. (See Area 1002 on the map.)

Scientific input on the importance of the area has come from at least two different perspectives. Wildlife biologists have determined that the coastal plain has the greatest concentration of wildlife and is the major calving ground for the Porcupine caribou herd. Studies of the migration of the Porcupine caribou herd show that they migrate from southern areas of ANWR and the adjacent Yukon Territory to the coastal plain at calving time. Typically 50 to 75 percent of caribou calves are born in the coastal plain each summer. The coastal plain is important because there are fewer predators (grizzly bears, wolves, and eagles) and the area provides abundant forage needed by pregnant and nursing female caribou.

Studies by petroleum geologists at the U.S. Geological Survey led them to estimate that there were about 10 billion barrels of recoverable oil in the coastal plain and adjacent waters of the Arctic Ocean. A comparable estimated figure for the rest of the United States is 120 billion barrels. It is important to differentiate these estimates from known oil deposits, known as proved reserves. The total proved reserves for the United States is about 28 billion barrels.

Migrating Caribou in ANWR

There are also international implications. The refuge borders Canada's Northern Yukon National Park. Many animals, particularly members of the Porcupine caribou herd, travel across the border on a regular yearly migration. The United States is obligated by treaty to protect these migration routes.

Thus, the status of ANWR is at the center of a struggle among competing interests. The major players are environmentalists, who seek to preserve this region as wilderness; the state of Alaska, which funds a major portion of its activities with dividends from oil production; Alaska

plants. (See table 9.2.) The United States has 100 operating nuclear power plants, about 23 percent of the world total. Figure 9.17 shows the locations of the nuclear power plants in North America.

In addition to the trend of building additional nuclear power plants, many nuclear power plants have been relicensed to operate for a longer period than their original design life. Most current nuclear power plants originally had a design lifetime of up to 40 years, but engineering assessments of many plants have established that many can operate safely and economically for much longer. In the United States, and elsewhere in the world, several reactors have been granted license renewals that extend their operating lives to 60 years. As additional plants reach the end of their 40-year licenses, it is likely that they will have their licenses extended as well. Extending the operating life of a plant reduces the cost of producing power because the decommissioning is delayed and no new power plant (nuclear or fossil fuel) needs to be built to replace the lost capacity.

9.6 The Nature of Nuclear Energy

To understand where nuclear energy comes from, it is necessary to review some aspects of atomic structure presented in chapter 4. All atoms are composed of a central region called the nucleus, which contains positively charged protons and neutrons that have no charge. Moving around the nucleus are smaller, negatively charged electrons. Since like charges repel one another, some force is required to hold the positively charged protons together in the nucleus. This is called the *nuclear force*. In most atoms, the various forces in the nucleus are balanced and the nucleus is stable. However, some isotopes of atoms are **radioactive;** that is, the nuclei of these atoms are unstable and spontaneously decompose. Neutrons, electrons, protons, and other larger particles are released during nuclear disintegration, along with a great deal of energy. The rate of decomposition is consistent for any given isotope. It is measured and expressed as **radioactive half-life,**

residents, who receive a dividend payment from oil revenues; oil companies that want to drill in the refuge; members of Congress, who must vote on any changes in the status of the refuge; and the president, who must sign or veto any legislation related to the refuge.

- Environmentalists support designating the entire refuge as wilderness and argue that none of the reserve should be developed when improvements in energy conservation could reduce the demand for oil. They maintain that drilling in the reserve will harm the habitat of millions of migratory birds, caribou, and polar bears.
- Alaska relies on oil for about 80 percent of its revenue and has no sales or income tax. Furthermore, each Alaskan citizen receives a yearly dividend check from a state fund established with proceeds from oil companies. Even so, some Alaskan citizens support drilling; oth-

ers oppose it. The Inupiat Eskimos who live along the north Alaskan coast mostly are in favor of drilling in ANWR. The Inupiat believe oil revenues and land-rental fees from oil companies will raise their living standards. The other Native American tribe in the region, the Gwich'in, who live on the southern fringe of the refuge, oppose drilling. They argue that the drilling will impact the caribou migration through the area every fall and thus affect their ability to provide food for their families.

- Oil companies have repeatedly stated that the oil reserves should be developed and that the oil can be recovered without endangering wildlife or the fragile Arctic ecosystem.
- Members of Congress are split on this issue. In recent years it has become a partisan issue, with most Republicans supporting drilling

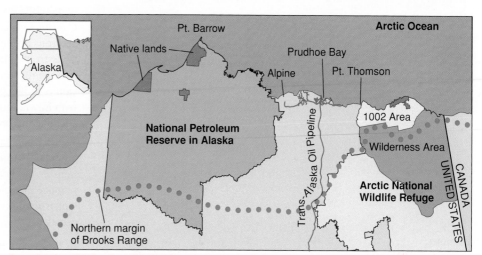

Source: Data from USGS Fact Sheet 0028-01: online report.

and most Democrats opposing drilling. Alaska's members of Congress support drilling. One of the arguments in support of drilling is a desire to reduce dependence on foreign oil. Every year a bill is introduced in Congress to allow drilling in ANWR.

- The U.S. president also has an important role to play. In 2002, President George W. Bush reconfirmed his support for drilling. However, no bill authorizing drilling passed Congress for him to sign. President Obama has expressed his opposition to drilling. If a bill passes Congress, he must decide whether to sign or veto it.

It is clear that the competing interests of environmentalists, Alaskans, the energy industry, and politicians will continue to keep ANWR in the headlines.

which is the time it takes for one-half of the radioactive material to spontaneously decompose. Table 9.3 lists the half-lives of several radioactive isotopes.

Nuclear disintegration releases energy from the nucleus as **radiation,** of which there are three major types:

Alpha radiation consists of a moving particle composed of two neutrons and two protons. Alpha radiation usually travels through air for less than a meter and can be stopped by a sheet of paper or the outer, nonliving layer of the skin.

Beta radiation consists of moving electrons released from nuclei. Beta particles travel more rapidly than alpha particles and will travel through air for a couple of meters. They are stopped by a layer of clothing, glass, or aluminum.

Gamma radiation is a type of electromagnetic radiation that does not consist of particles. Other forms of electromagnetic radiation are X rays, light, and radio waves. Gamma radiation can

pass through your body, several centimeters of lead, or nearly a meter of concrete.

When a radioactive isotope disintegrates and releases particles, it becomes a different kind of atom. For example, uranium-238 ultimately produces lead—but it goes through several steps in the process. (See figure 9.18.)

9.7 Nuclear Chain Reaction

In addition to releasing alpha, beta, and gamma radiation when they disintegrate, the nuclei of a few kinds of atoms release neutrons. When moving neutrons hit the nuclei of certain other atoms, they can cause those nuclei to split as well. An atom that has a nucleus that will split is said to be **fissionable** and the process

Table 9.2 Nuclear Reactor Statistics
October 2013

Region	Reactors Operable	Reactors Under Construction	Reactors Planned
World	*432*	*70*	*173*
United States	100	3	9
France	58	1	1
Japan	50	3	9
Russia	33	10	28
South Korea	23	5	6
India	20	7	18
Canada	19	0	2
China	17	30	59
United Kingdom	16	0	4
Ukraine	15	0	2
Sweden	10	0	0
Germany	9	0	0
Rest of World	62	11	35

Source: Data from World Nuclear Association.

of splitting is known as **nuclear fission.** If these splitting nuclei also release neutrons, they can strike the nuclei of other atoms, which also disintegrate, resulting in a continuous process called a **nuclear chain reaction.** (See figure 9.19.) Only certain kinds of atoms are suitable for the development of a nuclear chain reaction. The two materials commonly used in nuclear reactions are uranium-235 and plutonium-239. In addition, there must be a certain quantity of nuclear fuel (a critical mass) in order for a nuclear chain reaction to occur. It is this process that results in the large amounts of energy released from bombs or nuclear reactors.

9.8 Nuclear Fission Reactors

A **nuclear reactor** is a device that permits a sustained, controlled nuclear fission chain reaction. There are four important materials that are involved in producing a controlled nuclear chain reaction: the fuel, a moderator, control rods, and the core coolant.

- The *fuel* most commonly used in nuclear reactors is **uranium-235 (U-235).** When the

nucleus is split, two to three rapidly moving neutrons are released, along with large amounts of energy that can be harnessed to do work.

- *Moderators* are used to slow down the fast-moving neutrons so that they are more effective in splitting other U-235 nuclei and maintaining a chain reaction. The most commonly used moderators are water and graphite.
- *Control rods* contain nonfissionable materials that absorb the neutrons produced by fissioning uranium and prevent the neutrons from splitting other atoms. By moving the control rods into the reactor the number of neutrons available to cause fission is reduced and the reaction slows. If the control rods are withdrawn, more fission occurs, and more particles, radiation, and heat are produced.
- The *coolant* is needed to manage the large amount of heat produced within the nuclear reactor. The coolant is needed to transfer the heat away from the reactor core to the turbine to produce electricity. In most reactors the core coolant is water, which also serves as the moderator, but gases and liquid metals can also be used as coolants in special kinds of nuclear reactors.

In the production of electricity, a nuclear reactor serves the same function as burning a fossil fuel. It produces heat, which

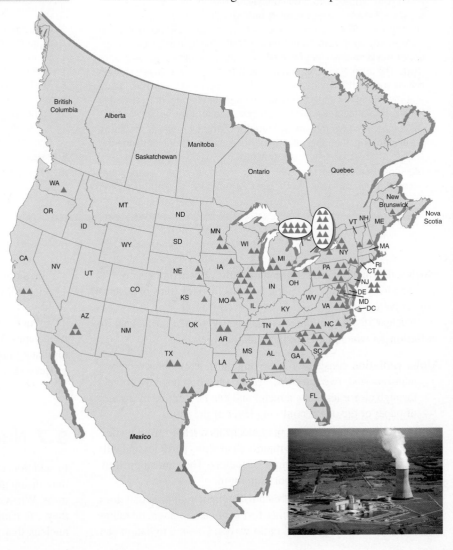

FIGURE 9.17 Distribution of Nuclear Power Plants in North America The United States has the largest number of nuclear power plants of any country in the world—100. Canada currently has 19 operating nuclear power plants, and Mexico has two.

Table 9.3 Half-Lives and Significance of Some Radioactive Isotopes

Radioactive Isotope	Half-life	Significance
Uranium-235	700 million years	Fuel in nuclear power plants
Plutonium-239	24,110 years	Nuclear weapons Fuel in some nuclear power plants
Carbon-14	5,730 years	Establish age of certain fossils
Americium-241	432.2 years	Used in smoke detectors
Cesium-137	30.17 years	Treat prostate cancer Used to measure thickness of objects in industry
Strontium-90	29.1 years	Power source in space vehicles Treat bone tumors
Cobalt-60	5.27 years	Sterilize food by irradiation Cancer therapy Inspect welding seams
Iridium-192	73.82 days	Inspect welding seams Treat certain cancers
Phosphorus-32	14.3 days	Radioactive tracer in biological studies
Iodine-131	8.06 days	Diagnose and treat thyroid cancer
Radon-222	3.8 days	Naturally occurs in atmosphere of some regions where it causes lung cancers
Radon-220	54.5 seconds	Naturally occurs in atmosphere of some regions where it causes lung cancers

converts water to steam to operate a turbine that generates electricity. After passing through the turbine, the steam must be cooled, and the water is returned to the reactor to be heated again. Various types of reactors have been constructed to furnish heat for the production of steam. They differ in the moderator used, in how the reactor core is cooled, and in how the heat from the core is used to generate steam. The three most common kinds of nuclear reactors are boiling-water reactors, pressurized-water reactors, and heavy-water reactors.

Boiling-water reactors use water as both a moderator and a reactor-core coolant. Steam is formed within the reactor and transferred directly to the turbine, which turns to generate electricity. A disadvantage of the boiling-water reactor is that the steam passing to the turbine must be treated to remove any radiation. Even then, some radioactive material is left in the steam; therefore, the generating building must be shielded. About 20 percent of the nuclear reactors in the world are boiling-water reactors.

Pressurized-water reactors also use water as a moderator and reactor-core coolant but the water is kept under high pressure so that steam is not allowed to form in the reactor. (See figure 9.20.) A secondary loop transfers the heat from the pressurized water in the reactor to a steam generator. The steam is used to turn the turbine and generate electricity. Such an arrangement reduces the risk of radiation in the steam but adds to the cost of construction by requiring a secondary loop for the steam

Isotope of Element	Type of Radiation	Half-Life
Uranium-238 (U-238) Protons = 92 Neutrons = 146		4.5 billion years
	Alpha (2 protons and 2 neutrons)	
Thorium-234 (Th-234) Protons = 90 Neutrons = 144		24.5 days
	Beta (electron)	
Protactinium-234 (Pa-234) Protons = 91 Neutrons = 143		1.14 minutes
	Beta (electron)	
Uranium-234 (U-234) Protons = 92 Neutrons = 142		233,000 years
	Alpha (2 protons and 2 neutrons)	
Thorium-230 (Th-230) Protons = 90 Neutrons = 140		83,000 years
	Alpha (2 protons and 2 neutrons)	
Radium-226 (Ra-226) Protons = 88 Neutrons = 138		1,590 years
	Alpha (2 protons and 2 neutrons)	
Radon-222 (Rn-222) Protons = 86 Neutrons = 136		3.825 days
	Alpha (2 protons and 2 neutrons)	
Polonium-218 (Po-218) Protons = 84 Neutrons = 134		3.05 minutes
	Alpha (2 protons and 2 neutrons)	
Lead-214 (Pb-214) Protons = 82 Neutrons = 132		26.8 minutes
	Beta (electron)	
Bismuth-214 (Bi-214) Protons = 83 Neutrons = 131		19.7 minutes
	Beta (electron)	
Polonium-214 (Po-214) Protons = 84 Neutrons =130		0.00015 seconds
	Alpha (2 protons and 2 neutrons)	
Lead-210 (Pb-210) Protons = 82 Neutrons = 128		22 years
	Beta (electron)	
Bismuth-210 (Bi-210) Protons = 83 Neutrons = 127		5 days
	Beta (electron)	
Polonium-210 (Po-210) Protons = 84 Neutrons = 126		140 days
	Alpha (2 protons and 2 neutrons)	
Lead-206 (Pb-206) Protons = 82 Neutrons = 124		Stable

FIGURE 9.18 Radioactive Decay Path of U-238 When an unstable isotope disintegrates, it ultimately leads to a stable isotope of a different element. In the process it is likely to go through several unstable isotopes of other elements before it becomes a stable element.

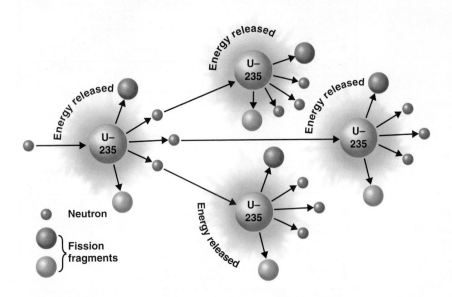

FIGURE 9.19 Nuclear Chain Reaction When a neutron strikes the nucleus of a U-235 atom, energy is released, and several fission fragments and neutrons are produced. The newly released neutrons may strike other atoms of U-235, causing their nuclei to split with the release of additional neutrons and energy. This series of events is called a *nuclear fission chain reaction.*

generator. About 60 percent of the nuclear reactors in the world are pressurized-water reactors. Most of the reactors currently under construction are of this type.

Heavy-water reactors are a third type of reactor that uses water as a coolant. This type of reactor was developed by Canadians and uses water that contains the hydrogen isotope deuterium in its molecular structure as the reactor-core coolant and moderator. Since the deuterium atom is twice as heavy as the more common hydrogen isotope, the water that contains deuterium weighs slightly more than ordinary water. Heavy-water reactors are similar to pressurized-water reactors in that they use a steam generator to convert regular water to steam in a secondary loop. The major advantage of a heavy-water reactor is that naturally occurring uranium isotopic mixtures serve as a suitable fuel while other reactors require that the amount of U-235 be enriched to obtain a suitable fuel. This is possible because heavy water is a better neutron moderator than is regular water. Since it does not require enriched fuel, the cost of producing fuel for a heavy-water reactor is less than that for other reactors. About 10 percent of nuclear reactors are of this type.

There are two types of reactors that are not currently popular but may become important in the future.

Gas-cooled reactors were developed by atomic scientists in the United

Kingdom. Carbon dioxide serves as a coolant for a graphite-moderated core. As in the heavy-water reactor, natural isotopic mixtures of uranium are used as a fuel. However, this is not a popular type of reactor. China is the only country that plans to build one in the near future.

Nuclear breeder reactors are nuclear fission reactors that form new nuclear fuel as they operate to produce electricity. Because the process requires fast-moving neutrons, water cannot be used as a moderator because it slows the neutrons too much and most models of breeder reactors function without a moderator. Because it is necessary to move heat away from the reactor core very efficiently, most breeder reactors use liquid metal (often liquid sodium) as a core coolant. Thus, they are often called *liquid metal fast-breeder reactors.* When a fast-moving neutron hits a nonfissionable uranium-238 (U-238) nucleus and is absorbed, an atom of fissionable **plutonium-239 (Pu-239)** is produced. Remember that U-235 is a nuclear fuel and U-238 is not. Thus, a breeder reactor converts a nonfuel (U-238) into a fuel (Pu-239). (See figure 9.21.)

During the early stages of the development of nuclear power plants, breeder reactors were seen as the logical step after nuclear fission development because they would reduce the need for uranium, which is a nonrenewable material. However, most breeder reactors are considered experimental and because they produce plutonium-239, which can be used to produce nuclear

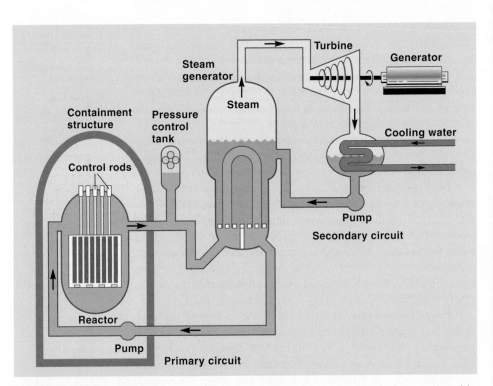

FIGURE 9.20 Pressurized-Water Reactor In a pressurized-water reactor, the heat generated by nuclear fission raises the temperature of the water that serves as the moderator and core coolant. Since the reactor is in a sealed container, the water does not produce steam. The superheated water transfers the heat to a secondary loop that produces steam that turns the turbine to produce electricity.

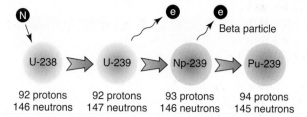

Beta particle

U-238 → U-239 → Np-239 → Pu-239

| 92 protons | 92 protons | 93 protons | 94 protons |
| 146 neutrons | 147 neutrons | 146 neutrons | 145 neutrons |

FIGURE 9.21 Formation of Pu-239 in a Breeder Reactor When a fast-moving neutron (N) is absorbed by the nucleus of a U-238 atom, a series of reactions result in the formation of Pu-239 from U-238. Two Intermediate atoms are U-239 and Np-239, which release beta particles (electrons) from their nuclei. (A neutron in the nucleus can release a beta particle and become a proton.) This series of reactions is important because, while U-238 is stable and is not a nuclear fuel, the addition of a neutron makes it unstable and leads to the production of Pu 239, which is fissionable and can serve as a nuclear fuel.

weapons, they are politically sensitive. Most countries have discontinued their breeder reactor programs following accidents or political decisions. The United Kingdom. Germany, the United States, and France have breeder reactors that are currently not in operation. Russia, China, India, and Japan have operating breeder reactors, and India is currently building a new breeder reactor.

9.9 The Nuclear Fuel Cycle

To appreciate the consequences of using nuclear fuels to generate energy, it is important to understand the nuclear fuel cycle, which follows the process from the mining of uranium to the disposal of the waste from power plants.

1. **Mining and milling** of uranium ore are the first steps in the nuclear fuel cycle. (See figure 9.22.) Low-grade uranium ore is obtained by underground or surface mining. The ore contains about 0.2 percent uranium by weight. After it is mined, the ore goes through a milling process. It is crushed and treated with a solvent to concentrate the uranium. Milling produces yellow-cake, a material containing 70 to 90 percent uranium oxide. About 70 percent of the world's production of uranium comes from mines in Australia, Kazakhstan, Russia, South Africa, Canada, and the United States.

2. **Enrichment** of the U-235 content is necessary because naturally occurring uranium contains about 99.3 percent nonfissionable U-238 and only 0.7 percent fissionable U-235. This concentration of U-235 is not high enough for most types of reactors. Since the masses of the isotopes U-235 and U-238 vary only slightly, and the chemical differences are very slight, enrichment is a difficult and expensive process. The enrichment process involves the use of centrifuges to separate the two isotopes of uranium by their slight differences in mass. Enrichment increases the U-235 content from 0.7 percent to 3 percent.

3. **Fuel fabrication** involves converting the enriched material into a powder, which is then compacted into pellets about the size of a pencil eraser. These pellets are sealed in metal fuel rods about 4 meters (13 feet) in length, which are loaded into a reactor.

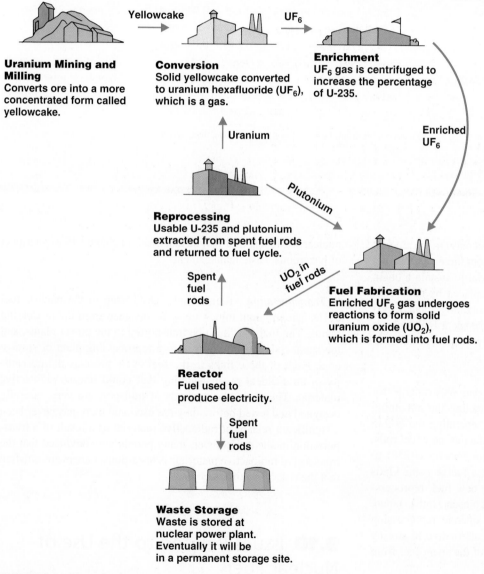

Uranium Mining and Milling
Converts ore into a more concentrated form called yellowcake.

Conversion
Solid yellowcake converted to uranium hexafluoride (UF₆), which is a gas.

Enrichment
UF₆ gas is centrifuged to increase the percentage of U-235.

Yellowcake → UF₆ → Enriched UF₆

Uranium

Plutonium

Reprocessing
Usable U-235 and plutonium extracted from spent fuel rods and returned to fuel cycle.

Spent fuel rods

UO₂ in fuel rods

Fuel Fabrication
Enriched UF₆ gas undergoes reactions to form solid uranium oxide (UO₂), which is formed into fuel rods.

Reactor
Fuel used to produce electricity.

Spent fuel rods

Waste Storage
Waste is stored at nuclear power plant. Eventually it will be in a permanent storage site.

FIGURE 9.22 Steps in the Nuclear Fuel Cycle The process of obtaining nuclear fuel involves mining, extracting the uranium from the ore, concentrating the U-235, fabricating the fuel rods, installing and using the fuel in a reactor, and disposing of the waste. Some countries reprocess the spent fuel as a way of reducing the amount of waste they must deal with.
Source: U.S. Department of Energy.

Focus On

Measuring Radiation

There are several different kinds of radiation measurements for different purposes. The table shows some of the common units and their values.

The amount of radiation absorbed by the human body is a way to quantify the potential damage to tissues from radiation. The **absorbed dose** is the amount of energy absorbed by matter. It is measured in grays or rads. However, the damage caused by alpha particles is 20 times greater than that caused by beta particles or gamma rays. Therefore, a dose equivalent is used. The **dose equivalent** is the absorbed dose times a quality factor. The units for dose equivalents are seiverts or rems. The quality factor for beta and gamma radiation is 1. Therefore, the dose equivalent is the same as the absorbed dose. The quality factor for alpha radiation is 20. Therefore, the dose equivalent for alpha radiation is 20 times the absorbed dose.

Radiation Measurement Units

What Is Measured	International Scientific Units	U.S. Commonly Used Units	Application
Number of nuclear disintegrations	becquerel (Bq) 1 Bq = 1 disintegration/ second	curie (Ci) 1 Ci = 37 billion disintegration/second 1 Ci = 37 billion Bq	Quantify the strength of a radiation source
Absorbed dose	gray (Gy) 1 Gy = 1 joule/kilogram of matter	rad 1 rad = 0.01 joule/ kilogram of matter 1 rad = 0.01 Gy	Quantify the amount of energy absorbed
Dose equivalent	sievert (Sv) 1 Sv = Gy X quality factor*	rem 1 rem = rad X quality factor* 1 rem = 0.01 Sv	Quantify the potential biological effect of a dose

*For beta and gamma radiation, the quality factor = 1. For alpha radiation, the quality factor = 20.

---

4. **Use of the fuel** in a reactor results in a decrease in the amount of U-235 in the fuel rods over time. After about three years, a fuel rod does not have enough radioactive material to sustain a chain reaction, and the spent fuel rods must be replaced by new ones. The spent rods are still very radioactive, containing about 1 percent U-235 and 1 percent plutonium. Because spent fuel rods are radioactive, they must be managed carefully to prevent environmental damage and risks to health.

5. **Reprocessing or storage** are the two options available for spent fuel rods. Reprocessing involves extracting the remaining U-235 and plutonium from the spent fuel, separating the U-235 from the plutonium, and using both to manufacture new fuel rods. The plutonium is mixed with spent uranium (mostly U-238) to form a mixed oxide fuel that can be used as a fuel in some kinds of nuclear power plants. Besides providing new fuel, reprocessing reduces the amount of nuclear waste. At present, India, Japan, Russia, France, and the United Kingdom operate reprocessing plants that reprocess spent fuel rods as an alternative to storing them as a nuclear waste. About 30 percent of the spent fuel from the past has been reprocessed.

Those countries that do not reprocess nuclear waste have made the decision to store it. Initially, the waste is stored onsite at the nuclear power plant, but the long term plan is to bury the waste in stable geologic formations. These and other issues related to

nuclear waste disposal will be discussed in chapter 19 as an aspect of hazardous waste management.

6. **Transportation** is involved in each step in the nuclear fuel cycle. The uranium mines are some distance from the processing plants. The fuel rods must be transported to the power plants, and the spent rods must be moved to a reprocessing plant or storage area. Each of these links in the fuel cycle presents the possibility of an accident or mishandling that could release radioactive material. Therefore, the methods of transport are very carefully designed and tested before they are used and there has never been a significant release of radioactive material as a result of a transportation incident. However, many people are convinced that the transport of radioactive materials is hazardous. Others are satisfied that the risks are extremely small.

9.10 Issues Related to the Use of Nuclear Fuels

Most of the concerns about the use of nuclear fuels relate to the danger associated with the radiation produced by the fuel and the waste products of its use.

The Biological Effects of Ionizing Radiation

When an alpha or beta particle or gamma radiation interacts with atoms, it can dislodge electrons from the atoms and cause the formation of ions. Therefore, these kinds of radiation are called **ionizing radiation.** (X rays are also a form of ionizing radiation, although they are not formed as a result of nuclear disintegration.) When ionization occurs in living tissue it can result in damage to DNA or other important molecules in cells. The degree and kind of damage vary with the kind of radiation, the amount of radiation, the duration of the exposure, and the types of cells irradiated.

Because ionizing radiation affects DNA, it can cause mutations, which are changes in the genetic messages within cells. Mutations can cause two quite different kinds of problems. Mutations that occur in the ovaries or testes can form mutated eggs or sperm, which can lead to abnormal offspring. Care is usually taken to shield these organs from unnecessary radiation. Mutations that occur in other tissues of the body may manifest themselves as abnormal tissue growths known as cancer. Two common cancers that are strongly linked to increased radiation exposure are leukemia and breast cancer. Because mutations are essentially permanent, they may accumulate over time. Therefore, the accumulated effects of radiation over many years may result in the development of cancer later in life. Focus On: Measuring Radiation describes how radiation exposure is measured.

The effects of large doses (1,000 to 1 million rems) are clearly evident because people become ill and die. However, demonstrating known harmful biological effects from smaller doses is much more difficult. (See table 9.4.) Moderate doses (10 to 100 rems) are known to increase the likelihood of cancer and birth defects. The higher the dose, the higher the incidence of abnormality. Lower doses may cause temporary cellular changes, but it is difficult to demonstrate long-term effects.

Current research is trying to assess the risks associated with repeated exposure to low-level radiation. Thus, the effects of low-level, chronic radiation generate much controversy. Some people feel that all radiation is harmful, that there is no safe level, and that special care must be taken to prevent exposure. Others feel that there is a threshold level of exposure below which there are no biological effects. One problem is that it is likely that there are delayed effects of radiation exposure. Since it is assumed that radiation affects DNA, the effect may not be evident at the time of exposure but would show up later when the affected gene becomes active. This complicates our attempts to link certain health problems such as cancers with previous exposures to radiation. It should be pointed out that no one experiences zero exposure; the average person is exposed to 0.2 to 0.3 rem per year from natural and medical sources.

Radiation Protection

Time, distance, and shielding are the basic principles of radiation protection. The basic idea behind these three principles is that the total cumulative dose should be minimized.

- *Time* is important because the longer one is exposed to a source of radiation, the more radiation is absorbed.
- *Distance* is important because the farther a person is from the source, the less likely that person will be hit by radiation. In addition, alpha and beta radiation will only travel a short distance through air.
- *Shielding* is important because it can stop radiation and prevent a person from being exposed. Because gamma radiation can travel many meters through air and can penetrate the body, shielding is very important. Water, lead, and concrete are common materials used for shielding from gamma radiation. Another kind of shielding is needed to prevent materials that produce radiation from entering the body. Even though alpha and beta radiation are easily stopped, if radioactive isotopes that emit alpha or beta radiation are ingested or inhaled,

Table 9.4 Radiation Effects

Source or Benchmark	Dose	Biological Effects
Nuclear bomb blast or accidental exposure in a nuclear facility	100,000 rems/incident	Immediate death
Nuclear accident or accidental	10,000 rems/incident	Coma, death in 1–2 days
exposure to X rays	1,000 rems/incident	Death in 2–3 weeks
	800 rems/incident	100% death eventually
	500 rems/incident	50% survival with good medical care
	100 rems/incident	Increased probability of leukemia
	50 rems/incident	Changes in numbers of blood cells observed
	10 rems/incident	Early embryos may show abnormalities
X ray of intestine	1 rem/procedure	Damage or effects difficult to demonstrate
Upper limit for occupationally exposed persons	5 rems/year	
Upper limit for release from nuclear facilities that are not nuclear power plants	0.5 rem/year	
Natural background radiation	0.2–0.3 rem/year	
Upper limit for exposure of general public to radiation above background	0.1 rem/year	
Upper limit for release from nuclear power plants	0.005 rem/year	

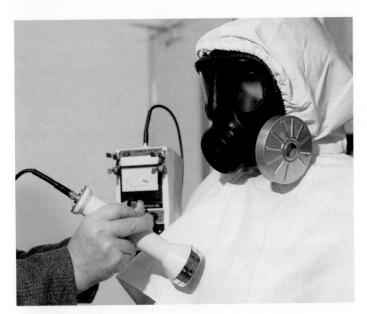

FIGURE 9.23 **Protective Equipment** Persons who work in an area that is subject to radiation exposure must take steps to protect themselves. This worker is wearing protective clothing and a respirator. The surface of his protective suit is being checked with a Geiger counter to determine if he has any radioactive dust particles on his clothing.

the radiation they emit can damage the cells they are in contact with. Thus, shielding from isotopes involves such devices as masks, gloves, and protective clothing. (See figure 9.23.)

Because in most cases people cannot sense ionizing radiation, protection requires specialized sensors. These can be as simple as film badges that record the amount of radiation hitting them. These kinds of devices measure *past exposure* to radiation, which can be useful in determining whether a person should avoid *future exposure*. This is most important for people who work in environments that have a radiation hazard. Because there are limits to the amount of radiation exposure an employee can receive, monitoring allows for changes in job assignments or procedures that limit total exposure.

Reactor Safety

There are currently 432 operating nuclear power plants and about an additional 150 which have operated in the past but have been shut down. Nearly all nuclear power plants have operated without serious accidents. However, three accidents generated concern about nuclear power plant safety: Three Mile Island in 1979, Chernobyl in 1986, and Fukushima Dai-ichi in 2011.

Three Mile Island

The Three Mile Island nuclear plant is located in the Susquehanna River near Middletown, Pennsylvania. On March 28, 1979, the main pump that supplied cooling water to the reactor broke down. At this point, an emergency coolant should have flooded the reactor and stabilized the temperature. The coolant did start to flow into the reactor core, but a pressure relief valve was stuck in the open position so that water also was flowing from the core. In addition, there was no sensor to tell the operator if the reactor was flooded with coolant. Relying on other information, the operator assumed the reactor core was flooded and overrode the automatic emergency cooling system. Without the emergency coolant, the reactor temperature rose rapidly. The control rods eventually stopped fission in the reactor, but because of the loss of coolant, a partial core meltdown had occurred. Thus, the accident was caused by a combination of equipment failures, lack of appropriate information to the operator, and decisions by the operator.

However, in retrospect, the containment structure worked as designed and prevented the release of radioactive materials from the core. Later that day, radioactive steam was vented into the atmosphere. The dose of radiation that would have been received by a person in the area was estimated to have been about 0.001 rem. This compares to normal background radiation of 0.2–0.3 rem.

The crippled reactor was eventually defueled in 1990 at a cost of about $1 billion. It has been placed in monitored storage until the companion reactor, which is still operating, reaches the end of its useful life. (See figure 9.24.) At that time, both reactors will be decommissioned.

Chernobyl

Chernobyl is a small city in Ukraine north of Kiev that became infamous in the spring of 1986, when it became the site of the world's largest nuclear power plant accident.

FIGURE 9.24 **Three Mile Island** The damaged nuclear power plant (on the left) at Three Mile Island is in monitored storage alongside its active companion reactor (on the right).

At 1 A.M. on April 25, 1986, at Chernobyl Nuclear Power Station-4, a test was begun to measure the amount of electricity that the still-spinning turbine would produce if the steam were shut off. This was important information since the emergency core cooling system required energy for its operation, and the coasting turbine could provide some of that energy until another source became available.

During the experiment, operators violated six important safety rules. They shut off all automatic warning systems, automatic shutdown systems, and the emergency core cooling system for the reactor.

As the test continued, the power output of the reactor rose beyond its normal level and continued to rise. The operators activated the emergency system that was designed to put the control rods back into the reactor and stop the fission. But it was too late. The core had already been deformed, and the control rods would not fit properly; the reaction could not be stopped. In 4.5 seconds, the energy level of the reactor increased 2,000 times. The cooling water in the reactor converted to steam and blew the 1,000-metric ton (1,102-ton) concrete roof from the reactor, and the graphite that was part of the reactor core caught fire.

In less than 10 seconds, Chernobyl became the scene of the world's worst nuclear power plant accident. (See figure 9.25.) It took 10 days to bring the burning reactor under control. The immediate consequences were 37 fatalities; 500 persons hospitalized, including 237 with acute radiation sickness; and 116,000 people evacuated. Of the evacuees, 24,000 received high doses of radiation. Currently, some of these people are experiencing health problems attributable to their radiation exposure. In particular, children or fetuses exposed to fallout are showing increased frequency of thyroid cancer because of exposure to radioactive iodine 131 released from Chernobyl.

Fukushima Dai-ichi

On March 11, 2011, a magnitude 9 earthquake 130 km off the northeast coast of Japan triggered the automatic shutdown of several nuclear reactors in the vicinity of the earthquake. The subsequent tsunami flooded the site at Fukushima Dai-ichi (dai-ichi means 1 in this context). This caused several problems. The heat exchangers were damaged, power to the site was cut off, and the diesel generators designed to provide power in an emergency were flooded and stopped operating. The reactors at Fukushima Dai-ichi were older, boiling-water reactors that required electrical power to pump cooling water through them to remove heat during the shutdown process. At the time of the earthquake and tsunami, there were six reactors in operation and three were shut down for refueling or maintenance. Each of the six reactors also had a spent fuel storage site adjacent to the reactor.

The loss of the ability to cool the reactors and their adjacent spent fuel storage sites resulted in explosions and fires that were caused by the release of hydrogen from the overheated reactors and one of the spent fuel storage sites. At one point, seawater was used to supply cooling water, but as power was restored and the pumping mechanisms repaired or replaced, freshwater was again used for cooling. By the end of April 2011, the reactors and spent fuel storage sites were considered stable, but 4 of the 6 reactor units were so badly damaged that they will need to be decommissioned. This will take several years.

The explosions, fires, and leaks in the cooling system released radiation into the atmosphere and seawater. People living within 20 km of the plant were urged to evacuate and take iodine supplements to reduce the likelihood that radioactive iodine-131 would accumulate in the thyroid gland. Crops and milk from the affected area were banned from sale to reduce the likelihood of exposure to airborne radioactive materials. Since iodine-131 has a half-life of 8 days, its levels fall quickly. However, two other radioactive isotopes, cesium-134 and cesium-137, have half-lives of 2 years and 30 years, respectively. Thus, they remain in the environment for a longer period. The Japanese government established an evacuation zone around the damaged nuclear site shortly after the incident and established monitoring sites to determine when it will be safe for residents to return. Measurements of the level of radioactivity in seawater in the vicinity of the plant showed only slightly elevated levels.

Employees and contractors working to control the damaged plants received higher than normal exposures. Thirty exceeded the original 100 millisievert limit and two workers exceeded a higher limit (250 millisieverts) authorized to allow workers to deal with the emergency. Currently, the situation is stable and it is planned to remove the spent fuel rods stored at the site first, then to begin to remove the damaged fuel rods in the reactors. Ultimately the three damaged reactors will be decommissioned in 30-40 years.

FIGURE 9.25 The Accident at Chernobyl An uncontrolled chain reaction in the reactor of unit 4 resulted in a series of explosions and fires. See the circled area in the photograph.

Going Green

Returning a Nuclear Plant Site to Public Use

Big Rock Point nuclear power plant was the fifth nuclear power plant built in the United States. It was a small (67-megawatt) boiling-water reactor. When the plant was shut down in 1997, it was the longest operating (35 years) nuclear power plant in the country and had operated safely for that entire period. During its last 20 years of operation, Big Rock Point operated without a lost time accident. The plant was licensed in 1962 and, in its early years, was used for research related to kinds of fueling and related issues. In 1965, its primary purpose became that of producing electricity, although for 10 years during the time it was producing electricity, it was also used to produce colbalt-60 for medical purposes.

In 1997, because of Big Rock Point's small size and the need for some equipment upgrades, it was determined to be uneconomic and was shut down. Decommissioning began almost immediately. The employees of the plant were retrained to work with the contractor responsible for the decommissioning. Because of its small size, the reactor vessel was removed and shipped to a low-level radioactive waste facility and the remainder of the plant was decontaminated and demolished. In 2007, the U.S. Nuclear Regulatory Commission released the cleaned-up site of 175 hectares (435 acres) with a mile of shoreline on Lake Michigan for unrestricted use. A small portion of the former nuclear plant site was reserved as storage for spent fuel rods. That function will cease if a nuclear waste facility is built to accept spent nuclear fuel. The decommissioning process took nearly nine years and cost $390 million.

Following each of these accidents it is understandable that the role of nuclear power as an energy source is reexamined. For about 25 years, the accidents at Three Mile Island and Chernobyl dampened interest in building new nuclear power plants. The Fukushima Dai-ichi accident caused many countries to reassess the safety of many of their older nuclear power plants, particularly those that require pumps to circulate cooling water during shutdown. However, the continuous safe operation of the most current plants, new designs that have passive safety systems, and the demand for a form of energy that does not release carbon dioxide has altered thinking by many governments. It appears that some older plants may be shut down but that those currently being built will be completed.

Terrorism

After the terrorist attacks on New York and Washington, D.C., on September 11, 2001, fear arose regarding nuclear plants as potential targets for terrorist attacks.

The consensus by nuclear experts is that damage to a nuclear power plant by an aircraft would not significantly damage the containment building or reactor and that normal emergency and containment functions would prevent the release of radioactive materials.

Since most spent nuclear fuel is stored at the nuclear plant, these facilities are a concern as being vulnerable to terrorist attacks. Again, analysis by nuclear experts leads to the conclusion that, although damage to these facilities would result in some radioactive material being released, it would be localized to the plant site. The most difficult contamination to deal with would be through the dispersal of airborne radioactive material into the natural environment.

Probably the greatest terrorism-related threat involving radioactive material is from dirty bombs. Dirty bombs are also known as Radiological Dispersal Devices (RDDs). They are not true nuclear weapons or devices but are simply a combination of conventional explosives and radioactive materials. The explosion of such a device is designed to scatter radioactive material about the environment to cause disruption and create panic rather than to kill large numbers of people. The most easily obtainable radioactive materials that could be used in fashioning such a bomb would be low-level nuclear materials from medical facilities and similar institutions. Highly radioactive materials from power plants or nuclear weapons facilities are more difficult to obtain but would scatter more dangerous nuclear material.

Nuclear Waste Disposal

The disposal of nuclear waste from power plants is a major concern. There are really only two alternatives: reprocessing the waste from fuel or placing the waste in a secure location. Several nations use reprocessing to reduce the amount of nuclear waste. Others rely on temporary storage at nuclear plants. Ultimately, these wastes will need to be placed in long-term storage underground. Issues related to disposal of nuclear waste will be covered in depth in chapter 19, where hazardous materials are discussed.

Decommissioning Nuclear Power Plants

All industrial facilities have a life expectancy—that is, the number of years they can be profitably operated. The life expectancy for an electrical generating plant, whether fossil fuel or nuclear, is about 30 to 40 years, after which time the plant is taken out of service. With a fossil-fuel plant, demolition is relatively simple and quick. A wrecking ball and bulldozers reduce the plant to rubble, which is trucked off to a landfill. The only harm to the environment is usually the dust raised by the demolition.

Demolition of a nuclear plant is not so simple. In fact, nuclear plants are not demolished; they are decommissioned. **Decommissioning** involves removing the fuel, cleaning surfaces,

Issues & Analysis

Drilling for Oil in Deep Water

In April 2010, the *Deepwater Horizon* oil rig caught fire and sank. Safety devices that were supposed to prevent the release of oil in the case of an accident failed and it was about three months before the flow of oil was stopped. Over that period about 10,000 metric tons of oil was released, making it the largest oil spill on record. The depth of the water at the well site was about 1.6 kilometers (1 mile). Thus, efforts to control the spill required the use of special remote devices and encountered many problems that delayed the sealing of the well.

The immediate response by the Department of Interior was to stop all drilling in the Gulf of Mexico; fishing was also curtailed. This met with considerable resistance from the state governments affected and oil companies, and was challenged in the courts. However, the moratorium was put into effect.

Several procedures were initiated to reduce the impact of the oil on the marine and coastal environment. Dispersants were used to break up the oil into small globules with the hope that microorganisms would more easily break down the oil. Booms were used to contain some of the oil, pump it into ships, and prevent it from reaching shore. Oil on the ocean surface was burned to reduce the amount of oil. It appears that these efforts greatly reduced the impact of the oil spill, although there are on-going studies to assess the damage.

In October 2010, the moratorium on drilling in the Gulf of Mexico was lifted and oil companies were permitted to resume drilling after meeting more stringent permit requirements.

However, the Gulf oil spill had an effect on planned exploration in other coastal areas along the east coast of the United States from Florida to Delaware. The Obama administration had opened these areas to oil exploration just prior to the Gulf oil spill. Proponents of drilling in these areas maintain that oil from these areas would reduce dependence on foreign oil. In addition, the coastal states involved would receive financial benefits. In November 2010, the federal government reversed the earlier

plan and delayed expansion of offshore oil drilling in offshore areas other than the Gulf of Mexico for seven years. Coastal states and oil companies have argued against the ban.

- If the cost of gasoline rose to $5.00 per gallon, would you support offshore drilling?
- Should the economic interests of the coastal states be a part of the equation for deciding to drill offshore?
- Are there ways to reduce dependence on oil that would make off-shore drilling unnecessary?

The burning Deepwater Horizon oil drilling rig

and permanently preventing people from coming into contact with the contaminated buildings or equipment.

The decommissioning of a plant is a two-step process:

Stage 1 involves removing fuel rods and water used in the reactor and properly storing or disposing of them. This removes 99 percent of the radioactivity. The radioactive material remaining consists of contaminants and activated materials. Activated materials are atoms that have been converted to radioactive isotopes as a result of exposure to radiation during the time the plant was in operation. These materials must be dealt with before the plant can be fully decommissioned.

Stage 2 leads to the final disposition of the facility. There are three options to this second stage of the decommissioning process.

1. Decontaminate and dismantle the plant as soon as it is shut down.

2. Secure the plant for many years to allow radioactive materials that have a short half-life to disintegrate and then dismantle

the plant. (However, this process should be completed within 60 years.)

3. Entomb the contaminated portions of the plant by covering the reactor with reinforced concrete and placing a barrier around the plant. (Currently this option is only considered suitable for small research facilities.)

Today, about 100 commercial nuclear power plants, 45 experimental (prototype) reactors, and 250 research reactors in the world have been shut down and are in various stages of being decommissioned.

Recent experience indicates that the cost for decommissioning a large plant will be between $200 million and $400 million, about 5 percent of the cost of generating electricity. Although the mechanisms vary among countries, the money for decommissioning is generally collected over the useful life of the plant.

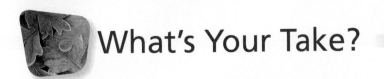

What's Your Take?

The United States has 100 operating nuclear power plants and generates about 20 percent of its electricity from nuclear power. Yet it seeks to prevent some other countries, such as North Korea and Iran, from developing nuclear power capabilities. Should the United States be able to prevent other countries from developing nuclear power?

Choose one side and develop arguments to support your point of view.

Summary

The major energy sources are fossil fuels that are nonrenewable and will eventually be used up. A resource is a naturally occurring substance of use to humans, a substance that can potentially be extracted. Reserves are known deposits from which materials can be extracted profitably with existing technology under present economic conditions. The amount of a reserve changes as new deposits are discovered, new technology is applied, and economic conditions change.

Coal is the world's most abundant fossil fuel. It formed as a result of plant material being buried and modified over time. Several qualities of coal are determined by the relative amounts of water and carbon present.

Oil and natural gas were formed from microorganisms that were buried, compressed, and modified.

Coal is obtained by either surface mining or underground mining. Problems associated with coal extractions are disruption of the landscape due to surface mining and subsidence due to underground mining. Black lung disease, waste heaps, water and air pollution, and acid mine drainage are additional problems. Oil and natural gas are obtained by drilling into geologic formations that contain the oil or natural gas. As oil becomes less readily available, there has been increased exploration and drilling in more difficult sites such as the arctic and the floor of the ocean. In addition, more expensive secondary and tertiary recovery methods have been developed to extract a greater percentage of the oil from the deposit. Oil spills are the major environmental issue related to the use of oil. Natural gas has the lowest environmental impact of the three forms of fossil fuels. The primary environmental concern is preventing explosions.

Nuclear power plants provide about 11 percent of the world's electrical energy. As the cost of oil has increased and there is concern about climate change, many countries have plans to build additional nuclear power plants.

Nuclear fission is the splitting of the nucleus of the atom that releases particles (alpha, beta, neutrons, etc.) and energy. If concentration of radioactive material is high enough, it will support a nuclear chain reaction. In a nuclear chain reaction, the neutrons released by the disintegration of an atom strike other atoms and cause them to split, also releasing neutrons. The resulting energy can be used for a variety of purposes. The splitting of U-235 in a nuclear reactor can be used to heat water to produce steam that generates electricity. Various kinds of nuclear reactors have been constructed, including boiling-water reactors, pressurized-water reactors, heavy-water reactors, gas-cooled reactors, and experimental breeder reactors. All reactors contain a core with fuel, a moderator to control the rate of the reaction, and a cooling mechanism to prevent the reactor from overheating.

The nuclear fuel cycle involves mining and enriching the original uranium ore, fabricating it into fuel rods, using the fuel in reactors, and reprocessing or storing the spent fuel rods. The fuel and wastes must also be transported. At each step in the cycle, there is danger of exposure. During the entire cycle, great care must be taken to prevent accidental releases of nuclear material.

The biological effects of exposure to radiation occur because the various forms of radiation cause certain molecules in tissues to ionize, which makes them very reactive. These very reactive molecules interact with other important biological molecules and alter them so these molecules cannot perform their normal function. A primary concern is the change that occurs to DNA which can cause genetic mutations and cancer. Extremely high doses of radiation directly kill cells by causing major molecular changes. Simple methods of protecting people from radiation are to limit the amount of time people are exposed, keep people at a distance from the source of the radiation, and provide shielding materials around radiation sources.

Although the accidents at Three Mile Island in the United States, Chernobyl in Ukraine, and Fukushima in Japan raised concerns about the safety of nuclear power plants, rising energy prices and concerns about climate change have stimulated increased building of nuclear power plants in many countries and many more are in the planning stages. The disposal of nuclear waste is expensive and controversial. Long-term storage in geologically stable regions is the most commonly supported option. Russia, Japan, and the United Kingdom operate nuclear reprocessing facilities as an option for reducing the amount of nuclear waste that must ultimately be placed in long-term storage.

Acting Green

1. Use the chart at the right to calculate your monthly carbon footprint. The average carbon dioxide produced per person in the United States is 1.7 tons of CO_2 per month (20 tons per year).

   Note: If there are several persons in your household, divide the total energy used in each category by the number of people in the household to determine your individual carbon footprint.

2. Contact your local electric utility or visit its website and determine what percent of the electricity produced comes from nuclear power, coal-fired power plants, and natural gas-fired power plants.

3. The United States has 100 electricity-producing nuclear power plants. Identify the one that is closest to you geographically. An Internet search should allow you to find this information.

Energy Use	Tons of CO_2
Kilowatt-hours of electricity $\times$ 0.0006 =	
Therms of natural gas $\times$ 0.00591 =	
Gallons of heating oil $\times$ 0.01015 =	
Gallons of gasoline $\times$ 0.0087 =	

Review Questions

1. Name the three most important sources of energy.
2. Distinguish between reserves and resources.
3. Describe three factors that can cause the amount of an oil reserve to increase.
4. Describe the geologic processes that resulted in the formation of coal.
5. Describe the differences between lignite, bituminous, and anthracite coal.
6. Describe the processes that resulted in the formation of oil and natural gas.
7. What regions of the world have the largest reserves of coal? Of oil? Of natural gas?
8. List three environmental impacts of the use of coal.
9. What are secondary and tertiary oil recovery methods? Why is their use related to the price of oil?
10. What is the most common environmental problem associated with the extraction and transportation of oil?
11. What environmental advantage does natural gas have over oil and coal?
12. What environmental concern has caused people to be more accepting of nuclear power?
13. What are the products of the nuclear disintegration of a radioactive isotope?
14. What is a nuclear chain reaction?
15. Describe how a nuclear power plant generates electricity.
16. Name the steps in the nuclear fuel cycle.
17. How does radiation cause damage to organisms?
18. List the three primary methods of protecting people from damaging radiation.
19. What happened at Chernobyl, Three Mile Island, and Fukushima? Why did it happen?
20. What are the major environmental problems associated with the use of nuclear power?
21. What happens during Stage 1 of the decommissioning of a nuclear power plant?
22. What options are available during Stage 2 of the decommissioning of a nuclear power plant?

Critical Thinking Questions

1. Coal-burning electric power plants in the Midwest have contributed to acid rain in the eastern United States. Other energy sources would most likely be costlier than coal, thereby raising electricity rates. Should citizens in eastern states be able to pressure utility companies in the Midwest to change the method of generating electricity? What mechanisms might be available to make these changes? How effective are these mechanisms?

2. Recent concerns about global warming have begun to revive the nuclear industry in the United States. Do you think nuclear power should be used instead of coal for generating electricity? Why?

3. Some states allow consumers to choose an electric supplier. Would you choose an alternative to nuclear or coal even if it cost more?

4. Given what you know about the economic and environmental costs of different energy sources, would you recommend that your local utility company use nuclear power or coal to supplement electric production? What criteria would you use to make your recommendation?

To access additional resources for this chapter, please visit ConnectPlus® at connect.mheducation.com. There you will find interactive exercises, including Google Earth™, additional Case Studies, and SmartBook™, an adaptive eBook that integrates our LearnSmart® adaptive learning technology.

Sunlight provides a continuous source of energy to Earth. Sunlight is converted to the energy in biomass, which is the world's most commonly used renewable energy source.

OBJECTIVES

After reading this chapter, you should be able to:

- Distinguish renewable energy sources from nonrenewable sources.
- Recognize that renewable energy sources currently provide about 13 percent of world energy.
- Recognize that fuelwood is a major source of energy in many parts of the less-developed world and that fuelwood shortages are common.
- Identify industries that typically use the wastes they produce to provide energy.
- Describe technical and economic factors that must be considered when burning municipal solid waste to produce energy.

- Identify negative environmental effects of using crop residues to provide energy.
- Describe the economic factors that must be taken into account when evaluating the use of crop residues or energy plantations to provide energy.
- List four processes used in the production of energy from biomass.
- Describe four environmental issues related to the use of biomass to provide energy.
- Describe environmental issues related to the development of hydroelectric power.

- Describe how active and passive solar heating designs differ.
- Describe two methods used to generate electricity from solar energy.
- Describe how wind, geothermal, and tidal energy are used to produce electricity.
- Recognize that wind, geothermal, and tidal energy can be developed only in areas with the proper geologic or geographical features.
- Recognize that energy conservation can significantly reduce our need for additional energy sources.

Energy Return on Investment

The use of energy by society is a very complex process. This makes evaluating energy alternatives difficult and requires some way of comparing them. One way is to measure the amount of energy that an energy source produces and the amount of energy needed to produce the energy. All energy sources require an input of energy to develop their potential. For example, in order to produce a barrel of oil, a considerable amount of energy went into developing the well to produce the oil. Energy was used to explore for the source of the oil. Energy was used to drill the well to get to the oil. Energy was used to pump the oil to the surface. So if one does the energy accounting it is possible to measure the amount of energy invested to get energy out. This is referred to as *Energy Return On Investment* (EROI).

$$EROI = \frac{Energy\ gained}{Energy\ required\ to\ get\ the\ energy}$$

It is also important to understand that you can measure EROI at several points in the process of producing and using energy.

1. One common way to begin an accounting of energy is to measure the EROI at the point of production—at the oil well, wind turbine, or coal mine.

2. However, some energy sources must be modified before they can be used by a consumer. Oil must be refined, coal must be crushed and washed, and electricity at a wind turbine must be stepped up to a higher voltage before it can be transmitted over lines.

3. Furthermore there are energy costs of getting the energy source to the consumer. Natural gas must be pumped to the user, gasoline must be delivered to the gas station, coal must be shipped to the consumer, and electrical lines must be maintained to send energy from source to consumer.

4. Finally there is an energy cost to using energy. Roadways must be built and maintained to allow cars to travel, blowers on furnaces are needed to disperse the heat through a building, and replacements must be manufactured for appliances that wear out.

When all of these costs are taken into account it is likely that an EROI of 3:1 at the production site (oil well, etc.) is necessary to have a net positive energy balance for the use of an energy source by society.

The table provides information about the range of EROIs gleaned from several sources.

Energy Source	Energy out/Energy in (EROI)
Fossil fuels	
Oil and gas 1930	100+
Oil and gas 1970	30
Oil and gas 2005	11 to 18
Discoveries 1970	8
Natural gas 2005	10
Coal (mine mouth)	80
Tar sands	2–4
Tight oil	5
Nuclear	**5–15**
Renewable	
Hydroelectric	100+
Wind turbines	18
Flat plate solar collectors	1.9
Concentrating solar collectors	1.6
Photovoltaic	6.8
Ethanol from sugar cane	0.8–10
Ethanol from corn	0.8–1.6
Biodiesel	1.3

Data from Murphy, David J., and Hall, Charles A. S. *Year in review—EROI or energy return on (energy) invested.* Annals of the New York Academy of Sciences 1185 (2010) 102–118.

There are several conclusions that can be derived from these data.

- More recent discoveries and development of oil and natural gas have much lower EROIs than early discoveries.

- Unconventional sources of oil (tar sand, tight oil) have a lower EROI than traditional sources.

- Hydroelectricity, wind turbines, and photovoltaics have the best EROI of renewables.

10.1 The Status of Renewable Energy

The burning of fossil fuels (oil, natural gas, and coal) and electricity from nuclear power provide about 87 percent of the energy used in the world. The burning of fossil fuels is also responsible for most of the human-caused carbon dioxide emissions. In the last 10 years, energy consumption has increased by about 30 percent and we should expect continued substantial increases in the demand for energy into the future.

Renewable energy is provided by processes that replenish themselves or are continuously present as a feature of the solar system. The sun is the primary source of renewable energy. In terms of world energy usage, biomass is the primary renewable energy source. The process of photosynthesis converts sunlight energy into plant biomass.

$$\text{Carbon dioxide} + \text{Water} + \text{Sunlight energy} \longrightarrow \text{Biomass (Chemical energy)} + \text{Oxygen}$$

This energy is stored in the organic molecules of the plant as wood, starch, oils, or other compounds. Any form of biomass—plant, animal, alga, or fungus—can be traced back to the energy of the sun. Since biomass is constantly being produced, it is a form of renewable energy. Solar, wind, geothermal, and tidal energy are renewable energy sources because they are continuously available.

There is currently a great deal of interest in renewable forms of energy. There are several factors that stimulate this interest. Since fossil fuels are nonrenewable, as they become scarce the price rises. When the price of fossil fuels increases, the economics of renewable energy sources improves and there is an economic incentive to invest in renewable energy. When fossil fuels are cheap, many renewable forms of energy cannot compete economically. However, as the price of fossil fuels increases, many forms of renewable energy become economically viable. A third factor is the concern about climate change, which is driven by carbon dioxide emissions, primarily from the burning of fossil fuels. In general, renewable energy sources do not add to carbon dioxide emissions.

Currently, renewable energy sources—biomass, hydroelectricity, wind turbines, solar energy, geothermal energy, and tidal energy—supply about 13 percent of the world's total energy. Biomass accounts for about 10 percent of the energy used in the world, since firewood and other plant materials are the primary source of energy in much of the developing world.

Hydroelectric power accounts for about 2.3 percent and the remaining renewable technologies account for about 1 percent. (See figure 10.1.) Some optimistic studies suggest that these sources could provide half of the world's energy needs by 2050. It is unlikely that that will occur, but renewable sources certainly will become much more important as fossil-fuel supplies become more expensive.

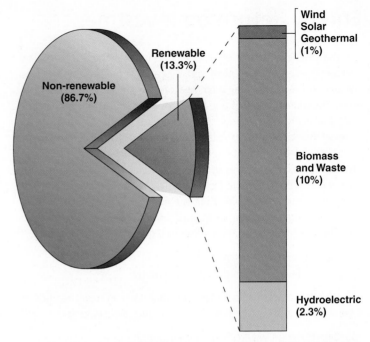

FIGURE 10.1 Renewable Energy as a Share of Total Energy Consumption (World 2011) Of the energy consumed in the world, about 87 percent is from nonrenewable fossil fuels and nuclear power. Renewable energy sources provide 13 percent, and the burning of biomass and waste accounts for over 75 percent of all renewable energy consumed. Source: Data from International Energy Agency.

requirements have been met by the combustion of fossil fuels. Biomass, however, is still the predominant form of energy used by people in the less-developed countries. For example, in the region of Africa, nearly 50 percent of the energy supply is obtained from biomass. India obtains about 25 percent and Bangladesh obtains about 28 percent of its energy from biomass.

Major Types of Biomass

There are several distinct sources of biomass energy: fuelwood, municipal and industrial wastes, agricultural crop residues and animal waste, and energy plantations.

Fuelwood Because of its bulk and low level of energy compared to equal amounts of coal or oil, wood is not practical to transport over a long distance, so most of it is used locally. In less-developed countries, wood has been the major source of fuel for centuries. In fact, wood is still the primary source of energy for nearly half of the world's population. In these regions, the primary use of wood is for cooking. In much of the less-developed world, cooking is done over open fires. Using fuel-efficient stoves instead of open fires could reduce these energy requirements by 50 percent. Improving efficiency would protect wood resources, reduce the time or money needed to obtain firewood, and improve the health of people because they would breathe less wood smoke. (See figure 10.2.)

Even developed countries that have abundant forest resources, such as the United States, Canada, Norway, Finland, and Sweden, obtain 4 to 20 percent of their energy from fuelwood.

10.2 Major Kinds of Renewable Energy

While many consider renewable energy to be the answer to energy supply and climate change problems, there are still many technical, economic, and cultural challenges to overcome before renewable energy will meet a significant percentage of humans' energy demands. This section will examine the major sources of renewable energy, and look at their potential and challenges.

Biomass Conversion

Biomass fulfilled almost all of humankind's energy needs prior to the Industrial Revolution. All biomass is traceable back to green plants that convert sunlight into plant material through photosynthesis. As recently as 1850, 91 percent of total U.S. energy consumption was biomass in the form of wood. Since the Industrial Revolution, the majority of the developed world's energy

(a)

(b)

FIGURE 10.2 Wood Use for Cooking Many people in the developing world use wood as a primary fuel for cooking. Reliance on wood is a major cause of deforestation. (a) In many parts of the world such as Africa, cooking is often done over open fires. This is a very inefficient use of fuel. (b) The use of simple mud stoves as in this situation in India greatly increases efficiency.

Waste Wastes of various kinds are a major source of biomass and other burnable materials produced by society. Certain industries, such as lumber mills, paper mills, and plants that process sugar from sugarcane, use biomass as a raw material for their products and in the process produce wastes such as sawdust, wood scrap, waste paper, or bagasse (sugarcane stalks), which are burnable. These industries typically burn these wastes to provide energy for their operations. Municipal solid waste is also a source of energy. About 80 percent of municipal waste is combustible and over 60 percent is derived from biomass. Plastics, textiles, rubber, and similar materials contribute the rest of the burnable portion of trash. (See figure 10.3.) However, to use municipal solid waste to produce energy requires that the waste be sorted to separate the burnable organic material from the inorganic material. The sorting is done most economically by those who produce the waste, which means that residents and businesses must separate their trash into garbage, burnable materials, glass, and metals. The trash must be gathered by compartmentalized collection trucks.

The burning of municipal solid waste to produce energy makes economic sense only when the cost of waste disposal is taken into account. In other words, although energy from solid waste is expensive, one can deduct the avoided landfill costs from the cost of producing energy from waste. Where landfill costs are high, municipal waste-to-energy plants make economic sense. In the United States, about 12 percent of solid waste is burned in 86 plants, resulting in about 2,500 megawatts of electricity. Europe and Japan have much less available land and have placed restrictions on landfills. Thus, these countries have a much higher rate of burning of solid waste. Countries in Western Europe have over 400 waste-to-energy plants. Japan burns about 80 percent of its waste and Germany burns nearly all of its waste that is not recyclable.

Crop residues and animal wastes As we seek alternative sources of energy, crop residues have been identified as possibilities. In many parts of the world, the straw and stalks left on the field are collected and used to provide fuel for heat and cooking. Animal wastes are also used for energy. Animal dung is dried and burned or processed in anaerobic digesters to provide a burnable gas.

However, there are several impediments to this development. Crop residues are bulky and have a low energy-to-weight ratio. Thus, harvesting crop residues to burn or convert to ethanol makes sense

(a) Sawdust and wood scrap can be used to produce energy

(b) Paper mills burn waste

(c) Trash can be burned to provide energy

FIGURE 10.3 Energy from Waste Many segments of the forest products industry, such as lumber and paper mills, use the waste they produce as a source of energy for their industrial process. Municipal solid waste also contains large amounts of burnable material that can be used to generate energy.

Focus On

Biomass Fuels and the Developing World

Although most of the world uses fossil fuels as energy sources, much of the developing world relies on *biomass* as its source of energy. The biomass can be wood, grass, agricultural waste, or dung. According to the United Nations, 2 billion people (30 percent of the world's population) use biomass as fuel for cooking and heating dwellings. In developing countries, nearly 40 percent of energy used comes from biomass. In some regions, however, the percentage is much higher. For example, in sub-Saharan Africa, fuelwood provides about 80 percent of energy consumed. Worldwide, about 60 percent of wood removed from the world's forests is used for fuel.

This dependence on biomass has several major impacts:

- Often women and children must walk long distances and spend long hours collecting firewood and transporting it to their homes.

- Because the fuel is burned in open fires or inefficient stoves, smoke contaminates homes and affects the health of the people. The World Health Organization estimates that in the developing world, 40 percent of acute respiratory infections are associated with poor indoor air quality related to burning biomass. A majority of those who become ill are women and children because the children are in homes with their mothers who spend time cooking food for their families.

- Often the fuel is harvested unsustainably. Thus, the need for an inexpensive source of energy is a cause of deforestation. Furthermore, deforested areas are prone to soil erosion.

- When dung or agricultural waste is used for fuel, it cannot be used as an additive to improve the fertility or organic content of the soil. Thus, the use of these materials for fuel negatively affects agricultural productivity.

Nepali woman carrying brushwood.

only if the travel distance from field to processing plant is short. Removal of crop residues from fields has several negative agricultural impacts. It leads to increased erosion, since the soil is more exposed, and reduces the benefits provided by organic matter in the soil, since the crop residues are not incorporated into the soil. Finally, the production of ethanol from cellulose is currently not economical.

Energy plantations Many crops can be grown for the express purpose of energy production. Crops that have been used for energy include forest plantations, sugarcane, corn, sugar beets, grains, kelp, palm oil, and many others. Two main factors determine whether a crop is suitable for energy use. Good energy crops have a very high yield of dry material per unit of land (dry metric tons per hectare). A high yield reduces land requirements and lowers the cost of producing energy from biomass. Similarly, the

amount of energy that can be produced from a biomass crop must be more than the amount of energy required to grow the crop. In some circumstances, such as the heavily mechanized corn farms of the U.S. Midwest, the amount of energy in ethanol produced from corn is not much greater than the energy used for tractors, to manufacture fertilizer, and to process the grain into ethanol. (See Issues and Analysis—Does Corn Ethanol Fuel Make Sense?)

Biomass Conversion Technologies

In order to use biomass as a source of energy, it must be burned. Often it is important to convert the biomass into a different form that is easier to transport and use. There are several technologies capable of making the energy of biomass available for use. These include direct combustion and cogeneration, ethanol production, anaerobic digestion, and pyrolysis.

Direct combustion   The most common way that biomass and waste are used for energy production is by burning them to produce heat or electricity. In much of the developing world, the primary use of energy derived from biomass is as fires to provide heat for heating homes and cooking.

Large-scale operations are used to power industrial processes or to generate electricity. Large biomass electric-power generation systems have lower efficiencies than comparable fossil-fuel systems because of the higher moisture content of biomass. However, using the biomass as a component of a coal-fired power plant improves the efficiency. Furthermore, if a biomass-fired power plant can be used to provide both heat and electricity (cogeneration system), the economics and energy efficiency improve significantly. This is how biomass is used in the wood products, paper, and sugar industries.

Biofuels production   Ethanol can be produced by fermenting sugars. The sugars can be obtained directly from plants or may be produced by converting starch or cellulose of plants to sugars.

Typically, sugar or starch is extracted from the biomass crop by crushing and mixing with water and yeast and then keeping the mixture warm in large tanks called fermenters. The yeast breaks down the sugar and converts it to ethanol and carbon dioxide. A distillation process is required to remove the water and other impurities from the dilute alcohol product. The low price of sugar coupled with the high price of oil has prompted Brazil to use its large crop of sugarcane to produce ethanol. Ethanol is sold in a variety of mixtures for automobile fuel, from 100 percent ethanol to 20 percent ethanol and 80 percent gasoline. In total, ethanol provides about 50 percent of Brazil's automobile fuel.

In the United States, corn is used for ethanol production and then blended with gasoline. Most gasoline in the United States contains some ethanol (up to 10 percent) because it improves engine efficiency, reduces air pollution, and reduces the likelihood that water will freeze in gas lines. E85 is a fuel that is 85 percent ethanol and 15 percent gasoline. It can be used in later model cars that are known as flex-fuel vehicles. (See figure 10.4.)

Biodiesel can be produced from the oils in a variety of crops, including soybeans, rapeseed, and palm oil as well as animal fats. These raw materials need to be modified chemically before they can be used as fuel. Currently, about 2 percent of the diesel fuel consumed in the world is biodiesel. Europe leads the world in production of biodiesel fuel with about 50 percent of the total world production.

FIGURE 10.4 Biofuels Biofuels (E85 and biodiesel) are available for purchase in many parts of the world.

Anaerobic digestion   Anaerobic digestion involves the decomposition of wet and green biomass or animal waste through bacterial action in the absence of oxygen. This process produces a mixture, consisting primarily of methane and carbon dioxide, known as biogas. The most commonly used technology involves small digesters on farms that generate gas for use in the home or for farm-related activities. Millions of small methane digesters are in use in countries like China, India, and Korea. (See figure 10.5.) Large, industrial size farming operations also use anaerobic disgesters to process animal waste and capture biogas. (See figure 10.6.) Similarly many sewage treatment plants capture biogas. In both instances the methane collected can be used to provide heat or run machinery. Anaerobic digestion also occurs in landfills. In many landfills the methane gas produced eventually escapes into the atmosphere. However, the gas can be extracted by inserting perforated pipes into the landfill. In this way, the gas will travel through the pipes, under natural pressure, to be used as an energy source, rather than simply escaping into the atmosphere to contribute to greenhouse gas emissions. Some newer landfills have even been designed to encourage anaerobic digestion, which reduces the volume of the waste and provides a valuable energy by-product.

Pyrolysis   Pyrolysis is a process for converting solid biomass to a more useful fuel. Biomass is heated or partially burned in

What's Your Take?

Many states require electric utilities to "generate" a certain amount of the electricity they produce from renewable sources. To meet their requirements, many electric utilities simply buy the "renewable" electricity from another company and do not invest in technologies that are renewable. Should they be allowed to do this? Choose one side or the other and develop arguments to support your point of view.

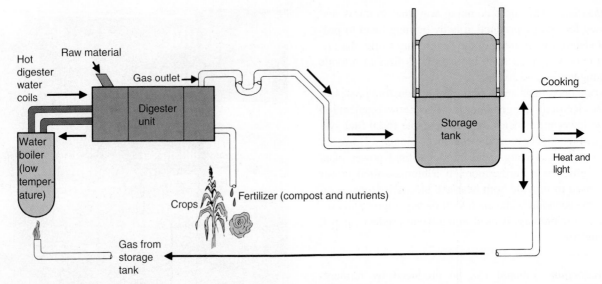

FIGURE 10.5 Methane Digester In the digester unit, anaerobic bacteria convert animal waste into methane gas. This gas is then used as a source of fuel. The sludge from this process serves as a fertilizer. In many less-developed countries, this type of digester has the advantages of providing a source of energy and a supply of fertilizer and managing animal wastes, which helps reduce disease.

FIGURE 10.6 Anaerobic Bioreactor This bioreactor is used to produce methane from animal manure on a large dairy farm.

FIGURE 10.7 Desertification The demand for fuelwood in many regions has resulted in the destruction of forests. This is a major cause of desertification.

an oxygen-poor environment to produce a hydrocarbon-rich gas mixture, an oil-like liquid, and a carbon-rich solid residue (charcoal), which has a higher energy density than the original fuel. In developing countries, charcoal kilns are simply mounds of wood or wood-filled pits in the ground that are set afire and then covered with earth. The smoldering fire slowly converts the wood to charcoal.

Gasification is a form of pyrolysis, carried out with more air and at high temperatures, to optimize the gas production. The resulting gas, known as producer gas, is a mixture of carbon monoxide, hydrogen, and methane, together with carbon dioxide and nitrogen. The gas is more versatile than the original solid biomass, and it can be used as a source of heat or used in internal combustion engines or gas turbines to produce electricity.

Environmental Issues

Although the use of biomass and waste as an energy source is often thought to be environmentally benign, it has many significant environmental impacts.

Habitat and biodiversity loss It is estimated that throughout the world there are 1.3 billion people who cannot obtain enough wood or must harvest wood at a rate that exceeds its growth. This has resulted in the destruction of much forest land in Asia and Africa and has hastened the rate of desertification in these regions. (See figure 10.7.)

The Renewable Fuel Mandate

Scientific studies have provided a great deal of information about how energy can be produced from biomass. An international body of scientists and policymakers has also provided information about the impact of rising CO_2 concentrations in the atmosphere and the potential for climate change. The production of ethanol from sugars and starches is well understood. Theoretically it should also be possible to produce ethanol from cellulose, which is the primary structural molecule of plants. Using all these bits of scientific information along with a desire to reduce dependence on oil imports and stimulate agriculture and alternative energy industries, Congress passed two bills—the Energy Policy Act of 2005 and the Energy Independence and Security Act of 2007—that mandated specific amounts of renewable fuels in gasoline. Although these two laws addressed many energy-related issues, the following items relate to the renewable fuel mandate.

- Subsidies were provided to build ethanol plants.
- Different kinds of renewable fuels were identified—corn-based ethanol, cellulosic ethanol, biodiesel, and advanced fuels. Each fuel must demonstrate that it produces less greenhouse gases than gasoline.
- Gasoline producers were required to blend specific amounts of renewable fuels into gasoline or pay a fine.
- A schedule of increasing amounts of each of the kinds of renewable fuels was required in gasoline from 2005 to 2020. (See graph.)
- A tax credit of $0.45 per gallon of ethanol was provided to blenders who incorporated ethanol into their gasoline.

Unintended Consequences

Between 2005, when the Energy Policy Act of 2005 was passed, and 2012, the amount of corn used for ethanol production increased to 40 percent of corn production.

- This resulted in increases of 16 percent in the amount of land planted to corn and 240 percent increase in the price of corn between 2005 and 2012.
- Farmers took erodible farmland out of conservation programs and planted corn because they could make more money by planting corn than they could from agriculture programs that paid them to remove marginal lands from production.
- Since the additional demand for corn to make ethanol caused the price to rise, farmers who needed to feed corn to livestock had to pay higher prices, resulting in increased costs of meat and dairy products.
- Since farmers planted more corn and less of other crops, the prices of other crops rose because of a lower supply.

Companies that produce gasoline experienced problems.

- They cannot add more than 10 percent ethanol to gasoline because many vehicles could be damaged by higher amounts of ethanol. Thus

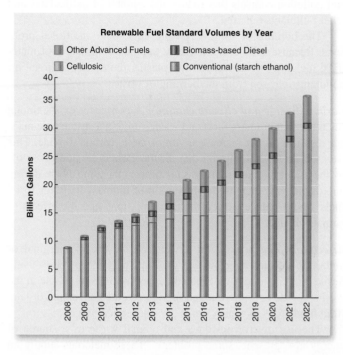

Source: Data from DOE Alternative Fuels Data Center.

they are limited in the amount of ethanol they can put into gasoline and have not met mandated amounts since 2011.
- They were required to begin using cellulosic ethanol in 2010 but were unable to do so because it is not available.
- Because of these two problems they have paid fines for not meeting the mandated amounts. They appealed the fines and requested that the mandated quantities of ethanol be reduced.

Initially there was a tax credit of $0.45/gallon of ethanol to blenders of gasoline for using ethanol in gasoline. This subsidy expired in 2011, which meant that the cost of ethanol increased. At the same time gasoline blenders were limited in the amount of ethanol they could purchase.

- The increased price of ethanol was passed along to consumers, although prices for gasoline are so volatile that it is difficult to see.
- Although flex fuel vehicles can use E85 fuel, which is 85 percent ethanol and 15 percent gasoline, they do not do so because the cost per mile driven is higher for E85. (See Issues and Analysis at the end of this chapter for details.)
- Several major producers of ethanol have gone bankrupt.

Another issue associated with biomass energy is the loss of biodiversity. Destroying natural ecosystems to plant sugarcane, grains, palm oil, or other plants can reduce the biodiversity of a region. The plantations lack the complexity of a natural ecosystem and are susceptible to widespread damage by pests or disease.

Air pollution Burning wood is a source of air pollution. Often the people in developing countries use wood in open fires or poorly designed, inefficient stoves. This results in the release of high amounts of smoke (particulate matter) and other products of incomplete combustion, such as carbon monoxide and hydrocarbons, which contribute to ill health and death.

Respiratory illnesses are particularly common among women and children who spend the most time in the home. Even in the developed world, air pollution from the burning of biomass is a problem. Some cities, such as London, England, have a total ban on burning wood. Vail, Colorado, permits only one wood-burning stove per dwelling. Many areas require woodstoves to have special pollution controls that reduce the amount of particulates and other pollutants released.

The burning of solid waste presents some additional problems. Because solid waste is likely to contain a mixture of materials, including treated paper and plastic, there are additional air pollutants released from the burning of waste.

Carbon dioxide and climate change   A consensus exists among scientists that biomass fuels and wastes used in a sustainable manner result in no net increase in atmospheric carbon dioxide. This is based on the assumption that all the carbon dioxide given off by the use of biomass fuels was taken from the atmosphere during photosynthesis. Increased substitution of biomass fuels for fossil fuels would reduce carbon dioxide emissions and the effects of climate change.

Effects on food production   Although the use of marginal or underutilized land to grow energy crops may make sense, using fertile cropland does not. Since there are millions of people in the world who do not have enough food to eat, the conversion of land from food crops to energy crops presents ethical issues.

The use of crop residues and animal waste as a source of energy also presents some problems. These materials supply an important source of organic matter and soil nutrients for farmers. This is particularly true among subsistence farmers in the developing world. They cannot afford fertilizer and rely on these materials to maintain soil fertility. However, they also need energy. Thus, they must make difficult decisions about how to use this biomass resource.

Hydroelectric Power

People have long used water to power a variety of machines. Some early uses of water power were to mill grain, saw wood, and run machinery for the textile industry. Flowing water has energy that can be captured and turned into electricity. This is called hydroelectric power, or *hydropower.*

At present, hydroelectricity provides about 16 percent of the world's electricity. In some areas of the world, hydroelectric power is the main source of electricity. In South and Central America, 65 percent of the electricity comes from hydroelectric power. Norway gets 95 percent of its electricity and over 50 percent of all its energy from hydroelectricity. Canada gets 59 percent and the United States gets about 8 percent of its electricity from hydroelectric plants.

Technology for Obtaining Hydropower

The most common type of hydroelectric power plant uses a dam on a river to store water in a reservoir. (See figure 10.8.) Water released from the reservoir flows through a turbine, spinning it, which in turn activates a generator to produce electricity. Most of

(a) Kuroyon Dam, Japan

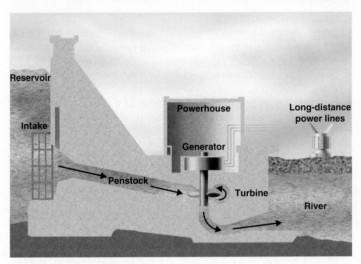

(b) Hydroelectric power plant

FIGURE 10.8 Hydroelectric Power Plant (a) The water impounded in this reservoir is used to produce electricity. In addition, this reservoir serves as a means of flood control and provides an area for recreation. (b) This figure shows how a hydroelectric dam produces electricity.
Source: (b) Tennesese Valley Authority.

this hydroelectricity comes from large dams, which rank among humanity's greatest engineering feats. Table 10.1 lists the locations and sizes of the largest hydroelectric facilities. In 2006, China completed construction of the largest hydroelectric dam in the world, the Three Gorges Dam, on the Yangtze River. It has a generating capacity of 22,500 megawatts. Although most hydroelectricity comes from large plants (1,000s of megawatts) built on large reservoirs, small (less than 10 megawatts) plants are also important contributors. They can be built on small rivers, and in some areas of the world where the streams have steep gradients and a constant flow of water, hydroelectricity may be generated without a reservoir. China has over 80,000 small stations, and the United States has nearly 1,500. In addition, microhydroelectric power systems (less than 1 megawatt) can be built in remote places to supply electricity for local needs such as a home, ranch, or village.

Potential for Additional Hydropower

Over the past ten years, the energy furnished by hydroelectricity worldwide increased by nearly 40 percent. The potential for developing hydroelectric power is best in mountainous regions and large river valleys. Future increases in hydroelectric power will come mainly from the development of large plants (1,000s of megawatts) on reservoirs.

Some areas of the world, such as Canada, the United States, Europe, and Japan, have already developed about 50 percent of their hydroelectric potential. Most of the potential for development of new hydroelectric power facilities is in Africa, Asia, South America, Eastern Europe, and Russia.

Environmental Issues

Although hydroelectric power is a renewable energy source and does not emit air or water pollutants, there are still environmental consequences associated with developing hydroelectric facilities. The most obvious impact of hydroelectric dams is the flooding of vast areas of land, much of it previously forested or used for agriculture. The size of reservoirs created can be extremely large. The James Bay project on the Le Grande River in northern Quebec has submerged over 13,000 square kilometers (5,000 square miles) of land. The construction of the Three Gorges Dam in China inundated 153 towns and 4,500 villages and caused the displacement of over a million people. In addition, numerous archeological sites were submerged and the nature of the scenic canyons of the Three Gorges was changed.

Dams and reservoirs also greatly alter watersheds. Damming a river alters the normal flow of the river and changes the quality of water (temperature, amount of particulate matter, oxygen content, etc.) in the river downstream of the dam. Dams also alter the migration patterns of fish and often prevent fish from migrating upstream to spawn. These impacts can be reduced by requiring that dams release enough water to maintain minimum flows downstream of a dam and by creating fish ladders that allow fish to move upstream past the dam. Silt, normally carried downstream to the lower reaches of a river, is trapped by a dam and deposited on the bed of the reservoir. This silt slowly fills a reservoir, decreasing the amount of water that can be stored and used for electrical generation. The river downstream of the dam is also deprived of silt, which normally fertilizes the river's floodplain during high-water periods.

The flooded area behind a new dam contains vegetation that decomposes. The bacteria involved in decomposition have two negative effects. They release carbon dioxide and some of them can convert mercury that is naturally in the soil into methylmercury that can accumulate in fish. This poses a health hazard to those who depend on these fish for food. It is thought that both of these problems are temporary and will be reduced once the flooded vegetation is decomposed.

Solar Energy

The sun is often mentioned as the ultimate answer to the world's energy problems. It provides a continuous supply of energy that far exceeds the world's demands. In fact, the amount of energy received from the sun each day is 600 times greater than the amount of energy produced each day by all other energy sources combined. However, solar energy, in all its forms, provides less than one percent of energy needs. The major problems with solar energy are that, by nature, it is both intermittent and diffuse. It is available only during the day when it is sunny, and it is spread out over the entire Earth, falling on many places like the oceans where it is difficult to collect. All systems that use solar energy must store energy or use supplementary sources of energy when sunlight is not available. Because of differences in the availability of sunlight, some parts of the world are more suited to the use of solar energy than others.

Solar energy is utilized in three ways:

1. In a passive heating system, the sun's energy is converted directly into heat for use at the site where it is collected.
2. In an active heating system, the sun's energy is converted into heat, but the heat must be transferred from the collection area to the place of use.
3. The sun's energy also can be used to generate electricity by heating water to turn turbines or by using photovoltaic cells.

Passive Solar Systems

Anyone who has walked barefoot on a sidewalk or blacktopped surface on a sunny day has experienced the effects of passive solar heating. In a **passive solar system,** light energy is transformed to heat energy when it is absorbed by a surface. Homes and buildings can be designed to use passive solar energy for heating and lighting, which reduces the need for nonrenewable energy sources. (See figure 10.9.)

In the Northern Hemisphere, the south side of a building always receives the most sunlight. Therefore, buildings designed for passive solar heating usually have large south-facing windows.

Table 10.1 World's Largest Hydroelectric Plants

Name	Country	Rated Capacity Megawatts
Three Gorges Dam	China	22,500
Itaipu	Brazil/Paraguay	14,000
Guri (Simón Bolívar)	Venezuela	10,200
Tucuruí	Brazil	8,370
Grand Coulee	United States	6,809
Longtan Dam	China	6,426
Sayano Shushenskaya	Russia	6,400
Krasnoyarskaya	Russia	6,000
Robert-Bourassa	Canada	5,616
Churchill Falls	Canada	5,429
Bratskaya	Russia	4,500
Laxiwa	China	4,200
Xiaowan	China	4,200

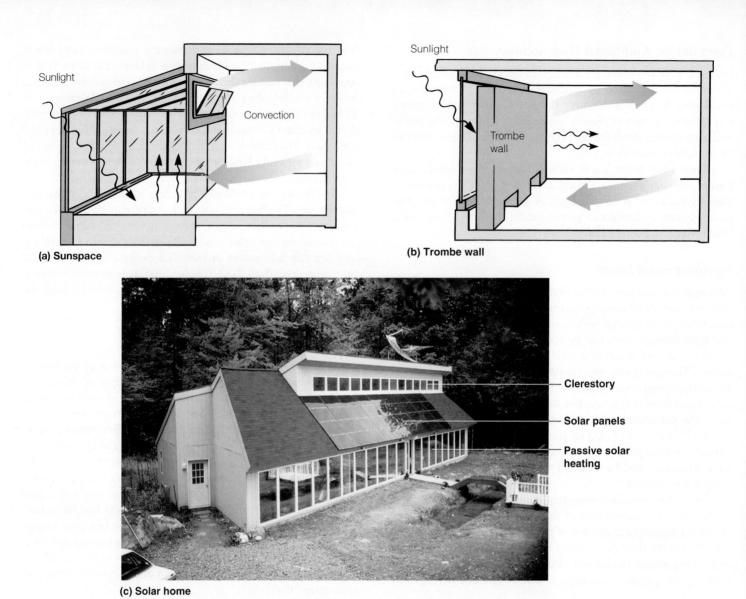

(a) Sunspace

(b) Trombe wall

(c) Solar home

Clerestory

Solar panels

Passive solar heating

FIGURE 10.9 Passive Solar Designs (a) A sunspace is like a greenhouse attached to a house. The heat captured in the sunspace can be transferred to the living space. (b) A trombe wall is a heat-absorbing body placed near a window that reradiates heat to the house when the sun goes down. (c) This house shows several features that make use of solar energy. The clerestory at the top allows sunlight to enter and provide lighting. The south-facing windows allow sunlight to warm the house, and the solar panels on the roof capture additional solar energy.

Materials that absorb and store the sun's heat can be built into the sunlit floors and walls. The floors and walls heat up during the day and slowly release heat at night, when the heat is needed most. This passive solar design feature is called *direct gain.*

Other passive solar heating designs involve sunspaces or trombe walls. A *sunspace* (which is much like a greenhouse) is built on the south side of a building. As sunlight passes through glass, it warms the sunspace. Proper ventilation allows the heat to circulate into the building. A *trombe wall* is a very thick, south-facing wall painted black and made of a material that absorbs a lot of heat. A pane of glass installed a few centimeters in front of the wall, helps hold in the heat. The wall heats up slowly during the day; then, as it cools gradually during the night, it gives off its heat inside the building.

Many passive solar systems also provide daylighting. *Daylighting* is simply the use of natural sunlight to brighten a building's interior, reducing the need for electricity to light the interior of a building. To lighten north-facing rooms and upper levels, a clerestory—a row of windows near the peak of the roof— is often used along with an open floor plan inside that allows the light to bounce throughout the building.

Of course, too much solar heating and daylighting can be a problem during hot summer months. There are design features that can help keep passive solar buildings cool in the summer. For instance, overhangs can be designed to shade windows when the sun is high in the summer. Sunspaces can be closed off from the rest of the building. And a building can be designed to use fresh-air ventilation in the summer.

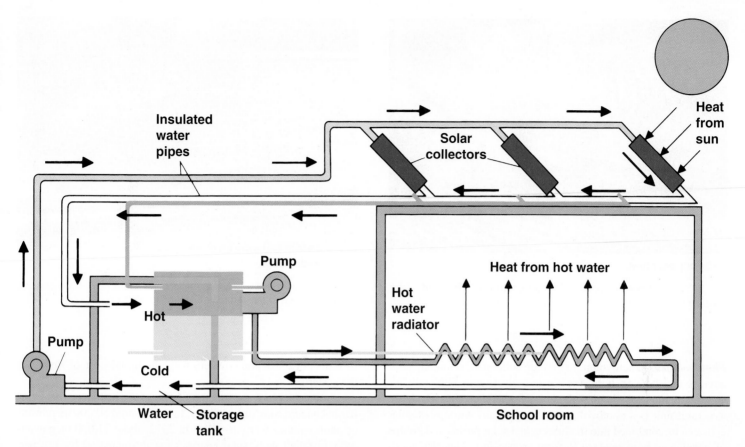

FIGURE 10.10 An Active Solar Heating Design An active solar system requires a solar collector, a pump, a heat storage system, and a system of pipes to convey the heat from one place to another.

Active Solar Systems

An **active solar system** requires a solar collector, a pump, and a system of pipes to transfer the heat from the site of production to the area to be heated. (See figure 10.10.)

Active solar collector systems are most commonly used to provide heat energy for water heaters, pools, and homes.

Because sunlight is intermittent, active solar heating systems require heat storage mechanisms and usually also require conventional energy sources to provide energy when solar energy is inadequate. Rock, water, or specially produced products are used to store heat during the day, and the storage medium releases heat when the sun is not shining.

An active solar space heating system makes economic sense if it can offset considerable amounts of heating energy from conventional systems over the life of the building or the system. Active solar systems are most easily installed in new buildings, but in some cases they can be installed in existing structures. A major consideration in the use of an active solar system is the initial cost of installation.

Solar-Generated Electricity

Solar energy can be used to generate electricity in two different ways. It can be used to create steam that is used to run a turbine

similar to that of a conventional power plant, or photovoltaic cells can be used to generate electricity directly from sunlight.

Conventional electric generation To produce electricity using a turbine, the energy from the sun must be collected and concentrated to heat water to make steam. There are basically two designs used. One design, called a *solar power tower*, uses mirrors to focus sunlight at a central point that raises the temperature and allows for the production of steam. All current facilities are small (5–20 megawatts). Several large (50–400 megawatt) solar power tower facilities are under construction in the United States, Israel, and South Africa.

Currently, the most successful commercial design is the parabolic trough, which can heat oil in pipes to 390°C (734°F). This heat can be transferred to water, which is turned into steam that is used to run conventional electricity-generating turbines. The 354-megawatt Solar Energy Generating System (SEGS) in the Mojave Desert in California is the largest solar electric generation facility in the world. A 64-megawatt Nevada Solar One plant opened in 2007 and the 280-megawatt Solana Generating Station came online in Arizona in 2013. There are about 12 parabolic trough plants under construction and due to go online between 2014 and 2017. Figure 10.11 shows examples of a power tower and a parabolic trough.

(a) Solar power tower

(b) Parabolic trough

FIGURE 10.11 Conventional Solar Generation of Electricity (a) A power tower produces steam by focusing sunlight at a central point. (b) A parabolic trough heats oil that transfers heat to water and converts it to steam. In both cases, the steam is used to turn a turbine and produce electricity.

Photovoltaics Photovoltaics (PV) are solid-state semiconductor devices that convert light directly into electricity. Photovoltaics are usually made of silicon with traces of other elements. The basic structural unit is a photovoltaic cell (solar cell). Groups of solar cells can be combined into modules called solar panels and groups of panels can be connected to form arrays. Individual PV modules are low-voltage, direct current (DC) devices, but connecting arrays of PV modules produces higher voltages. In some homes and other applications, the electricity is used as direct current and requires motors and other appliances that can use direct current. In these applications, a battery storage system is required to provide energy during nighttime hours. However, since most homes use alternating current, a device called an inverter is needed to convert direct current to alternating current. This allows the home to use alternating current from the grid, when needed, and to supply electricity to the grid when the solar panels are producing more electricity than is needed by the home.

Thin-film photovoltaics use layers of semiconductor materials only a few micrometers thick. Furthermore, they can be folded or formed into a variety of shapes, which has made it possible for photovoltaics to double as rooftop shingles, roof tiles, building facades, or the glazing for skylights or atria. However, they are less efficient at producing electricity than traditional solar cells.

Three factors drive the photovoltaic industry: cost of the solar installation, efficiency of the system, and government policy. As the cost of the system is reduced and efficiency increases, the price per kilowatt-hour of electricity falls. Currently, the cost of installation is falling and efficiency is increasing. Thus, the price of electricity from photovoltaics is falling as the cost of generating electricity from coal and nuclear power is rising. Electricity from photovoltaics is now less expensive in some places than electricity from the grid. However, the cost of installing a photovoltaic system on a house is expensive. Many governments and some utility companies offer subsidies for installing photovoltaic systems in order to assist homeowners with the initial cost of installing the system.

In recent years, the amount of PV power installed worldwide has been increasing dramatically. In 2010, about 10,000 megawatts of photovoltaics were installed. In 2012, about 31,000 megawatts were installed. U.S. solar energy capacity increased by 138 percent between 2011 and 2012. Much of this growth is attributable to government subsidies. Many European countries have pledged to reduce their dependence on fossil fuels and subsidize renewable energy installations. Germany, Spain, Japan, China, and the United States lead in the amount of photovoltaics installed. Figure 10.12 shows typical applications of photovoltaic technology.

Environmental Issues

Since solar energy is renewable, it has minimal environmental impact. However, the manufacture of the silicon or other materials that make up the units requires large amounts of energy. Thermal or photovoltaic power plants require large amounts of land to position their mirrors or solar collectors. The SEGS system in California covers 6.4 km^2 (2.5 mi^2). The installation of photovoltaics or water heating systems on buildings does not require additional space and is often incorporated into the design of the building.

Wind Energy

For centuries, wind has been used to move ships, grind grains, pump water, and do other forms of work. In more recent times, wind has been used to generate electricity. (See figure 10.13.)

In 2008, the U.S. Department of Energy published a report that stated that it was technically feasible to generate 20 percent of electricity in the United States from wind by 2030. Some areas are better suited for producing wind energy than others.

(a) Photovoltaic shingles being installed

(b) Solar panels

(c) Photovoltaic power plant

FIGURE 10.12 Photovoltaic Applications Photovoltaic devices can be used in many ways. (a) Photovoltaic shingles can be installed on roofs. (b) Solar panels can be installed on the surfaces of homes and other buildings. (c) Large numbers of solar panels can be combined to serve as power plants that supply the electrical grid.

FIGURE 10.13 Wind Energy Fields of wind-powered generators can produce large amounts of electricity.

Figure 10.14 shows the wind energy potential of regions within the United States. However, location can be a problem. Although places such as the Dakotas have the strongest winds, they are remote from energy-using population centers, and large losses in the amount of electricity would occur as it is transmitted through electric lines.

Because winds are variable, so is the amount of energy generated by each wind turbine. This means that electrical energy from wind must be coupled with other, more reliable sources of energy.

Since the technology to generate electricity from wind is relatively easy to install, sizable increases in capacity occur each year. In 2010, there were about 200,000 megawatts of install capacity worldwide. By 2012, that had increased by 20 percent to 238,000 megawatts. The United States, Germany, and China constitute about 60 percent of the installed capacity. Although there has been rapid development of new wind power capacity, it is important to recognize that the total electrical energy produced by wind today is less than 1 percent of total worldwide energy consumption.

Environmental Issues

Wind generators do have some negative effects. The moving blades are a hazard to birds and bats. Newer windmills, however, have slower-moving rotors that many birds such as the golden eagle find easier to avoid. In addition, some people consider the sight of a large number of wind generators to be visual pollution.

Geothermal Energy

Geothermal energy is obtained in two different ways. In geologically active areas where hot magma approaches the surface, the heat from the underlying rock can be used to heat water. The heated water can then be used directly either to heat buildings or to generate electricity by way of a steam turbine. (See figure 10.15.)

The United States produces about 30 percent of the world's geothermally generated electricity. The Pacific Gas and Electric Company (PG&E) has been producing electricity from geothermal energy since 1960. PG&E's complex of generating units located north of San Francisco is the largest in the world and provides 700 megawatts of power, enough for 700,000 households, or 2.9 million people. However, to put this in perspective, geothermal

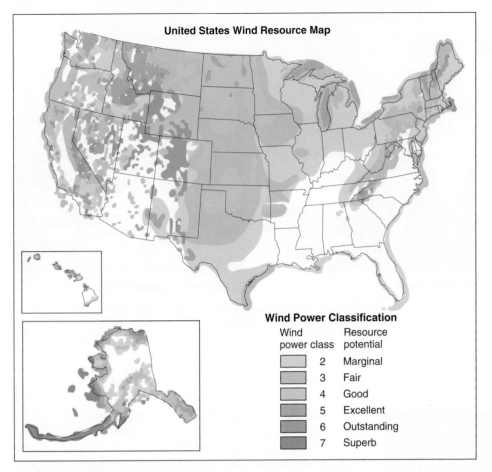

United States Wind Resource Map

Wind Power Classification

Wind power class	Resource potential
2	Marginal
3	Fair
4	Good
5	Excellent
6	Outstanding
7	Superb

FIGURE 10.14 Wind Energy Potential This map ranks regions of the United States in terms of their potential to supply electricity from wind energy.
Source: U.S. Department of Energy.

FIGURE 10.15 Geothermal Power Plant Steam obtained from geothermal wells is used to produce electricity.

electricity accounts for less than 1 percent of total electricity consumption in the United States. Other countries that produce significant amounts of geothermal electricity are the Philippines, Italy, Mexico, Japan, New Zealand, Indonesia, and Iceland. In Iceland, half of the geothermal energy is used to produce electricity and half is used for heating. In the capital, Reykjavik, all of the buildings are heated with geothermal energy at a cost that is less than 25 percent of what it would be if oil were used.

It is also possible to use heat pumps to obtain geothermal energy from areas that are not geologically active. All objects contain heat energy, which can be extracted from and transferred to other locations. Geothermal heat pump systems utilize a closed loop of underground pipes. A water–antifreeze solution is circulated through the pipes and heat is extracted from the solution and transferred to the building in the same way a refrigerator moves heat from the inside to the outside. Typically the amount of heat energy harvested is three to four times the amount of electrical energy used to run the system.

Environmental Issues

The use of geothermal energy from geologically active areas creates some environmental problems. The steam often contains hydrogen sulfide gas, which has the odor of rotten eggs and is an unpleasant form of air pollution. (The sulfides from geothermal sources can be removed.) The minerals in the steam corrode pipes and equipment, causing maintenance problems. The minerals are also toxic to fish if wastewater is discharged into local bodies of water.

Tidal Power

Tides are caused by the gravitational force exerted by the moon and the sun. The magnitude of the gravitational attraction between two objects depends on the masses of the objects and the distance between them. The moon exerts a larger gravitational force on the Earth because, although it is much smaller in mass than the sun, it is a great deal closer than the sun. This force of attraction causes the oceans, which make up 71 percent of the Earth's surface, to bulge along an axis pointing toward the moon. (There is also a bulge on the side of the Earth farthest from the moon because the moon is pulling the Earth away from the water on its surface.) Tides are produced by the rotation of the Earth beneath this bulge in its watery coating, resulting in the rhythmic rise and fall of water levels that can be observed along coasts. Thus, there are two high and two low tides each day. When the sun, moon, and Earth are in a line, the combined effects of sun and moon generate higher tides.

Certain coastal regions experience higher tides than others. This is a result of the amplification of tides caused by local geographical features such as bays and inlets. To produce practical amounts of power, a difference between high and low tides of at least 5 meters (16 feet) is required. About 40 sites around the world have this tidal range. The higher the tides, the more electricity can be generated from a given site and the lower the cost of electricity produced. Due to the constraints just described, it has been estimated that the potential to produce electricity from tides is only about 60,000 megawatts.

Technology for Obtaining Tidal Energy

The technology required to convert tidal energy into electricity is very similar to that used in traditional hydroelectric power plants. The first requirement is a dam or "barrage" across a tidal bay or estuary. Building such dams is expensive. Therefore, the best tidal sites are those where a bay has a narrow opening, thus reducing the length of dam required. At certain points along the dam, gates and turbines are installed. When the difference in the elevation of the water on the two sides of the barrage is adequate, the gates are opened. As the water flows from the high side to the low side of the dam, the flowing water causes turbines to spin and produce electricity.

Although the technology required to harness tidal energy is well established, tidal power is expensive, and only one major tidal generating station is in operation. This is a 240-megawatt facility (1 megawatt = 1 million watts) at the mouth of the La Rance river estuary on the northern coast of France (a large coal or nuclear power plant generates about 1,000 megawatts of electricity). The La Rance generating station has been in operation since 1966 and has been a very reliable source of electricity for France. (See figure 10.16.) Elsewhere, there is a 20-megawatt facility at Annapolis Royal in Nova Scotia, Canada, a 0.4-megawatt tidal

power plant near Murmansk, Russia, and a 0.5-megawatt facility on Jangxia Creek in the East China Sea. A 252-megawatt facility is currently being built in South Korea and other sites throughout the world are being evaluated for their potential.

Environmental Issues

Tidal energy is a renewable source of electricity and does not contribute to climate change. However, changing tidal flows by damming a bay or estuary could result in negative impacts on aquatic and shoreline ecosystems, as well as affecting navigation and recreation.

Studies undertaken to identify the environmental impacts of tidal power have determined that each site is different and the impacts depend greatly on local geography. Local tides changed only slightly due to the La Rance barrage, and the environmental impact has been negligible, but this may not be the case for all other sites. In all cases the barriers and turbines will affect the migration of fish and other marine species.

10.3 Energy Conservation

The amount of energy used by a society is determined by many factors. These include the level of economic development, the cost of energy, societal expectations, and government policies related to energy use.

Many cultural or lifestyle factors have been shaped by the availability of relatively low-cost energy. Large homes, outdoor lighting, large lawns, home entertainment centers, automobile travel, and many "labor-saving" devices (leaf blowers, riding lawn mowers, dishwashers, etc.) use large amounts of energy. If the cost of energy were higher, people would be likely to make different choices about what is essential and would evaluate energy efficiency more carefully.

There is typically a relationship between the cost of an item and its energy efficiency. Often, poorly designed, energy-inefficient buildings and machines can be produced inexpensively. The short-term cost (purchase price) is low, but the long-term cost for upkeep and energy utilization is high. Typically, the cost of more efficient buildings or machines is higher, but the difference in initial price is made up by savings in energy cost over several years. This is known as the payback period.

It is clear that the United States and Canada have about twice the per capita energy consumption of other countries with similar economic status. (See chapter 8.) Therefore, it seems plausible that energy conservation measures should be able to substantially lower per person energy consumption. Energy conservation can be thought of as a "source" of energy, since it reduces energy demands and thus makes it easier to meet future energy needs. In addition, it saves money for the consumer.

Some conservation technologies are sophisticated and require substantial investment, while others are simple and cost little to achieve. (See figure 10.17.) For example, highly efficient compact fluorescent light bulbs that can be used in regular incandescent fixtures give the same amount of light for 25 percent of the energy, last about 6 times longer, and produce less heat. LED bulbs

FIGURE 10.16 Tidal Generating Station La Rance River Estuary Power Plant in France is the world's largest tidal electricity-generating station.

Going Green

Hybrid and Electric Vehicles

In many metropolitan areas, emissions from automobiles are a major contributor to air-quality problems. An increase in the efficiency with which the chemical energy of fuel is converted to the motion of automobiles would greatly reduce air pollution. Because sources of fossil fuels are limited, they will become less available and greater efficiency will extend the limited supplies. There has been an ongoing process of modifying vehicles to improve performance and fuel efficiency. These include: reducing vehicle weight by using lighter materials, streamlining the shape to reduce air resistance, using better tires, and many other modifications. To further increase energy efficiency, more innovative techniques are required.

A great deal of energy is needed to get a vehicle to begin moving from a stop (accelerate), and it takes an equal amount of energy to stop a vehicle once it is moving (decelerate). Internal combustion engines operate most efficiently when they are running at a specific speed (rpm) and work at less than peak efficiency when the vehicle is accelerating or decelerating. Thus, using an internal combustion engine to accelerate is not efficient and contributes significantly to air pollution. When the brakes are applied to stop the vehicle, the kinetic energy possessed by the moving vehicle is converted to heat in the braking system. So the energy that has just been used to accelerate the vehicle is lost as heat when the vehicle is brought to a stop. Furthermore, in metropolitan areas, an automobile sits in traffic with its engine running a significant amount of the time. Thus, the stop-and-go traffic common in city driving provides conditions that significantly reduce the efficient transfer of the chemical energy of fuel to the kinetic energy of turning wheels.

Since electric motors are highly efficient, they are used in several kinds of vehicles. In full electric vehicles batteries are the only source of energy and must be plugged into an external source of electricity to be charged. In hybrid electric vehicles electric motors provide the energy to start the car moving (accelerate) and when the car is being driven at low speed. An internal combustion engine provides the extra power needed to reach higher speeds on the highway. In conventional vehicles, a great deal of energy is wasted while the car idles in traffic or at stoplights. In hybrid vehicles, only the electric motors powered by batteries are used in stop-and-go traffic, so no fuel is used. The battery is recharged when the car switches back to the gas engine. The battery is also recharged by a process called *regenerative braking,* in which the kinetic energy of the car is captured during braking and stored as electrochemical energy in the battery. Since the internal combustion engine is only used to assist at high speeds, it is smaller, weighs less, and consumes less fuel than engines of conventional vehicles. Today, nearly all automobile manufacturers include one or more hybrid vehicles in their product mix.

Full electric (plug-in) vehicles have also become important. Currently, they are used primarily as commuter cars, since most have a range of 130 km (80 mile) or less on a full charge. They also generally require several hours to achieve a full charge of the battery. Although they use no fuel and produce no emissions during operation, the electricity used to charge them comes from a power plant. If the power plant is fired by coal or natural gas, fuel consumption and emissions do occur.

The Chevrolet Volt is an electric vehicle that also has a gasoline engine. However, instead of using the gasoline engine to propel the vehicle, the engine powers a generator that provides electricity for the electric motor. So it is similar to a hybrid vehicle in many ways.

(a)

(b)

(c)

FIGURE 10.17 Energy Conservation The use of (a) fluorescent light bulbs, (b) energy-efficient appliances, and (c) low-emissive glass could reduce energy consumption significantly.

are even better. They use about 20 percent of the energy and last 25 times longer than incandescent bulbs. Since lighting accounts for 14 percent of U.S. electricity consumption, widespread use of these lights could significantly reduce energy consumption. Low-emissive glass for windows can reduce the amount of heat entering a building, while allowing light to enter. The use of this glass in new construction and replacement windows could have a major impact on the energy picture. Many other technologies, such as automatic dimming devices or automatic light-shutoff devices, are being used in new construction. Installing energy-saving windows and doors, replacing inefficient heating systems and appliances, or upgrading insulation in buildings would save a great deal of energy but require a substantial investment. Conversely, reducing the temperature in buildings during the winter or increasing the temperature in the summer costs little and would reduce the energy needed to provide heat and air conditioning.

Government Incentives

The shift to more efficient use of energy needs encouragement. Some people may be convinced to reduce energy use by moral arguments but most people are not likely to make changes in their lifestyle unless there are significant penalties for wasteful use or rewards for reducing energy consumption. In recent years, the U.S. government has passed laws or established policies that will reduce average energy consumption. Most of these initiatives provide economic rewards in the form of tax incentives or impose economic penalties (taxes and fines). The following are examples of government actions designed to improve energy efficiency.

- Imposition of higher fuel economy standards on the manufacture of automobiles and trucks. The new standard established by the Obama administration set a fuel efficiency of 54.5 mpg by 2025.
- Tax incentives for those who upgrade insulation, windows, doors, heating and cooling systems, and other appliances.
- Phase out by 2014 of most uses of incandescent light bulbs, which are very inefficient. Although there were politically motivated efforts to undermine the law, most retail establishments only provide higher efficiency bulbs.
- Established higher energy efficiency standards for appliances.
- Investment in more efficient electricity distribution.
- Improvement in high speed rail transportation.

Electric utilities are also part of the energy conservation picture. Many electric utilities have energy conservation programs that help their customers reduce their energy needs. Since the building of new power plants is expensive, electric utilities have an incentive to keep electricity use from increasing faster than it would if people used the energy inefficiently.

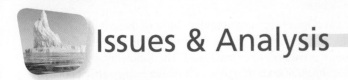
Does Corn Ethanol Fuel Make Sense?

To evaluate the feasibility of using ethanol as fuel it is necessary to do some accounting from two points of view: energy gain and economic competitiveness.

Energy Analysis

Many studies have assessed the amount of energy necessary to produce ethanol. Currently in North America, ethanol is made by fermenting the starch contained in corn. It is essentially the same process as that used to make beer and wine. Therefore, the energy input to make ethanol includes fuel used by farmers to till, plant, harvest, and transport corn. In addition, fertilizer, herbicides, and other agricultural chemicals require energy to produce and apply. Finally, the fermenting of corn to make ethanol and the distillation of the ethanol from a dilute solution both require energy. Various attempts have been made to account for all of these energy inputs and the results range from zero net energy gain to a net gain of 30 percent. If it takes 1 unit of energy to produce about 1.3 units of energy as ethanol, the net gain is 0.3 units.

Flex-fuel vehicles have engines and other components designed to use high concentrations of ethanol as fuel. The most commonly available mixture is E85 fuel, which contains 85 percent ethanol and 15 percent gasoline. Since ethanol has only 67.2 percent of the energy of gasoline, a vehicle will travel fewer miles on a gallon of ethanol or E85 fuel than on a gallon of gasoline. (See chart.)

Gallons Needed to Provide Equivalent Energy		
Gasoline	Ethanol	E85 fuel
1	1.49	1.42

In the United States, ethanol is most commonly added to gasoline (up to 10 percent) to increase the octane of the gasoline/ethanol mixture. This mixture can be used in unmodified engines and improves engine performance, resulting in less air pollution.

The amount of ethanol produced in 2013 was about 50 billion liters (13 billion gallons). If the amount of energy needed to produce the ethanol is subtracted, the net gain in energy is equivalent to about 15 billion liters (4 billion gallons) of ethanol. This is less than 3 percent of the amount of gasoline used in 2013.

Economic Analysis

Since ethanol only contains 67.2 percent and E85 only contains 72 percent of the energy of gasoline, they must cost less than gasoline to be competitive. Thus, in order for E85 to be more economical than regular gasoline, its cost must only be 72 percent of the price of gasoline. Or, in other words, it must cost 28 percent less than regular gasoline. Although E85 costs less than regular gasoline, it does not cost 28 percent less and thus is more expensive to use than regular gasoline.

Price/Gallon to Provide Equivalent Energy per Dollar		
Gasoline	Ethanol	E85 fuel
$3.00	$2.00	$2.15

What Do You Think?

- Because it is a renewable fuel, is it important to provide ethanol as a fuel?
- Should the public be required to use ethanol fuel?
- Should the production of ethanol be subsidized?

Summary

About 13 percent of the world's energy comes from renewable energy sources of which 10 percent is from biomass. Fuelwood is a minor source of energy in industrialized countries but is the major source of fuel in many less-developed nations. Biomass can be burned to provide heat for cooking or to produce electricity, or it can be converted to alcohol or used to generate methane. In some communities, solid waste is burned to reduce the volume of the waste and also to supply energy. The use of biomass to provide energy in the less-developed world contributes to habitat and biodiversity loss and contributes to air pollution because the fuel is often burned in inefficient stoves or over open fires.

Hydroelectric power provides about 16 percent of the world's electricity and can be increased significantly in many parts of the world. However, its development requires flooding areas and thus may require the displacement of people. Other environmental effects of developing hydroelectric plants are: altering the habitat for fish and other organisms and altering stream flow. Solar energy can be collected and used in either passive or active heating systems and can also be used to generate electricity in two different ways. Mirrors can be used to concentrate sunlight to produce steam which can be used to power a turbine. Photovoltaic cells can be used to produce electricity. Lack of a constant supply of sunlight is solar energy's primary limitation. Wind power may be used to generate electricity but may require wide, open areas and a large number of wind generators. The use of geothermal and tidal energy is determined by specific geologic and geographic features of the land. Therefore, the use of these sources of energy is limited.

Energy conservation can reduce energy demands without noticeably changing standards of living. However, the public typically needs incentives to make changes that do not provide immediate benefits. Therefore, most energy conservation programs are associated with economic incentives, which take the form of added costs (fines, taxes) or subsidies (tax deductions).

Acting Green

1. Contact your local electric utility or visit its website and determine what percentage of the electricity produced comes from fossil fuels, nuclear energy, hydroelectric, or other renewable energy sources.

2. Does your electric utility offer an opportunity to purchase green energy? If so, what does it cost?

3. Develop a list of ten ways you could reduce energy use. Implement five of them.

Review Questions

1. What are the general characteristics of renewable energy sources?
2. What percent of world energy comes from renewable energy sources?
3. What renewable energy source provides the majority of renewable energy?
4. List industries that typically make use of the waste they produce to provide themselves with energy.
5. Why is burning of municipal waste to produce energy more common in Europe than in North America?
6. How is the biofuel ethanol produced?
7. List three negative environmental impacts of using biomass to provide energy.
8. What are negative environmental impacts of developing hydroelectric power?
9. Compare a passive solar heating system with an active solar heating system.
10. Describe two different ways sunlight is used to make electricity.
11. List two reasons people oppose additional wind energy development.
12. List three energy conservation techniques.

Critical Thinking Questions

1. Imagine you are an official with the Department of Energy and are in the budgeting process for alternative energy research. Decide where you would invest money and explain why you made your choice. What do you think the political repercussions of your decision would be? Why?

2. Do you believe that large dam projects like the Three Gorges Dam project in China are, on the whole, beneficial? Do you believe they are not beneficial? What alternatives would you recommend? Why?

3. Energy conservation is one way to decrease dependence on fossil fuels. What are some things you can do at home, work, or school that would reduce fossil-fuel use and save money?

4. What alternative energy resources that the text has outlined are most useful in your area? How might these be implemented?

To access additional resources for this chapter, please visit ConnectPlus® at connect.mheducation.com. There you will find interactive exercises, including Google Earth™, additional Case Studies, and SmartBook™, an adaptive eBook that integrates our LearnSmart® adaptive learning technology.

One of the measures of biodiversity is the number of different species in an area. This photograph shows a leopard overlooking migrating wildebeests and zebras on the African plains of the Serengeti in Tanzania, Africa.

OBJECTIVES

After reading this chapter, you should be able to:

- Recognize that humans significantly modify natural ecosystems.
- State the major causes of biodiversity loss.
- Give examples of genetic diversity, species diversity, and ecosystem diversity.
- Describe the values of biodiversity.
- Identify causes of desertification.

- Describe the Red List of Threatened Species.
- Understand why biodiversity in freshwater ecosystems is under great threat from human actions.
- Describe why salmon runs in the Pacific Northwest have been in decline for more than a century.

- Describe the role of endangered species legislation and the biodiversity treaty.
- Describe techniques that foster the sustainable use of wildlife and fisheries resources.

Bio-Prospecting and Medicine—The Value of Biodiversity

Many of the world's major medicines are indirectly derived from biodiversity. The process of looking for potentially valuable genetic resources and biochemical compounds in nature is called *bio-prospecting.* Nearly half of all cancer drugs approved by the U.S. Food and Drug Administration from 1945 to 2012 were developed from natural products or derivatives of natural products. In 2012, 43 percent of the world's top-selling drugs were either obtained directly from or derived from natural products. Many of the drugs that people are familiar with are derived from natural products. Aspirin, for example, was originally derived from willow bark. Taxol, used in cancer treatment, was derived from the yew tree.

A number of plants and animals that are known to hold significant opportunities for medical research are threatened with extinction. Bears, for example, neither lose bone mass nor excrete urine while hibernating. Studying this ability could potentially provide insights into osteoporosis and means of dealing with renal failure. Six of the eight bear species are listed in the International Union for Conservation of Nature's 2012 Red List as threatened with extinction. The southern gastric brooding frog went extinct before researchers could investigate the properties of the substance it used to inhibit acid and enzyme secretions and protect its young that were raised in its stomach. It could have offered relief to sufferers of peptic ulcers.

The relationship between biodiversity, ecosystems, and human health goes beyond just the production of natural-product-based drugs. Studies have shown that in areas of extreme poverty, where ecosystem services such as access to fresh water, healthy soils, or pollinators are compromised, so too is human health. Poor nutrition, caused by increasingly degraded ecosystems, is linked with high rates of illness and disease. In addition, evidence is mounting that patterns of disease will be significantly shifted by the world's changing biodiversity.

Approximately 80 percent of people in developing countries rely on traditional medicines, the majority of which are derived from plants. Many medicinal plants are at risk of extinction, particularly in those locations where people are most dependent on them for health care and income.

Increasing the motivation of the pharmaceutical industry to value biodiversity and the active ingredients derived from it for drug production will help provide a greater economic incentive to conserve biodiversity. Incentives could include demonstration of the links between product innovation and natural-product-based drug discovery and underwriting elements of the bio-prospecting process in countries like Brazil and Kenya. Without the help of such incentives it is estimated that, at the current plant and animal extinction rate, one major drug will be lost every two years—a valuable opportunity lost.

"Rosy Periwinkle (Catharanthus roseus) harvested by Malagasy woman, leaves and flowers used for Anti-cancer medicine, Madagascar."

11.1 Biodiversity Loss and Extinction

Biodiversity is a broad term that is used to describe the diversity of genes and species and ecosystems in a region. Biodiversity is lost when populations are greatly reduced in size, when a species becomes extinct, or when ecosystems are destroyed or greatly modified. **Extinction** is the death of a species—the elimination of all the individuals of a particular kind. Extinction is a natural and common event in the long history of biological evolution. However, as we will see later in this section, extinction and loss of biodiversity are a major consequence of human domination of the earth. Over the past few hundred years, humans are estimated to have increased the extinction rate by a factor of 1,000 to 10,000 times above rates typical over the planet's history. About one-eighth of bird species, one-fourth of mammal species, one-third of amphibian species, and one-half of turtle species are threatened. Over 10 percent of the world's coral reefs have been lost. Nearly

60 percent of the remaining reefs are threatened by human activity. Mangrove forests, which are found in swampy areas near the ocean, are being reduced by over 1 percent per year. About 25 percent of the global land surface has been converted to raising crops. Approximately 60 percent of the ecosystem services (closely linked to biodiversity) are being degraded or used unsustainably. These include the maintenance of freshwater; the survival of fishery stocks; air and water purification; and the regulation of regional and local climate, natural hazards, and pests.

Currently, the most rapid changes affecting biodiversity are taking place in developing countries. This means that the harmful effects of biodiversity loss and the degradation of ecosystem services are borne disproportionately by the poor. Climate change and excessive nutrient loading are two major factors affecting biodiversity that are expected to become more severe in the future. Better protection of biodiversity and natural assets will require coordinated efforts across all levels of government, business, and international institutions. The productivity of ecosystems depends on policy choices in investment, trade, subsidy, taxation, and regulation, among others.

You may want to review chapter 5 before you begin this chapter so that you can apply ecological concepts to the issues regarding human impact on biodiversity.

Kinds of Organisms Prone to Extinction

Complete extinction occurs when all the individuals of a species are eliminated. Besides complete extinction, we commonly observe *local extinctions* of populations. Although not as final, a local extinction indicates that the future of the species is not encouraging. Furthermore, as a population is reduced in size, some of the genetic diversity in the population is likely to be lost.

Studies of modern local extinctions suggest that certain kinds of species are more likely than others to become extinct. Table 11.1 lists the major factors that affect extinction. First, species that have small populations of dispersed individuals (low population density) are more prone to extinction because successful breeding is more difficult for them than it is for species that have large populations of relatively high density. Second, organisms in small, restricted areas, such as islands, are also prone to extinction because an environmental change in their locale can eliminate the entire species at once. Organisms scattered over large areas are much less likely to be negatively affected by one event. Third, specialized organisms are more likely to become extinct than are generalized ones. Since specialized organisms rely on a few key factors in the environment, anything that negatively affects these factors could result in their extinction, whereas generalists can use alternate resources. Finally, some kinds of organisms, such as carnivores at higher trophic levels in food chains, typically have low populations but also have low rates of reproduction compared to their prey species.

Rabbits, raccoons, and rats are good examples of animals that are not likely to become extinct soon. They have high population densities and a wide geographic distribution. In addition, they have high reproductive rates and are generalists that can live under a variety of conditions and use a variety of items as food.

The cheetah is much more likely to become extinct because it has a low population density, is restricted to certain parts of Africa, has low reproductive rates, and has very specialized food habits. It must run down small antelope, in the open, during daylight, by itself. Similarly, the entire wild whooping crane species consists of about 375 individuals restricted to small winter and summer ranges that must have isolated marshes. (Captive and experimental populations bring the total number to slightly over 500 individuals.) In addition, their rate of reproduction is low. About one young bird is successfully raised by every two mated pairs.

Extinction as a Result of Human Activity

At one time, a human was just another consumer somewhere in the food chain. Humans fell prey to predators and died as a result of disease and accidents just like other animals. The simple tools they used would not allow major changes in their surroundings, so these people did not have a long-term effect on their surroundings.

As human populations grew, and as their tools and methods of using them became more advanced, the impact that a single human could have on his or her surroundings increased tremendously. Although in many ecosystems fires were natural events, the use of fire by humans to capture game and to clear land for gardens could destroy climax communities and return them to earlier successional stages more frequently than normal.

As technology advanced, wood was harvested for fuel and building materials, land was cleared for farming, streams were dammed to provide water power, and various mineral resources were exploited to provide energy and build machines. These modifications allowed larger human populations to survive, but always at the expense of previously existing ecosystems. (See figure 11.1.)

Today, with over 7 billion people on Earth, nearly all of the Earth's surface has been affected in some way by human activity. One of the major impacts of human activity has been to reduce biodiversity.

Genetic Diversity

Genetic diversity is a term used to describe the number of different kinds of genes present in a population or a species. High genetic diversity indicates that there are many different kinds of genes present and that individuals within the population will have different structures and abilities. Low genetic diversity indicates that nearly all the individuals in the

Table 11.1 Probability of Becoming Extinct

Most Likely to Become Extinct	Least Likely to Become Extinct
Low population density	High population density
Found in small area	Found over large area
Specialized niche	Generalized niche
Low reproductive rates	High reproductive rates

Most likely to become extinct

Least likely to become extinct

(a) (b) (c)

FIGURE 11.1 Changes in the Ability of Humans to Modify Their World As technology has advanced, the ability of people to modify their surroundings has increased significantly. (a) When humans lacked technology, they had only minor impacts on the natural world. (b) The agricultural revolution resulted in many of the suitable parts of the Earth being converted to agriculture. (c) Modern agricultural technology allows major portions of the Earth to be transformed into agricultural land.

population have the same characteristics. Several things influence the genetic diversity of a population.

1. *Mutations* are changes in the genetic information of an organism. Mutations introduce new genetic information into a population by modifying genes that are already present. Most of the mutations we observe are harmful, but occasionally a mutation results in a new valuable characteristic. For example, at some time in the past, mutations occurred in the DNA of certain insect species that made some individuals tolerant to the insecticide DDT, even though the chemical had not yet been invented. These mutant characteristics remained very rare in these insect populations until DDT was used. Then, this characteristic became very valuable to the insects that carried it. Because insects that lacked the ability to tolerate DDT died when they came in contact with DDT, more of the DDT-tolerant individuals were left to reproduce the species, and therefore, the DDT-tolerance became much more common in these populations. The evolution of DDT resistance is similar to the evolution of resistance in human antibiotic medication.

   If we think about the evolution of organisms, it is clear that incredible numbers of mutations were required for the evolution of complex plants, animals, and fungi from their single-celled ancestors. In the final analysis, all the characteristics shown by a species are the result of mutations.

2. *Migration* of individuals of a species from one population to another is also an important way to alter the genetic diversity of a population. Most species consist of many separate populations that are adapted to local environmental conditions. Whenever an organism leaves one population and enters another, it subtracts its genetic information from the population it left and adds it to the population it joins. If the migrating individual contains rare characteristics, it may significantly affect the genetic diversity of both populations. The extent of migration need not be great. As long as genes are entering or leaving a population, genetic diversity will change.

3. *Sexual reproduction* is another process that influences genetic diversity. Although the process of sexual reproduction does

not create new genetic information, it tends to generate new *genetic combinations* when the genetic information from two individuals mixes during fertilization, generating a unique individual. This doesn't directly change the genetic diversity of the population, but the new member may have a unique combination of characteristics so superior to those of other members of the population that the new member will be much more successful in producing offspring and will influence the genetic diversity of future generations.

4. *Population size* is a very important factor related to genetic diversity. The smaller the population, the less genetic diversity it can contain and the fewer the variations in the genes for specific characteristics. In addition, random events often can significantly alter the genetic diversity in small populations, with rare characteristics being lost from the population. For example, consider a population of a species of rose that consists of 100 individuals in which 95 of them have red flowers and five have white flowers. The death of five plants with red flowers would not change genetic diversity very much. However, the death of the five plants with white flowers could eliminate the characteristic from the population.

5. *Selective breeding* also can affect the genetic diversity of a species. Domesticated plants and animals have been modified over many generations by our choosing certain desired characteristics. Undesirable characteristics were eliminated, and desirable ones selected. One of the consequences of this process is a loss of genetic diversity. (See figure 11.2.) Many modern domesticated organisms are very different from their wild ancestors and could not survive without our help. Often when genetic diversity is reduced, whether by reduced population size or selective breeding, deleterious genes for a characteristic may be present in both parents. When this occurs and the deleterious genes from both parents are passed to the offspring, the survival of the offspring may be jeopardized.

Species Diversity

Species diversity is a measure of the number of different species present in an area. Some localities naturally have high species

(a) Genetically diverse Indian corn

(b) Genetically uniform field corn

FIGURE 11.2 Genetic Diversity Genetic diversity is a measure of the number of kinds of genes in a population. There is more genetic variety displayed in the multicolored and various shaped seeds of Indian corn than the yellow, uniform seeds of field corn.

(a) Pine plantation—low species diversity

(b) Desert ecosystem—high species diversity

FIGURE 11.3 Species Diversity (a) Forest plantations have low species diversity—the dominant vegetation consists of a single species of tree. (b) This desert has several species making up the dominant vegetation. How many different species can you count?

diversity, while others have low species diversity. (See figure 11.3.) For example, it is well known that tropical rainforests have very high species diversity, while Arctic regions have low species diversity. Some people find it useful to distinguish two kinds of species diversity. One approach is to simply count the number of different kinds of species in an area. This is often referred to as *species richness*. Another way to look at species diversity is to take into account the number of different taxonomic categories of the species present. This can be called *taxonomic richness*. A region with many different taxonomic categories of organisms (frogs, birds, mammals, insects, pine trees, mosses) would have a higher species diversity than one that had fewer taxonomic categories (grasses, insects, birds, mammals).

Measuring species diversity is not easy. For one thing, we do not know how many species there are. Estimates of the actual number of species range from a few million to 100 million. About 1.4 million species have been described, but each year, new species are discovered, and in some groups, such as the bacteria, there may be millions of species yet unnamed. Another problem associated with measuring species diversity is the difficulty of discovering all the species in an area. Many species are naturally rare, and others live in difficult-to-reach places or are active only at certain times. For example, an examination of the kinds of organisms in the tops of trees in tropical forests led to the discovery of many species that were previously unknown. Table 11.2 gives current data about the number of different kinds of species in specific taxonomic groups.

Several factors are known to influence the species diversity of a particular location.

1. *The geologic and evolutionary history* of a region impacts its species diversity. As mentioned earlier, tropical rainforests

Table 11.2 Numbers of Described Species by Taxonomic Group

Taxonomic Group	Approximate Number of Described Species
Insects	950,000
Plants	270,000
Non-insect arthropods (spiders, mites, crustaceans, etc.)	113,000
Fungi	100,000
Roundworms	80,000
Mollusks	70,000
Protozoa	31,000
Fish	29,300
Algae	27,000
Flatworms	25,000
Earthworms and related organisms	12,000
Sponges	10,000
Birds	9,900
Bacteria, blue-green algae	9,000
Jellyfish, corals, comb jellies	9,000
Reptiles	8,240
Starfish, sea urchins, sand dollars, etc.	7,000
Amphibians	5,900
Mammals	5,400
Rotifers	1,800
Archaea	260
Other	27,000
Total	**1,800,800**

naturally have greater species diversity than polar regions. Perhaps this is due to relatively recent climatic events. Twenty thousand years ago, glaciers covered much of northern Europe, the more northerly parts of North America, and parts of Asia. The glaciers would have drastically reduced the number of species present, and species would have migrated into the region as the glaciers receded. This same kind of drastic change did not take place in tropical regions; therefore, they retain more species.

2. *Migration* can introduce new species to an area where they were not present previously. While it is easy to see how such introductions can increase species diversity, some invading species actually result in a reduction in species diversity because the species originally present were unable to compete with the invaders. For example, the introduction of the zebra mussel to North America has had the effect of reducing the population size of many native mussels. Many of these native mussels are endangered.

3. *The size of the area* being considered also affects species diversity. In general, the larger the area being considered, the larger the species diversity. This is a natural consequence of including more kinds of spaces and the organisms that are adapted to them. For example, if species diversity were being measured in a desert, the diversity would increase greatly if a stream or water hole were included in the area under investigation.

4. *Human activity* has a great effect on the species diversity of a region. When humans exploit an area, they convert natural ecosystems to human-managed agricultural, forest, aquaculture, and urban ecosystems. They harvest certain species for their use. They specifically eliminate species that compete with more desirable species. They introduce species that are not native to the area. These topics will be discussed in greater depth later in the chapter.

Ecosystem Diversity

Ecosystem diversity is a measure of the number of kinds of ecosystems present in an area. Many regions of the world appear to be quite uniform in terms of the kinds of ecosystems present. For example, large parts of central Australia, North Africa, and southwestern United States and adjacent Mexico are deserts. While there are general similarities (low rainfall, thorny woody plants, and animals that can survive on little water), each of these deserts is different and has specific organisms typical to the region. Furthermore, within each of these deserts are local regions where water is available. These areas may include rivers, springs, or rocky outcrops that collect and hold water. These locations support species of organisms not found elsewhere in the desert. In many other cases, natural events such as hurricanes, fires, floods, or volcanic eruptions may have destroyed the original vegetation, resulting in patches of early successional stages that contribute much to the diversity of organisms present.

In addition, local topographic conditions may create patches of the landscape that differ from the prevailing type. Ridges typically have different vegetation on north- and south-facing slopes. (See figure 11.4.) Differences in soil type will support unique vegetation. Rolling terrain may provide pockets that contain temporary ponds that support different vegetation types. Each of these unique localities also can have animals that are specialized to the locality. Since many microhabitats exist within a region, the larger the area studied, the more the kinds of ecosystems present.

11.2 The Value of Biodiversity

Although most would agree that maintaining biodiversity is a good thing, a need exists to place a value on preserving biodiversity so that governments and individuals can apply logic to the daily decisions that affect biodiversity. There are many different ways to assign value to biodiversity. Some involve understanding the ecological roles played by organisms, others involve cold economic analysis, and still others stem from ethical considerations.

FIGURE 11.4 Ecosystem Diversity This location shows high ecosystem diversity. There are at least four different ecosystem types visible; the lake, the deciduous trees surrounding the lake, the conifer trees in the middle, and low-growing vegetation on the mountainside.

Biological and Ecosystem Services Values

Our species is totally dependent on the diversity of organisms on Earth. It is important to recognize that each organism is involved in a vast network of relationships with other organisms. Symbiotic nitrogen-fixing bacteria live in the roots of certain plants. Soil-building organisms live on the dead organic matter provided by plants and animals. Animals eat plants or other animals. It is impossible to have an organism function optimally unless it has its supporting cast of players that are part of the ecosystem. Although it is impossible to think of every possible role for each species on Earth, experts have identified some broad categories of services provided by ecosystems and the organisms that make them up. This list describes ecosystem services that are not part of our economic system—we do not directly pay for them. Examining this list will help us recognize how important ecosystem services are to our lives.

Nutrient Cycling

Carbon, nitrogen, phosphorus, and many other chemical elements are cycled through ecosystems by a complex array of bacteria and other organisms. Nitrogen-fixing bacteria are particularly valuable for providing nitrogen in a usable form to plants.

Cultural Uses

People use the natural world for many kinds of nonconsumptive uses. Enjoyment of landscapes and individual organisms, scientific study and other educational activities, and the spiritual significance of specific places and things are all examples of cultural uses of ecosystems.

Water Regulation and Supply

Intact soil and vegetation slow the flow of water and allow water to penetrate the soil and recharge aquifers. These processes make water available for agricultural, industrial, and domestic uses. In areas with intact ecosystems, water is released slowly so that intact ecosystems are less prone to damage from drought. New York City found that it could provide water to its residents less expensively by protecting the watershed from which the water comes, rather than building expensive water purification plants to clean water from local rivers.

Disturbance Regulation and Erosion Control

Colonization of disturbed sites—caused by fires, floods, windstorms, landslides, or human actions—by plants and animals heals the scars and prevents continued damage. Furthermore, intact ecosystems provide flood and erosion control. Mangrove forests, marshes, and other wetlands protect shorelands from erosion. The network of roots in forested areas and grasslands ties the soil together and protects watersheds.

Waste Treatment

Decomposer organisms recycle both natural and human-produced organic wastes. Excess nutrients are removed by organisms and pollutants are removed from air, soil, and water and converted to less harmful materials.

Food and Raw Materials

In many parts of the world, people are involved in a subsistence economy and rely directly on ecosystems for food and raw materials. They harvest wild plants and animals as food and medicine and use plants to provide food for livestock, building materials, and firewood. The UN Food and Agriculture Organization estimates that over 60 percent of all wood harvested worldwide is burned as fuel. Much of this is collected by individuals and used by them for heating homes and cooking.

Atmospheric and Climate Services

Many atmospheric gases are cycled between organisms and the atmosphere. In particular, removal of carbon dioxide during photosynthesis is important in controlling the warming of the planet. Many countries have planted trees to help remove carbon dioxide from the air as a response to concerns about climate change. Other pollutants such as nitrogen and sulfur compounds are also modified by organisms, particularly when they come in contact with soil. Ozone in the upper atmosphere protects from ultraviolet light.

Recreation

Natural areas provide important recreational opportunities for an increasingly urban population. Camping, hiking, kayaking, fishing, hunting, ecotourism, and sight-seeing provide exercise and enjoyment.

Biological Control Services

All organisms are involved in a complex set of interrelationships with other kinds of organisms. Some of these relationships are harmful. We label organisms that cause harm to us or our domesticated plants and animals as *pests*. However, every pest also has organisms that cause it harm—dragonflies, swallows, and bats eat mosquitoes that carry disease and are annoying; ladybird beetles eat aphids; and cats are frequently kept to control rodent populations on farms.

Pollination Services

Many different kinds of insects are pollinators that are extremely important to the successful fruiting of plants. The careless elimination of these beneficial insects by the broad use of insecticides can negatively affect agricultural production. Recent declines in honeybee populations have highlighted the value of this pollination service.

Habitat/Refuges

Refuges and other protected areas provide places that protect species of concern, serve as nursery sites for specific species, or provide temporary stopping places for migratory species. Ducks Unlimited, an organization that supports waterfowl hunting, uses money provided by its members to protect wetland nesting habitats for ducks and geese.

Genetic Resources

If we cause the extinction of a potentially useful organism, we have lost the opportunity to use it for our own ends. Most of the wild ancestors of our most important food grains, such as maize (corn), wheat, and rice, are thought to be extinct. Over 50 percent of the most common drugs used to control and cure disease are derived from plants and animals and new ones are discovered each year.

Soil Formation

The weathering of rock provides new mineral material for the building of soil. Bacteria, fungi, tiny animals, and the roots of plants are involved in building soil by breaking down organic matter, incorporating it into the mineral part of the soil, and creating a loose texture to the soil. All terrestrial ecosystems, including agricultural and commercial forest lands, rely on the soil-forming

services provided by these organisms. Our food supply depends on the protection and management of the soil and the organisms that assist in soil building.

Assigning Value to Ecosystem Services

What are Mother Nature's life-support services worth? In one sense, their value is infinite. The Earth's economies would soon collapse without fertile soil, freshwater, breathable air, and an amenable climate. But "infinite" too often translates to "zero" in the equations that guide land use and policy decisions. Practitioners in the young field of ecological economics believe more concrete numbers are required to help nations avoid unsustainable economic choices that degrade both their natural resources and the vital services that healthy natural ecosystems generate.

All of these services can be converted into monetary terms, since it takes money to purify water, purchase land, and buy and plant trees. Since choices between competing uses for ecosystems often are determined by financial values assignable to ecosystems, many environmental thinkers have begun to try to put a value on the many services provided by intact, functioning ecosystems. Obviously, this is not an easy task, and many will belittle these initial attempts to put monetary values on ecosystem services, but it is an important first step in forcing people to consider the importance of ecosystem services when making economic decisions about how ecosystems should be used. Table 11.3 presents approximate values for ecosystem services assigned by a panel of experts including ecologists, geographers, and economists. The study assigned an estimated value of $33 trillion per year, which many consider to be low. The current world gross national product is about $50 trillion per year. Therefore, the "free" services of ecosystems must not be overlooked when decisions are made about land use and how natural resources should be managed.

Table 11.3 Estimates of Various Ecosystem Services

Ecosystem Services	Value (trillion $US)
Soil formation	17.1
Recreation	3.0
Nutrient cycling	2.3
Water regulation and supply	2.3
Climate regulation (temperature and precipitation)	1.8
Habitat	1.4
Flood and storm protection	1.1
Food and raw materials production	0.8
Genetic resources	0.8
Atmospheric gas balance	0.7
Pollination	0.4
All other services	1.6
Total value of ecosystem services	**33.3**

Direct Economic Values

There are three major economic sectors that are highly dependent on the ecological health of the planet: agriculture, forestry, and fisheries. Each contributes many billions of dollars to the global economy. The Food and Agriculture Organization of the United Nations estimates, based on exports, that agriculture contributes about US $670 billion to the world economy, while the harvest and trade in forest products is responsible for about US $200 billion and fisheries resources are responsible for about US $80 billion yearly. However, since much of this production is consumed in the country where it is produced—the products are not exported— the real economic value is much greater than the value of exports. For example, less than 40 percent of fisheries products are traded internationally, so the real value is greater than US $80 billion.

It is also important to understand that in many parts of the world, many agricultural, forestry, and fisheries products never enter the formal economy of a region, so it is difficult to assign specific monetary value to them. For example, firewood, fish, game meat, and other wild foods are often consumed directly by the people who harvest them, sold at local markets, or bartered for other goods and services. (See figure 11.5.)

There are also markets for specialty products such as medicinal plants, ivory, skins, and many other natural products. There are many historical examples of valued species that were or are being harvested beyond their capacity to reproduce. The harvesting of ivory was steadily reducing the number of elephants, and the harvesting of alligators for their hides was reducing alligator numbers. In both cases, regulation of the harvest has preserved the resource. Alligator populations have rebounded and they are again being harvested in the United States. Many countries in Africa maintain that elephant populations are so large that harvesting of elephants for meat and ivory should be allowed as well.

When looking at the economic value of biodiversity it is also important to consider the potential value of biodiversity. For example, many current drugs are derived from organisms. How many potential drugs are yet undiscovered? Protecting biodiversity can be thought of as protecting future economically important resources.

Ethical Values

Many would argue that a case can be made that all species have an intrinsic value and a fundamental right to exist without being needlessly eliminated by the unthinking activity of the human species. This is an ethical position that is unrelated to social or economic considerations. According to those who support this philosophical position, extinction by itself is not bad, but human-initiated extinction is. This contrasts with the philosophical position that humans are simply organisms that have achieved a preeminent position on Earth, and therefore, extinctions we cause are no different from extinctions caused by other forces.

Others would argue that experiencing natural landscapes and processes is an important human right. The beauty of a forested hillside in full autumn color, the graceful movements of deer, the raw power of a grizzly bear, and the wonder shown by a child when it discovers a garter snake or a butterfly are all emotional events. There is a value in their just being, so that they can be observed.

FIGURE 11.5 Informal Economy It is difficult to measure the real economic value of the agricultural, forestry, and fisheries industries because many products, like these fish, are sold or bartered in local markets.

The ethical positions held by a person are typically shaped by experience. In 1800, 94 percent of North Americans lived in rural settings and had daily encounters with plants and animals in a natural setting. Furthermore, there was a basic understanding of how natural ecosystems operated. They milked their own cows, slaughtered animals for meat, and observed the effect of weather, seasons, and other cycles of change on the natural systems of which they were a part. Although they often changed ecosystems to provide for agriculture, they had a basic understanding of how ecosystems function. Today, about 80 percent of North Americans live in urban settings and have limited contact and experience with these natural processes. This same dynamic is at work throughout the world. Today, over 50 percent of the world population is urban.

Because urban dwellers have little personal experience with naturally functioning ecosystems, they have a different set of ethical values about what natural ecosystems provide. The values people place on biodiversity are important, since they shape the way political and social institutions respond to threats to biodiversity.

Focus On

The Serengeti Highway Route

The Tanzanian government is planning to create a highway across the Serengeti. The controversial project is proposed as a way of increasing trade and helping stimulate economic growth. The proposed new highway could have a severe impact on traditional animal migration routes and impact delicate ecosystems in the African plains. The negative impact on the annual migration of the wildebeest and zebras could potentially decimate the big herds from their present number of an estimated 1.5 million animals to as few as 200,000. This has raised serious concerns among tourism operators in both Kenya and Tanzania.

In 2010 scientists from 32 different countries signed a petition urging Tanzania to find an alternate route around the Serengeti National Park, rather than going straight through it. The scientists believe that the construction of the road "would result in severe, negative, irreversible impacts on migratory animals." The scientists also said, "the proposed road cuts through a critical wilderness area that is essential to the migration. The type of road surface matters little. The migration itself could easily collapse, with a devastating effect on all wildlife, the grasslands, and the entire ecosystem."

An organization called Save the Serengeti was created and works with scientists, conservation organizations, and the travel industry to try to resolve the issues regarding the proposed highway. The Tanzanian government, however, seems intent on moving ahead with the plan. As of 2013 the Serengeti highway issue was still being heard before the East African Court of Justice.

How would you address this conflict?

Is there a compromise position or is compromise not always an option?

How would a highway affect this migration?

11.3 Threats to Biodiversity

Efforts to preserve biodiversity involve a tension between the desire to use biotic resources and a wish to maintain biodiversity. The value of exploiting a resource can be measured in economic terms. Farmland, lumber, or animal products can be given a measurable monetary value by the economic marketplace. On the other hand, it is often difficult to put an economic value on the preservation of biodiversity and the environmental services provided by organisms. Therefore, environmentalists often must rely on ethical or biological arguments to make their points. Often the decisions that are made involve a compromise that allows some utilization of a resource while preserving some of the biodiversity.

Five major human impacts threaten to reduce biodiversity: habitat loss, overexploitation, introduction of exotic species, predator and pest control activities, and climate change.

Habitat Loss

Habitat loss occurs when human activities result in the conversion of natural ecosystems to human-dominated systems. The resulting changes eliminate or reduce the numbers of species that were a part of the original ecosystem.

The International Union for Conservation of Nature and Natural Resources (IUCN) estimates that about 80 percent to 90 percent of threatened species are under threat because of habitat loss or fragmentation. Similarly, habitat loss and fragmentation are thought to be a major cause of past extinctions. The primary activities that result in habitat loss are farming, forestry, grazing by livestock, modification of aquatic habitats, and conversion to urban and industrial landscapes.

Often the destruction of natural habitats is not total but leaves patches of relatively unaltered habitat interspersed with human-modified landscapes. However, scientists studying the effects of this activity on forest birds have noticed that as the forest is reduced to small patches, many species of birds disappear from the area.

Conversion to Agriculture

About 40 percent of the world's land surface has been converted to cropland and permanent pasture. Typically, the most productive natural ecosystems (forests and grasslands) are the first to be modified by human use and the most intensely managed. Since nearly all agricultural practices involve the removal of the original vegetation and substitution of exotic domesticated crops and animals, the loss of biodiversity is significant. For example, much of the Mediterranean and Middle East once supported extensive forests that were converted to agriculture but now consist of dry scrubland. In Europe, little of the original forest is left. In North

America, the eastern deciduous forests were reduced, and almost all the original prairie in the United States has been converted to agricultural land, resulting in the loss of some species.

As the human population grows, it needs more space to grow food. Current pressures to modify the environment are greatest in areas that have high population density. (See figure 11.6.) Today, agricultural land is being pushed to feed more people, and its

wise use is essential to the health and welfare of the people of the world. Chapters 13 and 14 take a close look at patterns of use of agricultural land.

Forestry Practices

The history and current status of the world's forests are well known. The economic worth of the standing timber can be assessed, and the importance of forests for wildlife and watershed protection can be given a value. Originally, almost half of the United States, three-fourths of Canada, almost all of Europe, and significant portions of the rest of the world were forested. The forests were removed for fuel, for building materials, to clear land for farming, and just because they were in the way. These forest-destroying activities are often called **deforestation.** This activity destroyed much of the forested land and returned other forests to an earlier successional stage, which resulted in the loss of certain animal and plant species that required mature forests for their habitat. In general, all continents still contain significant amounts of forested land, although most of these forest ecosystems have been extensively modified by human activity. Figure 11.7 shows that temperate parts of the world (particularly North America, Europe, and China, which have large forested areas) are maintaining or increasing their forests, while tropical forests are being lost. The changes in China are recent and involve extensive reforestation projects.

FIGURE 11.6 Conversion of Forest to Agriculture This farm in the mountains of Costa Rica was once covered by forest.

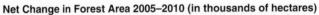

Net Change in Forest Area 2005–2010 (in thousands of hectares)

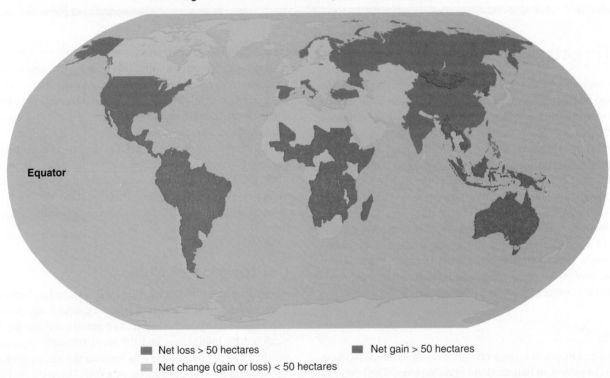

Equator

■ Net loss > 50 hectares   ■ Net gain > 50 hectares
■ Net change (gain or loss) < 50 hectares

FIGURE 11.7 Changes in Forest Area Areas shown in red have experienced a net loss in forest area of greater than 50,000 hectares over the years shown while areas shown in brown have experienced a net gain of more than 50,000 hectares. Areas shown in brown are increasing at a rate of greater than 0.5 percent per year. Areas shown in green are maintaining their forests. It is obvious that most tropical forests in South America, Africa, and Australia are being lost, while nontropical areas are holding their own or growing.
Source: Data from the Food and Agriculture Organization of the United Nations.

Science, Politics, & Policy

The Economic and Political Value of Biodiversity

Humanity was plagued by smallpox for nearly 3,000 years. By the mid eighteenth century, smallpox, with a mortality rate of 30 percent, was killing some 400,000 Europeans every year. It was not until 1796 that Edward Jenner demonstrated the effectiveness of cowpox to induce immunity against smallpox. And in 1979 smallpox became the first disease to be eradicated by humanity and its ingenious use of genetic material. Cowpox, a nuisance virus to cows and milkmaids, had saved millions of lives.

At its heart, valuing biodiversity is about valuing the fabric of life and solutions—particularly genetic—to our problems: curing diseases, preserving genetic diversity for plants and animals we use for food, and recognizing the value of organisms that provide their ecosystem services such as pest control, pollination, and climate regulation. But, with 8.7 million species on Earth (give or take 1.3 million), around 90 percent of which are undiscovered, can we save them all from a fast-growing world economy?

Economists have addressed this question for years from a variety of perspectives. In each case the costs of conserving a species, or its functional unit (such as an ecosystem), are weighed against the economic benefits, calculated by determining the value of using a species or the loss of value that occurs if a species is not used. The real answers, however, are not determined by economists, but by society and its political leaders.

For example the intrinsic (or existence) value of a species—how much we care about it even if we do not need or use it—can overcome other considerations, and convert a species' otherwise dismal fate into a hopeful future. An example of this is the panda, which is mostly useless to our economies and has fairly high conservation costs, but which can contribute significantly to our (and their) well-being just by its continued existence.

The real value of biodiversity is evidently enormous, and extends well beyond the economists human-centric focus; yet the value we as a society are attributing to its conservation is rather small. While we recognize current problems of resource use, are today's problems a good indicator of the kinds of problems we will face in the future? Given where our economies are currently taking us, in the future we will have far fewer resources and live in a less stable world. Even if we believe a more precautionary approach is needed if we are to explicitly value our future generations, how precautionary should we be? For example, some countries have begun storing genetic material (such as seed banks) for future problems. While this may be a noble idea, is it enough and is it really just a Band-Aid approach to preserving biodiversity? Given the high rate of biodiversity loss, the United Nations declared 2010 the International Year of Biodiversity to highlight the problem. The decade 2011–2020 has become the United Nations Decade on Biodiversity.

Are we doing enough to address the loss of biodiversity? Does society believe that our declining reserve of biodiversity is enough to solve the problems of the future? If we get this wrong, are we willing to pay the price of another smallpox? Are policymakers up to this task?

Modern forest-management practices involve a compromise that allows economic exploitation while maintaining some of the biodiversity of the forest. Logging activities are often selective and preferentially remove certain species of trees. This causes a shift in the species diversity of plants. Logging also destroys the habitat for many kinds of animals that require mature stands of timber. The pine martin, grizzly bear, and cougar all require forested habitat that is relatively untouched by human activity. The removal of mature forests reduces their population size and consequently the genetic diversity of their populations.

For example, today in Australia, a squirrel-sized marsupial known as a numbat is totally dependent on old forests that have termite- and ant-infested trees. Termites are numbats' major source of food, and the hollow trees and limbs caused by the insects' activities provide numbats with places to hide. Clearly, tree harvesting will have a negative impact on this species, which is already in danger of extinction. (See figure 11.8.)

In addition to serving as habitats for many species of plants and animals, forests provide many other ecosystem services. Forested areas modify the climate, reduce the rate of water runoff, protect soil from erosion, and provide recreational opportunities. Because trees transpire large quantities of water and shade the soil, their removal often leads to a hotter, drier climate. Trees and other plants hold water on their surfaces, thus reducing the rate of runoff. Slowing runoff also allows more water to sink into

FIGURE 11.8 A Specialized Marsupial—the Numbat This small marsupial mammal requires termite- and ant-infested trees for its survival. Termites serve as food, and the hollow limbs and logs provide hiding places. Loss of old-growth forests with diseased trees will lead to the numbat's extinction.

the soil and recharge groundwater resources. Therefore, removal of trees results in more rapid runoff so that flooding and soil erosion are more common. Soil particles wash into streams, where they cause siltation. The loss of soil particles reduces the soil's fertility. The particles that enter streams may cover spawning sites and reduce the biodiversity of fish populations. If the trees along the stream are removed, the water in the stream will be warmed by increased exposure to sunlight. This may also negatively affect fish populations.

In many parts of the world, the construction of logging roads increases access to the forest and results in colonization by peasant "squatters" who seek to clear the forest for agriculture. The roads also permit poachers to have greater access to wildlife in the forest. Finally, the "wilderness" nature of the area is destroyed, which results in a loss of value for many who like to visit mature forests for recreation.

Environmental Implications of Various Forest Harvesting Methods

One of the most controversial logging practices is **clear-cutting.** (See figure 11.9.) As the name implies, all of the trees in a large area are removed. This is a very economical method of harvesting, but it exposes the soil to significant erosive forces. If large blocks of land are cut at one time, it may slow the reestablishment of forest and have significant effects on wildlife. On some sites with gentle slopes, clear-cutting is a reasonable method of harvesting trees, and environmental damage is limited. This is especially true if a border of undisturbed forest is left along the banks of any streams in the area. The roots of the trees help to stabilize the stream banks and retard siltation. Also, the shade provided by the trees helps to prevent warming of the water, which might be detrimental to some fish species.

Clear-cutting can be very destructive on sites with steep slopes or where regrowth is slow. Under these circumstances, it may be possible to use **patchwork clear-cutting.** With this method, smaller areas are clear-cut among patches of untouched forest. This reduces many of the problems associated with clear-cutting and can also improve conditions for species of game animals that flourish in successional forests but not in mature forests.

For example, deer, grouse, and rabbits benefit from a mixture of mature forest and early-stage successional forest.

Clear-cut sites where natural reseeding or regrowth is slow may need to be replanted with trees, a process called **reforestation.** Reforestation is especially important for many of the conifer species, which often require bare soil to become established. Many of the deciduous trees will resprout from stumps or grow quickly from the seeds that litter the forest floor, so reforestation is not as important in deciduous forests.

Selective harvesting of some species of trees is also possible but is not as efficient or as economical as other methods, from the point of view of the harvesters. It allows them, however, to take individual, mature, high-value trees without completely disrupting the forest ecosystem. In many tropical forests, high-value trees, such as mahogany, are often harvested selectively. However, there may still be extensive damage to the forest by the construction of roads and to noncommercial trees by the felling of the selected species.

Special Concerns About Tropical Deforestation

Tropical forests have greater species diversity than any other terrestrial ecosystem. The diverse mixture of tree species requires harvesting techniques different from those traditionally used in northern temperate forests. In addition, because tropical soils have low fertility and are highly erodible, tropical forests are not as likely to regenerate after logging as are temperate forests. Currently, few tropical forests are being managed for long-term productivity; they are being harvested on a short-term economic basis only, as if they were nonrenewable resources. Worldwide tropical forests are being lost at a rate of about 0.6 percent per year. (See figure 11.10.)

Several concerns are raised by tropical deforestation. First, the deforestation of large tracts of tropical forest is significantly reducing the species diversity of the world. Second, because tropical forests very effectively trap rainfall and prevent rapid runoff

FIGURE 11.9 Clear-Cutting Clear-cutting removes all of the trees in a large area and exposes the soil to erosive forces. This shows a clear-cut in British Columbia, Canada.

FIGURE 11.10 Tropical Deforestation The deforestation of tropical regions leads to great reduction in biodiversity.

and the large amount of water transpired from the leaves of trees tends to increase the humidity of the air, the destruction of these forests also can significantly alter climate, generally resulting in a hotter, more arid climate. Third, high rainfall coupled with the nature of the soils results in the deforested lands being easily eroded. Finally, people have become concerned about preserving the potential of forests to trap carbon dioxide. As they carry on photosynthesis, trees trap large amounts of carbon dioxide. This may help to prevent increased carbon dioxide levels that contribute to global warming. (See chapter 16 for a discussion of global warming and climate change.)

Another complicating factor is that the human population is growing rapidly in tropical regions of the world. More people need more food, which means that forestland will be converted to agriculture and the value of forest for timber, fuel, watershed protection, wildlife habitat, biodiversity, and carbon dioxide storage will be lost.

Plantation Forestry

Many forest products companies manage forest plantations in the same way farmers manage crops. They plant single-species, even-aged forests of fast-growing hybrid trees that have been developed in the same way as high-yielding agricultural crops. Fertilizer is applied if needed, and weeds and pests are controlled. Often controlled fires and aerial application of pesticides are used to control competing species and pests. Such forests have low species diversity and are not as valuable for wildlife and other uses as are more natural, mixed species forests. Furthermore, the trees planted in many managed forests may be exotic species. *Eucalyptus* trees from Australia have been planted in South America, Africa, and other parts of the world, and most of the forests in northern England and Scotland have a mixture of native pines and imported species from the European mainland and North America.

In these intensively managed forests, some single-species plantations mature to harvestable size in 20 years, rather than in the approximately 100 years typical for naturally reproducing mixed forests. However, the quality of the lumber products is reduced. In many of these forests, clear-cutting is the typical method of harvest, and the cutover area is immediately replanted. (See figure 11.11.)

Rangeland and Grazing Practices

Rangelands consist of the many arid and semiarid lands of the world that support grasses or a mixture of grasses and drought-resistant shrubs. These lands are too dry to support crops but are often used to raise low-density populations of domesticated or semidomesticated animals. In some cases, the animals are maintained on permanent open ranges, while in others, nomadic herds are moved from place to place in search of suitable grazing. Usually, the animals are introduced species not native to the region. Sheep, cattle, and goats that are native to Europe and Asia have been introduced into the Americas, Australia, New Zealand, and many areas of Africa.

FIGURE 11.11 Forest Plantation This *Eucalyptus* forest plantation in South Africa shows both mature trees and young trees in the foreground. *Eucalyptus* species are not native to South Africa.

The conversion of rangelands to grazing by domesticated animals has major impacts on biodiversity. In an effort to increase the productivity of rangelands, management techniques may specifically eliminate certain species of plants that are poisonous or not useful as food for the grazing animals, or specific grasses may be planted that are not native to the area. In some cases, native animals are reduced if they are a threat to livestock because they are predators or because they may spread disease to the livestock. In addition, the selective eating habits of livestock tend to reduce certain species of native plants and encourage others.

Because rainfall is low and often unpredictable, it is important to regulate the number of livestock on the range. In areas where the animals are on permanent pastures, the numbers of animals can be adjusted to meet the capacity of the range to provide forage. In many parts of the world, nomadic herders simply move their animals from areas where forage is poor to areas that have better forage. This is often a seasonal activity that involves the movement of animals to higher elevations in the summer or to areas where rain has fallen recently. (See figure 11.12.)

In many parts of the world where human population pressures are great, overgrazing is a severe problem. As populations increase, desperate people attempt to graze too many animals on the land. They also cut down the trees for firewood. If overgrazed, many plants die, and the loss of plant cover subjects the soil to wind erosion, resulting in a loss of fertility, which further reduces the land's ability to support vegetation. The cutting of trees for firewood has a similar effect, but it is especially damaging because many of these trees are legumes, which are important in nitrogen fixation. Their removal further reduces soil fertility. This severe overuse of the land results in conversion of the land to a more desertlike ecosystem. This process of converting arid and semiarid land to desert because of improper use by humans is called **desertification.** Desertification can be found throughout the world but is particularly prevalent in northern Africa and parts of Asia, where rainfall is irregular and unpredictable, and where

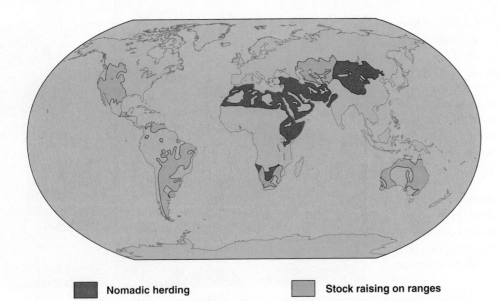

■ Nomadic herding   ■ Stock raising on ranges

FIGURE 11.12 Use of Rangelands The arid and semiarid regions of the world will not support farming without irrigation. In many of these areas, livestock can be raised. Permanent ranges occur where rainfall is low but regular. Nomadic herders can utilize areas that have irregular, sparse rainfall.

many people are subsistence farmers or nomadic herders who are under considerable pressure to provide food for their families. (See figure 11.13.)

Habitat Loss in Aquatic Ecosystems

Freshwater ecosystems are especially rich in species. The IUCN estimates that there are about 45,000 species that rely on freshwater ecosystems for their survival. River systems are linear and provide a variety of different habitats from their headwaters to their mouths. Different species are found in the cooler headwaters than at the warmer mouth of a river. Furthermore, each river system is isolated from others. Therefore, species have evolved to match the peculiar characteristics of each river system. Consequently, there is a high degree of *endemism* (local species found nowhere else) in river systems. Similarly, many lakes are isolated and show a high degree of endemism. Lake Victoria at one time perhaps had 300 species of fish found nowhere else in the world.

Biodiversity in freshwater ecosystems is under great threat from human actions.

- Rivers have been channelized to facilitate navigation, which increases the rate of flow in the river and reduces the amount of shallow water habitat.
- Dams have been built to provide for flood control and provide power, which changes the flow of the river and the temperature of the water.

- Pollution from industries, cities, and farms enters lakes and streams making the water unsuitable for sensitive species.
- Exotic species have been purposely and accidentally introduced. For example, the introduction of the Nile perch into Lake Victoria led to the extinction of about 200 species of native fish and the zebra mussel in the United States is responsible for the endangerment of many species of native mussels.
- Fish and other biotic resources have been overexploited.
- Withdrawal of water for municipal, industrial, and irrigation purposes has resulted in rivers drying up. The Aral Sea has shrunk in size and increased in salinity as a result of water withdrawals to irrigate farmland. Rivers such as the Rio Grande and Colorado in the United States and the Yellow River in China regularly dry up before they reach the sea as a result of water withdrawals for municipal and irrigation uses.
- Deforestation and agricultural activities lead to siltation and warming of waters.

The magnitude of the problem is illustrated in table 11.4.

Habitat loss is also a problem in aquatic systems. In marine ecosystems, much of the harvest is restricted to shallow parts of the ocean where bottom-dwelling fish can be easily harvested. The typical method used to harvest bottom-dwelling fish and shellfish involves the use of trawls, which are nets that are dragged along the bottom. These nets capture various species, many of which are not commercially valuable. The trawls disturb

Table 11.4 Species Threatened with Extinction in Freshwater Ecosystems

Group of Organisms	Percent of Species
Freshwater turtles (worldwide)	50%
Amphibians (worldwide)	32
Fish (worldwide)	20
Fish (Europe)	38
Fish (Madagascar)	54
Fish (East Africa)	54
Fish (United States)	37
Mussels (United States)	69
Crayfish (United States)	51
Stoneflies (United States)	43
Dragonflies and damselflies (United States)	18

Data Sources: IUCN; U.S. data from http://www.natureserve.org.

Risk of Human-Induced Desertification

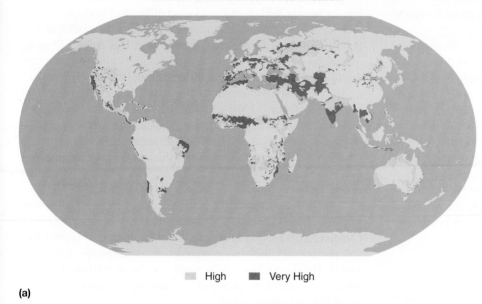

High ■ Very High ■

(a)

(b)

FIGURE 11.13 Desertification (a) Areas of the world where desertification is prevalent. (b) Arid and semiarid areas can be converted to deserts by overgrazing or unsuccessful farming practices. The loss of vegetation increases erosion by wind and water, increases the evaporation rate, and reduces the amount of water that infiltrates the soil. All of these conditions encourage the development of desertlike areas. Source: U.S. Department of Agriculture.

the seafloor and create conditions that make it more difficult for the fish populations to recover. In addition, 25 percent of the catch typically consists of species that have no commercial value. These are discarded by being thrown overboard. However, they are usually dead, and their removal further alters the ecological nature of the seafloor. Some people have even advocated that the trawl should be banned as a fishing technique because of the damage done to the ocean bottom.

Freshwater lakes, streams, and rivers are modified for navigation, irrigation, flood control, or power production purposes, all of which may alter the natural ecosystem and change the numbers or kinds of aquatic organisms present. These topics are discussed in greater detail in chapter 15.

Salmon runs, endemic to the Pacific Northwest, have been in decline for more than a century. As early as the late 1800s, hatchery augmentation was deemed necessary to maintain viable populations in the Columbia River basin. In the early 1900s, both Oregon State and Washington State imposed fishing season closures and gear-type regulations for the salmon fishery. Pacific salmon have disappeared from approximately 40 percent of the rivers they used to inhabit in Washington, Oregon, Idaho, and California. The decline of Pacific salmon has led to the listing of 28 salmon and steelhead stocks as either endangered or threatened under the Endangered Species Act (ESA). Causes of the salmon's decline, such as overharvest, poor hatchery management, logging, mining, dams, irrigation, grazing, and urban and industrial development, have been recognized. The specific cause-and-effect relationship between human activities and a negative impact on a salmon population, however, is often hard to establish.

In the Pacific Northwest, the extensive development of dams to provide power and aid navigation has made it nearly impossible for adult salmon to migrate upstream to spawn and difficult for young fish to migrate downstream to the ocean. Fish ladders and other techniques have not been successful in allowing the fish to pass the dams. The only solution to the problem is to remove or greatly modify several of the dams.

Conversion to Urban and Industrial Uses

About 4.3 percent of U.S. land is developed as urban centers, industrial sites, and the transportation infrastructure that allows for the movement of people and products about the country. Although this is a relatively small percentage of the total land area of the United States, urban areas are the most heavily affected by human activity. Many such areas are covered with impermeable surfaces that prevent plant growth and divert rainfall to local streams and rivers. In addition, streams and other natural features are altered to serve the needs of the people. Biodiversity is drastically reduced, and only the most adaptable organisms can survive in such settings.

Many industrial sites are associated with urban centers, although some industries such as mining and oil and natural gas production may be located far from urban centers. Their impact, however, is similar to that of an urban center. The land is altered in such a way that the natural ecosystems are destroyed.

Overexploitation

Overexploitation occurs when humans harvest organisms faster than the organisms are able to reproduce. Overexploitation has driven some organisms to extinction and threatens many others.

According to the IUCN, overexploitation is responsible for over 30 percent of endangered species of animals and about 8 percent of plants.

Organisms are harvested for a wide variety of purposes. Animals of all kinds are killed and eaten as a source of protein. We use organisms for a variety of purposes in addition to food. Many plants and animals are used as ornaments. Flowers are picked, animal skins are worn, and animal parts are used for their purported aphrodisiac qualities. In the United States, many species of cactus are being severely reduced because people like to have them in their front yards. In other parts of the world, rhinoceros horn is used to make dagger handles or is powdered and sold as an aphrodisiac. Because some people are willing to pay huge amounts of money for these products, unscrupulous people are willing to break the law and poach these animals for the quick profit they can realize.

Overfishing of Marine Fisheries

The United Nations estimates that 70 percent of the world's marine fisheries are being overexploited or are being fully exploited and are in danger of overexploitation as the number of fishers increases. Figure 11.14 shows both the amount of fish captured and the amount produced by aquaculture. While the amount of fish captured by fishers has remained essentially constant since 1987, the amount produced by fish farming has increased. It is important to note that more effective capture methods and increased fishing

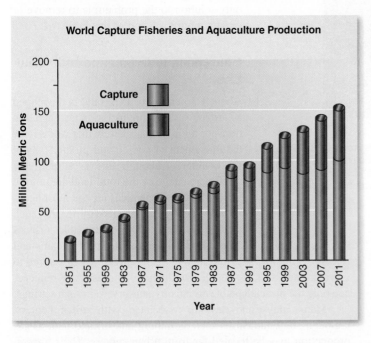

FIGURE 11.14 Trends in World Fish Production The amount of fish captured increased steadily until about 1987. Since then, the amount has remained essentially constant. This indicates that the world fisheries are being exploited to their capacity. Aquaculture has continued to increase so the total production continues to increase. These data have been called into question by many because it is thought that China has grossly overreported its fish production statistics. This would mean that the capture fisheries may actually have been declining in recent years.
Source: Data from Food and Agriculture Organization of the United Nations.

effort were required to maintain the production levels for the capture of marine fish. A study conducted by the National Oceanic and Atmospheric Administration showed that 40 stocks of fish populations were subject to overfishing in U.S. waters. Among the stocks being overfished are cod in the Northeast, red snapper in the Gulf of Mexico, and Pacific bluefin tuna off the West Coast. The study also indicated that 21 stocks have been rebuilt to healthy levels since 2000, and three key stocks in the Northeast— Georges Bank haddock, Atlantic pollock, and spiny dogfish— reached healthy levels in 2010. Overall about 16 percent of all fish stocks in 2010 were subject to overfishing in U.S. waters, meaning they were being fished at too high a level for what the population can sustain over time.

Another indication that marine fishery resources are being overexploited is the change in the kinds of fish being caught. The commercial fishing industry has been attempting to market fish species that previously were regarded as unacceptable to the consumer. These activities are the result of reduced catches of desired species. Examples of "newly discovered" fish in this category are monkfish and orange roughy.

Aquaculture

Fish farming (aquaculture) is becoming increasingly important as a source of fish production. (See figure 11.15.) Salmon farming has been particularly successful. This involves raising various species of salmon in "pens" in the ocean, which allows the introduction of food and other management techniques to achieve rapid growth of the fish. Norway, Chile, Canada, and Scotland are the leading countries in salmon production. The production of salmon from fish farms has increased rapidly. In 1988, less than 20 percent of the salmon sold were from fish farms, compared to over 65 percent in 2004. During the same period of time, the production of wild-caught salmon has been relatively constant. However, the farming method of producing fish is not without its environmental effects.

Raising fish in such concentrated settings results in increased nutrients in the surrounding water from uneaten food and the wastes the fish release. This can cause local algal blooms that negatively affect other fisheries, such as shellfish. Many of the salmon species raised in farms are not native to the waters in which they are raised. Inevitably, some of these fish escape. The introduction of exotic species can have a negative impact on native species. Some people even worry about the genetic stocks of fish that are raised. The introduction of new genetic stocks that escape and interbreed with wild fish can alter the genetic makeup of the wild populations. On an economic front, the increase in the amount of farmed salmon is reducing the prices salmon fishers receive for their catch.

Currently, about 60 percent of all aquaculture production is from freshwater systems, and production is growing rapidly. Aquaculture of freshwater species typically involves the construction of ponds, which allow the close management of the fish, shrimp, or other species. This can be done with a very low level of technology and is easily accomplished in underdeveloped areas of the world.

The environmental impacts of freshwater aquaculture are similar to those of aquaculture in marine systems. Nutrient overloads

(a) Aquaculture in ocean

(b) Aquaculture in freshwater

FIGURE 11.15 Aquaculture Photo (a) shows aquaculture pens in the ocean in British Columbia, Canada. Fish reared in these pens are fed and managed like livestock. Photo (b) shows a freshwater fish pond. In many parts of the world such ponds provide important sources of protein for the local populace.

from concentrations of fish can pollute local bodies of water, and the escape of exotic species may harm native species. In addition, freshwater aquaculture involves the conversion of land to a new use. Often the lands involved are mangrove swamps or other wetlands that many people feel should be protected. Regardless of environmental concerns, the productivity of freshwater aquaculture has the potential to provide for the protein needs of a growing population, so it is likely to continue to increase at a rapid rate.

Unsustainable Harvest of Wildlife and Plants

Wildlife may be harvested for a variety of reasons. Primary among them is the need for food in much of the world. Any animal that can be captured or killed is used for food. Meat from wild animals is often referred to as **bush meat,** since it is gathered from the "bush," a common term for the wild. Estimates about the extent of the problem vary, but there is wide agreement that the harvest is unsustainable. One estimate by the Wildlife Conservation Society is that about 70 percent of wildlife species in Asia and Africa and about 40 percent of species in Latin America are being hunted unsustainably. It is clear that this practice is causing local extinctions of certain species of wildlife. Furthermore, endangered species such as chimpanzees and gorillas are often harvested.

Trade in bush meat and other animal products has become a major problem for several reasons. Hunting of wildlife is a part of all subsistence cultures, so it is considered a normal activity. Poor people can earn money by harvesting and selling wild animals for meat or other purposes. Many kinds of wildlife are considered delicacies and are highly prized for the home and restaurant trade. In many parts of the developing world, regulations regarding the preservation of wildlife are poorly developed or widely ignored. The roads associated with logging operations in tropical forests give access to a greater part of the forest. Modern technology (guns and artificial lights) allows more efficient location and killing of animals.

The harvest of living animals for the pet and aquarium trade is a significant problem. Many kinds of birds, reptiles, and fish are threatened because they are desired as pets. Generally, it is not the isolated taking of an individual organism that is the problem but, rather, the method of capture, which may have far-reaching effects. For example, many nests are destroyed when nestling birds are taken. In the case of tropical marine fishes, toxins are often used to stun the fish, and the few valuable individuals are recovered while many of the others die. Furthermore, since much of the market in exotic pets involves transporting animals long distances, the mortality rate among those captured is high.

Wildlife species also are hunted because parts of the animal may have particular value. For example, ivory and animal skins are highly valued in many cultures as art objects or clothing. Parts of some animals are thought to have particular medicinal properties, and others are hunted because of their use in traditional cultural practices. For example, rhinoceros horn is thought to be an aphrodisiac, and some cultures prize rhinoceros horn as handles for knives. These beliefs lead to a brisk trade in the desired species.

Traditional medicine makes use of a great number of different plants and animals. (See figure 11.16.) Since about 80 percent of the world's population uses traditional cultural medicine as opposed to modern Western-style medical treatments, harvesting of medicinal plants and animals is a significant problem in much of the world. Many of the medicines used in these practices are derived from endangered plants and animals. Harvest of some species of plants has led to reductions in populations to the point that the species are endangered. For example, both the Asian and American ginseng plants have been greatly reduced because of their use in traditional medicine. Tiger bone and rhinoceros horn are important for a variety of traditional medicinal uses. The trade in these products puts a significant pressure on populations of these animals.

Introduction of Exotic Species

Globalization is responsible for spreading thousands of invasive alien species around the world. These species are not only the

FIGURE 11.16 Traditional Medicine Traditional medicine for sale in Mukuru slum in Nairobi, Kenya.

second greatest threat to native biodiversity but can also be harmful to human health. Exotic diseases are on the rise because of increased transportation and encroachment of humans into previously remote ecosystems. Introduced birds, rodents, and insects including mosquitoes and fleas can all serve as vectors and reservoirs of human diseases. Shipping moves over 80 percent of the world's commodities and transfers more than 10 billion metric tons of ballast water internationally each year. Many unwanted organisms take hold in foreign locations such as those that cause harmful algal blooms. Invasive species have long-term public health implications and incur huge costs to national economies. And efforts to control this growing menace can bring new problems. Pesticides applied to treat a particular pest species, for example, could pollute soil and freshwater supplies.

Introduction of exotic species can also have a significant effect on biodiversity because exotic species often kill or directly compete with native species and drive them to extinction.

The introduction of exotic species occurs for a variety of reasons. Some introductions are purposeful, while others are accidental. The reasons for purposeful introductions are quite varied. Most agriculturally important plants and animals are introductions. Other organisms were introduced because people had a fondness for a particular plant or animal. Many plants were introduced for horticultural purposes. Some introduced species are easily controlled. For example, most agriculturally important plants do not grow in the wild. Other introductions are accidental and arrived as stowaways on imported materials. Not all exotic species become problems; however, some organisms disperse broadly and displace native species.

The IUCN estimates that about 30 percent of birds and 15 percent of plants are threatened because they are unable to successfully compete against invasive exotic species. On many islands, the introduction of rats has had a major impact on nesting birds because the rats eat the eggs and kill the nestlings. The introduction of cats and foxes into Australia has resulted in the reduction of populations of many species of native marsupials. Cats and foxes are efficient predators of the native wildlife and caused the extinction of many kinds of native mammals on the Australian mainland.

The introduction of diseases has also had considerable impact on American forests. Two fungal diseases of trees have significantly changed the nature of American forests. Chestnut blight essentially eliminated the American chestnut, and Dutch elm disease has greatly reduced the numbers of American elms.

Various kinds of insects have also had an effect on the structure of ecosystems. The gypsy moth has spread throughout North America. Its larvae eat the leaves of forest trees and have significantly altered forest ecosystems because mature oaks are more likely to die as a result of defoliation than some other forest trees. The Asian longhorned beetle is a recent arrival to North America. It was discovered in New York in 1996. Its larvae feed on the wood of deciduous trees and often cause their death. Asian longhorned beetles probably enter North America inside solid-wood packing material from China. Although currently confined to a few locations in the eastern United States and Canada, the beetle has the potential to do great damage to the deciduous forest of the region, and both the U.S. and Canadian governments are taking action to control the problem. The only method of controlling the infestations involves the destruction of infected trees.

Certain species of fire ants are causing problems. The red imported fire ant is an invasive pest that has spread to many areas of the world, particularly the United States, Australia, the Philippines, and China, and is feared for the severity of its multiple stings and bites. Some people are allergic to the ant's venom; in some cases, the allergy may be severe enough to be fatal. Millions of dollars are spent every year on medical treatment and control efforts.

The European wasp is a native of Europe, North Africa, and temperate Asia but has spread since the early 1950s to North America, South Africa, South America, New Zealand, and Australia. The wasp can be a serious pest, aggressively defending its nest and swarming out to attack if disturbed. Its sting is painful, and multiple stings, or a sting in the throat, can be dangerous.

Freshwater ecosystems also have been greatly affected by accidental and purposeful introductions. The zebra mussel and Eurasian milfoil are two accidental introductions that have caused great problems. Zebra mussels were first discovered in Lake St. Clair near Detroit about 1985. They were most likely released from ballast water from European ships. The zebra mussel has spread through much of the eastern United States and adjacent Canada in the Great Lakes region and the Mississippi River watershed. It has three major impacts. It clogs intake pipes for water treatment plants and other industrial users. It establishes colonies on the surface of native mussels, often resulting in their death. And it has altered freshwater ecosystems by filtering much plankton from the water and allowing more aquatic plants to grow.

Eurasian milfoil was first discovered in Washington, D.C., in the 1940s but has since spread to much of the United States and southern Canada. Although its introduction may have been on purpose, its spread has occurred by accidental transfer by boats, wildlife, and drifting. It forms thick mats in many bodies of water that can interfere with boat traffic. Figure 11.17 shows examples of these exotic organisms.

(a) Adult gypsy moths

(b) Gypsy moth caterpillar

(c) Asian longhorned beetle

(d) Zebra mussel

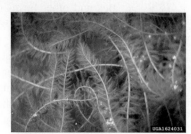

(e) Eurasian milfoil

FIGURE 11.17 Invasive Species The routes by which each of these species entered the United States is different. (a, b) The gypsy moth was purposely brought to the United States and accidentally escaped. (c) The Asian longhorned beetle appeared to have arrived in packing crates. (d) The zebra mussel arrived in ballast water from Europe, and (e) the Eurasian milfoil arrived as an aquarium plant.

The introduction of exotic fish species also has greatly affected naturally occurring freshwater ecosystems. The Great Lakes, for example, have been altered considerably by the accidental and purposeful introduction of fish species. The sea lamprey, smelt, carp, alewife, brown trout, and several species of salmon are all new to this ecosystem. (See figure 11.18.) The sea lamprey is parasitic on lake trout and other species and nearly eliminated the native lake trout population. Controlling the lamprey problem requires the use of a very specific larvicide that kills the immature lamprey in the streams. This technique works because mature lamprey migrate upstream to spawn, and the larvae spend several years in the stream before migrating downstream into the lake. With the partial control of lamprey, lake trout populations have been increasing. (See figure 11.19.) Recovery of the lake trout is particularly desirable, since at one time it was an important commercial species.

Another accidental introduction to the Great Lakes, the alewife, a small fish of little commercial or sport value, became a problem during the 1960s, when alewife populations were so great that they died in large numbers and littered beaches. Various species of salmon were introduced about this time in an attempt to control the alewife and replace the lake trout population, which had been depressed by the lamprey. While this salmon introduction has been an economic success and has generated millions of dollars for the sport fishing industry, it may have had a negative impact on some native fish such as the lake trout, with which the salmon probably compete. Salmon also migrate upstream, where they disrupt the spawning of native fish. Most species of salmon die after spawning, which causes a local odor problem.

Other more recent arrivals in the Great Lakes include a clam, the zebra mussel, and two fishes, the round goby and river ruffe. All of these probably arrived in ballast water from fresh or brackish water bodies in Europe. These exotic introductions are increasing rapidly and changing the mix of species present in Great Lakes waters.

Control of Predator and Pest Organisms

The systematic killing of certain organisms because they interfere with human activities also results in reduced biodiversity. Many large predators have been locally exterminated because they preyed on the domestic animals that humans use for food. Mountain lions and grizzly bears in North America have been reduced to small, isolated populations, in part because they were hunted to reduce livestock loss. Tigers in Asia and the lion and wolf in Europe were reduced or eliminated for similar reasons.

Some extinctions of pest species are considered desirable. Most people would not be disturbed by the

Lake trout

Native	Introduced
Brook lamprey	Sea lamprey
Lake trout Brook trout Whitefish Herring Ciscoes	Brown trout Rainbow trout Pink salmon Coho salmon Chinook salmon Atlantic salmon
Suckers	
Chubs Shiners	Carp
Catfish Bullheads	
	Smelt Alewife
White bass	White perch
Smallmouth bass Rock bass	Sunfish
Yellow perch Walleye	Ruffe

Brown trout

Smelt

Yellow perch

FIGURE 11.18 Native and Introduced Fish Species in the Great Lakes The Great Lakes have been altered considerably by the introduction of many non-native fish species. Some were introduced accidentally (lamprey, alewife, and carp), and others were introduced on purpose (salmon, brown trout, rainbow trout).

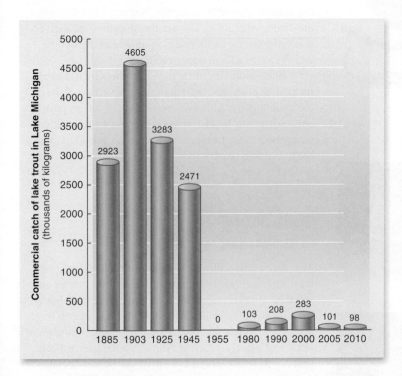

FIGURE 11.19 The Impact of the Lamprey on Commercial Fishing The lamprey entered the Great Lakes in 1932. Because it is an external parasite on lake trout, it had a drastic effect on the population of lake trout in the Great Lakes. As a result of programs to prevent the lamprey from reproducing, the number of lamprey has been reduced somewhat, and the lake trout population is recovering with the aid of stocking programs.

extinction of malaria parasite, mosquitoes, rats, or fleas. In fact, people work hard to drive some species to extinction. For example, in the *Morbidity and Mortality Weekly Report* (October 26, 1979), the U.S. Centers for Disease Control triumphantly announced that the virus that causes smallpox was extinct in the human population after many years of continuous effort to eliminate it.

At one time, it was thought that populations of game species could be increased substantially if predators were controlled. In Alaska, for example, the salmon-canning industry claimed that bald eagles were reducing the salmon population, and a bounty was placed on eagles. From 1917 to 1952 in Alaska, 128,000 eagles were killed for bounty money. This theory of predator control to increase populations of game species has not proven to be valid in most cases, however, since the predators do not normally take the prime animals anyway. They are more likely to capture sick or injured individuals not suitable for game hunting.

Although at one time they were thought to be helpful, bounties and other forms of predator management have been largely eliminated in North America. In fact, the pendulum has swung the other way, and predator control is not considered to be cost-effective in most cases. One exception to this general trend is the hunting and trapping of wolves in Alaska and Canada. One rationale for the taking of wolves is that they kill moose and caribou. Since many of the people who live in these areas rely on wild game as a significant part of their food source, controlling wolf populations is politically popular. However, the number of wolves taken is regulated.

The control of cowbird populations has been used to enhance the breeding success of the Kirtland's warbler. Cowbirds lay their eggs in the nests of other birds, including Kirtland's warblers. When the cowbird egg hatches, the hatchling pushes the warbler chicks out of the nest. Thus, trapping and killing of cowbirds in the vicinity of nesting Kirtland's warblers has been used to enhance the reproduction of this endangered species.

Many species may require refuges where they are protected from competing introduced species or human interference. Many native rangeland species of wildlife benefit from the exclusion of introduced grazing animals such as cattle and sheep because the absence of grazing livestock allows a more natural grassland community to be reestablished. (See figure 11.20.)

Climate Change

As scientists and the public have come to recognize that the climate is changing, the role of climate change on the survival of species has become an issue. Many species exist near the limit of their physiological tolerance. A slight change in the temperature may push them over the brink. Three groups of organisms appear to be greatly affected by climate change: amphibians, corals, and Arctic species. The International Union for Conservation of Nature and Natural Resources (IUCN) estimates that about 30 percent of amphibian species are threatened because of the effects of climate change. Many species of frogs are threatened by a fungal disease that has become common in the last few decades. Many people think there is a link between a warming planet and the prevalence of the disease. Corals are declining throughout the world. Many scientists suspect that warming water is encouraging diseases that are causing the death of corals.

Arctic and Antarctic species are adapted to cold temperatures and often rely on ice cover. The melting of sea ice in both the Arctic and Antarctic is changing migration patterns and food availability. The listing of polar bears as vulnerable by the IUCN and as a threatened species by the U.S. Fish and Wildlife Service

FIGURE 11.21 Climate Change Affects Arctic Animals The polar bear was listed as a threatened species as a result of global warming. Warmer waters result in less sea ice and bears have greater difficulty capturing their primary prey—seals. Note the remains of a kill on the ice behind the bear.

is recognition that breakup of ice cover in the Arctic oceans is affecting the ability of polar bears to obtain food. The bears use seals as a primary food source and wait at breathing holes or at the edge of ice flows to capture seals when they come up to breath. They also stalk seals that are resting on ice flows. With more open water, seals do not need to rely as much on breathing holes and bears are less successful at obtaining food. Furthermore, bears must swim greater distances to reach ice flows. (See figure 11.21.)

In the Antarctic, loss of sea ice has resulted in some populations of penguins migrating southward and appears to be related to a reduction in the amount of krill (small crustaceans), which are at the base of the Antarctic food web.

Climate change is also affecting non-native plant species. Since the beginning of the twentieth century, the growing season in many areas of North America has expanded by about two weeks. Many non-native species are changing their flowering schedule in concert with the longer growing season. For example, purple loosestrife, a European import, blooms 24 days earlier in the eastern United States than it did a century ago. Invasive species of plants are by nature highly flexible and respond to unusual environments more quickly than do natives. With the help of climate change, invasives also reap the benefits that come with early blooming, such as shading out competitors and capturing a larger share of nutriments, water, or pollinators.

11.4 What is Being Done to Preserve Biodiversity?

Efforts to preserve biodiversity involve a variety of approaches. Foremost among them is the need to understand the life history of organisms so that effective measures can be taken to protect species from extinction. Several international organizations work to prevent the extinction of organisms. The IUCN has assessed more

FIGURE 11.20 Habitat Protection The area on the left side of the fence has been protected from cattle grazing. This area provides a haven for many native species of plants and animals that cannot survive in heavily grazed areas.

than 70,000 species for its Red List and lists over 21,000 species as threatened with extinction. The IUCN classifies species threatened with extinction into three categories: critically endangered, endangered, and vulnerable. Threatened species are unlikely to survive if the conditions threatening their extinction continue. These organisms need action by people to preserve them, or they will become extinct.

The IUCN publishes updates to its Red List of Threatened Species. In a recent study of 25,000 vertebrates, 41 percent of amphibians are threatened, as are 25 percent of mammals, 22 percent of reptiles, 13 percent of birds, 33 percent of cartilaginous fish such as sharks, and 15 percent of bony fish such as southern bluefin tuna.

While vertebrates make up just 3 percent of all species, they are vital to the food chain and the overall ecosystem, and much more data are available for them. Ecologist Edward O. Wilson has described vertebrates as the "backbone" of diversity.

Although the IUCN is a highly visible international conservation organization, it has very little power to effect change. It generally seeks to protect species in danger by encouraging countries to complete inventories of plants and animals within their borders. It also encourages the training of plant and animal biologists within the countries involved. (There is currently a critical shortage of plant and animal biologists who are familiar with the organisms of the tropics.) The IUCN also encourages the establishment of preserves to protect species in danger of extinction.

Legal Protection

International efforts to preserve biodiversity have involved various activities of the United Nations. Most countries of the world have ratified the Convention on Biological Diversity (CBD), commonly known as the international biodiversity treaty. The United States is one of five countries that have not ratified the convention. The United States signed the treaty but it was not ratified by the U.S. Senate. Some of the key components of the Convention on Biological Diversity are that the signatory countries will:

- Develop national strategies for the conservation and sustainable use of biological diversity.
- Identify components of biological diversity important for its conservation.

- Monitor biological diversity.
- Identify activities that have adverse impacts on the conservation and sustainable use of biological diversity.
- Establish a system of protected areas.
- Rehabilitate and restore degraded ecosystems and promote the recovery of threatened species.
- Develop or maintain necessary legislation for the protection of threatened species and populations.
- Integrate consideration of the conservation and sustainable use of biological resources into national decision making.

When governments adopted the Convention on Biological Diversity in 1992 and its Cartagena Protocol on Biosafety in 2000, they acknowledged the critical importance of conserving biodiversity and sustainability using its components for meeting the food, health, and other needs of a growing world population. Specific reference to human health in the texts of the CBD and the Cartagena Protocol is made in relation to the risks associated with the use and release of living modified organisms resulting from modern biotechnology. In addition, the convention calls on parties to identify and monitor species which are of medicinal, agricultural, or other economic value as well as those used in medical research. As such, the convention offers a useful framework for integrating human health and biodiversity policy and increasing awareness of the critical links between the two.

Awareness and concern about loss of biodiversity are high in many of the developed countries of the world. However, these parts of the world have few undisturbed areas and many vulnerable species have already been eliminated. The potential for the loss of biodiversity is greatest in tropical, developing countries. Many biologists estimate that there may be as many species in the tropical rainforests of the world as in the rest of the world combined. Unfortunately, loss of biodiversity is not a high priority for the general public in developing countries, even though their national governments have ratified the biodiversity treaty. This difference in level of concern is understandable since the developed world has surplus food, higher disposable income, and higher education levels, while people in many of the less-developed countries, where population growth is high, are most concerned with immediate needs for food and shelter, not with long-term issues such as species extinction. The developed world should not be

What's Your Take?

The United Nations stated in 2006 that the target of significantly reducing the rate of loss in biodiversity by 2020 would not be met without strong science and effective governance mechanisms. Some scientists estimate that the current rate of extinction is a thousand times greater than at any other time in the course of humanity's development. The director of the UN environment program stated that, "if we fail to demonstrate measurable success by 2020, political commitment will be undermined, public interest will be lost, investment in biodiversity research and management will be reduced, environmental institutions will be further weakened." Do you feel this is an accurate statement, or do you feel it is "alarmist" in tone? Defend your position.

Focus On

Millennium Ecosystem Assessment Report and the Millennium Declaration

In 2005 a major study called the Millennium Ecosystem Assessment Synthesis and Natural Resources Report was published. It involved input from about 1,300 of the world's leading experts and a partnership among several international organizations including the United Nations, the World Bank, and the International Union for the Conservation of Nature (IUCN). The study was conducted over four years and involved experts from 95 countries. The report is recognized by governments as a mechanism to meet part of the assessment needs of four international environmental treaties—the UN Convention on Biological Diversity, the Ramscar Convention on Wetlands, the UN Convention to Combat Desertification, and the Convention on Migratory Species. Following are highlights of the main findings of the report.

- Humans have changed ecosystems more rapidly and extensively in the last 50 years than in any other period. This was done largely to meet rapidly growing demands for food, freshwater, timber, fiber, and fuel. More land was converted to agriculture since 1945 than in the eighteenth and nineteenth centuries combined. Experts say that this resulted in a substantial and largely irreversible loss in diversity of life on Earth, with 10 to 30 percent of the mammal, bird, and amphibian species currently threatened with extinction.

- Ecosystem changes that have contributed substantial net gains in human well-being and economic development have been achieved at growing costs in the form of degradation of other services. Only three ecosystem services have been enhanced in the last 50 years—increases in crop, livestock, and aquaculture production. Two services—capture fisheries and fresh water—are now well beyond levels that can sustain current, much less future, demands.

- The degradation of ecosystem services could grow significantly worse during the first half of this century. In all the plausible futures explored by the scientists, they project progress in eliminating hunger, but at a slower rate than needed to halve the number of people suffering from hunger by 2015. Experts warn that changes in ecosystems such as deforestation influence the abundance of human pathogens such as malaria and cholera, as well as increasing the risk from emerging new diseases.

The major conclusion of this assessment is that it lies within the power of human societies to ease the strains we are putting on the nature services of the planet, while continuing to use them to bring better living standards to all. Achieving this, however, will require major changes in the way nature is treated at every level of decision making and new forms of cooperation between government, business, and civil society.

Values Underlying the Millennium Declaration

The Millennium Declaration—which outlines 60 goals for peace; development; the environment; human rights; the vulnerable, hungry, and poor; Africa; and the United Nations—is founded on a core set of values described as follows:

"We consider certain fundamental values to be essential to international relations in the twenty-first century. These include:

- **Freedom.** Men and women have the right to live their lives and raise their children in dignity, free from hunger and from the fear of violence, oppression, or injustice. Democratic and participatory governance based on the will of the people best assures these rights.

- **Equality.** No individual and no nation must be denied the opportunity to benefit from development. The equal rights and opportunities of women and men must be assured.

- **Solidarity.** Global challenges must be managed in a way that distributes the costs and burdens fairly in accordance with basic principles of equity and social justice. Those who suffer or who benefit least deserve help from those who benefit most.

- **Tolerance.** Human beings must respect one other, in all their diversity of belief, culture, and language. Differences within and between societies should be neither feared nor repressed, but cherished as a precious asset of humanity. A culture of peace and dialogue among all civilizations should be actively promoted.

- **Respect for nature.** Prudence must be shown in the management of all living species and natural resources, in accordance with the precepts of sustainable development. Only in this way can the immeasurable riches provided to us by nature be preserved and passed on to our descendants. The current unsustainable patterns of production and consumption must be changed in the interest of our future welfare and that of our descendants.

- **Shared responsibility.** Responsibility for managing worldwide economic and social development, as well as threats to international peace and security, must be shared among the nations of the world and should be exercised multi-laterally. As the most universal and most representative organization in the world, the United Nations must play the central role."[1]

The Millennium Ecosystem Assessment Report is available at http://www.maweb.org/en/Article.aspx?id=58.

1.United Nations General Assembly, "United Nations Millennium Declaration," Resolution 55/2, United Nations A/RES/55/2, 8 September 2000.

Going Green

Consumer Choices Related to Biodiversity

Decisions you make as a consumer can have an important impact on biodiversity. There are many organizations and businesses that provide information about how purchasing decisions affect biodiversity. The following is a sampling.

The Earth Island Institute certifies dolphin-safe tuna products and maintains a website that lists companies that agree to adhere to several criteria, including having an independent observer onboard fishing vessels.

The Forest Stewardship Council certifies wood and paper products that are produced from forests that are managed according to ten key management principles, one of which is the maintenance of biodiversity in management of the forest resource. The organization maintains a website that lists companies and products that meet their criteria.

The Rainforest Alliance provides certification for a variety of products and services related to rainforests. These include products such as shade-grown coffee, cocoa, and ecotourism.

The Marine Aquarium Council certifies that organisms involved in the aquarium trade are produced in a sustainable fashion.

All of these organizations have logos that identify their product as being produced in a manner that is protective of biodiversity.

---

held blameless, however, because our activities may indirectly be involved in extinction. When we purchase lumber, agricultural products, and fish from the less-developed world at low prices, we often are indirectly encouraging people in the developing world to overexploit their resources and are thus affecting the rates of extinction.

In the United States, the primary action related to the preservation of biodiversity involved the passage of the U.S. Endangered Species Act in 1973. This legislation designates species as endangered or threatened and gives the federal government jurisdiction over any species designated as endangered. **Endangered species** are those that have such small numbers that they are in immediate jeopardy of becoming extinct. (See figure 11.22.) **Threatened species** could become extinct if a critical factor in their environment were changed. About 1,350 U.S. species and subspecies have been designated as endangered or threatened by the Office of Endangered Species of the Department of the Interior. The Endangered Species Act directs that no activity by a governmental agency should lead to the extinction of an endangered species and that all governmental agencies must use whatever measures are necessary to preserve these species.

Since one key to preventing extinctions is preservation of the habitat required by the endangered species, many U.S. governmental agencies and private organizations have purchased sensitive habitats or have managed areas to preserve suitable habitats for endangered species. Since the protection of a species requires the preservation of its habitat, inevitably landowners face losing the use of their land. Many lawsuits have been filed by landowners claiming that the federal government has taken their land from them by preventing them from using it for development purposes.

As a result of these pressures, the rules were changed to allow some of the land to be used if a species conservation plan is in place. Another modification to the Endangered Species Act occurred in 1978, when Congress amended it so that exemptions to the act could be granted for federally declared major disaster areas or for national defense, or by a seven-member Endangered Species Review Committee. Because this group has the power to sanction the extinction of an organism, it has been nicknamed the "God Squad." If the committee found that the economic benefits of a project outweighed the harmful ecological effects, it would exempt a project from the Endangered Species Act.

The designation of a species as endangered or threatened is biased toward large organisms that are relatively easy to identify. Many groups of organisms are very poorly known, so it is difficult to assess whether specific species are endangered or not. Table 11.5 compares the number of existing species in different taxonomic categories and the numbers and percentages considered endangered. Most endangered or threatened species are birds, mammals, certain categories of plants, some insects (particularly butterflies), a few mollusks, and fish. Bacteria, fungi, most insects, and many other inconspicuous organisms rarely show up on endangered species lists, even though they play vital roles in the nitrogen and carbon cycles and as decomposers.

The Convention on International Trade in Endangered Species of Wild Fauna and Flora (CITES) is an international agreement among governments. Its aim is to ensure that international trade in specimens of wild animals and plants does not threaten their survival. Currently, there are over 170 member countries who agree to limit export and import of certain species in order to protect the populations in their countries of origin. There are about 5000 species of animals and 28,000 species of plants that are protected by CITES against overexploitation through international trade.

Sustainable Management of Wildlife Populations

Many kinds of terrestrial wild animals are managed for their value as game animals, to protect them from extinction, or for other purposes. Several techniques, some of which result in ecosystem modifications, are used to enhance certain wildlife populations. These techniques include habitat analysis and management, population assessment and management, and establishing refuges.

Habitat Analysis and Management

Managing a particular species requires an understanding of the habitat needs of that species. An animal's habitat must provide the

Blackfooted ferret

Galápagos tortoise

Mission blue butterfly

Green pitcher plant

Giant panda

Ocelot

Nene (Hawaiian goose)

Pine Barrens treefrog

FIGURE 11.22 Endangered Species Endangered species are those present in such low numbers that they are in immediate danger of becoming extinct. These organisms are examples of endangered species.

Table 11.5 Endangered and Threatened Species by Taxonomic Group

Taxonomic Group	Approximate Number of Described Species in the World	Number of U.S. Species Endangered or Threatened	IUCN Threatened List	Percent of Total Species on IUCN Threatened List
Insects	950,000	57	623	0.07%
Plants	270,000	742	8,438	3.1%
Invertebrates other than insects or mollusks	257,800	36	507	0.2%
Mollusks	81,000	145	979	1.2%
Fishes	29,300	138	1,201	4.0%
Birds	9,900	90	1,217	12.0%
Reptiles	8,240	40	422	5.0%
Amphibians	5,900	23	1,808	29.0%
Mammals	5,400	84	1,094	20.0%
Bacteria, algae, fungi and protozoa	163,000	2	18	0.0%

Sources: Data from the IUCN Red List of Threatened Species and The U.S. Fish and Wildlife Service website.

following: food, water, and cover. **Cover** refers to any set of physical features that conceals or protects animals from the elements or enemies. Several kinds of cover are important and include places where the animal can rest and raise young, where it can escape from enemies, and where it is protected from the elements.

Once the critical habitat requirements of a species are understood, steps can be taken to alter the habitat to improve the success of the species. The habitat modifications made to enhance the success of a species are known as **habitat management.** The endangered Kirtland's warbler builds nests near the ground in dense stands of young jack pine. The density is important since their nests are vulnerable to predators and are better hidden in dense cover. Jack pine is a fire-adapted species that releases seeds following forest fires and naturally reestablishes dense stands following fire. Therefore, planned fires to regenerate new dense jack pine stands is a technique that has been used to provide appropriate habitat for this endangered bird.

Habitat management also may take the form of encouraging some species of plants that are the preferred food of the game species. For example, habitat management for deer may involve encouraging the growth of many young trees, saplings, and low-growing shrubs by cutting the timber in the area and allowing the natural regrowth to supply the food and cover the deer need. Both forest management and deer management may have to be integrated in this case because some other species of animals, such as squirrels, will be excluded if mature trees are cut, since they rely on the seeds of trees as a major food source.

Population Assessment and Management

Population management is another important activity that requires planning. Species of game animals are often managed so that they do not exceed the carrying capacity of their habitat. Wildlife managers use several techniques to establish and maintain populations at an appropriate level. Population censuses are used to check the population regularly to see if it is within acceptable limits. These census activities include keeping records of the number of animals killed by hunters, recording the number of singing birds during the breeding season, counting the number of fecal pellets, direct counting of large animals from aircraft, and a variety of other techniques.

Since wildlife management often involves the harvesting of animals by hunting for sport and for meat, regulation of hunting activity is an important population management technique. Seasons are usually regulated to ensure adequate reproduction and provide the largest possible healthy population during the hunting season. Hunting seasons usually occur in the fall so that surplus animals are taken before the challenges of winter. Winter taxes the animal's ability to stay warm and is also a time of low food supplies in most temperate regions. A well-managed wildlife resource allows for a large number of animals to be harvested in the fall and still leaves a healthy population to survive the winter and reproduce during the following spring. (See figure 11.23.)

In some cases, when a population of a species is below the desired number, organisms may be artificially introduced. Introductions are most often made when a species is being reintroduced to an area where it had become extinct. Wolves have been successfully reintroduced to Yellowstone National Park and large tracts of wooded land in the northern Midwest (Michigan, Wisconsin, Minnesota). Similarly, an exchange between the state of Michigan and the province of Ontario reintroduced moose to Michigan and turkeys to parts of Ontario.

Some game species are non-native species that have been introduced for sport hunting purposes. The ring-necked pheasant was originally from Asia but has been introduced into Europe, Great Britain, and North America. All of the large game animals of New Zealand are introduced species, since there were none in

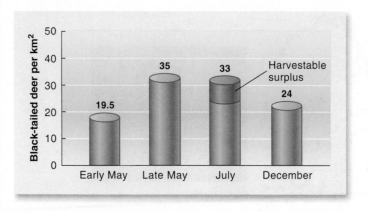

FIGURE 11.23 Managing a Wildlife Population The seasonal changes in this population of black-tailed deer are typical of many game species. The hunting season is usually timed to occur in the fall so that surplus animals will be harvested before winter, when the carrying capacity is lower.

Source: Data from R. D. Taber and R. F. Dasmann, "The Dynamics of Three Natural Populations of the Deer *Odocoileus hemionus columbianus*," *Ecology* 38, no. 2 (1957): 233–46.

the original biota of these islands. Several species of deer have been introduced into Europe from Asia. Many of these are raised in deer parks, where the animals are similar to free-ranging domestic cattle. They may be hunted for sport or slaughtered to provide food. Introductions of exotic species were once common. However, today it is recognized that these introductions often result in the decreased viability of native species or in the introduced species becoming pests.

Special Issues with Migratory Animals

Waterfowl (ducks, geese, swans, rails, etc.) present some special management problems because they are migratory. **Migratory birds** can fly thousands of kilometers and can travel north in the spring to reproduce during the summer months and return to the south when cold weather freezes the ponds, lakes, and streams that serve as their summer homes. (See figure 11.24.)

Because many waterfowl nest in Canada and the northern United States and winter in the southern United States and Mexico, an international agreement among Canada, the United States, and Mexico is necessary to manage and prevent the destruction of this wildlife resource. Habitat management has taken several forms. In Canada, where much of the breeding occurs, government and private organizations such as Ducks Unlimited have worked to prevent the draining of small ponds and lakes that provide nesting areas for the birds. In addition, new impoundments have been created where it is practical. Because birds migrate southward during the fall hunting season, a series of wildlife refuges provide resting places, food, and protection from hunting. In addition, these refuges may be used to raise local populations of birds. During the winter, many of these birds congregate in the southern United States. Refuges in these areas are important overwintering areas where the waterfowl can find food and shelter.

A similar management problem exists in Africa and Eurasia. Many birds that rely on wetlands migrate between Eurasia and Africa. In 1995, an international Agreement on the Conservation of African-Eurasian Waterbirds was developed. Its development was a complicated process, since it involves about 40 percent of the world's land surface and includes 117 countries. Several important countries have signed the agreement, but many have yet to do so.

In sub-Saharan Africa, many species of wildlife migrate across large expanses of land that may involve several countries. Often different countries have differing rules for harvest, and often there are barriers to the movement of migratory animals. For example, fences are commonly used to designate the boundaries between nations. A relatively new concept of transboundary parks has been developed to accommodate the movements of these animals. One typical action that occurs with the formation of transboundary parks across national boundaries is the removal of structures that interfere with the movement of the animals.

Sustainable Management of Fish Populations

One of the major problems associated with the management of marine fisheries resources is the difficulty in achieving agreement on limits to the harvest. The oceans do not belong to any country; therefore, each country feels it has the right to exploit the resource wherever it wants. Since each country seeks to exploit the resource for its advantage without regard for the sustainable use of the resource, international agreements are required to manage the resource. Such agreements are difficult to achieve because of political differences and disregard for the nature of the resource. (See Common Property Resource Problems—The Tragedy of the Commons in chapter 3.)

For several reasons, the most productive areas of the ocean are those close to land. In shallow water, the entire depth (water column) is exposed to sunlight. Plants and algae can carry on photosynthesis, and biological productivity is high. The nutrients washed from the land also tend to make these waters more fertile. Furthermore, land masses modify currents that bring nutrients up from the ocean bottom. Many of the commercially important fish and other seafood species are bottom-dwellers, but fishing for them at great depths is not practical. Therefore, fishing pressure is concentrated on areas where the water is shallow and relatively nutrient rich. Countries have taken control over the waters near them by establishing a 200-nautical-mile limit (approximately 300 kilometers) within which they control the fisheries. This has not solved conflicts, however, as neighboring countries dispute the fishing practices used in waters they both claim. Canada and the United States have had continuing arguments over the management of the lobster fishery and argued as well over the North Atlantic cod fishery until its collapse in the early 1990s.

In recent years, several countries have designated areas within their territorial waters where fishing is not allowed. Several studies suggest that these reserves are effective ways to protect a portion of the fishery resource and serve to repopulate surrounding areas that are open to fishing. This approach is particularly effective for fish species that live on the bottom or are associated with specific structures on the bottom.

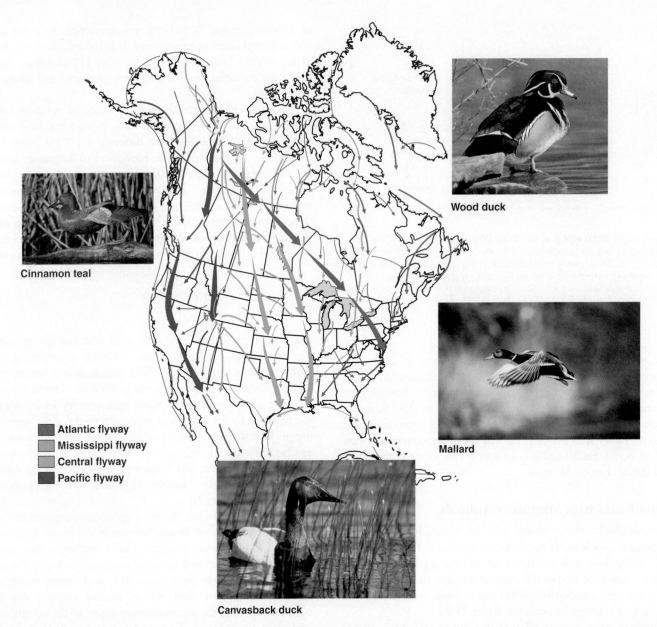

Wood duck

Cinnamon teal

Mallard

Canvasback duck

■ Atlantic flyway
■ Mississippi flyway
■ Central flyway
■ Pacific flyway

FIGURE 11.24 Migration Routes for North American Waterfowl Migratory waterfowl follow traditional routes when they migrate. These have become known as the Atlantic, Mississippi, Central, and Pacific flyways. Many of these waterfowl are hatched in Canada, migrate through the United States, and winter in the southern United States or Mexico.

Although freshwater fisheries face some of the same problems as marine fisheries, they are typically easier to regulate because rivers, lakes, and streams are often contained within a particular country. Although people in many places in the world harvest freshwater fish for food, the quantities are relatively small compared to those produced by marine fisheries. Furthermore, the management of the freshwater fish resource is much more intense since the bodies of water are smaller and human populations have greater access to the resource. Freshwater fish management in much of the world includes managing for both recreational and commercial food purposes.

The fisheries resource manager must try to satisfy two interest groups. Both sports fishers and those who harvest for commercial purposes must adhere to regulations. However, the regulations usually allow the commercial fisher to use different harvesting methods, such as nets. In much of North America, freshwater fisheries are primarily managed for sport fishing. The management of freshwater fish populations is similar to that of other wild animals. Fish require cover, such as logs, stumps, rocks, and weed beds, so they can escape from predators. They also need special areas for spawning and raising young. These might be a gravel bed in a stream for salmon, a sandy area in a lake for bluegills or bass,

Focus On

The California Condor

The California condor (*Gymnogyps californianus*) is thought to have been adapted to feed on the carcasses of large mammals found in North America during the Ice Age. With the extinction of the large, Ice Age mammals, the condors' major food source disappeared. By the 1940s, their range had shrunk to a small area near Los Angeles, California. Further fragmentation of their habitat caused by human activity, death by shooting, and death by eating animals containing lead shot reduced the wild population to about 17 animals by 1986.

A low reproductive potential makes it difficult for the species to increase in numbers. They do not become sexually mature until six years of age, and females typically lay one egg every two years. Because of concerns about the survival of the species, in 1987, all the remaining wild condors were captured to serve as breeding populations. The total population was 27 individuals. The plan was to raise young condors in captivity, leading to a large population that would ultimately allow for the release of animals back into the wild. Captive breeding is a very involved activity. Extra eggs can be obtained by removing the first egg laid and incubating it artificially. The female will lay a second egg if the first is removed. Raising the young requires careful planning. Although the young are fed by humans, they must not associate food with humans. This would result in inappropriate behaviors in animals eventually released to the wild. Therefore, puppets that resemble the parent birds are used to feed the young birds.

These efforts increased the number of offspring produced per female and resulted in a captive population of 54 individuals by 1991, when two condors were released into the wild north of Los Angeles. Six more were released in 1992. In 1996, a second population of condors was introduced in Arizona north of the Grand Canyon near the Utah border. The goal is to have two populations of up to 150 individuals each. In addition to natural mortality from predation, there has been unanticipated mortality from contact with power lines and lead poisoning from ingesting lead shot with food. To help reduce these mortality problems, the birds that are currently being released go through a period of training in which they are taught to avoid power lines. In addition, several condors have been recaptured to be treated for lead poisoning. These will eventually be released back to the wild.

By 2013, the population had increased to 435 birds, with 237 living in the wild and 198 in captivity. Since 2002, condors have been released in Baja, California, in conjunction with the Mexican government. There are 11 condors in Mexico. In 2007, for the first time, a pair laid an egg. However, the egg disappeared from the nest.

FIGURE 11.25 Monitoring Fish Populations A freshwater fisheries biologist studies stunned trout in a river in Montana. These fish were raised in a hatchery and released in the river to enhance recreational fishing and tourism.

or a marshy area for pike. A freshwater fisheries biologist tries to manipulate some of the features of the habitat to enhance them for the desired species of game fish. This might take the form of providing artificial spawning areas or cover. Regulation of the fishing season so that the fish have an opportunity to breed is also important.

In addition to these basic concerns, the fisheries biologist pays special attention to water quality. Whenever people use water or disturb land near the water, water quality is affected. (See figure 11.25.) For example, toxic substances kill fish directly, and organic matter in the water may reduce the oxygen. The use of water by industry or the removal of trees lining a stream warms the water and makes it unsuitable for certain species. Poor watershed management results in siltation, which covers spawning areas, clogs the gills of young fish, and changes the bottom so food organisms cannot live there. The fisheries biologist is probably as concerned about what happens outside the lake or stream as what happens within it.

Issues & Analysis

The Problem of Image

When we think of endangered species, we almost always visualize a mammal or a bird. In North America, we identify wolves, grizzly bears, various whales, bald eagles, whooping cranes, and similar species as endangered. We rarely think about clams, fish, plants, or insects. Because certain species are able to grab the attention of the public, they are called charismatic species. In addition, most of the charismatic species are carnivores at high trophic levels. Groups of people will organize to save the whales, whooping cranes, elephants, or osprey, but little interest is generated to save the Tar River spiny mussel, cave crayfish, or San Diego fairy shrimp. The graph shows the percentage of species in selected groups that are in various categories of concern in the United States. By far the most vulnerable category of organisms consists of freshwater species of

mussels (clams), crayfish, fish, and stoneflies. The least vulnerable are birds and mammals, yet they capture most of the public's interest and are highlighted by the U.S. Fish and Wildlife Service on its endangered species website.

- What factors cause us to rank birds and mammals higher than clams and crayfish?
- Should we spend as much money to save the Tar River spiny mussel as we are spending on wolves or California condors?
- Since money is limited, how would you decide which species to spend limited resources on?

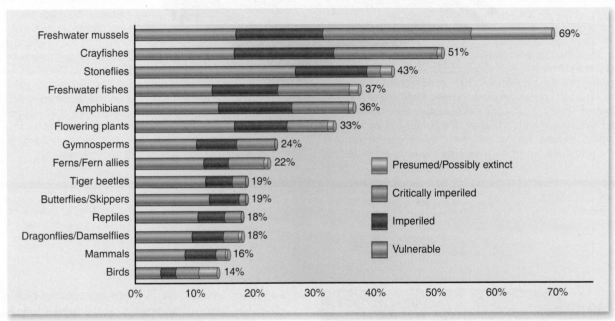

Source: Data from Natureserve, Precious Heritage.

Summary

The loss of biodiversity has become a major concern. Biodiversity can be measured in several ways. Genetic diversity is reduced when populations are reduced in size. Species diversity is reduced when a species becomes extinct from a local population or worldwide. Ecosystems involve the interactions of organisms and their physical environment.

Concern about the loss of biodiversity is based on several ways of looking at the value of a species. Functioning ecosystems and their component organisms provide many valuable services that are often overlooked because they are not easily

measured in economic terms. For example, nitrogen-fixing bacteria provide nitrogen for crops, photosynthesis is responsible for food production, and decomposers recycle organic materials. Other biotic resources can be measured in economic terms. Meat and fish harvested from the wild have great economic value, many kinds of drugs are derived from plants and animals, and ivory or animal hides are frequently traded. Many people also consider the loss of biodiversity to be an ethical problem. They ask the question, Do humans have the right to drive other species to extinction?

The primary causes of biodiversity loss are habitat loss by humans converting ecosystems to agriculture and grazing, over-exploitation by harvesting species at unsustainable levels, the introduction of exotic species that disrupt ecosystems and compete with or prey on native organisms, purposeful killing of pest organisms and large predators, and climate change.

Protection of biodiversity typically involves two kinds of activities: legal protection by national laws and international agreements, and management of the use of species and ecosystems at sustainable levels. These management activities typically include the establishment of regulations for hunting, fishing, and other uses.

Acting Green

1. Do not plant exotic species as landscape plants. They may become pests or carry diseases that harm native plant species.
2. Never release plants, fish, or other animals into a body of water unless they came out of that body of water.
3. Do not purchase exotic pets. Some become pests while others are harvested in unsustainable ways.
4. Participate in work parties to eliminate exotic, invasive species. Call your local nature center, Audubon club, Nature Conservancy, or similar organization.
5. Look for labels on products that indicate they were produced in a sustainable manner.

Review Questions

1. Name three ways humans directly alter ecosystems.
2. Why is the impact of humans greater today than at any time in the past?
3. Describe three factors that influence the genetic diversity of a population.
4. Describe three major causes of the loss of biodiversity.
5. What are the major causes of biodiversity loss in marine ecosystems?
6. List three problems associated with forest exploitation.
7. What is desertification? What causes it?
8. List three key components of the Convention on Biological Diversity (biodiversity treaty).
9. List six techniques utilized by wildlife managers.
10. What special problems are associated with waterfowl management?
11. What is extinction, and why does it occur?
12. List three examples of ecosystem services provided by biological resources.
13. List three actions that can be taken to prevent extinctions.
14. Describe the relationship between habitat needs of a species and the overall survival success of the species.
15. Describe the role of the Red List of Threatened Species in species conservation.

Critical Thinking Questions

1. Perhaps 98 percent to 99 percent of all species that have ever existed are extinct. Nearly all went extinct long before humans arrived on the scene. Why should we be concerned about the extinction of organisms today?
2. Would you support clearing of forests and plowing of grasslands that are ecologically important in order to support agriculture in countries that have significant hunger? Where do you draw the line between preserving ecosystems and human interest?
3. A Pacific Northwest Native American tribe has been permitted to hunt an endangered whale species to preserve its traditional culture and uphold its treaty rights. Now there is fear that other countries will hunt the whales, too. Do you think the tribe should be denied its rights? Why or why not?
4. Pharmaceutical companies are helping some developing countries to preserve their rainforests so these companies can look for organisms with possible pharmacological value. How do you feel about these arrangements? What limits, if any, would you place on the pharmaceutical companies? Why?

To access additional resources for this chapter, please visit ConnectPlus® at connect.mheducation.com. There you will find interactive exercises, including Google Earth™, additional Case Studies, and SmartBook™, an adaptive eBook that integrates our LearnSmart® adaptive learning technology.

CHAPTER OUTLINE

Land-use planning involves understanding the geologic characteristics of a place when determining its use. These homes in a slum of Rio de Janeiro should never have been built on a steep hillside prone to landslides.

OBJECTIVES

After reading this chapter, you should be able to:

- Recognize that humans alter the environment and that a growing population makes land-use planning necessary.

- Explain why most major cities are located on rivers, lakes, or the ocean.

- Describe the historical factors that led people to move from rural areas to urban areas in the late 1800s.

- State social and economic factors that encourage people to leave the city and move to the suburbs.

- Explain why floodplains and wetlands are often mismanaged.

- Describe physical, social, and land- use characteristics typical of urban sprawl.

- Describe tract development, leapfrog development, and ribbon sprawl.

- State how lifestyle and economic factors, along with poor planning, encouraged suburban sprawl.

- Describe societal problems associated with poorly planned urban growth.

- List common land-use planning principles.

- Describe common ways used by governments to implement good land use.

- State common urban problems that require special attention from urban planners.

- List common characteristics of a smart growth urban planning program.

- List the federal agencies involved in determining land use on federal land and their primary responsibilities.

- Describe different kinds of users of public land and how their desires result in conflict.

Oregon's Statewide Land Use Planning Program

Oregon's Statewide Land Use Planning Program is perhaps the best known in the United States. Created in 1973, the plan coordinates city and county planning agencies to ensure consistent, environmentally sound land-use practices across the state. Through a combination of zoning laws that protect forest and farm land, establish urban growth boundaries, and provide for some development, Oregon has served as a model for other states looking to reduce costs and pollution and stop urban sprawl.

Before 1973, land-use planning in Oregon was a wholly local matter. Some cities and counties planned and zoned their lands, while others did not. While some of these plans were effective, some were not. Some of the plans were coordinated with those of other jurisdictions, while some were not. This mosaic of differing approaches to planning promised to plague the state with uncoordinated development and urban sprawl, threatening Oregon's scenic beauty and contributing to environmental degradation.

Under the plan, each city adopted an urban growth boundary that marks the limits of urban development. In addition, each county created strong conservation zoning laws on its farmlands and forestlands. The goals of the plan are:

- To conserve farmland, forestland, coastal resources, and other natural resources vital to the state's economy and environment.
- To encourage efficient development and provision of public services.
- To coordinate the planning activities of local governments and state and federal agencies.
- To enhance the state's economy.
- To establish and maintain a planning process that is open and accessible to Oregon's citizens.

Since its inception, Oregon's land-use planning system has had a great deal of success in protecting private forest, farm, and rangeland from development. By 2013, despite rapid population growth since the law was enacted, the vast majority of private land—98 percent—remained in the same use as it was in 1974.

Some concerns regarding the law have arisen over the years. One of these is the rise of low-density residential land uses, especially in forest and range areas in central Oregon. Several hundred thousand hectares of low-density residential land uses were developed since 1974. Although it is less visible, low-density development can cause substantial degradation to natural habitat and wetlands. Another concern is that regionally the degree of development is not balanced. Western Oregon, especially the Portland metro area and the North Willamette Valley, led the way in resource land lost to development, losing over a tenth of their respective resource lands.

There have been challenges to the Oregon law, and there have been concerns raised about certain types of development. The consensus in the state, however, is that the 1973 law has been largely successful at meeting Oregonians' goals of protecting valuable farms, forests, and working rangeland.

Oregon's statewide land use planning program provides for both development and conservation.

12.1 The Need for Planning

Human development impacts upon all facets of the environment. While most apparent in urbanized areas, there is literally no place on the globe that does not bear some evidence of the impact of humankind. Most of this change occurred as people converted the land to agriculture and grazing, but, in our modern world, significant amounts have been covered with buildings, streets, highways, and other products of society. In many cases, cities became established before there was an understanding of the challenges presented by the location. When these cities grew and technology and society changed, the shortcomings of the locations became apparent. For example, Los Angeles and Mexico City have severe air pollution problems because of their geographic location and climate, Venice and New Orleans are threatened by high sea levels, and San Francisco and Tokyo are subject to earthquakes. Currently, most land-use decisions are still based primarily on economic considerations or the short-term needs of a growing population rather than on careful analysis of the capabilities and unique values of the land and landscape. Each piece of land has specific qualities based on its location and physical makeup. Some is valued for the unique species that inhabit it, some is valued for its scenic beauty, and some has outstanding potential for agriculture or urban uses. Since land and the resources it supports (soil, vegetation, elevation, nearness to water, watersheds) are not being created today (except by such phenomena as volcanoes and river deltas), land should be considered a nonrenewable resource.

Once land has been converted from natural ecosystems or agriculture to intensive human use, it is generally unavailable for other purposes. As the population of the world grows, competition for the use of the land will increase, and systematic land-use planning will become more important. Furthermore, as the population of the world becomes more urbanized and cities grow, urban planning becomes critical.

Almost all population growth in the next 30 years will be concentrated in urban areas, and most of this growth will occur in less-developed countries.

12.2 Historical Forces That Shaped Land Use

Today, most of the North American continent has been significantly modified by human activity. In the United States, about 52 percent of the land is used for crops and livestock, about 44 percent is forests and natural areas, and nearly 4 percent is used intensively by people in urban centers and as transportation corridors. Canada is 54 percent forested and wooded and uses only 8 percent of its land for crops and livestock. Less than 1 percent of the land is in urban centers and transportation corridors. A large percentage of its remaining land is wilderness in the north.

This pattern of land use differs greatly from the original conditions experienced by the early European colonists who immigrated to the New World. The first colonists converted only small portions of the original landscape to farming, manufacturing, and housing, but as the population increased, more land was converted to agriculture, and settlements and villages developed into towns and cities. Although most of this early development was not consciously planned, it was not haphazard.

Waterways and Development

If you look at a map of North America, you see that the major cities are located on rivers, lakes, and oceans. (See figure 12.1.) Waterways were the primary method of transportation, which allowed exploration and the development of commerce in the early European settlement of North America. Thus, early towns were usually built near rivers, lakes, and oceans. Typically, cities developed as far inland as rivers were navigable. Where abrupt changes in elevation caused waterfalls or rapids, goods being transported by boat or barge needed to be offloaded, transported around the obstruction, and loaded onto other boats. Cities often developed at these points. Buffalo, New York, on the Niagara River, which connects Lake Erie to Lake Ontario, and Sault Sainte Marie, Ontario, on the Saint Marys River, which connects Lake Superior with Lake Huron, are examples of such cities. In addition to transportation, bodies of water provided drinking water, power, and waste disposal for growing villages and towns. Those towns and villages with access to waterways that provided easy transportation could

readily receive raw materials and distribute manufactured goods. Some of these grew into major industrial or trade centers. Without access to water, St. Louis, Montreal, Chicago, Detroit, Vancouver, and other cities would not have developed. The availability of other natural resources, such as minerals, good farmland, or forests, was also important in determining where villages and towns were established. Industrial development began on the waterfront, since water supplied transportation, waste disposal, and power. As villages grew into towns and cities, large factories replaced small gristmills, sawmills, and blacksmith shops. The waterfront became a center of intense industrial activity. As industrial activity increased in the cities, people began to move from rural to urban centers for the job opportunities these centers presented.

The Rural-to-Urban Shift

North America remained essentially rural until industrial growth began in the last third of the 1800s and the population began a trend toward greater urbanization. (See figure 12.2.) Several forces led to this rural-to-urban transformation. First, the Industrial Revolution led to improvements in agriculture that required less farm labor at the same time industrial jobs became available in the city. Thus, people migrated from the farm to the city. The average person was no longer a farmer but, rather, a factory worker, shopkeeper, or clerk (with a regular paycheck) living in a tenement or tiny apartment near where he or she worked. This pattern of rural-to-urban migration occurred throughout North America, and it is still occurring in developing countries today. A second factor that affected the growth of cities was the influx of immigrants from Europe. Although some became farmers, many of these new citizens settled in towns and cities, where jobs were available. A third reason for the growth of cities was that they offered a greater variety of cultural, social, and artistic opportunities than did rural communities. Thus, cities were attractive for cultural as well as economic reasons. San Antonio, Texas, and Las Vegas, Nevada, are examples of cities that developed around unique cultural attractions.

Migration from the Central City to the Suburbs

During the early stages of industrial development, there was little control of industry activities, so the waterfront typically became a polluted, unhealthy, undesirable place to live. As roads and rail transport became available, anyone who could afford to do so moved away from the original, industrial city center. The more affluent moved to the outskirts of the city, and the development of suburban metropolitan regions began. Thus, the agricultural land surrounding the towns was converted to housing. Most cities originally had good farmland near them, since the floodplain near rivers typically has a deep, rich soil and agricultural land adjacent to the city was one of the factors that determined whether the city grew or not. This was true because until rapid land transportation systems became well developed, farms needed to be close to the city so that farmers with horses and wagons could transport their produce to the markets in the city. This rich farmland adjacent to

The Scioto River, Columbus, Ohio   Boston, Massachusetts

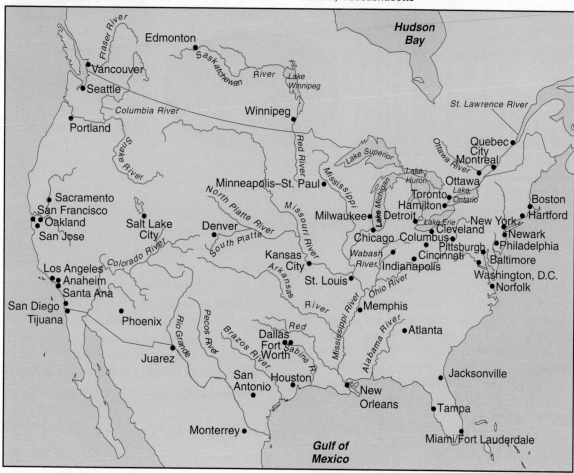

FIGURE 12.1 Water and Urban Centers Note that most of the large urban centers are located on water. Water is an important means of transportation and was a major determining factor in the growth of cities. The urban centers shown have populations of 1 million or more (except Winnipeg and Quebec City).

the city was ideal for the expansion of the city. As the population of the city grew, demand for land within the city increased. As the price of land in the city rose, people and businesses began to look for cheaper land farther away from the city. As cities continued to grow, certain sections within each city began to deteriorate. Industrial activity continued to be concentrated near water in the city's center. Industrial pollution and urban crowding turned the cores of many cities into undesirable living areas. In the early

1900s, people who could afford to leave began to move to the outskirts. This trend continued after World War II, in the 1940s, 1950s, and 1960s, as a strong economy and government policies that favored new home purchases (tax deductions and low-interest loans) allowed more people to buy homes.

Developers and real estate agents were quick to respond and to help people acquire and convert agricultural land to residential or commercial uses. Land was viewed as a commodity

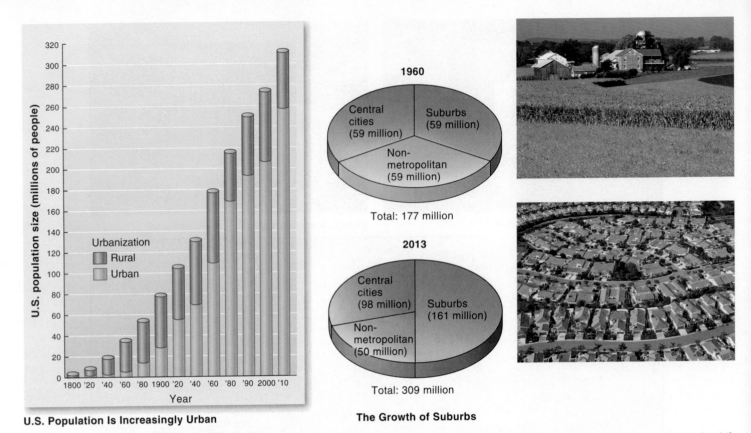

U.S. Population Is Increasingly Urban

The Growth of Suburbs

1960

Central cities (59 million)
Suburbs (59 million)
Non-metropolitan (59 million)

Total: 177 million

2013

Central cities (98 million)
Suburbs (161 million)
Non-metropolitan (50 million)

Total: 309 million

FIGURE 12.2 Rural-to-Urban Population Shift In 1800, the United States was essentially a rural country. Industrialization in the late 1800s began the shift to an increasing urban population. In 2013, 84 percent of the U.S. population was urban or suburban.

to be bought and sold for a profit, rather than as a nonrenewable resource to be managed. As long as money could be made by converting agricultural land to other purposes, it was impossible to prevent such conversion. There were no counteracting forces strong enough to prevent it. The conversion of land around cities in North America to urban uses destroyed many natural areas that people had long enjoyed. The Sunday drive from the city to the countryside became more difficult as people had to drive farther to escape the ever-growing suburbs.

Characteristics of Suburbs

In 1950, about 60 percent of the urban population lived in the central city; by 2013, this number was reduced to about 27 percent as blocks of homes were added to the periphery of the city. This unplanned suburban growth had become known as urban sprawl. **Urban sprawl** has been defined as the spread of low-density, auto-dependent development on rural land outside compact urban centers. Sprawl is typically characterized by the following:

- Excessive land consumption because of low population densities in comparison with older city centers
- Dependence on automobiles and poor availability of public transport
- Fragmented open space with wide gaps between developed and undeveloped areas

- Lack of choice in housing types and prices
- Streets in residential areas typically show branching patterns and often include cul-de-sacs
- Segregation from one another of commercial, industrial, single-family housing, and multiple-family housing areas
- Commercial buildings surrounded by large parking lots
- Lack of public spaces and community centers

Patterns of Urban Sprawl

There are several patterns of urban sprawl. (See figure 12.3.)

One type of growth involves the development of exclusive, wealthy suburbs adjacent to the city. These homes are usually on large individual lots in the more pleasing geographic areas surrounding the city. Often they are located along water, on elevated sites, or in wooded settings. A second development pattern is tract development. **Tract development** is the construction of similar residential units over large areas. Initially, these tracts are often separated from each other by farmland. This discontinuous pattern of isolated housing tracts is often called leapfrog development.

Roads that link new housing to the central city and other suburbs are constructed or improved, which stimulates the development of a form of urban sprawl along transportation routes. This is referred to as **ribbon sprawl** and usually consists of commercial

Focus On

Megacities

A **megacity** is usually defined as a metropolitan area with a total population in excess of ten million people. Some definitions also set a minimum level for population density of at least 2,000 persons per square kilometer.

As of 2013, there were 24 megacities in the world. The largest of these are the metropolitan areas of Tokyo, Delhi, Mexico City, New York, and Shanghai; each of these has a population in excess of 20 million inhabitants. Tokyo is the largest metropolitan area, while Shanghai is the largest city proper.

In 1800, only 3 percent of the world's population lived in cities, a figure that rose to 47 percent by the end of the twentieth century. In 1950, there were 83 cities with populations exceeding one million; by 2013, this number had risen to 498. The United Nations forecasts that today's urban population of 3.4 billion will rise to nearly 5 billion by 2030, when three out of five people will live in cities.

In the decade of 1950–1960, 60 percent of the growth of megacities was in the developing world. Between 2000 and 2013, the developing world accounted for 90 percent of the growth. Out of the 28 biggest cities on Earth, only six are in the developed world. The most rapid growth is taking place in countries that still have large rural hinterlands and relatively young populations. These mostly poor places—most with median incomes between Dhaka at $3,000 per capita and Bangkok at $23,000—will continue to grow, at least until their populations begin to see the results of decreasing birthrates.

Projections indicate that urban growth over the next 25 years will be in developing countries. One billion people, almost one-seventh of the world's population, now live in shanty towns. In many poor countries overpopulated slums exhibit high rates of disease due to unsanitary conditions and malnutrition. By 2030, over 2 billion people in the world will be living in slums. Over 90 percent of the urban population of Ethiopia, Malawi, and Uganda, three of the world's most rural countries, now live in slums.

Many migrate to the cities because they feel they will have greater access to social services and other cultural benefits than are available in rural areas. Many also feel there are more employment opportunities. However, the increase in the urban population is occurring so rapidly that

View of laundry hung out to dry in a slum in Mumbai, India.

it is very difficult to provide the services needed by the population, and jobs are not being created as fast as the urban population is growing. Thus, many of the people arriving in urban centers live in poverty on the fringes of the city in shantytowns that lack water, sewer, and other services. Often these shantytowns are constructed on land, without permission, only a short distance from affluent urban dwellers. Because the poor lack safe drinking water and sewer services, they pollute the local water sources, and disease is common. Because they burn wood and other poor-quality fuels in inefficient stoves, air pollution is common. The additional people also create traffic problems of staggering proportions. Because the shantytowns are unplanned developments, there is also a loss of open space for parks and other recreational areas. The table lists the ten megacities with the largest rate of population increase. All are in the developing world.

Table A Fastest Growing Megacities in the World (Urban areas with more than 10 million residents)

Rank	Country	Urban Area	Population Estimate	Growth (2000–2010)
1	Pakistan	Karachi	20.8 million	80.5%
2	China	Shenzhen	12.5 million	56.1%
3	Nigeria	Lagos	12.1 million	48.2%
4	China	Beijing	18.3 million	47.6%
5	Thailand	Bangkok	14.6 million	45.2%
6	Bangladesh	Dhaka	14.4 million	45.2%
7	China	Guangzhou	17.8 million	43.0%
8	China	Shanghai	21.8 million	40.1%
9	India	Delhi	22.9 million	39.2%
10	Indonesia	Jakarta	26.8 million	34.6%

(a) Wealthy Suburbs

(b) Ribbon Sprawl

(c) Tract Development

(d) Leapfrog Development

FIGURE 12.3 Types of Urban Sprawl There are several types of sprawl shown in these photos. (a) Large lots with isolated houses are typical of wealthy suburbs adjacent to the city. (b) Ribbon sprawl develops as a commercial strip along highways. (c) Tract development results in neighborhoods consisting of large numbers of similar houses on small lots with branching streets and cul-de-sacs. Note the absence of sidewalks. (d) Many housing tracts are relatively isolated from one another, which is a pattern called leapfrog development.

and industrial buildings that line each side of the highway that connects housing areas to the central city and shopping and service areas. Ribbon sprawl results in high costs for the extension of utilities and other public services. It also makes the extent of urbanization seem much larger than it actually is, since the undeveloped land is hidden by the storefronts that face the highway.

As suburbs continued to grow, cities (once separated by farmland) began to merge, and it became difficult to tell where one city ended and another began. This type of growth led to the development of regional cities. Although these cities maintain their individual names, they are really just part of one large urban area called a **megalopolis.** (See figure 12.4.) The eastern seaboard of the United States, from Boston, Massachusetts, to Washington, D.C., is an example of a continuous city. In the midwestern United States, the area from Milwaukee, Wisconsin, to Chicago, Illinois, is a further example. Other examples are London to Dover in England, the Toronto-Mississauga region of Canada, and the southern Florida coast from Miami northward.

FIGURE 12.4 Regional Cities in the United States and Canada The lights in this satellite image show population concentrations. Many cities have merged with their neighbors to form huge regional cities. Major urban regions are the northeast coast of the United States (Boston to Washington, D.C.), the region south of the Great Lakes (Milwaukee to Chicago and Detroit to Cleveland), the east coast of Florida from Jacksonville to Miami, the Toronto and Montreal regions of Canada, and the west coast of California (San Francisco area and Los Angeles to San Diego).

12.3 Factors That Contribute to Sprawl

Early in the twenty-first century, many, including government policymakers, are beginning to look at ways to prevent continued sprawling urban development. In order to change the current situation, it is important to understand the powerful social, economic, and policy forces that encourage sprawl.

Lifestyle Factors

One of the factors that has supported urban sprawl is the relative wealth of the population. This wealth is reflected in material possessions, two of which are automobiles and homes. In 2013, there were about 247 million passenger cars in the United States. This equates to one car for every person of driving age. In addition, about two-thirds of the population live in family-owned homes.

With this level of wealth, people are free to choose where and how they want to live. Many are attracted to a lifestyle that includes low-density residential settings, with large homes and yards, isolated from the problems of the city. They are also willing to drive considerable distances to live in these settings. Thus, a contributing factor to urban sprawl is the high rate of automobile ownership, which allows for ease of movement between home, the workplace, and shopping.

Economic Factors

Several economic forces operate to encourage sprawl development. First of all, it is less expensive to build on agricultural and other nonurban land than it is to build within established cities. The land is less expensive, the regulations and permit requirements are generally less stringent, and there are fewer legal issues to deal with. An analysis of building costs in the San Francisco Bay area of California determined that it costs between 25 percent and 60 percent more to build in the city than it does to build in the suburbs.

Several tax laws also contributed to encouraging home ownership. The interest on home mortgages is deductible from income taxes, and, in the past, people could avoid paying capital gains taxes on homes they sold if they bought another home of equal or higher value.

Typically, businesses and industries are attracted to newly developed regions and are able to influence local politicians and civic leaders to support rezoning to allow for commercial development and demand tax-exempt status for a period. However, once these businesses are in place they require infrastructure improvements (roads, sewer, water, parking, etc.) that encourage further development in the area.

Planning and Policy Factors

Many planning and policy issues have contributed to sprawl development. First of all, until recently, little coordinated effort has been given to planning how development should occur in metropolitan areas. There are several reasons for this. First of all, most metropolitan areas include hundreds of different political jurisdictions (the New York City metropolitan area includes about 700 separate government units and involves the states of New York, New Jersey, and Connecticut). It is very difficult to integrate the activities of these separate jurisdictions in order to achieve coordinated city planning. In addition, it is very difficult for a small local unit of government to see the "big picture," and many are unwilling to give up their autonomy to a regional governmental body.

Local zoning ordinances have often fostered sprawl by prohibiting the mixing of different kinds of land use. Single-family housing, multiple-family housing, commercial, and light and heavy industry were restricted to specific parts of the community. In addition, many ordinances specify minimum lot sizes and house sizes. This tends to result in a decentralized pattern of development that is supported by the heavy use of automobiles and the roads and parking facilities necessary to support them.

In addition, many government policies actually subsidize the development of decentralized cities. For example, in many cases, developers and the people who buy the homes and businesses they build are able to avoid paying for the full cost of extending services (water, sewer, electricity, police protections, fire protection, schools, and road improvements) to new areas. The local unit of government picks up the cost of these improvements, and the cost is divided among all the taxpayers rather than just those who will benefit. The roads that are needed to support new developments are usually paid for with federal and state monies, so again the cost is not borne by those who benefit most but by the taxpayers of the state or nation. Furthermore, federal and state governments have not supported public transport in an equivalent way.

12.4 Problems Associated with Unplanned Urban Growth

Poor planning for the growth of urban areas has resulted in a variety of problems that can be organized into several broad categories.

Transportation Problems

Most urban areas experience continual problems with transportation. This is primarily because, as cities grew, little thought was given to how people were going to move around and through the city. Furthermore, when housing patterns and commercial sectors changed, transportation mechanisms had to be changed to meet the shifting needs of the public. This often involved the abandonment of old transportation corridors and the establishment of new ones.

A strong automobile-based bias exists in the culture in the United States. In Los Angeles, for example, 70 percent of surface area in the city is dedicated in some way to the automobile (roads, parking garages, etc.), compared with only 5 percent devoted to parks and open space. The reliance on the automobile as the primary method of transportation has required the constant building of new highways. U.S. Department of Transportation (DOT) data state that about 25 percent of major streets and highways in urban areas are in poor condition and that state and local governments spend about $1 trillion a year on highway and street construction. Paradoxically, the establishment of new transportation corridors stimulates growth in the areas served, and the transportation corridors soon become inadequate.

According to the DOT, the average person in the United States travels about 260 kilometers (about 160 miles) per week in a car. The DOT, also states that the average person in a large metropolitan area spends more than 40 hours per year stuck in traffic delays.

Because little attention was paid to planning for public transportation as cities grew, it is difficult to establish mass transit in many cities. The dispersed nature of the suburbs and the need to find corridors in which to build mass transit requires the taking of land from current uses. Furthermore, it requires diverting funding from the building and maintenance of roads and highways that are currently underfunded. (See figure 12.5.)

FIGURE 12.5 Traffic and Suburbia Traffic congestion is a common experience for people who work in cities but live in the suburbs. The popularity of automobiles and the desire to live in the suburbs are closely tied.

Death of the Central City

Currently, less than 10 percent of people work in the central city. When people leave the city and move to the suburbs, they take their purchasing power and tax payments with them. Therefore, the city has less income to support the services needed by the public. When the quality of services in urban centers drops, the quality of life declines, the flight from the city increases, and a downward spiral of decay begins. An additional problem is the decline of the downtown business district. When shopping malls are built to accommodate the people in the suburbs, the downtown business district declines. Because people no longer need to come to the city to shop, businesses in the city center fail or leave, which deprives the remaining residents of basic services. They must now travel greater distances to satisfy their basic needs. The poor and elderly are often left behind in the city center, with reduced access to services.

High Infrastructure and Energy Costs

Infrastructure includes all the physical, social, and economic elements needed to support the population. Whenever a new housing or commercial development occurs on the outskirts of the city, municipal services must be extended to the area. Sewer and water services, natural gas and electric services, schools and police stations, roads and airports—all are needed to support this new population. Extending services to these new areas is much more costly than supplying services to areas already in the city because most of the basic infrastructure is already present in the city.

Energy costs are also high because of low energy efficiency. Energy efficiency is low for several reasons. First of all, automobiles are the least energy-efficient means of transporting people from one place to another. Second, the separation of blocks of homes from places of business and shopping requires that additional distances be driven to meet basic life needs. Third, congested traffic routes result in hours being spent in stop-and-go traffic, which wastes much fuel. Finally, single-family homes require more energy for heating and cooling than multifamily dwellings.

Loss of Open Space and Farmland

Although urban areas are dominated by buildings, an important difference between a pleasant and an unfriendly urban setting is the presence of open space. Open fields, parks, boulevards, and similar land uses allow people to visually escape from the congestion of the city. Unplanned urban growth does not take this important factor into account and the open space that remains often consists of small fragments. Consequently, in order to provide green space in many older cities and suburbs, buildings must be torn down, and disused spaces must be renovated into parks and other open space at great expense.

Currently, land in the United States is being converted to urban uses at a rate of over 400,000 hectares (1 million acres) per year. About one-third of this land is prime agricultural land, but forest, rangeland, and desert habitats are also converted to housing. Land that is flat, well drained, accessible to transportation, and close to cities is ideal farmland. However, it is also prime development land. Areas that once supported crops now support housing developments, shopping centers, and parking lots. (See figure 12.6.)

Loss of Forest Habitat

Loss of Agricultural Land

Loss of Rangeland Habitat

FIGURE 12.6 Loss of Open Space and Farmland As urban areas expand there is loss of open space and the loss of habitat for wildlife. In addition, in many cases the land was previously used for agriculture.

Regardless of the kind of land that is converted, the previous use of the land is lost and a common outcome is the loss of available open space and the loss and fragmentation of wildlife habitat. Because of concerns about the loss of farmland, several states have established programs that provide protection to farmers who do not want to sell their land to developers. The programs may require farmers, in return for lower taxes on the land, to put their land in a conservation easement that prevents them or future owners from using the land for anything other than farming.

Air and Water Pollution Problems

Most of the large industrial sources of air pollution in North America have been contained. However, reliance on the automobile as the primary method of transportation has resulted in significant air pollution problems in many cities. Automobile engines release carbon monoxide, hydrocarbons, and nitrogen oxides that interact to produce photochemical smog, which continues to degrade the air quality of many cities. (See figure 12.7a.) (See chapter 16 for a detailed discussion of urban air quality.) A simple solution to this problem is a centralized, efficient public transportation system. However, this is difficult to achieve with a highly dispersed population.

The infrastructure needed to support automobile travel (streets, highways, parking lots, etc.) is impervious to water. A typical shopping mall has a paved parking lot that is four times larger than the space taken up by the building. Because rainwater and snow melt cannot soak into the soil through the asphalt and cement, it is all channeled into drainage systems that carry the water directly to local water sources. Thus, the runoff from paved parking lots, streets, and highways carries pollutants from automobiles (oil, coolant, pieces of rubber) into local streams and ponds. (See figure 12.7b.)

Because the water cannot soak into the soil, urban centers with a large impervious surface area experience flash floods following heavy rains or rapid snow melt.

Floodplain Problems

Because most cities were established along water, many cities are located in areas called *floodplains*. **Floodplains** are the low areas near rivers and thus are subject to periodic flooding. Some floodplains may flood annually, while others flood less regularly. Floodplain elevations are often designated by the likelihood of flooding. An area in the 100-year floodplain would, on average, flood about once in a hundred years. However, this is a statistical concept and a location could experience a 100-year flood more or less frequently than once in a hundred years. In essence, these floodplain levels are really elevations above the river. Areas that are in the 500-year floodplain are at a higher elevation than those at the 100-year floodplain.

Because many towns and cities were established on rivers, and land near rivers is generally flat, this land was often developed for residential or commercial purposes even though it suffered periodic flooding. A better use of these areas is for open space or recreation, yet developers continue to build houses and light industry

(a) Automobiles and Smog

(b) Storm Water Runoff

FIGURE 12.7 Urban Sprawl Contributes to Air and Water Pollution (a) The dominance of the automobile as a mode of transportation produces air pollutants which in some cases cause smog. (b) The impermeable surfaces of roads, streets, and parking lots prevent water from soaking into the ground and the contaminants from these surfaces are washed into storm drains.

on them. Usually, when the land in a floodplain is developed for residential or commercial use, flood-control structures (retaining walls, levees, dikes, dams, retention basins, etc.) are built to prevent the periodic flooding natural to the area. Frequently, federal or state tax monies are used to build these flood-control structures and to repair the damage that results from floods. Thus, the cost of building these structures is typically spread over the population in general and the full cost is not paid for by the people who receive the service.

Floods are natural phenomena. Contrary to popular impressions, no evidence supports the premise that floods are worse today than they were 100 or 200 years ago, except perhaps on small, isolated watersheds. What has increased is the economic loss from the flooding. There is little doubt, however, that urban flood-control methods (walls, barriers, and levees) have had detrimental downstream effects during high-water periods because preventing the water from spreading into the floodplain forces increased amounts of water to be channeled to areas downstream. In 1993, during an extensive flooding event that involved both the Mississippi and Missouri Rivers, the U.S. Army Corps of Engineers estimated over $1.5 billion in damage to residential and commercial property. (See figure 12.8.) In 2010, the damage caused by the flood in Nashville, Tennessee, was estimated at $1.9 billion. Sometimes there are no alternatives to floodplain development—it is not possible to move an entire city to a higher elevation—but too often risks are simply ignored. Because flooding causes loss of life and property, floodplains should no longer be developed for uses other than agriculture and recreation.

Many communities have enacted **floodplain zoning ordinances** to restrict future building in floodplains. Although such ordinances may prevent further economic losses, what happens to individuals who already live in floodplains? Floodplain building ordinances usually allow current residents to remain. Relocation, usually at a financial loss, is the only alternative. Such situations are unfortunate; perhaps proper planning in the future will prevent these problems.

Wetlands Misuse

Since access to water was and continues to be important to industrial development, many cities are located in areas with extensive wetlands. **Wetlands** are areas that periodically are covered with water. They include swamps, tidal marshes, coastal areas, and estuaries. Some wetlands, such as estuaries and marshes, are permanently wet, while others, such as many swamps, have standing water during only part of the year. Many wetlands may have standing water for only a few weeks a year, often in the spring of the year when the snow melts. Because wetlands breed mosquitoes and are sometimes barriers to the free movement of people, they have often been considered useless or harmful. Most of them have been drained, filled, or used as dumps. Many modern cities have completely covered over extensive wetland areas and may even have small streams running under streets, completely enclosed in concrete. Not including Alaska, the United States has lost about 53 percent of its wetlands from pre-European settlement to the present. That equates to a loss from 89 million to 42 million hectares (220–100 million acres). The current loss rate is about 50,000 hectares (124,000 acres) per year. There are also 68 million hectares (170 million acres) of wetlands in Alaska. These are mostly peatlands, with almost no loss in the past 200 years.

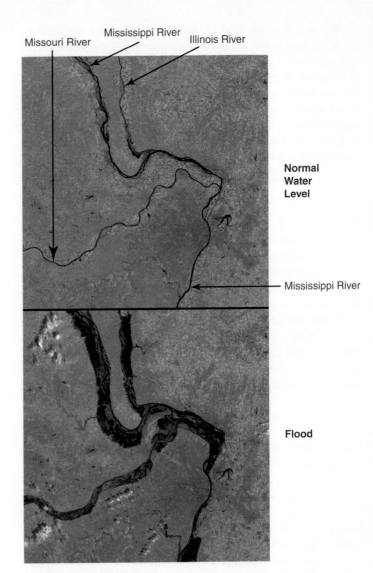

FIGURE 12.8 Flooding in Floodplain These two maps show where the Illinois River, Mississippi River, and Missouri River join. In the summer of 1993, the rivers occupied much of the low-lying floodplain causing extensive flooding.

Each kind of wetland has unique qualities and serves as a home to many kinds of plants and animals. Because most wetlands receive constant inputs of nutrients from the water that drains from the surrounding land, they are highly productive and excellent places for aquatic species to grow rapidly. Wetlands are frequently critical to the reproduction of many kinds of animals. Many fish use estuaries and marshes for spawning. Wetlands also provide nesting sites for many kinds of birds and serve as critical habitats for many other species. Waterfowl hunters and commercial and sport fisheries depend on these habitats to produce and protect the young of the species they harvest. Human impact on wetlands has severely degraded or eliminated these spawning and nursery habitats.

Besides providing a necessary habitat for fish and other organisms, wetlands provide natural filters for sediments and run-off. This filtration process allows time for water to be biologically

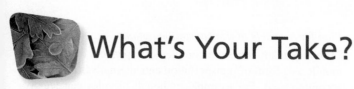

What's Your Take?

In a 2006 article in the *Washington Post,* the architect and professor of architecture Roger Lewis addressed the rebuilding of New Orleans following Hurricane Katrina. "Why," he asks, "do we stubbornly refuse to acknowledge that there are places on the Earth's surface—wetlands and floodplains, seismically active regions, arid deserts, steep hillsides and cliffs—where erecting cities endangers not only humans, but also the natural environment?" Develop a position paper that either supports or refutes Lewis's question as it pertains to the rebuilding of New Orleans or other places that have suffered great damage from natural disasters such as the New Jersey coast or areas around Los Angeles that are affected by wildfires and mud slides.

cleaned before it enters larger bodies of water, such as lakes and oceans, and reduces the sediment load carried by runoff. Wetlands also protect shorelines from erosion. When they are destroyed, the natural erosion protection provided by wetlands must be replaced by costly artificial measures, such as breakwalls.

Geology and Resource Limitations

The geologic status of an area must also be considered in land-use decisions. Building cities on the sides of volcanoes or on major earthquake-prone faults has led to much loss of life and property. Building homes and villages on unstable hillsides or in areas subject to periodic fires is also unwise. Yet, every year more houses slide down California hillsides and wildfires consume homes throughout the dry West.

Another problem in some locations is lack of water. Southern California and metropolitan areas in Arizona must import water to sustain their communities. Wise planning would limit growth to whatever could be sustained by available resources. As the population of cities in such locations continues to grow, the strain on regional water resources will increase.

Water-starved cities often cause land-use dilemmas far from the city boundaries. Supplying water and power to cities often involves the construction of dams that flood valleys that may have significant agricultural, scenic, or cultural value. See chapter 15 on water use for a more extensive development of this topic.

Aesthetic Issues

Unpleasant odors, disagreeable tastes, annoying sounds, and offensive sights can be aggravating. Yet it is difficult to get complete agreement on what is acceptable and when some aesthetic boundary has been crossed that is unacceptable. Furthermore, many useful activities generate stimuli that are offensive while the activity itself may be essential or at least very useful. Many of these do not harm us physically but may be harmful from an aesthetic point of view.

Unplanned or poorly planned development can place housing developments downwind of dairy operations or industries that produce odors. Building subdivisions next to major highways or airports exposes residents to noise. In some cases, barriers are built to reduce the volume of the noise. The visual aspects of a place can also be influenced by the surroundings. Although it is done, placing homes near landfills, junkyards, or industry does not make much sense.

12.5 Land-use Planning Principles

Many of the problems outlined in the previous sections can be avoided with proper land-use planning. **Land-use planning** is a process of evaluating the needs and wants of the population, the characteristics and values of the land, and various alternative solutions to the use of a particular land surface before changes are made. Planning land use brings with it the need to examine the desires of many competing interests. The economic and personal needs of the population are a central driving force that requires land-use decisions to be made. However, the unique qualities of particular portions of the land surface may prevent some uses, poorly accommodate others, but be highly suitable for still others. For example, the floodplain beside a river is unsuitable for building permanent structures; it can easily accommodate recreational uses such as parks, but it may be most useful as a nature preserve. Agricultural land near cities can be easily converted to housing but may be more valuable for growing fruits and vegetables that are needed by the people of the city. This is particularly true when agricultural land is in short supply near urban centers. To make good land-use decisions, each piece of land must be evaluated and one of several competing uses assigned to it. When land-use decisions are made, the decision process usually involves the public, private landowners, developers, government, and special interest groups. Each interest has special wants and will argue that its desires are most important. Since this is the case, people must have principles and processes to guide land-use decisions, irrespective of their personal wants and interests.

A basic rule should be to make as few changes as possible, but when changes are suggested or required, several things should be considered.

1. *Evaluate and record any unique geologic, geographic, or biologic features of the land.*

   Some land has unique features that should be preserved because of their special value to society. The Grand Canyon,

Yellowstone National Park, and many wilderness areas have been set aside to preserve unique physical structures, scenic characteristics, special ecosystems, or unusual organisms. On a more local level, a stream may provide fishing opportunities near a city, or land may have excellent agricultural potential that should take precedence over other uses. New York City purchased and protects a watershed that provides water for the city at a much lower cost than treating water from the Hudson River.

2. *Preserve unique cultural or historic features.*

Some portions of the landscape, areas within cities, and structures have important cultural, historic, or religious importance that should not be compromised by land-use decisions. In many cities, historic buildings have been preserved and particular sections set aside as historic districts. Sacred sites, many battlefields, and places of unique historic importance are usually protected from development.

3. *Conserve open space and environmental features.*

It must be recognized that open space and natural areas are not unused, low-value areas. Many studies of human behavior have shown that when people are given choices, they will choose settings that provide a view of nature or that allow one to see into the distance. Some have argued that this is a deep-seated biological need, while others suggest that it is a culturally derived trait. Regardless of its origin, urban planners know that access to open space and natural areas is an important consideration when determining how to use land. Therefore, it makes sense to protect open space within and near centers of population.

4. *Recognize and calculate the cost of additional changes that will be required to accommodate altered land use.*

Whenever land use is altered, additional modifications will be required to accommodate the change. For example, when a new housing development is constructed, schools and other municipal services will be required, roads will need to be improved, and the former use of the land is lost. Frequently, the cost of these changes is not borne by the developer or the homeowner but becomes the responsibility of the entire community; everyone pays for the additional cost through tax increases. In many areas, basic services, such as provision of water, are severe problems. It makes no sense to build new housing when there is not enough water to support the existing population.

5. *Plan for mixed housing and commercial uses of land in proximity to one another.*

One of the major problems associated with development in North America is segregation of different kinds of housing from one another and from shopping and other service necessities. Mixing various kinds of uses together (single-family housing, apartments, shopping and other service areas, and offices) allows easier connection between uses without reliance on the automobile. Walking and biking become possible when these different uses are within a short distance of one another.

6. *Plan for a variety of transportation options.*

Currently, many areas are completely dependent on the automobile as a form of transportation and alternatives are poorly accommodated. For example, although bicycles can legally be ridden on roads and streets, it is generally unsafe because of the high speeds of vehicular traffic and poor road surface conditions near the edge of the pavement. Special bike lanes are available in some areas, but this is rare in much of North America. Walking is a healthy, pleasant way to get from one task to another. However, crossing wide, busy streets is difficult, many areas lack sidewalks, and related service areas are often far apart. Clustering housing and service areas allows people to walk to places of business. Clustering service businesses—rather than stringing them out along a highway—also makes it easier to plan bus and rail routes to allow people to get from one place to another without relying on automobiles.

7. *Set limits and require managed growth with compact development patterns.*

Much unplanned growth occurs because there is no plan or the land-use plan is not enforced. One very effective tool that promotes efficient use of the land is to establish an urban growth limit for a municipality. An **urban growth limit** establishes a boundary within which development can occur. Development outside the boundary is severely restricted. One of the most important outcomes of setting urban growth boundaries is that a great deal of planning must precede the establishment of the limit. This lets all in the community know what is going on and can allow development to occur in logical stages that do not stress the community's ability to supply services. This mechanism also stimulates higher density uses of the urban land.

8. *Encourage development within areas that already have a supportive infrastructure so that duplication of resources is not needed.*

Because all development of land for human activity requires that services be provided, it makes sense that housing and commercial development occur where the infrastructure is already present. Infrastructure includes electric, phone, sewer, water, and transportation systems and service industries, such as shopping, banking, restaurants, hotels, and entertainment. It also includes schools, hospitals, and police protection. If development occurs far from these services, it is very costly to extend or duplicate those services in the new location. Furthermore, all large cities and most smaller ones have vacant lots or abandoned buildings that have outlived their usefulness or are vacant because of changing business or housing needs. This land is already urban and close to municipal services, and the buildings can be readily renovated or demolished and replaced. Many will argue that this is too expensive, but if other options are discouraged, these spaces will be used, and their redevelopment can be important in revitalizing inner-city spaces. Figure 12.9 provides examples of how these principles can be used.

(a) Protect Historic and Cultural Resources

(b) Protect Open Space

(c) Reuse Abandoned Urban Space

(d) Plan for Pedestrian and Bicycle Travel

FIGURE 12.9 Land-Use Planning Principles These photos illustrate some of the issues that must be faced when making land-use decisions. (a) Protect historic resources (b) protect open space (c) reuse abandoned urban space (d) plan for pedestrian and bicycle travel (e) encourage mixing service businesses with housing.

(e) Encourage Mixed Housing and Business

12.6 Mechanisms for Implementing Land-Use Plans

Land-use planning is the construction of an orderly list of priorities for the use of available land. Developing a plan involves gathering data on current use and geological, biological, and sociological information. From these data, projections are made about what human needs will be. All of the data collected are integrated with the projections, and each parcel of land is evaluated and assigned a best use under the circumstances. There are basically three components that contribute to the successful implementation of a land-use development plan: land-use decisions can be assigned to a regional governmental body, the land or its development rights can be purchased, and laws or ordinances can be used to regulate land use.

Establishing State or Regional Planning Agencies

State and regional planning is often more effective than local land-use planning because watersheds, groundwater resources, and many other important geological, geographic, and habitat characteristics cross local political boundaries. In addition, a regional approach is also likely to prevent duplication of facilities and lead to greater efficiency. For example, airport locations should be based on a regional plan that incorporates all local jurisdictions.

State or regional planning bodies are also more likely to have the financial resources to hire professional planners to assist in the planning process. The first state to develop a comprehensive statewide land-use program was Hawaii. During the early 1960s, much of Hawaii's natural beauty was being destroyed to build houses and apartments for the increasing population. The landscape that attracted tourists was being destroyed to provide hotels and supermarkets for them. Local governments had failed to establish and enforce land-use controls. Consequently, in 1961, the Hawaii State Land-Use Commission was founded. This commission designated all land as urban, agricultural, or conservational. Each parcel of land could be used only for its designated purpose. Other uses were allowed only by special permit. To date, the record for Hawaii's action shows that it has been successful in controlling urban growth and preserving the islands' natural beauty, even though the population continues to grow. (See figure 12.10.)

Often it is difficult to establish regional planning agencies because local units of government, private landowners, and developers do not want to give up land-use decisions to a regional body. However, many states have recognized the value of regional planning and have passed legislation to encourage or mandate regional planning. (See table 12.1.) Some have passed legislation dealing with special types of land use. Examples include wetland preservation, coastal zone management, floodplain protection, and scenic and historic site preservation. Others have required that local units of government establish land-use plans and state the criteria that must be included.

FIGURE 12.10 Hawaii Land-Use Plan Hawaii was the first state to develop a comprehensive land-use plan. This development in an agricultural area was stopped when the plan was implemented in the 1960s.

Restricting Use

Many kinds of land-use restrictions involve some form of zoning. **Zoning** is a common type of land-use regulation that designates specific areas within a community for certain kinds of land use. Common designations are agricultural, commercial, residential, recreational, environmental, and industrial. (See figure 12.11.)

Zoning has had both positive and negative effects on good land-use planning. Most people would agree that it is important to protect important historic, scenic, or cultural sites. Zoning can do this effectively. However, in many ways certain zoning processes contributed to the urban sprawl described earlier in this chapter. Many communities practiced single-use zoning that required certain areas within the community to be designated for particular uses. Certain areas were set aside for commercial development, others for housing, and others for industrial development. This segregation of uses, along with the dominance of the automobile as a form of transportation, contributed to the sprawl we see today.

Zoning decisions become very complicated when people must weigh many competing uses and economic issues related to land use.

Most local zoning boards are elected or appointed and often lack specific training in land-use planning. As a result, zoning regulations are frequently made by people who see only the short-term economic gain and not the possible long-term loss. Often the land is simply zoned so that its current use is sanctioned and is rezoned when another use appears to have a higher short-term value to the community. Even when well-designed land-use plans exist, they are usually modified to encourage local short-term growth rather than to provide for the long-range needs of the community. The public needs to be alert to variances from established land-use plans, because once the plan is compromised, it becomes easier to accept future deviations that may not be in the best interests of the community. Many times, individuals who make zoning decisions are real-estate agents, developers, or local businesspeople. These individuals wield significant local political power and are not always unbiased in their decisions. Concerned citizens must try to combat special interests by attending zoning commission meetings and by participating in the planning process.

Another approach to land-use planning is to limit the amount of new development that can take place within the community. New development requires additional services, alters the community structure, and creates new traffic patterns. Therefore, many cities have recognized that they must regulate the growth of their communities and have enacted various laws and ordinances that restrict development. Many involve a growth management plan that restricts the areas that may be developed or places limits on how many new buildings can be constructed. (See table 12.2.)

Table 12.1 Examples of State Land-Use Planning Legislation

State	Legislation	Significant Elements
California	California Coastal Act (1976)	Generally defines the coastal zone as extending 3 miles out to sea and 1,000 yards inland and requires permits from the California Coastal Commission or an approved local body for development within the zone.
Florida	Omnibus Growth Management Act (1985)	Established a list of goals for the state and requires that communities develop comprehensive plans compatible with state goals.
Georgia	The Georgia Planning Act (1989)	Requires cities and counties to prepare a comprehensive plan and update their plan every ten years.
Hawaii	Hawaiian Land-use Law (1961)	Requires that all land in Hawaii be classified into one of four districts (urban, agricultural, rural, and conservation), each of which is subject to different procedures and standards for managing land uses.
Massachusetts	Cape Cod Commission Act (1989)	Established a regional planning commission for Cape Cod.
New York	Adirondack Park Agency Act (1971)	Adirondack State Park includes both privately held lands (about 50%) and those owned by governmental units. The Adirondack Park Agency performs long-range planning for the park and reviews plans of private landowners to determine if their plans are compatible with the park's plan.
Vermont	Growth Management Act (1988)	Established state planning goals. Regional commissions are required to adopt regional plans that meet state goals and to act as links between towns and state and local government.
Washington	Growth Management Act (1990)	Required the state's largest and fastest growing counties to adopt growth management comprehensive plans that address land-use and transportation planning, concentrate new growth in urban areas, and protect natural resources and environmentally important areas. It established 13 planning goals to guide the preparation of local plans and regulations.

FIGURE 12.11 Zoning Most communities have a zoning authority that designates areas for particular use. This sign indicates that decisions have been made about the "best" use for the land.

Table 12.2 Examples of Local Growth Management Actions

Action	Specific Requirements	Examples of Locations
Limit Building Permits	Limit the number of building permits	Livermore, CA; Boulder, CO; Westminster, CO
	Limits building permits to 500; permits awarded based on evidence of planning	Petaluma, CA
	Established a cap on growth at 40,000 dwellings in 1972	Boca Raton, FL
	The ordinance was overturned by the U.S. Supreme Court	
Initiate Growth Management Planning Process	Establish a growth management plan	Boulder, CO
	Adopted growth management plan	San Diego, CA
	Established regional urban growth areas	Larimer County, CO
	Established growth management controls	Montgomery County, MD
	Growth management plan to provide affordable housing	Palm Beach County, FL
	Established growth limits based on environmental carrying capacity The city is adjacent to the Sanibel Island National Wildlife Refuge	Sanibel, FL
	Eliminated rural zoning by establishing county-wide planning commission	Hardin County, KY
Require Infrastructure Improvements	Development not allowed until municipal services are established in the area	Ramapo, NY
	Public facilities must be developed along with new housing	Broward County, FL
Limit Development to Certain Areas	Development is limited to the area within city boundaries	Minneapolis, MN

12.7 Special Urban Planning Issues

Urban areas present a large number of planning issues. Transportation, open space, and improving the quality of life in the inner city are significant problems.

Urban Transportation Planning

A growing concern of city governments is to develop comprehensive urban transportation plans. While the specifics of such plans might vary from region to region, urban transportation planning usually involves four major goals:

1. Conserve energy and land resources.
2. Provide efficient and inexpensive transportation within the city, with special attention to people who are unable to drive, such as many elderly, young, handicapped, and financially disadvantaged persons.
3. Provide suburban people opportunities to commute efficiently to and from the city.
4. Reduce urban pollution.

Since automobiles are an important means of travel, transportation corridors and parking facilities must be included in any urban transportation plan. This is particularly true in North America where automobile travel dominates. Governments in North America encourage automobile use by financing highways and expressways, by maintaining a cheap energy policy, and by withdrawing support for most forms of mass transportation. Thus, they encourage automobile transportation with hidden subsidies (highway construction and cheap gasoline) but maintain that rail and bus transportation should not be subsidized. North Americans will seek alternatives to private automobile use only when the cost of fuel, the cost of parking, or the inconvenience of driving becomes too high.

However, many urban planners recognize that the automobile's disadvantages (pollution, land devoted to freeways and parking, congestion, noise, etc.) may outweigh its advantages, so some cities, such as Toronto, London, San Francisco, and New York, have attempted to encourage the use of public transport and dissuade automobile use by making it more costly or less convenient (few places to park, high fees for parking, special fees to drive in parts of the city, restricting access during certain parts of the day, etc.). (See Science, Politics, & Policy . . . Reducing Automobile Use in Cities, chapter 8.)

The major urban mass transit systems are railroads, subways, trolleys, and buses. In many parts of the world, mass transportation is extremely efficient and effective. However, in the United States, where the automobile is the primary method of transportation, mass transportation systems are often underfunded and difficult to establish. There are several reasons for this. Some are real and some are perceived.

1. One impediment to the development of mass transit is the dispersed housing patterns typical of many urban areas. Mass transit is most economical along heavily populated routes.

2. In many smaller cities, mass transit is less convenient than the automobile. This is less true in large metropolitan areas where traffic congestion and parking problems are common.

3. Trains and buses are often crowded and uncomfortable during peak travel times.

4. There is a general perception that mass transit is expensive to build and operate compared to highways and freeways. This is very difficult to evaluate. The cost of building and maintaining highways is probably comparable to that of building and maintaining track for passenger trains.

Urban Open Space and Recreation Planning

About 80 percent of the population of North America live in urban areas. These urban dwellers value open space because it breaks up the sights and sounds of the city and provides a place for recreation. Inadequate land-use planning in the past often ignored the value of open space. Until recently, creating a new park within a city was considered an uneconomical use of the land, but people are now beginning to realize the need for parks and open spaces.

Some cities recognized the need for open space a long time ago and allocated land for parks. (See figure 12.12.) London, Toronto, and Perth, Australia, have centrally located and well-used parks. New York City set aside approximately 200 hectares (500 acres) for Central Park in the late 1800s. Boston has developed a park system that provides a variety of urban open spaces. Other cities have not dealt with this need for open space because they have lacked either the foresight or the funding. Large urban centers are discovering that it is important to provide adequate, low-cost recreational opportunities coveniently located to people in neighborhoods. Some of these opportunities take the form of commercial establishments, such as bowling centers, amusement parks, and theaters. Others must be subsidized by the community. (See figure 12.13.) The responsibility for playgrounds, organized recreational activities, and open space typically resides in an arm of the municipal government known as the parks and recreation department. Cities spend millions of dollars to develop and maintain recreation programs. Often, conflict arises over the allocation of financial and land resources. These are closely tied because open land is scarce in urban areas, and it is expensive. Riverfront property is ideal for park and recreational use, but it is also prime land for industry, commerce, or high-rise residential buildings. Although conflict is inevitable, many metropolitan areas are beginning to see that recreational resources may be as important as economic growth for maintaining a healthy community.

An outgrowth of the trend toward urbanization is the development of **nature centers.** In many urban areas, so little natural area is left that the people who live there need to be given opportunities to learn about nature. Nature centers are basically teaching institutions that provide a variety of ways for people to learn about and appreciate the natural world. Zoos, botanical gardens, and some urban parks, combined with interpretative centers, also provide recreational experiences. Nature centers are usually located near urban centers, in places where some appreciation of the natural processes and phenomena can be developed. They may be operated by municipal governments, school systems, or other nonprofit organizations.

Redevelopment of Inner-City Areas

A common problem in many older cities is inner-city decay, which results from the migration of people and businesses from the inner city to the suburbs. Many old industrial sites sit vacant. Businesses moved to suburban malls. The quality of housing has declined. Services have been reduced. To improve the quality of life of the residents of the city, special efforts must be made to revitalize the city. Although activities that will improve the quality of life in the inner city vary from city to city, several land-use processes can help.

One problem that has plagued industrial cities is vacant industrial and commercial sites. Many of these buildings have

Washington Square Park, San Francisco

Quiet Open Space

Central Park, New York City

FIGURE 12.12 Urban Open Space Urban open space takes many forms. Some are elaborate parks with many different kinds of space and uses. Others are simply quiet places where people can read or exercise.

FIGURE 12.13 Urban Recreation In urban areas, recreation often takes the form of sports programs, playgrounds, and walking. Most cities recognize the need for such activities and facilities and develop extensive recreation programs for their citizens.

matched to the intended use of the site. Although an old industrial site with specific contamination problems may not be suitable for housing, it may be redeveloped as a new industrial site, since access to the contamination can be controlled. An old industrial site that has soil contamination may be paved to provide parking.

Another important focus in urban redevelopment is the remodeling of abandoned commercial buildings for shopping centers, cultural facilities, or high-density housing. Chattanooga, Tennessee, has achieved a reputation for revitalizing its inner-city area. The process of revitalization involved extensive planning activities that included the public, public and private funding of redevelopment activities, establishment of an electric bus system to alleviate air pollution, renovation of existing housing, redevelopment of old warehouses into a shopping center, and incorporation of a condemned bridge over the Tennessee River into a portion of a park that is also an important pedestrian connection between a residential area and the downtown business district.

Smart Growth Urban Planning

Though they typically support growth, many cities and communities are questioning the practice of abandoning older communities while converting adjacent open space and prime agricultural lands to urban use. There is an economic cost to both abandoning infrastructure in the city and rebuilding it farther out. They are also questioning the dispersed nature of suburban areas, which relies on automobiles as the major method of transportation.

Smart growth urban planning is a reaction to the twin problems of deteriorating central cities and dysfunctional suburban development. In central cities and older suburbs, these problems include deteriorating infrastructure, poor schools, and a shortage of affordable, quality housing. In newer suburban areas, these problems may include increased traffic congestion, declining air quality, and the loss of open space. Smart growth is an approach that argues that these problems are two sides of the same coin and that the neglect of our central cities is fueling the growth and related problems of the suburbs.

A smart growth approach recognizes that growth will occur but seeks to direct the growth to enhance the quality of the urban environment to make it more "livable" while accommodating larger numbers of people. Livability suggests, among other things, that the quality of our built environment and how well we preserve the natural environment both directly affect our quality of life. In order to accomplish the goal of livability, the cultural, economic, environmental, and social aspects of urban areas must be taken into account. The smart growth network has developed a series of guidelines about smart growth approaches to urban planning.

- Preserve open space, farmland, natural beauty, and critical environmental areas.
- Direct development toward existing urban areas. This encourages the reuse of abandoned or poorly used urban space and discourages sprawl.
- Take advantage of compact building design. Buildings can be designed so that more people can be housed in the same space. Multistory buildings have a smaller footprint than a

remained vacant because the cost of cleanup and renovation is expensive. Such sites have been called **brownfields.** Many of these sites involved environmental contamination, and since the EPA required that they be cleaned up to a pristine condition, no one was willing to do so. A new approach to utilizing these sites is called **brownfields development.** This involves a more realistic approach to dealing with the contamination at these sites. Instead of requiring complete cleanup, the degree of cleanup required is

Science, Politics, & Policy

Urban Farming in Detroit

Detroit, Michigan, is a shrinking city. Nearly 2 million people once lived in Detroit. Fewer than 800,000 remain today. Consequently, about 10,000 hectares (25,000 acres), nearly 30 percent of the city, consist of abandoned properties. Detroit has more than 200,000 vacant parcels—almost half of them residential plots—that generate no significant tax revenue and cost more to maintain than the city can afford. Finding new uses for this land has become one of the most pressing challenges for a city that lost a quarter of its population in the past decade. One plan to reutilize these properties is to develop urban farms on vacant lots.

Several organizations have evaluated the feasibility of a Detroit urban farm project. A study by Michigan State University stated that vacant property could supply a large proportion of the fruits and vegetables needed by the population in the area. This would involve a variety of traditional farming techniques including intensive farming, such as greenhouses that extend the growing season. The American Institute of Architects said that Detroit was well suited to become a pioneer in urban agriculture at a commercial scale. A private company, Hantz Farms, has plans to develop about 85 hectares (200 acres) of underutilized vacant lands and abandoned properties for agricultural uses. Hantz Farms plans to buy about 2,300 parcels. They want to plant oak trees, then maybe fruit orchards and hydroponic vegetables. Cities such as Cleveland, Milwaukee, and Buffalo, New York, have reclaimed some vacant land through small-scale farming projects. But no other major city is dealing with as much empty space as Detroit.

Large-scale farming in Detroit still faces a number of legal, political and logistical challenges, including concerns about soil quality, the price of the land and the impact on neighbors. Some argue that the soil is too toxic to raise healthy crops. The question also is asked about how easy it would be to convince buyers that the produce would be safe to eat. Because lead was used in gasoline and paints, lead contamination of soil is a concern in older cities. In order to assure that soils are safe, it is necessary to test for lead and other contaminants. In some cases, it may make sense to remove the top layer of contaminated soil and replace it.

Local property owners are also concerned that they may be forced to relocate to accommodate the large, contiguous area desired for the farming project.

Other local residents already are involved in local garden projects. They fear that financial support for community gardens will disappear if large farms are developed.

Finally, there are issues of zoning. Agriculture brings with it smells, sounds, and problems with livestock. Thus, the community must develop zoning ordinances that regulate agricultural use of the land so that it is compatible with the urban areas that surround it.

In 2012 Michigan State University agreed to invest $1.5 million to establish an urban-agriculture research center in Detroit. The mayor of Detroit stated, "the goal of the research center would be to demonstrate that innovation based on urban food production can create new businesses and jobs, return idle land to productivity and grow a more environmentally sustainable and economically vital city." A representative from the university said, "intensive indoor urban agriculture is going to be the global solution (for food security) to cities like Mexico City and Sao Paolo, Brazil. Detroit can be an innovator in urban agriculture by developing indoor farming systems that have zero energy loss and 90 percent water efficiency, and then manufacture equipment for those systems and sell them worldwide." What are your thoughts? Is large-scale urban farming an alternative use of land in selective cities? Do the pluses of such a concept outweigh the concerns or is it the other way around?

Urban farming in Detroit. This urban farm was previously a street of abandoned homes.

single-story building with the same floor space. Apartment and condominium buildings are even more efficient in the use of land. These approaches reduce the need to develop new land for housing.

- Create a range of housing opportunities and choices. Housing should provide a variety of styles to accommodate people with different lifestyles, desires, and income levels.

- Foster distinctive, attractive communities with a strong sense of place. Attention to the design of buildings and their relationship with open space and cultural attractions is important in creating a pleasing urban setting.

- Mix land uses. Create neighborhoods with a mixture of housing, commercial services, and parks so that residents do not need to drive somewhere for shopping, dining, or recreation.

- Create walkable neighborhoods. This could involve controlling vehicle access in certain areas, and providing sidewalks, pathways, pedestrian bridges, and other ways to separate vehicle traffic from pedestrians.

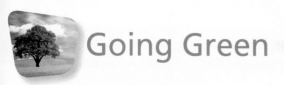

Going Green

Using Green Building Techniques in Urban Planning

Green building practices involve a variety of architectural design, land-use practice, and resource conservation techniques in the construction of buildings and landscapes. These practices can be an important consideration in land-use decision making and can be applied to both renovation of existing sites and new construction.

Green building practices can include high-tech elements such as automatic control of lighting, high-efficiency heat pumps, and photovoltaic cell arrays. They can also include low-tech practices such as orienting buildings to take advantage of sunlight, reusing wastewater, using permeable surfaces in parking lots to allow water to soak into the soil, collecting rainwater for irrigation, and recycling materials from demolition.

Another aspect of green design is to pay attention to the aesthetics of a place. Often what is pleasing to the eye has additional benefits. Trees provide shade but also reduce the amount of heat gained by the surfaces of buildings and parking lots. This is translated into reduced need for cooling a building. Water that runs off buildings or parking lots can be used to provide ponds or other landscape features and at the same time reduce the amount of water entering storm drains.

The Green Building Council estimates that green building reduces energy use by 30 percent, carbon emissions by 35 percent, and water use by 40 percent and reduces the cost of getting rid of waste by 50–90 percent. Although the green building market is expected to grow rapidly in the next decade, a substantial shift from the status quo is needed to make these practices the norm. A particularly significant problem in advancing a green building ethic is the need for developers and planners to recognize the difference between short-term and long-term costs. Many green technologies and designs are expensive but pay off over a period of years due to reduced operating costs. As long as buildings are seen as "temporary," many developers will opt for the least expensive way to build and will ignore the long-term cost.

But things are changing and green building approaches are increasingly seen as part of comprehensive urban development programs that can reduce energy, waste disposal, and wastewater management costs while at the same time providing a more livable city.

- Provide a variety of transportation choices including walking, biking, and easy access to public transportation.
- Encourage community and stakeholder collaboration in development decisions. Because changes in land use involve owners and users of current lands and buildings, owners and users need to be involved in discussions about the purpose and costs of proposed changes.
- Make development decisions predictable, fair, and cost effective. In order to obtain community support, people must be able to see that the proposed changes will be fair to those involved—current users, landowners, and business owners—and that the cost is reasonable.

An example of the principles of smart growth being applied is Suisun City, California. In 1989, the *San Francisco Chronicle* rated Suisun City, a town of 25,000 people between San Francisco and Sacramento, the worst place to live in the Bay Area. At that time, Suisun City's historic Main Street was a strip of boarded-up storefronts and vacant lots. Several blocks away, an oil refinery sat at the head of the polluted, silt-laden Suisun Channel. Today, Suisun's harbor is filled with boats and lined with small businesses. A train and bus station that connects the city to the rest of northern California is located nearby. The town is diverse, walkable, and picturesque. Its crime rate is low and its housing affordable. This dramatic change occurred because Suisun City's residents, businesses, and elected officials agreed on a common vision for their town's future. Cleaning up the polluted Suisun Channel and making the waterfront a focal point of their town was a common goal. The citizens also wanted to reestablish the historic Main Street as a social and retail gathering place. It was recognized that by encouraging tax-generating commercial development such as retail shops and restaurants along Main Street and the waterfront, municipal finances would be strengthened.

The Mizner Park redevelopment in Boca Raton, Florida, is a good example of the use of current space to meet new purposes. A failed shopping mall was torn down and the site was used to provide high-density housing, an outdoor performance space, a central tree-lined boulevard with pedestrian walkways in the middle, wide sidewalks that front retail shops, restaurants, office space, and entertainment venues. It has become a new community meeting place in the middle of a large urban area.

12.8 Federal Government Land-use Issues

The U.S. government owns and manages about 30 percent of U.S. land. These lands are generally specified for certain uses which include military reserves, national parks, Indian reservations, forests, and rangelands. They are managed by several government agencies including the Forest Service, the Bureau of Land Management, the National Park Service, and the Fish and Wildlife Service. The National Park Service and the Fish and Wildlife Service are primarily involved in preserving lands for their designated purposes. However, the Forest Service and the Bureau of Land Management manage land for multiple uses. The Forest Service is responsible for managing the harvesting of trees as well as hunting and other recreational uses. In many areas of the west, they also manage grazing rights. The Bureau of Land Management controls huge territories in the west, primarily rangelands, and lease much of this land to ranchers and mining interests. However, they are also involved in managing recreational use of their lands.

The 1960 Multiple Use Sustained Yield Act divided the use of national forests into four categories: wildlife habitat preservation, recreation, lumbering, and watershed protection. This act was designed to encourage both economic and recreational use of the forests. However, specific users of this public land are often in conflict, particularly recreational users with timber harvesters.

Today, one of the major uses for public lands is outdoor recreation. Many people want to use the natural world for recreational purposes because nature can provide challenges that may be lacking in their day-to-day lives. Whether the challenge is hiking in the wilderness, hunting or fishing, exploring underwater, climbing mountains, or driving a vehicle through an area that has no roads, these activities offer a sense of adventure. (See table 12.3.) All of these activities use the out-of-doors but not in the same way. Conflicts develop because some activities cannot occur in the same place at the same time. For example, wilderness camping and backpacking often conflict with off-road vehicles; and hunting and fishing activities conflict with grazing rights or wilderness activities.

A basic conflict exists between those who prefer to use motorized vehicles and those who prefer to use muscle power in their recreational pursuits. (See figure 12.14.) This conflict is particularly strong because both groups would like to use the same public

Table 12.3 Number of People Who Participated in Selected Outdoor Recreational Activities in 2013

Activity	Percent of Population over 7 Years of Age Participating
Exercise walking	33
Swimming	27
Bicycling	24
Fishing	15
Camping	18
Golf	14
Hiking	10
Running	8
Hunting	6
Backpacking	4
Skiing	4

Source: Data from *Statistical Abstract of the United States,* 2013.

land. Both have paid taxes, and both feel that it should be available for them to use as they wish. Finally, as more rangelands and forests have had vehicular access controlled or eliminated, those who

FIGURE 12.14 Conflict Over Recreational Use of Land and Water The people who use land or water for motorized and nonmotorized activities are often antagonists over the allocation for recreational use.

want to use public lands for motorized recreation have become upset.

Land-use conflicts also arise between business interests and recreational users of public lands. Federal and state governments give special use permits to certain users of public lands. Many ski resorts in the West make use of public lands. Grazing is also an important use of federal land. Based on "Animal Use Months" established by the Bureau of Land Management or the Forest Service, ranchers are allowed to graze cattle on certain public lands. Technically, failure to comply can mean a loss of grazing rights. However, since the establishment of regulations is highly political, many maintain that the political influence of ranchers allows them to use a public resource without adequately compensating the government. In addition, the regulatory agencies are understaffed and find it difficult to adequately regulate the actions of individual ranchers. As a result, some lands are overgrazed. Many people who want to use the publicly owned rangelands for outdoor recreation resent the control exercised by grazing interests. On the other hand, ranchers resent the intrusion of hikers and campers on land they have traditionally controlled.

A particularly sensitive issue is the designation of certain lands as wilderness areas. Obviously, if an area is to be wilderness, human activity must be severely restricted. This means that the vast majority of Americans will never see or make use of it. Many people argue that this is unfair because they are paying

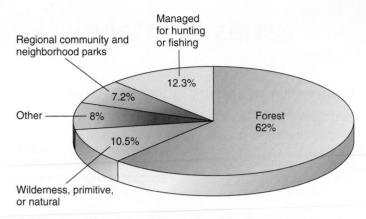

FIGURE 12.15 U.S. Federal Recreational Lands Of the approximately 108 million hectares (267 million acres) of federal recreational lands in the United States, approximately 10 percent is designated as wilderness, primitive, or natural.

taxes to provide recreation for a select few. Others argue that if everyone were to use these areas, their charm and unique character would be destroyed and that, therefore, the cost of preserving wilderness is justifiable. Areas designated as wilderness make up a very small proportion of the total public land available for recreation. (See figure 12.15.)

Summary

Historically, waterways served as transportation corridors that allowed for the exploration of new land and for the transport of goods. Therefore, most large urban centers began as small towns located near water. Water served the needs of the towns in many ways, especially as transportation. Several factors resulted in the shift of the population from rural to urban. These included the Industrial Revolution, which provided jobs in cities, and the addition of foreign immigrants to the cities. As towns became larger, the farmland surrounding them became suburbs surrounding industrial centers. Unregulated industrial development in cities resulted in the degradation of the waterfront and stimulated the development of suburbs around the city as people sought better places to live and had the money to purchase new homes. The rise in automobile ownership further stimulated the movement of people from the cities to the suburbs.

Many problems have resulted from unplanned growth. The building of new housing, commercial, and industrial sites on farmland results in the loss of valuable agricultural land. Floodplains and wetlands are often mismanaged. Loss of property and life results when people build on floodplains. Wetlands protect our shorelines and provide a natural habitat for fish and wildlife. Transportation problems and lack of open space are also typical in many large metropolitan areas.

Land-use planning involves gathering data, projecting needs, and developing mechanisms for implementing the plan. Good land-use planning should include assessment of the unique

geologic, geographic, biological, and historic and cultural features of the land; the costs of providing additional infrastructure; preservation of open space; provision for a variety of transportation options; a mixture of housing and service establishments; redevelopment of disused urban land; and establishment of urban growth limits. Establishing regional planning agencies, purchasing land or its development rights, and enacting zoning ordinances are ways to implement land-use planning. The scale of local planning is often not large enough to be effective because problems may not be confined to political boundaries. State or regional planning units can afford professional planners and are better able to withstand political and economic pressures. A growing concern of urban governments is to develop comprehensive urban transportation plans that seek to conserve energy and land resources, provide efficient and inexpensive transportation and commuting, and help to reduce urban pollution. Urban areas must also provide recreational opportunities for their residents and seek ways to rebuild decaying inner cities.

Federal governments own and manage large amounts of land; therefore, national policy must be developed. This usually involves designating land for particular purposes, such as timber production, grazing land, parks, or wilderness. The recreational use of public land often requires the establishment of rules that prevent conflict between potential users who have different ideas about what appropriate uses should be. Often federal policy is a compromise between competing uses and land is managed for multiple uses.

Smart Communities' Success Stories

Cool Cities—Michigan

The Cool Cities initiative launched by the Michigan governor is an urban redevelopment endeavor aimed at attracting creative young workers to the state's cities. A series of forums throughout the state—and an online survey—collected input on ways to revitalize cities, strengthen the economy, and make Michigan a magnet state for new job creation and business expansion.

Learning Corridor—Hartford, Connecticut

Trinity College helped revitalize its neighborhood by building a $175 million "Learning Corridor." The development combines a neighborhood school, job training facility, playing fields, and open space to revitalize what had been an urban area in serious decline. The "Learning Corridor" has become a model for other communities that want to avoid "school sprawl" that pushes neighborhood schools to suburbs accessible only by car.

Civano—Tucson, Arizona

The 700-hectare (1,700-acre) neighborhood development integrates housing with shopping, workplace, school, and civic facilities so that most places are within walking distance of one another. Its buildings incorporated solar design and water conservation features.

Stapleton Development Corporation—Boulder, Colorado

This is the redevelopment of the former Stapleton International Airport site, near Denver. The project involves constructing a community of urban villages, employment centers, and greenways that focus on sustainability, environmental preservation, and economic and social development.

Jackson Meadow—Marine on St. Croix, Minnesota

This 64-home residential development focuses on ecological land use. Jackson Meadow has a "commitment to create a sustainable environment that respects the unique nature of this special place." The development is combined with more than 100 hectares (250 acres) of open space for people and wildlife and includes a communally constructed wetlands and natural ponds for water filtration and runoff.

Grand Valley Metropolitan Council—Grand Rapids, Michigan

Grand Valley Metropolitan Council (GVMC) is an alliance of governmental units in the Grand Rapids, Michigan, metropolitan area that are appointed to plan for growth and development, improve the quality of community life, and coordinate governmental services. Through joint planning they coordinate land-use planning with transportation and other systems.

The Vail Environmental Strategic Plan—Vail, Colorado

This plan describes a program that was adopted to maintain and improve environmental quality in the Vail Valley and to ensure the prolonged economic health of the region. Efforts include monitoring and improving air and water quality, preserving open space, and protecting the area's wildlife.

Community Greens—Arlington, Virginia

This concept promotes more livable cities through the creation of parks that are collectively owned and managed by the neighbors whose homes and backyards, decks, patios, and balconies enclose the green.

Examples of parks include redeveloping alleys as green space and action by developers to create a common backyard green space when they redevelop older urban communities. These parks provide accessible and safe play spaces for children, shared space for neighbors to meet, and increased property values.

After reading these examples:

- Can you identify characteristics of your community that could be considered "smart"?
- Can you think of innovative ways to help your community grow smarter?
- What barriers would need to be overcome to make your community a "smart community"?

Acting Green

1. Visit your city's website and locate its land-use planning page. What are the city's major rules associated with land use?
2. As you travel to class, observe changes that are occurring to land use. What is being lost with the change? What is being gained from the change?
3. Attend a local meeting of the planning or zoning board.
4. Be an advocate and user of public transportation in your community.
5. Become active in an organization that is promoting good land use in your community.
6. Visit several of the parks or green spaces in your community. Visit them during different times of the year.

Review Questions

1. List three reasons why land-use planning is necessary.
2. Why did urban centers develop near waterways?
3. List three factors that encouraged people to move from rural farms to cities in the 1800s.
4. List three factors that encouraged people to leave the cities for suburbs.
5. List three physical and three social consequences of suburban sprawl.
6. Describe tract development, ribbon sprawl, and leapfrog development.
7. What is a megalopolis?
8. Describe how each of the following factors contributed to the development of suburban sprawl: automobiles, home ownership, tax laws, local zoning ordinances, federal government policies.
9. State three consequences of the dominance of the automobile as a means of transport in urban areas.
10. What characteristics of suburbs contribute to a loss of sense of community?
11. What characteristics of suburbs contribute to high infrastructure and high energy costs?
12. How does the development of suburbs around a city contribute to loss of open space, and air and water pollution problems?
13. What land uses are suitable on floodplains?
14. Describe three values provided by wetlands.
15. Why is an understanding of the geology and resource base of an area important in land-use planning?
16. List eight principles that should be involved in good land-use planning.
17. What role do state and regional planning, purchasing of land, and use restrictions play in implementing land-use plans?
18. Describe three common land-use issues cities must deal with.
19. List ten common smart growth principles.
20. List four major federal agencies that manage federal lands and their primary responsibilities.
21. Give examples of conflict over the use of federally owned property.

Critical Thinking Questions

1. Choose the city where you live. Interview local residents and look at old city maps. What did the city look like 75 years ago? What were the city's boundaries? Where did people do their shopping? How did they get around? How does this compare with the current situation in the city?
2. What historical factors brought members of your family to the city? How does this compare to the factors that are currently contributing to the growth of cities in the developing world?
3. Consider the outer rim of the city closest to you. Which, if any, of the problems associated with unplanned growth are associated with your city? What factors make them a problem? What do you think can be done about them?
4. There has been tremendous development in the arid West of the United States over the past few decades, creating demands for water. How should these demands be met? Should there be limits to this type of development? What kinds of limits, if any?
5. Imagine you are a U.S. Forest Service supervisor who is creating a ten-year plan that is in the public comment stage. What interests would be contacting you? What power would each interest have? How would you manage the competing interests of timber, mining, grazing, and recreation or those between motorized and nonmotorized recreation? What values, beliefs, and perspectives help you to form your recommendations?
6. Imagine that you lived in an area of the country that has the potential to be named a wilderness area. What conflicts do you think would arise from such a declaration? Who might be some of the antagonists? Which perspective do you think is most persuasive? How would you answer the objections of the other perspective?

Soil and Its Uses

Soil is a resource that supports plants, which are the base of most food chains. We should think of soil as a valuable resource that should be preserved.

CHAPTER OUTLINE

OBJECTIVES

After reading this chapter, you should be able to:

- Describe how the study of soil science developed.
- Describe the geologic processes that build and erode the Earth's surface.
- List the physical, chemical, and biological factors involved in soil formation.
- Explain the importance of humus to soil fertility.
- Differentiate between soil texture and soil structure.

- Explain how texture and structure influence soil atmosphere and soil water.
- Explain the role of living organisms in soil formation and fertility.
- Describe the various layers in a soil profile.
- Describe the processes of soil erosion by water and wind.
- Explain how contour farming, strip farming, terracing, waterways,

windbreaks, and conservation tillage reduce soil erosion.
- Understand that the misuse of soil reduces soil fertility, pollutes streams, and requires expensive remedial measures.
- Explain how land not suited for cultivation may still be productively used for other purposes.

The Living Soil

Many words have been written about the living soil. Many were scientific and factual, some were emotional and moving, others creative. But in generation after generation, it is important to bring attention back to the central theme—the soil sustains all life on Earth. Without the soil, nothing lives. Healthy soils support healthy environments, and healthy environments support healthy life. When you are taking your next walk (away from concrete) think of the millions of beneficial organisms that are going through their daily routine of eating, breathing, living, and dying in the soil beneath your feet. One cup of fertile soil may contain as many bacteria as there are people on Earth.

We eat the food, drink the water, breathe the air, and enjoy the views, but only a few of us walk the fields and forests on a regular basis and understand what those lands need from us in order to sustain the living soil. The fate of Earth's land and waters will be determined by how people use them to meet their daily needs. Only if we can satisfy today's needs without reducing the opportunity for coming generations to meet their needs will we meet the test of sustainability.

Most of us do not live on the land or work with it daily. But we can seek opportunities to involve ourselves in a soil-healing effort somewhere in our community. If we are suburbanites, we can compost household and yard waste and apply only the proper rates of fertilizers, pesticides, and herbicides. If we are farmers, we can adopt conservation tillage and other conservation practices. If we raise livestock, we can create an effective nutrient management system that turns mountains of manure into a useful byproduct. We can also make our voices heard in the public forum. By becoming active in local political activities we can communicate our concerns to political decision makers. You may find ways to work within your church or community organization to guide mission and outreach efforts toward long-term solutions such as those that protect and build soil quality around the world.

In the final analysis, each of us can take some kind of active role in soil stewardship. That role will take many forms, but it must have the effect of building, restoring, and improving the world that we touch.

13.1 The Study of Soil as a Science

Soil science is the study of soil as a natural resource on the surface of the Earth. The study includes soil formation; classification and mapping; physical, chemical, biological, and fertility properties of soils; and these properties in relation to the use and management of soils. While the field of soil science is relatively new, the impact of soils on civilization is timeless.

For six millennia we have tilled, drained, and irrigated soils for agriculture. For even longer, humans have used soils as a construction material. Four thousand years ago, the Chinese were classifying soils according to their productivity and using those classes as a basis for tax assessment. Throughout early human history, classification systems were developed to distinguish among the soils that are important to human lives.

The study of soil as a science is a recent advancement. The fields of chemistry, geography, and geology began to emerge following the Renaissance, chemistry being accepted as a science in the seventeenth century. In the eighteenth century, geographers were classifying and mapping soils in Great Britain. In 1862, Friedrich Fallou coined the term *pedology* for the scientific study of soils. In 1876, an interdisciplinary commission was organized in Russia to study the chernozioum soil. Through this collaboration, the geologist-geographer Vasili Dokuchaev developed the fundamentals of soil investigation. In 1941, Hans Jennys presented the conceptual equation $S = f(cl, o, r, p, t, \ldots)$, where (S) is a function of climate (cl), organisms (o), relief or topography (r), parent material (p), time (t), and unspecified factors (…), one of which includes human activities. His equation synthesized the concepts of the time and formed a paradigm of soil science that is still followed today. In order to understand soil and its uses, it is important to begin with the basic geologic processes that formed the Earth.

13.2 Geologic Processes

We tend to think of the Earth as being stable and unchanging until we recognize such events as earthquakes, volcanic eruptions, floods, and windstorms changing the surface of the places we live. There are forces that build new land and opposing forces that tear it down. Much of the building process involves shifting of large portions of the Earth's surface known as *plates*.

The Earth is composed of an outer crust, a thick layer of plastic mantle, and a central core. The **crust** is an extremely thin, less dense, solid covering over the underlying mantle. The **mantle** is a layer that makes up the majority of the Earth and surrounds a small core made up primarily of iron. The outermost portion of the mantle, adjacent to the crust, is solid. Collectively, the crust and solid outer mantle are known as the **lithosphere.** Just below the outer mantle is a thin layer known as the **asthenosphere,** which is capable of plastic flow. Below the asthenosphere, the mantle is more solid. The core consists primarily of iron and nickel and has a solid center and a liquid outer region. (See figure 13.1.)

Plate tectonics is the concept that the outer surface of the Earth consists of large plates composed of the crust and the outer portion of the mantle and that these plates are slowly moving over the surface of the liquid outer mantle. The heat from the Earth causes slow movements of the outer layer of the mantle similar to what happens when you heat a liquid on the stove, only much slower. The movements of the plates on this plastic outer layer of the mantle are independent of each other. Therefore, some of the plates are pulling apart from one another, while others are colliding.

Where the plates are pulling apart from one another, the liquid mantle moves upward to fill the gap and solidifies. Thus new crust is formed from the liquid mantle. Approximately half of the surface of the Earth has been formed in this way in the past 200 million years.

If plates are pulling apart on one portion of the Earth, they must be colliding elsewhere. Where plates collide, several things can happen. (See figure 13.2.) Often, one of the plates slides under the other and is melted. When this occurs frequently some of the liquid mantle makes its way to the surface and volcanoes are formed, which results in the formation of mountains. The west coasts of North and South America have many volcanoes and mountain ranges where the two plates are colliding. The volcanic activity adds new material to the crust. When a collision occurs between two plates under the ocean, the volcanoes may eventually reach the surface and form a chain of volcanic islands, such as can be seen in the Aleutian Islands and many of the Caribbean Islands. When two continental plates collide, neither plate slides under the other and the crust buckles to form mountains. The Himalayan, Alp, and Appalachian mountain ranges are thought to have formed from the collision of two continental plates. All of these movements of the Earth's surface are associated with earthquakes. The movements of the plates are not slow and steady sliding movements but tend to occur in small jumps. However, what is a small movement between two plates on the Earth is a huge movement for the relatively small structures and buildings produced by humans, so these small movements can cause tremendous amounts of damage.

These building processes are counteracted by processes that tend to make the elevated surfaces lower. Gravity provides a force that tends to wear down the high places. Moving water and ice (glaciers) and wind assist in the process; however, their

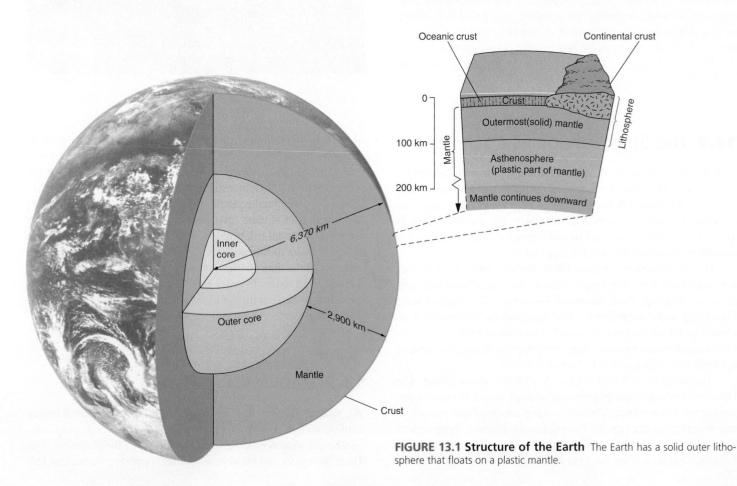

FIGURE 13.1 Structure of the Earth The Earth has a solid outer lithosphere that floats on a plastic mantle.

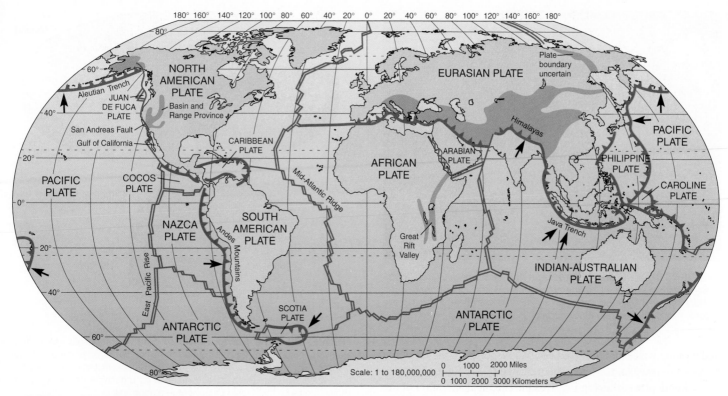

FIGURE 13.2 **Tectonic Plates** The plates that make up the outer surface of the Earth move with respect to one another. They pull apart in some parts of the world and collide in other parts of the world.

effectiveness is related to the size of the rock particles. Several kinds of **weathering** processes are important in reducing the size of particles that can then be dislodged by moving water and air. **Mechanical weathering** results from physical forces that reduce the size of rock particles without changing the chemical nature of the rock. Common causes of mechanical weathering are changes in temperature that tend to result in fractures in rock, the freezing of water into ice that expands and tends to split larger pieces of rock into smaller ones, and the actions of plants and animals.

Because rock does not expand evenly, heating a large rock can cause it to fracture, so that pieces of the rock flake off. These pieces can be further reduced in size by other processes, such as the repeated freezing of water and thawing of ice. Water that has seeped into rock cracks and crevices expands as it freezes, causing the cracks to widen. Subsequent thawing allows more water to fill the widened cracks, which are enlarged further by another period of freezing. Alternating freezing and thawing breaks large rock pieces into smaller ones. (See figure 13.3.) The roots of plants growing in cracks can also exert enough force to break rock.

The physical breakdown of rock is also caused by forces that move and rub rock particles against each other (abrasion). For example, a glacier causes rock particles to grind against one another, resulting in smaller fragments and smoother surfaces. These particles are deposited by the glacier when the ice melts. In many parts of the world, the parent material from which soil is formed consists of glacial deposits. Wind and moving water also cause small particles to collide, resulting in further weathering.

The smoothness of rocks and pebbles in a stream or on the shore is evidence that moving water has caused them to rub together, removing their sharp edges.

Wind and moving water also remove small particles and deposit them at new locations, exposing new surfaces to the weathering process. For example, the landscape of the Grand Canyon in the southwest United States was created by a combination of wind

FIGURE 13.3 **Physical Fragmentation by Freezing and Thawing** The crack in the rock fills with water. As the water freezes and becomes ice, it expands. The pressure of the ice enlarges the crack. The ice melts, and water again fills the crack. The water freezes again and widens the crack. Alternate freezing and thawing splits the rock into smaller fragments.

Soil and Its Uses   293

FIGURE 13.4 An Eroded Landscape This landscape in Grand Canyon National Park was created by the action of wind and moving water. The particles removed by these forces were deposited elsewhere and may have become part of the soil in that new location.

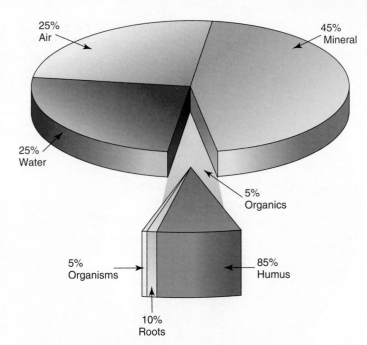

FIGURE 13.5 The Components of Soil Although soils vary considerably in composition, they all contain the same basic components: mineral material, air, water, and organic material. The organic material can be further subdivided into humus, roots, and other living organisms. The percentages shown are those that would be present in a good soil.

and moving water that removed easily transported particles, while rocks more resistant to weathering remained. (See figure 13.4.)

Chemical weathering involves the chemical alteration of the rock in such a manner that it is more likely to fragment or to be dissolved. Some small rock fragments exposed to the atmosphere may be oxidized; that is, they combine with oxygen from the air and chemically change to different compounds. Other kinds of rock may combine with water molecules in a process known as *hydrolysis*. Often, the oxidized or hydrolyzed molecules are more readily soluble in water and, therefore, may be removed by rain or moving water. Rain is normally slightly acidic, and the acid content helps dissolve rocks.

Because of gravity, the prevailing movement of particles is from high elevations to lower ones. This process of loosening and redistributing particles is known as **erosion.** Wind can move sand and dust and can cause the wearing away of rocky surfaces by sandblasting their surfaces. Glaciers can move large rocks and cause their surfaces to be rounded by being rubbed against each other and the surface of the Earth. Moving water transports much material in streams and rivers. In addition, wave action along the shores of lakes and the coasts of oceans constantly wears away and transports particles.

13.3 Soil and Land

The geologic processes just discussed are involved in the development of both soil and land; however, soil and land are not the same. **Land** is the part of the world not covered by the oceans. **Soil** is a thin covering over the land consisting of a mixture of minerals, organic material, living organisms, air, and water that together support the growth of plant life. The proportions of the soil components vary with different types of soils, but a typical, "good" agricultural soil is about 45 percent mineral, 25 percent air,

25 percent water, and 5 percent organic matter. (See figure 13.5.) This combination provides good drainage, aeration, and organic matter. Farmers are particularly concerned with soil because the nature of the soil determines the kinds of crops that can be grown and which farming methods must be employed. Urban dwellers should also be concerned about soil because its health determines the quality and quantity of food they will eat. If the soil is so abused that it can no longer grow crops or if it is allowed to erode, degrading air and water quality, both urban and rural residents suffer.

13.4 Soil Formation

Soils and their horizons differ from one another, depending on how and when they are formed. Soil scientists use five soil factors to explain how soils form and to help them predict where different soils may occur. The scientists also allow for additions and removal of soil material and for activities and changes within the soil that continue each day.

Soil Forming Factors

Parent material Few soils weather directly from the underlying rocks. These "residual" soils have the same general chemistry as the original rocks. More commonly, soils form in materials that have moved in from elsewhere. Materials may have moved a great distance or only a short distance. Windblown "loess" is common in the Midwest. It buries "glacial till" in many areas. Glacial till is material ground up and moved by glaciers. The material in which soils form is called **parent material.** In the lower part of the soils, these materials may be relatively unchanged from when they were deposited by moving water, ice, or wind.

Sediments along rivers have different textures, depending on whether the stream moves quickly or slowly. Fast-moving water leaves gravel, rocks, and sand. Slow-moving water and lakes leave fine textured material (clay and silt) when sediments in the water settle out.

Climate   Soils vary depending on the climate. Temperature and moisture amounts cause different patterns of weathering and leaching. Wind redistributes sand and other particles especially in arid regions. The amount, intensity, timing, and kind of precipitation influence soil formation. Seasonal and daily changes in temperature affect moisture effectiveness, biological activity, rates of chemical reactions, and kinds of vegetation.

Topography   Slope and aspect affect the moisture and temperature of soil. Steep slopes facing the sun are warmer, just like the south-facing side of a house. Steep soils may be eroded and lose their topsoil as they form. Thus, they may be thinner than the more nearly level soils that receive deposits from areas upslope. Deeper, darker colored soils may be expected on the bottomland.

Biological factors   Plants, animals, microorganisms, and humans affect soil formation. Animals and microorganisms mix soils and form burrows and pores. Plant roots open channels in the soils. Different types of roots have different effects on soils. Grass roots are "fibrous" near the soil surface and easily decompose, adding organic matter. Microorganisms affect chemical exchanges between roots and soil. Humans can mix the soil so extensively that the soil material is again considered parent material.

The native vegetation depends on climate, topography, and biological factors plus many soil factors such as soil density, depth, chemistry, temperature, and moisture. Leaves from plants fall to the surface and decompose on the soil. Organisms decompose these leaves and mix them with the upper part of the soil. Trees and shrubs have large roots that may grow to considerable depths.

Time   Time for all these factors to interact with the soil is also a factor. Over time, soils exhibit features that reflect the other forming factors. Soil formation processes are continuous. Recently deposited material, such as the deposition from a flood, exhibits no features from soil development activities. The previous soil surface and underlying horizons become buried. Terraces above the active floodplain, while genetically similar to the floodplain, are older land surfaces and exhibit more development features.

These soil forming factors continue to affect soils even on "stable" landscapes. Materials are deposited on their surface, and materials are blown or washed away from the surface. Additions, removals, and alterations are slow or rapid, depending on climate, landscape position, and biological activity.

A combination of physical, chemical, and biological events acting over time is responsible for the formation of soil. Soil building begins with the fragmentation of the parent material, which consists of ancient layers of rock or more recent geologic deposits from lava flows or glacial activity. The kind and amount of soil developed depend on the kind of parent material present, the plants and animals present, the climate, the time involved, and the slope of the land. As was discussed earlier, the breakdown of parent material is known as weathering. The climate and chemical nature of the rock material greatly influence the rate of weathering. Similarly, the size and chemical nature of the particles have a great impact on the nature of the soil that will develop in an area.

The role of organisms in the development of soil is also very important. The first organisms to gain a foothold in this modified parent material also contribute to soil formation. Lichens often form a pioneer community that grows on the surface of rocks and traps small particles. The decomposition of dead lichens and other organic matter releases acids that chemically alter the underlying rock, causing further fragmentation. The release of chemicals from the roots of plants causes further chemical breakdown of rock particles.

The organic material resulting from the decay of plant and animal remains is known as **humus.** It is a very important soil component that accumulates on the surface and ultimately becomes mixed with the top layers of mineral particles. This material contains nutrients that are taken up by plants from the soil. Humus also increases the water-holding capacity and the acidity of the soil so that inorganic nutrients, which are more soluble under acidic conditions, become available to plants. Humus also tends to stick other soil particles together and helps to create a loose, crumbly soil that allows water to soak in and permits air to be incorporated into the soil. Compact soils have few pore spaces, so they are poorly aerated, and water has difficulty penetrating, so it runs off.

Burrowing animals, soil bacteria, fungi, and the roots of plants are also part of the biological process of soil formation. One of the most important burrowing animals is the earthworm. One hectare (2.47 acres) of soil may support a population of 500,000 earthworms that can process as much as 9 metric tons of soil a year. These animals literally eat their way through the soil, resulting in further mixing of organic and inorganic material, which increases the amount of nutrients available for plant use. They often bring nutrients from the deeper layers of the soil up into the area where plant roots are more concentrated, thus improving the soil's fertility. Soil aeration and drainage are also improved by the burrowing of earthworms and other small soil animals, such as nematodes, mites, pill bugs, and tiny insects. Small soil animals also help to incorporate organic matter into the soil by collecting dead organic material from the surface and transporting it into burrows and tunnels. When the roots of plants die and decay, they release organic matter and nutrients into the soil and provide channels for water and air.

Fungi and bacteria are decomposers and serve as important links in many mineral cycles. (See chapter 5.) They, along with animals, improve the quality of the soil by breaking down organic material to smaller particles and releasing nutrients.

Climate and time are also important in the development of soils. In general, extremely dry or cold climates develop soils very slowly, while humid and warm climates develop them more rapidly. Cold and dry climates have slow rates of accumulation of organic matter needed to form soil. Furthermore, chemical weathering proceeds more slowly at lower temperatures and in the absence of water. Under ideal climatic conditions, soft parent material may develop into a centimeter (less than 1/2 inch) of soil within 15 years. Under poor climatic conditions, a hard parent material may require hundreds of years to develop into that much soil. In any case, soil formation is a slow process. (See table 13.1.)

The amount of rainfall and the amount of organic matter influence the pH of the soil. In regions of high rainfall, basic ions such as calcium, magnesium, and potassium are leached from the soils, and more acid materials are left behind. In addition, the decomposition of organic matter tends to increase the soil's acidity. Soil pH is important, since it influences the availability of nutrients, which affects the kinds of plants that will grow, which affects the amount of organic matter added to the soil. Since calcium, magnesium, and potassium are important plant nutrients, their loss by leaching reduces the fertility of the soil. Excessively acidic soils also cause aluminum ions to become soluble, which in high amounts are toxic to many plants. (See the discussion of acid rain in chapter 16.) Most plants grow well in soils with a pH between 6 and 7, although some plants such as blueberries and potatoes grow well in acidic soils. In most agricultural situations, the pH of the soil is adjusted by adding chemicals to the soil. Lime can be added to make soils less acid, and acid-forming materials such as sulfates can be added to increase acidity.

13.5 Soil Properties

Soil properties include soil texture, structure, atmosphere, moisture, biotic content, and chemical composition. **Soil texture** is determined by the size of the mineral particles within the soil. The largest soil particles are gravel, which consists of fragments larger than 2.0 millimeters in diameter. Particles between 0.05 and 2.0 millimeters are classified as sand. Silt particles range from 0.002 to 0.05 millimeter in diameter, and the smallest particles are clay particles, which are less than 0.002 millimeter in diameter.

Large particles, such as sand and gravel, have many tiny spaces between them, which allow both air and water to flow through the soil. Water drains from this kind of soil very rapidly, often carrying valuable nutrients to lower soil layers, where they are beyond the reach of plant roots. Clay particles tend to be flat and are easily packed together to form layers that greatly reduce the movement of water through them.

However, rarely does a soil consist of a single size of particle. Various particles are mixed in many different combinations, resulting in many different soil classifications. (See figure 13.6.) An ideal soil for agricultural use is a **loam,** which combines the good aeration and drainage properties of large particles with the nutrient-retention and water-holding ability of clay particles.

Table 13.1 In the Time It Took to Form 1 Inch of Soil

2012	Superstorm Sandy devastates parts of the northeast United States
2011	World population reaches 7 billion
2005	Hurricane Katrina devastates U.S. Gulf Coast region
2001	Terrorist attacks in U.S. kill thousands
1989	Berlin Wall torn down signaling an end to the Cold War
1975	U.S. Armed Forces withdraw from Vietnam
1973	Congress passes Endangered Species Act
1970	First "Earth Day" observed
1969	American astronaut Neil Armstrong walks on moon
1964	Rev. Martin Luther King, Jr., receives Nobel Peace Prize in Oslo, Norway
1963	President John F. Kennedy assassinated in Dallas, Texas
1961	Soviet cosmonaut Yuri Gagarin becomes first human to orbit Earth
1953	Korean War ends
1952	Dr. Jonas Salk successfully tests polio vaccine
1945	World War II ends
1935	U.S. Soil Conservation Service established
1934	Drought leads to severe dust storms in "Dust Bowl" region of Great Plains
1918	World War I ends
1903	Orville Wright makes first successful flight in self-propelled airplane
1877	Thomas Edison invents phonograph
1875	Alexander Graham Bell invents telephone
1838	Samuel F. B. Morse develops code for electric telegraph systems
1829	Louis Braille publishes system of writing for the blind
1804	Lewis and Clark begin "Corps of Discovery" expedition to find a water route across North America
1789	French Revolution begins with attack on the Bastille
1776	Continental Congress adopts Declaration of Independence
1741	Astronomer Anders Celcius introduces Centigrade temperature scale
1732	George Washington born
1724	Daniel Gabriel Fahrenheit, inventor of mercury thermometer, devises Fahrenheit temperature scale
1689	Peter the Great becomes tsar of Russia
1650	World population estimated at 500 million
1631	First newspaper, Gazette de France, published in Paris
1620	Pilgrim leaders at Plymouth Colony establish a governing authority through Mayflower Compact
1564	Shakespeare and Galileo born
1513	Juan Ponce de Leon becomes first European to set foot in Florida
1512	Michelangelo finishes painting Sistine Chapel ceiling
1507	Martin Waldseemuller, German geographer, first to call the New World "America"

Adapted from the NRCS poster PI-06, 2006.

Soil structure is different from its texture. **Soil structure** refers to the way various soil particles clump together. The particles in sandy soils do not attach to one another; therefore, sandy soils have a granular structure. The particles in clay soils tend to stick to one another to form large aggregates. Other soils that have a mixture of particle sizes tend to form smaller aggregates. A good soil is **friable,** which means that it crumbles easily. The soil structure and its moisture content determine how friable a soil is. Sandy soils are very friable, while clay soils are not. If clay soil is worked when it is too wet, it can stick together in massive blocks that will be difficult to break up.

A good soil for agricultural use will crumble and has spaces for air and water. In fact, the air and water content depends on the presence of these pore spaces. (See figure 13.7.) In good soil, about one-half to two-thirds of the spaces contain air after the excess water has drained. The air provides a source of oxygen for plant root cells and all the other soil organisms. The relationship between the amount of air and water is not fixed. After a heavy rain, most of the spaces may be filled with water and less oxygen is available to plant roots and other organisms. If some of the excess water does not drain from the soil, the plant roots may die from lack of oxygen. They are literally drowned. On the other hand, if there is not enough soil moisture, the plants wilt from lack of water. Soil moisture and air are also important in determining the numbers and kinds of soil organisms.

Protozoa, nematodes, earthworms, insects, algae, bacteria, and fungi are typical inhabitants of soil. (See figure 13.8.) The role of protozoa in the soil is not firmly established, but they seem to act as parasites and predators on other forms of soil organisms and, therefore, help to regulate the populations of those organisms. Nematodes, which are often called wireworms or roundworms, may aid in the breakdown of dead organic matter. Some nematodes are parasitic on the roots of plants. Insects and other soil arthropods contribute to the soil by forming burrows and consuming and fragmenting organic materials, but they are also major crop pests that feed on plant roots. Several kinds of bacteria are able to fix nitrogen from the atmosphere. Algae carry on photosynthesis and are consumed by other soil organisms. Bacteria and fungi are particularly important in the decay and recycling of materials. Their chemical activities change complex organic materials into simpler forms that can be used as nutrients by plants. For example, some of these microorganisms can convert the nitrogen contained in the protein component of organic matter into ammonia or nitrate, which are nitrogen compounds that can be utilized by plants. The amount of nitrogen produced varies with the type of organic matter, type of microorganisms, drainage, and temperature. Finally, it is important to recognize that the soil contains a complicated food chain in which all organisms are subject to being consumed by others. All of these organisms are active within distinct layers of the soil, known as the soil profile.

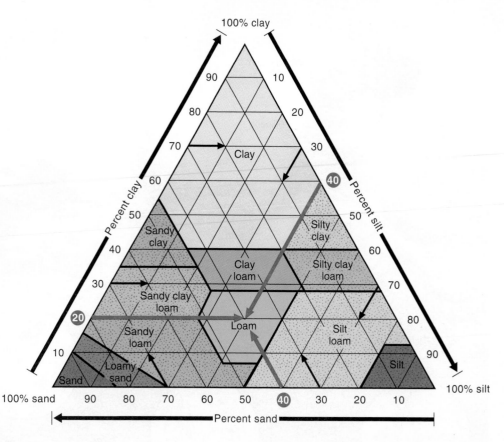

FIGURE 13.6 **Soil Texture** Texture depends on the percentage of clay, silt, and sand particles in the soil. A loam soil has the best texture for most crops. As shown in the illustration, if a soil were 40 percent sand, 40 percent silt, and 20 percent clay, it would be a loam.
Source: Data from U.S. Soil Conservation Service.

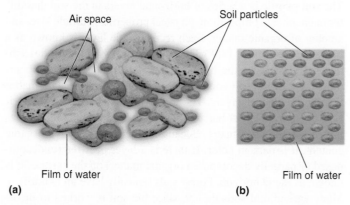

FIGURE 13.7 **Pore Spaces and Particle Size** (a) Soil that is composed of particles of various sizes has spaces for both water and air. The particles have water bound to their surfaces (represented by the colored halo around each particle), but some of the spaces are so large that an air space is present. (b) Soil composed of uniformly small particles has less space for air. Since roots require both air and water, the soil in (a) would be better able to support crops than would the soil in (b).

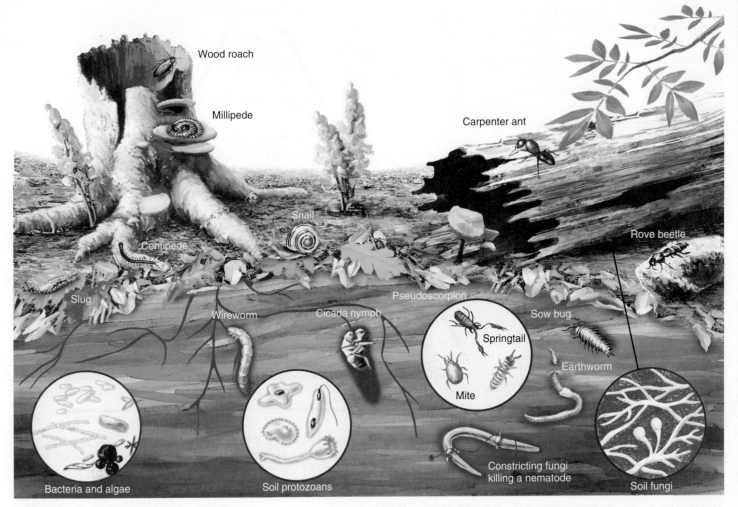

FIGURE 13.8 **Soil Organisms** All of these organisms occupy the soil and contribute to it by rearranging soil particles, participating in chemical transformation, and recycling dead organic matter.

13.6 Soil Profile

The **soil profile** is a series of horizontal layers in the soil that differ in chemical composition, physical properties, particle size, and amount of organic matter. Each recognizable layer is known as a **horizon.** (See figure 13.9.) There are several systems for describing and classifying the horizons in soils. In general, the uppermost layer of the soil contains more nutrients and organic matter than do the deeper layers. The top layer is known as the *A* horizon, or topsoil. The *A* horizon consists of small mineral particles mixed with organic matter. Because of the relatively high organic content, it is dark in color. If there is a layer of **litter** (undecomposed or partially decomposed organic matter) on the surface, it is known as the *O* horizon. Forest soils typically have an *O* horizon. Many agricultural soils do not, since the soil is worked to incorporate surface crop residue. As the organic matter decomposes, it becomes incorporated into the *A* horizon. The thickness of the *A* horizon may vary from less than a centimeter (less than 1/2 inch) on steep mountain slopes to over a meter (over 40 inches) in the rich grasslands of central North America. Most of the living organisms and nutrients are found in the *A* horizon. As water moves down through the *A* horizon, it carries dissolved organic matter and minerals to lower layers. This process is known as **leaching.** Because of the leaching away of darker materials such as iron compounds, a lighter-colored layer develops below the *A* horizon that is known as the *E* horizon. Not all soils develop an *E* horizon. This layer usually contains few nutrients because water flowing down through the soil dissolves and transports nutrients to the underlying *B* horizon. The *B* horizon, often called the subsoil, contains less organic material and fewer organisms than the *A* horizon. However, it contains accumulations of nutrients that were leached from higher levels. Often, clay minerals that are leached from the topsoil are deposited in this layer. Because nutrients are deposited in this layer, the *B* horizon in many soils is a valuable source of nutrients for plants, and such subsoils support a well-developed root system. Because the amount of leaching depends on the available rainfall, grasslands soils, which develop under low rainfall, often have a poorly developed *B* horizon, while soils in woodlands that receive higher rainfall usually have a well-developed *B* horizon.

The area below the subsoil is known as the *C* horizon, and it consists of weathered parent material. This parent material

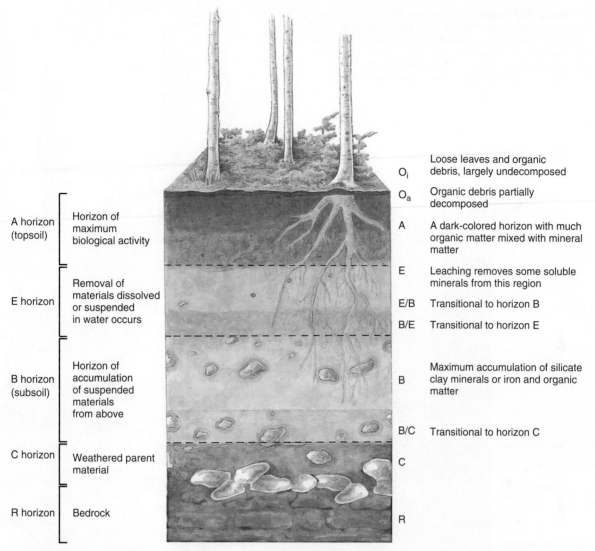

	O_i	Loose leaves and organic debris, largely undecomposed
	O_a	Organic debris partially decomposed
A horizon (topsoil) — Horizon of maximum biological activity	A	A dark-colored horizon with much organic matter mixed with mineral matter
E horizon — Removal of materials dissolved or suspended in water occurs	E	Leaching removes some soluble minerals from this region
	E/B	Transitional to horizon B
	B/E	Transitional to horizon E
B horizon (subsoil) — Horizon of accumulation of suspended materials from above	B	Maximum accumulation of silicate clay minerals or iron and organic matter
	B/C	Transitional to horizon C
C horizon — Weathered parent material	C	
R horizon — Bedrock	R	

FIGURE 13.9 Soil Profile A soil has layers that differ physically, chemically, and biologically. The top layer is known as the *A* horizon and contains most of the organic matter. Organic matter that collects on the surface is known as the *O* horizon, which can be subdivided into an undecomposed top layer (*O$_i$*) and a partially decomposed bottom layer (*O$_a$*). Many soils have a light-colored *E* horizon below the *A* horizon. It is light in color because dark-colored materials are leached from the layer. The *B* horizon accumulates minerals and particles as water carries dissolved minerals downward from the *A* and *E* to the *B* horizon. The *B* horizon is often called the subsoil. Below the *B* horizon is a *C* horizon of weathered parent material.

contains no organic materials, but it does contribute to some of the soil's properties. The chemical composition of the minerals of the *C* horizon helps to determine the pH of the soil. If the parent material is limestone, the soil will tend to neutralize acids; if the parent material is granite rock, the soil will not be able to do so. The characteristics of the parent material in the *C* horizon may also influence the soil's rate of water absorption and retention. Ultimately, the *C* horizon rests on bedrock, which is known as the *R* horizon.

Soil profiles and the factors that contribute to soil development are extremely varied. Over 15,000 separate soil types have been classified in North America, but the three major classifications are grassland, forest, and desert. (See figure 13.10.) Most of the cultivated land in the world can be classified as either grassland soil or forest soil.

Because the amount of rainfall in grassland areas is relatively low, it does not penetrate into the soil layers very far. Most of the roots of the grasses and other plants remain near the surface, and little leaching of minerals from the topsoil to deeper layers occurs. Since the roots of the plants rot in place when the grasses die, a deep layer of topsoil develops. This lack of leaching also results in a thin layer of subsoil, which is low in mineral and organic content and supports little root growth.

Forest soils develop in areas of more abundant rainfall. Water moves down through the soil so that deeper layers of the soil have a great deal of moisture. The roots of the trees penetrate to this layer and extract the water they need. The leaves and other plant parts that fall to the soil surface form a thin layer of organic matter on the surface. This organic matter decomposes and mixes with the mineral material of the top layers of the soil. The water that moves through the soil tends to carry material from the topsoil to the subsoil, where many of the roots of the plants are located. One of the materials that accumulates in the *B* horizon is clay. In some

FIGURE 13.10 Major Soil Types There are thousands of different soil types, but many of them can be classified into three broad categories. Soils formed in the grasslands have a deep *A* horizon. The shallow *B* horizon does not have sufficient nutrients to support root growth. In forest soils, the *A* horizon is thinner, and leaching transfers many nutrients to the *B* horizon. Thus, roots are found in both the *A* and *B* horizons. Desert soils have very thin *A* horizons.

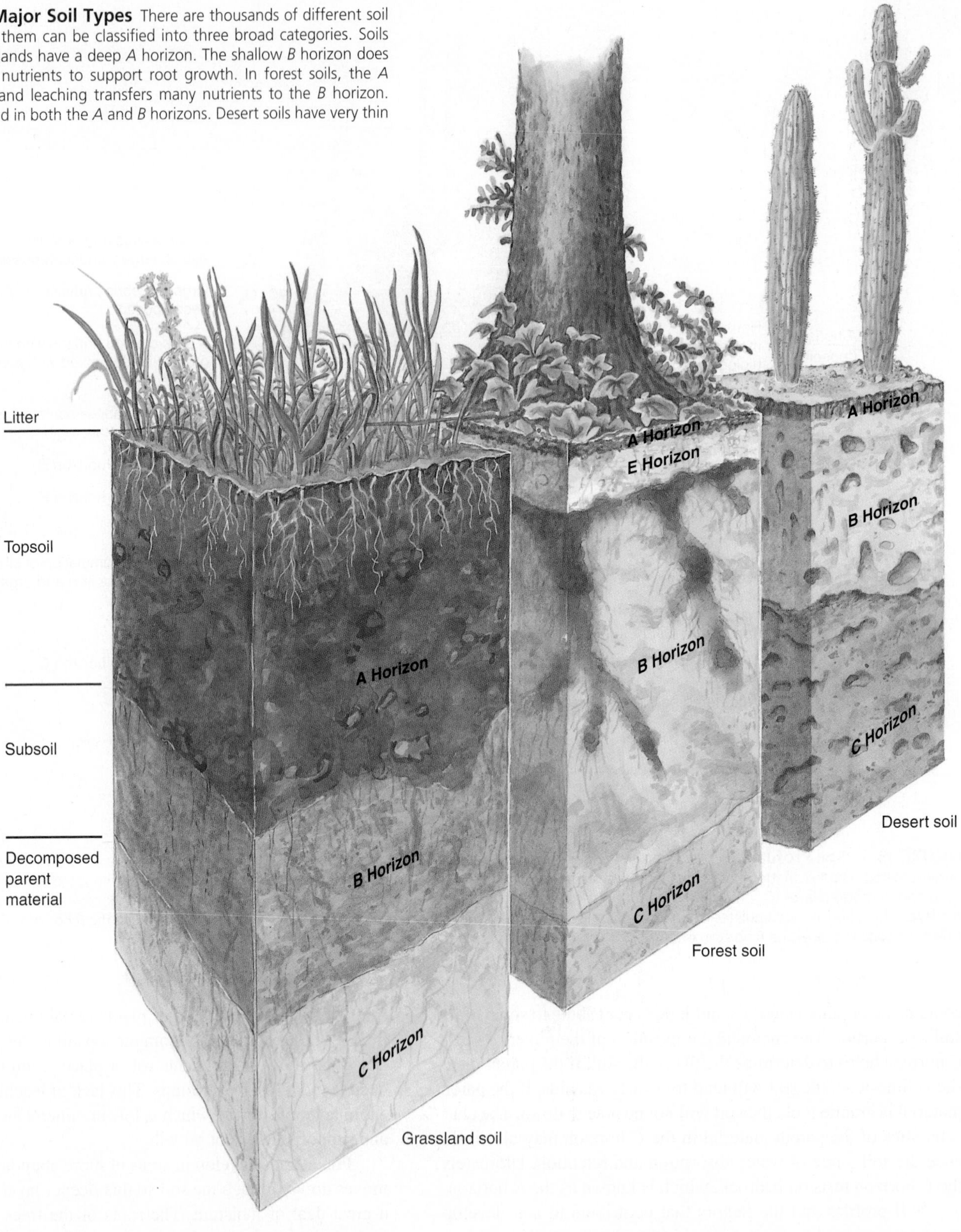

Litter

Topsoil

Subsoil

Decomposed
parent
material

A Horizon

B Horizon

C Horizon

Grassland soil

A Horizon

E Horizon

B Horizon

C Horizon

Forest soil

A Horizon

B Horizon

C Horizon

Desert soil

soils, particularly forest soils, clay or other minerals may accumulate and form a relatively impermeable "hardpan" layer that limits the growth of roots and may prevent water from reaching the soil's deeper layers.

Desert soils have very poorly developed horizons. Since there is little rainfall, deserts do not support a large amount of plant growth, and much of the soil is exposed. Therefore, little organic matter is added to the soil, and little leaching of materials occurs from upper layers to lower layers. Since much of the soil is exposed to wind and water erosion, much of the organic material and smaller particles are carried away by wind or by flash floods when it rains.

Science, Politics, & Policy

Organic Crops, Healthy Soil, and Policy Debates

A study conducted in 2010 by Washington State University compared the fruit and soil properties of organic and conventional strawberry farms. The study analyzed 31 chemical and biological soil properties, and the taste and nutrition of three strawberry varieties. The study found the organic farms produced more flavorful and nutritious berries while leaving the soil healthier and with more genetically diverse soil microbes.

All the farms in the study were in California, which produces 90 percent of the strawberries in the United States. California is also the center of an ongoing debate about the use of soil fumigants. Conventional farms in the study use the ozone-depleting methyl bromide as a soil fumigant to control pests. Methyl bromide is slated to be replaced by the highly toxic methyl iodide although the use of methyl iodide is opposed by health advocates and more than 50 Nobel laureates. California elected officials are asking the EPA to reconsider its approval of methyl iodide.

Other findings of the study included:

- The organic strawberries had significantly higher antioxidant activity and concentrations of ascorbic acid.
- The organic strawberries had longer shelf life.
- The organic strawberries had more dry matter, or, "more strawberry in the strawberry."

The study also found the soils on organic farms were superior in several key chemical and biological properties. These include higher nitrogen and micronutrient content, microbial biomass, and enzyme activity, and the ability of the soil to sequester carbon.

Another study conducted by the U.S. Department of Agriculture (USDA) found that organic farming improved soil organic matter more than no-till farming. This was contrary to conventional thinking that organic farming endangered soil because it relies on tillage and cultivation to kill weeds. However, the USDA study showed that the addition of organic matter in manure and cover crops as normal activities on organic farms more than offset losses from tillage.

These two studies are representative of growing research in the area of organic farming and soil health and quality. Congress has been involved in several aspects of the debate. The USDA, as directed by Congress, has established guidelines for determining what can be marketed as "organic." Congress is also lobbied by advocates of both organic and conventional farming. Can you identify examples of organizations on both sides of this policy debate? Have the positions of groups changed over the past decade? If so, why do you think this happened?

In cold, wet climates, typical of the northern parts of Europe, Russia, and Canada, there may be considerable accumulations of organic matter, since the rate of decomposition is reduced. The extreme acidity of these soils also reduces the rate of decomposition. Hot, humid climates also tend to have poorly developed soil horizons because the organic matter decays very rapidly, and soluble materials are carried away by the abundant rainfall.

Because tropical rainforests support such a vigorous growth of plants and an incredible variety of plant and animal species, it is often assumed that tropical soils must be very fertile. Consequently, many people have tried to raise crops on tropical soils. It is possible to grow certain kinds of crops that are specially adapted to tropical soils, but raising of most traditional crop species is not successful. To understand why this is so, it is important to understand the nature of tropical rainforest soils. Two features of the tropical rainforest climate have a great influence on the nature of the soil. High temperature results in rapid decomposition

of organic matter, so that the soils have very little litter and humus. High rainfall tends to leach nutrients from the upper layers of the soil, leaving behind a soil that is rich in iron and aluminum. The high iron content results in a reddish color for most of these soils. Because the nutrients are quickly removed, these soils are very infertile. Furthermore, when the vegetation is removed, the soil is quickly eroded.

In addition to the differences caused by the kind of vegetation and rainfall, topography influences the soil profile. (See figure 13.11.) On a relatively flat area, the topsoil formed by soil-building processes will collect in place and gradually increase in depth. The topsoil formed on rolling hills or steep slopes is often transported down the slope as fast as it is produced. On such slopes, the accumulation of topsoil may not be sufficient to support a cultivated crop. The topsoil removed from these slopes is eventually deposited in the flat floodplains. These regions serve as collection points for topsoil that was

FIGURE 13.11 The Effect of Slope on a Soil Profile The topsoil formed on a large area of the hillside is continuously transported down the slope by the flow of water. It accumulates at the bottom of the slope and results in a thicker *A* horizon. The resulting "bottomland" is highly productive because it has a deep, fertile layer of topsoil, while the soil on the slope is less productive.

produced over extensive areas. As a result, these river-bottom and delta regions have a very deep topsoil layer and are highly productive agricultural land.

13.7 Soil Erosion

Erosion is the wearing away and transportation of soil by water, wind, or ice. Concerns about soil degradation and increased runoff from loss of tree cover date as far back as Plato writing about Attica in the fourth century B.C.:

> . . . there are remaining only the bones of the wasted body, as they may be called, as in the case of small islands, all the richer and softer parts of the soil having fallen away, and the mere skeleton of the land being left. But in the primitive state of the country, its mountains were high hills covered with soil, and the plains, as they are termed by us, of Phelleus were full of rich earth, and there was abundance of wood in the mountains. Of this last the traces still remain, for although some of the mountains now only afford sustenance to bees, not so very long ago, there were still to be seen roofs of timber cut from trees growing there, which were of a size sufficient to cover the largest houses; and there were many other high trees, cultivated by man and bearing abundance of food for cattle. Moreover, the land reaped the benefit of the annual rainfall, not as now losing the water which flows off the bare earth into the sea . . .

Soil erosion takes place everywhere in the world, but some areas are more exposed than others. Erosion occurs wherever grass, bushes, and trees are disappearing. Deforestation and desertification both leave land open to erosion. In deforested areas, water washes down steep, exposed slopes, taking the soil with it. The Grand Canyon of the Colorado River, the

floodplains of the Nile in Egypt, the little gullies on hillsides, and the deltas that develop at the mouths of rivers all attest to the ability of water to move soil. Anyone who has seen muddy water after a rainstorm has observed soil being moved by water. The force of moving water allows it to carry large amounts of soil. While erosion is a natural process, it is greatly accelerated by agricultural practices that leave the soil exposed. Each year, the Mississippi River transports over 325 million metric tons of soil from the central regions of North America to the Gulf of Mexico. This is equal to the removal of a layer of topsoil approximately 1 millimeter (0.04 inch) thick from the entire region. Although the rate of erosion varies from place to place, movement of soil by water occurs in every stream and river in the world. Dry Creek, a small stream in California, has only 500 kilometers (310 miles) of mainstream and tributaries; however, each year, it removes 180,000 metric tons of soil from a 340-square-kilometer (130-square-mile) area. In desertified regions, exposed soils, cleared for farming, building, or mining, or overgrazed by livestock, simply blow away. Wind erosion is most extensive in Africa and Asia. Blowing soil not only leaves a degraded area behind but can bury and kill vegetation where it settles. It will also fill drainage and irrigation ditches. When high-tech farm practices are applied to poor lands, soil is washed away and chemical pesticides and fertilizers pollute the runoff. Every year, erosion carries away far more topsoil than is created, primarily because of agricultural practices that leave the soil exposed. (See figure 13.12.)

Worldwide, erosion removes about 25.4 billion metric tons of soil each year. In Africa, soil erosion has reached critical levels, with farmers pushing farther onto deforested hillsides. In Ethiopia, for example, soil loss occurs at a rate of between 1.5 billion and 2 billion cubic meters (53–70 billion cubic feet) a year, with some 4 million hectares (about 10 million acres) of highlands considered irreversibly degraded. In Asia, in the eastern hills of Nepal, 38 percent of the land area is fields that have been abandoned because the topsoil has washed away. In the Western Hemisphere, Ecuador is losing soil at 20 times the acceptable rate.

According to the International Fund for Agricultural Development (IFAD), traditional labor-intensive, small-scale soil conservation efforts that combine maintenance of shrubs and trees with crop growing and cattle grazing work best at controlling erosion. In parts of Pakistan, a program begun by IFAD in 1980 to control rainfall runoff, erosion, and damage to rivers from siltation has increased crop yields and livestock productivity by 20 percent to 30 percent.

Badly eroded soil has lost all of the topsoil and some of the subsoil and is no longer productive farmland. Most current agricultural practices lose soil faster than it is replaced. Farming practices that reduce erosion, such as contour farming and terracing, are discussed in the Soil Conservation Practices section of this chapter.

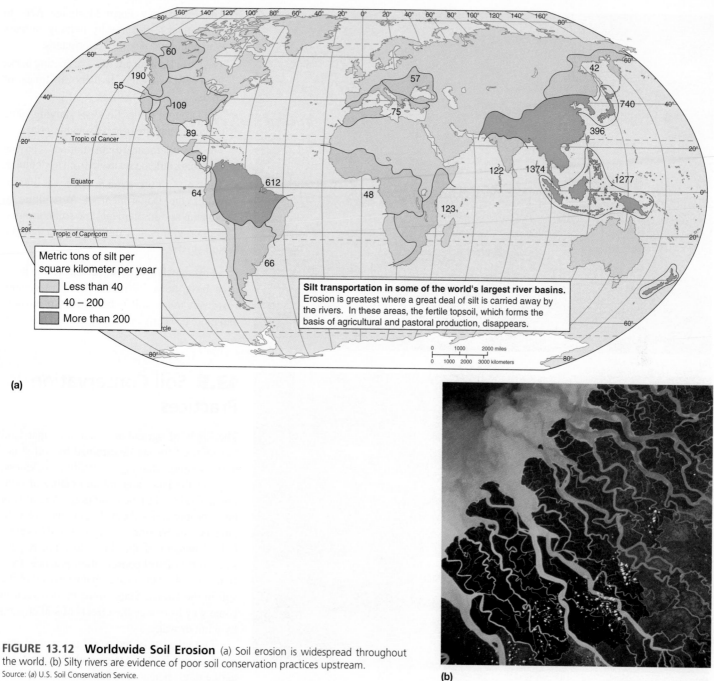

(a)

(b)

FIGURE 13.12 Worldwide Soil Erosion (a) Soil erosion is widespread throughout the world. (b) Silty rivers are evidence of poor soil conservation practices upstream.
Source: (a) U.S. Soil Conservation Service.

Wind is also an important mover of soil. Under certain conditions, it can move large amounts. (See figure 13.13a.) Wind erosion may not be as evident as water erosion, since it does not leave gullies. Nevertheless, it can be a serious problem. Wind erosion is most common in dry, treeless areas where the soil is exposed. In the Sahel region of Africa, much of the land has been denuded of vegetation because of drought, overgrazing, and improper farming practices. This has resulted in extensive wind erosion of the soil. (See figure 13.13b.) In the Great Plains region of North America, there have been four serious periods of wind erosion since European settlement in the 1800s. If this area receives less than 30 centimeters (12 inches) of rain per year, there is not enough moisture to support crops. When this occurs for several years in a row, it is called a drought. Farmers plant crops, hoping for rain. When the rain does not come, they plow their fields again to prepare them for another crop. Thus, the loose, dry soil is left exposed, and wind erosion results. Because of the large amounts of dust in the air during those times, the region is known as the Dust Bowl. During the 1930s, wind destroyed 3.5 million hectares (over 8.5 million acres) of farmland and seriously damaged an additional 30 million hectares (75 million acres) in the Dust Bowl.

In the United States in the late 1920s, good crop yields and high prices for wheat encouraged a rapid increase in the cropped area. When drought hit in the following decade, there was

(a)

(b)

FIGURE 13.13 Wind Erosion (a) The dry, unprotected topsoil from this field is being blown away. The force of the wind is capable of removing all the topsoil and transporting it several thousand kilometers. (b) The semiarid region just south of the Sahara Desert is in an especially vulnerable position. The rainfall is unpredictable, which often leads to crop failure. In addition, population pressure forces people in this region to try to raise crops in marginal areas. This often results in increased wind erosion.

catastrophic soil erosion that drove many farmers from the land. By 1940, 2.5 million people had left the Great Plains.

During the 1930s, the U.S. government responded with a comprehensive package of measures, both to give short-term economic relief and provide for long-term agricultural research and development. Examples of these initiatives include:

- The Emergency Farm Mortgage Act—to prevent farm closures by helping farmers who could not pay their mortgages.
- The Farm Bankruptcy Act—restricting banks from dispossessing farmers in times of crisis.
- The Drought Relief Service—buying cattle in emergency areas at reasonable prices.
- The Resettlement Administration— buying land that could be set aside from agriculture.
- The Soil Conservation Service was created within the Department of Agriculture to develop and implement new soil conservation programs. It also undertook a national soil survey.

Fortunately, many soil conservation practices have been instituted that protect soil. However, much soil is being lost to erosion, and more protective measures should be taken.

13.8 Soil Conservation Practices

The kinds of agricultural activities that land can be used for are determined by soil structure, texture, drainage, fertility, rockiness, slope of the land, amount and nature of rainfall, and other climatic conditions. A relatively large proportion—about 20 percent—of U.S. land is suitable for raising crops. However, only 2 percent of that land does not require some form of soil conservation practice. (See figure 13.14.) This means that nearly all of the soil in the United States must be managed in some way to reduce the effects of soil erosion by wind or water.

Not all parts of the world are as well supplied as the United States with land that has agricultural potential. (See table 13.2.) For example, worldwide, approximately 11 percent of the land surface is suitable for crops, and an additional 24 percent is in permanent pasture. In the United States, about 20 percent is cropland, and 25 percent is in permanent pasture. Contrast this with the continent of Africa, in which only 6 percent is suitable for crops and 29 percent can be used for pasture. Canada has only 5 percent suitable for crops and 3 percent for pasture. Europe has the highest percentage of cropland with 30 percent, but it has only 17 percent in permanent pasture.

Since very little land is left that can be converted to agriculture, we must use what we have wisely. A study by the Food and Agricultural Organization states that up to 40 percent of the world's agricultural lands are seriously affected by soil degradation. The World Resources Institute stated that land degradation

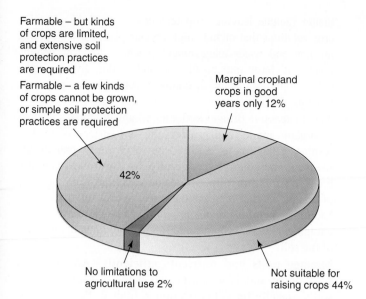

Farmable – but kinds of crops are limited, and extensive soil protection practices are required

Farmable – a few kinds of crops cannot be grown, or simple soil protection practices are required

Marginal cropland crops in good years only 12%

42%

No limitations to agricultural use 2%

Not suitable for raising crops 44%

FIGURE 13.14 U.S. Land Used for Agricultural Purposes Only 2 percent of the land in the United States can be cultivated without some soil conservation practices. This 2 percent is primarily flatland, which is not subject to wind erosion. On 42 percent of the remaining land, some special considerations for protecting the soil are required, or the kinds of crops are limited. Twelve percent of the land is marginal cropland that can provide crops only in years when rainfall and other conditions are ideal. Forty-four percent is not suitable for cultivating crops but may be used for other purposes, such as cattle grazing or forests.

Table 13.2 Percentage of Land Suitable for Agriculture

Country	Percent Cropland	Percent Pasture
World	11.0	26.0
Africa	6.3	28.8
Egypt	2.8	5.0
Ethiopia	12.7	40.7
Kenya	7.9	37.4
South Africa	10.8	66.6
North America	13.0	16.8
Canada	4.9	3.0
United States	19.6	25.0
South America	6.0	28.3
Argentina	9.9	51.9
Venezuela	4.4	20.2
Asia	15.2	25.9
China	10.3	30.6
Japan	12.0	1.7
Europe	29.9	17.1

Source: Data from *World Resources,* 2011.

affects around 70 percent of the world's rangelands, 40 percent of rain-fed agricultural lands, and 30 percent of irrigated lands. According to the United Nations Environment Programme, an estimated 500 million hectares (1.24 billion acres) of land in Africa have been affected by soil degradation since about 1950, including as much as 65 percent of agricultural land.

Africa south of the Sahara is the only remaining region of the world where per capita food production has remained stagnant over the past 40 years. About 180 million Africans—up 100 percent since 1970—do not have access to sufficient food to lead healthy and productive lives.

Depletion of soil fertility is a major cause of low per capita food production in Africa. Over decades, small-scale farmers have removed large quantities of nutrients from their soils without using sufficient quantities of manure or fertilizer to replenish the soil.

A soil fertility replenishment approach has been developed during the past decade using resources naturally available in Africa. These included the use of nitrogen-fixing leguminous trees and biomass transfer of the leaves of nutrient-accumulating shrubs.

Leguminous trees are interplanted into a young maize (corn) crop and allowed to grow as fallows during dry seasons. The quantities of nitrogen captured are similar to those applied as fertilizers by commercial farmers to grow maize in developed countries. After the wood is harvested from the trees, nitrogen-rich leaves, pods, and green branches are hoed into the soil before maize is planted at the start of a subsequent season. This litter decomposes with the tree roots, releasing nitrogen and other nutrients to the soil.

The transfer of the leaf biomass of nutrient-accumulating shrubs from roadsides and hedges into cropped fields also adds nutrients and can double maize yields without fertilizer additions. This organic source of nutrients is effective because it also adds other plant nutrients, particularly potassium and micronutrients. Because of the high labor requirements for cutting and carrying the biomass to fields, this is profitable with high-value crops such as vegetables but not with relatively low-valued maize.

While the global expansion of agricultural area has been modest in recent decades, intensification has been rapid, as irrigated area increased and fallow time decreased to produce more output per hectare.

Many techniques can protect soil from erosion while allowing agriculture. Some of the more common methods are discussed here. Whenever soil is lost by water or wind erosion, the topsoil, the most productive layer, is the first to be removed. When the topsoil is lost, the soil's fertility decreases, and larger amounts of expensive fertilizers must be used to restore the fertility that was lost. This raises the cost of the food we buy. In addition, the movement of excessive amounts of soil from farmland into streams has several undesirable effects. First, a dirty stream is less aesthetically pleasing than a clear stream. Second, too much sediment in a stream affects the fish population by reducing visibility, covering spawning sites, and clogging the gills of the fish. Fishing may be poor because of unwise farming practices hundreds of kilometers upstream. Third, the soil carried by a river is eventually deposited somewhere. In many cases, this soil must be removed by dredging to clear shipping channels. We pay for dredging with our tax money, and it is a very expensive operation.

For all of these reasons, proper soil conservation measures should be employed to minimize the loss of topsoil. Figure 13.15 contrasts poor soil conservation practices with proper soil protection. When soil is not protected from the effects of running water, the topsoil is removed and gullies result. This can be prevented by slowing the flow of water over sloping land.

Soil Quality Management Components

- **Enhance organic matter:** Whether soil is naturally high or low in organic matter, adding new organic matter every year is perhaps the most important way to improve and maintain soil quality. Regular additions of organic matter improve soil structure, enhance water- and nutrient-holding capacity, protect soil from erosion and compaction, and support a healthy community of soil organisms. Practices that increase organic

(a)

(b)

FIGURE 13.15 Poor and Proper Soil Conservation Practices (a) This land is losing its topsoil because there is little vegetation to protect it from erosion. (b) This rolling farmland shows strip contour farming to minimize soil erosion by running water. It should continue indefinitely to be productive farmland.

matter include leaving crop residues in the field, choosing crop rotations that include high-residue plants, using optimal nutrient and water management practices to grow healthy plants with large amounts of roots and residue, growing cover crops, applying manure or compost, using low- or no-tillage systems, and mulching.

- **Avoid excessive tillage:** Reducing tillage minimizes the loss of organic matter and protects the soil surface with plant residue. Tillage is used to loosen surface soil, prepare the seedbed, and control weeds and pests. But tillage can also break up soil structure, speed the decomposition and loss of organic matter, increase the threat of erosion, destroy the habitat of helpful organisms, and cause compaction. New equipment allows crop production with minimal disturbance of the soil.

- **Manage pests and nutrients efficiently:** An important function of soil is to buffer and detoxify chemicals, but soil's capacity for detoxification is limited. Pesticides and chemical fertilizers have valuable benefits, but they also can harm nontarget organisms and pollute water and air if they are mismanaged. Nutrients from organic sources also can pollute when misapplied or overapplied. Efficient pest and nutrient management means testing and monitoring soil and pests; applying only the necessary chemicals, at the right time and place to get the job done; and taking advantage of nonchemical approaches to pest and nutrient management such as crop rotations, cover crops, and manure management.

- **Prevent soil compaction:** Compaction reduces the amount of air, water, and space available to roots and soil organisms. Compaction is caused by repeated traffic, heavy traffic, or traveling on wet soil. Deep compaction by heavy equipment is difficult or impossible to remedy, so prevention is essential.

- **Keep the ground covered:** Bare soil is susceptible to wind and water erosion and to drying and crusting. Ground cover protects soil, provides habitats for larger soil organisms, such as insects and earthworms, and can improve water availability. Ground can be covered by leaving crop residue on the surface or by planting cover crops. In addition to ground cover, living cover crops provide additional organic matter, and continuous cover and food for soil organisms. Ground cover must be managed to prevent problems with delayed soil warming in spring, diseases, and excessive buildup of phosphorus at the surface.

- **Diversify cropping systems:** Diversity is beneficial for several reasons. Each plant contributes a unique root structure and type of residue to the soil. A diversity of soil organisms can help control pest populations, and a diversity of cultural practices can reduce weed and disease pressures. Diversity across the landscape can be increased by using buffer strips, small fields, or contour strip cropping. Diversity over time can be increased by using long crop rotations. Changing vegetation across the landscape or over time increases not only plant diversity but also the types of insects, microorganisms, and wildlife that live on a farm.

Contour Farming

Contour farming, which is tilling at right angles to the slope of the land, is one of the simplest methods for preventing soil erosion. This practice is useful on gentle slopes and produces a series of small ridges at right angles to the slope. (See figure 13.16.) Each ridge acts as a dam to prevent water from running down the incline. This allows more of the water to soak into the soil. Contour farming reduces soil erosion by as much as 50 percent and, in drier regions, increases crop yields by conserving water.

Strip Farming

When a slope is too steep or too long, contour farming alone may not prevent soil erosion. However, a combination of contour and strip farming may work. **Strip farming** is alternating strips of closely sown crops such as hay, wheat, or other small grains with strips of row crops such as corn, soybeans, cotton, or sugar beets. (See figure 13.17.) The closely sown crops retard the flow of water, which reduces soil erosion and allows more water to be absorbed into the ground. The type of soil, steepness, and length of slope dictate the width of the strips and determine whether strip or contour farming is practical.

Terracing

On very steep land, the only practical method of preventing soil erosion is to construct terraces. **Terraces** are level areas constructed at right angles to the slope to retain water and greatly reduce the amount of erosion. (See figure 13.18.) Terracing has been used for centuries in nations with a shortage of level farmland. The type of terracing seen in figure 13.18a requires the use

FIGURE 13.16 Contour Farming Tilling at right angles to the slope creates a series of ridges that slows the flow of the water and prevents soil erosion. This soil conservation practice is useful on gentle slopes.

FIGURE 13.17 Strip Farming On rolling land, a combination of contour and strip farming prevents excessive soil erosion. The strips are planted at right angles to the slope, with bands of closely sown crops, such as wheat or hay, alternating with bands of row crops, such as corn or soybeans.

(a)

(b)

FIGURE 13.18 Terraces Since the construction of terraces requires the movement of soil and the protection of the steep slope between levels, terraces are expensive to build. (a) The terraces seen here are extremely important for people who live in countries that have little flatland available. They require much energy and hand labor to maintain but make effective agricultural use of the land without serious erosion. (b) This modification of the terracing concept allows the use of the large farm machines typical of farming practices in Canada, Europe, and the United States.

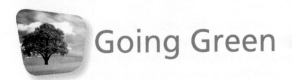

Going Green

Green Landscaping

Traditional landscaping and current landscape maintenance practices, while frequently meeting human needs and aesthetics, often have harmful environmental impacts. These impacts include:

- Gasoline-powered landscape equipment (mowers, trimmers, blowers) account for over 5 percent of urban air pollution in North America.
- Residential application of pesticides is typically at a rate 20 times that of farm applications per acre.
- Yard wastes (mostly grass clippings) comprise 20 percent of municipal solid waste collected and most still ends up in landfills.
- A lawn has less than 10 percent of the water absorption capacity of a natural woodland—often a reason for suburban flooding.

Natural or native landscaping attempts to balance our needs and aesthetics with those of the environment. Natural landscaping provides habitat for local and migratory animals, conserves native plants, and improves water quality. Landowners benefit by reducing the time and expense of mowing, watering, fertilizing and treating lawn and garden areas. Natural landscaping can also be used to address problems such as erosion, poor soils, steep slopes, and poor drainage. The principles of natural landscaping include:

- Protect existing natural areas to the greatest extent possible (woodlands and wetlands, stream corridors, and meadows).
- Select regionally native plants to form the backbone of the landscape.

- Reduce use of turf; instead, install woodland, meadow, or other natural plantings.
- Reduce use of pesticides.
- Reduce use of power landscape equipment. Shrinking the size of the lawn and planting appropriate native species in less formal arrangements will reduce the need for extensive use of power equipment. Once you begin to explore and experiment with native plants, you will soon discover that many of these plants go beyond just replacing worn out selections in your yard. Native plants will eventually reduce your labor and maintenance costs, provide habitat for wildlife, and create a sense of place.
- Avoid use of invasive exotics that outcompete native plants and result in declines in biodiversity. Examples include: Norway maples, kudzu, purple loosestrife, and multiflora rose.
- Practice soil and water conservation. Stabilize slopes with natural plantings, mulch around plants, and install drought tolerant species.
- Compost and mulch on site to eliminate solid waste. Generate your own mulch—a soil additive can replace the need for most fertilizers.

A growing number of communities are developing municipal compost facilities. These communities offer curbside pickup of "organic green waste," including yard waste, food scraps, and biodegradable products. Often the resulting compost is available to community members for their garden use.

Does your community offer curbside pickup of organic green waste and make the compost available for your use?

---

of small machines and considerable hand labor and is not suitable for the mechanized farming typical in much of the world. The modification shown in figure 13.18b allows for the use of large farm machines. Terracing is an expensive method of controlling erosion, since it requires moving soil to construct the level areas, protecting the steep areas between terraces, and constant repair and maintenance. Many factors, such as length and steepness of slope, type of soil, and amount of precipitation, determine whether terracing is feasible.

Waterways

Even with such soil conservation practices as contour farming, strip farming, and terracing, farmers must often provide protected channels for the movement of water. **Waterways** are depressions on sloping land where water collects and flows off the land. When not properly maintained, these areas are highly susceptible to erosion. (See figure 13.19.) If a waterway is maintained with a permanent sod covering, the speed of the water is reduced, the

(a)

(b)

FIGURE 13.19 Protection of Waterways Prevents Erosion (a) An unprotected waterway has been converted into a gully. (b) A well-maintained waterway is not cultivated; a strip of grass retards the flow of water and protects the underlying soil from erosion.

roots tend to hold the soil particles in place, and soil erosion is decreased.

Windbreaks

Contour farming, strip farming, terracing, and maintaining waterways are all important ways of reducing water erosion, but wind is also a problem with certain soils, particularly in dry areas of the world. Wind erosion can be reduced if the soil is protected. The best protection is a layer of vegetation over the surface. However, the process of preparing soil for planting and the method of planting often leave the soil exposed to wind. **Windbreaks** are plantings of trees or other plants that protect bare soil from the full force of the wind. Windbreaks reduce the velocity of the wind, thereby decreasing the amount of soil that it can carry away. (See figure 13.20.)

(a)

(b)

FIGURE 13.20 **Windbreaks** (a) In sections of the Great Plains, trees provide protection from wind erosion. The trees along the road protect the land from the prevailing winds. (b) In this field, temporary strips of vegetation serve as windbreaks.

In some cases, rows of trees are planted at right angles to the prevailing winds to reduce their force, while in other cases, a kind of strip farming is practiced in which strips of hay or grains are alternated with row crops that leave large amounts of the soil exposed. In some areas of the world, the only way to protect the soil is to not cultivate it at all but to leave it in a permanent cover of grasses.

13.9 Conventional Versus Conservation Tillage

Conventional tillage methods in much of the world require the extensive use of farm machinery to prepare the soil for planting and to control weeds. Typically, a field is plowed and then disked or harrowed one to three times before the crop is planted. The plowing, which turns the soil over, has several desirable effects: any weeds or weed seeds are buried, thus reducing the weed problem in the field. Crop residue from previous crops is incorporated into the soil, where it will decay faster and contribute to soil structure. Nutrients that had been leached to deeper layers of the soil are brought near the surface. And the dark soil is exposed to the sun so that it warms up faster. This last effect is most critical in areas with short growing seasons. In many areas, fields are plowed in the fall, after the crop has been harvested, and the soil is left exposed all winter.

After plowing, the soil is worked by disks or harrows to break up any clods of earth, kill remaining weeds, and prepare the soil to receive the seeds. After the seeds are planted, there may still be weed problems. Farmers often must cultivate row crops to kill the weeds that begin to grow between the rows. Each trip over the field costs the farmer money, while at the same time increasing the amount of time the soil is exposed to wind or water erosion.

In recent years, several new systems of tillage have developed as innovations in chemical herbicides and farm equipment have taken place. These tilling practices protect the soil by leaving the crop residue on the soil surface, thus reducing the amount of time it is exposed to erosion forces. **Reduced tillage** is a method that uses less cultivation to control weeds and to prepare the soil to receive seeds but generally leaves 15 percent to 30 percent of the soil surface covered with crop residue after planting. **Conservation tillage** methods further reduce the amount of disturbance to the soil and leave 30 percent or more of the soil surface covered with crop residue following planting. Selective herbicides are used to kill unwanted vegetation before planting the new crop and to control weeds afterward. Several variations of conservation tillage are used:

1. Mulch tillage involves tilling the entire surface just before planting or as planting is occurring.
2. Strip tillage is a method that involves tilling only in the narrow strip that is to receive the seeds. The rest of the soil and the crop residue from the previous crop are left undisturbed.

It has been stated by some soil scientists that the only way to have successful action against soil erosion on a large scale is to implement land management practices that increase production capacity over a large area. This belief is counter to another school of thought that soil protection is best served by smaller organic farms. Develop an argument in support of one of these contradictory attitudes on soil conservation and farming practices.

3. Ridge tillage involves leaving a ridge with the last cultivation of the previous year and planting the crop on the ridge with residue left between the ridges. The crop may be cultivated during the year to reduce weeds.

4. No-till farming involves special planters that place the seeds in slits cut in the soil that still has on its surface the residue from the previous crop.

Both reduced tillage and conservation tillage methods reduce the amount of time and fuel needed by the farmer to produce the crop and, therefore, represent an economic savings. By 2000, over half of the cropland in the United States was being farmed using reduced or conservation tillage methods. (See table 13.3.) For many kinds of crops, yields are comparable to those produced by conventional tillage methods.

Other positive effects of reduced tillage, in addition to reducing erosion, are:

1. The amount of winter food and cover available for wildlife increases, which can lead to increased wildlife populations.

2. Since there is less runoff, siltation in streams and rivers is reduced. This results in clearer water for recreation and less dredging to keep waterways open for shipping.

3. Row crops can be planted on hilly land that cannot be converted to such crops under conventional tilling methods. This allows a farmer to convert low-value pastureland into cropland that gives a greater economic yield.

4. Since fewer trips are made over the field, petroleum is saved, even when the petrochemical feedstocks necessary to produce the herbicides are taken into account.

5. Two crops may be grown on a field in areas that had been restricted to growing one crop per field per year. In some areas, immediately after harvesting wheat, farmers have planted soybeans directly in the wheat stubble.

6. Because conservation tillage reduces the number of trips made over the field by farm machinery, the soil does not become compacted as quickly.

However, there are also some drawbacks to conservation tillage methods:

1. The residue from previous vegetation may delay the warming of the soil, which may, in turn, delay planting some crops for several days.

2. The crop residue reduces evaporation from the soil and the upward movement of water and soil nutrients from deeper layers of the soil, which may retard the growth of plants.

3. The accumulation of plant residue can harbor plant pests and diseases that will require more insecticides and fungicides. This is particularly true if the same crop is planted repeatedly in the same location.

Conservation tillage is not the complete answer to soil erosion problems but may be useful in reducing soil erosion on well-drained soils. It also requires that farmers pay close attention to the condition of the soil and the pests to be dealt with.

In 1972, 12 million hectares (30 million acres) in the United States were under some form of conservation tillage. About 1.3 million hectares (3.2 million acres) were being farmed using no-till methods. By 1992, this had risen to 60 million hectares (150 million acres) of conservation-tilled land, of which 6.2 million hectares (15 million acres) were being farmed using no-till methods. In 2013, 93 percent of U.S. cropland was in some form of reduced tillage practice.

Table 13.3 Comparison of Various Tillage Methods

Tillage Method	Fuel Use (Liters per Hectare)	Time Involved (Hours per Hectare)
Conventional plowing	8.08	3.00
Reduced tillage	5.11	2.20
Mulch tillage	4.64	2.07
Ridge tillage	4.12	2.25
No tillage	2.19	1.21

Source: Data from University of Nebraska, Institute of Agriculture and Natural Resource.

13.10 Protecting Soil on Nonfarmland

Each piece of land has characteristics, such as soil characteristics, climate, and degree of slope, that influence the way it can be used. When all these factors are taken into consideration, a proper use can be determined for each portion of the planet. Wise planning and careful husbandry of the soil are necessary if the land and its soil are to provide food and other necessities

Focus On

Land Capability Classes

Not all land is suitable for raising crops or urban building. Such factors as the degree of slope, soil characteristics, rockiness, erodibility, and other characteristics determine the best use for a parcel of land. In an attempt to encourage people to use land wisely, the U.S. Soil Conservation Service has established a system to classify land-use possibilities. The table shows the eight classes of land and lists the characteristics and capabilities of each. The photo shows how these classes can coexist.

Unfortunately, many of our homes and industries are located on type I and II land, which has the least restrictions on agricultural use. This does not make the best use of the land. Zoning laws and land-use management plans should consider the land-use capabilities and institute measures to ensure that land will be used to its best potential.

- Can you provide examples from your community where you feel land has not been used properly?

- In your opinion, should there be more or less land-use planning?

- What are some of the consequences of building or farming on land that is not suitable?

	Land Class	Characteristics	Capability	Special Conservation Measures
Land suitable for cultivation	I	Excellent, flat, well-drained land	Cropland	Normal good practices adequate
	II	Good land; has minor limitations, such as slope, sandy soil, or poor drainage	Cropland Pasture	Strip cropping Contour farming
	III	Moderately good land with important limitations of soil, slope, or drainage	Cropland Pasture Watershed	Contour farming Strip cropping Terraces Waterways
	IV	Fair land with severe limitations of soil, slope, or drainage	Pasture Orchards Urban Industry Limited cropland	Crops on a limited basis Contour farming Strip cropping Terraces Waterways
Land not suitable for cultivation	V	Use for grazing and forestry; slightly limited by rockiness, shallow soil, or wetness	Grazing Forestry Watershed Urban Industry	No special precautions if properly grazed or logged; must not be plowed
	VI	Moderate limitations for grazing and forestry because of moderately steep slopes		Grazing or logging may be limited at times
	VII	Severe limitations for grazing and forestry because of very steep slopes vulnerable to erosion		Careful management is required when used for grazing or logging
	VIII	Unsuitable for grazing and forestry because of steep slope, shallow soil, lack of water, or too much water		Not to be used for grazing or logging; steep slope and lack of soil present problems

Phytoremediation—Using Plants to Clean Up Polluted Soil

Polluted soil poses a severe problem for both ecosystem health and land development. Because soil lies at the confluence of many systems, soil pollution can be spread to other parts of the natural environment. Groundwater, for instance, percolates through the soil and can carry the soil pollutants into streams, rivers, wells, and drinking water. Erosion can create the same problem. Plants growing on polluted soil may contain harmful levels of pollutants themselves. Dust blown from polluted soil can be inhaled directly. In an urban setting polluted soil makes valuable open land unusable for parks or commercial development.

Despite the benefits of cleaning polluted soil, remediation often fails to take place because of the cost and effort of the work. Both soil minerals and soil pollutants carry small electric charges that can cause each to bond with the other, thus making polluted soil very hard to clean. Soil is also a dense medium. This causes excavation of polluted soil for off-site treatment or disposal to be very expensive because of the time, labor, and heavy machinery necessary to do the job. Therefore, cheaper on-site remediation techniques have been the focus of much attention. One of the most interesting and promising of these techniques is phytoremediation.

Phytoremediation is the use of specialized plants to clean up polluted soil. While most plants exposed to high levels of soil toxins will die, researchers have discovered that certain plants are resistant, and an even smaller group actually thrive. Both groups of plants are of interest, but the thriving plants show a particular potential for remediation because it has been shown that some of them actually transport and accumulate high levels of soil pollutants within their bodies. These plants are called **hyperaccumulators.**

Hyperaccumulators are being used to help clean up heavy metal polluted soil. For example, mustard greens are being used to remove lead from polluted areas in Boston, and hydroponically grown sunflowers are being used to take radioactive metals out of the water surrounding the Chernobyl nuclear power plant.

Most heavy metals are also essential plant nutrients, so plants have the ability to take up the metals. However, on polluted soil, the levels of heavy metals are often hundreds of times greater than normal and can be toxic to plants. Hyperaccumulators actually prefer these high concentrations. Essentially, hyperaccumulators are acting as natural vacuum cleaners, sucking pollutants out of the soil and depositing them in their leaves and shoots. Removing the metals is as simple as pruning or cutting the hyper-accumulators above the ground, not excavating tons of soil.

Resistant, but not hyperaccumulating, plants also have a role in phytoremediation. Organic toxins, those that contain carbon such as the hydrocarbons found in gasoline and other fuels, can be broken down by microbial processes. Resistant plants can thrive on sites that are often too toxic for other plants to grow. They in turn give the microbial processes the boost they need to remove organic pollution more quickly from the soil.

There are many limitations to phytoremediation. It is a slow process that may take many growing seasons before an adequate reduction of pollution is seen. Also, hyperaccumulators can be a pollution hazard themselves. For instance, animals can eat the metal rich plants and cause the toxins to enter the food chain. If the concentration of metals in the plants is thought to be high enough to cause toxicity, there must be a way to segregate the plants from humans and wildlife. Additionally, phytoremediation is still a new practice and its effectiveness in cleaning up various toxins compared to conventional means of treatment is not always known. However, with more research and practice phytoremediation could well increase.

- Do you feel that phytoremediation has a future in cleaning up polluted soil?
- Are there polluted sites in your community where you think phytoremediation could be utilized?

of life. Not all land is suitable for crops or continuous cultivation. Some has highly erodible soil and must never be plowed for use as cropland, but it can still serve other useful functions. With the use of appropriate soil conservation practices, much of the land not usable for crops can be used for grazing, wood production, wildlife production, or scenic and recreational purposes. Figure 13.21 shows land that is not suitable for cultivation because it is too arid but that, if it is used properly, can provide grazing for cattle or sheep.

The land shown in figure 13.22 is not suitable for either crops or grazing because of the steep slope and thin soil. However, it is still a valuable and productive piece of land, since it can be used to furnish lumber, wildlife habitats, and recreational opportunities.

FIGURE 13.21 Noncrop Use of Land to Raise Food As long as it is properly protected and managed, this land can produce food through grazing, but it should never be plowed to plant crops because the topsoil is too shallow and the rainfall is too low.

FIGURE 13.22 Forest and Recreational Use Although this land is not capable of producing crops or supporting cattle, it furnishes lumber, a habitat for wildlife, and recreational opportunities.

Summary

The surface of the Earth is in constant flux. The movement of tectonic plates results in the formation of new land as old land is worn down by erosive activity. Soil is an organized mixture of minerals, organic material, living organisms, air, and water. Soil formation begins with the breakdown of the parent material by such physical processes as changes in temperature, freezing and thawing, and movement of particles by glaciers, flowing water, or wind. Oxidation and hydrolysis can chemically alter the parent material. Organisms also affect soil building by burrowing into and mixing the soil, releasing nutrients, and decomposing.

Topsoil contains a mixture of humus and inorganic material, both of which supply soil nutrients. The ability of soil to grow crops is determined by the inorganic matter, organic matter, water, and air spaces in the soil. The mineral portion of the soil consists of various mixtures of sand, silt, and clay particles.

A soil profile typically consists of the *O* horizon of litter; the *A* horizon, which is rich in organic matter; an *E* horizon from which materials have been leached; the *B* horizon, which accumulates materials leached from above; and the *C* horizon, which consists of slightly altered parent material. Forest soils typically have a shallow *A* horizon and an *E* horizon and a deep, nutrient-rich *B* horizon with much root development. Grassland soils usually have a thick *A* horizon containing most of the roots of the grasses. They lack an *E* horizon and have few nutrients in the thin *B* horizon.

Soil erosion is the removal and transportation of soil by water or wind. Proper use of such conservation practices as contour farming, strip farming, terracing, waterways, windbreaks, and conservation tillage can reduce soil erosion. Misuse reduces the soil's fertility and causes air- and water-quality problems. Land unsuitable for crops may be used for grazing, lumber, wildlife habitats, or recreation.

Acting Green

1. Work with a local group to plant trees and native vegetation along rivers and streams.
2. Compost your biodegradable home and yard waste.
3. Support natural or native landscaping in your community and college.
4. Plant a garden—get some soil under your fingernails!
5. Find out what kind of soil is present where you live. Dig a small hole. Is the soil clay, sand, or rubble from a previous building?

Review Questions

1. How are soil and land different?
2. Name the five major components of soil.
3. Describe the process of soil formation.
4. Name five physical and chemical processes that break parent material into smaller pieces.
5. In addition to fertility, what other characteristics determine the usefulness of soil?
6. How does soil particle size affect texture and drainage?
7. Describe a soil profile.
8. Define erosion.
9. Describe three soil conservation practices that help to reduce soil erosion.
10. Besides cropland, what are other possible uses of soil?

Critical Thinking Questions

1. Minimum tillage soil conservation often uses greater amounts of herbicides to control weeds. What do you think about this practice? Why?
2. As populations grow, should we try to bring more land into food production, or should we use technology to aid in producing more food on the land we already have in production? What are the trade-offs?
3. Given what you know about soil formation, how might you explain the presence of a thick *A* horizon in soils in the North American Midwest?
4. Why should nonfarmers be interested in soil conservation?
5. Imagine that you are a scientist hired to consult on a project to evaluate land-use practices at the edge of a small city. The area in question has deep ravines and hills. What kinds of agricultural, commercial, and logging practices would you recommend in this area to help preserve the environment?
6. Look at your own community. Can you see examples of improper land use (urban or rural)? What are the consequences of these land-use practices? What recommendations would you make to improve land use?

To access additional resources for this chapter, please visit ConnectPlus® at connect.mheducation.com. There you will find interactive exercises, including Google Earth™, additional Case Studies, and SmartBook™, an adaptive eBook that integrates our LearnSmart® adaptive learning technology.

chapter 14

Agricultural Methods and Pest Management

Agricultural use of land results in the modification of the natural world to allow for the growth of crops. Often pest species are controlled with the use of pesticides.

OBJECTIVES

After reading this chapter, you should be able to:

- Describe shifting agriculture.
- Describe situations that require labor- intensive farming practices.
- Explain how mechanization encouraged monoculture farming.
- List the advantages and disadvantages of monoculture farming.
- Explain why chemical fertilizers are used.
- Describe the difference between macronutrients and micronutrients.
- Explain why soil organic matter is important.

- Explain why modern agriculture makes extensive use of pesticides.
- Differentiate between persistent pesticides and nonpersistent pesticides.
- List four problems associated with pesticide use.
- Explain why certain pesticides became biomagnified in food chains.
- Explain why pesticides continue to be used.
- List characteristics typical of a sustainable approach to agriculture.

- Explain why integrated pest management depends on a complete knowledge of the pest's life history.
- Describe techniques used in integrated pest management.
- Describe how a genetically modified organism is created.
- Describe the two most common kinds of genetic modifications introduced into crop plants.
- Describe the economic and social changes typically required when using sustainable agriculture techniques.

The Challenge for Agriculture—Feed More People

Everyone needs agriculture. Agriculture feeds our entire population. In the developing world, agriculture contributes significantly to economic growth and helps in poverty reduction. Agriculture also has one of the highest potentials for reducing carbon emissions and helping vulnerable people adapt to climate change.

Increasing population With a predicted 9.7 billion people on Earth by 2050 (an increase of 36 percent over 2013), agricultural production will have to increase to meet new demands, for food for people, feed for livestock, and fuel for machines. Agriculture must not only meet demands—it must also do so while minimizing its environmental footprint and creating sustainable livelihoods for farmers and others.

Increasing demand With higher incomes, emerging middle classes in developing countries can afford to consume more fruits and vegetables and in particular more meat, which requires much more water and land to produce.

Land resources Land degradation—whether in the form of desertification, deforestation, overgrazing, salinization, or soil erosion—poses a serious threat to long-term food security, especially since arable land is already scarce in Asia and cultivating additional land in Latin America and Africa would come at high environmental and infrastructure costs. Worldwide some 1.3 billion people live on fragile lands with limited possibility for increased agricultural production.

Water availability Agriculture accounts for 70–80 percent of global freshwater consumption, and since other demands for water are expected to increase much faster, the amount available for irrigation will see only a minimal increase. About one-third of the world's people live in water-scarce areas and nearly one-quarter of the world's gross domestic product is produced in those areas; the wider repercussions of an increase in water scarcity go beyond the agricultural sector. Further, it is estimated that the agricultural sector is responsible for up to 30 percent of global greenhouse emissions, which directly contribute to climate change.

Rural poverty Many of the people who live with food insecurity are the rural poor in developing countries. Therefore, we should not ignore the potential of agriculture to achieve the goals of a secure food supply, poverty reduction through improved rural livelihoods, and environmental sustainability through reduced footprint of production and climate change adaptation.

In order to face these challenges, agricultural development will have to focus on the following:

Innovation, research, and education Agriculture is a knowledge-intensive sector. Investment must be made in research and development and in training farmers to use new technology to improve productivity.

Enhancing sustainability The world will need to produce more with less to meet demand and reduce its environmental footprint. This will require more efficient use of land and water resources and a reduction in losses that occur between the farm and consumer. Even developed countries experience losses of food during harvest, transport, and distribution to stores, restaurants, and households.

Reducing poverty The rural poor are locked into a cycle of poverty that makes it difficult for them to make changes in traditional farming practices. They often lack access to profitable markets because they do not have available transportation or methods of preserving their produce. Government investments in transportation infrastructure, marketing mechanisms, and food preservation methods would reduce waste and increase incomes for farmers. Because farmers have improved incomes they are likely to be willing and can afford to try new crops or farming practices.

14.1 The Development of Agriculture

Our early ancestors obtained food from nature by hunting and gathering. The development of agriculture involved manipulating the natural environment to produce the kinds of foods humans want and allowed for an increase in the size of the human population. The history of the development of agriculture has involved various kinds of innovations. One of the simplest is shifting agriculture, also known as "slash and burn" agriculture.

Shifting Agriculture

Shifting agriculture involves cutting down and burning the trees and other vegetation in a small area of forest. (See figure 14.1.) Burning releases nutrients that were tied up in the biomass and allows a few crops to be raised before the nutrients in the soil are depleted.

FIGURE 14.1 Shifting Agriculture In many areas of the world where the soils are poor and human populations are low, crops can be raised by disturbing small parts of the ecosystem followed by several years of recovery. The burning of vegetation releases nutrients that can be used by crops for one or two years before the soil is exhausted. The return of the natural vegetation prevents erosion and repairs the damage done by temporary agricultural use.

Once the soil is no longer suitable for raising crops (within two or three years), the site is abandoned and, over time the area returns to forest through the process of succession.

In some parts of the world with poor soil and low populations, shifting agriculture is still used successfully. This method is particularly useful on thin, nutrient-poor tropical soils and on steep slopes. Tropical areas with high rainfall are nutrient poor because the rain washes the nutrients from the soil. The small size of the openings in the forest and their temporary existence prevent widespread damage to the soil, and erosion is minimized. While this system of agriculture is successful when human population densities are low, it is not suitable for large, densely populated areas. When populations become too large, the size and number of the garden plots increase and the time between successive uses of the same plot of land decreases. When a large amount of the forest is disturbed and the time between successive uses is decreased, the forest cannot return and repair the damage done by the previous use of the land, and the nature of the forest is changed.

The traditional practices of the people who engage in small-scale, shifting agriculture have been developed over hundreds of years and often are more effective for their local conditions than other methods of gardening. Typically, these gardens are planted with a mixture of plants, a system known as **polyculture.** Mixing plants together in a garden often is beneficial, since shade-requiring species may be helped by taller plants, or nitrogen-fixing legumes may provide nitrogen for species that require it. In addition, mixing species may reduce insect pest problems because some plants produce molecules that are natural insect repellents. The small, isolated, temporary nature of the gardens also reduces the likelihood of insect infestations. While today we see this form of agriculture practiced most commonly in tropical areas, it is important to note that many Native American cultures used shifting agriculture and polyculture in temperate areas.

Labor-Intensive Agriculture

In areas of the world with better soils, more intense forms of agriculture developed that involved a great deal of manual labor to till, plant, and harvest the crop. This style of agriculture is still practiced in much of the developing world today. Three factors favor labor-intensive farming: (1) the growing site does not allow for mechanization, (2) the kind of crop requires much hand labor, or (3) the economic condition of the people does not allow them to purchase the tools and machines used for mechanized agriculture.

Crops or terrain that requires that fields be small discourages mechanization, since large tractors and other machines cannot be used efficiently on small, oddly shaped fields. Many mountainous areas of the world fit into this category. Many densely populated countries have numerous small farms of 1 to 2 hectares (a hectare is about the size of a soccer field) that can be effectively managed with human labor, supplemented by that of draft animals and a few small gasoline-powered engines. (See figure 14.2.) In addition, in the less-developed regions of the world, the cost of labor is low, which encourages the use of hand labor rather than relatively expensive machines to do planting, weeding, and other activities.

In addition, some crops require such careful handling in planting, weeding, or harvesting that large amounts of hand labor are required. The planting of paddy rice and the harvesting of many fruits and vegetables are examples. However, the primary reason for labor-intensive farming is economic.

The mechanized farming typical of much of the developed world requires large tracts of land. These must either be purchased or created by forming larger cooperative farms from many small units. Even if social and political obstacles to such large landholdings could be overcome, there is still the problem of obtaining the necessary capital to purchase the machines. Large parts of the developing world, including much of Africa and many areas in Asia, are in this situation. About 60 percent of the population of Africa and southern Asia (India, Pakistan, Bangladesh, Afghanistan, Indonesia, Thailand, etc.) is rural. Many of these people are engaged in nonmechanized agriculture.

Hand labor required

Small irregular fields

FIGURE 14.2 Labor-Intensive Agriculture In many of the less-developed countries of the world, the extensive use of hand labor allows for impressive rates of production with a minimal input of fossil fuels and fertilizers. This kind of agriculture is also necessary in areas that have only small patches of land suitable for farming.

Mechanized Monoculture Agriculture

Mechanized agriculture is typical of North America, Australia, New Zealand, and much of South America and Europe. In large measure, machines and fossil-fuel energy replace the energy formerly supplied by human and animal muscles. Mechanization requires large expanses of fairly level land for the machines to operate effectively. In addition, large tracts of land must be planted in the same crop for efficient planting, cultivating, and harvesting, a practice known as **monoculture.** (See figure 14.3.)

Energy and Mechanized Agriculture

Mechanized agriculture has substituted the energy stored in petroleum products for the labor of humans. For example, in the United States in 1913, it required 135 hours of labor to produce 2,500 kilograms (5,500 pounds) of corn. In 1980, it required only about 15 hours of labor to produce the same amount of corn. The energy

FIGURE 14.3 Mechanized Monoculture This wheat field is an example of monoculture, a kind of agriculture that is highly mechanized and requires large fields for the efficient use of machinery. In this kind of agriculture, machines and fossil-fuel energy have replaced the energy of humans and draft animals.

supplied by petroleum products replaced the equivalent of 120 hours of labor. Energy is needed for tilling, planting, harvesting, and pumping irrigation water. The manufacture of fertilizer and pesticides also requires the input of large amounts of fossil fuels, both as a source of energy for the industrial process and as raw material from which these materials are made. For example, about 5 metric tons of fossil fuel are required to produce about 1 metric ton of fertilizer. In addition, the pesticides used in mechanized agriculture require the use of oil as a raw material. Since the developed world depends on oil to run machines and manufacture fertilizer and pesticides to support its agriculture, any change in the availability or cost of oil will have a major impact on the world's ability to feed itself.

Issues with Monoculture

Even though mechanized monoculture is an efficient method of producing food, it is not without serious drawbacks.

- *Protecting the soil*—Traditionally, mechanized farming practices removed much of the organic matter each year when the crop was harvested. This tended to reduce soil organic matter. As agricultural scientists and farmers have recognized the need to improve the organic matter in soils, many farmers have been leaving increased amounts of organic matter after harvest, or they specifically plant a crop that is later plowed under to increase the soil's organic content. These soil improvement practices also help to reduce soil erosion.

    Soil erosion is most severe when soil is left uncovered by vegetation or unharvested plant parts. Because of problems with erosion, many farmers are now using methods that reduce the time the fields are left bare. They use no-till farming and other techniques that reduce erosion. (See chapter 13.)

- *Loss of genetic diversity*—To ensure that a crop can be planted, tended, and harvested efficiently by machines, farmers plant fields with seeds that are genetically identical. These special seeds can ensure that all the plants germinate at the same time, resist the same pests, ripen at the same time, and grow to the same height. These are valuable characteristics to the farmer, but the plants in the field have little genetic variety. When all the farmers in an area plant the same genetic varieties, pest control becomes a serious problem. If diseases or pests begin to spread, the magnitude of the problem becomes devastating because all the plants have the same characteristics and, thus, are susceptible to the same diseases. If genetically diverse crops are planted or crops are rotated from year to year, this problem is not as great.

- *Reliance on fertilizer and pesticides*—Because farm equipment is expensive, farmers tend to specialize in a few crops. This means that the same crop may be planted in the same field several years in a row. This lack of crop rotation may deplete certain essential soil nutrients, thereby requiring special attention be paid to soil chemistry. Fertilizer must be used to replace the soil nutrients needed by plants. In addition, planting the same crop repeatedly encourages the growth of insect and fungus pest populations because they have a huge food supply at their disposal. This requires the frequent use of insecticides and fungicides or other methods of pest control.

14.2 Fertilizer and Agriculture

Fertilizers are valuable because they replace the soil nutrients removed by plants. Some of the chemical building blocks of plants, such as carbon, hydrogen, and oxygen, are easily replaced by carbon dioxide from the air and water from the soil, but others are less easily replaced. The three primary soil nutrients often in short supply are nitrogen, phosphorus, and potassium compounds. They are often referred to as **macronutrients** and are the common ingredients of chemical fertilizers.

Certain other elements are necessary in extremely small amounts and are known as **micronutrients.** Examples are boron, zinc, and manganese. As an example of the difference between macronutrients and micronutrients, harvesting 1 metric ton of potatoes removes 10 kilograms (22 pounds) of nitrogen (a macronutrient) but only 13 grams (0.03 pound) of boron (a micronutrient).

The replacement of macronutrients and micronutrients is important because when the crop is harvested, the chemical elements that are a part of the crop are removed from the field. Since many of those elements originated from the soil, they need to be replaced if another crop is to be grown.

Various experts estimate that approximately 25 percent of the world's crop yield can be directly attributed to the use of chemical fertilizers. The use of fertilizer has increased significantly over the last few decades and is projected to increase even more. (See figure 14.4.) However, since fertilizer production relies on energy from fossil fuels, the price and availability of chemical fertilizers are strongly influenced by world energy prices. If the price of oil increases, the price of fertilizer goes up, and so does the cost of food. This is felt most acutely in parts of the world where people are poor since the farmers are unable to buy fertilizer and crop yields fall accordingly.

Although chemical fertilizers replace inorganic nutrients, they do not replace soil organic matter. Organic material is important because it modifies the structure of the soil, preventing compaction and maintaining pore space, which allows water and air to move to the roots. The decomposition of organic matter produces humus, which helps to maintain proper soil chemistry because it tends to loosely bind many soil nutrients and other molecules and modifies the pH so that nutrients are not released too rapidly. Soil bacteria and other organisms use organic matter as a source of energy. Since these organisms serve as important links in the carbon and nitrogen cycles, the presence of organic matter is important to their function. Thus, total dependency on chemical fertilizers usually reduces the amount of organic matter and can change the physical, chemical, and biological properties of the soil.

As water moves through the soil, it dissolves soil nutrients (particularly nitrogen compounds) and carries them into streams and lakes, where they may encourage the growth of unwanted plants and algae. This is particularly true when fertilizers are applied at the wrong time of the year, just before a heavy rain, or

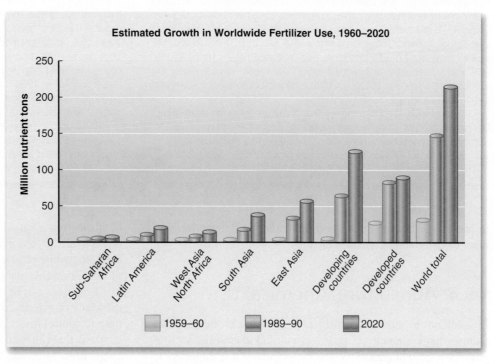

FIGURE 14.4 Increasing Fertilizer Use The use of fertilizer is increasing rapidly. Growth in use will be most rapid in the developing countries over the next 20 years.

Science, Politics, & Policy

Regulation of Pesticides

Countries in the developed world have established careful regulations for the sale and use of pesticides. In the United States, the Environmental Protection Agency requires careful studies of the effectiveness and possible side effects of each pesticide licensed for use. It has banned several pesticides from further use because new information suggested that they were not as safe as originally thought. For example, it banned the use of dinoseb, a herbicide, because tests by the German chemical company Hoechst AG indicated that dinoseb causes birth defects in rabbits. Other studies indicate that dinoseb causes sterility in rats. Similarly, in 1987, Velsicol Chemical Corporation signed an agreement with the EPA to stop producing and distributing chlordane in the United States. Chlordane had been banned previously for all applications except for termite control. After it was shown that harmful levels of chlordane could exist in treated homes, its use was finally curtailed. Also in 1987, the Dow Chemical Company announced that it would stop producing the controversial herbicide 2,4,5-T. In both of these cases, new pesticides were developed to replace the older types, so that discontinuing them did not result in an economic hardship for farmers or other consumers.

Perhaps the best-known regulation of a pesticide is that of DDT. The accumulated scientific evidence of the negative effects of DDT caused it to be banned for all agricultural use in the United States in 1972. Although the ban was challenged in the courts, it was upheld. The results of this ban are seen in the reduced levels of DDT in the environment and in the recovery of populations of such organisms as eagles and pelicans.

The United Nations Environment Program was asked to ban DDT from use throughout the world in 1999. This was because DDT was showing up worldwide in the bodies of many kinds of animals that were great distances from sources of DDT and was present in the breast milk of women where DDT was used. This resulted in a great deal of protest from many public health professionals concerned about controlling malaria, since DDT is still widely used in many parts of the world to kill mosquitoes that spread malaria. They feared that a total ban of DDT would result in less effective control of mosquitoes and increased deaths from malaria. DDT was subsequently banned for agricultural use worldwide under the Stockholm Convention but its use in disease vector control continues and remains controversial.

In the 1990s, the EPA required that many pesticides be reregistered. During this process, manufacturers needed to justify the continued production of their products. In many cases, companies did not want to spend the money to go through the registration process; the product was then

withdrawn from the market. In other cases, the variety of uses was modified. For example, chlorpyrifos, one of the most widely used insecticides, had its use altered. Through negotiations with the producers, its use in homes and schools was eliminated, but it was allowed in most agricultural settings.

Today, before a company can sell or distribute any pesticide in the United States, the EPA must review studies on the pesticide to determine that it will not pose unreasonable risks to human health or the environment. Once the EPA has made that determination, it will license or register that pesticide for use in accordance with label directions. Before allowing a pesticide to be used on a food commodity the EPA sets limits on how much of a pesticide may be used on food during growing and processing, and how much can remain on the food. The Food Quality Protection Act of 1996 further stated that the public's overall exposure to pesticides (through food, water, and home environments) must be considered when making decisions to set standards for pesticide use on food.

Many in the corporate world feel that we overregulate pesticides in the United States. On the other side there are strong supporters of stricter regulations. What are your thoughts?

---

in such large amounts that the plants cannot efficiently remove them from the soil before they are lost.

14.3 Agricultural Chemical Use

In addition to chemical fertilizers, mechanized monoculture requires large amounts of other agricultural chemicals, such as pesticides, growth regulators, and preservatives. It is important to have an understanding of basic terms used to discuss the various chemicals employed to control pests. A **pesticide** is any chemical used to kill or control populations of unwanted fungi, animals,

or plants, often called **pests.** The term *pest* is not scientific but refers to any organism that is unwanted. Insects that feed on crops are pests, while others, such as bees, are beneficial for pollinating plants. Unwanted plants are generally referred to as **weeds.**

Pesticides can be subdivided into several categories based on the kinds of organisms they are used to control. **Insecticides** are used to control insect populations by killing them. Unwanted fungal pests that can weaken plants or destroy fruits are controlled by **fungicides.** Mice and rats are killed by **rodenticides,** and plant pests (weeds) are controlled by **herbicides.** Since pesticides do not kill just pests but can kill a large variety of living things, including humans, these chemicals might be more appropriately called

biocides. A perfect pesticide is one that kills or inhibits the growth of only the specific pest organism causing a problem. The pest is often referred to as the **target organism.** However, most pesticides are not very specific and kill many **nontarget organisms** as well. For example, most insecticides kill both beneficial and pest species, rodenticides kill other animals as well as rodents, and most herbicides kill a variety of plants, both pests and nonpests.

Many of the older pesticides were very stable and remained active for long periods of time. These are called **persistent pesticides.** Pesticides that break down quickly are called **nonpersistent pesticides.**

Insecticides

If insects are not controlled, they consume a large proportion of the crops produced by farmers. In small garden plots, insects can be controlled by manually removing them and killing them. However, in large fields, this is not practical, so people have sought other ways to control pest insects.

In addition, many insects harm humans because they spread diseases, such as sleeping sickness, bubonic plague, and malaria. Mosquitoes are known to carry over 30 diseases harmful to humans. Currently, the World Health Organization estimates that there are about 250 million cases of malaria and about 800,000 people die of the disease each year. Most of these deaths are children. Malaria is one of the top five causes of death in children. The discovery of chemicals that could kill insects was celebrated as a major advance in the control of disease and the protection of crops.

Nearly 3,000 years ago, the Greek poet Homer mentioned the use of sulfur to control insects. For centuries, it was known that natural plant products could repel or kill insect pests. Plants with insect-repelling abilities were planted with crops to help control the pests. Nicotine from tobacco, rotenone from tropical legumes, and pyrethrum from chrysanthemums were extracted and used to control insects. In fact, these compounds are still used today. However, because plant products are difficult to extract and apply and have short-lived effects, other compounds were sought. In 1867, the first synthetic inorganic insecticide, Paris green, was formulated. It was a mixture of acetate and arsenide of copper and was used to control Colorado potato beetles.

The first synthetic organic insecticide to be used was DDT [1,1,1-trichloro-2,2-*bis*-(p-chlorophenyl)ethane]. DDT was originally thought to be the perfect insecticide. It was inexpensive, long-lasting, relatively harmless to humans, and very deadly to insects. After its discovery, it was widely used in agriculture and to control disease-carrying insects. During the first ten years of its use (1942–52), DDT is estimated to have saved five million lives, primarily because of its use in controlling disease-carrying mosquitoes.

However, scientists began to recognize several problems associated with the use of DDT. They documented that insect populations that were repeatedly subjected to spraying by DDT developed resistance to it and larger doses were required to kill the pest insects. Ecologists and crop scientists did studies on how the death of nontarget organisms affected surrounding ecosystems and natural predators and parasites of pest insects. This was a particular problem because DDT is a persistent chemical. Studies of the breakdown of DDT in the environment showed that it had a long half-life

(the amount of time required for half of the chemical to decompose). In temperate regions of the world, DDT has a half-life of 10 to 15 years. This means that if 1,000 kilograms (2,200 pounds) of DDT were sprayed over an area, 500 kilograms (1,100 pounds) would still be present in the area 10 to 15 years later; 30 years from the date of application, 250 kilograms (550 pounds) would still be present. The half-life of DDT varies depending on soil type, temperature, the kinds of soil organisms present, and other factors. In tropical parts of the world, the half-life may be as short as six months.

Finally, it was discovered that because it is persistent, DDT tends to accumulate and reach higher concentrations in older animals and in animals at higher trophic levels. The problem was particularly acute in species of birds such as eagles, pelicans, and cormorants that were predators on fish. Scientists documented that as DDT levels in the birds increased, they produced eggs with thin shells, which were easily broken in the nest. Consequently, the populations of these birds dropped precipitously.

Since DDT was introduced, over 60,000 different compounds that have potential as insecticides have been synthesized. However, most of these have never been put into production because cost, human health effects, or other drawbacks make them unusable. Several categories of these compounds have been developed. Three that are currently used are chlorinated hydrocarbons, organophosphates, and carbamates.

Chlorinated Hydrocarbons

Chlorinated hydrocarbons are a group of pesticides of complex, stable structure that contain carbon, hydrogen, and chlorine. DDT was the first such pesticide manufactured, but several others have been developed. Other chlorinated hydrocarbons are chlordane, aldrin, heptachlor, dieldrin, and endrin. It is not fully understood how these compounds work, but they are believed to affect the nervous systems of insects, resulting in their death.

One of the major characteristics of these pesticides is that they are very stable chemical compounds. This is both an advantage and a disadvantage. They can be applied once and be effective for a long time. However, since they do not break down easily, they tend to accumulate in the soil and in the bodies of animals in the food chain. Thus, they affect many nontarget organisms, not just the original target insects.

Because of their negative effects, most of the chlorinated hydrocarbons are no longer used in most parts of the world. DDT, aldrin, dieldrin, toxaphene, chlordane, and heptachlor have been banned in the United States and many other developed countries. The use of DDT was prohibited in the United States in the early 1970s. Aldrin, dieldrin, heptachlor, and chlordane have also been prohibited from use on crops, although heptachlor and chlordane were still used for termite control until recently. In 1987, Velsicol Chemical Corporation agreed to stop selling chlordane in the United States. However, many developing countries still use chlorinated hydrocarbons for insect control to protect crops and public health. Because of their persistence and continued use in many parts of the world, chlorinated hydrocarbons are still present in the food chain, although the level of contamination has dropped. These molecules continue to enter parts of the world where their use has been banned through the atmosphere and as trace contaminants of imported products.

Organophosphates and Carbamates

Because of the problems associated with persistent insecticides, nonpersistent insecticides that decompose to harmless products in a few hours or days were developed. However, like other insecticides, these are not species-specific; they kill beneficial insects as well as harmful ones. Although the short half-life prevents the accumulation of toxic material in the environment, it is a disadvantage for farmers, since more frequent applications are required to control pests. This requires more labor and fuel and, therefore, is more expensive.

Both **organophosphates** and **carbamates** work by interfering with the ability of the nervous system to conduct impulses normally. Under normal conditions, a nerve impulse is conducted from one nerve cell to another by means of a chemical known as a neurotransmitter. One of the most common neurotransmitters is acetylcholine. When this chemical is produced at the end of one nerve cell, it causes an impulse to be passed to the next cell, thereby transferring the nerve message. As soon as this transfer is completed, an enzyme known as cholinesterase destroys acetylcholine, so the second nerve cell in the chain is stimulated for only a short time. Organophosphates and carbamates interfere with cholinesterase, preventing it from destroying acetylcholine. This results in nerve cells being continuously stimulated, causing uncontrolled spasms of nervous activity and uncoordination that result in death.

Although these pesticides are less persistent in the environment than are chlorinated hydrocarbons, they are generally much more toxic to humans and other vertebrates because they affect the nervous system of all animals—not just insects. Persons who apply such pesticides must use special equipment and should receive special training because improper use can result in death. Since organophosphates interfere with cholinesterase more strongly than do carbamates, they are considered more dangerous and, for many applications, have been replaced by carbamates.

Common organophosphates are malathion, parathion, and diazinon. Malathion is widely used for such projects as mosquito control, but parathion is a restricted organophosphate because of its high toxicity to humans. Diazinon is widely used in gardens. Carbaryl, propoxur, and aldicarb are examples of carbamates. Carbaryl (Sevin) is widely used in home gardens and in agriculture to control many kinds of insects. Propoxur is not used on crops but is used to control insects around homes and farms. Aldicarb is a restricted-use insecticide that is used primarily on cotton, soybeans, and peanuts. It has been associated with groundwater contamination and has been discontinued for some uses, such as control of insects in potatoes.

Herbicides

Weeds are plants we do not want to have growing in a particular place. Weed control is extremely important for agriculture because weeds take nutrients and water from the soil, making them unavailable to the crop species. In addition, weeds may shade the crop species and prevent it from getting the sunlight it needs for rapid growth. At harvest time, weeds reduce the efficiency of harvesting machines. Also, weeds generally must be sorted from the crop before it can be sold, which adds to the time and expense of harvesting.

Traditionally, farmers have expended much energy trying to control weeds. Initially, weeds were eliminated with manual labor and the hoe. Tilling the soil also helps to control weeds. Once the crop is planted, row crops such as corn or sugar beets may be cultivated to remove weeds from between the rows. All of these activities are expensive in terms of time and fuel. Selective use of herbicides (commonly called weed killers) can have a tremendous impact on a farmer's profits.

Herbicides are used heavily in genetically modified crops. In the United States in 2012, 2.1 billion kilograms (4.6 billion pounds) of pesticides were used on three crops; corn, soybeans, and cotton. Ninety percent were herbicides. In addition to use on crops, herbicides are widely used to control unwanted vegetation along power-line rights-of-way, railroad rights-of-way, highways, and to control weeds in lawns.

Many of the recently developed herbicides can be very selective if used appropriately. Some are used to kill weed seeds in the soil before the crop is planted, while others are used after the weeds and the crop begin to grow. In some cases, a mixture of herbicides can be used to control several weed species simultaneously. Figure 14.5 shows the effects of using a herbicide that kills weeds in a soybean field.

Several major types of herbicides are in current use. One type is synthetic plant-growth regulators that mimic natural-growth regulators known as **auxins.** Two of the earliest herbicides were of this type: 2,4-dichlorophenoxyacetic acid (2,4-D) and 2,4,5- trichlorophenoxyacetic acid (2,4,5-T). When applied to broad-leaf plants, these chemicals disrupt normal growth, causing the death of the plant. 2,4-D has been in use for about 50 years and is still one of the most widely used herbicides. Many newer herbicides have other methods of action. Some disrupt the photosynthetic activity of plants, causing their death. Others inhibit enzymes, precipitate proteins, stop cell division, or destroy cells directly. Depending on the concentration of the herbicide used, some are toxic to all plants, while others are very selective as to which species of plants they affect. One such herbicide is diuron. In proper

FIGURE 14.5 The Effect of Herbicides The weeds in this photograph have been treated with herbicides. The soybeans are unaffected and grow better without competition from weeds.

concentrations and when applied at the appropriate time, it can be used to control annual grasses and broad-leaf weeds in over 20 different crops. However, at higher concentrations, it kills all vegetation in an area. Fenuron is an herbicide that kills woody plants. In low concentration, it is used to control woody weed plants in cropland. In high concentrations, it is used on noncroplands, such as power-line rights-of-way. (See figure 14.6.)

Atrazine is widely used to control broad-leaf and grassy weeds in corn, sorghum, sugarcane, pineapple, and other crops, and on Christmas tree and conifer reforestation plantings. Glyphosate is a broad-spectrum, nonselective, systemic herbicide used to control annual and perennial plants, including grasses, sedges, broad-leaved weeds, and woody plants. It can be used on noncropland as well as on a great variety of crops.

Fungicides and Rodenticides

Fungus pests can be divided into two categories. Some are natural decomposers of organic material, but when the organic material being destroyed happens to be a crop or other product useful to humans, the fungus is considered a pest. Other fungi are parasites on crop plants; they weaken or kill the plants, thereby reducing the yield. Fungicides are used as fumigants (gases) to protect agricultural products from spoilage, as sprays and dusts to prevent the spread of diseases among plants, and as seed treatments to protect seeds from rotting in the soil before they have a chance to germinate. Methylmercury is often used on seeds to protect them from spoilage before germination. However, since methylmercury is extremely toxic to humans, these seeds should never be used for food. To reduce the chance of a mix-up, treated seeds are usually dyed a bright color.

Like fungi, rodents are harmful because they destroy food supplies. In addition, they can carry disease and damage crops in the field. In many parts of the world, such as India, the government pays a bounty to people who kill rats, because this is an inexpensive way to protect the food supply. Several kinds of rodenticides have been developed to control rodents. One of the most widely used is warfarin, a chemical that causes internal bleeding in animals that consume it. It is usually incorporated into a food substance so that rodents eat warfarin along with the bait. Because it is effective in all mammals, including humans, it must be used with care to prevent nontarget animals from having access to the chemical. As with many kinds of pesticides, some populations of rodents have become tolerant of warfarin, while others avoid baited areas. In many cases, rodent problems can be minimized by building storage buildings that are rodent-proof, rather than relying on rodenticides.

Other Agricultural Chemicals

In addition to herbicides, other agrochemicals are used for special applications. For example, a synthetic auxin sprayed on cotton plants before harvest causes the leaves to drop off. This makes harvesting easier because the leaves do not clog the machine that picks the cotton.

NAA (naphthaleneacetic acid) is used by fruit growers to prevent apples from dropping from the trees and being damaged. This chemical can keep the apples on the trees for up to 10 extra days, which allows for a longer harvest period and fewer lost apples.

FIGURE 14.6 Herbicide Use to Maintain Rights-of-Way Power-line rights-of-way are commonly maintained by herbicides that kill the woody vegetation, which might grow so tall that it interferes with the power lines.

FIGURE 14.7 Chemical Loosening of Cherries to Allow Mechanical Harvesting By using chemicals and machinery, this farmer can rapidly harvest the cherry crop. This practice reduces the amount of labor required to pick the cherries but requires the application of chemicals to loosen the fruit.

Under other conditions, it may be valuable to get fruit to fall more easily. Cherry growers use ethephon to promote loosening of the fruit so that the cherries will fall more easily from the tree when shaken by the mechanical harvester. This method lowers the cost of harvesting the fruit. (See figure 14.7.)

14.4 Problems with Pesticide Use

A perfect pesticide would have the following characteristics:

1. It would be inexpensive.
2. It would affect only the target organism.
3. It would have a short half-life.
4. It would break down into harmless materials.

However, the perfect pesticide has not been invented. Many of the more recently developed pesticides have fewer drawbacks than the early pesticides, but none is without problems.

Focus On

The Dead Zone of the Gulf of Mexico

Each summer, a major "dead zone" of about 18,000 square kilometers (7,000 square miles) develops in the Gulf of Mexico off the mouth of the Mississippi River. This dead zone contains few fish and bottom-dwelling organisms. It is caused by low oxygen levels (hypoxia) brought about by the rapid growth of algae and bacteria in the nutrient-rich waters. The nutrients can be traced to the extensive use of fertilizer in the major farming areas of the central United States, farming areas drained by the Mississippi River and its tributaries. About 1.6 million metric tons of nitrogen—mostly fertilizer runoff from midwestern farms—flows out of the Mississippi River every year.

The hypoxia problem begins when nitrogen and other nutrients wash down the Mississippi River to the Gulf of Mexico, where they trigger a bloom of photosynthetic bacteria and algae that are consumed by protozoa and tiny animals. The dead cells and fecal matter of the organisms then fall to the seafloor. As growing colonies of decomposer bacteria digest this waste, they consume dissolved oxygen faster than it can be replenished. The flow of oxygen-rich water from the Mississippi cannot rectify the problem, because differences in temperature and density cause the warm freshwater to float above the colder, salty ocean water.

Crustaceans, worms, and any other animals that cannot swim out of the hypoxic zone die.

According to the EPA, nutrient pollution has degraded more than half of U.S. estuaries. The National Research Council named nutrient pollution and the sustainability of fisheries as the most important problems facing the U.S. coastal waters in the next decade.

Gulf of Mexico fisheries, which generate some $2.8 billion a year in revenue, are one potential casualty of hypoxia. Hypoxia can block crucial migration of shrimp, which must move from inland nurseries to feed and spawn offshore. In other places in the world, such as the Black and Baltic Seas, hypoxia has been responsible for the collapse of some commercial fisheries.

Several approaches could reduce the amount of hypoxia-causing nitrogen released into the Mississippi River Basin. These include:

- Reduce use of nitrogen-based fertilizers and improve storage of manure. Reduce runoff from feedlots.

- Plant perennial crops instead of fertilizer-intensive corn and soybeans on 10 percent of the acreage.

- Remove nitrogen and phosphorus from domestic wastewater.

- Restore 2 million to 4 million hectares (5–10 million acres) of wetlands, which absorb nitrogen runoff.

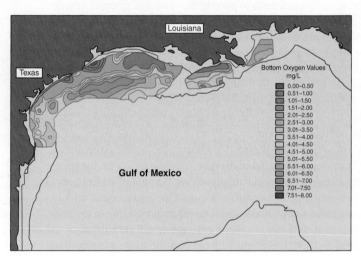

Urban and agricultural discharge into the Mississippi River Basin includes nutrients such as nitrogen and phosphorus that are very important to the rapid growth of algae and bacteria.
Source: Image on left and data for map from NOAA.

Persistence

Although the trend has been away from using persistent pesticides in North America and much of the developed world, some are still allowed for special purposes, and they are still in common use in other parts of the world. Because of their stability, these chemicals have become a long-term problem. Persistent pesticides become attached to small soil particles, which are easily moved by wind or water to any part of the world. Persistent pesticides and other pollutants have been discovered in the ice of the poles and are present in detectable amounts in the body tissues of animals, including humans, throughout the world. Thus, chemicals originally sprayed

to control mosquitoes in Africa or to protect a sugarcane field in Brazil may be distributed throughout the world.

Bioaccumulation and Biomagnification

A problem associated with persistent chemicals is that they may accumulate in the bodies of animals. If an animal receives small quantities of persistent pesticides or other persistent pollutants in its food and is unable to eliminate them, the concentration within the animal increases. This process of accumulating higher and higher amounts of material within the body of an animal is called

bioaccumulation. Many of the persistent pesticides and their breakdown products are fat soluble and build up in the fat of animals. When affected animals are eaten by a carnivore, these toxins are further concentrated in the body of the carnivore, causing disease or death, even though lower-trophic-level organisms are not injured. This phenomenon of acquiring increasing levels of a substance in the bodies of higher-trophic-level organisms is known as **biomagnification.**

The well-documented case of DDT is an example of how biomagnification occurs. DDT is not very soluble in water but dissolves in oil or fatty compounds. When DDT falls on an insect or is consumed by the insect, the DDT is accumulated in the insect's fatty tissue. Large doses kill insects but small doses do not, and

their bodies may contain as much as one part per billion of DDT. This is not very much, but it can have a tremendous effect on the animals that feed on the insects.

If an aquatic habitat is sprayed with a small concentration of DDT or receives DDT from runoff, small aquatic organisms may accumulate a concentration that is up to 250 times greater than the concentration of DDT in the surrounding water. These organisms are eaten by shrimp, clams, and small fish, which are, in turn, eaten by larger fish. DDT concentrations of large fish can be as much as 2,000 times the original concentration sprayed on the area. What was a very small initial concentration has now become so high that it could be fatal to animals at higher trophic levels. Eagles, osprey, cormorants, and pelicans are particularly susceptible species. (See figure 14.8.)

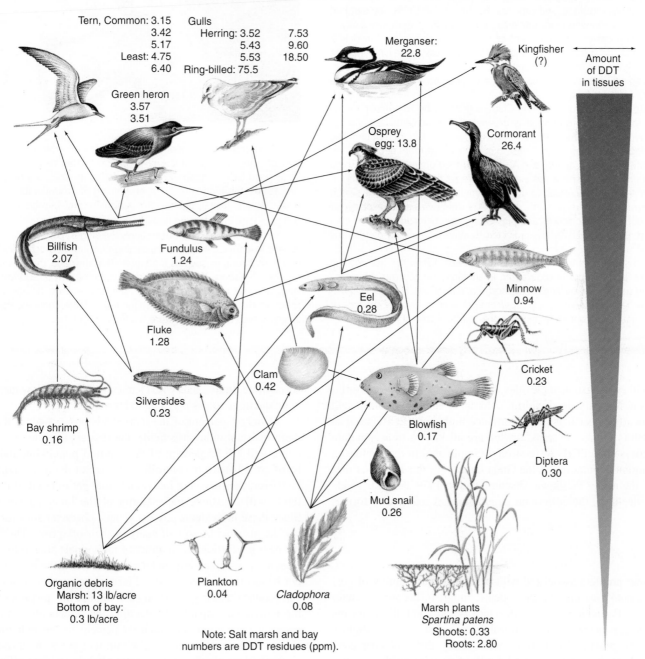

FIGURE 14.8 The Biomagnification of DDT All the numbers shown are in parts per million (ppm). A concentration of one part per million means that in a million equal parts of the organism, one of the parts would be DDT. Notice how the amount of DDT in the bodies of the organisms increases as we go from producers to herbivores to carnivores. Because DDT is persistent, it builds up in the top trophic levels of the food chain.

Focus On

Economic Development and Food Production in China

When the economic conditions of a people improve, the change is reflected in the quality of their diet. In particular, meat becomes a bigger part of the diet. Therefore, people feed some of their grains to animals and consume animal products such as meat, fish, eggs, and butter. They eat more vegetables and use more edible oils. In short, they move a bit higher up the food chain.

A sizable percentage of China's 1.3 billion citizens are moving up the food chain. In the early 1980s, the typical urban Chinese diet consisted of rice, porridge, and cabbage. By the middle 1990s, the diet had dramatically changed to include meat, eggs, or fish at least once a day.

China has about 19 percent of the world's population but only 7 percent of the arable land; in other words, it has less arable land than the United States but more than four times the population. China's farm sector is simply unable to keep up with the surging demand for what people regard as better food.

Total meat consumption in China is growing by 10 percent a year; feed for animal consumption is growing by 15 percent. Demand for poultry, which requires 2 to 3 kilograms of feed per kilogram of bird, has doubled in five years. China's new diet will make it more dependent on the United States, Canada, and Australia for feed grains. By 1997, China had gone from being a net exporter of grain to importing 16 million metric tons. This increased to 60 million metric tons in 2010. The switch in corn is even more dramatic. Until 2009 China was a net exporter of corn but with the chickens and pigs eating so much corn, China began importing corn in 2010. It is projected to import 15 million metric tons by 2015.

To support its increased agricultural needs, China has become the world's largest importer of fertilizer. In addition, the government is converting some marginal lands in the north for agriculture. Even so, its farmland is declining by at least 0.5 percent per year because of the loss of fertile, multiple-cropped farmland in the southern coastal provinces.

As in much of the world, irrigation is responsible for a major portion of agricultural productivity in China. In the Yellow River Valley's large grain belt, irrigation projects tripled crop yields during the 1950s and 1960s, resulting in severe overpumping of groundwater. Today, the water table is falling, aquifers are vanishing, and farmers now have to compete with industry and households for water.

China is only a part—but certainly the biggest part—of a larger story. What's true for China is also happening throughout much of Asia. Large, populous countries such as Indonesia, Thailand, and the Philippines are rapidly urbanizing. They are gaining purchasing power and losing farmland, and the people are adding more animal proteins and processed foods to their diets. Given the magnitude of the changes, the world's ability to feed itself in the future is difficult to predict.

Other persistent molecules are known to behave in similar fashion. Mercury, lead, aldrin, chlordane, and other chlorinated hydrocarbons, such as polychlorinated biphenyls (PCBs) used as insulators in electric transformers, are all known to accumulate in ecosystems. PCBs were strongly implicated in the decline in cormorant populations in the Great Lakes. For this and other reasons, the use of PCBs was discontinued in 1979. As PCB levels have declined, the cormorant population has returned to former levels.

Pesticide Resistance

Another problem associated with pesticides is the ability of pest populations (insects, weeds, rodents, fungi) to become resistant to them. Pesticide resistance develops because not all organisms within a given species are identical. Each individual has a slightly different genetic composition and slightly different characteristics. If an insecticide is used for the first time on the population of a particular insect pest, it kills all the individuals that are susceptible. Individuals with characteristics that allow them to tolerate the insecticides may live to reproduce.

If only 5 percent of the individuals possess genes that make them resistant to an insecticide, the first application of the insecticide will kill 95 percent of the exposed population and so will be of great benefit in controlling the size of the insect population. However, the surviving individuals that are tolerant of the insecticide will constitute the majority of the breeding populations. Since these individuals possess genetic characteristics for tolerating the insecticide, so will many of their offspring. Therefore, in the next generation, the number of individuals able to tolerate the insecticide will increase, and the second use of the insecticide will not be as effective as the first. Since some species of insect pests can produce a new generation each month, this process of selecting individuals capable of tolerating the insecticide can result in resistant populations in which 99 percent of the individuals are able to tolerate the insecticide within five years. As a result, that particular insecticide is no longer as effective in controlling insect pests, and increased dosages or more frequent spraying may be necessary. Over 500 species of insects have populations resistant to insecticides.

For example, a study was done on populations of houseflies found in poultry facilities where chickens were housed and managed to produce eggs. The insecticide Cyfluthrin was used regularly to control fly populations in these facilities. Poultry facilities provide excellent conditions for developing resistance, since the flies spend their entire lives within the facility and are subjected to repeated applications of insecticide. Table 14.1 shows that repeated exposure to the insecticide Cyfluthrin resulted in resistance in the poultry facility housefly population and that a population not previously exposed was not resistant. The standard dose that killed susceptible houseflies did not kill flies from the resistant population. Even at a dosage of 100 times the standard dose, nearly 40 percent of the houseflies in the resistant, poultry facility population, survived.

Just as insects develop resistance to insecticides, weeds develop resistance to herbicides, and fungi develop resistance to fungicides. There are now over 300 examples of weeds that have developed resistance to one or more herbicides and about 100 examples of resistant fungi.

Effects on Nontarget Organisms

Most pesticides are not specific and kill beneficial species as well as pest species. In agriculture, there is less concern about the effects of herbicides on nontarget organisms because a herbicide is chosen that does not harm the desired crop plant, and generally all other plants in the field are competing pests. However, with insecticides, several problems are associated with the effects on nontarget organisms. The use of insecticides can harm populations of birds, mammals, and harmless or beneficial insects. In general, insecticides that harm vertebrates are restricted in their use. However, it is difficult to apply insecticides in such a way that only the harmful species are affected. When beneficial insects are killed by insecticides, pesticide use can be counterproductive. If an insecticide kills predator and parasitic insects that normally control the pest insects, there are no natural checks to control the population growth of the pest species. Additional applications of insecticides are necessary to prevent the pest population from rebounding to levels even higher than the initial one. Once the

decision is made to use pesticides, it often becomes an irreversible tactic, because stopping their use would result in rapid increases in pest populations and extensive crop damage.

An associated problem is that the use of insecticides may change the population structures of the many species present so that a species that was not previously a problem becomes a serious pest. For example, when synthetic organic insecticides came into common use with cotton in the 1940s, the bollworm and tobacco budworm became major pests, because nontarget insects that were parasites and predators on the bollworm and tobacco budworm were eliminated. In the mid-1990s, a similar situation developed when repeated use of malathion killed predator insects and allowed beet army worms to become a major pest in tropical and semitropical areas of the United States. The repeated use of insecticides caused a different pest problem to develop.

Human Health Concerns

Short-term and long-term health effects to persons applying pesticides and the public that consumes pesticide residues in food are also concerns. If properly applied, most pesticides can be used with little danger to the applicator. However, in many cases, people applying pesticides are unaware of how they work and what precautions should be used in their application. About one-third of pesticides are used in developing countries. In many parts of the developing world, regulations regarding use of pesticides may be poorly enforced. In addition, farmers often are not able to understand the caution labels on the packages, do not have access to or use appropriate protective gear, or use poorly functioning equipment. Therefore, many incidences of acute poisoning occur each year. In most cases, the symptoms disappear after a period free from exposure. Estimates of the number of poisonings are very difficult to obtain, since many go unreported, but in the United States, pesticide poisonings requiring medical treatment are in the thousands per year. The World Health Organization estimates that each year, there are between 1 million and 5 million acute pesticide poisonings and that about 20,000 deaths occur from pesticide poisonings.

For most people, however, the most critical health problem related to pesticide use is inadvertent exposure to very small

Table 14.1 Insecticide Resistance of Houseflies to the Insecticide Cyfluthrin

	Houseflies from a Normal Population	Houseflies from a Poultry Facility Population		
Concentration (ng/cm^2)	8.3	8.3 (1×)	83 (10×)	830 (100×)
Percent survival	0	100	90	38

Source: Data from Scott, Jeffrey G., et al., "Insecticide Resistance in Houseflies from Caged-Layer Poultry Facilities," *Pest Management Science* 56 (2000): 147–153.

quantities of pesticides. Many pesticides have been proven to cause mutations, produce cancers, or cause abnormal births in experimental animals. Studies of farmers who were occupationally exposed to pesticides over many years show that they have higher levels of certain kinds of cancers than the general public. There also are questions about the effects of chronic, minute exposures to pesticide residues in food or through contaminants in the environment.

Although the risk of endangering health by consuming tiny amounts of pesticides is very small compared to other risk factors, such as automobile accidents, smoking, or poor eating habits, many people find pesticides unacceptable and seek to prohibit their sale and use.

14.5 Why Are Pesticides So Widely Used?

If pesticides have so many drawbacks, why are they used so extensively? There are three primary reasons. First, the use of pesticides has increased the amount of food that can be grown in many parts of the world. On a worldwide basis, pests consume approximately 35 percent of crops. Farmers, grain-storage operators, and the food industry continually seek to reduce this loss. A retreat from dependence on pesticides would certainly reduce the amount of food produced. Agricultural planners in most countries are not likely to suggest changes in pesticide use that would result in malnutrition and starvation for many of their inhabitants.

The economic value of pesticides is the second reason they are used so extensively. The cost of pesticides is more than offset by the savings that result from less need to till fields and increased yields resulting in greater profits for the farmer. If it were not economical, farmers would not use pesticides. In addition, the production and distribution of pesticides is big business. Companies that have spent millions of dollars developing a pesticide are going to advertise the value of their product to the farmer and encourage its use.

A third reason for extensive pesticide use is that many health problems are currently impossible to control without pesticides. This is particularly true in areas of the world where insecticides are used to control insect-borne diseases. Malaria, yellow fever, dengue fever, sleeping sickness, chagas disease, and several other insect-borne diseases kill and disable millions of people each year, mostly in the developing world. It is hard to argue against the use of insecticides in these cases when the benefit far outweighs the risk and widespread public health consequences would occur if insecticides were not used.

14.6 Alternatives to Conventional Agriculture

Before the invention of synthetic fertilizers, herbicides, fungicides, and other agrochemicals, farmers typically raised a variety of crops as well as livestock. Most farms had cows, chickens, and pigs to provide milk, eggs, and meat and horses or mules to provide power. Farmers used a variety of techniques to maintain the fertility of their fields. Crop rotation involved planting a field to a crop, then replacing it the next year with a different kind of crop. Because animals were an important part of the farming operation, grain and hay, which were fed to the animals, were an important part of the rotation. Animal manure and crop rotation provided soil nutrients; a mixture of crops prevented regular pest problems; and manual labor killed insects and weeds. With the development of mechanization, larger areas could be farmed, draft animals (horses, mules, and oxen) were no longer needed to pull farm equipment, and many farmers changed from mixed agriculture, in which several kinds of crops and animals were important ingredients, to monoculture. Chemical fertilizers replaced manure as a source of soil nutrients, and crop rotation was no longer as important, since hay and grain were no longer grown for draft animals, cattle, and other farm animals. The larger fields of crops such as corn, wheat, and cotton presented opportunities for pest problems to develop, and chemical pesticides were used to "solve" this problem.

Today, many people feel current agricultural practices are not sustainable and that we should look for ways to reduce reliance on fertilizer and pesticides, protect soil and water resources, and protect wildlife and other environmental resources, while ensuring good yields and controlling the pests that compete with us for the food that we raise. The approaches to solving this problem involve a continuum of farming practice from highly technical intensive industrial operations to small-scale organic farming. Some of the terms used to describe these approaches are: *alternative agriculture, sustainable agriculture,* and *organic agriculture.*

Alternative agriculture is the broadest term. It includes all nontraditional agricultural methods and encompasses sustainable agriculture, organic agriculture, alternative uses of traditional crops, alternative methods for raising crops, and producing crops for industrial use.

Sustainable agriculture involves modifications to conventional farming practices to reduce reliance on fertilizers and pesticides and protect agricultural and natural systems while producing adequate, safe food in an economically viable manner.

Organic agriculture is distinguished by methods that do not involve the use of artificial fertilizers, chemical growth regulators, antibiotics, pesticides, and genetically modified organisms. Thus, most organic farms are mixed farms with livestock and raise a variety of crops, which are involved in a crop rotation.

In general, a sustainable approach to agriculture is characterized by several broad goals:

- Conserve soil and water resources.
- Reduce use of fertilizer and pesticides.
- Promote biodiversity in the farming operation and in the surrounding natural ecosystems.
- Sustain the economic viability of farm operations.
- Enhance the quality of life for farmers and society as a whole.

The following sections will explore in more detail some of the techniques involved in meeting these goals.

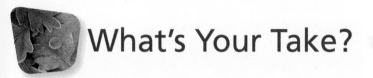

What's Your Take?

Techniques for Protecting Soil and Water Resources

Conventional farming practices have several negative effects on soil and water. Soil erosion is a problem throughout the world but can be controlled by a variety of changes in the way the land is used. Many of these techniques, such as contour farming, terracing, and reduced tillage were discussed in chapter 13. However, two other problems are also important: compaction of the soil and reduction in soil organic matter. Several changes in agricultural production methods can help to reduce these problems. Reducing the number of times farm equipment travels over the soil will reduce the degree of compaction. Leaving crop residue on the soil and incorporating it into the soil reduces erosion and increases soil organic matter. In addition, introducing organic matter into the soil makes compaction less likely.

Agriculture is the largest nonpoint source of water pollution. Fertilizer runoff stimulates aquatic growth and degrades water resources; pesticides can accumulate in food chains; and groundwater resources can be contaminated by fertilizer, pesticides, or animal waste. Reducing or eliminating these sources of contamination would protect natural ecosystems and reduce a threat to human health. Fertilizer runoff can be lessened by reducing the amount of fertilizer applied and the conditions under which it is applied. Applying fertilizer as plants need it will ensure that more of it is taken up by plants and less runs off. Increased organic matter in the soil also tends to reduce runoff. More careful selection, timing, and use of pesticides would decrease the extent to which these materials become environmental contaminants.

Precision agriculture is a new technique that addresses many of these concerns. With modern computer technology and geographic information systems, it is now possible, based on the soil and topography, to automatically vary the chemicals applied to the crop at different places within a field. Thus, less fertilizer is used, and it is used more effectively.

True organic agriculture that uses neither chemical fertilizers nor pesticides is the most effective in protecting soil and water resources but it requires several adjustments in the way in which farming is done.

Integrated Pest Management

Integrated pest management uses a variety of methods to control pests rather than relying on pesticides alone. Integrated pest management is a technique for establishing pest control strategies that depends on a complete understanding of all ecological aspects of the crop and the particular pests to which it is susceptible. It requires information about the metabolism of the crop plant, the biological interactions between pests and their predators or parasites, the climatic conditions that favor certain pests, and techniques for encouraging beneficial insects. It may involve the selective use of pesticides. Much of the information necessary to make integrated pest management work goes beyond the knowledge of the typical farmer. The metabolic and ecological studies necessary to pinpoint weak points in the life cycles of pests can usually be carried out only at universities or government research institutions. These studies are expensive and must be completed for each kind of pest, since each pest has a unique biology.

Several methods are employed in integrated pest management. These include disrupting reproduction, using beneficial organisms to control pests, developing resistant crops, modifying farming practices, and selectively using pesticides.

Disrupting Reproduction

In some species of insects, a chemical called a **pheromone** is released by females to attract males. Males of some species of moths can detect the presence of a female from a distance of up to 3 kilometers (nearly 2 miles). Since many moths are pests, synthetic odors can be used to control them. Spraying an area with the pheromone confuses the males and prevents them from finding females, which results in a reduced moth population the following year. In a similar way, a synthetic sex attractant molecule known as Gyplure is used to lure gypsy moths into traps, where they become stuck. Since the females cannot fly and the males are trapped, the reproductive rate drops, and the insect population may be controlled.

Another technique that reduces reproduction is male sterilization. In the southern United States and Central America, the screwworm fly weakens or kills large grazing animals, such as cattle, goats, and deer. The female screwworm fly lays eggs in open wounds on these animals, where the larvae feed. However, it was discovered that the female mates only once in her lifetime. Therefore, the fly population can be controlled by raising and releasing large numbers of sterilized male screwworm flies. Any female that mates with a sterile male fails to produce fertilized eggs and cannot reproduce. The screwworm fly has been eliminated from the United States and northern Mexico, and much of Central America may become free of them as well. In 1988, the screwworm fly was accidentally introduced into Libya with a

South American cattle shipment. The sterile male technique was used to successfully eradicate the fly.

Using Beneficial Organisms to Control Pests

The manipulation of predator-prey relationships can also be used to control pest populations. For instance, the ladybird beetle, commonly called a ladybug, is a natural predator of aphids and scale insects. (See figure 14.9a.) Artificially increasing the population of ladybird beetles reduces aphid and scale populations. In California during the late 1800s, scale insects on orange trees damaged the trees and reduced crop yields. The introduction of a species of ladybird beetle from Australia quickly brought the pests under control. Years later, when chemical pesticides were first used in the area, so many ladybird beetles were accidentally killed that scale insects once again became a serious problem. When pesticide use was discontinued, ladybird beetle populations rebounded, and the scale insects were once again brought under control. (See figure 14.9b.)

Herbivorous insects can also be used to control weeds. Purple loosestrife (*Lythrum salicaria*) is a wetlands plant accidentally introduced from Europe in the mid-1800s. It takes over sunny wetlands and eliminates native vegetation such as cattails, thereby eliminating many species that rely on cattails for food, nesting places, or hiding places. (See figure 14.10.) The plant exists in most states and provinces in the United States and Canada. In Europe, the plant is not a pest because there are several insect species that attack it in various stages of the plant's life cycle. Since purple loosestrife is a European plant, a search was made to identify candidate species of insects that would control it. The criteria were that the insects must live only on purple loosestrife and not infest other plants and must have the capacity to do major damage to purple loosestrife. Five species of beetles have been identified that will attack the plant in various ways. After extensive studies to determine if introductions of these insects from Europe would be likely to cause other problems, several were selected as candidates to help control purple loosestrife in the United States and Canada. Some combination of these beetles has been released in a large proportion of the states and provinces infested with purple loosestrife. Two species (*Galerucella calmariensis* and *Galerucella pusilla*) feed on the leaves, shoots, and flowers of the newly growing purple loosestrife; one (*Hylobius transversovittatus*) has larvae that feed on the roots. Two other species (*Nanophyes brevis* and *Nanophyes marmoratus*) feed on the flowers. This multipronged attack by several species of beetles has been effective in reducing populations of purple loosestrife.

Naturally occurring pesticides found in plants also can be used to control pests. For example, marigolds are planted to reduce the number of soil nematodes, and garlic plants are used to check the spread of Japanese beetles.

Developing Resistant Crops

Farmers have been involved in manipulating the genetic makeup of their plants and animals since these organisms were first domesticated. Initially, farmers either consciously or accidentally chose to plant specific seeds or breed certain animals that had specific desirable characteristics. In the past, if pests devastated a field of

(a) Ladybird beetle eating an aphid

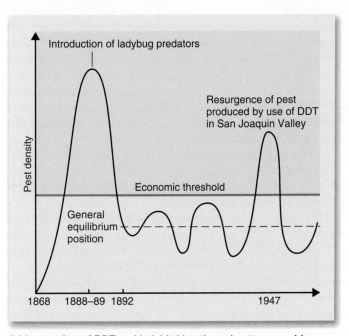

(b) Interaction of DDT and ladybird beetle and cottony cushion scale populations

FIGURE 14.9 Insect Control with Natural Predators (a) The ladybird beetle is a predator of many kinds of pest insects, including aphids. (b) In 1889, the introduction of ladybird beetles (ladybugs) brought the cottony cushion scale under control in the orange groves of the San Joaquin Valley. In the 1940s, DDT reduced the ladybird beetle population, and the cottony cushion scale population increased. Stopping the use of DDT allowed the ladybird population to increase, reducing the pest population and allowing the orange growers to make a profit.
Source: (b) V. M.Stern, et al., *Hilgardia*, 29: 93, 1959.

crops and a few plants stayed alive and healthy, the seeds from these healthy plants were used to generate the next crop. Thus, the beneficial factors that made the plants resistant were transferred to the next generation, making the new generation of crops slightly more resistant to the same pests. This resulted in local varieties with particular characteristics. When the laws of genetics began to be understood in the early 1900s, scientists began to make

(b) *Galerucella calmariensis*

(c) *Galerucella pusilla*

FIGURE 14.10 Biological Control of Purple Loosestrife (a) Purple loosestrife is a European plant that invades wetlands and prevents the growth and reproduction of native species of plants. Several kinds of European beetles that feed on purple loosestrife have been released as biological control agents. (b, c) Two species (*Galerucella calmariensis* and *Galerucella pusilla*) feed on the leaves, shoots, and flowers of the newly growing purple loosestrife. (d) Another species (*Hylobius transversovittatus*) has larvae that feed on the roots. It appears that these insect species are slowing the spread of purple loosestrife.

(a) **Purple loosestrife**

(d) *Hylobius transversovittatus*

precise crosses between carefully selected individuals to enhance the likelihood that their offspring would have certain highly desirable characteristics. This led to the development of hybrid seeds and specific breeds of domesticated animals. Controlled plant and animal breeding resulted in increased yields and better disease resistance in domesticated plants and animals. These activities are still the major driving force for developing improved varieties of domesticated organisms.

Modifying Farming Practices

Modification of farming practices can often reduce the impact of pests. In some cases, all crop residues are destroyed to prevent insect pests from finding overwintering sites. For example, shredding and plowing under the stalks of cotton in the fall reduces overwintering sites for boll weevils and reduces their numbers significantly, thereby reducing the need for expensive insecticide applications. Many farmers are also returning to crop rotation, which tends to prevent the buildup of specific pests that typically occurs when the same crop is raised in a field year after year.

Selective Use of Pesticides

Pesticides can also play a part in integrated pest management. Identifying the precise time when the pesticide will have the greatest effect at the lowest possible dose has these advantages: It reduces the amount of pesticide used and may still allow the parasites and predators of pests to survive. Such precise applications often require the assistance of a trained professional who can correctly identify the pests, measure the size of the population, and time pesticide applications for maximum effect. In several instances, pheromone-baited traps capture insect pests from fields, and an assessment of the number of insects caught can be a guide to when insecticides should be applied.

Integrated pest management will become increasingly popular as the cost of pesticides rises and knowledge about the biology of specific pests becomes available. However, as long as humans raise crops, there will be pests that will outwit the defenses we develop. Integrated pest management is just another approach to a problem that began with the dawn of agriculture.

Genetically Modified Crops

When the structure of DNA was discovered and it was determined that the DNA of organisms could be manipulated, an entirely new field of plant and animal breeding arose. **Genetic engineering** or **biotechnology** involves inserting specific pieces of DNA into the genetic makeup of organisms. The organism with the altered genetic makeup is known as a **genetically modified organism.** The DNA inserted could be from any source, even an entirely different organism.

In agricultural practice, two kinds of genetically modified organisms have received particular attention. One involves the insertion of genes from a specific kind of bacterium called *Bacillus thuringiensis israeliensis* (Bti). The bacterial genes produce a material that causes the destruction of the lining of the gut of insects that eat it. It is a natural insecticide that is a part of the plant. To date, these genes have been inserted into the genetic makeup of several crop plants, including corn and cotton.

A second kind of genetic engineering involves inserting a gene for herbicide resistance into the genome of certain crop plants. The value of this to farmers is significant. For example, a farmer can plant cotton with very little preparation of the field to rid it of weeds. When both the cotton and the weeds begin to grow, the field is sprayed with a specific herbicide that will kill the weeds but not harm the cotton because it contains genes that allow it to resist the effects of the herbicide.

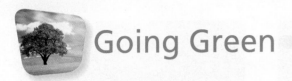

Going Green

Sustainability and Lawn Care

The typical lawn is a monoculture of a particular species of grass. Like all monocultures, it requires the expenditure of time, energy, and money to control pests and provide nutrients for growth. And wherever green grass grows there was once habitat—a forest, prairie, wetland, or even a desert. This is why many gardeners concerned about disappearing wilderness and wildlife declines are trying to grow the habitat back. We can apply the same sustainability principles employed by farmers to how we maintain our lawns. One simple change is to eliminate or reduce the size of the grassy area of a yard that requires constant attention. Replacing grassy areas with ground cover plants, rock gardens, or native ecosystems can conserve water, reduce the need for fertilizer and pesticides, and increase biodiversity. In some cases, particularly in desert areas, it is possible to replace lawns with native vegetation, which makes watering and other management practices unnecessary. Lawn mowing can also be more sustainable. Allowing the grass to get at least 8 centimeters (3 inches) tall before cutting provides several advantages. The taller grass shades the soil, which reduces evaporation of water from the soil and reduces the invasion of some weeds that need bright sunlight to germinate and grow. Allowing grass clippings to stay on the lawn and decompose adds organic matter and returns soil nutrients that would be removed if the grass clippings are bagged.

New varieties of slow growing and drought-resistant grasses are available that require little or no watering and grow slowly so that the lawn needs to be mowed less frequently. In some cases, the seeds can be sown into the current lawn and will slowly replace the original grass species.

Test the soil for its nutrient content and pH. The primary soil nutrients are potassium, phosphorus, and nitrogen. Most fertilizers are a mixture that includes these three nutrients. If the soil has adequate nutrients, the excess ends up being washed from the soil and becomes a source of water pollution. Soil pH greatly influences the kind of plants that will grow and also affects the amount of nutrients released in the soil. The tests can be performed by the agricultural extension service in your area, many lawn care businesses, and with kits you can purchase. Although there is a cost for doing the tests, it also costs money to buy fertilizer that you may not need.

Leaves from trees and other plant materials are a resource, not a waste. These materials can be composted or used as mulch around trees and other plantings. If the volume of leaves is not too great, they may simply be chopped up during lawn mowing and will incorporate into the soil.

Reduced size of lawn and increased biodiversity

Large lawn with low biodiversity

Table 14.2 lists several kinds of traits that have been modified by genetic engineering and the kinds of crops involved. The use of genetically modified crops has become extremely important worldwide. Three crops are particularly important: corn (maize), soybeans, and cotton. In the United States, at least 90 percent of cotton, soybeans, and corn grown are genetically modified.

Table 14.2 Some Genetically Modified Crops and Their Altered Traits

Modified Trait	Crop
Input Traits	
Herbicide resistance	Sugar beet, soybean, corn, canola, cotton, flax, alfalfa, sugarcane, rice, radish
Insect resistance	Tomato, corn, potato, cotton
Insect and herbicide resistance	Corn, cotton
Virus resistance	Zucchini, papaya
Male sterile	Corn, canola, radish
Output Traits	
Modified oil	Soybean, canola
Delayed fruit ripening	Tomato, cantaloupe
Provitamin A enriched	Rice
Iron fortification	Rice
Beta-carotene, lycopene enriched	Tomato
Detoxification of mycotoxins	Corn
Detoxification of cyanogens	Cassava
Caffeine-free	Coffee beans
Vitamin E enriched	Canola
High sucrose content	Sugarcane
Modified starch	Potato
High protein content	Potato

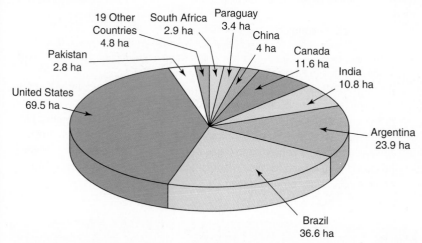

FIGURE 14.11   Millions of Hectares of GMO Crops In 2012, genetically modified crops were grown in 28 countries on 170.3 million hectares (421 million acres) of land. About 40 percent was grown in the United States.

Source: Data for U.S. from USDA, 2013; data for world from International Service for the Acquisition of Agri-Biotech Applications 2012.

Worldwide 81 percent of soybeans and cotton, 35 percent of corn, and 30 percent of rapeseed (canola) are genetically modified.

The United States accounts for 40 percent of all genetically modified crops planted globally. (See figure 14.11.) About 50 percent of the cropland planted in the United States is to genetically modified crops; primarily corn, soybeans, and cotton.

Many groups oppose the use of genetically modified organisms. They argue that this technology is going a step too far, that no long-term studies have been done to ensure their safety, that there are dangers we cannot anticipate, and that if such crops are grown, they should be labeled so that the public knows when they are consuming products from genetically modified organisms. In 2013 the citizens of Washington State narrowly defeated a ballot proposal that would have required that foods containing ingredients from genetically engineered plants be labeled as such. The Washington State effort is part of an ongoing fight by those opposed to genetically engineered crops to push for labeling. In 2013, 26 states introduced labeling legislation. Connecticut passed GMO labeling legislation in 2013, but it will not go into effect until four other New England states pass labeling laws. Maine passed legislation that will not go into effect until five other states, or any amount of states with a total population of 20 million, enact similar laws. The European Union (EU) resisted the purchase of genetically modified grains from the United States. This resulted in farmers having to segregate their stores of grains so that they can guarantee to an EU buyer that a crop is not genetically modified and at the same time sell a genetically modified crop to other buyers. However, attitudes are changing and some EU countries are now growing genetically modified crops and relaxing their resistance to the purchase of genetically modified crops.

Supporters argue that all plant and animal breeding involves genetic manipulation and that this is just a new kind of genetic manipulation. A great deal of evidence exists that genes travel between species in nature and that genetic engineering simply makes a common, natural process more frequent. Over the next 20 to 30 years, scientists hope to use biotechnology to produce high-yield plant strains that are more resistant to insects and disease, thrive on less fertilizer, make their own nitrogen fertilizer, do well in slightly salty soils, withstand drought, and use solar energy more efficiently during photosynthesis. These new kinds of genetically modified organisms will continue to be developed and tested, and the political arguments about their appropriateness will continue as well.

Genetically altered foods are prevalent in the United States and the developed world. More than 60 percent of the foods we purchase from the supermarket today have ingredients derived from genetically modified crops. Most of these are either from corn or soybeans, which are the bases for numerous ingredients manufactured for the food industry, including starch, oils, proteins, and other ingredients.

Economic and Social Aspects of Sustainable Agriculture

Crop rotation is an effective way to enhance soil fertility, reduce erosion, and control pests. The use of nitrogen-fixing legumes, such as clover, alfalfa, beans, or soybeans, in crop rotation increases soil nitrogen but places other demands on the farmer. For example, it typically requires that cattle be a part of the farmer's operation in

Issues & Analysis

What Is Organic Food?

It is important to have standards that organic foods must meet and to label the foods so that the consumer can clearly identify what is organic and what is not. All producers of organic products must be certified.

Since 2002, organic certification in the United States has taken place under the authority of the USDA National Organic Program, which accredits organic certifying agencies and oversees the regulatory process. Certification includes annual submission of an organic system plan and inspection of farm fields and processing facilities by independent state or private organizations accredited by the USDA. Inspectors verify that organic practices such as long-term soil management, buffering between organic farms and neighboring conventional farms, and recordkeeping are being followed.

The following tables describe the major requirements for organic food production and the labeling requirements for different levels of organic content in foods.

- Many people feel there is no difference in food quality between foods that are labeled organic and those that are not. Do you think the nutritional quality of organic foods differs from that of conventionally produced foods?
- The rules permit small amounts of pesticide residues. Why do you think these are allowed?
- The rule that animals must have access to the outdoors has been challenged. Do you think being outdoors changes the quality of the meat, milk, or eggs produced?

Table A Requirements for Organic Certification

Organic Crops	Organic Livestock
Land must not have prohibited substances applied to it for at least three years. Use of genetic engineering, ionizing radiation, and sewage sludge is prohibited. No chemical fertilizer can be used. Only animal and crop wastes and tillage methods may be used to provide fertility. Synthetic pesticides must not be used, but pesticides made from plants and other natural sources may be used.	Animals for slaughter must be raised under organic management from the last third of gestation, or no later than the second day of life for poultry. Livestock must be fed products that are 100 percent organic. Vitamin and mineral supplements are permitted. No antibiotics or hormones may be used. Animal welfare conditions require that: 1. Animals have access to the outdoors. (They may be temporarily confined only for reasons of health, safety, the animal's stage of production, or protection of soil or water quality.) 2. Preventive health management practices must be employed, including the use of vaccines to keep animals healthy. 3. Sick or injured animals must be treated; however, animals treated with a prohibited medication (antibiotic, hormone, etc.) may not be sold as organic.
Farms and handling operations that sell less than $5,000 annually of organic agricultural products are exempt from certification. But if they comply with all other national standards they may label their products as organic.	A dairy herd can be converted to organic production if 80 percent organically produced feed is used for nine months, followed by three months of 100 percent organically produced feed.

Table B Labeling Requirements for Organic Foods

Requirements	Label Statement Permitted
Products must contain only organically produced ingredients (excluding water and salt).	"100 percent organic"
Products must consist of at least 95 percent organically produced ingredients (excluding water and salt). Products with pesticide residues up to 5 percent of the EPA pesticide tolerance level may be sold as organic.	"Organic"
Processed products that contain at least 70 percent organic ingredients.	"Made from organic ingredients"
Processed products that contain less than 70 percent organic ingredients cannot use the term *organic* anywhere on the label but may list the individual ingredients that are organic.	Label may list those ingredients that are organic

order to make use of forage crops (clover and alfalfa) and provide organic fertilizer for subsequent crops. Crop rotation also requires a greater investment in farm machinery, since certain crops require specialized equipment. Also, the raising of cattle requires additional expenditures for feed supplements and veterinary care.

But no matter how elegant the system or how accomplished the farmer, no agriculture is sustainable if it's not also profitable and able to provide a healthy family income and a good quality of life. Critics say that sustainable or organic farming cannot produce the amount of food required for today's population and that it can be economically successful only in specific cases. Proponents disagree and stress that when the hidden costs of soil erosion and pollution are included, organic agriculture or some modification of it is a viable alternative approach to conventional means of food production. Furthermore, farmers are willing to accept lower yields because they do not have to pay for expensive chemical fertilizers and pesticides. In addition, farmers often receive premium prices for products that are organically grown. Thus, even with lower yields, they can still make a profit.

Many sustainable practices lend themselves to smaller, family-scale farms. From a social point of view, this implies that all members of the family participate in some way in the economic unit. This lifestyle often involves three generations. Smaller family farms that produce a variety of products tend to find their best niches in local markets, often selling directly to consumers. Thus, marketing skills are important to the success of many small sustainable farming businesses.

Perhaps sustainable agricultural practices can help address the major food issues facing an increasingly hungry planet. About one out of seven of the world's people are hungry periodically because of lack of food. This is true although world food production is capable of feeding everyone. Some countries have surplus food while others do not have enough. Long-term solutions to feeding the world's hungry are complex and are unlikely to be found without dealing with the following:

- *Increasing population.* A simple solution to the food problem is to have fewer people to feed. However, the population is projected to increase and is growing most rapidly in parts of the world where food is already in short supply.
- *Poverty.* When a significant proportion of the people of the world live on a few dollars a day, the price of food is a critical factor in determining whether a family will go hungry.
- *Crop subsidies and trade barriers.* Governments protect their farmers with subsidies and trade barriers. Government subsidies are common in Europe and North America. Providing subsidies to farmers allows them to produce food for an artificially low price. Thus, subsidized farmers can sell their products for less than unsubsidized farmers in other countries and it makes it uneconomic for unsubsidized farmers to continue farming. Trade barriers are used to protect a country's own farmers or to put pressure on another country. Both subsidies and trade barriers hinder the development of farming in poorer nations and make them more dependent on purchasing food.
- *Higher fuel costs.* Since much of the surplus food in the world is grown in the developed world with mechanized agriculture, an increase in the price of fuel adds to the cost of food.
- *Production of biofuels.* Increased oil prices have encouraged the diversion of crops to the production of biofuels.
- *Increasing demand for protein.* The rise in global wealth has created a greater demand for protein. Animal protein requires that crops be fed to animals with a loss in the total amount of energy, which reduces total food available.

Summary

Although small slash-and-burn garden plots are common in some parts of the world, most of the food in the world is raised on more permanent farms. In countries where population size is high and money is in short supply, much of the farming is labor-intensive, using human labor for many of the operations necessary to raise crops. However, much of the world's food is grown on large, mechanized farms that use energy rather than human muscle for tilling, planting, and harvesting crops, and for producing and applying fertilizers and pesticides.

Monoculture involves planting large areas of the same crop year after year. This causes problems with plant diseases, pests, and soil depletion. Although chemical fertilizers can replace soil nutrients that are removed when the crop is harvested, they do not replace the organic matter necessary to maintain soil texture, pH, and biotic richness.

Mechanized monoculture depends heavily on the control of pests by chemical means. Persistent pesticides are stable and persist in the environment, where they may biomagnify in ecosystems. Consequently, many of the older persistent pesticides have been quickly replaced by nonpersistent pesticides that decompose much more quickly and present less of an environmental hazard. However, most nonpersistent pesticides are more toxic to humans and must be handled with greater care than the older persistent pesticides.

Pesticides can be divided into several categories based on the organism they are used to control. Insecticides are used to control insects, herbicides are used for plants, fungicides for fungi, and rodenticides for rodents. Because of the problems of persistence, biomagnification, resistance of pests to pesticides, and effects on human health, many people are seeking pesticide-free alternatives to raising food.

Many people feel that current mechanized agriculture is not sustainable and support the following kinds of changes in agricultural practice: reduced use of fertilizer and pesticides, increased biodiversity by raising a variety of crops, improved soil and water conservation techniques, and improved economic and lifestyle changes for farmers. These modifications include: crop rotation, integrated pest management, a mixture of crops and animals in the farming operation, and the use of animal manure for fertilizer.

Genetically modified organisms have become important in world agriculture. Crop plants (particularly corn, soybeans, and cotton) have been genetically modified to include genes that make the plants resistant to insect attack and to certain herbicides. Herbicide resistance is important because the crop plant can be sprayed with a herbicide and remain unaffected, while weeds in the field are killed.

Acting Green

1. Reduce or eliminate the use of pesticides and herbicides on your yard or garden.
2. Support local or regional producers of sustainable agricultural products.
3. Visit your local "farmers market." Discuss his or her marketing strategy with one of the vendors.
4. Visit a grocery store and identify three produce items that were grown within your state or region.
5. Plant your own organic garden even if it is only a small container in your apartment.
6. Inventory all pesticides in your home. Take those not needed to the next household hazardous waste disposal in your community.
7. Plant a butterfly garden.

Review Questions

1. How does the practice of shifting agriculture provide nutrients for the growth of agricultural products?
2. Why is abandonment of fields important in a shifting agriculture system?
3. List conditions that make labor-intensive farming necessary.
4. How were the invention of machines and the development of monoculture linked?
5. Why are fertilizers used? What problems are caused by fertilizer use?
6. List three advantages and three disadvantages of large-scale mechanized monoculture.
7. What are micronutrients? How do they differ from macronutrients?
8. List two functions of soil organic matter.
9. Describe why pesticides are commonly used in mechanized agriculture.
10. How do persistent and nonpersistent pesticides differ?
11. What is biomagnification? What problems does it cause?
12. Describe how some populations of pests become resistant to pesticides.
13. How do sustainable farming practices differ from conventional mechanized monoculture?
14. Explain why a complete knowledge of the biology of a pest is important in using integrated pest management.
15. Describe three techniques used to control pests that do not involve the use of pesticides.
16. How is a genetically modified organism different from other organisms?
17. What characteristics have been introduced into insect-resistant and herbicide-resistant genetically modified crops?
18. List three reasons many sustainable farms are small family enterprises.

Critical Thinking Questions

1. If you were a public health official in a developing country, would you authorize the spraying of DDT to control mosquitoes that spread malaria? What would be your reasons?
2. Look at table 14.1. What caused the changes in the effectiveness of the insecticide? If you were an agricultural extension agent, what alternatives to pesticides might you recommend?
3. Imagine that you are a scientist examining fish in Lake Superior and you find toxaphene in the fish you are studying. Toxaphene was used primarily in cotton farming and has been banned since 1982. How can you explain its presence in these fish?
4. Are the risks of pesticide use worth the benefits? What values, beliefs, and perspectives lead you to this conclusion?
5. Do you think that current agricultural practices are sustainable? Why or why not? What changes in agriculture do you think will need to happen in the next 50 years?
6. Imagine you are an EPA official who is going to make a recommendation about whether an agricultural pesticide can remain on the market or should be banned. What are some of the facts you would need to make your recommendation? Who are some of the interest groups interested in the outcome of your decision? What arguments might they present regarding their positions? What political pressures might they be able to bring to bear on you?
7. Why are few consumers demanding alternative methods of crop production, and why are farmers not using those methods?

 |ENVIRONMENTAL SCIENCE

To access additional resources for this chapter, please visit ConnectPlus® at connect.mheducation.com. There you will find interactive exercises, including Google Earth™, additional Case Studies, and SmartBook™, an adaptive eBook that integrates our LearnSmart® adaptive learning technology.

Freshwater is essential to maintain life for most terrestrial plants and animals. Human use of water for domestic, industrial, and agricultural use often requires diversion of water. While we must manage water use, we should also preserve the scenic and other aesthetic values of water as well.

OBJECTIVES

After reading this chapter, you should be able to:

- Explain how water is cycled through the hydrologic cycle.
- Explain the significance of groundwater, aquifers, and runoff.
- Explain how land use affects infiltration and surface runoff.
- List the various kinds of water use and the problems associated with each.
- List the problems associated with water impoundment.
- List the major sources of water pollution.
- Define biochemical oxygen demand (BOD).

- Differentiate between point and nonpoint sources of pollution.
- Explain how heat can be a form of pollution.
- Differentiate among primary, secondary, and tertiary sewage treatments.
- Describe some of the problems associated with stormwater runoff.
- List sources of groundwater pollution.
- Explain how various federal laws control water use and prevent misuse.
- List the problems associated with water-use planning.

- Explain the rationale behind the federal laws that attempt to preserve certain water areas and habitats.
- List the problems associated with groundwater mining.
- Explain the problem of salinization associated with large-scale irrigation in arid areas.
- List the water-related services provided by local governments.
- Explain why wars are fought over water.

Who Owns the Water?

We take water for granted. We assume it is our right to have water; yet water is a resource. It is not always available where we need it, or when we need it. Rivers do not follow political boundaries—they flow through states and over international borders. There are endless demands for water ranging from agriculture to manufacturing. There is also the ecosystem that depends on water getting downstream. It has been said that oil was the fluid of the last century where there was a lot of turmoil, but water is the fluid of this century.

So what are the legal rights in the United States when it comes to water? Basically there are two doctrines that govern surface water rights in the United States—one for the West and one for the East.

The riparian doctrine applies to the eastern United States. Under the riparian doctrine if you live close to a river or lake you have "reasonable rights" to use that water. The western United States uses the prior appropriation doctrine. As people started exploring the West and looking for water for agriculture and mining, there was a need to move water away from the rivers. People wanted a claim to water but often lived too far away from a river for the riparian doctrine to make any sense. Thus, the prior appropriation was devised.

Prior appropriation allocates water rights based on who started using the water first. So if you are first in time, you are first in rights. This was based on a permitting system. Whoever had the first permit had the water rights. Recently, however, the courts have stated that first use is not strictly based on the first permit but rather on who was actually there first. This interpretation has led to Native American tribes in many areas claiming water rights since they were the first users of the local water. Two rivers—the Klamath and Chattahoochee—on opposite sides of the United States highlight the growing disputes over water rights.

In an arid area like the Klamath River Basin, there is often not enough water available for everyone who has a right to use it. And the person with the oldest water rights gets all the water they are entitled to first. For decades, groups in the basin have contested who holds the senior water rights. The state reached a decision in 2013 in favor of the tribes, and state regulators told ranchers near the headwaters of the Klamath River in Oregon to shut off their irrigation pumps. The state said the shutdown was necessary to protect treaty rights of Native American tribes who live downstream. The water shut-off, however, jeopardizes a multimillion-dollar cattle ranching industry. The river system also provides water for salmon and suckerfish. The water shut-off came after the Klamath Indian Tribe said they needed more water left in the river system to protect their fishery.

The tribes are currently in negotiations with the ranchers to come up with a better plan to share water in dry years as long as the salmon and suckerfish are protected.

Water management may seem like it would be more of a problem in the arid western United States, but it is truly a national issue. The U.S. Army Corps of Engineers operates hundreds of reservoirs in the United States. These reservoirs impound trillions of liters of water, regulating the flows of multistate river systems. Since the dams were constructed over the past decades, the demands on the water stored behind them in reservoirs has changed.

That is the case for the Chattahoochee River. It flows 800 kilometers from the Blue Ridge Mountains in northeast Georgia through metropolitan Atlanta, down through Alabama and Florida and then into Apalachicola Bay in the Gulf of Mexico. As metro Atlanta expanded it took more water out of the Lake Lanier reservoir, on the Chattahoochee, which meant that those living downstream, in Alabama and Florida, had less water. Runoff from Atlanta also diminishes the quality of the river water, and low water levels affect the flow of nutrients down river. Since the Corps of Engineers operates the four federal dams on the river, it found itself in the middle of the conflict. This resulted in over 20 years of court battles. Interstate compacts were created to try to resolve the conflict, and the issue went all the way to the U.S. Supreme Court. The river basin does not follow political boundaries, and thus there is a great deal of tension over how much water is going to be available for everybody to do all the things that they need to do with the river.

An underlying issue is the need to emphasize water conservation and water efficiency throughout the entire river system. We need to also start thinking about how we are growing, both in rural and urban areas, and whether there is enough water available to sustain that growth.

Fog on the Chattahoochee River

15.1 The Global Water Challenge

In New Delhi, India, citizens wake in the morning to the sound of a megaphone announcing that freshwater will be available only for the next one hour. People then rush to fill the bathtub and other receptacles with water to last the day. New Delhi's endemic short-falls of water occur largely because water managers decided several years ago to divert large amounts from upstream rivers and reservoirs to irrigate crops.

In arid Phoenix, Arizona, citizens often wake to the sounds of sprinklers watering suburban lawns and golf courses. Although Phoenix sits amid the Sonoran Desert, it enjoys a virtually unlimited water supply. Politicians there have allowed irrigation water to be shifted away from farming operations to cities and suburbs, while permitting recycled wastewater to be employed for landscaping and other nonpotable applications.

In China, almost a quarter of all surface water remains so polluted that it is unfit even for industrial use, while less than half of total supplies are drinkable. Despite tougher regulations over the last decade, China has struggled to rein in the thousands of small paper mills, cement factories, and chemical plants discharging industrial waste directly into the country's waterways. These three stories provide a look at some of the critical water challenges we are facing in the world today.

According to the World Meteorological Organization (WMO), if current projections for global population growth and resource availability remain unchanged, about 34 countries will experience serious difficulties in obtaining water supplies by 2025. Currently, about 29 countries are suffering from moderate to severe water shortages. The number of people living in countries experiencing water shortages will increase from around 132 million (in 1990) to about 653 million by the year 2025, which will represent about 8 percent of the world's population.

Global freshwater resources are threatened by rising demands from many areas. Growing populations need more water for drinking, hygiene, sanitation, food production, and industry. Climate change, meanwhile, is expected to contribute to more droughts. Policymakers must figure out how to supply water without degrading the natural ecosystems that provide it. Existing low-tech approaches can help reduce scarcity, as can ways to increase supplies, such as improved methods to desalinate water. Governments at all levels must start setting policies and making investments in infrastructure for water conservation. And, finally, individuals need to know how much water they use and how that amount used can be reduced. It has been said that we will find subsitute forms of energy that are not fossil-fuel based, but we will never find a subsitute for water.

15.2 The Water Issue

Water in its liquid form is the material that makes life possible on Earth. All living organisms are composed of cells that contain at least 60 percent water. Furthermore, their metabolic activities take place in a water solution. Organisms can exist only where they have access to adequate supplies of water. Water is also unique because it has remarkable physical properties. Water molecules are polar; that is, one part of the molecule is slightly positive and the other is slightly negative. Because of this, the water molecules tend to stick together, and they also have a great ability to separate other molecules from each other. Water's ability to act as a solvent and its capacity to store heat are a direct consequence of its polar nature. These abilities make water extremely valuable for societal and industrial activities. Water dissolves and carries substances ranging from nutrients to industrial and domestic wastes. A glance at any urban sewer will quickly point out the importance of water in dissolving and transporting wastes. Because water heats and cools more slowly than most substances, it is used in large quantities for cooling in electric power generation plants and in other industrial processes. Water's ability to retain heat also modifies local climatic conditions in areas near large bodies of water. These areas do not have the wide temperature-change characteristic of other areas.

For most human as well as some commercial and industrial uses, the quality of the water is as important as its quantity. Water must be substantially free of dissolved salts, plant and animal waste, and bacterial contamination to be suitable for human consumption. The oceans, which cover approximately 70 percent of the Earth's surface, contain over 97 percent of its water. However, saltwater cannot be consumed by humans or used for many industrial processes. Freshwater is free of the salt found in ocean waters. Of the freshwater found on Earth, only a tiny fraction is available for use. (See figure 15.1.) Unpolluted freshwater that is suitable for drinking is known as **potable water.** Early human migration routes and settlement sites were influenced by the availability of drinking water. At one time, clean freshwater supplies were considered inexhaustible. Today, despite advances in drilling, irrigation, and purification, the location, quality, quantity, ownership, and control of potable waters remain important concerns.

To understand how water can become scarce, consider that the human population numbered under 2 billion in 1900, surpassed 6 billion by 2000, and is surging toward 9 billion by 2050. And we're not the only living organisms on this planet that need freshwater to survive.

The global water supply, by comparison, is finite. Whether it's in the form of ice, vapor, steam, or liquid, the amount of water on Earth today is about the same as the water that slaked the planet's thirst a million years ago. Only a very thin slice of that supply is the liquid freshwater that we depend upon for drinking, washing, irrigation, manufacturing, energy, and more.

The world's total water supply is estimated at about 1.38 billion cubic kilometers (333 million cubic miles). Of this, about 97.5 percent is saltwater and about 2.5 percent is freshwater. Most of the freshwater (78 percent) is locked up in glaciers and ice caps, about 21 percent is groundwater in deep sediments or soil, and less than 1 percent is surface water. About 87 percent of surface freshwater is in lakes, 11 percent in wetlands, and 2 percent in rivers.

Some areas of the world have abundant freshwater resources, while others have few. In addition, demand is increasing for freshwater for industrial, agricultural, and personal needs.

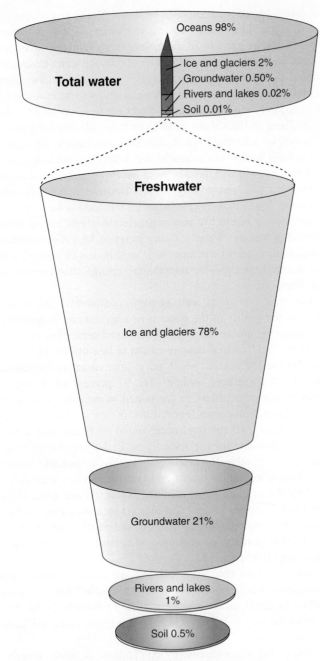

Oceans 98%

Total water

Ice and glaciers 2%
Groundwater 0.50%
Rivers and lakes 0.02%
Soil 0.01%

Freshwater

Ice and glaciers 78%

Groundwater 21%

Rivers and lakes 1%

Soil 0.5%

FIGURE 15.1 Freshwater Resources Although water covers about 70 percent of the Earth's surface, over 97 percent is saltwater. Of the less than 3 percent that is freshwater, only a tiny fraction is available for human use.

past decade and that without large economic investments in safe drinking-water supplies, the rate of increase will continue.

Unfortunately, the outlook for the world's freshwater supply is not very promising. According to studies by the United Nations and the International Joint Commission (a joint U.S.-Canadian organization that studies common water bodies between the two nations), many sections of the world are currently experiencing a shortage of freshwater, and the problem will only intensify.

Water, used by households, agriculture, and industry, is clearly the most important good provided by freshwater systems. Humans now withdraw about one-fifth of the world's rivers' base flow (the dry-weather flow or the amount of available water in rivers most of the time), but in river basins in arid or populous regions the proportion can be much higher. This has implications for the species living in or dependent on these systems, as well as for human water supplies. Between 1900 and 2010, withdrawals increased by a factor of more than six, which is greater than twice the rate of population growth.

Water supplies are distributed unevenly around the world, with some areas containing abundant water and others a much more limited supply. In water basins with high water demand relative to the available runoff, water scarcity is a growing problem. Many experts, governments, and international organizations are predicting that water availability will be one of the major challenges facing human society in the twenty-first century and that the lack of water will be one of the key factors limiting development.

Figure 15.2 shows areas of the world experiencing water stress. Water experts define areas where per capita water supply drops below 1,700 m^3/year as experiencing water stress—a situation in which disruptive water shortages can frequently occur. The map shows that by 2025, assuming current consumption patterns continue, at least 3.5 billion people—or 43 percent of the world's projected population—will live in water-stressed river basins. Of these, 2.4 billion will live under high water stress conditions. This per capita water supply calculation, however, does not take into account the coping capabilities of different countries to deal with water shortages. For example, high-income countries that are water scarce may be able to cope to some degree with water shortages by investing in desalination or reclaimed wastewater.

Several river basins, including the Volta in Ghana, Nile in Sudan and Egypt, Tigris and Euphrates in the Middle East, Narmada in India, and the Colorado River basin in the United States, will go from having more than 1,700 m^3 to less than 1,700 m^3 of water per capita per year. Another 29 basins will descend further into scarcity by 2025, including the Indus in Pakistan, Huang He in China, Seine in France, Balsas in Mexico, and Rio Grande in the United States.

Increasing scarcity, competition, and arguments over water in the first quarter of the twenty-first century could dramatically change the way we value and use water and the way we mobilize and manage water resources. Furthermore, changes in the amount of rain from year to year result in periodic droughts for some areas and devastating floods for others. However, rainfall is needed to regenerate freshwater and, therefore, is an important link in the cycling of water.

Shortages of potable freshwater throughout the world can also be directly attributed to human abuse in the form of pollution. Water pollution has negatively affected water supplies throughout the world. In many parts of the developing world, safe drinking water is scarce. The World Health Organization estimates that about 25 percent of the world's people do not have access to safe drinking water. Even in the economically advanced regions of the world, water quality is a major issue. According to the United Nations Environment Programme, 5 million to 10 million deaths occur each year from water-related diseases. The illnesses include cholera, malaria, dengue fever, and dysentery. The United Nations also reports that these illnesses have been increasing over the

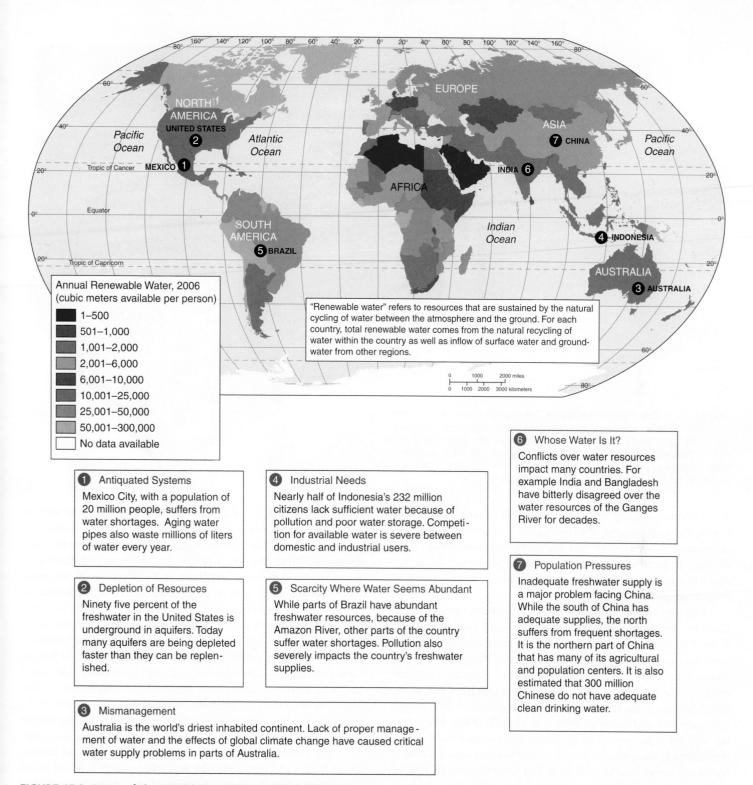

Annual Renewable Water, 2006
(cubic meters available per person)

- ■ 1–500
- ■ 501–1,000
- ■ 1,001–2,000
- ■ 2,001–6,000
- ■ 6,001–10,000
- ■ 10,001–25,000
- ■ 25,001–50,000
- ■ 50,001–300,000
- □ No data available

"Renewable water" refers to resources that are sustained by the natural cycling of water between the atmosphere and the ground. For each country, total renewable water comes from the natural recycling of water within the country as well as inflow of surface water and groundwater from other regions.

❶ Antiquated Systems

Mexico City, with a population of 20 million people, suffers from water shortages. Aging water pipes also waste millions of liters of water every year.

❷ Depletion of Resources

Ninety five percent of the freshwater in the United States is underground in aquifers. Today many aquifers are being depleted faster than they can be replenished.

❸ Mismanagement

Australia is the world's driest inhabited continent. Lack of proper management of water and the effects of global climate change have caused critical water supply problems in parts of Australia.

❹ Industrial Needs

Nearly half of Indonesia's 232 million citizens lack sufficient water because of pollution and poor water storage. Competition for available water is severe between domestic and industrial users.

❺ Scarcity Where Water Seems Abundant

While parts of Brazil have abundant freshwater resources, because of the Amazon River, other parts of the country suffer water shortages. Pollution also severely impacts the country's freshwater supplies.

❻ Whose Water Is It?

Conflicts over water resources impact many countries. For example India and Bangladesh have bitterly disagreed over the water resources of the Ganges River for decades.

❼ Population Pressures

Inadequate freshwater supply is a major problem facing China. While the south of China has adequate supplies, the north suffers from frequent shortages. It is the northern part of China that has many of its agricultural and population centers. It is also estimated that 300 million Chinese do not have adequate clean drinking water.

FIGURE 15.2 Areas of the World Experiencing Water Stress The map shows areas of the world where people are experiencing a shortage of water. By 2025, assuming current consumption patterns continue, at least 3.5 billion people—or about 43 percent of the world's projected population—will live in water-stressed river basins.

Source: United Nations, *Population Division, UNDP,* and *AQUASTAT-FAQ;* Population Reference Bureau.

15.3 The Hydrologic Cycle

All water is locked into a constant recycling process called the **hydrologic cycle.** (See figure 15.3.) Two important processes involved in the cycle are the evaporation and condensation of water. Evaporation involves adding energy to molecules of a liquid so that it becomes a gas in which the molecules are farther apart. Condensation is the reverse process in which molecules of a gas give up energy, get closer together, and become a liquid. Solar energy provides the energy that causes water to evaporate from the

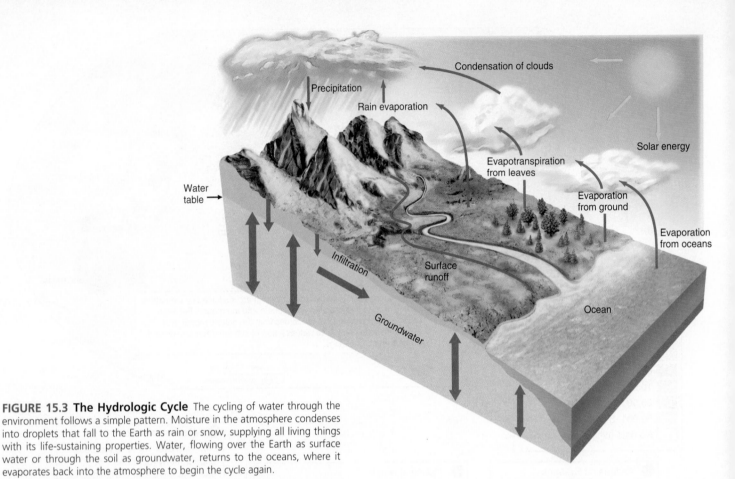

FIGURE 15.3 The Hydrologic Cycle The cycling of water through the environment follows a simple pattern. Moisture in the atmosphere condenses into droplets that fall to the Earth as rain or snow, supplying all living things with its life-sustaining properties. Water, flowing over the Earth as surface water or through the soil as groundwater, returns to the oceans, where it evaporates back into the atmosphere to begin the cycle again.

ocean surface, the soil, bodies of freshwater, and the surfaces of plants. The water evaporated from plants comes from two different sources. Some is water that has fallen on plants as rain, dew, or snow. In addition, plants take up water from the soil and transport it to the leaves, where it evaporates. This process is known as **evapotranspiration.**

The water vapor in the air moves across the surface of the Earth as the atmosphere circulates. As warm, moist air cools, water droplets form and fall to the land as precipitation. Although some precipitation may simply stay on the surface until it evaporates, most will either sink into the soil or flow downhill and enter streams and rivers, which eventually return the water to the ocean. Surface water that moves across the surface of the land and enters streams and rivers is known as **runoff.** Water that enters the soil and is not picked up by plant roots moves slowly downward through the spaces in the soil and subsurface material until it reaches an impervious layer of rock. The water that fills the spaces in the substrate is called **groundwater.** It may be stored for long periods in underground reservoirs.

The porous layer that becomes saturated with water is called an **aquifer.** An aquifer is an underground layer of gravel, sand, or permeable rock that holds groundwater that can be extracted by wells. There are three basic kinds of aquifers: unconfined, semiconfined, and confined. (See figure 15.4.) An **unconfined aquifer** usually occurs near the land's surface where water enters the aquifer from the land above it. The top of the layer saturated with water is called the **water table.** The lower boundary of the aquifer

is an impervious layer of clay or rock that does not allow water to pass through it. Unconfined aquifers are replenished (recharged) primarily by rain that falls on the ground directly above the aquifer and infiltrates the layers below. The water in such aquifers is at atmospheric pressure and flows in the direction of the water table's slope, which may or may not be similar to the surface of the land above it. Above the water table and below the land surface is a layer known as the **vadose zone** (also known as the unsaturated zone or zone of aeration) that is not saturated with water.

A **confined aquifer** is bounded on both the top and bottom by layers that are impervious to water and is saturated with water under greater-than-atmospheric pressure. An impervious confining layer is called an **aquiclude.** If water can pass in and out of the confining layer, the layer is called an **aquitard** and the aquifer is known as a **semiconfined aquifer.** Both confined and semiconfined aquifers are primarily replenished by rain and surface water from a recharge zone (the area where water is added to the aquifer) that may be many kilometers from where the aquifer is tapped for use. If the recharge area is at a higher elevation than the place where an aquifer is tapped, water will flow up the pipe until it reaches the same elevation as the recharge area. Such wells are called **artesian wells.** If the recharge zone is above the elevation of the top of the well pipe, it is called a flowing artesian well because water will flow from the pipe.

The nature of the substrate in the aquifer influences the amount of water the aquifer can hold and the rate at which water moves through it. **Porosity** is a measure of the size and number of

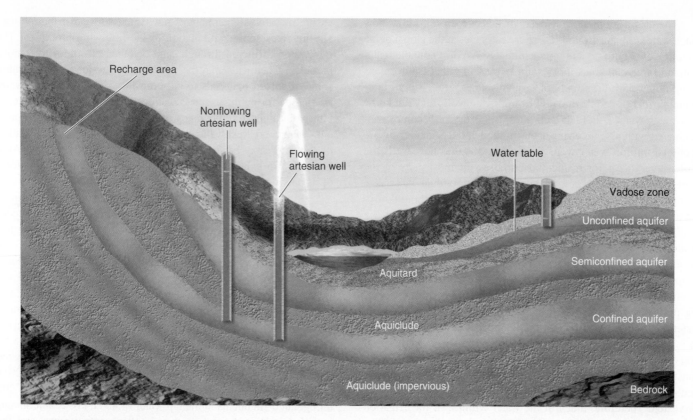

FIGURE 15.4 Aquifers and Groundwater Groundwater is found in the pores in layers of sediment or rock. The various layers of sediment and rock determine the nature of the aquifer and how it can be used.

the spaces in the substrate. The greater the porosity, the more water it can contain. The rate at which water moves through an aquifer is determined by the size of the pores, the degree to which they are connected, and any cracks or channels present in the substrate. The rate at which the water moves through the aquifer determines how rapidly water can be pumped from a well per minute.

15.4 Human Influences on the Hydrologic Cycle

The world's oceans are the primary regulator of global climate and an important sink for greenhouse gases. At continental, regional, and ocean basin scales, the water cycle is being affected by long-term changes in climate, threatening human security. These changes are affecting Arctic temperatures and sea and land ice, including mountain glaciers. They also affect ocean salinity and acidification, sea levels, precipitation patterns, extreme weather events, and, possibly, the ocean's circulatory regime. The trend to increase urbanization and tourism development has considerable impacts on coastal ecosystems. The socioeconomic consequences of all these changes are potentially immense.

Human activities can significantly impact evaporation, runoff, and infiltration. When water is used for cooling in power plants or to irrigate crops, the rate of evaporation is increased. Water impounded in reservoirs also evaporates rapidly. This rapid evaporation can affect local atmospheric conditions. Runoff and the rate of infiltration also are greatly influenced by human activity. Removing the vegetation by logging or agriculture increases runoff and decreases infiltration. Because there is more runoff, there is more erosion of soil. Urban complexes with a high percentage of impervious, paved surfaces have increased runoff and reduced infiltration. A major concern in urban areas is providing ways to carry stormwater away rapidly. This involves designing and constructing surface waterways and storm sewers. Often cities have significant flooding problems when heavy rains overtax the ability of their stormwater management systems to remove excess water. Many cities combine their storm-sewer water with wastewater at their treatment plants, which can cause serious pollution problems after heavy rains. The treatment plants cannot process the increased volume of water and must discharge the combined untreated wastewater and sewer water into the receiving body of water.

Cities also have problems supplying water for industrial and domestic use. In many cases, cities rely on surface water sources for their drinking water. The source may be a lake or river, or an impoundment that stores water. Over one-third of the world's largest cities rely on large areas of protected forests to maintain a steady flow of clean drinking water. For example, New York City, with a population of over 8 million people, relies on a 998 square kilometer (385 square mile) watershed protected by the Catskill State Park for 90 percent of its drinking water. In addition, aquifers are extremely important in supplying water. Many large urban areas in the United States depend on underground water for their water supply. This groundwater supply can be tapped as long as it is not used faster than it can be replaced.

There is a problem, however, in that we do not know how much water is in the ground. There are estimates at the global or the regional level but they are only estimates. We do know the water demand in certain areas. For example we know that Californians pump about 55 billion liters of groundwater a year. But an individual farmer or people drawing from private wells may not know how much water is in their well until it runs dry or becomes contaminated with arsenic or nitrogen. In the United States, there are about 16 million water wells, and about 500,000 new wells are drilled every year for residential purposes. In most parts of the world, including much of the United States, individual well usage is not metered, and anyone can pump groundwater without notifying an authority. Few places measure agricultural water use.

There are several ways to monitor water use from surface and groundwater sources. Water withdrawals are measurements of the amount of water taken from a source. This water may be used temporarily and then returned to its source and used again. For example, when a factory withdraws water from a river for cooling purposes, it returns most of the water to the river; thus, the water can be used later. Water that is incorporated into a product or lost to the atmosphere through evaporation or evapotranspiration cannot be reused in the same geographic area and is said to be consumed. Much of the water used for irrigation is lost to evaporation and evapotranspiration or is removed with the crop when it is harvested. Therefore, much of the water withdrawn for irrigation is consumed.

15.5 Kinds of Water Use

Water use varies considerably around the world, depending on availability of water and degree of industrialization. However, use can be classified into four broad categories: (1) domestic use, (2) agricultural use, (3) industrial use, and (4) in-stream use. It is important to remember that some uses of water are consumptive, while others are nonconsumptive. Freshwater availability and use, as well as the conservation of aquatic resources, are key to human well-being. The quality and quantity of surface and groundwater resources, and life-supporting ecosystem services, are being jeopardized by the impacts of population growth, rural-to-urban migration, rising wealth and resource consumption, and by climate change. If present trends continue, 1.8 billion people will be living in countries or regions with absolute water scarcity by 2025, and two-thirds of the world population could be subject to water stress.

Domestic Use of Water

Over 90 percent of the water used for domestic purposes in North America is supplied by municipal water systems, which typically include complex, costly storage, purification, and distribution facilities. Many rural residents, however, can obtain safe water from untreated private wells. Nearly 37 percent of municipal water supplies come from wells.

Regardless of the water source (surface or groundwater), water supplied to cities in the developed world is treated to ensure its safety. Treatment of raw water before distribution usually involves some combination of the following processes. The raw water is filtered through sand or other substrates to remove particles. Chemicals may be added to the water that will cause some dissolved materials to be removed. Then, before the water is released for public use, it is disinfected with chlorine, ozone, or ultraviolet light to remove any organisms that might still be present. Where no freshwater is available, expensive desalinization of saltwater may be the only option available. The United Arab Emirates, for example, uses desalinated water for over 90 percent of its needs.

Domestic activities in highly developed nations require a great deal of water. This domestic use includes drinking, air conditioning, bathing, washing clothes, washing dishes, flushing toilets, and watering lawns and gardens. On average, each person in a North American home uses about 400 liters (about 100 gallons) of **domestic water** each day. About 69 percent is used as a solvent to carry away wastes (nonconsumptive use; laundry, bathing, toilets, and dishes), about 29 percent is used for lawns and gardens (consumptive use), and only a tiny amount (about 2 percent) is actually consumed by drinking or in cooking. (See figure 15.5.) Yet all water that enters the house has been purified and treated to make it safe for drinking. However, a study conducted in 2011 of the drinking water in the United States found more than 200 unregulated chemicals in the tap water of 45 states. Contaminants included industrial solvents, weed killers, refrigerants, and the rocket fuel component perchlorate. The pollutants come from agriculture, factory discharges, consumer products, urban runoff, and wastewater treatment plants. Water utilities in the United States currently spend more than $50 billion a year to treat drinking water and over $200 million a year to protect source waters and prevent pollution from sources such as urban runoff. Natural processes cannot cope with the highly concentrated wastes typical of a large urban area. The unsightly and smelly wastewater also presents a potential health problem, so cities and towns must treat it before returning it to a local water source. Until recently, the cost of water in almost every community has been so low that

Water Use by a Typical North American Family of Four

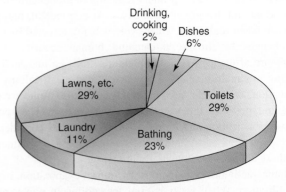

FIGURE 15.5 Urban Domestic Water Uses Over 150 billion liters (40 billion gallons) of water are used each day for urban domestic purposes in North America. (Nonconsumptive uses are shown in blue and red, and consumptive uses are shown in green and brown.)

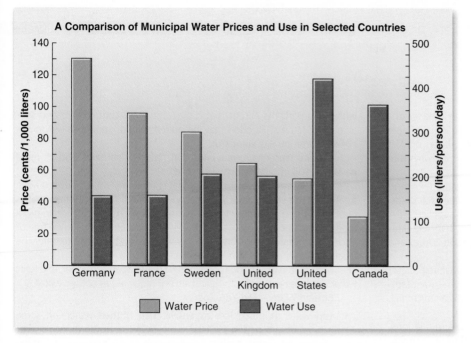

A Comparison of Municipal Water Prices and Use in Selected Countries

FIGURE 15.6 Water Use Decreases as Water Price Increases A general correlation exists between the amount of water that is used and its price.

there was very little incentive to conserve, but shortages of water and increasing purification costs have raised the price of domestic water in many parts of the world. (See figure 15.6.)

The price of turning on the tap varies greatly from city to city. In the United States, San Diego's water is the costliest ($1.65 per 400 liters) (100 gallons) due to the fact that 90 percent is pumped in from northern California and the Colorado River. New York City residents pay $0.80 for the same amount of water. Copenhagen, Denmark, has the most expensive rates in Europe at $3.43 per 400 liters. Despite water shortages, China has kept water fees low to stall inflation at $0.20 per 400 liters. Rates in China are now rising to promote conservation. Many people believe that we would conserve more water if the rates were higher—and the cost is generally increasing around the world. The question remains, however, as to how to make clean water affordable for the world's poorest citizens.

Although domestic use of water is a relatively small component of the total water-use picture, urban growth has created problems in the development, transportation, and maintenance of quality water supplies. (See figure 15.7.) In regions experiencing rapid population growth, such as Asia, domestic use is expected to increase sharply. Many cities in China are setting quotas on water use that are enforced by higher prices for larger users. In the coastal city of Dalian, a family that uses more than 8,000 liters (2,113 gallons) of water per month will pay four times more than a family that conserves water. In North America, more than 36 states expect a water crisis in the next ten years. Fast-growing cities in the West, especially those in arid areas, are experiencing water shortages. Demand for water in urban areas sometimes exceeds the immediate supply, particularly when the supply is local surface water. This is especially true during the summer, when water demand is high and precipitation is often low. Many communities have

begun public education campaigns designed to help reduce water usage. In addition, at the federal level, the U.S. Environmental Protection Agency (EPA) is in the process of developing a national consumer labeling program of water-efficient products modeled after its highly successful Energy STAR program. Mexico City, with a population of nearly 20 million and minimal access to surface water, has one of the most serious water management problems in the world. The water supply for the Mexico Basin is severely stressed by water management practices that do not ensure a sustainable supply of clean water. Since extractions from the Valley of Mexico aquifer began nearly 100 years ago, groundwater levels have dropped significantly. As the city's population continues to grow, managers have been forced to look at alternative sources of water from more than 160 kilometers (100 miles) away.

In addition to encouraging the public to conserve water, municipalities need to pay attention to losses that occur within the distribution system. Leaking water pipes and mains account for significant losses of water. Even in the developed world, such losses may be as high as 20 percent. Poorer countries typically exceed this, and some

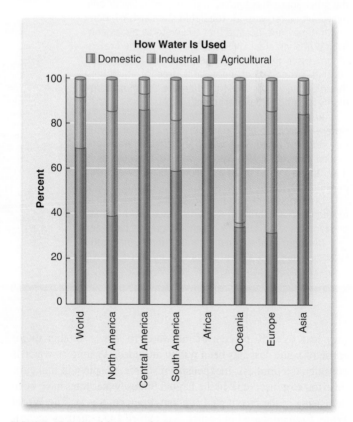

FIGURE 15.7 World Uses of Water Domestic, industrial, and agricultural uses dominate the allocation of water resources. However, there is considerable variety in different parts of the world in how these resources are used.
Source: Data from World Resources Institute, 2012.

Focus On

The Bottled Water Boom

Consumers spend billions of dollars on bottled water, making it the fastest-growing drink of choice in many parts of the world. Some people drink bottled water as an alternative to other beverages; others drink it because they prefer its taste or because their tap water is not always safe to drink.

In the United States, the Environmental Protection Agency (EPA) and the Food and Drug Administration (FDA) set drinking water standards. EPA sets standards for tap water provided by public water suppliers; FDA sets standards for bottled water based on EPA standards. FDA regulates bottled water as a packaged food under the Federal Food, Drug, and Cosmetic Act.

Any bottled water sold in interstate commerce in the United States, including products that originate overseas, must meet federal standards. Bottled water must meet FDA standards for physical, chemical, microbial, and radiological contaminants.

Bottlers use standard identifiers, prescribed by FDA regulations, to describe their water, but the meanings may be different than you expect. The terms refer to both the geological sources of the water and the treatment methods applied to the water. The terms do not necessarily describe the geographic location of the source or determine its quality. For instance, *spring water* can be collected at the point where water flows naturally to the Earth's surface or from a borehole that taps into the underground source. Other terms used on the label about the source, such as *glacial water* and *mountain water,* are not regulated standards of identity and may not indicate that the water is necessarily from a pristine area. Likewise, the term *purified* refers to processes that remove chemicals and pathogens. Purified water is not necessarily free of microbes, though it might be. The following are the most commonly used bottled water terms:

Artesian water, groundwater, spring water, well water—water from an underground aquifer that may or may not be treated. Well water and artesian water are tapped through a well. Spring water is collected as it flows to the surface or via a borehole. Groundwater can be either:

Distilled water—steam from boiling water that is recondensed and bottled. Distilling water kills microbes and removes water's natural minerals, giving it a flat taste.

Drinking water—water intended for human consumption that is sealed in bottles or other containers with no ingredients except that it may optionally contain safe and suitable disinfectants. Fluoride may be added within limitations set in the bottled water quality standards.

Mineral water—groundwater that naturally contains 250 or more parts per million of total dissolved solids.

Purified water—water that originates from any source but has been treated to meet the U.S. *Pharmacopeia* definition of purified water. Purified water is essentially free of all chemicals (it must not contain more than 10 parts per million of total dissolved solids) and may also be free of microbes if treated by reverse osmosis. Purified water may alternately be labeled according to how it is treated.

Carbonated water, soda water, seltzer water, sparkling water, and tonic water are considered soft drinks and are not regulated as bottled water.

Bottled Water—Did You Know?

- Tap water is much less expensive than bottled water—in some cases, you are paying for little more than the bottle itself. At least 25 percent (some experts say as much as 40 percent) of bottled water is nothing more than processed tap water.

- There is an environmental cost—it takes three to four times the amount of water in the bottle just to make the plastic for the bottle, and that's not including how much oil is used and how much carbon dioxide is created when the water is shipped to the store.

- Buy domestic—if you cannot break the bottled water habit, look for a brand that has not been shipped across the world. The less distance the water has to travel, the fewer greenhouse gases are produced.

- 28 billion—the number of plastic water bottles purchased in the United States annually, of which only 16 percent are recycled. 1.5 million barrels—the amount of oil used to make those bottles. $11 billion— the amount that people in the United States spend on bottled water annually.

- 2.5 million—the number of plastic water bottles that people in the United States discard hourly.

In an effort to encourage recycling and discourage consumption, the city of Chicago became the first major U.S. city to put a 5-cent tax on bottled water. Many other cities are debating similar measures.

may lose over 50 percent of the water to leaks. Another major cause of water loss has been public attitudes. As long as water is considered a limitless, inexpensive resource, people will make little effort to conserve it. In the United States, water rates have been increasing in the recent past at about 4 percent a year. The World Water Council, based in France, has ranked 147 countries according to the efficiency of their water use. It ranked the United States last—as the most wasteful and least efficient. One explanation for this is that, historically, water in the United States has been very cheap. For example, the Germans pay $1.82 for a cubic meter of water, the French $1.12, and the British $1.34. In the United States, the cost is 55 cents, or roughly $15 a month for the average family of four. As the cost of water rises and attitudes toward water change, so will usage and efforts to conserve. For example, the California Urban Water Conservation Council, together with the EPA, has launched a highly innovative virtual home tour that allows Internet users to click on different areas of the H_2OUSE floorplan for facts and advice on saving water.

In 1990, 30 states reported "water-stress" conditions. In 2011, the number was 46. Taking measures at home to conserve water not only saves money, it also is of benefit to the greater community.

Saving water at home does not require a significant cost outlay. Although there are water-saving appliances and water conservation systems such as rain barrels and on-demand water heaters that are more expensive, the bulk of water saving methods can be achieved at little cost. (See figure 15.8.) For example, 75 percent

of water used indoors is in the bathroom, and 25 percent of this is for the toilet. Installing a ULF (ultra-low flush) toilet and low-flow showerheads and aerators can significantly reduce water use. When buying low-flow aerators, be sure to read the label for the actual "gpm" (gallons per minute) rating. Often, retailers promote "low-flow," rated at 2.5 gpm, which is the top of the low-flow spectrum. This may be needed for the kitchen sink, but a 1.5 gpm aerator will work fine for the bathroom sink and most water outlets. By using water-saving features you could reduce your in-home water use by 35 percent.

Agricultural Use of Water

In North America, groundwater accounts for about 37 percent of the water used in agriculture and surface water about 63 percent. **Irrigation** is the major consumptive use of water in most parts of the world and accounts for about 80 percent of all the water consumed in North America. About 500 billion liters per day (134 billion gallons per day or 150 million acre-feet per year) are used in irrigation in the United States. The amount of water used for irrigation and livestock continues to increase throughout the world. Future agricultural demand for water will depend on the cost of water for irrigation; the demand for agricultural products, food, and fiber; governmental policies; the development of new technology; and competition for water from a growing human population.

Since irrigation is common in arid and semiarid areas, local water supplies are often lacking, and it is often necessary to transport water great distances to water crops. This is particularly true in the western United States, where about 14 million hectares (35 million acres) of land are irrigated.

Four methods of irrigation are commonly used. Surface or flood irrigation involves supplying the water to crops by having the water flow over the field or in furrows. This requires extensive canals and is not suitable for all kinds of crops. Spray irrigation involves the use of pumps to spray water on the crop. Trickle irrigation uses a series of pipes with strategically placed openings so that water is delivered directly to the roots of the plants. Subirrigation involves supplying water to plants through underground pipes. Often this method is used where soils require draining at certain times of the year. The underground pipes can be used to drain excess water at one time of the year and supply water at others. Each of these methods has its drawbacks and advantages as well as conditions under which it works well.

Construction and maintenance of irrigation structures, such as dams, canals, pipes, and pumps, are expensive. Costs for irrigation water have traditionally been low, since many of the dams and canals were constructed with federal assistance, and farmers have often used water wastefully. As competition has grown between urban areas and agriculture for scarce water resources, there has been pressure to raise the cost of water used for irrigation. Increasing the cost of water will stimulate farmers to conserve, just as it does homeowners. Another way to reduce the demand for irrigation water is to reduce the quantity of water-demanding crops grown in dry areas or change from high water-demanding to lower water-demanding crops. For example, wheat or soybeans

Water Savings Guide

Conservative use will save water		Normal use will waste water
Wet down, soap up, rinse off 15 liters (4 gal)	Shower	Regular shower 95 liters (25 gal)
May we suggest a shower?	Tub bath	Full tub 135 liters (36 gal)
Minimize flushing Each use consumes 20–25 liters (5–7 gal) New toilets use 6 liters (1.6 gal)	Toilet	Frequent flushing is very wasteful
Fill basin 4 liters (1 gal)	Washing hands	Tap running 8 liters (2 gal)
Fill basin 4 liters (1 gal)	Shaving	Tap running 75 liters (20 gal)
Wet brush, rinse briefly 2 liters (1/2 gal)	Brushing teeth	Tap running 38 liters (10 gal)
Take only as much as you require	Ice	Unused ice goes down drain
Repair leaks	Leaks	A small drip wastes 95 liters (25 gal) per week

(a)

(b)

FIGURE 15.8 Conserving Water (a) Minor changes in the way people use water could significantly reduce domestic water use. Note: 1 gallon equals approximately 3.785 liters. (b) The photo shows a billboard in central Florida encouraging citizens to conserve water.

require less water than do potatoes or sugar beets. It is also becoming increasingly important to modify irrigation practices to use less water. (See figure 15.9.) For example, the use of trickle irrigation and some variations of spray irrigation use water more efficiently than the more traditional flood irrigation methods.

Many forms of irrigation require a great deal of energy. This is particularly true when pumps are used to deliver the water to the crop. It is estimated that 40 percent of the energy devoted to agriculture in Nebraska is used for irrigation. Increasing energy costs may force some farmers to reduce or discontinue irrigation. In addition, much of western Nebraska relies on groundwater for irrigation, and the water table is dropping rapidly. If a water shortage develops, land values will decline. Land use and water use are interrelated and cannot be viewed independently. When farmers abandoned their homesteads in Kansas, Nebraska, and Texas during the 1930s Dust Bowl, they had no idea they stood atop part of the High Plains aquifer (Ogallala aquifer), one of North America's most abundant underground reservoirs. Two decades later, water pumped from the aquifer transformed the region into the breadbasket of the United States. Today, wells pumping water from the aquifer nourish a multibillion-dollar farm economy. But the withdrawal of water has greatly surpassed the aquifer's rate of natural replenishment from precipitation. In fewer than a hundred years, the aquifer will not be able to support the current rate of irrigation, thus threatening the region's agricultural base. In the Salinas Valley, along the central coast of California, farmers are irrigating their fields with recycled water from the Monterey Regional Water Pollution Control Agency's (MRWPCA) local wastewater treatment plant. Because seawater had entered the groundwater as far as 10 kilometers (6 miles) inland, well water was becoming too salty for agricultural use. MRWPCA, aware that this problem may be caused by overpumping of groundwater, in the 1980s began studying the possibility of irrigating with treated municipal wastewater. The MRWPCA conducted a study that showed that crops irrigated with recycled water were not contaminated with pathogens. California then approved the plan, which required an upgrade of the local wastewater treatment plant. Sixty kilometers (40 miles) of pipeline were constructed to deliver the water to some 75 users throughout the system, which has the capacity to provide 24 billion liters (19,500 acre-feet) of water per year and irrigate 4,900 hectares (12,000 acres) of coastal farmland. Such systems may become a common solution to the problem of deteriorating and diminishing groundwater supplies in California and perhaps throughout the world.

Industrial Use of Water

Industrial water use accounts for nearly half of total water withdrawals in the United States, about 70 percent in Canada, and about 23 percent worldwide. Since most industrial processes involve heat exchanges, 90 percent of the water used by industry is for cooling and is returned to the source, so only a small amount is actually consumed. Industrial use accounts for less than 20 percent of the water consumed in the United States. For example, electric-power generating plants use water to cool steam so that it changes back into water. Many industries, especially power

(a)

(b)

(c)

FIGURE 15.9 Types of Irrigation Many arid areas require irrigation to be farmed economically. (a) Surface or flood irrigation uses irrigation canals and ditches to deliver water to the crops. The land is graded so that water flows from the source into the fields. Water is siphoned from a canal into ditches between rows of crops. (b) Spray irrigation uses a pump to spray water into the air above the plants. (c) Trickle irrigation conserves water by delivering water directly to the roots of the plants but requires an extensive network of pipes.

plants, actually can use saltwater for cooling purposes. About 30 percent of the cooling water used by power plants is saltwater. If the water heated in an industrial process is dumped directly into a watercourse, it significantly changes the water temperature. This affects the aquatic ecosystem by increasing the metabolism of the organisms and reducing the water's ability to hold dissolved oxygen.

Industry also uses water to dissipate and transport waste materials. In fact, many streams are now overused for this purpose, especially in urban centers. The use of watercourses for waste dispersal degrades the quality of the water and may reduce its usefulness for other purposes, as well as directly harming fish and wildlife. This is especially true if the industrial wastes are toxic. In 2010, a dam in Hungary holding back a vast reservoir of toxic red sludge from an alumina manufacturing plant burst, killing four people and injuring 120. An estimated 700,000 cubic meters (24 million cubic feet) of toxic red sludge flowed over seven towns causing the evacuation of hundreds of local residents.

Historically, industrial waste and heat were major causes of pollution. However, most industrialized nations have passed laws that severely restrict industrial discharges of wastes or heated water into watercourses. In the United States, the federal government's role in maintaining water quality began in 1972, with the passage of the Federal Water Pollution Control Act (PL 92-500). This act provided federal funds and technical assistance to strengthen local, state, and interstate water-quality programs. The act (and subsequent amendments in 1977, 1981, 1987, and 1993) is commonly referred to as the "Clean Water Act." The Clean Water Act is a comprehensive and technically rigorous piece of legislation. It seeks to protect the waters of the United States from pollution. To do this, the act specifically regulates pollutant discharges into "navigable waters" by implementing two concepts: setting water quality standards for surface water and limiting effluent discharges into the water. The policy objectives of the Clean Water Act are to restore and maintain the "chemical, physical and biological integrity of the nation's waters." Enforcement of this act has been extremely effective in improving surface-water quality. However, many countries in the developing world have done little to control industrial pollution, and water quality is significantly reduced by careless use.

In-Stream Use of Water

In-stream water use does not remove water but makes use of it in its channels and basins. Therefore, all in-stream uses are nonconsumptive. Major in-stream uses of water are for hydroelectric power, recreation, and navigation. Although in-stream uses do not remove water, they may require modification of the direction, time, or volume of flow and can negatively affect the watercourse.

Electricity from hydroelectric power plants is an important energy resource. Presently, hydroelectric power plants produce about 13 percent of the total electricity generated in the United States. (See figure 15.10.) They do not consume water and do not add waste products to it. However, the dams needed for the plants have definite disadvantages, including the high cost of construction and the resulting destruction of the natural habitat in streams

(a)

(b)

FIGURE 15.10 Dams Interrupt the Flow of Water The flow of water in most large rivers is controlled by dams. Most of these dams provide electricity. In addition, they prevent flooding and provide recreational areas. (a) Large dams, however, destroy the natural river system, while smaller dams (b) do not always have a significant impact.

and surrounding lands. The sudden discharge of impounded water from a dam can seriously alter the downstream environment. If the discharge is from the top of the reservoir, the stream temperature rapidly increases. Discharging the colder water at the bottom of the reservoir causes a sudden decrease in the stream's water temperature. Either of these changes is harmful to aquatic life. The impoundment of water also reduces the natural scouring action of a flowing stream. If water is allowed to flow freely, the silt accumulated in the river is carried downstream during times of high water. This maintains the river channel and carries nutrient materials to the river's mouth. But if a dam is constructed, the silt is deposited behind the dam, eventually filling the reservoir. Many other dams were constructed to control floodwaters. While dams reduce flooding, they do not eliminate it. In fact, the building of a dam often encourages people to develop the floodplain. As a result, when flooding occurs, the loss of property and lives may be greater.

Science, Politics, & Policy

Water Wars

It is a given in developed countries that nearly everyone has access to clean water. But roughly 1.1 billion people around the world have no such reliable access. The result is both grim and predictable: the lack of clean water leads directly to a higher incidence of preventable waterborne diseases such as cholera and dysentery, which kill 2.2 million a year.

Many conflicts are caused or inflamed by water scarcity. The conflicts from Chad to Darfur, Sudan, to Ethiopia, to Somalia and its pirates, and across to Yemen, Iraq, Pakistan, and Afghanistan, lie in a great arc of arid lands where water scarcity is leading to failed crops, dying livestock, extreme poverty, desperation and death from contaminated water.

Extremist groups like the Taliban find ample recruitment possibilities in such impoverished communities. Governments lose their legitimacy when they cannot guarantee their populations most basic needs: safe drinking water, stable food crops, and fodder and water for the animal herds on which communities depend for their livelihoods.

Water supplies are increasingly under stress in large parts of the world, especially in the world's arid regions. Rapidly intensifying water scarcity reflects bulging populations, depletion of groundwater, waste and pollution, and the increasingly harmful effects of climate change.

The consequences of water scarcity are dire: drought and famine, loss of livelihood, the spread of waterborne diseases, forced migrations, and even open conflict. Practical solutions will include many components, including better water management, improved technologies to increase the efficiency of water use, and new investments undertaken jointly by governments, the business sector, and civic organizations.

The precise nature of the water crisis will vary, with different pressure points in different regions. For example, Pakistan, an already arid country, will suffer from a rapidly growing population, which has grown from 42 million in 1950 to a projected 363 million by 2050. To compound that concern farmers are relying on groundwater that is being depleted by overpumping. In addition, the Himalayan glaciers that feed Pakistan's rivers may melt by 2050 due to climate change.

Solutions to the water crisis will have to be found at all levels from community concerns to global issues such as climate change. Solutions will require partnerships between government, business, and civil society. Such partnerships, while critical, can be difficult to develop because these different sectors of society often have little or no experience in working with each other and may not trust each other.

Lasting solutions to water challenges will require a broad range of expert knowledge about climate, ecology, farming, population, engineering, economics, politics, and local culture. It will also be necessary to bring together scientific, political, and business leaders from countries facing water scarcity issues to look at common solutions to the problems. In short, the global society will have to cooperate in ways it has not been accustomed to. Is the world up to the challenge?

The massive flooding along the Missouri River in 2011 is a case in point. The Missouri River is the longest river in the United States. Beginning in southwestern Montana in the Rocky Mountains it flows north then generally southeast across the heart of the United States, ending at the Mississippi River, just to the north of St. Louis, Missouri, a total length of 4,023 kilometers (2,500 miles). In 2011 the Missouri River saw one of the worst flooding events in United States history. The flooding was related to the record snowfall in the Rocky Mountains of Montana and Wyoming along with near record spring rainfall in central and eastern Montana. In order to prevent structural damage caused by overflow the six major dams along the Missouri River released record amounts of water during the spring and summer of 2011. This release of large quantities of water resulted in severe flooding in many agricultural areas, towns and cities along the river.

The construction of the dams to tame the "Mighty Mo" during the last century encouraged development and farming in what had been the natural floodplains of the Missouri River. The public along the river had become dependent on constructed levees

to protect against flooding in the absence of the river's natural floodplains.

Because dams create lakes that have a large surface area, evaporation is increased. In arid regions, the amount of water lost can be serious. This is particularly evident in hot climates. Furthermore, flow is often intermittent below the dam, which alters the water's oxygen content and interrupts fish migration. The populations of algae and other small organisms are also altered. Because of all these impacts, dam construction requires careful planning.

Dam construction often creates new recreational opportunities because reservoirs provide sites for boating, camping, and related recreation. (See figure 15.11.) However, these opportunities come at the expense of a previously free-flowing river. Some recreational pursuits, such as river fishing, are lost. Sailing, water-skiing, swimming, fishing, and camping all require water of reasonably good quality. Water is used for recreation in its natural setting and often is not physically affected. Even so, it is necessary to plan for recreational use, because overuse or inconsiderate use can degrade water quality.

FIGURE 15.11 Recreational Opportunities Created by Dams Over 4 million people enjoy fishing and boating on Lake Havasu, located on the border between California and Arizona. Lake Havasu was created in 1938 by the building of Parker Dam on the Colorado River. The lake's surface area is 19,300 acres (7,800 ha). The primary purpose of the dam and lake is to store water for pumping into two aqueducts serving several western states.

Most major rivers and large lakes are used for navigation. North America currently has more than 40,000 kilometers (25,000 miles) of commercially navigable waterways. These waterways must have sufficient water depth to ensure the passage of ships and barges. Canals, locks, and dams are used to ensure that adequate depths are provided. Often, dredging is necessary to maintain the proper channel depth. Dredging can resuspend contaminated sediments. Another problem is determining where to deposit the contaminated sediments when they are removed from the bottom. In addition, the flow within the hydrologic system is changed, which, in turn, affects the water's value for other uses.

Most large urban areas rely on water to transport resources. During recent years, the inland waterway system in the United States has carried about 10 percent of goods such as grain, coal, ore, and oil. In North America, expenditures for the improvement of the inland waterway system have totaled billions of dollars.

In the past, almost any navigation project was quickly approved and funded, regardless of the impact on other uses. Today, however, such decisions are not made until ecological impacts are analyzed.

15.6 Kinds and Sources of Water Pollution

Water quality degradation from human activities continues to harm human and ecosystem health. Three million people die from waterborne diseases each year in developing countries—the majority are children under the age of five. Pollutants of primary concern include microbial pathogens and excessive nutrient loads. Water contaminated by microbes remains the greatest single cause of human illness and death on a global scale. High nutrient loads lead to eutrophication of downstream and coastal waters and loss of beneficial human uses. Pollution from land sources, particularly agriculture and urban run-off, needs urgent action by governments and the agricultural sector. Pesticide pollution and suspended sediments are also hard to control.

Water pollution occurs when something enters water that changes the natural ecosystem or interferes with water use by segments of society. In an industrialized society, maintaining completely unpolluted water in all drains, streams, rivers, and lakes is probably impossible. (See table 15.1.) But we can evaluate the water quality of a body of water and take steps to preserve or improve its quality by eliminating sources of pollution. Some pollutants seriously affect the quality and possible uses of water. In general, water pollutants can be divided into several broad categories.

Toxic chemicals or acids may kill organisms and make the water unfit for human use. If these chemicals are persistent, they may bioaccumulate in individual organisms and biomagnify in food chains.

Dissolved organic matter is a significant water pollution problem because it decays in the water. As the microorganisms naturally present in water break down the organic matter, they use up available dissolved oxygen from the water. If too much dissolved oxygen is removed, aquatic organisms die. The amount of oxygen required to decay a certain amount of organic matter is called the **biochemical oxygen demand (BOD).** (See figure 15.12.) Measuring the BOD of a body of water is one way to determine how polluted it is. If too much organic matter is added to the water, all of the available oxygen will be used up. Then, anaerobic (not requiring oxygen) bacteria begin to break down wastes. Anaerobic respiration produces chemicals that have a foul odor and an unpleasant taste and generally interfere with the well-being of humans.

Disease-causing organisms are a very important pollution problem in most of the world. Untreated or inadequately treated human or domesticated animal waste is most often the source of these organisms. In the developed world, sewage treatment and drinking-water treatment plants greatly reduce this public health problem.

Nutrients are also a pollution problem. Additional nutrients in the form of nitrogen and phosphorus compounds from fertilizer, sewage, detergents, and animal waste increase the rate of growth of aquatic plants and algae. However, phosphates and nitrates are generally present in very limited amounts in unpolluted freshwater and, therefore, are a limiting factor on the growth of aquatic plants and algae. (A **limiting factor** is a necessary material that is in short supply, and because of the lack of it, an organism cannot reach its full potential growth. See chapter 5.) Thus, when phosphates or nitrates are added to the surface water, they act as a fertilizer and promote the growth of undesirable algae populations. The excessive growth of algae and aquatic plants due to added nutrients is called **eutrophication.** Algae and larger aquatic plants may interfere with the use of the water by fouling boat propellers, clogging water-intake pipes, changing the taste and odor of water, and causing the buildup of organic matter on the bottom. As this organic matter decays, oxygen levels decrease, and fish and other aquatic species die.

Physical particles also can negatively affect water quality. Particles alter the clarity of the water, can cover spawning sites, act as abrasives that injure organisms, and carry toxic materials.

Determining acceptable water quality involves economic considerations. Removing the last few parts per million of some

Table 15.1 Sources and Impacts of Selected Pollutants

Pollutant	Source	Effects on Humans	Effects on Aquatic Ecosystem
Acids	Atmospheric deposition; mine drainage; decomposing organic matter	Reduced availability of fish and shellfish; increased heavy metals in fish	Death of sensitive aquatic organisms; increased release of trace metals from soils, rock, and metal surfaces, such as water pipes
Chlorides	Runoff from roads treated for removal of ice or snow; irrigation runoff; brine produced in oil extraction; mining	Reduced availability of drinking water supplies; reduced availability of shellfish	At high levels, toxic to freshwater organisms
Disease-causing organisms	Dumping of raw and partially treated sewage; runoff of animal wastes from feedlots	Increased costs of water treatment; death and disease; reduced availability and contamination of fish, shellfish, and associated species	Reduced survival and reproduction of aquatic organisms due to disease
Elevated temperatures	Heat trapped by cities that is transferred to water; unshaded streams; solar heating of reservoirs; warm-water discharges from power plants and industrial facilities	Reduced availability of fish	Elimination of cold-water species of fish and shellfish; less oxygen; heat-stressed animals susceptible to disease; inappropriate spawning behavior
Heavy metals	Atmospheric deposition; road runoff; discharges from sewage treatment plants and industrial sources; creation of reservoirs; acidic mine effluents	Increased costs of water treatment; disease and death; reduced availability and healthfulness of fish and shellfish; biomagnification	Lower fish population due to failed reproduction; death of invertebrates leading to reduced prey for fish; biomagnification
Nutrient enrichment	Runoff from agricultural fields, pastures, and livestock feedlots; landscaped urban areas; dumping of raw and treated sewage and industrial discharges; phosphate detergents	Increased water treatment costs; reduced availability of fish, shellfish, and associated species; color and odor associated with algal growth; impairment of recreational uses	Algal blooms occur; death of algae results in low oxygen levels and reduced diversity and growth of large plants; reduced diversity of animals; fish kills
Organic molecules	Runoff from agricultural fields and pastures; landscaped urban areas; logged areas; discharges from chemical manufacturing and other industrial processes; combined sewers	Increased costs of water treatment; reduced availability of fish, shellfish, and associated species; odors	Reduced oxygen; fish kills; reduced numbers and diversity of aquatic life
Sediment	Runoff from agricultural land and livestock feedlots; logged hillsides; degraded streambanks; road construction; and other improper land use	Increased water treatment costs; reduced availability of fish, shellfish, and associated species; filling of lakes, streams, and artificial reservoirs and harbors, requiring dredging	Covering of spawning sites for fish; reduced numbers of insect species; reduced plant growth and diversity; reduced prey for predators; clogging of gills and filters
Toxic chemicals	Urban and agricultural runoff; municipal and industrial discharges; leachate from landfills and mines; atmospheric deposits	Increased costs of water treatment; increased risk of certain cancers; reduced availability and healthfulness of fish and shellfish	Reduced growth and survivability of fish eggs and young; fish diseases; death of carnivores due to biomagnification in the food chain

Source: Data, in part, from World Resources Institute.

materials from the water may not significantly improve water quality and may not be economically justifiable. This is certainly true of organic matter, which is biodegradable. However, radioactive wastes and toxins that may accumulate in living tissue are a different matter. Vigorous attempts to remove these materials are often justified because of the materials' potential harm to humans and other organisms.

Sources of pollution are classified as either point sources or nonpoint sources. When a source of pollution can be readily identified because it has a definite source and place where it enters the water, it is said to come from a **point source.** Municipal and industrial discharge pipes are good examples of point sources. Diffuse pollutants, such as from agricultural land and urban paved surfaces, acid rain, and runoff, are said to come from **nonpoint sources** and are much more difficult to identify and control. Initial attempts to control water pollution were focused on point sources of pollution, since these were readily identifiable and economic pressure and adverse publicity could be brought to bear on

companies that continued to pollute from point sources. In North America, most point sources of water pollution have been identified and are regulated.

Nonpoint sources of water pollution are being addressed, but this is much more difficult to do, since regulating many small, individual human acts is necessary.

There are many things that you can do to protect surface and ground waters from nonpoint source pollution.

- Be aware that many chemicals commonly used around the home are toxic. Use nontoxic substitutes wherever possible.
- Buy chemicals only in the amount you expect to use, and apply them only as directed.
- Take unwanted household chemicals to hazardous waste collection centers; do not pour them down the drain.
- Never pour unwanted chemicals on the ground. Soil cannot destroy most chemicals, and they may eventually contaminate runoff.

- Use water-based products whenever possible.
- When landscaping your yard, select native plants that have low requirements for water, fertilizers, and pesticides.
- Test your soil before applying fertilizers. Over-fertilization is a common problem, and the excess can leach into groundwater or contaminate rivers or lakes.

Municipal Water Pollution

Municipalities are faced with the double-edged problem of providing suitable drinking water and disposing of wastes. These wastes consist of stormwater runoff, wastes from industry, and wastes from homes and commercial establishments. Wastes from homes consist primarily of organic matter from garbage, food preparation, cleaning of clothes and dishes, and human wastes. Human wastes are mostly undigested food material and a concentrated population of bacteria, such as *Escherichia coli* and *Streptococcus faecalis*. These particular bacteria normally grow in the large intestine (colon) of humans and are present in high numbers in the feces of humans; therefore, they are commonly called **fecal coliform bacteria.** Fecal coliform bacteria are also present in the wastes of other warm-blooded animals, such as birds and mammals. Low numbers of these bacteria in water are not harmful to healthy people. However, because they can be easily identified, their presence in the water is used to indicate the amount of pollution from the fecal wastes of humans and other warm-blooded animals. The numbers of these types of bacteria present in water are directly related to the amount of fecal waste entering the water.

When human wastes are disposed of in water systems, potentially harmful bacteria from humans may be present in amounts too small to detect by sampling. Even in small numbers, these harmful bacteria may cause disease epidemics. It is estimated that some 1.5 million people in the United States become ill each year from infections caused by fecal contamination. In 1993, for example, a protozoan pathogen called *Cryptosporidium* was identified in the Milwaukee, Wisconsin, public water system. This resulted in over 400,000 people becoming ill and at least 100 deaths. The total costs of such diseases amount to billions of dollars per year in the United States alone. In earthquake ravished Haiti, a cholera outbreak began in 2010 due to contaminated drinking water. By 2011 over 2,000 Haitians had died from cholera. The greater the amount of wastes deposited in the water, the more likely it is that there will be populations of disease-causing bacteria. Therefore, the presence of fecal coliform bacteria is used as an indication that other, more harmful organisms may be present as well.

Wastewater from cleaning dishes and clothing contains some organic material along with the soap or detergent, which helps to

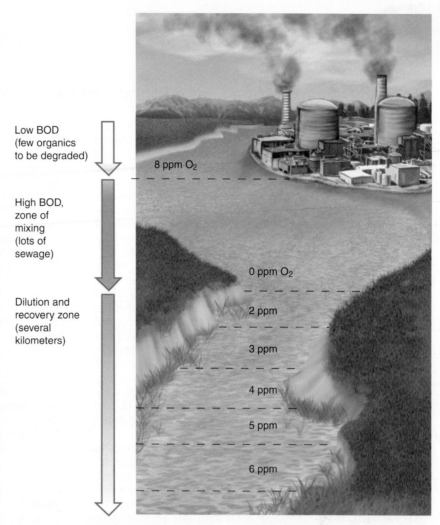

Low BOD (few organics to be degraded)

8 ppm O_2

High BOD, zone of mixing (lots of sewage)

0 ppm O_2

Dilution and recovery zone (several kilometers)

2 ppm

3 ppm

4 ppm

5 ppm

6 ppm

FIGURE 15.12 Effect of Organic Wastes on Dissolved Oxygen Sewage contains a high concentration of organic materials. When these are degraded by organisms, oxygen is removed from the water. This is called the biochemical oxygen demand (BOD). An inverse relationship exists between the amount of organic matter and oxygen in the water. The greater the BOD, the more difficult it is for aquatic animals to survive and the less desirable the water is for human use. The more the organic pollution, the greater the BOD.

separate the contaminant from the dishes or clothes. Soaps and detergents are useful because one end of the molecule dissolves in dirt or grease and the other end dissolves in water. When the soap or detergent molecules are rinsed away by the water, the dirt or grease goes with them.

At one time, many detergents contained phosphates as a part of their chemical makeup, which contributed to eutrophication. However, because of the environmental effects of phosphate on aquatic environments, since 1994 most major detergent manufacturers in North America and other developed countries have eliminated phosphates from most of their formulations. Today, the majority of phosphate entering water in North America is from human waste and runoff from farm fields and livestock operations. A study conducted by the Toxic Substances Hydrology Program of the U.S. Geological Survey (USGS) revealed that a broad range of chemicals found in residential, industrial, and agricultural wastewaters commonly occurs at low concentrations downstream from areas of intense urbanization and animal production. The

Focus On

Growing Demands for a Limited Supply of Water in the West

As one of the fastest-growing population centers in the United States, Las Vegas has a daily growing demand for water. Every month newcomers arrive to retire or find jobs, meaning the already swollen population could double in 20 to 30 years. This is forcing Las Vegas to look for water as far away as the Snake Valley (see map).

In 2005 the city started to file for groundwater rights in counties hundreds of kilometers away, setting off a water war that could be repeated across the parched but popular southwest United States. It is argued that if Las Vegas is successful in its claims, it could upset a complex web of aquifers that run as far away as California's Death Valley and western Utah.

That Las Vegas has real water concerns cannot be denied. The city exceeded the capacity of its own groundwater field several decades ago and currently is 90 percent dependent on a limited allotment from the Colorado River—an allotment it is fast outgrowing. That is what has driven the city to petition for water rights in the outlying counties, but if the history of Western development has shown one thing, it's that this kind of water shopping can go seriously wrong. In the early part of the twentieth century, Los Angeles famously—and secretly—bought up thousands of hectares of land in California's Owens Valley and then began to drain the surface and subsurface water for the city's use. After decades of pumping, a dozen Owens Valley springs have dried up, and water tables in places are too low to support once-abundant native grasses and shrubs.

Equally controversial plans such as Las Vegas's are developing in other Western cities. After over a decade of drought, Lake Powell is less than half full. Water flows into Lake Powell, which is between Utah and Arizona, from the Rocky Mountains via the Colorado River. More than 30 million people in seven states depend on the Colorado for water to grow crops, fuel power plants, and keep cities such as Las Vegas alive. While Las Vegas gets most of its water from Lake Mead, if water doesn't flow from Lake Powell, the amount of water in Lake Mead is diminished. The decade-long drought has severely slowed the flow of water into Lake Powell.

In 2013 the Bureau of Reclamation cut, by 10 percent, the amount of water people in the southwestern United States could take from Lake Powell. As states and counties fight over their allotment of water in the coming years, hydroelectric plants (including the one on the Hoover Dam) could become idle, and farmers are potentially looking at reduced crop production.

Few urban areas are more vulnerable to water shortages than Las Vegas, which has been referred to as the dryest big city in the United States. Las Vegas takes 90 percent of its water from Lake Mead, although Nevada gets by far the smallest share of water among the seven states that border the Colorado—just 2 percent of the total. Las Vegas has worked to conserve water, paying residents to replace lawns with desert-appropriate landscaping. The city's overall water use has dropped since 2002, even as population and visitor numbers have continued to rise.

If the rest of the Southwest United States can use its water more efficiently, it should have enough for decades. One solution could involve diverting more of the Colorado River's water away from agriculture—which claims 85 percent of the supply—in favor of the region's thirsty cities. That would be challenging politically, but something has to give. While Lake Mead has shrunk to about 40 percent of capacity, the immense reservoir still contains about 14 trillion liters of water (3.6 trillion gallons). But the dry sky above and the rock all around reinforce the inescapable fact that this land was a desert, is a desert, and always will be a desert. When the explorer J.C. Ives visited the present location of the Hoover Dam in 1857, he declared the land "worthless," adding, "There is nothing there to do but leave." Today's residents are hoping there is another choice.

Water Fountain in Las Vegas, Nevada

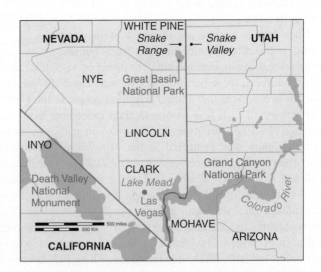

Going Green

From Toilet Water to Tap Water

Wastewater has long been recycled for agricultural and industrial purposes. But only a few places in the world, including Singapore, have been bold enough to add it to the drinking supply. The technology—if it gains wider public acceptance—could change the face of potential water-rights disputes in places such as the United States, Vietnam, and Egypt.

No federal law governs recycled wastewater for drinking, but states such as California require additional testing for contaminants.

Despite the not-so-pleasant nickname—"toilets to tap"—given to the technology, only about 10 percent of household wastewater typically comes from toilets, while the rest comes from showers, sinks, and laundry machines.

Treated wastewater typically doesn't go directly into the tap, but is piped into the ground, lakes, or reservoirs. It mixes with water from other sources, and may be cleaned further to meet drinking water standards before being funneled to consumers' taps—often months later.

Roughly 884 million people—1 of every 8 in the world—still lack access to safe drinking water, according to the World Health Organization and UNICEF. Meanwhile, water usage has increased by more than twice the rate of the world's population growth during the past century.

In Windhoek, Namibia, purifying wastewater to drink has been a way of life for decades. The drinking supply of Fairfax Water, which serves 1.7 million in Northern Virginia, outside Washington, D.C., has included recycled sewage water since the 1970s. About 5 percent of the area's drinking water now comes from purified sewage.

Parts of Orange County, California, began purifying sewer water in the 1970s, but the water district has significantly ramped up capacity so that recycled wastewater now supplies up to one-fifth of the daily water demand of the 2.4 million people within the area. Singapore has built advanced wastewater recycling facilities in less than a decade to meet almost one-third of its daily water needs.

Only in the past few years, however, has treatment technology improved to the point where a growing number of municipalities and countries are considering adopting wastewater purification programs.

To make the water potable, Singapore and Orange County use several steps. First, during a process called microfiltration, the water passes through a membrane with tiny holes—hundreds of times smaller than a human hair—that trap bacteria. It then undergoes a reverse-osmosis process in which it's pushed through a second, semi-permeable membrane that blocks salt, viruses, and pharmaceuticals. Finally, the water is zapped with high-intensity ultraviolet light and hydrogen peroxide to destroy any trace organics.

Yet, because of the controversy surrounding this type of water recycling, public acceptance has become as crucial as politics or cost in whether it's implemented.

In Toowoomba, Australia, about 100 kilometers (60 miles) west of Brisbane, residents soundly defeated a 2006 proposal to add recycled wastewater to the drinking supply, despite the area's perpetual water shortage.

Singapore, though, has become known for its efficiency in implementing wastewater recycling. By 2060, it expects its purified wastewater to accommodate half the nation's water demand. It has given out 19 million bottles of its so-called NEWater to athletic groups and at community events. And it's drawn more than 800,000 visitors to its visitors' center for wastewater purification demonstrations.

All water, to some extent, is recycled. River water often is treated and used by one city, then waste may be funneled into pipes and discharged downstream. Some environmental groups are hoping that as knowledge about wastewater recycling grows, so will its adoption across North America. What are your thoughts? Would you have difficulty drinking "recycled" wastewater? In truth, don't you drink such water already?

---

chemicals include human and veterinary drugs (including antibiotics), natural and synthetic hormones, detergent metabolites, plasticizers, insecticides, and fire retardants. The report found that in addition to caffeine, the most frequently detected compounds were cholesterol and coprostanol, which is a by-product of DEET, a common insect repellent. The compounds found in the water are sold on supermarket shelves and are in virtually every medicine cabinet and broom closet, as well as farms and factories. Although they are flushed or rinsed down the drain every day, they do not disappear.

In a study of 139 streams throughout the United States, the USGS found that one or more of these chemicals was in 80 percent of the streams sampled. Half of the streams contained seven or more of these chemicals, and about one-third contained ten or more. This was the first national-scale examination of streams for these organic wastewater contaminants. The chemicals identified largely escape regulation and are not removed by municipal wastewater treatment.

In the 1970s scientists began detecting pharmaceutical residue in waterways, but in an era when rivers were choking on industrial sludge, traces of drugs seemed a small matter. It would take until the 1990s for that view to change. That was when pharmaceutical estrogens, principally from birth control pills, began showing up in the water too, leading to male fish with female-like sex organs. Scarily, it did not take much estrogen to affect the fish—just 5 or 6 nanograms, or billionths of a gram, per liter of lake water.

Agricultural Water Pollution

Agricultural activities are the primary cause of water pollution problems. Excessive use of fertilizer results in eutrophication in many aquatic habitats because precipitation carries dissolved nutrients (nitrogen and phosphorus compounds) into streams and lakes. In addition, groundwater may become contaminated with fertilizer and pesticides. The exposure of land to erosion results in increased amounts of sediment being added to watercourses. Runoff from animal feedlots carries nutrients, organic matter, and bacteria. Water used to flush irrigated land to get rid of excess salt in the soil carries a heavy load of salt that degrades the water body. And the use of agricultural chemicals results in contamination of sediments and aquatic organisms. One of the largest water pollution problems is agricultural runoff from large expanses of open fields. See chapter 13 for a general discussion of methods for reducing runoff and soil erosion.

Farmers can reduce runoff in several ways. One is to leave a zone of undisturbed, permanently vegetated land, called a conservation buffer, near drains or streambanks. This retards surface runoff because soil covered with vegetation tends to slow the movement of water and allows the silt to be deposited on the surface of the land rather than in the streams. This can be costly because farmers may need to remove valuable cropland from cultivation. One goal of the Clean Water Action Plan is to establish 3.2 million kilometers (2 million miles) of conservation buffer strips. Another way to retard runoff is keep the soil covered with a crop as long as possible. Careful control of the amount and the timing of fertilizer application can also reduce the amount of nutrients lost to streams. This makes good economic sense because any fertilizer that runs off or leaches out of the soil is unavailable to crop plants and results in less productivity.

Industrial Water Pollution

Factories and industrial complexes frequently dispose of some or all of their wastes into municipal sewage systems. Depending on the type of industry involved, these wastes contain organic materials, petroleum products, metals, acids, toxic materials, organisms, nutrients, or particulates. Organic materials and oil add to the BOD of the water. The metals, acids, and specific toxic materials need special treatment, depending on their nature and concentration. In these cases, a municipal wastewater treatment plant will require that the industry pretreat the waste before sending it to the wastewater treatment plant. If this is not done, the municipal sewage treatment plants must be designed with their industrial customers in mind. In most cases, cities prefer that industries take care of their own wastes. This allows industries to segregate and control toxic wastes and design wastewater facilities that meet their specific needs.

Since industries are point sources of pollution, they have been relatively easy to identify as pollution sources, have been vigorously regulated, and have responded to mandates that they clean up their effluent. Most companies, when they remodel their facilities, include wastewater treatment as a necessary part of an industrial complex. However, some older facilities continue to pollute. These companies discharge acids, particulates, heated water, and noxious gases into the water. While industrial water pollution in the industrialized world has been significantly regulated, in much of the developing world this is not the case, and many lakes, streams, and harbors are severely polluted with heavy metals and other toxic materials, organic matter, and human and animal waste.

A special source of industrial water pollution is mining. By its very nature, mining disturbs the surface of the Earth and increases the chances that sediment and other materials will pollute surface waters. Hydraulic mining is practiced in some countries and involves spraying hillsides with high-pressure water jets to dislodge valuable ores. Often, chemicals are used to separate the valuable metals from the ores, and the waste from these processes is released into streams as well. Water that drains from current or abandoned coal mines is often very acidic. Pyrite is a mineral associated with many coal deposits. It contains sulfur, and when exposed to weathering, the sulfur reacts with oxygen, resulting in the formation of sulfuric acid. In addition, fine coal-dust particles are suspended in the water, which makes the water chemically and physically less valuable as a habitat. Dissolved ions of iron, sulfur, zinc, and copper also are present in mine drainage. Control involves containing mine drainage and treating it before it is released to surface water. Although federal legislation was passed in the 1990s requiring backfilling and land restoration after a mine is closed down, issues of responsibility and compliance with the law persist.

Thermal Pollution

Amendments to the Federal Water Pollution Control Act of 1972 mandated changes in how industry treats water. Industries were no longer allowed to use water and return it to its source in poor condition. One of the standards regulates the temperature of the water that is returned to its source. Because many industries use water for cooling, thermal pollution can be a problem. **Thermal pollution** occurs when an industry removes water from a source, uses the water for cooling purposes, and then returns the heated water to its source.

Power plants heat water to convert it into steam, which drives the turbines that generate electricity. For steam turbines to function efficiently, the steam must be condensed into water after it leaves the turbine. This condensation is usually accomplished by taking water from a lake, stream, or ocean to absorb the heat. This heated water is then discharged. The least expensive and easiest method of discharging heated water is to return the water to the aquatic environment, but this can create problems for aquatic organisms. Although an increase in temperature of only a few degrees may not seem significant, some aquatic organisms are very sensitive to minor temperature changes. Many fish are triggered to spawn by increases in temperature, while others may be inhibited from spawning if the temperature rises. For example, lake trout will not spawn in water above 10°C (50°F). If a lake has a temperature of 8°C (46°F), the lake trout will reproduce, but an increase of 3°C (5°F) would prevent spawning and result in this species' eventual elimination from that lake. Another problem associated with elevated water temperature is that it results in a decrease in the amount of oxygen dissolved in the water.

Ocean estuaries are very fragile. The discharge of heated water into an estuary may alter the type of plants present. As a result, animals with specific food habits may be eliminated because the warm water supports different food organisms. The entire food web in the estuary may be altered by only slight temperature increases.

Cooling water used by industry does not have to be released into aquatic ecosystems. Today in the industrialized world, most cooling water is not released in such a way that aquatic ecosystems are endangered. Three other methods of discharging the heat are commonly used. One method is to construct a large shallow pond. Hot water is pumped into one end of the pond, and cooler water is removed from the other end. The heat is dissipated from the pond into the atmosphere and substrate.

A second method is to use a cooling tower. In a cooling tower, the heated water is sprayed into the air and cooled by evaporation. The disadvantage of cooling towers and shallow ponds is that large amounts of water are lost by evaporation. The release of this water into the air can also produce localized fogs.

The third method of cooling, the dry tower, does not release water into the atmosphere. In this method, the heated water is pumped through tubes, and the heat is released into the air. This is the same principle used in an automobile radiator. The dry tower is the most expensive to construct and operate.

Marine Oil Pollution

Marine oil pollution has many sources. One source is accidents, such as oil-drilling blowouts or oil tanker accidents. The *Exxon Valdez,* which ran aground in Prince William Sound, Alaska, in 1989, released over 42 million liters (11 million gallons) of oil and affected nearly 1,500 kilometers (930 miles) of Alaskan coastline. The event had a great effect on the algae and animal populations of the Sound, and the economic impact on the local economy was severe. A U.S. National Oceanic and Atmospheric Administration study estimates that 50 percent of the oil biodegraded on beaches or in the water; 20 percent evaporated; 14 percent was recovered; 12 percent is at the bottom of the sea, mostly in the Gulf of Alaska; 3 percent lies on shorelines; and less than 1 percent still drifts in the water column. River otter, seabird, and bald eagle populations recovered to prespill numbers by 1992. Long-term impacts on reproduction and susceptibility to disease for many marine species are still under study. This points out the tremendous resilience of natural ecosystems to respond to and recover (within limits) from disastrous events.

Although accidents such as the *Exxon Valdez* and the deep water horizon spill in 2010 are spectacular events, much more oil is released as a result of small, regular discharges from other, less-visible sources. Nearly two-thirds of all human-caused marine oil pollution comes from three sources: (1) runoff from streets, (2) improper disposal of lubricating oil from machines or automobile crankcases, and (3) intentional oil discharges that occur during the loading and unloading of tankers. With regard to the latter, pollution occurs when the tanks are cleaned or oil-contaminated ballast water is released. Oil tankers use seawater as ballast to stabilize the craft after they have discharged their oil. This oil-contaminated water is then discharged back into the ocean when the tanker is refilled. In addition to human-caused oil pollution, oil naturally seeps into the water from underlying oil deposits in many places.

As the number of offshore oil wells and the number and size of oil tankers have grown, the potential for increased oil pollution has also grown. Many methods for controlling marine oil pollution have been tried. Some of the more promising methods are recycling and reprocessing used oil and grease from automobile service stations and from industries, and enforcing stricter regulations on the offshore drilling, refining, and shipping of oil. As a result of oil spills from shipping tankers, an international agreement was reached in 1992 that required that all new oil tankers be constructed with two hulls—one inside the other. Such double-hulled vessels would be much less likely to rupture and spill their contents. Today, approximately 25 percent of oil tankers are double hulled.

Groundwater Pollution

A wide variety of activities, some once thought harmless, have been identified as potential sources of groundwater contamination. In fact, possible sources of human-induced groundwater contamination span every facet of social, agricultural, and industrial activities. (See figure 15.13.) Once groundwater pollution has occurred, it is extremely difficult to remedy. Pumping groundwater and treating it is very slow and costly, and it is difficult to know when all of the contaminated water has been removed. A much better way to deal with the issue of groundwater pollution is to work very hard to prevent the pollution from occurring in the first place.

Major sources of groundwater contamination include:

1. Agricultural products. Pesticides contribute to unsafe levels of organic contaminants in groundwater. Seventy-three different pesticides have been detected in the groundwater in Canada and the United States. Accidental spills or leaks of pesticides pollute groundwater sources with 10 to 20 additional pesticides. Other agricultural practices contributing to groundwater pollution include animal-feeding operations, fertilizer applications, and irrigation practices.

2. Underground storage tanks. For many years in North America, a large number of underground storage tanks containing gasoline and other hazardous substances have leaked. Four liters (1 gallon) of gasoline can contaminate the water supply of a community of 50,000 people. A major program of replacing leaking underground storage tanks recently was completed in the United States. However, the effects of past leaks and abandoned tanks will continue to be a problem for many years.

3. Landfills. Even though recently constructed landfills have special liners and water collection systems, approximately 90 percent of the landfills in North America have no liners to stop leaks to underlying groundwater, and 96 percent have no system to collect the leachate that seeps from the landfill. Sixty percent of landfills place no restrictions on the waste accepted, and many landfills are not inspected even once a year.

4. Septic tanks. Poorly designed and inadequately maintained septic systems have contaminated groundwater with nitrates, bacteria, and toxic cleaning agents. Over 20 million septic tanks are in use in the United States, and up to a third have been found to be operating improperly.

5. Surface impoundments. Over 225,000 pits, ponds, and lagoons are used in North America to store or treat wastes. Seventy-one percent are unlined, and only 1 percent use a plastic or other synthetic, nonsoil liner. Ninety-nine percent of these impoundments have no leak-detection systems. Seventy-three percent have no restriction on the waste placed

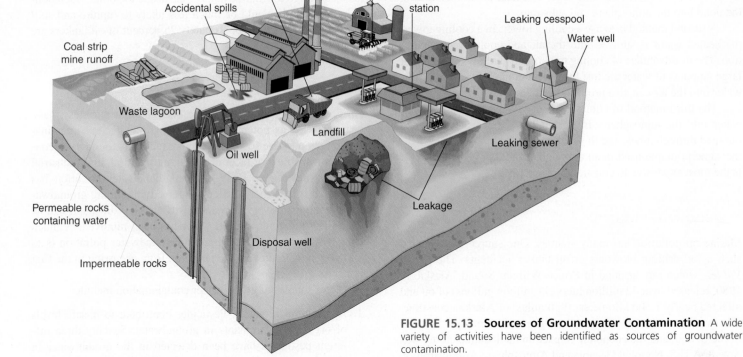

FIGURE 15.13 Sources of Groundwater Contamination A wide variety of activities have been identified as sources of groundwater contamination.

in the impoundment. Sixty percent are not even inspected annually. Many of these ponds are located near groundwater supplies.

Other sources of groundwater contamination include mining wastes, salting for controlling road ice, land application of treated wastewater, open dumps, cemeteries, radioactive disposal sites, urban runoff, construction excavation, fallout from the atmosphere, and animal feedlots.

15.7 Water-Use Planning Issues

In the past, wastes were discharged into waterways with little regard for the costs imposed on other users by the resulting decrease in water quality. Furthermore, as the population has grown and the need for irrigation and domestic water has intensified, in many parts of the world there has not been enough water to satisfy everyone's needs. With today's increasing demands for high-quality water, unrestrained waste disposal and unlimited withdrawal of water could lead to serious conflicts about water uses, causing social, economic, and environmental losses at both local and international levels. (Table 15.2 summarizes some of the areas of controversy involving water throughout the world.)

Metropolitan areas must deal with a variety of issues and maintain an extensive infrastructure to provide three basic water services:

1. Water supply for human and industrial needs
2. Wastewater collection and treatment
3. Stormwater collection and management

Water sources must be identified and preserved for use. Some cities obtain all their municipal water from groundwater and must have a thorough understanding of the size and characteristics of the aquifer they use. Some cities, such as New York City, obtain potable water by preserving a watershed that supplies the water needed by the population. Other cities have abundant water in the large rivers that flow by them but must deal with pollution problems caused by upstream users. In many places where water is in short supply, municipal and industrial-agricultural needs for water conflict.

Water for human and industrial use must be properly treated and purified. It is then pumped through a series of pipes to consumers. After the water is used, it flows through a network of sewers to a wastewater treatment plant, where it is treated before it is released. Maintaining the infrastructure of pipes, pumps, and treatment plants is expensive.

Metropolitan areas must also deal with great volumes of excess water during storms. This water is known as **stormwater runoff.** Because urban areas are paved and little rainwater can be absorbed into the ground, managing stormwater is a significant problem. Cities often have severe local flooding because the water is channeled along streets to storm sewers. If these sewers are overloaded or blocked with debris, the water cannot escape and flooding occurs.

The Water Quality Act of 1987 requires that municipalities obtain permits for discharges of stormwater runoff so that nonpoint sources of pollution are controlled. In the past, many cities had a single system to handle both sewage and stormwater runoff. During heavy precipitation or spring thaws, the runoff from streets could be so large that the wastewater treatment plant could not handle the volume. The wastewater was then diverted directly into the receiving body of water without being treated.

Table 15.2 International Water Disputes

Rivers/Lakes	Countries Involved	Issues
Asia		
Brahmaputra, Ganges, Farakka	Bangladesh, India, Nepal	Alluvial deposits, dams, floods, irrigation, international quotas
Mekong	Cambodia, Laos, Thailand, Vietnam	Floods, international quotas
Salween	Tibet, China (Yunan), Burma	Alluvial deposits, floods
Middle East		
Euphrates, Tigris	Iraq, Syria, Turkey	International quotas, salinity levels
West Bank Aquifer, Jordan, Litani, Yarmuk	Israel, Jordan, Lebanon, Syria	Water diversion, international quotas
Africa		
Nile	Mainly Egypt, Ethiopia, Sudan	Alluvial deposits, water diversion, floods, irrigation, international quotas
Lake Chad	Nigeria, Chad	Dam
Okavango	Namibia, Angola, Botswana	Water diversion
Europe		
Danube	Hungary, Slovak Republic	Industrial pollution
Elbe	Germany, Czech Republic	Industrial pollution, salinity levels
Meuse, Escaut	Belgium, Netherlands	Industrial pollution
Szamos (Somes)	Hungary, Romania	Water allocation
Tagus	Spain, Portugal	Water allocation
Americas		
Colorado, Rio Grande	United States, Mexico	Chemical pollution, international quotas, salinity levels
Great Lakes	Canada, United States	Pollution
Lauca	Bolivia, Chile	Dams, salinity
Paran	Argentina, Brazil	Dams, flooding of land
Cenepa	Ecuador, Peru	Water allocation

jurisdictions (governmental and bureaucratic areas) that divide responsibility for management of basic water services. The Chicago metropolitan area is a good example. This area is composed of six counties and approximately 2,000 local units of government. It has 349 separate water-supply systems and 135 separate wastewater disposal systems. Efforts to implement a water-management plan, when so many layers of government are involved, can be complicated and frustrating.

To meet future needs, urban, agricultural, and national interests will need to deal with a number of issues, such as the following:

- Increased demand for water will generate pressure to divert water to highly populated areas or areas capable of irrigated agriculture.

- Increased demand for water will force increased treatment of wastewater and reuse of existing water supplies.

- In many areas where water is used for irrigation, evaporation of water from the soil over many years results in a buildup of salt in the soil. When the water used to flush the salt from the soil is returned to a stream, the quality of the water is lowered.

- In some areas, wells provide water for all categories of use. If the groundwater is pumped out faster than it is replaced, the water table is lowered.

- In coastal areas, seawater may intrude into the aquifers and ruin the water supply.

- The demand for water-based recreation is increasing dramatically and requires high-quality water, especially for activities involving total body contact, such as swimming.

Because of these new requirements, some cities have created areas in which to store this excess water until it can be treated. This is expensive and, therefore, is done only if federal or state funding is available. Many cities have also gone through the expensive process of separating their storm sewers from their sanitary sewers. A good example of the long and costly process of separation of sewers is Portland, Oregon. By 2011 Portland had completed only half of a 20-year project to separate its storm and sanitary sewer systems. The final costs are estimated to exceed the project's $1 billion budget.

Providing water services is expensive. We must understand that water supplies are limited. We must also understand that water's ability to dilute and degrade pollutants is limited and that proper land-use planning is essential if metropolitan areas are to provide services and limit pollution.

In pursuing these objectives, city planners encounter many obstacles. Large metropolitan areas often have hundreds of local

Water Diversion

Water diversion is the physical process of transferring water from one area to another. The aqueducts of ancient Rome are early examples of water diversion. Thousands of diversion projects have been constructed since then. New York City, for example, receives 90 percent of its water supply from the Catskill Mountains (Schoharie and Ashokan Reservoirs) and from four reservoirs collecting water from tributaries to the Delaware River west of the Catskills. About 10 percent of its supply comes from the Croton watershed east of the Hudson River. The most distant watershed serving New York City is some 200 kilometers (125 miles) away. Los Angeles is another example. It began importing water from the Owens Valley, 400 kilometers (250 miles) to the north, as early as 1913.

While diversion is seen as a necessity in many parts of the world, it often generates controversy. An example of this is the

It has been stated that while water resources have rarely, if ever, been the sole source of violent conflict or war, the next major war could be a water war. This is in response to the growing pressure on natural resources that is being experienced throughout the world in the context of increasing demand. With the very high numbers of international watercourses that are shared between countries, water and its use are undoubtedly a cause of tension and often strain relations between countries. Water is a security concern for many countries. Develop a position paper on the likelihood of a future "water war" and where you feel it could develop.

Garrison Diversion Unit in North Dakota, which was originally envisioned as a way to divert water from the Missouri River system for irrigation. (See figure 15.14.) The initial plan to irrigate portions of the Great Plains was developed during the Dust Bowl era of the 1930s. The Federal Flood Control Act of 1944 authorized the construction of the Garrison Diversion Unit. However, one of the original intents of the plan—to divert water from the Missouri River to the Red River—has not been met. Since the Red River flows north into Canada, the project requires international cooperation. The Canadian government has concerns about the effects on water quality of releasing additional water into the Red River. There are also concerns about environmental consequences. Portions of wildlife refuges, native grasslands, and waterfowl breeding marshes would be damaged or destroyed. Some states also have expressed concern about diverting water from the Missouri River, since any water diverted is not available for those downstream on the Missouri River.

While the plan has been modified several times, two sections of canal and a pumping station at Lake Sakakawea have been completed but not used. Proponents continue to push for legislation to complete the connection between the two canals that would allow water to be diverted to the Red River for municipal use. The original intent of diverting water for irrigation has been eliminated from the most recent proposal.

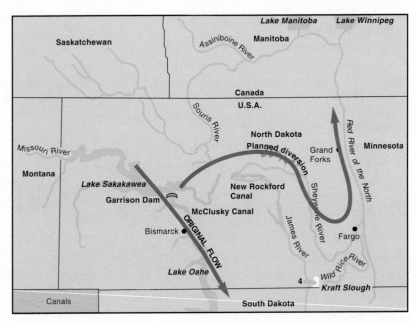

FIGURE 15.14 The Garrison Diversion Unit The original intent of this plan was to divert water from the Missouri River to the McClusky Canal and the Sheyenne River and eventually into Lake Winnepeg by way of the Red River. In the process, additional land could be irrigated, and growing populations in the Red River Valley could be served with adequate water. Opposition has stalled the project, and only portions of the canal system and a pumping station have been completed.

One major consequence of diverting water for irrigation and other purposes is that the water bodies downstream of the diversion are deprived of their source of water. This often has major ecological consequences. Water levels in Lake Chad in Africa are falling due to drought and increased demand for irrigation water. This has affected the populations of fish and other wildlife. In Mexico, 2,500 kilometers (1,550 miles) of rivers have dried up because the water was diverted for other purposes, resulting in the extinction of 15 species of fish and threatening half of the remaining species.

As people recognize the significance of wildlife habitat, plans are being worked out to balance societal needs with the need for water to maintain wildlife habitat. For example, Mono Lake in the Sierra Mountains in central California began to shrink in 1941 when much of the water that fed the lake was diverted to supply Los Angeles. Since Mono Lake has no outlet, its size is determined by the balance between the water flowing into the lake and evaporation from the lake surface. Consequently, the water level in the lake dropped 13 meters (43 feet), the lake's volume decreased by half, and its salinity doubled. These changes resulted in a loss of habitat important for many ducks and other waterfowl. In 1994, an agreement was reached to increase the amount of water flowing to the lake so that the level of the lake would rise. The plan is to increase the level of the lake by 5 meters (17 feet) over a 20- to 30-year period.

In New South Wales, Australia, an area known as the Macquarie Marshes was affected by water diversion. The original extent of the marshes was reduced by about 50 percent because the Macquarie River, which feeds the marsh, was dammed to provide irrigation water. In the mid-1990s, because of concerns about the loss of wildlife, an agreement was reached to provide additional water to the area to maintain the marsh and protect the breeding areas of waterfowl.

Table 15.3 Percent of Sewage Treated in Selected Areas

Area	Percent
North America	90
Europe	72
Mediterranean Sea	30
Caribbean Basin	Less than 10
Southeast Pacific	Almost zero
South Asia	Almost zero
South Pacific	Almost zero
West and Central Africa	Almost zero

Source: Data from World Resources Institute, 2010.

Wastewater Treatment

Because water must be cleaned before it is released, most companies and municipalities in the developed world maintain wastewater treatment facilities. The percentage of sewage that is treated, however, varies greatly throughout the world. (See table 15.3.) Treatment of sewage is usually classified as primary, secondary, or tertiary. **Primary sewage treatment** is mostly a physical process that removes larger particles by filtering water through large screens and then allowing smaller particles to settle in ponds or lagoons. Water is removed from the top of the settling stage and released either to the environment or to a subsequent stage of treatment. If the water is released to the environment, it does not have any sand or grit; but it still carries a heavy load of organic matter, dissolved salts, bacteria, and other microorganisms. The microorganisms use the organic material for food, and as long as there is sufficient oxygen, they will continue to grow and reproduce. If the receiving body of water is large enough and the organisms have enough time, the organic matter will be degraded. In crowded areas, where several municipalities take water and return it to a lake or stream within a few kilometers of each other, primary water treatment is not adequate and major portions of the receiving body of water are affected.

Secondary sewage treatment is a biological process that usually follows primary treatment. It involves holding the wastewater until the organic material has been degraded by the bacteria and other microorganisms. Secondary treatment facilities are designed to promote the growth of microorganisms. To encourage this action, the wastewater is mixed with large quantities of highly oxygenated water, or the water is aerated directly, as in a trickling filter system or an activated-sludge system. In a **trickling filter system,** the wastewater is sprayed over the surface of rock or other substrate to increase the amount of dissolved oxygen. The rock also provides a place for a film of bacteria and other microbes to attach so they are exposed simultaneously to the organic material in the water and to oxygen. These microorganisms feed on the dissolved organic matter and small suspended particles, which then become incorporated into their bodies as part of their cell structure. The bodies of the microorganisms are larger than the dissolved and suspended organic matter, so this process concentrates

the organic wastes into particles that are large enough to settle out. This mixture of organisms and other particulate matter is called **sewage sludge.** The sludge that settles consists of living and dead microorganisms and their waste products.

In **activated-sludge sewage treatment** plants, the wastewater is held in tanks and has air continuously bubbled through it. The sludge eventually is moved to settling tanks where the water and sludge can be separated. To make sure that the incoming wastewater has appropriate kinds and amounts of decay organisms, some of the sludge is returned to aeration tanks, where it is mixed with incoming wastewater. This kind of process uses less land than a trickling filter. (See figure 15.15.) Both processes produce a sludge that settles out of the water.

The sludge that remains is concentrated and often dewatered (dried) before disposal. Sludge disposal is a major problem in large population centers. In the San Francisco Bay area, 2,500 metric tons of sludge are produced each day. Most of this is carried to landfills and lagoons, and some is composted and returned to the land as fertilizer. Some municipalities incinerate their sludge. In other areas, if the sewage sludge is free of heavy metals and other contaminants that might affect plant growth or the quality and safety of food products, it is applied directly to agricultural land as a fertilizer and soil conditioner.

In North America and much of the developed world, wastewater receives both primary and secondary sewage treatment. While the water has been cleansed of its particles and dissolved organic matter, it still has microorganisms that might be harmful. Therefore, the water discharged from these sewage treatment plants must be disinfected. The least costly method of disinfection is chlorination. However, many people feel that the use of chlorine should be discontinued, since chlorination may be responsible for the creation of chlorinated organic compounds that are harmful. Therefore, other methods of killing microorganisms are being explored. The chemical ozone also kills microorganisms and has been substituted for chlorine by some facilities. Ultraviolet light and ultrasonic energy can also be used. Vancouver, Washington, for example, has been using ultraviolet light treatment since 1998 as a final treatment before returning water to the Columbia River. However, chlorine is inexpensive and very effective, so it continues to be the primary method used.

A growing number of larger sewage treatment plants use additional processes called tertiary sewage treatment. **Tertiary sewage treatment** involves a variety of techniques to remove inorganic nutrients left after primary and secondary treatments. (See table 15.4.) The tertiary treatment of municipal wastewater is often used to remove phosphorus and nitrogen that could increase aquatic plant growth. Some municipalities are using natural or constructed wetlands to serve as tertiary sewage treatment systems. In other cases, the effluent from the treatment facility is used to irrigate golf courses, roadside vegetation, or cropland. The vegetation removes excess nutrients and prevents them from entering streams and lakes where they would present a pollution problem. Tertiary treatment of industrial and other specialized wastewater streams is very costly because it requires specific chemical treatment of the water to eliminate specific problem materials. Many industries maintain their own wastewater facilities and design specific tertiary treatment processes to match the specific nature of their waste products.

(a)

(b)

(c)

FIGURE 15.15 **Primary and Secondary Wastewater Treatment**
Primary treatment is physical; it includes filtrating and settling of wastes. Photograph (a) is of a settling tank in which particles settle to the bottom. Secondary treatment is mostly biological and includes the concentration of dissolved organics by microorganisms. Two major types of secondary treatment are trickling filter and activated-sludge methods. Photograph (b) shows a trickling filter system, and photograph (c) shows an activated-sludge system.

expand as well. Because treated wastewater is often located close to its customers, the cost of conveying this water supply source can be much less than that of other water supply options. A number of communities and water providers use treated wastewater for direct and indirect reuse. For example, El Paso, Texas, Northwest Reclaimed Water Project provides more than a million liters of reclaimed water per year to schools, parks, a golf course, and multifamily housing developments and residential customers for irrigation.

The amount of existing supply related to water reuse is based on the amount of water that can be produced with current permits and

As water has become scarce in many parts of the world, people have looked at the reuse of wastewater for other purposes. Ultimately, it is possible to have a closed loop system for domestic water in which the output of the wastewater plant becomes the input for the drinking water supply. Water reuse is the use of water that has been already used beneficially once. There are two types of water reuse: direct reuse and indirect reuse. Direct reuse is the use of effluent from a wastewater treatment plant that is piped directly from the plant to the place where it is used. For example, treating wastewater and piping it to a golf course for irrigation is direct water reuse. Indirect reuse is the use of water, usually treated effluent, that is placed back into a river or stream and then diverted further downstream to be used again. An example of indirect reuse is treating wastewater, discharging the effluent into a river, and using the river to transport it downstream where a golf course diverts the water from the river for irrigation.

Reuse is a promising source of additional water in the future. As municipal water supply and wastewater facilities expand to support a growing customer base, the volume of treated wastewater can be expected to

Table 15.4 Tertiary Treatment Methods

Kind of Tertiary Treatment	Problem Chemicals	Methods
Biological	Phosphorus and nitrogen compounds	1. Large ponds are used to allow aquatic plants to assimilate the nitrogen and phosphorus compounds from the water before the water is released.
		2. Columns containing denitrifying bacteria are used to convert nitrogen compounds into atmospheric nitrogen.
Chemical	Phosphates and industrial pollutants	1. Water can be filtered through calcium carbonate. The phosphate substitutes for the carbonate ion, and the calcium phosphate can be removed.
		2. Specific industrial pollutants, which are nonbiodegradable, may be removed by a variety of specific chemical processes.
Physical	Primarily industrial pollutants	1. Distillation
		2. Water can be passed between electrically charged plates to remove ions.
		3. High-pressure filtration through small-pored filters
		4. Ion-exchange columns

existing infrastructure. As the amount of effluent and the need for additional supplies of water increase, water reuse will have an important role in meeting future water supply needs. However, policy issues related to permitting and environmental flows will have to be addressed before the full potential of this water management strategy is realized.

Salinization

Another water-use problem results from **salinization,** an increase in salinity caused by growing salt concentrations in soil. This is primarily a problem in areas where irrigation has been practiced for several decades. As water evaporates from soil or plants extract the water they need, the salts present in all natural waters become concentrated. Since irrigation is most common in hot, dry areas that have high rates of evaporation, there is generally an increase in the concentration of salts in the soil and in the water that runs off the land. (See figure 15.16.) Every river increases in salinity as it flows to the ocean. The salinity of the Colorado River water increases 20 times as it passes through irrigated cropland between Grand Lake in north-central Colorado and the Imperial Dam in southwest Arizona. The problem of salinity will continue to increase as irrigation increases.

Groundwater Mining

Groundwater mining means that water is removed from an aquifer faster than it is replaced. When this practice continues for a long time, the water table eventually declines. Groundwater mining is common in areas of the western United States and throughout the world. In North America, it is a particular problem due to growing cities and increasing irrigation. In aquifers with little or no recharge, virtually any withdrawal constitutes mining, and sustained withdrawals will eventually exhaust the supply. This problem is particularly serious in communities that depend heavily on groundwater for their domestic needs.

Groundwater mining can also lead to problems of settling or subsidence of the ground surface. Removal of the water allows the ground to compact, and large depressions may result. For example, in the San Joaquin Valley of California, groundwater has been withdrawn for irrigation and cultivation since the 1850s, and groundwater levels have fallen over 100 meters (300 feet). More than 1,000 hectares (approximately 2,500 acres) of ground have subsided, some as much as 6 meters (20 feet). Currently, the ground surface in that area is sinking 30 centimeters (12 inches) per year. London, Mexico City, Venice, Houston, and Las Vegas are some other cities that have experienced subsidence as a result of groundwater withdrawal. Table 15.5 lists some estimated amounts of groundwater depletion in certain areas of the United States and the world. As people recognize the severity of the problem, public officials are beginning to develop water conservation plans for their cities. Albuquerque, New Mexico, which relies on groundwater for its water supply, has an extensive public education program to encourage people to reduce their water consumption. Since grass demands water, people are encouraged to use desert plants for landscaping or collect rainwater to water their lawns. Finding and correcting leaks, reducing the amount of water used in bathing, and recycling water from swimming pools are other conservation strategies.

An average of 85 billion gallons (320 billion liters) of groundwater are withdrawn daily in the United States. More than 90 percent

FIGURE 15.16 Salinization As water evaporates from the surface of the soil, the salts it was carrying are left behind. In some areas of the world, this has permanently damaged cropland.

of these withdrawals are used for irrigation, public supply (deliveries to homes, businesses, industry), and self-supplied industrial uses. Irrigation is the largest use, accounting for about two-thirds of the amount. The percentage of total irrigation withdrawals provided by groundwater increased from 23 percent in 1950 to 44 percent in 2010. Groundwater provides about half of the drinking water in the United States with nearly all those in rural areas reliant upon groundwater.

Rapid expansion in groundwater use occurred between 1950 and 1975 in many industrial nations and subsequently in much of the developing world. The intensive use of groundwater for irrigation in arid and semi-arid countries has been called a "silent revolution" as millions of independent farmers worldwide have chosen to become increasingly dependent on the reliability of groundwater resources, reaping abundant social and economic benefits but with limited management controls by government water agencies. Perhaps as many as 2 billion people worldwide depend directly upon groundwater for drinking water. The dependence on groundwater for drinking water is particularly high in Europe, where about 75 percent of the drinking-water supply is obtained from groundwater.

Groundwater mining poses a special problem in coastal areas. As the fresh groundwater is pumped from wells along the coast, the saline groundwater moves inland, replacing fresh groundwater with unusable saltwater. This process, called **saltwater intrusion,** is shown in figure 15.17. Saltwater intrusion is a serious problem in heavily populated coastal areas throughout the world.

Achieving an acceptable trade-off between groundwater use and the long-term effects of that use is a central theme in the evolving concept of groundwater sustainability.

Groundwater sustainability is commonly defined in a broad context as the development and use of groundwater resources in a manner that can be maintained for an indefinite amount of time without causing unacceptable environmental, economic, or social consequences. Groundwater sustainability management strategies are composed of a small number of general approaches, including:

- Use of sources of water other than local groundwater, by shifting the local source of water (either completely or in part) from groundwater to surface water or importing water from outside the local water-system boundaries (the California Central Valley and Houston have implemented these approaches).

Table 15.5 Groundwater Depletion in Major Regions of the World

Region/Aquifer	Estimates of Depletion
California	Groundwater overdraft exceeds 1.7 billion cubic meters (60 billion cubic feet) per year. The majority of the depletion occurs in the Central Valley, which is referred to as the vegetable basket of the United States.
Southwestern United States	In parts of Arizona, water tables have dropped more than 120 meters (400 feet). Projections for parts of New Mexico indicate that water tables will drop an additional 22 meters (70 feet) by 2020.
High Plains aquifer system, United States	The Ogallala aquifer underlies nearly 20 percent of all the irrigated land in the United States. To date, the net depletion of the aquifer is in excess of 350 billion cubic meters (12 trillion cubic feet), or roughly 15 times the average annual flow of the Colorado River. Most of the depletion has been in the Texas High Plains, which witnessed a 26 percent decline in irrigated land from 1979 to 1989. Current depletion is estimated to be in excess of 13 billion cubic meters (450 billion cubic feet) a year.
Mexico City and Valley of Mexico	Use exceeds natural recharge by 60 to 85 percent, causing land subsidence and falling water tables.
African Sahara	North Africa has vast nonrecharging aquifers where current depletion exceeds 12 billion cubic meters (425 billion cubic feet) a year.
India	Water tables are declining throughout much of the most productive agricultural land in India. In parts of the country, groundwater levels have declined 90 percent during the past two decades.
North China	The water table underneath portions of Beijing has dropped 40 meters (130 feet) during the past 40 years. A large portion of northern China has significant groundwater overdraft.
Arabian Peninsula	Groundwater use is nearly three times greater than recharge. At projected depletion rates, exploitable groundwater reserves could be exhausted within the next 50 years. Saudi Arabia depends on nonrenewable groundwater for roughly 75 percent of its water. This includes irrigation of 2 million to 4 million metric tons of wheat per year.

- Control or regulation of groundwater pumping through implementation of guidelines, policies, taxes, or regulations by water management authorities.
- Conservation practices, techniques, and technologies that improve the efficiency of water use, often accompanied by public education programs on water conservation.
- Reuse of wastewater (grey water) and treated wastewater (reclaimed water) for nonpotable purposes such as irrigation of crops, lawns, and golf courses.
- Desalinization of brackish groundwater or treatment of otherwise impaired groundwater to reduce dependency on fresh groundwater sources.

Preserving Scenic Water Areas and Wildlife Habitats

Some bodies of water have unique scenic value. To protect these resources, the way in which the land adjacent to the water is used must be consistent with preserving these scenic areas.

The U.S. Federal Wild and Scenic Rivers Act of 1968 established a system to protect wild and scenic rivers from development. All federal agencies must consider the wild, scenic, or recreational values of certain rivers in planning for the use and development of the rivers and adjacent land. The process of designating a river or part of a river as wild or scenic is complicated. It often encounters local opposition from businesses dependent on growth. Following reviews by state and federal agencies, rivers may be designated as wild and scenic by action of either Congress or the secretary of the interior. Sections of over 150 streams comprising about 12,000 kilometers (7,700 miles) in the United States have been designated as wild or scenic.

Many unique and scenic shorelands have also been protected from future development. Until recently, estuaries and shorelands have been subjected to significant physical modifications, such as

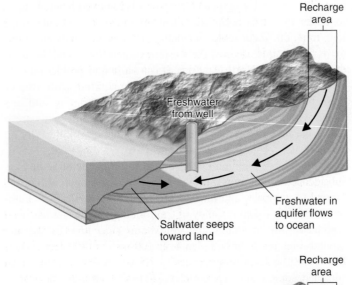

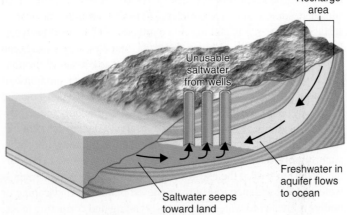

FIGURE 15.17 Saltwater Intrusion When saltwater intrudes on fresh groundwater, the groundwater becomes unusable for human consumption and for many industrial processes.

dredging and filling, which may improve conditions for navigation and construction but destroy fish and wildlife habitats. Recent actions throughout North America have attempted to restrict the development of shorelands. Development has been restricted in some particularly scenic areas, such as Cape Cod National Seashore in Massachusetts and the Bay of Fundy in the Atlantic provinces of Canada.

Historically, poorly drained areas were considered worthless. Subsequently, many of these wetlands were filled or drained and used for building sites. At the time of European settlement, the area that is now the conterminous United States contained an estimated 89 million hectares (221 million acres) of wetlands. Over time, wetlands have been drained, dredged, filled, leveled, and flooded to the extent that less than half of the original wetland acreage remains. Figure 15.18 shows the most recent causes of wetlands loss.

Today, 95 percent of the remaining 20 million hectares (50 million acres) of wetlands in the United States are inland freshwater wetlands. The remaining 5 percent are in saltwater estuarine environments. Freshwater forested wetlands make up the single largest category of all wetlands in the conterminous United States. Until recently, wetlands were commonly thought of as wastelands, and many were filled, dredged, and developed for purposes such as industry, housing, and agriculture. Today, we know that wetlands serve important functions, including flood protection, filtering sediments and pollutants, erosion protection, and water storage for release in times of drought. They provide a vital link in the chain of life, offering habitat and food for many species of plants, animals, and microscopic organisms. In addition, they provide economic benefits and opportunities for recreation, education, and research.

Freshwater wetland ecosystems include ponds, marshes, seasonally flooded meadows, and riparian areas. These systems filter pollutants, assist in flood control, and provide breeding habitat for fish, birds, insects, and amphibians.

Wetlands are among the most biologically productive ecosystems in the worlds—similar to rainforests in the diversity of species they support. The U.S. Fish and Wildlife Service estimates that as many as 43 percent of threatened and endangered species rely directly or indirectly on wetlands for their survival. The loss of wetlands habitat ultimately results in declines of wildlife and fish populations.

Diking, dredging, agriculture, and urbanization have been the primary causes of loss of wetlands. Increase in floods, reduction in water flow, and declining wildlife populations are, in part, the result of wetlands destruction.

Even a small percentage of wetlands in watersheds can be critical. Wetlands can slow and store floodwater. Studies show that flood peaks may be as much as 80 percent higher in watersheds without wetlands than in similar basins with large wetland areas. In fact, keeping even a few wetlands in a watershed has a disproportionately positive effect on reducing flood flows. A watershed with as little as five percent of its area in wetlands would have a storm flood peak that is 50 percent lower than if there were no wetlands.

In addition to the direct loss of wetlands from filling or draining, there are many indirect influences that affect wetland functions. The greatest sources of harm in developing areas are the conversion of forests and the creation of impervious surfaces (pavements, roofs, etc.). These changes cause greater fluctuations in water levels in the winter and lower levels in summer.

The introduction of non-native species such as spartina (cordgrass) in estuarine wetlands and purple loosestrife or reed canary grass in freshwater wetlands presents additional risks to the health of the wetlands. These plant species can take over a wetland and grow so densely that no other plants can survive; that, in turn, affects the fish and wildlife that depend on the native plants for food and cover. (See figure 15.19.) Coastal estuarine

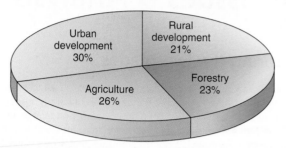

Causes of Wetlands Loss (1986–2010)

Urban development 30%
Rural development 21%
Forestry 23%
Agriculture 26%

FIGURE 15.18 Wetlands Conversion The loss of wetlands occurs because people convert wetlands to other uses. Conversion to urban and rural housing and other infrastructure accounts for just over 50 percent of the wetlands loss. Draining wetlands for agriculture accounts for an additional 26 percent. Source: U.S. Geological Survey.

- Filter toxic wastes, excess nutrients, sediments, and other pollutants
- Help prevent erosion
- Reduce flooding by storing stormwater

- Reduce storm damage by absorbing waves
- Provide feeding and resting spots for migratory waterfowl

- Provide food and habitat for other aquatic species

- Provide nursery sites for the young of a number of species, including oysters, clams, crabs, and shrimp

FIGURE 15.19 The Value of Wetlands Wetlands are areas covered with water for most of the year that support aquatic plant and animal life. Wetlands can be either fresh- or saltwater and may be isolated potholes or extensive areas along rivers, lakes, and oceans. We once thought of wetlands as only a breeding site for mosquitoes. Today, we are beginning to appreciate their true value.

Restoring the Everglades

Everglades National Park is a unique, subtropical, freshwater wetland visited by about a million people per year. This unique ecosystem exists because of an unusual set of conditions. South Florida is a nearly flat landscape with a slight decrease in elevation from the north to the south. In addition, the substrate is a porous limestone that allows water to flow through rather easily. Originally, water drained in a broad sheet from Lake Okeechobee southward to Florida Bay. This constant flow of water sustained a vast, grassy wetland interspersed with patches of trees.

After Everglades Park was established in 1947, about 800,000 hectares (2 million acres) of wetlands to the north were converted to farms and urban development. South Florida boomed. Some 4.5 million people currently live in the horseshoe crescent around the Everglades region, and new residents arrive each day. Conversion of land to agriculture and urban development required changes in the natural flow of water. Dams, drainage canals, and water diversion supported and protected the human uses of the area but cut off the essential, natural flow of freshwater to the Everglades. Changes in the normal pattern of water flow to the Everglades resulted in periods of drought and a general reduction in the size of this unique wetland region. Populations of wading birds in the southern Everglades fell drastically as their former breeding and nesting areas dried up. Other wildlife, such as alligators, Florida panthers, snail kites, and wood storks, also were negatively affected because drought reduces suitable habitat during parts of the year and the animals are forced to congregate around the remaining sources of water. The quality of the water is also important. The original wetland ecosystem was a nutrient-poor system. The introduction of nutrients into the water from agricultural activities encouraged the growth of exotic plants that replaced natural vegetation.

As people recognized that the key element necessary to preserve the Everglades was a constant, reliable source of clean freshwater, several steps were taken to modify water use to preserve the Everglades. The Kissimmee River, which flows into Lake Okeechobee, was channelized into an arrow-straight river in the late 1960s. This project destroyed the

marshes and allowed nutrients from dairy farms and other agricultural activities to pollute the lake and Everglades Park, to which the water from Lake Okeechobee eventually flowed. To help alleviate this problem, in 1990, the U.S. Army Corps of Engineers began to return the Kissimmee to its natural state, with twisting oxbow curves and extensive wetlands. This allows the plants in the natural wetlands to remove much of the nutrient load before it enters the lake. To reduce the likelihood of further development near the park, in 1989, Congress approved the purchase of 43,000 hectares (100,000 acres) for an addition to the east section of the park. Florida obtained an additional 60,000 hectares (150,000 acres) as an additional buffer zone for the park.

For several years, the U.S. Army Corps of Engineers and the South Florida Water Management District cooperated in the development of a comprehensive restoration plan that was finished in 2000. The development of the plan involved the input of scientists, politicians, various business interests, and environmentalists. Key components of the plan are:

1. Developing facilities to store surface water and pump water into aquifers so that it can be released when needed

2. Developing wetlands to treat municipal and agricultural runoff so that nutrient loads are reduced

3. Using clean wastewater to recharge aquifers and supply water to wetlands in the Miami area

4. Reducing the amount of water lost through levees and redirecting water to the Everglades

5. Removing barriers to the natural flow of water through the Everglades

The plan received strong support from Congress in the fall of 2001, when it allocated $1.4 billion to begin implementing the plan. It will require many more billions of dollars and up to 30 years to accomplish all aspects of the plan, but if the plan continues to be implemented over the next few decades, the Everglades ecosystem will be restored to a more stable condition and will have a more hopeful future.

zones and adjoining sand dunes also provide significant natural flood control. Sand dunes act as barriers and absorb damaging waves caused by severe storms. In recent years, public appreciation of the ecological, social, and economic values of wetlands has increased substantially. The increased awareness of how

much wetland acreage has been lost or damaged since the time of European settlement and the consequences of those losses has led to the development of many federal, state, and local wetland protection programs and laws.

In 2008, an agreement was entered into between the state of Florida and U.S. Sugar, the largest producer of cane sugar in the United States. Under the agreement, the state of Florida will purchase U.S. Sugar for $1.7 billion. U.S. Sugar will continue operations for six years and will then transfer to Florida 187,000 acres (760 km²). The newly acquired land will remain undeveloped to allow it to be restored to its pre-drainage state. The land will be used to reestablish a part of the connection between Lake Okeechobee and the Everglades through a managed system of water storage and treatment and, at the same time, safeguard the St. Lucie and Caloosahatchee rivers and estuaries. The land acquisition would also prevent massive amounts of nutrient phosphorus from entering the Everglades every year. Phosphorus is used as a fertilizer for sugar production. Phosphorus runoff pollutes the water to 20 times the tolerable level, endangering native wildlife.

Phosphorus changes the chemistry of the water and destroys the microbial populations, an essential source of food for many aquatic organisms, which then do not flourish. As one result, 90 percent of the wading birds in the Everglades have disappeared and 68 species of plants and animals are either endangered or threatened.

- What do you think?
- Is it possible to restore the Everglades?
- Where should the funding come from to support the restoration plan?
- Should development in the Everglades be curtailed even more?
- Should there be stronger laws regarding phosphorus?

Everglades wading birds once numbered more than a million, but pollution and drought brought on by water diversion have decimated their number.

Construction threatens to destroy the fragile ecosystem of the Everglades.

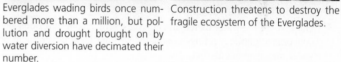

Summary

Water is a renewable resource that circulates continually between the atmosphere and the Earth's surface. The energy for the hydrologic cycle is provided by the sun. Water loss from plants is called evapotranspiration. Water that infiltrates the soil and is stored underground in the tiny spaces between rock particles is called groundwater, as opposed to surface water that enters a river system as runoff. There are two basic kinds of aquifers. Unconfined aquifers have an impervious layer at the bottom and receive water that infiltrates from above. The top of the layer of water is called the water table. A confined aquifer is sandwiched between two impervious layers and is often under pressure. The recharge area may be a great distance from where the aquifer is tapped for use.

The way in which land is used has a significant impact on rates of evaporation, runoff, and infiltration.

The four human uses of water are domestic, agricultural, instream, and industrial. Water use is measured by either the amount withdrawn or the amount consumed. Domestic water is in short supply in many metropolitan areas. Most domestic water is used for waste disposal and washing, with only a small amount used for drinking. The largest consumptive use of water is for agricultural irrigation. Major in-stream uses of water are for hydroelectric power, recreation, and navigation. Most industrial uses of water are for cooling and for dissipating and transporting waste materials.

Major sources of water pollution are municipal sewage, industrial wastes, and agricultural runoff. Nutrients, such as nitrates and phosphates from wastewater treatment plants and agricultural runoff, enrich water and stimulate algae and aquatic plant growth. Organic matter in water requires oxygen for its decomposition and, therefore, has a large biochemical oxygen demand (BOD). Oxygen depletion can result in fish death and changes in the normal algal community, which leads to visual and odor problems.

Point sources of pollution are easy to identify and resolve. Nonpoint sources of pollution, such as agricultural runoff and mine drainage, are more difficult to detect and control than those from municipalities or industries.

Thermal pollution occurs when an industry returns heated water to its source. Temperature changes in water can alter the kinds and numbers of plants and animals that live in it. The methods of controlling thermal pollution include cooling ponds, cooling towers, and dry cooling towers.

Wastewater treatment consists of primary treatment, a physical settling process; secondary treatment, biological degradation of the wastes; and tertiary treatment, chemical treatment to remove specific components. Two major types of secondary wastewater treatments are the trickling filter and the activated-sludge sewage methods.

Groundwater pollution comes from a variety of sources, including agriculture, landfills, and septic tanks. Marine oil pollution results from oil drilling and oil-tanker accidents, runoff from streets, improper disposal of lubricating oil from machines and car crankcases, and intentional discharges from oil tankers during loading or unloading.

Reduced water quality can seriously threaten land use and in-place water use. In the United States and other nations, legislation helps to preserve certain scenic water areas and wildlife habitats. Shorelands and wetlands provide valuable services as buffers, filters, reservoirs, and wildlife areas. Water management concerns of growing importance are groundwater mining, increasing salinity, water diversion, and managing urban water use. Urban areas face several problems, such as providing water suitable for human use, collecting and treating wastewater, and handling stormwater runoff in an environmentally sound manner. Water planning involves many governmental layers, which makes effective planning difficult.

Acting Green

1. Plant native and/or drought-tolerant grasses, ground covers, shrubs, and trees. Once established, they do not need water as frequently and usually will survive a dry period without watering. Group plants together based on similar water needs.

2. Install a new water-saving toilet.

3. Get involved in water management issues. Voice your questions and concerns at public meetings conducted by your local government or water management district.

4. Conserve water because it is the right thing to do. Don't waste water just because someone else is paying for it, such as when you are staying at a hotel.

5. Support projects that will lead to an increased use of reclaimed wastewater for irrigation and other uses.

6. Encourage your friends and neighbors to be part of a water-conscious community. Promote water conservation in community newsletters, on bulletin boards, and by example. Encourage your friends, neighbors, and co-workers to "do their part."

7. Efforts to reduce water consumption are expanding rapidly on university and college campuses. Three such programs are underway at Duke University in North Carolina, Princeton University in New Jersey, and Cuyamaca College in California.

A Duke initiative included the dispensing of 5,000 low-water-flow showerheads to faculty, staff, and off-campus students. It is estimated that each 1.5-gallons-per-minute showerhead will save an estimated 7,300 gallons of water annually, compared to a standard 2.5-gallons-per-minute fixture.

Residence halls at Princeton are having older toilets replaced with new dual flushing systems designed to save water. The new toilets allow users to push the flush handle one way to use less water for liquid waste and another way to release more water for solid waste.

Cuyamaca College is promoting water conservation in the southern California landscape through exhibits and programs that educate and inspire the public. One project is the Water Conservation Garden. The 5-acre garden has displays that showcase water conservation in themed gardens, such as a native plant garden and a vegetable garden, as well as how-to displays, such as mulch and irrigation exhibits. What is your campus doing to conserve water?

Review Questions

1. Describe the hydrologic cycle.
2. Distinguish between withdrawal and consumption of water.
3. What are the similarities between domestic and industrial water use? How are they different from in-stream use?
4. How is land use related to water quality and quantity? Can you provide local examples?
5. What is biochemical oxygen demand? How is it related to water quality?
6. How can the addition of nutrients such as nitrates and phosphates result in a reduction of the amount of dissolved oxygen in the water?
7. Differentiate between point and nonpoint sources of water pollution.
8. How are most industrial wastes disposed of? How has this changed over the past 25 years?
9. What is thermal pollution? How can it be controlled?
10. Describe primary, secondary, and tertiary sewage treatment.
11. What are the types of wastes associated with agriculture?
12. Why is stormwater management more of a problem in an urban area than in a rural area?
13. Define groundwater mining.
14. How does irrigation increase salinity?
15. What are the three major water services provided by metropolitan areas?

Critical Thinking Questions

1. Leakage from freshwater distribution systems accounts for significant losses. Is water so valuable that governments should require systems that minimize leakage to preserve the resource? Under what conditions would you change your evaluation?
2. Do nonfarmers have an interest in how water is used for irrigation? Under what conditions should the general public be involved in making these decisions along with the farmers who are directly involved?
3. Should the United States allow Mexico to have water from the Rio Grande and the Colorado River, both of which originate in the United States and flow to Mexico?
4. Do you believe that large-scale hydroelectric power plants should be promoted as a renewable alternative to power plants that burn fossil fuels? What criteria do you use for this decision?
5. What are the costs and the benefits of the proposed Garrison Diversion Unit? What do you think should happen with this project?
6. How might you be able to help save freshwater in your daily life? Would the savings be worth the costs?
7. Look at the hydrologic cycle in figure 15.3. If global warming increases the worldwide temperature, how should increased temperature directly affect the hydrologic cycle?

|ENVIRONMENTAL SCIENCE

To access additional resources for this chapter, please visit ConnectPlus® at connect.mheducation.com. There you will find interactive exercises, including Google Earth™, additional Case Studies, and SmartBook™, an adaptive eBook that integrates our LearnSmart® adaptive learning technology.

Air Quality Issues

The atmosphere contains gases that are constantly mixed by wind caused by uneven heating of the Earth's surface by the sun. One of the ways we recognize this mixing process is in changes in weather and wind patterns brought about by the movement of air masses. This photo shows the pattern of clouds associated wit Hurricane Sandy in 2012.

OBJECTIVES

After reading this chapter, you should be able to:

- List the gases that make up most of the atmosphere.
- Recognize that the atmosphere consists of several distinguishable layers.
- Recognize that the atmosphere can accept and disperse significant amounts of pollutants.
- List the major sources and effects of the six criteria air pollutants.
- Recognize that air pollutant standards are set by governments and organizations.
- Distinguish between primary and secondary air pollutants.

- Recognize that there is a special category of air pollutants known as hazardous air pollutants.
- Describe how photochemical smog is formed and how it affects humans.
- Describe how climate and geography contribute to photochemical smog.
- Describe a thermal inversion.
- List the actions taken by the EPA as a result of the Clean Air Act and the effect that these actions have had on air quality.
- Describe the primary molecules involved in producing acid rain and their sources.

- Describe the effects of acid rain on structures, terrestrial ecosystems, and aquatic ecosystems.
- Describe the link between chlorofluorocarbon use and ozone depletion.
- Describe the effects of ozone depletion.
- Describe how the problem of ozone depletion has been addressed.
- Describe how sources of air pollution in economically developed countries differ from those in newly developing countries.
- Identify noise and how it is measured.

Improvements in Air Quality in Mexico City

In 1992 the United Nations declared that Mexico City had the worst air pollution in the world. It had only 8 days during the year that were considered good. Several factors contributed to the problem.

- Geography—Since the city is at the bottom of a bowl surrounded by mountains, air pollutants tend to accumulate. Because it is at an elevation of 2,240 meters (7,350 feet) the amount of oxygen in the air is less than at sea level. This results in less complete combustion of fuels for cars, factories, and homes, which results in more pollutants.
- Population—The city and the surrounding suburbs have a population of over 20 million people. The population requires electricity, transportation, jobs, and services. Meeting the needs of the population requires the burning of fuels that cause air pollution. The emissions from 6 million motor vehicles are major contributors to the air pollution in the city.
- Economic Activity—Mexico City is the major industrial and business center of Mexico. In terms of economic activity, Mexico City is one of the top ten cities in the world. About half of Mexico's industrial output is produced in the Mexico City region. Important industrial activities include textiles, chemicals, furniture, plastics and metals, electronics assembly, and the production of pharmaceutical products. Many factories were heavy polluters.

The city and federal governments have approached the air pollution problem from several points of view.

- Industrial Activity—The industrial sector was required to make several changes.

  - Industries were required to reduce pollution or relocate.
  - Industries and power plants were required to use fuels with reduced sulfur content.
  - Many industries switched from oil to natural gas, which results in less pollution.

- Automotive Changes—Several changes to transportation have resulted in lower pollution.

  - No-lead gasoline has been required since 1997.
  - Automobile fuels were blended to burn cleaner at the elevation of the city.
  - Inspections of vehicles were required to verify the effectiveness of pollution-control devices, and it has been made difficult for people to circumvent the inspection.
  - Vapor recovery mechanisms for volatile organic compounds (VOCs) have been installed at gasoline stations and distribution centers.

- Other Activities—Although changes to industry and automobiles are responsible for the major changes in air pollution, a few other actions may have had an impact.

  - Improvements in public transportation (buses, light rail, taxis) resulted in less air pollution.
  - Public information about air pollution encouraged people to make changes that reduce the problem.

These actions have significantly improved air quality in Mexico City. In 2012 Mexico City had 248 good air days. Compared to 8 good air days in 1992, that is a major improvement in 20 years.

16.1 The Atmosphere

The atmosphere (air) is composed of 78.1 percent nitrogen, 20.9 percent oxygen, and a number of other gases such as argon, carbon dioxide, methane, and water vapor that total about 1 percent. Most of the atmosphere is held close to the Earth by the pull of gravitational force, so it gets less dense with increasing distance from the Earth. Throughout the various layers of the atmosphere, nitrogen and oxygen are the most common gases present, although the molecules are farther apart at higher altitudes.

The atmosphere is composed of four layers. (See figure 16.1.)

- The *troposphere* extends from the Earth's surface to about 10 kilometers (about 6.2 miles) above the Earth. It actually varies from about 8 to 18 kilometers (5–11 miles), depending on the position of the Earth and the season of the year. The temperature of the troposphere declines by about 6°C (11°F) for every kilometer above the surface. The troposphere contains most of the water vapor of the atmosphere and is the layer in which weather takes place.

- The *stratosphere* extends from the top of the troposphere to about 50 kilometers (about 31 miles) and contains most of the ozone. The ozone is in a band between 15 and 30 kilometers

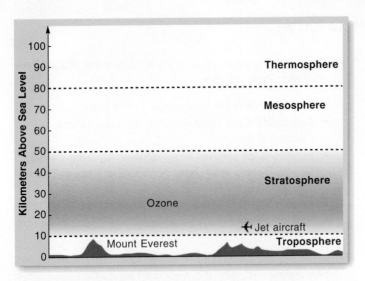

FIGURE 16.1 The Atmosphere The atmosphere is divided into the troposphere, the relatively dense layer of gases close to the surface of the Earth; the stratosphere, more distant with similar gases but less dense; the mesosphere; and the thermosphere. Weather takes place in the troposphere, and the important ozone layer is present in the stratosphere.

(9–19 miles) above the Earth's surface. Because the ozone layer absorbs sunlight, the upper layers of the stratosphere are warmer than the lower layers.

- The *mesosphere* is a layer with decreasing temperature from 50 to 80 kilometers (31–50 miles) above the Earth.

- The *thermosphere* is a layer with increasing temperature that extends to about 300 kilometers (186 miles) above the Earth's surface.

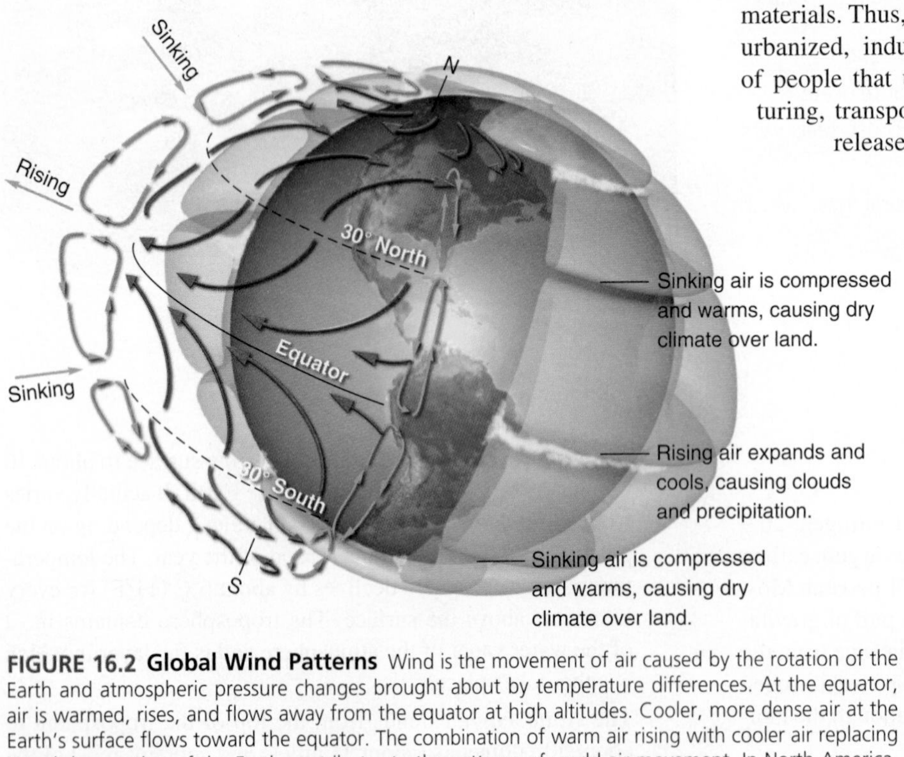

FIGURE 16.2 Global Wind Patterns Wind is the movement of air caused by the rotation of the Earth and atmospheric pressure changes brought about by temperature differences. At the equator, air is warmed, rises, and flows away from the equator at high altitudes. Cooler, more dense air at the Earth's surface flows toward the equator. The combination of warm air rising with cooler air replacing it and the rotation of the Earth contributes to the patterns of world air movement. In North America, most of the winds are westerlies (from the west to the east).

Even though gravitational force keeps the majority of the air near the Earth, it is not static. As air absorbs heat from the Earth, it expands and rises. When its heat content is radiated into space, the air cools, becomes more dense, and flows toward the Earth. As the air circulates vertically due to heating and cooling, it also moves horizontally over the surface of the Earth because the Earth rotates on its axis. The combination of all air movements creates the wind and weather patterns characteristic of different regions of the world. (See figure 16.2.)

16.2 Pollution of the Atmosphere

Pollution is any addition of matter or energy that degrades the environment for humans and other organisms. Because human actions are the major cause of pollution we can do something to prevent it. There are several natural sources of gases and particles that degrade the quality of the air, including material emitted from volcanoes, dust from wind erosion, and gases from the decomposition of dead plants and animals. Since these events are not human-induced, we cannot do much to control them. However, automobile emissions, chemical odors, factory smoke, and similar materials are considered air pollution and will be the focus of this chapter.

The problem of air pollution is directly related to the number of people living in an area and the kinds of activities in which they are involved. When a population is small and its energy use is low, the impact of people is minimal. The pollutants released into the air are diluted, carried away by the wind, washed from the air by rain, or react with oxygen in the air to form harmless materials. Thus, the overall negative effect is slight. However, our urbanized, industrialized civilization has dense concentrations of people that use large quantities of fossil fuels for manufacturing, transportation, and domestic purposes. These activities release large quantities of polluting by-products into our environment.

Gases or small particles released into the atmosphere are likely to stay near the Earth due to gravity. We do not get rid of them; they are just diluted and moved out of the immediate area. In industrialized urban areas, pollutants cannot always be diluted sufficiently before the air reaches another city. The polluted air from Chicago is further polluted when it reaches Gary, Indiana, supplemented by the wastes of Detroit and Cleveland, and finally moves over southeastern Canada and New England to the ocean. While not every population center adds the same kind or amount of waste, each adds to the total load carried.

A good example to illustrate the effects of population centers on pollution levels involves the production of ground-level ozone. Ground-level ozone is a by-product of automobile usage. (This topic will be dealt with in great detail later in the chapter.) Although ozone in the upper atmosphere is valuable in protecting the Earth from ultraviolet light, ground-level ozone can severely damage lung tissue. The maps in figure 16.3 show the population centers and peak values for ground-level ozone and fine particulate matter for September 2, 2010. This was a particularly bad day, but it points out the regional nature of air pollution problems and that pollution is related to population density.

Air pollution causes health problems. In the industrialized nations of the world, attention to air pollution problems has improved air quality substantially. However, in the developing world, air pollution is a major problem. Many of the megacities of the developing world have extremely poor air quality. The World Health Organization estimates that urban air pollution accounts for over 3 million deaths annually. Deaths from air pollution occur primarily among the elderly, the infirm, and the very young. The causes of this air pollution are open fires, large numbers of poorly maintained motor vehicles, and poorly regulated industrial plants. Not only does poor air quality in such cities increase the death rate, but the general health of the populace is lowered. Approximately 20 to 30 percent of all respiratory diseases appear to be caused by air pollution.

16.3 Categories of Air Pollutants

Around the world, five major types of substances are released directly into the atmosphere in their unmodified forms in sufficient quantities to pose a health risk and are called **primary air pollutants.** They are carbon monoxide, volatile organic compounds (hydrocarbons), particulate matter, sulfur dioxide, and oxides of nitrogen. Primary air pollutants may interact with one another in the presence of sunlight to form new compounds such as ozone that are known as **secondary air pollutants.** Secondary air pollutants also form from reactions with substances that occur naturally in the atmosphere. In addition, the U.S. Environmental Protection Agency (EPA) has established air quality standards for six principal air pollutants, which are called the **criteria air pollutants.**

The criteria air pollutants are carbon monoxide (CO), particulate matter (PM), sulfur dioxide (SO_2), nitrogen dioxide (NO_2), lead (Pb), and ozone (O_3). Four of these pollutants (CO, SO_2, NO_2, and Pb) are emitted directly from a variety of sources. Ozone is not directly emitted but is formed when nitrogen dioxide, other oxides of nitrogen, and volatile organic compounds

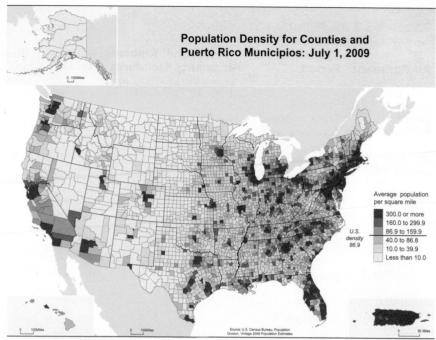

Population Density for Counties and Puerto Rico Municipios: July 1, 2009

Average population per square mile

	300.0 or more
	160.0 to 299.9
U.S. density 86.9	86.9 to 159.9
	40.0 to 86.8
	10.0 to 39.9
	Less than 10.0

Source: U.S. Census Bureau, Population Division, Vintage 2009 Population Estimates

(a) Population Density

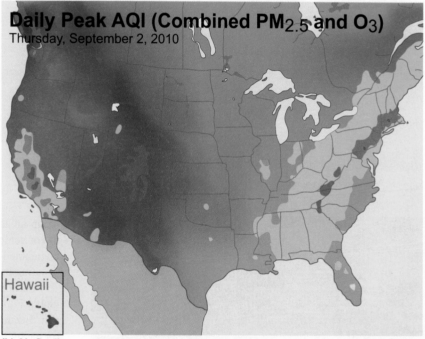

Daily Peak AQI (Combined PM$_{2.5}$ and O$_3$)
Thursday, September 2, 2010

Hawaii

(b) Air Quality

FIGURE 16.3 Air Pollution and Population Centers In general the areas with the worst air quality are those with high population density. (a) This map shows the population density of counties in the United States. (b) This map shows the areas with poor air quality due to ozone on September 2, 2010. Note that in general the air quality is poorest in the most densely populated areas.

Sources: U.S. Environmental Protection Agency and U.S. Census Bureau.

(VOCs) react in the presence of sunlight. Particulate matter can be directly emitted—naturally occurring dust and dust from agriculture, construction, and industry—or it can be formed when emissions of nitrogen oxides, sulfur oxides (SO_x)—primarily SO_2 and SO_3—ammonia, organic compounds, and other gases react in the

Table 16.1 Air Pollutant Standards

Air Pollutant	Measurement Period	U.S. National Ambient Air Quality Standards	European Union Air Quality Standards
EPA Criteria Air Pollutants			
Carbon monoxide (CO)	8-hour average	9 ppm (10 mg/m³)	10 mg/m³
	1-hour average	35 ppm (40 mg/m³)	
Nitrogen dioxide (NO₂)	Annual mean	0.053 ppm (100 µg/m³)	40 µg/m³
	1-hour	0.1 ppm (190 µg/m³)	200 µg/m³
Ozone (O₃)	8-hour average	0.075 ppm (150 µg/m³)	120 µg/m³
Lead (Pb)	3-month average	0.15 µg/m³	
	Annual mean		0.5 µg/m³
Particulate matter (PM₁₀)	Annual mean		40 µg/m³
	24-hour average	150µg/m³	50 µg/m³
Particulate matter (PM₂.₅)	Annual mean	12 µg/m³	25 µg/m³
	24-hour average	35 µg/m³	
Sulfur dioxide (SO₂)	24-hour average		125 µg/m³
	1-hour average	0.075 ppm (200 µg/m³)	350 µg/m³
Other Common Air Pollutants			
Benzene	Annual mean	No standards set, but current levels are below 2.5 µg/m³	5 µg/m³
Volatile organic compounds		No standards set, but reductions are needed to reduce ground-level ozone	Polycyclic aromatic hydrocarbons and organic solvents regulated (1 ng/m³)
Mercury		New regulations will result in 90 percent reduction in mercury releases from power plants.	Industrial uses and releases of mercury are regulated

Source: U.S. Environmental Protection Agency and European Commission.

FIGURE 16.4 Carbon Monoxide The major source of carbon monoxide is the internal combustion engine, which is used to provide most of our transportation. The more concentrated the number of automobiles, the more concentrated the pollutants. Carbon monoxide concentrations of a hundred parts per million are not unusual in rush-hour traffic in large metropolitan areas. These concentrations are high enough to cause fatigue, dizziness, and headaches.

atmosphere. (See table 16.1.) In addition, certain compounds with high toxicity are known as **hazardous air pollutants** or **air toxics.**

Carbon Monoxide

Carbon monoxide (CO) is a chemical compound produced when carbon-containing materials such as gasoline, coal, wood, and trash are burned with insufficient oxygen. When carbon-containing compounds are burned with abundant oxygen present, carbon dioxide is formed ($C + O_2 \rightarrow CO_2$). When the amount of oxygen is restricted, carbon monoxide is formed ($2C + O_2 \rightarrow 2CO$) instead of carbon dioxide. Any process that involves the burning of fossil fuels has the potential to produce carbon monoxide. The single largest source of carbon monoxide is the automobile. (See figure 16.4.) About 60 percent of CO comes from vehicles driven on roads and 25 percent comes from vehicles not used on roads (earth movers and other heavy equipment, ATVs, snowmobiles, lawn tractors, etc.). Most of the remainder comes from other processes that involve burning (fires, power plants, industry, wood-burning stoves, burning leaves, etc.). Although increased fuel efficiency and the use of catalytic converters have reduced carbon monoxide emissions per kilometer driven, carbon monoxide remains a problem because the number of automobiles on the road and the number of kilometers driven have risen. In urban areas, as much as 90 percent of carbon monoxide is from motor vehicles. In many parts of the world, automobiles are poorly maintained and may have inoperable pollution control equipment, resulting in even greater amounts of carbon monoxide.

Carbon monoxide is dangerous because it binds to the hemoglobin in blood and makes the hemoglobin less able to carry oxygen. Because carbon monoxide remains attached to hemoglobin for a long time, even small amounts tend to accumulate and reduce the blood's oxygen-carrying capacity. Several hours of exposure to air containing only 0.001 percent of carbon monoxide can cause death. Carbon monoxide is most dangerous in enclosed spaces, where it is not diluted by fresh air entering the space. The amount of carbon monoxide produced in heavy traffic can cause headaches, drowsiness, and blurred vision. Cigarette smoking is also an important source of carbon monoxide because the smoker is inhaling it directly. A heavy smoker in congested traffic is doubly exposed and may experience severely impaired reaction time compared to a nonsmoking driver.

Fortunately, carbon monoxide is not a persistent pollutant. It readily combines with oxygen in the air to form carbon dioxide

$(2CO + O_2 \rightarrow 2CO_2)$. Therefore, the air can be cleared of its carbon monoxide if no new carbon monoxide is introduced into it. Control of carbon monoxide in the United States has been very good. The U.S. EPA reports that carbon monoxide levels decreased by about 75 percent between 1990 and 2012, and all communities now meet the standards set by the EPA. This was accomplished with a variety of controls on industry and, in particular, on motor vehicles. Catalytic converters reduce the amount of carbon monoxide released by vehicle engines, and specially formulated fuels that produce less carbon monoxide are used in many cities that have a carbon monoxide problem. Often these special fuels are only used in winter, when car engines run less efficiently and produce more carbon monoxide.

Particulate Matter

Particulate matter consists of tiny solid particles and liquid droplets dispersed into the atmosphere. The Environmental Protection Agency has set air pollution standards for coarse particles between 10 microns and 2.5 microns (**PM$_{10}$**) and for fine particles less than 2.5 microns (**PM$_{2.5}$**). A micron is one millionth of a meter. Many bacteria are about 1 micron in diameter. Most of the coarse particles (PM$_{10}$) are primary pollutants such as dust and carbon particles that are released directly into the air. Dust from natural wind erosion, road travel, agricultural activities, construction sites, industrial processes, and smoke particles from fires are the other primary sources of coarse particles. (See figure 16.5.) Fine particles (PM$_{2.5}$) are mostly secondary pollutants that form in the atmosphere from interactions of primary air pollutants. Sulfates and nitrates formed from sulfur dioxide and nitrogen oxides are examples. Other sources of fine particulates are fires and road dust.

Particulates cause problems ranging from the annoyance of reduced visibility and particulates settling on a backyard picnic table to serious health effects. Particulates (especially fine particulates) can irritate respiratory passages and accumulate deep in the lungs and reduce the ability of the lungs to exchange gases. Such lung damage usually occurs in people who are repeatedly exposed to large amounts of particulate matter on the job. Miners and others who work in dusty conditions are most likely to be affected. Droplets and solid particles can also serve as centers for the deposition of other substances from the atmosphere. As we breathe air containing particulates, we come in contact with concentrations of other potentially more harmful materials that have accumulated on the particulates. Sulfuric, nitric, and carbonic acids, which irritate the lining of our respiratory system, frequently are associated with particulates. Some particles, such as certain kinds of asbestos, are known to cause cancer.

According to the U.S. EPA, the amount of PM$_{10}$ pollution decreased by about 39 percent between 1990 and 2012. The EPA has been setting standards for PM$_{2.5}$ particles for a shorter period, and pollution by these particles decreased by about 33 percent between 2000 and 2012. The major region that still exceeds PM$_{10}$ standards includes parts of California, and southern Nevada and Arizona. Several population centers still exceed the standards set for PM$_{2.5}$.

Sulfur Dioxide

Sulfur dioxide (SO$_2$) is a compound of sulfur and oxygen that is produced when sulfur-containing fossil fuels are burned $(S + O_2 \rightarrow SO_2)$. There is sulfur in coal and oil because they were produced from the bodies of organisms that had sulfur as a component of some of their molecules. The sulfur combines with oxygen to form sulfur dioxide when fossil fuels are burned. Today, over 70 percent of sulfur dioxide released into the atmosphere is from power plants, primarily those that burn coal. (See figure 16.6.) Most of the rest comes from industrial processes and other activities that involve the burning of fossil fuels.

Sulfur dioxide has a sharp odor, irritates respiratory tissue, and aggravates asthmatic and other respiratory conditions. It also reacts with water, oxygen, and other materials in the air to form sulfur-containing acids. The acids can become attached to particles that, when inhaled, are very corrosive to lung tissue. These acid-containing particles are also involved in acid deposition, which is discussed in the Acid Deposition section of this chapter.

FIGURE 16.5 Particulate Matter The dust produced by vehicles is a major source of particulate matter. The process of extracting and loading coal into this truck also produces dust.

FIGURE 16.6 Sulfur Dioxide Coal-fired power plants are the primary source of sulfur dioxide.

Because the major sources of sulfur dioxide are power plants and they are easily monitored, much has been done to reduce the amount of sulfur dioxide released. In the United States between 1990 and 2012, SO_2 levels decreased about 72 percent and nearly all communities meet the SO_2 standards set by the U.S. EPA.

Nitrogen Dioxide

The burning of fossil fuels produces a mixture of nitrogen-containing compounds commonly known as **oxides of nitrogen** or **nitrogen oxides** (NO_x). These compounds are formed because the nitrogen and oxygen molecules in the air combine with one another when subjected to the high temperatures experienced during combustion. The two most commonly produced nitrogen oxide molecules are **nitrogen monoxide (NO)** and **nitrogen dioxide** (NO_2). In general, combustion reactions produce nitrogen monoxide ($N_2 + O_2 \rightarrow 2NO$), but nitrogen monoxide can be converted to nitrogen dioxide in the air ($2NO + O_2 \rightarrow 2NO_2$) to produce a mixture of NO and NO_2. Thus, NO_2 is, for the most part, a secondary pollutant.

Nitrogen dioxide is a reddish brown, highly reactive gas that causes respiratory problems, is responsible for much of the haze seen over cities, and is a component of acid precipitation. Nitrogen dioxide also is important in the production of the mixture of secondary air pollutants known as photochemical smog, which is discussed in the section on Photochemical Smog in this chapter.

The primary source of nitrogen oxides is the burning of fossil fuels—particularly in internal combustion engines. Automobiles account for about 35 percent, nonroad motorized equipment (construction equipment, farm equipment, lawn mowers, snowmobiles, etc.) account for about 20 percent, and electricity generation accounts for about 20 percent.

Catalytic converters significantly reduce the amount of nitrogen monoxide released from the internal combustion engine. (About 75 percent of the NO produced by an automobile engine is converted back into N_2 and O_2 by the car's catalytic converter.) However, the increase in the numbers of cars and kilometers driven offsets some of the gains attributable to catalytic converters. EPA rules that require cleaner burning fuel and require nonroad vehicles to reduce their releases of nitrogen oxides and sulfur have had a positive effect. (See figure 16.7.) There has been about a 46 percent decrease in nitrogen dioxide between 1990 and 2012. Although all communities now meet the EPA standards set for nitrogen oxides, nitrogen oxides remain a problem because they contribute to the development of photochemical smog and ground-level ozone.

Lead

Lead (Pb) is a metal that can enter the body when we inhale airborne particles or consume lead that was deposited on surfaces. Lead accumulates in the body and causes a variety of health effects, including mental retardation and kidney damage.

FIGURE 16.7 Nitrogen Dioxide Engineering changes to automobiles and the fuel they use have significantly reduced the amount of nitrogen dioxide and other oxides of nitrogen produced.

At one time, the primary source of airborne lead was from additives in gasoline. Lead was added to gasoline to help engines run more effectively. Recognition that lead emissions were hazardous resulted in the lead additives being removed from gasoline in North America and Europe. Today, nearly all countries have phased out lead as an additive to gasoline and nearly all of the gasoline produced worldwide is unleaded. Since leaded gasoline has been eliminated in much of the world, lead levels have fallen. In the United States, lead emissions peaked at about 258,000 tons per year. In 2007, they were about 1,300 tons per year—a reduction of 99.5 percent. Today, industrial sources such as metal smelters and manufacturers of batteries account for about 80 percent of lead emissions. The standards set by the U.S. EPA are met in all parts of the United States.

Another major source of lead contamination is lead paints. Many older homes have paints that contain lead, since various lead compounds are colorful pigments. Dust from flaking paint, remodeling, or demolition is released into the atmosphere. Although the amount of lead may be small, its presence in the home can result in significant exposure to inhabitants, particularly young children who chew on painted surfaces and often eat paint chips.

Volatile Organic Compounds

Volatile organic compounds (VOCs) are organic compounds that readily evaporate and become pollutants in the air. Since they are composed primarily of carbon and hydrogen, they are often referred to as **hydrocarbons.** The use of internal combustion engines accounts for about 44 percent of volatile organic compounds released into the air. The use of solvents contributes about 22 percent. (See figure 16.8.) VOCs are important in the processes that lead to the production of the secondary air pollutants found in smog. Smog will be discussed in the Photochemical Smog section of this chapter. Some VOCs are toxic and are known as hazardous air pollutants.

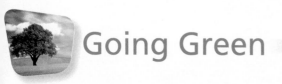

Going Green

Going Solvent-Free

Sometimes a good environmental change is so pervasive it is almost invisible. At one time volatile organic compounds (VOCs) were widely used in paints, printing inks, cleaners, and developing solutions for photographs. VOCs were also used as solvents and aerosol propellants. Today, powder coatings that go on dry and are cured with heat have replaced solvent-based paints in the manufacture of most household appliances, automobiles, and metal products. In other applications water-based paints have replaced solvent-containing oil-based paints. Most printing processes are now solvent free. The change from traditional photography, which required chemical solutions to develop pictures, to digital photography, which can be done with solvent-free printing, has eliminated many solvents. Many cleaning products and adhesives are now available as solvent-free versions. Even nail polish and art supplies for "oil" painting are available solvent-free.

Aerosol cans often contain products and propellants that are VOCs. They are marketed as convenient methods for applying substances to surfaces or dispersing materials into the environment. If the container warns that the contents are flammable, it is highly likely that it contains VOCs. There are several alternatives to aerosols. Pump spray containers can be used. Often the product—deodorant, cooking oil, paint, etc.—can be applied in other ways just as easily.

You as a consumer are in a position, by the kinds of purchases you make, to further reduce VOC releases. Do you really need a product in an aerosol can? Can you find a solvent-free alternative for a product you use? You can read the labels on the products you buy to find out if they are solvent-free.

FIGURE 16.8 Volatile Organic Compounds (VOCs) A major source of volatile organic compounds is consumer products that evaporate when they are used.

Many modifications to automobiles have significantly reduced the amount of volatile organic compounds entering the atmosphere. Recycling some gases through the engine so they burn rather than escape, increasing the proportion of oxygen in the fuel-air mixture to obtain more complete burning of the fuel, and using devices to prevent the escape of gases from the fuel tank and crankcase are three of these modifications. In addition, catalytic converters allow unburned organic compounds in exhaust gases to be oxidized more completely so that fewer volatile organic compounds leave the tailpipe. Reductions have also been achieved from industrial sources by requiring industries to account for their emissions and encouraging the substitution of nonvolatile compounds for volatile organic compounds. For example, the increased use of paints and coatings that do not require an organic solvent has substantially reduced the amount of VOCs released from industrial sources.

Emissions of VOCs have decreased by about 45 percent from 1990 to 2012.

Hazardous Air Pollutants

While all of the major air pollutants already discussed are environmental health concerns, hundreds of other dangerous chemical compounds that can cause harm to human health or damage the environment are purposely or accidentally released into the air. These compounds are collectively known as hazardous air pollutants (HAP) or air toxics. Pesticides are toxic materials that are purposely released to kill insects or other pests. Other dangerous materials are released as a result of consumer activities. Benzene in gasoline escapes when gasoline is put into the tank, and the use of some consumer products such as glues and cleaners releases toxic materials into the air. The majority of air toxics, however, are released as a result of manufacturing processes. Perchloroethylene is released from dry cleaning establishments, and toxic metals are released from smelters. The chemical and petroleum industries are the primary sources of hazardous air pollutants. Although air toxics are harmful to the entire public, their presence is most serious for people who are exposed on the job, since they are likely to be exposed to higher concentrations of hazardous substances over longer time periods. Chapter 19 deals with hazardous materials and the issues related to regulating their production, use, and disposal.

16.4 Photochemical Smog

The pollutants discussed in the previous sections are primary air pollutants that are released directly into the air. Secondary air pollutants are those that are created in the air from preexisting pollutants. Photochemical smog involves the production of secondary air pollutants.

Photochemical smog is a mixture of pollutants including ozone, aldehydes, and peroxyacetyl nitrates that result from the interaction of nitrogen dioxide, nitrogen monoxide, and volatile organic compounds with sunlight in a warm environment. (See figure 16.9.) The two most destructive components of photochemical smog are ozone and peroxyacetyl nitrates (PAN). **Ozone (O_3)** is a molecule that consists of three oxygen atoms bonded to one another. Ground-level ozone is a serious pollutant. However, in the upper atmosphere ozone serves to screen out ultraviolet light. Ultraviolet light can damage tissue and cause genetic mutations. The role of ozone in the upper atmosphere will be discussed in the section on Ozone Depletion later in this chapter. Both of these secondary pollutants are excellent oxidizing agents that will react readily with many other compounds, including those found in living things, causing destructive changes. Ozone is particularly harmful because it destroys chlorophyll in plants and injures lung tissue in humans and other animals. Peroxyacetyl nitrates, in addition to being oxidizing agents, are eye irritants.

How Smog Forms

For photochemical smog to develop, several ingredients are required. Nitrogen monoxide, nitrogen dioxide, and volatile organic compounds must be present, and sunlight and warm temperatures are important to support the chemical reactions involved. The chemical reactions that cause the development of photochemical smog are quite complicated, but a description of several key reactions will help you understand the process.

In most urban areas, nitrogen monoxide, nitrogen dioxide, and volatile organic compounds are present as by-products of the burning of fuel in vehicles and industrial processes. In the presence of sunlight, nitrogen dioxide breaks down to nitrogen monoxide and atomic oxygen:

$$NO_2 \xrightarrow{\text{Sunlight}} NO + O^*$$

Atomic oxygen is extremely reactive and will react with molecular oxygen in the air to form ozone:

$$O^* + O_2 \rightarrow O_3$$

Both ozone and atomic oxygen will react with volatile organic compounds to produce very reactive organic free radicals:

$$O^* \text{ or } O_3 + \begin{array}{c}\text{Organic} \\ \text{molecule}\end{array} \rightarrow \begin{array}{c}\text{Organic} \\ \text{free radical}\end{array}$$

Free radicals are very reactive and cause the formation of additional nitrogen dioxide from nitrogen monoxide:

$$NO + \begin{array}{c}\text{Free} \\ \text{radical}\end{array} \rightarrow NO_2 + \begin{array}{c}\text{Other organic} \\ \text{molecule}\end{array}$$

FIGURE 16.9 Photochemical Smog The interaction among hydrocarbons, oxides of nitrogen, and sunlight produces new compounds that are irritants to humans. The visual impact of smog is shown in this photograph taken in Xi'an, China, on a smoggy day.

This step is important because the presence of additional NO_2 results in the further production of ozone. The organic free radicals also react with nitrogen dioxide to form peroxyacetyl nitrates (PANs) and aldehydes. (Figure 16.10 summarizes these events.)

Human Activity and the Pattern of Smog Development

The development of photochemical smog in an area involves the interaction of climate, time of day, and motor vehicle emissions in the following manner:

1. During morning rush-hour traffic, emissions from automobiles cause an increase in the amounts of nitrogen monoxide (NO) and volatile organic compounds (VOCs) in the atmosphere.

2. The presence of NO and VOCs leads to an increase in the amount of NO_2. Sunlight and warming temperatures support these reactions. Later in the morning NO levels fall and NO_2 levels rise because NO is converted to NO_2.

3. Ozone, peroxyacetyl nitrates, and aldehydes are produced by chemical reactions between VOCs, NO, and NO_2. The levels of these compounds begin to rise by late morning and remain high throughout the middle of the day.

4. Since sunlight and warm temperatures support the production of photochemical smog, in the evening as the sun sets and temperatures fall, the production of ozone, PANs, and aldehydes lessens. In addition, ozone and other smog components react with their surroundings and are destroyed, so the destructive components of smog decline in the evening. Figure 16.11 shows how the concentration of these various molecules changes during the day.

The Role of Climate and Geography

While smog can develop in any area, some cities have greater problems because of their climate, the amount of automobile traffic, and geographic features. Cities with warm climates and those that have lots of sunlight are more prone to develop photochemical smog because the chemical reactions responsible for smog are supported by warm temperatures and sunlight. Similarly, smog is more likely to be a problem during summer months because of the higher temperatures and longer days.

Cities that are located adjacent to mountain ranges or in valleys have greater problems because the pollutants can be trapped by thermal inversions. Normally, air is warmer at the surface of the Earth and gets cooler at higher altitudes. (See figure 16.12a.) In some instances, a layer of warmer air may be above a layer of cooler air at the Earth's surface. This condition is known as a **thermal inversion.**

In cities located in valleys, as the surface of the Earth cools at night, the cooler air on the sides of the valley can flow down into the valley and create a thermal inversion. (See figure 16.12b.) As cool air flows into these valleys, it pushes the warm air upward. Similarly, in cities such as Los Angeles that have mountains to the east and the ocean to the west, cool air from the ocean may push in under a layer of warm air to create a thermal inversion. (See figure 16.12c.) In either case, the warm air becomes sandwiched between two layers of cold air and acts like a lid on the valley. The lid of warm air cannot rise because it is covered by a layer of cooler, denser air pushing down on it. It cannot move out of the area because of the mountains. Without normal air circulation, smog accumulates. Harmful chemicals continue to increase in concentration until a major weather change causes the lid of warm air to rise and move over the mountains. Then the underlying cool air can begin to circulate, and the polluted air can be diluted.

Smog problems could be substantially decreased by reducing the NO_X and VOCs associated with the use of internal combustion engines (perhaps eliminating them completely) or by moving population centers away from the valleys where thermal inversions occur. It is highly unlikely that population centers can be moved; however, it is possible to reduce the molecules that lead to the production of smog. Reformulating gasoline and diesel fuel, regulating emissions from nonroad vehicles, reducing VOCs from solvents, and installing devices on automobiles

Source	Necessary Conditions	Reactions Take Place in Atmosphere	Products
Primarily automobiles	Volatile organic compounds (VOC) present	VOC + O* or $O_3 \rightarrow$ Highly reactive organic radicals + $NO_2 \rightarrow$	Peroxyacetyl nitrates / Aldehydes
Primarily automobiles	Nitrogen monoxide (NO) present	NO + Organic radicals $\longrightarrow NO_2$	
From automobiles and formed from NO	Nitrogen dioxide (NO_2) present	$O^* + O_2 \rightarrow O_3$ (ozone) $\longrightarrow$ Ozone / $NO_2 \rightarrow$ NO + $\overset{.}{O}^*$ (atomic oxygen) / Sunlight	Ozone
Sun	Sunlight		
Sun (summer temperatures)	Heat	Reactions take place more rapidly at higher temperatures.	

FIGURE 16.10 Major Steps in the Development of Photochemical Smog Photochemical smog develops when specific reactants and conditions are present. The necessary reactants are volatile organic compounds, nitrogen monoxide, and nitrogen dioxide. The conditions that cause these compounds to react are warm temperatures and the presence of sunlight. When these conditions exist, ozone and other components of smog are formed as secondary pollutants.

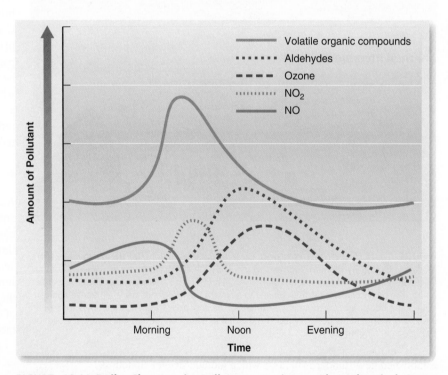

FIGURE 16.11 Daily Changes in Pollutants During a Photochemical Smog Incident The development of photochemical smog begins with a release of nitrogen oxides and volatile organic compounds associated with automobile use during morning traffic. As the sun rises and the day warms up, these reactants interact to form ozone and other secondary pollutants. These peak during the early afternoon and decline as the sun sets.

that reduce the amount of NO_2 and VOCs released have been beneficial. Between 1990 and 2012, levels of VOCs fell by 45 percent and levels of NO_2 fell by 46 percent. However, ozone levels fell by only about 12 percent over the same period and there are still many regions that do not meet EPA standards. The EPA

Normal Situation

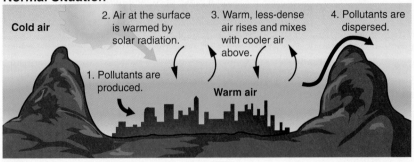

Cold air

2. Air at the surface is warmed by solar radiation.

3. Warm, less-dense air rises and mixes with cooler air above.

4. Pollutants are dispersed.

1. Pollutants are produced.

Warm air

(a)

Thermal Inversion

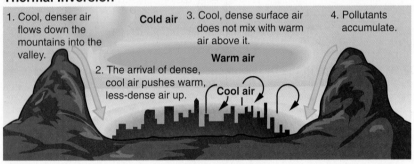

1. Cool, denser air flows down the mountains into the valley.

Cold air

3. Cool, dense surface air does not mix with warm air above it.

4. Pollutants accumulate.

Warm air

2. The arrival of dense, cool air pushes warm, less-dense air up.

Cool air

(b)

Thermal Inversion

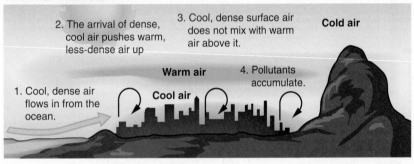

2. The arrival of dense, cool air pushes warm, less-dense air up

3. Cool, dense surface air does not mix with warm air above it.

Cold air

Warm air

4. Pollutants accumulate.

1. Cool, dense air flows in from the ocean.

Cool air

(c)

FIGURE 16.12 Thermal Inversion Under normal conditions (a) the air at the Earth's surface is heated by the sun and rises to mix with the cooler air above it. When a thermal inversion occurs (b and c), a layer of heavier cool air flows into a valley and pushes the warmer air up. The heavy cooler air is then unable to mix with the less-dense warm air above and cannot escape because of surrounding mountains. The cool air is trapped, sometimes for several days, and accumulates pollutants. If the thermal inversion continues, the levels of pollution can become dangerously high.

established new standards in 2008 which include stricter standards and new methods for meeting the standards that are currently being implemented.

16.5 Acid Deposition

Acid deposition is the accumulation of acid-forming particles on a surface. The acid-forming particles can be dissolved in rain, snow, or fog or can be deposited as dry particles. When dry particles are deposited, an acid does not actually form until these materials

mix with water. Even though the acids are formed and deposited in different ways, all of these sources of acid-forming particles are commonly referred to as **acid rain.**

Causes of Acid Precipitation

Acids in the atmosphere result from natural causes, such as vegetation, volcanoes, and lightning, and from human activities, such as the burning of coal and use of the internal combustion engine. (See figure 16.13.) These combustion processes produce sulfur dioxide (SO_2) and oxides of nitrogen (NO_X). Oxidizing agents, such as ozone, hydroxide ions, or hydrogen peroxide, along with water, are necessary to convert the sulfur dioxide or nitrogen oxides to sulfuric or nitric acid. Thus, acid precipitation is a secondary air pollutant.

Acid rain is a worldwide problem. Reports of high acid-rain damage have come from Canada, England, Germany, France, Scandinavia, and the United States. Rain is normally slightly acidic (pH between 5.6 and 5.7), since atmospheric carbon dioxide dissolves in water to produce carbonic acid. But acid rains sometimes have a concentration of acid a thousand times higher than normal. In 1969, New Hampshire had a rain with a pH of 2.1. In 1974, Scotland had a rain with a pH of 2.4. Currently, rain in much of the northeastern part of the United States and parts of Ontario has a pH between 4.6 and 5.1. This compares with levels in 1994, when pH readings were between 4.2 and 4.6. This is a substantial improvement brought about primarily by reductions in SO_2 emissions and, to a lesser extent, a reduction in NO_2 emissions.

Effects on Structures

Acid rain can cause damage in several ways. Buildings and monuments are often made from materials that contain limestone (calcium carbonate, $CaCO_3$), because limestone is relatively soft and easy to work. Sulfuric acid (H_2SO_4), a major component of acid rain, converts limestone to gypsum ($CaSO_4$), which is more soluble than calcium carbonate and is eroded over many years of contact with acid rain. (See figure 16.14.) Metal surfaces can also be attacked by acid rain.

Effects on Terrestrial Ecosystems

The effects of acid rain on terrestrial ecosystems are often subtle and difficult to quantify. A strong link has been established between acid rain and the decline of forests. In many parts of the world, acid rain is suspected of causing the death of many forests and reducing the vigor and rate of growth of others. (See figure 16.15.) In central Europe, many forests have declined significantly, resulting in the death of about 6 million hectares (14.8 million acres) of trees. Northeastern North America has been affected with significant tree death and reduction in vigor, particularly at higher elevations. Some areas have had 50 percent mortality of red spruce trees.

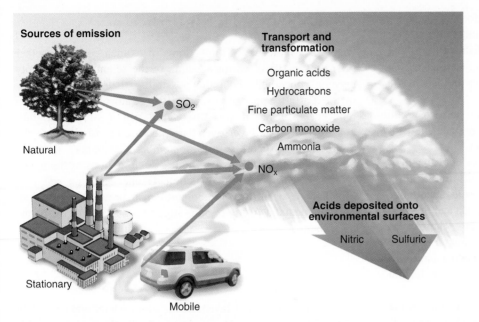

FIGURE 16.13 **Sources of Acid Deposition** Molecules from natural sources, and sulfur dioxide and nitrogen oxides from power plants and internal combustion engines react to produce acids responsible for acid deposition.

FIGURE 16.14 **Damage Due to Acid Deposition** Sulfuric acid (H_2SO_4), which is a major component of acid deposition, reacts with limestone ($CaCO_3$) to form gypsum ($CaSO_4$). Since gypsum is water soluble, it washes away with rain. The damage to this tombstone is the result of such acid reacting with the stone.

The deposition of acids causes major changes to the soils in areas where the soils are not able to buffer the additional acid. As soil becomes acidic, aluminum is released from binding sites and becomes part of the soil water, where it interferes with the ability of plant roots to absorb nutrients. A recent long-term study in New Hampshire strongly suggests that the many years of acid precipitation have reduced the amount of calcium and magnesium in the soil, which are essential for plant growth. Because there are no easy ways to replace the calcium even if acid rain were to stop, it would still take many years for the forests to return to health. Reduction in the pH of the soil may also change the kinds of bacteria in the soil and reduce the availability of nutrients for plants. While none of these factors alone would necessarily result in tree death, each could add to the stresses on the plant and may allow other factors, such as insect infestations, extreme weather conditions (particularly at high elevations), or drought, to further weaken trees and ultimately cause their death.

Effects on Aquatic Ecosystems

The effects of acid rain on aquatic ecosystems are much more clear-cut. In several experiments, lakes were purposely made acidic and the changes in the ecosystems recorded. The experiments showed that as lakes became more acidic, there was a progressive loss of many kinds of organisms. The food web becomes less complicated, many organisms fail to reproduce, and many others die. Most healthy lakes have a pH above 6. At a pH of 5.5, many desirable species of fish are eliminated; at a pH of 5, only a few starving fish may be found, and none is able to reproduce. Lakes with a pH of 4.5 are nearly sterile.

Several reasons account for these changes. Many of the early developmental stages of insects and fish are more sensitive to

FIGURE 16.15 **Forest Decline** Many forests, particularly at high elevations in northeastern North America, have shown significant decline, and dead trees are common.

acid conditions than are the adults. In addition, the young often live in shallow water, which is most affected by a flood of acid into lakes and rivers during the spring snowmelt. The snow and its acids have accumulated over the winter, and the snowmelt releases large amounts of acid all at once. Crayfish and other crustaceans need calcium to form their external skeletons. As the pH of the water decreases, the crayfish are unable to form new exoskeletons and so they die. Reduced calcium availability also results in the development of some fish with malformed skeletons. As mentioned earlier, increased acidity also results in the release of aluminum, which impairs the function of a fish's gills. About 14,000 lakes in Canada and 11,000 in the United States have been seriously altered by becoming acidic. Many lakes in Scandinavia are similarly affected.

The extent to which acid deposition affects an ecosystem depends on the nature of the bedrock in the area and the ecosystem's proximity to acid-forming pollution sources. (See figure 16.16.) Soils derived from igneous rock are not capable of buffering the effects of acid deposition, while soils derived from sedimentary rocks such as limestone release bases that neutralize the effects of acids. Because of this, eastern Canada and the U.S. Northeast are particularly susceptible to acid rain. These areas have a high proportion of granite rock and are downwind from the major air pollution sources of North America. Scandinavian countries have a similar geology and receive pollution from industrial areas in the United Kingdom and Europe.

Restrictions on power plant emissions and reductions in the release of nitrogen oxides from automobile exhaust have dramatically reduced acid rain. Figure 16.17 shows maps of typical pH readings of rain in 1994 and 2010. It is clear that the quality of the air has improved. Long-term monitoring of lakes and streams also shows improvement, but it will take many years before they return to normal.

16.6 Ozone Depletion

Ozone is a molecule of three atoms of oxygen bonded together (O_3). In the 1970s, various sectors of the scientific community became concerned about the possibility that the ozone layer in the Earth's upper atmosphere (the stratosphere) was being reduced. In 1985, it was discovered that a significant thinning of the ozone layer over the Antarctic occurred during the Southern Hemisphere spring (September–November); this area became known as the "ozone hole." Some regions of the ozone layer showed 95 percent depletion. Ozone depletion was also found to be occurring farther north than previously recorded. Measurements in Arctic regions suggest a thinning of the ozone layer there also.

Why Stratospheric Ozone is Important

The ozone in the outer layers of the atmosphere, approximately 15 to 35 kilometers (9–21 miles) from the Earth's surface, shields the Earth from the harmful effects of ultraviolet light radiation. Ozone absorbs ultraviolet light and is split into an oxygen molecule and an oxygen atom:

$$O_3 \xrightarrow{\text{Ultraviolet light}} O_2 + O$$

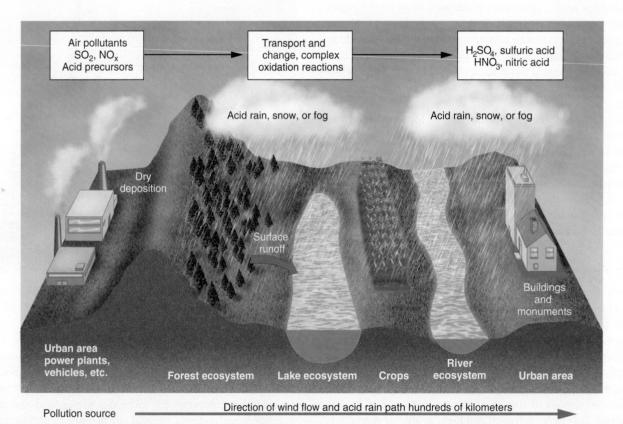

FIGURE 16.16 Factors That Contribute to Acid Rain Damage In any aquatic ecosystem, the following factors increase the risk of damage from acid deposition: (1) location downwind from a major source of pollution; (2) hard, insoluble bedrock with a thin layer of infertile soil in the watershed; (3) low buffering capacity in the soil of the watershed; (4) a low lake surface area-to-watershed ratio.

Source: Data from U.S. Environmental Protection Agency, *Acid Rain.*

Oxygen molecules are also split by ultraviolet light to form oxygen atoms:

$$O_2 \xrightarrow{\text{Ultraviolet light}} 2O$$

Recombination of oxygen atoms and oxygen molecules allows ozone to be formed again and to be available to absorb more ultraviolet light.

$$O_2 + O \rightarrow O_3$$

This series of reactions results in the absorption of about 99 percent of the ultraviolet light energy that comes from the sun and prevents it from reaching the Earth's surface. Less ozone in the upper atmosphere results in more ultraviolet light reaching the Earth's surface. Ultraviolet light is strongly linked to skin cancers and cataracts in humans and increased mutations in all living things.

Ozone Destruction

Chlorofluorocarbons are strongly implicated in the ozone reduction in the upper atmosphere. Chlorofluorocarbons and similar compounds can release chlorine atoms, which can lead to the destruction of ozone. Chlorine reacts with ozone in the following way to reduce the quantity of ozone present:

$$Cl + O_3 \rightarrow ClO + O_2$$
$$ClO + O \rightarrow Cl + O_2$$

These reactions both destroy ozone and reduce the likelihood that it will reform because atomic oxygen (O) is removed as well. It is also important to note that it can take 10 to 20 years for chlorofluorocarbon molecules to get into the stratosphere, and then they can react with the ozone for up to 120 years.

Actions to Protect the Ozone Layer

Since the 1970s, when chlorofluorocarbons were linked to the depletion of the ozone layer in the upper atmosphere, their use as propellants in aerosol cans has been banned in the United States,

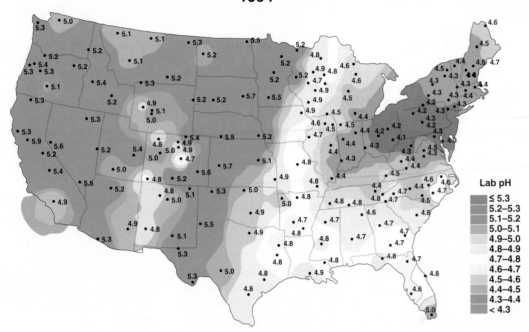

Hydrogen Ion Concentration as pH
1994

Lab pH
- ≤ 5.3
- 5.2–5.3
- 5.1–5.2
- 5.0–5.1
- 4.9–5.0
- 4.8–4.9
- 4.7–4.8
- 4.6–4.7
- 4.5–4.6
- 4.4–4.5
- 4.3–4.4
- < 4.3

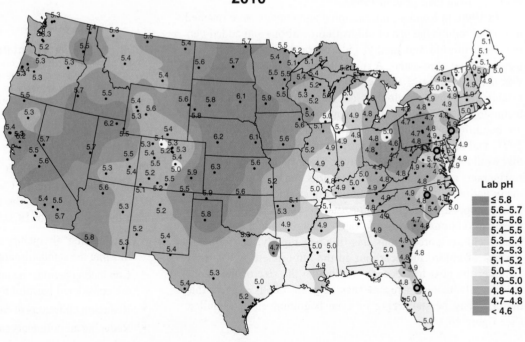

Hydrogen Ion Concentration as pH
2010

Lab pH
- ≤ 5.8
- 5.6–5.7
- 5.5–5.6
- 5.4–5.5
- 5.3–5.4
- 5.2–5.3
- 5.1–5.2
- 5.0–5.1
- 4.9–5.0
- 4.8–4.9
- 4.7–4.8
- < 4.6

FIGURE 16.17 Acid-Rain Improvements Regulations on emissions of sulfur dioxide and nitrogen oxides from power plants and automobiles has greatly reduced acid rain particularly in the northeastern United States where the problem was the greatest. Source: EPA and the National Atmospheric Deposition Program.

Canada, Norway, and Sweden, and the European Union agreed to reduce use of chlorofluorocarbons in aerosol cans. Today, aerosol cans do not contain chlorofluorocarbons.

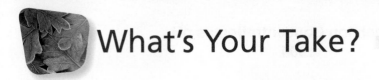

What's Your Take?

China's economy has grown continuously for over 30 years. However the country's environment and the Chinese people are paying a steep price. Air pollution is blamed for the premature death of over 1 million Chinese annually. Over the next decade Chinese consumers are expected to purchase hundreds of millions of automobiles that will add to the current air pollution problems. Chinese consumers argue that they want to enjoy the same standard of living as those in the West, and owning their own cars is part of that goal. Develop a position paper on car ownership from a Chinese citizen's perspective and another paper from the perspective of a citizen in Japan (which is downwind of China) or a citizen in North America.

In 1987, several industrialized countries, including Canada, the United States, the United Kingdom, Sweden, Norway, Netherlands, the Soviet Union, and West Germany, agreed to freeze chlorofluorocarbon and halon (used in fire extinguishers) production at current levels and reduce production by 50 percent by the year 2000. This document, known as the Montreal Protocol, was ratified by the U.S. Senate in 1988. Although the initial concerns about chlorofluorocarbons related to the problem of ozone depletion, efforts to reduce chlorofluorocarbons have been effective at removing a greenhouse gas as well. As a result of the 1987 Montreal Protocol, chlorofluorocarbon emissions dropped dramatically from their peak in 1988.

In 1990, in London, international agreements were reached to further reduce the use of chlorofluorocarbons and added carbon tetrachloride and methyl chloroform as chemicals whose use would be eliminated. A major barrier to these negotiations was the reluctance of the developed countries of the world to establish a fund to help less-developed countries implement technologies that would allow them to obtain refrigeration and air conditioning without the use of chlorofluorocarbons. In 1991, DuPont announced the development of new refrigerants that would not harm the ozone layer. These and other alternative refrigerants are now used in refrigerators and air conditioners in many nations, including the United States. In 1996, the United States stopped producing chlorofluorocarbons. As a result of these international efforts and rapid changes in technology, the use of chlorofluorocarbons has dropped rapidly, and concentrations of chlorofluorocarbons in the atmosphere will slowly fall over the next few decades. Although the size of the ozone hole has fluctuated in recent years, current trends suggest that the size may be stabilizing or even beginning to get smaller. (See figure 16.18.)

16.7 Control of Air Pollution

All of the air pollutants we have examined thus far are the result of human activity. That means their release into the atmosphere can be controlled. In the United States, implementation of the requirements of the Clean Air Act has been the primary means of controlling air pollution.

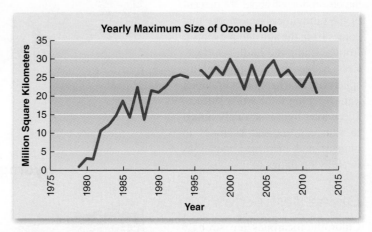

FIGURE 16.18 Size of Antarctic Ozone Hole The graph shows that the size of the Antarctic ozone hole has increased significantly between 1979 and 2000. The ban on the use of chlorofluorocarbons and other ozone-destroying chemicals appears to have had a positive effect. Since 2000, it appears that the size of the ozone hole has been stable or declining. Source: Data from NASA.

The Clean Air Act

What is commonly called the Clean Air Act is in reality several pieces of legislation. The original Clean Air Act of 1970 was amended in 1977 and again in 1990. The EPA has a number of responsibilities under the original act of 1970 and its amendments which include:

- Establishing air quality standards, developing strategies for meeting the standards, and ensuring that the standards are met.
- Conducting periodic reviews of the six criteria air pollutants that are considered harmful to public health and the environment.
- Reducing emissions of SO_2 and NO_X that cause acid rain.
- Reducing air pollutants such as PM, SO_X, and NO_X, which can reduce visibility across large regional areas, including many of the nation's most treasured parks and wilderness areas.
- Ensuring that sources of toxic air pollutants that may cause cancer and other adverse human health and environmental effects are well controlled and that the risks to public health and the environment are substantially reduced.
- Limiting the use of chemicals that damage the stratospheric ozone layer in order to prevent increased levels of harmful ultraviolet radiation.

Science, Politics, & Policy

A History of Mercury Regulations

Mercury is a metal that is liquid at room temperature. It has been known to be toxic for centuries. However, when mercury enters aquatic ecosystems it can be converted to methylmercury by the action of bacteria. The methylmercury becomes concentrated in aquatic food chains and becomes a human health problem when certain species of fish are eaten by humans. The toxic effects of methylmercury have been known since the 1930s. Since methylmercury impairs brain development, fetuses, infants, and children are particularly susceptible. Because of the level of methylmercury and other toxic materials in fish, every state has advisories against eating certain kinds of fish from certain locations.

Because of the known health effects of elemental mercury and methylmercury, the EPA successfully regulated several of the major sources of mercury emissions such as cement plants and those industries that use mercury in their manufacturing processes. The primary remaining source of mercury in the environment is the stack gases from power plants that burn coal or oil. Mercury from power plants is released into the atmosphere and enters watersheds, where it is converted to methylmercury and becomes incorporated into the bodies of organisms.

The George W. Bush administration in essence allowed power plants to be exempt from the Clean Air Act, which allowed power plants to avoid the cost of retrofitting their plants to reduce mercury emissions. In 2008, a lawsuit was filed by several environmental organizations in the U.S. Circuit Court of Appeals in Washington, D.C. The Appeals Court overturned the Bush Administration's mercury regulations and instructed the EPA to come up with a new rule. In 2009, an electric power industry group, Utility Air Regulatory Group, asked the Supreme Court to review the ruling by the Appeals Court, arguing that the Bush administration had legally decided not to regulate power plants under the Clean Air Act. The Supreme Court denied the request.

Ultimately, in 2011 the EPA announced new mercury standards for power plants (along with regulations on other hazardous materials). New and existing power plants must meet new stringent regulations that will ultimately reduce mercury emissions by over 90 percent. Because there are several kinds of power plants that use different kinds of coal and other fuels, the setting of rules was a complicated process. In general, new plants must meet much more stringent requirements than existing plants. Existing plants have up to 4 years to make the modifications needed to meet the regulations.

Implementation of the Clean Air Act and its amendments led to the establishment of a series of detailed control requirements to meet the goals of improving air quality:

1. All industries are required to obtain permits to release materials into the air.

2. All new and existing sources of air pollution are subject to national ambient air quality standards (NAAQS) established for sulfur dioxide (SO_2), nitrogen dioxide (NO_2), particulate matter (PM_{10} and $PM_{2.5}$), carbon monoxide (CO), ozone (O_3), and lead (Pb).

3. Newly constructed facilities are subject to more stringent control technology and permitting requirements than are pre-existing facilities.

4. Hazardous air pollutants are specifically identified (there are 187 substances identified) and regulated. Any source emitting 10 U.S. tons (9.1 metric tons) per year of any listed substance, or 25 U.S. tons (22.7 metric tons) of combined substances, is considered a major source and is subject to strict regulations.

5. Power plants are allowed to sell their sulfur dioxide release permits to other companies. This program encourages power plants that can easily reduce their emissions to do so. They can then sell their permits to other power plants that are having a more difficult time reducing emissions. The net result of this program has been a rapid reduction in sulfur dioxide emissions.

6. A program for the phaseout of ozone-depleting substances (chlorofluorocarbons [CFCs], halons, carbon tetrachloride, and methyl chloroform) was established.

7. In 2007, the U.S. Supreme Court ruled that greenhouse gases like CO_2 could be considered pollutants and gave the Environmental Protection Agency the power to regulate them under the Clean Air Act.

8. In 2011, the EPA established new regulations on the release of mercury and several other toxic substances. (See Science, Politics, & Policy.)

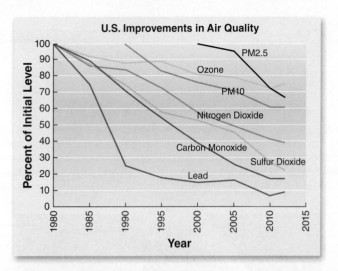

FIGURE 16.19 Improvement in Air Quality Many initiatives to improve air quality have been effective. Between 1980 and 2012, all of the major air pollutants have decreased significantly. The data for PM$_{10}$ are for the years 1990 to 2012 and the data for PM$_{2.5}$ are for the years 2000 to 2012, since the EPA did not begin collecting data on PM$_{10}$ until 1990 and PM$_{2.5}$ until 2000. The actual data for each pollutant were converted to the percent of the initial level for which data were available. This allowed all pollutants to be shown on the same graph in a meaningful way.

Source: U.S. Environmental Protection Agency.

Actions That Have Reduced Air Pollution

The improvement in air quality since the enactment of the Clean Air Act is a major success story. As a result of implementing the requirements of the Clean Air Act, a variety of pollution control mechanisms were effectively employed and air quality has improved significantly. (See figure 16.19.)

Motor Vehicle Emissions

Motor vehicles are the primary source of several important air pollutants: carbon monoxide, volatile organic compounds, and nitrogen oxides. In addition, ozone is a secondary pollutant of automobile use. Placing controls on the emissions from motor vehicles has resulted in a significant improvement in the air quality in North America, Western Europe, Japan, Australia, and New Zealand. Many engineering changes in automobiles have reduced the amount of VOCs that escape from the gas tank and crankcase. Modifications to the pumps at gas stations and to the filler pipes of cars have also reduced VOC emissions. Better fuel efficiency and specially blended fuels that produce less carbon monoxide and unburned organic compounds have also improved air quality. Catalytic converters reduced carbon monoxide, oxides of nitrogen, and volatile organic compounds in emissions and necessitated the use of lead-free fuel. This lead-free fuel requirement, in turn, greatly reduced the amount of lead (and other metal additives) in the atmosphere. Figure 16.19 shows that levels of carbon monoxide, nitrogen dioxide, and lead are down significantly. Ozone levels, which are dependent on VOC and NO$_2$ levels, have also fallen somewhat but still need improvement in some areas of the country—particularly California and the Northeast part of the United States.

Particulate Matter Emissions

Particulate matter comes from a variety of sources. Road dust is the major source of PM$_{10}$ particles. In addition, many industrial activities involve processes that produce dusts. Mining and other earth-moving activities, farming operations, and the transfer of grain or coal from one container to another all produce dust.

Fires (forest fires, grass fires, leaf burning, and fires from fireplaces and woodstoves, etc.) are a significant source of particulate matter—particularly PM$_{2.5}$. Burning of fossil fuels is another major source of particulate matter. Because of air quality regulations, industries have done much to reduce the amount of particulate matter released from the burning of fossil fuels. Various kinds of devices can be used to trap particles so they do not escape from the smokestack. These devices are very effective, and particulate matter from the burning of fossil fuels is greatly reduced. However, the smaller particles that form from gaseous emissions (SO$_2$, NO$_X$) are still a problem. Diesel engines are a significant source of particulate matter, and the gaseous emissions from motor vehicles contribute to the formation of smaller particles in the same way that industrial sources do. Recent changes in regulation of diesel engines will reduce the amount of particulate matter released.

While industrial activities, motor vehicle use, and land-use practices are major sources of particulate matter, many individual personal activities are also important. Many people in the world use wood as their primary source of fuel for cooking and heating. In developed countries such as the United States and Canada, some people use fireplaces and wood-burning stoves as a primary source of heat, but most use them for supplemental heat or for aesthetic purposes. However, the use of large numbers of wood-burning stoves and fireplaces can generate a significant air pollution problem. Some municipalities with air pollution problems regulate or ban the use of wood-burning stoves and fireplaces. High-efficiency wood-burning stoves significantly reduce the amount of particulate emissions. Figure 16.19 shows that the PM$_{10}$ level has decreased about 40 percent and the PM$_{2.5}$ level has decreased over 30 percent since data began to be collected.

Power Plant Emissions

The primary pollutants associated with electric power plants are particulates, sulfur dioxide (SO$_2$), nitrogen oxides (NO$_X$), and mercury. Most PM$_{10}$ emissions have been controlled with filters and other mechanical means. The control of sulfur dioxide requires more fundamental changes to the way electricity is produced. The EPA approached the problem by setting limits and allowing electric utilities to decide which options were the best for them. Electric utilities have used several strategies for meeting their emissions limits. Switching from a high-sulfur to a low-sulfur coal reduces the amount of sulfur dioxide released into the atmosphere by 66 percent. Chemical or physical treatment of coal to remove sulfur before it is burned can remove nearly 40 percent of the sulfur. Scrubbing the gases emitted from a smokestack is a third alternative. The technology is available, but, of course, these control devices are costly to install, maintain, and operate. Switching to natural gas, oil, or nuclear fuels reduces sulfur dioxide emissions even more. In recent years, the low price for natural gas has resulted in an increased use of natural gas by power plants.

Nitrogen oxides are produced wherever high temperature combustion occurs with air. Control of the release of nitrogen oxides typically involves removing the nitrogen oxides from the stack with chemical processes or using catalysts to encourage the breakdown of nitrogen oxides to nitrogen and oxygen. These processes are technical and expensive.

However, emissions of particulates, sulfur dioxide, and nitrogen oxides from power plants have been reduced significantly. Figure 16.19 shows that sulfur dioxide levels have fallen by nearly 80 percent. In some cases, this has resulted in an increased cost to consumers as the cost of improving emission control technology was passed along as a rate increase to the consumer.

Since the requirements to limit the release of mercury from power plants were established in 2011 and existing power plants have up to 4 years to meet the requirements, we will need to wait a few years to see improvements in mercury releases.

16.8 Air Pollution in the Developing World

Although much of the economically developed world has made great strides in improving air quality, this is not true in the less-developed countries of the world. In 2011 the World Health Organization published a study of the impact of outdoor air

pollution on world health. The World Health Organization estimated that worldwide there were about 1.3 million premature deaths that could be attributed to outdoor air pollution. Over 800,000 of those premature deaths were in Asia and the Eastern Mediterranean areas that include many of the developing economies of the world. They used data submitted by countries that stated the annual mean PM_{10} for cities. Some countries had data for many cities and others had few. Figure 16.20 shows the cities and the levels of PM_{10} pollution present. The map shows that much of the economically developed world has low PM_{10} pollution ($29 \mu g/m^3$ or less), while much of the developing world has much higher PM_{10} levels. In particular, the newly industrialized regions of Eastern Europe, India, China, and the Middle East have high levels of PM_{10} pollution.

Air pollution in much of the developing world is dominated by poorly regulated industry and transportation sections of the economy. It appears that the normal progress of economic development begins with the development of industry, which uses cheap labor and lax environmental regulations to manufacture products inexpensively. People and their governments are so eager to have the jobs and economic development that they are willing to accept the risks associated with poor environmental conditions. The pollution is not just the result of industry; since the people typically use solid fuels (coal, wood, dung) to heat their homes and cook their food, the smoke released from their homes adds to the particulate pollution. However, as people become better off economically, they demand better living conditions, which include better working conditions with fewer hours and higher pay, and an environment with less pollution. The opening story about changes in air pollution in Mexico City is an example of this change in attitude.

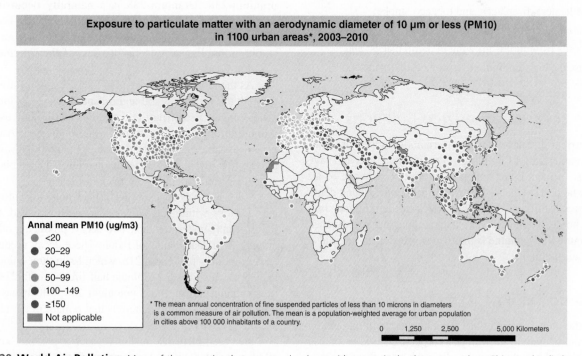

Exposure to particulate matter with an aerodynamic diameter of 10 μm or less (PM10) in 1100 urban areas*, 2003–2010

Annal mean PM10 (ug/m3)
- <20
- 20–29
- 30–49
- 50–99
- 100–149
- ≥150
- Not applicable

* The mean annual concentration of fine suspended particles of less than 10 microns in diameters is a common measure of air pollution. The mean is a population-weighted average for urban population in cities above 100 000 inhabitants of a country.

0   1,250   2,500   5,000 Kilometers

FIGURE 16.20 World Air Pollution Many of the countries that are experiencing rapid economic development such as China and India have high levels of air pollution.
Source: Data from World Health Organization.

16.9 Indoor Air Pollution

Even though we spend on average almost 90 percent of our time indoors, the movements to reduce indoor air pollution lag behind regulations governing outdoor air pollution. In the United States, the Environmental Protection Agency is conducting research to identify and rank the human health risks that result from exposure to individual indoor pollutants or mixtures of multiple indoor pollutants.

A growing body of scientific evidence indicates that the air within homes and other buildings can be more seriously polluted than outdoor air in even the largest and most industrialized cities. Many indoor air pollutants and pollutant sources are thought to have an adverse effect on human health.

Sources of Indoor Air Pollutants

Indoor air pollutants come from a variety of sources:

- Asbestos fibers from older ceiling and floor tiles and insulation;
- Formaldehyde, which is associated with many consumer products, including certain wood products and aerosols;
- Radon in certain parts of the world with the appropriate geology;
- Lead from lead-based paints in older homes;
- Volatile organic compounds such as airborne pesticide residues, perchloroethylene (associated particularly with dry cleaning), paradichlorobenzene (from mothballs and air fresheners), and gases from cleaning agents, polishes, and paints;
- Carbon monoxide and nitrogen dioxide from stoves, furnaces, kerosene heaters, fireplaces, and tobacco smoke;
- Particulate matter from burning materials;
- Mold spores from molds growing in damp places; and
- Other disease-causing or allergy-producing organisms.

Significance of Weatherizing Buildings

A recent contributing factor to the concern about indoor air pollution is the weatherizing of buildings to reduce heat loss and save on fuel costs. In most older homes, a complete exchange of air occurs every hour. This means that fresh air leaks in around doors and windows and through cracks and holes in the building. In a weatherized home, a complete air exchange may occur only once every five hours. Such a home is more energy efficient, but it also tends to trap air pollutants.

Secondhand Smoke

Smoking is the most important air pollutant source in the United States in terms of human health. The surgeon general estimates that 443,000 people in this country die each year from emphysema, heart attacks, strokes, lung cancer, or other diseases caused by tobacco smoking. Banning smoking probably would save more lives than would any other pollution-control measure.

Millions of nonsmoking people are exposed to environmental tobacco smoke (secondhand smoke) because they live and work in spaces where people smoke. The U.S. Environmental Protection Agency estimates that approximately 50,000 nonsmokers die each year of lung cancer or heart disease as a result of breathing air that contains secondhand smoke. About 18 percent of children are exposed to secondhand smoke because their parents smoke. They are much more likely to have respiratory infections than children not exposed.

In July 1993, the EPA recommended several actions to prevent people from being exposed to secondhand indoor smoke, including that:

- People not smoke in their homes or permit others to do so
- All organizations that deal with children have policies that protect children from secondhand smoke
- Every company have a policy that protects employees from secondhand smoke
- Smoking areas in restaurants and bars be placed so that the smoke will have little chance of coming into contact with nonsmokers

Today only 10 states do not have some kind of a statewide smoking ban in public places. Even in states that do not have legislation to protect people from secondhand smoke, many cities and other jurisdictions within those states do.

Radon

Radon-222, an inert radioactive gas with a half-life of 3.8 days, is one of the products formed during the breakdown of uranium-238. Uranium-238 is a naturally occurring element that makes up about 3 parts per million of the Earth's crust. Uranium-238 goes through 14 steps of decay before it becomes stable, nonradioactive lead-206. One of the intermediate steps is radon-222.

As the radon gas is formed in the rocks, it usually diffuses up through the rocks and soil and escapes harmlessly into the atmosphere. It can also diffuse into groundwater. Radon usually enters a home through an open space in the foundation. A crack in the basement floor or the foundation, the gap around a water or sewer pipe, or a crawl space allows the radon to enter the home. It may also enter in the water supply from wells.

Since radon is an inert gas, it does not enter into any chemical reactions within the body, but it can be inhaled. Once in the lungs, it may undergo radioactive decay, producing other kinds of atoms called "daughters" of radon. These decay products (daughters) of radon—plutonium 218, which has a three-minute half-life; lead 214, which has a 27-minute half-life; bismuth 214, which has a 20-minute half-life; and polonium 214, which has a millisecond half-life—are solid materials that remain in the lungs and are chemically active.

Increased incidence of lung cancer is the only known health effect associated with radon decay products. It is estimated that the decay products of radon are responsible for about 20,000 lung cancer deaths annually in the United States. This is about 10 percent of lung cancer deaths. The radon risk evaluation chart indicates that individuals exposed to greater than 4 picocuries per liter of air are at increased risk of developing lung cancer. (See table 16.2.)

About 10 percent of the homes in the United States have a potential radon problem. Figure 16.21 shows those regions of the United States that are likely to have elevated levels of radon. In addition, the Environmental Protection Agency and the U.S. surgeon general recommend that all Americans (other than those living in apartment buildings above the second floor) test their homes for radon. If the tests indicate radon levels at or above 4 picocuries per liter, the EPA recommends that the homeowner take action to lower the level. This is usually not expensive and consists of blocking the places where radon is entering or venting radon sources to the outside. People who are concerned about radon should contact their state's public health department or environmental protection agency.

Table 16.2 Radon Risk Evaluation Chart

Radon Level Picocuries/Liter (pCi/L)	Estimated Lung Cancer Deaths per 1,000 (Nonsmoker)	Estimated Lung Cancer Deaths per 1,000 (Smoker)	Comments
20 pCi/L	36/1,000	260/1,000	
10 pCi/L	18/1,000	150/1,000	
8 pCi/L	15/1,000	120/1,000	Lifetime risk of dying in auto accident 15/1,000
4 pCi/L	7/1,000	62/1,000	Should fix the problem
2 pCi/L	4/1,000	32/1,000	Reducing levels below 2 pCi/L is difficult
1.3 pCi/L	2/1,000	20/1,000	Average indoor radon level
0.4 pCi/L		3/1,000	Average outdoor radon level

16.10 Noise Pollution

Noise is referred to as unwanted sound. However, noise can be more than just an unpleasant sensation. Research has shown that exposure to noise can cause physical, as well as mental, harm. (See figure 16.22.) The loudness of the noise is measured by **decibels** (db). Decibel scales are logarithmic rather than linear. Thus, the change from 40 db (a library) to 80 db (a dishwasher or garbage disposal) represents a ten-thousandfold increase in sound loudness.

The frequency or pitch of a sound is also a factor in determining its degree of harm. High-pitched sounds are the most annoying. The most common sound pressure scale for high-pitched sounds is the A scale, whose units are written "dbA." Hearing loss begins with prolonged exposure (eight hours or more per day) to 80 or 90 dbA levels of sound pressure. Sound pressure becomes painful at around 140 dbA and can kill at 180 dbA. (See table 16.3.)

In addition to hearing loss, noise pollution is linked to a variety of other ailments, ranging from nervous tension headaches to neuroses. Research has also shown that noise may cause blood vessels to constrict (which reduces the blood flow to key body parts), disturbs unborn children, and sometimes causes seizures in epileptics. The U.S. Environmental Protection Agency has estimated that noise causes about 40 million U.S. citizens to suffer hearing damage or other mental or physical effects. Up to 64 million people are estimated to live in homes affected by aircraft, traffic, or construction noise.

The Noise Control Act of 1972 was the first major attempt in the United States to protect the public health and welfare from detrimental noise. This act also attempted to coordinate federal research and activities in noise control, to set federal noise emission standards

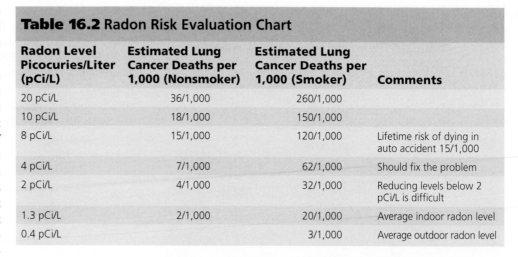

FIGURE 16.21 Generalized Geologic Radon Potential of the United States The EPA recommends that people check their homes for radon. Those regions of the country shown in pink have the highest likelihood of elevated radon levels.
Source: U.S. Geological Survey.

FIGURE 16.22 Hearing Protection People whose jobs require them to work with noisy machinery often wear devices to prevent hearing loss.

Table 16.3 Intensity of Noise

Source of Sound	Intensity in Decibels
Jet aircraft at takeoff	145
Pain occurs	140
Hydraulic press	130
Jet airplane (160 meters overhead)	120
Unmuffled motorcycle	110
Subway train	100
Farm tractor	98
Gasoline lawn mower	96
Food blender	93
Heavy truck (15 meters away)	90
Heavy city traffic	90
Vacuum cleaner	85
Hearing loss after long exposure	85
Garbage disposal unit	80
Dishwasher	65
Window air conditioner	60
Normal speech	60

for commercial products, and to provide information to the public. After the passage of the Noise Control Act, many local communities in the United States enacted their own noise ordinances. While such efforts are a step in the right direction, the United States is still controlling noise less than are many European countries. Several European countries have developed quiet construction equipment in conjunction with strongly enforced noise ordinances. The Germans and Swiss have established maximum day and night noise levels for certain areas. Regarding noise pollution abatement, North America has much to learn from European countries.

Issues & Analysis

Pollution, Policy, and Personal Choice

The air pollution problem that has shown the least improvement in North America is the control of ground-level ozone. Some geographic regions are more prone to developing conditions that generate ground-level ozone, and we know that the presence of nitrogen oxides and volatile organic compounds triggers the events leading to the development of unhealthy levels of ground-level ozone. Furthermore, it is clear that automobiles are the major source of the volatile organic compounds and nitrogen oxides.

It is also clear that a large proportion of the population is regularly exposed to unhealthy levels of ozone and that many millions of people become ill or have their activities restricted as a direct result of poor air quality.

There are several ways to attack this problem, but all of them involve restricting the freedom of people to use automobiles or restricting the kinds of automobiles they can drive. The following list describes some of the approaches used and their effects on the driving public.

1. *Restrict the kinds of cars that people can buy to less-polluting types.* Requirements that automobiles obtain better mileage with reduced emissions have successfully produced less pollution per mile driven. However, the number of miles driven is increasing, which negates some of the gains attributable to higher-mileage vehicles.

2. *Provide incentives to encourage fewer people to drive.* Many metropolitan areas have carpool lanes restricted to cars with two or more passengers. These lanes are largely unused. People choose to drive separately.

3. *Provide economic penalties for the use of automobiles.* In many cities of the world, people pay a fee to drive in parts of the city. In addition, tolls on roads and bridges are an economic incentive not to drive. However, this only reduces traffic if the fees are high and there are no "free" roads that people can use.

4. *Provide affordable, safe, and convenient alternative means of transport.* For every dollar spent on public transport in the United States, about $6 are spent on roads for cars.

These approaches to dealing with ground-level ozone raise several issues that affect personal choice.

- Should the use of automobiles be restricted to protect the health of people who are highly sensitive to ground-level ozone?

- Should governments determine what kinds of automobiles are allowed?

- Should fees, taxes, or assessments be used to increase the cost of using a car?

- A typical driver spends $1,000 or more per year for gasoline. Would you pay an annual fee of $1,000 to use public transport if it were available?

Summary

The atmosphere has a tremendous ability to disperse pollutants. Carbon monoxide, hydrocarbons, particulate matter, sulfur dioxide, and oxides of nitrogen compounds are the primary air pollutants. They can cause a variety of health problems. The U.S. Environmental Protection Agency established standards for six pollutants known as criteria air pollutants. They are carbon monoxide, nitrogen dioxide, sulfur dioxide, volatile organic compounds (hydrocarbons), ozone, and lead. The EPA also regulates hazardous air pollutants.

Photochemical smog is a secondary pollutant, formed when hydrocarbons and oxides of nitrogen are trapped by thermal inversions and react with each other in the presence of sunlight to form peroxyacetyl nitrates and ozone. Elimination of photochemical smog requires changes in technology, such as more fuel-efficient automobiles, special devices to prevent the loss of hydrocarbons, and catalytic converters to more completely burn hydrocarbons in exhaust gases.

Acid rain is caused by emissions of sulfur dioxide and oxides of nitrogen into the atmosphere, which form acids that are washed from the air when it rains or snows or settle as particles on surfaces. Direct effects of acid rain on terrestrial ecosystems are difficult to prove, but changes in many forested areas are suspected of being partly the result of additional stresses caused by acid rain. Recent evidence suggests that loss of calcium from the soil may be a major problem associated with acid rain. The effect of acid rain on aquatic ecosystems is easy to quantify. As waters become more acidic, the complexity of the ecosystem decreases, and many species fail to reproduce. The control of acid rain requires the use of scrubbers, precipitators, and filters—or the removal of sulfur from fuels. However, oxides of nitrogen are still a problem.

Significant impacts on our health, the vitality of forests and other natural areas, the distribution of freshwater supplies, and the productivity of agriculture are among the probable consequences of climate change. Chlorofluorocarbons also lead to the destruction of ozone in the upper atmosphere, which results in increased amounts of ultraviolet light reaching the Earth. Concern about the effects of chlorofluorocarbons has led to international efforts that have resulted in significant reductions in the amount of these substances reaching the atmosphere. Many commonly used materials release gases into closed spaces (indoor air pollution), where they cause health problems. The most important of these health problems are associated with tobacco smoking. Radon gas is also an important indoor air pollutant in certain parts of the world.

Acting Green

1. You can reduce automotive emissions by driving less or joining a ride carpool.
2. You can reduce emissions from power plants by reducing your electrical energy use.
3. Dispose of old CFC-containing air conditioners, refrigerators, freezers, and dehumidifiers appropriately.
4. Don't burn leaves and debris outdoors—let them decay naturally.
5. Use only water-based paints, which do not release VOCs.
6. Sit on a city street and record the two dominant sounds, two dominant sights, and two dominant smells. Visit a natural area in a park, nature center, or in the countryside. Make a similar list and compare your two lists.

Review Questions

1. Name the two primary gases in the atmosphere.
2. Describe two ways the gases in the troposphere differ from those in the stratosphere.
3. Describe two ways the atmosphere can get rid of pollutants.
4. List the five primary air pollutants commonly released into the atmosphere and their sources.
5. List the six criteria air pollutants, their sources, and their effects.
6. Define *secondary air pollutants* and give an example.
7. How is each of the following involved in the production of photochemical smog: volatile organic compounds, nitrogen oxides, thermal inversions, sunlight, automobiles, and ozone?
8. Why do some cities have greater problems with smog than others?
9. Describe three regulatory actions of the EPA that have significantly improved air quality and why they improved air quality.
10. What molecules produce acid rain and how are they produced?
11. What are the primary effects of acid rain on terrestrial and aquatic ecosystems?
12. Why is stratospheric ozone important?
13. What was done to protect stratospheric ozone?
14. What are the National Ambient Air Quality Standards?
15. Give an example of a hazardous air pollutant.
16. Explain why air pollution problems in economically developing countries are different from those in developed countries.
17. How does radon enter a home?
18. Why do buildings often have poor air quality?
19. Define noise.

Critical Thinking Questions

1. What could you do to limit the air pollution you create?
2. As a nation, the United States provides many subsidies to make energy cheap because policymakers feel that economic development depends on cheap energy. If these subsidies were withdrawn or taxes on energy were added, what effect would this have on your own energy consumption? Would you be willing to support high gasoline prices, in the $7 to $10-a-gallon range as in many European countries, if it would cut greenhouse gas emissions?
3. Why do you think air pollution is so much worse in developing countries than in developed countries? What should developed countries do about this, if anything?
4. What common indoor air pollutants are you exposed to? What can you do to limit this exposure?
5. Is it possible to have zero emissions of pollutants? What level of risk are you willing to live with?

To access additional resources for this chapter, please visit ConnectPlus® at connect.mheducation.com There you will find interactive exercises, including Google Earth™, additional Case Studies, and SmartBook™, an adaptive eBook that integrates our LearnSmart® adaptive learning technology.

Chapter 17

Climate Change:
A Twenty-first Century Issue

One of the obvious effects of climate change is the retreat of glaciers throughout the world. This photo shows the Bear Glacier in Kenai Fjords National Park in Alaska. The Bear Glacier retreated over 3 kilometers (about 2 miles) between 2002 and 2007.

CHAPTER OUTLINE

OBJECTIVES

After reading this chapter, you should be able to:

- Recognize that climate has changed in the geologic past.
- Understand that human activities can alter the atmosphere in such a way that these activities can change climate.
- List at least five kinds of evidence that support the concept that the Earth is getting warmer and that climate is changing.
- Describe how greenhouse gases cause the Earth to be warmer than it would

be if the greenhouse gases were not present.
- List the primary greenhouse gases and their sources.
- Describe the kinds of changes likely to occur as a result of climate change.
- Describe changes in the hydrologic cycle that are likely to occur with continued increases in temperature.
- Explain how increased carbon dioxide in the atmosphere is altering the ocean.

- State two factors that lead to the rise of sea levels with global warming.
- List three ways that an increased temperature of the Earth could affect human health.
- Explain how climate change is likely to change ecosystems and agriculture.
- Describe the kinds of actions that could reduce the release of greenhouse gases.

Bangladesh and Climate Change

Bangladesh is considered to be one of the countries most vulnerable to the effects of climate change. A combination of geography and climate place Bangladesh in the position of being highly susceptible to changes that could be brought about by continuing climate change. Seventy-five percent of the land is less than 10 meters above sea level. (Ten meters is about the height of a three-story building.) Three major rivers, the Ganges (Padma), Brahmaputra (Jamuna), and Meghna, and about 700 tributaries dominate the geography of the country and empty into the Bay of Bengal. About 80 percent of the land is flood plain. While the land area of Bangladesh is slightly less than 150,000 square kilometers (about the size of Iowa), the rivers that flow through Bangladesh drain an area of about 1.5 million square kilometers. This means that rainfall over an area ten times larger than the area of the country is ultimately funneled through the country of Bangladesh.

Bangladesh has a tropical monsoon climate with heavy rainfall during the June to November period and a warm, dry period from December to February. During the rainy season some parts of the country receive as much as 4,000 millimeters (160 inches) of rainfall. In addition, on average, one tropical cyclone per year impacts Bangladesh. The heavy rains and storm surges associated with cyclones result in widespread flooding. This combination of high rainfall and a geography dominated by rivers leads to annual flooding. This flooding typically affects 20 percent of the country, and major flooding can affect over 60 percent of the country.

A dense and rapidly growing population and a high degree of poverty make it difficult for people to adapt to a changing climate.

Bangladesh has a population of 157 million people and a population density of about 1,000 people per square kilometer (about 2,600 people per square mile). About 80 percent of the population lives in rural areas and about 60 percent are directly involved in agriculture. Furthermore about half of the population lives on US $1.25 per day.

The Bangladeshi government recognizes the potential of climate change to displace people and cause disruption to agriculture. In 2009 it developed a Bangladesh Climate Change Strategy and Action Plan, which seeks ways to reduce the impact of climate change on their people and economy.

17.1 Earth Is a Greenhouse Planet

The Earth is unique among the planets in having a temperature that allows for water to exist in a liquid form. Since all living things are primarily made up of water—the human body is over 50 percent water—and all early forms of life developed in the oceans, the presence of liquid water is essential for life as we know it. The Earth's temperature is determined by several factors: Earth's distance from the sun, changes in the energy output of the sun, and the presence of carbon dioxide in the atmosphere. Although Earth's orbit around the sun results in small changes in the distance between the Earth and sun and the sun's energy changes slightly on about an 11-year cycle, these differences appear to have little effect on Earth's temperature. However, several gases in the atmosphere are transparent to ultraviolet and visible light but absorb infrared radiation. These gases allow sunlight to penetrate the atmosphere and be absorbed by the Earth's surface. This sunlight energy is reradiated as infrared radiation (heat), which is absorbed by the greenhouse gases in the atmosphere. Because the effect is similar to what happens in a greenhouse (the glass allows light to enter but retards the loss of heat), these gases are called greenhouse gases, and the warming that occurs because of their presence is called the **greenhouse effect.** (See figure 17.1.)

Carbon dioxide is a greenhouse gas. Although the amount of carbon dioxide in the atmosphere is small, its effect is significant. According to NASA, if there were no carbon dioxide in the atmosphere, Earth's temperature would be about −18°C (0°F) instead of the current temperature of about 15°C (60°F). Thus the greenhouse effect caused by the presence of carbon dioxide in the atmosphere makes Earth suitable for life. Although the planet Venus is closer to the sun than Earth, it has a thick layer of clouds that tends to reflect the sun's energy. However, it has an atmosphere that is about 96 percent carbon dioxide. If there were no carbon dioxide in Venus's atmosphere, the surface temperature would be about −42°C (−44°F). However, the greenhouse effect caused by its carbon dioxide atmosphere results in a temperature of about 460°C (860°F). Clearly an increase in the amount of carbon dioxide in the atmosphere of any planet will cause the planet to be warmer.

17.2 Geologic Evidence for Global Warming and Climate Change

Although the phrases *global warming* and *climate change* are sometimes used interchangeably, they are really different aspects of the same problem. Global warming relates to an average increase in

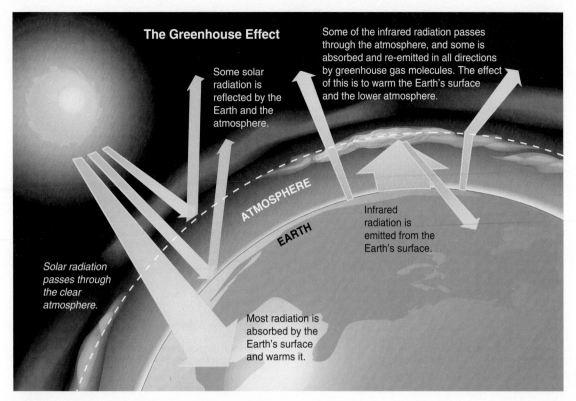

FIGURE 17.1 Greenhouse Effect Certain gases present in the atmosphere (carbon dioxide, methane, and others), known as greenhouse gases, allow light to reach the Earth from the sun but absorb infrared radiation from the Earth and reradiate some of it back to Earth. This results in a warming of the Earth. Without the greenhouse effect Earth would be much colder.

Source: Data from *Climate Change—State of Knowledge*. October 1997. Office of Science and Technology Policy, Washington, D.C.

temperature of the Earth's atmosphere, while climate change refers to the many other changes that come about because of global warming.

Like many new ideas in science, the idea that human activities could raise the Earth's temperature and change climate developed over more than a century. Geologic studies of Earth's history gave clues that climate had changed significantly over time. Some geologic periods were hotter than today and some were colder. Evidence of continental glaciers that covered large portions of North America, Europe, and Asia brought about a recognition that climate change had happened in the recent geologic past. (See figure 17.2.) The last ice age ended about 10,000 years ago. Studies of the remains of plants and animals showed that the climate present during the retreat of the glaciers was colder than present. In addition, it was recognized that the volume of water tied up in glaciers caused sea levels to fall. When scientists looked for events that could trigger changes in climate, they identified volcanos, changes in the sun's solar radiation, and a greenhouse effect. By the late 1800s the concept that atmospheric gases such as water vapor and carbon dioxide could "trap" heat was well established.

17.3 Growth in Knowledge of Climate Change

By the 1980s, scientists began to notice that the average temperature of the Earth was increasing and looked for causes for the change. That trend has continued to the present. (See figure 17.3.) It was clear that

FIGURE 17.2 Continental Glaciers Within the last 100,000 years, large glaciers covered much of northern North America and Europe. This is evidence that past climates were different from today's.

the Earth has had changes in its average temperature many times in the geologic past before humans were present. So, scientists initially tried to determine if the warming was a natural phenomenon or the result of human activity. Several gases such as carbon dioxide,

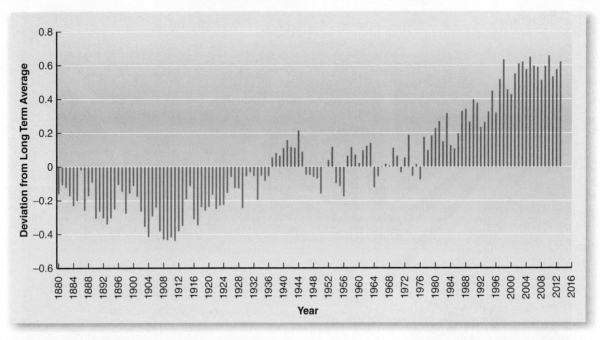

FIGURE 17.3 Changes in Average Global Temperature The graph shows how the average surface temperature of the Earth has deviated from the long-term average since 1880. Since 2001 the temperature has been over a half degree Celsius warmer than the long-term average.
Source: Data from NOAA Climate.gov.

chlorofluorocarbons, methane, and nitrous oxide are known as **greenhouse gases** because they let sunlight enter the atmosphere but slow the loss of heat from the Earth's surface. Evidence of past climate change going back as far as 160,000 years indicates a close correlation between the concentration of greenhouse gases in the atmosphere and global temperatures. Computer simulations of climate indicate that global temperatures will rise as atmospheric concentrations of greenhouse gases increase, and there are many other outcomes predicted by an increase in temperature. Since these predictions are based on computer models of climate, some scientists criticized them as being inaccurate and constructed from sketchy data.

Constructing computer models of climate is extremely difficult. The models must account for changes in the flow of heat between the atmosphere and oceans, winds and ocean currents, precipitation patterns, the impact of volcanic eruptions on climate, alterations in the amount of energy coming from the sun, albedo (amount of energy reflected from the Earth's surface), El Niño and El Niña, and many other factors. As with any new area of research, models improve as new information is discovered and better calculations are entered into the model.

In addition to improved models of climate, many areas of research have added data important to understanding the causes and effects of climate change. (See figure 17.4.) Some of these include:

- Records of the amount of carbon dioxide in the atmosphere that show steady increase in the amount of carbon dioxide.
- Studies of gas bubbles trapped in glaciers that indicate what the atmosphere was like before the time of the industrial revolution that began in the mid 1700s.
- Satellite photos that show how snow and ice conditions change.
- Migration behavior of terrestrial and marine animals that shows changes in the time of migration or the route followed.

- Ocean studies of CO_2 content, pH change, and other changes in chemistry.
- Changes in growing seasons.
- Physical measurements of the retreat of glaciers and thickness of ice sheets.
- Effects of increased carbon dioxide on photosynthesis.
- Wind patterns.
- Ocean currents.
- Effects of particulates from natural (wind erosion, volcanos) and human activities on climate.
- Sea level measurements.
- Frequency and strength of tropical storms.

Central to all the studies related to climate change was an understanding of the principal greenhouse gases and their contribution to warming.

17.4 Sources and Impacts of Principal Greenhouse Gases

The most important greenhouse gases are carbon dioxide (CO_2), methane (CH_4), chlorofluorocarbons (primarily CCl_3F and CCl_2F_2), and nitrous oxide (N_2O). Table 17.1 lists the relative contribution of each of these gases to the potential for global warming.

Carbon dioxide (CO_2) is the most abundant of the greenhouse gases and is responsible for about 64 percent of global warming. It occurs as a natural consequence of respiration and fermentation by organisms. However, much larger quantities are put into the atmosphere as a waste product of energy production.

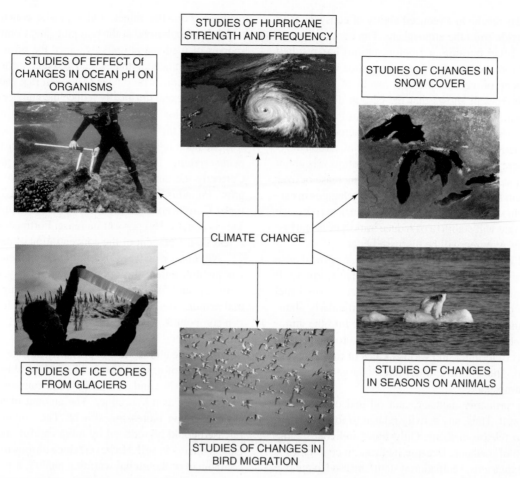

FIGURE 17.4 Research Related to Climate Change The recognition that climate change is a serious concern has led to the many areas of research related to the effects of climate change.

Table 17.1 Principal Greenhouse Gases

Greenhouse Gas	Pre-1750 Concentration (ppm)	2013 Concentration (ppm)	Contribution to Global Warming (percent)*	Principal Sources
Carbon dioxide (CO_2)	280	396.5	64	• Burning of fossil fuels • Deforestation
Methane (CH_4)	0.70	1.8	18	• Produced by bacteria in wetlands, rice fields, and guts of livestock • Release from fossil fuel use
Chlorofluorocarbons (CFCs)	Zero	0.00083	8	• Release from foams, aerosols, refrigerants, and solvents
Nitrous oxide (N_2O)	0.270	0.326	6	• Use of fertilizer and manure in agriculture • Burning of fossil fuels

Does not total 100 percent because a number of minor gases collectively contribute about 4 percent to the total warming.

Source: Data from NOAA Earth System Research Laboratory.

Coal, oil, natural gas, and biomass are all burned to provide heat and electricity for industrial processes, home heating, and cooking. The burning of any carbon-containing material results in the formation of carbon dioxide.

Another factor contributing to the increase in the concentration of carbon dioxide in the atmosphere is deforestation. Trees and other vegetation remove carbon dioxide from the air and use it for photosynthesis. Since trees live for a long time, they effectively tie up carbon in their structure. Cutting down trees to convert forested land to other uses (primarily agriculture), releases this carbon. Furthermore, agriculture relies primarily on plants that live one year and do not store carbon for long periods. Thus, a reduction in forests results in a reduced ability of ecosystems to remove carbon dioxide from the atmosphere. The combination of these factors (fossil-fuel burning and deforestation) has resulted in an increase in the concentration of carbon dioxide in the atmosphere. Measurement of carbon dioxide levels at the Mauna Loa Observatory in Hawaii shows that the carbon dioxide level increased from about 316 parts per million (ppm) in 1958 to about 396.5 ppm in 2013. (See figure 17.5.) This is an increase of about 25 percent. It is generally accepted that the amount of carbon dioxide in the atmosphere prior to the industrial revolution was about 280 ppm. The current concentration represents an increase of over 40 percent over preindustrial concentrations. Since changes in carbon dioxide levels in the atmosphere are due to human activity, we can make changes that will stabilize or reduce atmospheric carbon dioxide. The actions required will be discussed later.

Methane (CH₄) is the second most abundant greenhouse gas and is responsible for about 18 percent of global warming. It comes from biological sources and as a byproduct of fossil-fuel use. Several kinds of microorganisms that are particularly abundant in wetlands and rice paddies release methane into the atmosphere. Methane-releasing microorganisms are also found in large numbers in the guts of termites and various kinds of ruminant animals such as cattle. Thus, these animals release large amounts of methane into the atmosphere.

Natural gas is primarily methane, and oil and coal, contain some methane as well. Thus, any activity related to fossil fuel use has the potential to release methane. Oil drilling and coal mining result in the release of methane. Because methane in coal mines is often the cause of explosions, ventilation systems are used to remove methane from the mines. Oil typically contains some methane, which is often burned at the pumping site. Control of the biological sources of methane is unlikely, since the primary sources involve agricultural practices that would be very difficult to change. For example, nations would have to convert rice paddies to other forms of agriculture and drastically reduce the number of animals used for meat production. Neither is likely to occur, since food production in most parts of the world needs to be increased, not decreased.

Methane release from fossil fuel use can be reduced through better systems for preventing losses from production facilities. Currently the amount of methane in the atmosphere continues to grow. Preindustrial concentrations were about 700 parts per billion (ppb). Current concentrations are about 1,800 ppb. (See figure 17.6.) This is about a 160 percent increase. Fortunately its concentration is less than 1 percent of the concentration of carbon dioxide.

Nitrous oxide (N₂O), a minor component of the greenhouse gas picture, enters the atmosphere primarily through the use of fertilizers and fossil fuels. Nitrogen-containing fertilizers and animal manure used to improve agricultural production contain nitrogen compounds. Certain soil bacteria convert these compounds to nitrous oxide.

Since the atmosphere is primarily nitrogen and oxygen, when this mixture of gases is involved in the combustion of fossil fuels, the nitrogen (N_2) and oxygen (O_2) can combine to produce nitrous oxide (N_2O) as a byproduct. The amount of nitrous oxide in the atmosphere is increasing slowly. The amount of nitrous oxide released could be reduced by more careful management of nitrogen fertilizers in soil. However, since nitrous oxide is such a minor component of the global warming picture, it is not a priority item.

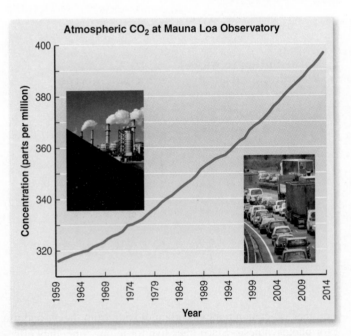

FIGURE 17.5 Changes in Atmospheric Carbon Dioxide Since the establishment of a carbon dioxide monitoring station at Mauna Loa Observatory in Hawaii, the amount of carbon dioxide in the atmosphere has increased every year.

Data from: NOAA Earth System Research Laboratory

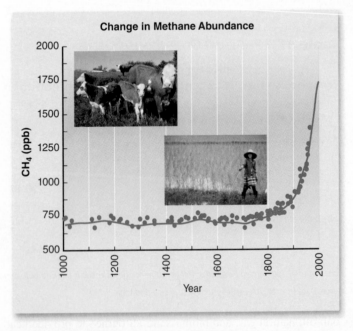

FIGURE 17.6 Changes in Atmospheric Methane Methane comes primarily from methane-producing microorganisms, which are common in the digestive tracts of ruminant animals like cattle and in wet soil like that found in swamps and rice paddies, although some is released from fossil fuel processing facilities.

Source: Data from UN Intergovernmental Panel on Climate Change.

Chlorofluorocarbons (CFCs) are also a minor component of the greenhouse gas picture and are synthetic compounds produced for particular uses. There are no natural sources of CFCs. CFCs were widely used as refrigerant gases in refrigerators and air conditioners, as cleaning solvents, as propellants in aerosol containers, and as expanders in foam products.

Although they are present in the atmosphere in minute quantities, they are extremely efficient as greenhouse gases (about 15,000 times more efficient at retarding heat loss than is carbon dioxide). Because chlorofluorocarbons are a major cause of ozone destruction, production and use of chlorofluorocarbons has been sharply reduced and scheduled to be eliminated by 2020. As a result of these actions the atmospheric concentrations of CFCs have begun to decline.

17.5 The Current State of Knowledge about Climate Change

Because of the complexity involved in climate studies and disagreements about the significance of global warming, in 1988 the United Nations Environment Programme and the World Meteorological Organization established the Intergovernmental Panel on Climate Change (IPCC) to study the issue and make recommendations. A main activity of the IPCC is to provide an assessment of the state of knowledge about climate change at regular intervals. The IPCC is organized into several working groups. Working Group I deals with the physical science that relates to climate change. Working Group II deals with the impacts of climate change. Working Group III deals with how to mitigate the effects of climate change.

Working Group I published its portion of the Fifth Assessment Report, *Climate Change 2013: The Physical Science Basis,* in September of 2013. Involved in writing portions of the report were 259 scientists from around the world. Over 600 people contributed material for the report. Their report restated several important conclusions from previous reports and added several new observations and conclusions:

1. Human activity is clearly influencing climate.
2. Increased concentrations of greenhouse gases, particularly carbon dioxide, are causing an increase in temperature.
3. Evidence of increased temperature is clear.

   - The average temperature of the Earth has increased between 0.65°C and 1.06°C (between1.2°F and 1.9°F) since 1880.
   - Amounts of spring snow and ice have decreased in the northern hemisphere. Snow is melting earlier in the year.
   - The number of cold days has decreased and the number of warm days has increased.
   - The arctic region is warming more than the rest of the world.
   - Permafrost (permanently frozen soil) temperatures have increased between 2°C and 3°C (3.6°F to 5.4°F), and the thickness of the permafrost layer and the area of the world that has a permafrost layer have decreased.

   - There has been a reduction in the area covered by arctic sea ice at the end of the summer season (September).
   - Greenland and Antarctic ice sheets have been losing mass.
   - Glaciers are melting.
   - The arrival of spring is earlier in many parts of the world.

4. Increased carbon dioxide in the atmosphere and increased temperature are affecting oceans.

   - About 28 percent of carbon dioxide emissions end up in the ocean.
   - Increased CO_2 dissolved in water has decreased pH by 0.1 pH unit. (A 26 percent increase in hydrogen ion concentration.)
   - About 90 percent of the additional energy added to the Earth has been stored in the oceans, resulting in an increase of about 0.44°C (0.8°F) in the temperature of the upper 75 meters of the oceans in the last 40 years.
   - Sea level has risen about 19 centimeters (7.5 inches) between 1901 and 2010.
   - The rate of sea level rise has been increasing and was about 3.2 millimeters/year (0.125 inches/year) from 1993 to 2010.

17.6 Consequences of Climate Change

It is important to recognize that although a small increase in the average temperature of the Earth may seem trivial, such an increase could set in motion changes that could significantly alter the climate of major regions of the world. Computer models suggest that rising temperature will lead to changes in the hydrologic cycle, sea level, human health, the survival and distribution of organisms, and the use of natural resources by people. Furthermore, some natural ecosystems or human settlements will be able to withstand or adapt to the changes, while others will not.

Poorer nations are generally more vulnerable to the consequences of global warming. These nations tend to be more dependent on economic activity that is climate-sensitive, such as subsistence agriculture, and lack the economic resources to adjust to the changes that global warming may bring. The Intergovernmental Panel on Climate Change has identified Africa as "the continent most vulnerable to the impacts of projected changes because widespread poverty limits adaptation capabilities."

Working Group II of the IPCC published its portion of the Fifth Assessment Report, *Climate Change 2014: Impacts, Adaptation, and Vulnerability,* in March of 2014. The report lists eight major risks due to climate change. The authors are highly confident that these risks are real.

1. Risk of death or harm from coastal flooding
2. Risk to health and livelihoods from inland flooding
3. Risk of severe weather disrupting infrastructure and public services
4. Risk of death and illness due to extreme heat
5. Risk of food insecurity due to warming, drought, or flooding
6. Risk to agricultural productivity due to a shortage of water for irrigation and drinking

Focus On

Doubters, Deniers, Skeptics, and Ignorers

Whenever a new theory, concept, or discovery occurs in science there is a healthy period of sorting through the new facts to find the underlying truths. Some people are slow to accept new ideas. Others are concerned that the new ideas may disrupt current culture. Others may find it hard to go against established religious dogma or cultural norms. For example, when the theory of natural selection was published, it created an uproar. Some people could not see evolutionary change take place in their lifetime and rejected the idea. Others felt that the idea was contrary to Biblical teaching and refused to consider the idea. Some still do today. However, over time and as mountains of new information piled up, the concept of evolution through natural selection became mainstream and is now central to all biological thinking.

Similarly when the theory of continental drift was first formulated it was rejected by most geologists. However, as concepts for how such movements could occur developed and measurements of distances between land masses became more accurate, the concept became central to understanding phenomena such as earthquakes, volcanoes, and the formation of mountains.

The concept of climate change is currently in the middle of a sorting process, but the supporting data are piling up. However, some individuals have powerful reasons for denying the validity of the concept or at least raising doubts about its authenticity.

- Energy companies are in the business of providing fossil fuels, which when burned release carbon dioxide—the most important greenhouse gas. Since much of business and industry involves the use of energy, they see any change in energy policy as a threat to their bottom line. A carbon tax would raise the price of fuels. Requiring the use of renewable energy presents an unknown that makes them nervous. Since the real problems caused by climate change are in the future and very difficult to characterize except in generalities, and most businesses function on a time line of ten years or less, it is difficult for many in business and industry to make the difficult changes that would reduce greenhouse gas emissions. This is particularly true if some wish to make changes but will be at an economic disadvantage if their competitors do not invest in the things that will reduce emissions. There is even an international aspect to this thinking. If the U.S. government imposes a carbon tax that raises the price of energy and China and India do not, then U.S. businesses will be at an economic disadvantage.

- Political conservatives typically are against increased government regulation. Any effort to reduce greenhouse gas emissions will require some kind of government regulation. Thus, there is a fundamental divide between those who would institute rules that would reduce greenhouse gas emissions and those who detest government regulation.

- Some conservative religious groups reject the concept of evolution and set up a false dichotomy between religion and science. They have an inherent distrust of science and often reject scientific ideas as being anti-religion.

- Others simply choose to ignore the problems associated with climate change and are not making any changes in the way they think and act.

For all of these reasons, some individuals seek to distort the facts that support the idea of climate change and deny that climate change is a problem. Their arguments take several directions.

- Some use the idea that climate change is a natural event that has always occurred—just look at the geological record.

- Others accept the facts that support the idea that climate is changing but do not accept that it could be the result of human activity.

- Some selectively choose data collected during specific periods that allow them to cast doubt on climate science while ignoring the long-term trend. If you look at the last ten years of temperature data, you can make the argument that the average temperature has been declining. However this ignores the fact that these ten years were some of the warmest on record.

---

7. Risk of loss of marine and coastal ecosystem services
8. Risk of loss of terrestrial and freshwater ecosystem services

The following sections provide details about the nature of the risks and the probable negative effects of climate change.

Disruption of the Hydrologic Cycle

In a fundamental way the hydrologic cycle (evaporation, precipitation, water flow, groundwater, etc.) is driven by energy. The primary source of energy is the temperature of the Earth, which is determined by the input of energy from the sun and the heat-trapping effect of greenhouse gases. Thus, a change in the Earth's temperature is expected to change weather and climate. Weather includes short-term activities such as; temperature changes, rain and snow events, winds, clouds, and other factors. Climate is the long- term average of weather patterns. Thus, if climate is changing we should expect changes in weather patterns also. Higher temperatures result in increased evaporation, which will cause some areas to become drier, while the increased moisture in the atmosphere will result in greater rainfall in other areas. This is likely to cause droughts in some areas and flooding in others. In those areas where evaporation increases more than precipitation, soil will become drier, lake levels will drop, and rivers will carry less water. Lower river flows and lake levels will impair navigation, hydroelectric power generation, and water quality, and reduce the supplies of water available for agricultural, residential, and industrial uses. In areas that receive more rainfall, flooding will occur, soil will become more moist, and the flow of water through rivers, lakes, and streams will increase. In addition, it is predicted that storms will be more severe and major tropical storms (variously called hurricanes, typhoons, tropical cyclones) will be more severe.

Snowfall patterns are also expected to change with some areas receiving more snow and others less. In some cases, because of higher temperatures, rainfall will increase as snowfall diminishes. In addition, with higher temperatures snow will melt earlier

What's Your Take?

Climate change will increase the incidence of flooding in several ways. Rising sea levels will threaten low-lying coastal areas. Intense storms will cause coastal storm surges that can affect areas not normally considered to be in a flood plain. Heavy rain storm events will cause streams and rivers to rise and flood areas adjacent to water courses. Most insurance policies do not cover flood damage. The federal government has made flood insurance available for those in areas at risk of floods; however, many people do not purchase the flood insurance because it is expensive. When flooding occurs victims seek help from government sources and charitable organizations. One way to reduce the cost of flooding is to prevent people from building in areas that are likely to flood. This could be done by federal or state laws or local ordinances. Draw up a law or ordinance that would reduce flood damage to homes and businesses. List at least three criteria that would be used to prevent building in flood-prone areas and justify your selection of each criterion.

---

in the year or may not accumulate at all. In California's Central Valley, for example, melting snow provides much of the summer water supply; warmer temperatures would cause the snow to melt earlier and thus reduce summer supplies, even if rainfall increased during the spring.

Although it is difficult to interpret changes in short-term weather patterns with respect to the general concept of climate change, there are some indications that weather patterns are changing. For example, recent data suggest that the amount of spring snow has decreased and that snow is melting earlier in the year, hurricanes are more powerful, there are more hot days and fewer cold days, and spring is arriving earlier. Additional information collected over the next few years will clarify if these weather changes are permanent or temporary.

Rising Sea Level

A warmer Earth will result in rising sea levels for two different reasons. When water increases in temperature, it expands and takes up more space. In addition, higher temperatures are causing the melting of glaciers, which adds more water to the oceans. Rising sea level erodes beaches and coastal wetlands, inundates low-lying areas, and increases the vulnerability of coastal areas to flooding from storm surges and intense rainfall. (See figure 17.7.)

FIGURE 17.7 Flooding from Hurricane Sandy A combination of increasing sea level and a storm surge from Hurricane Sandy caused extensive flood damage in New York and New Jersey in October of 2012.

By 2100, the IPCC projects sea level to rise by 26 to 98 centimeters (10 to 39 inches). A 50-centimeter (20-inch) sea-level rise will result in substantial loss of coastal land in North America, especially along the southern Atlantic and Gulf coasts, which are subsiding and are particularly vulnerable. Many coastal cities would be significantly affected by an increase in sea level. Worldwide about 600 million people live in low-lying coastal areas. The land area of some island nations and countries such as Bangladesh would change dramatically as flooding occurred. The oceans will continue to expand for several centuries after the Earth's average air temperature stops increasing.

Health Effects

Climate change will impact human health in a variety of ways.

Heat affects health The most direct effect of climate change is the impact of hotter temperatures. Extremely hot temperatures increase the number of people who die (of various causes) on a given day. For example, people with heart problems are vulnerable because the cardiovascular system must work harder to keep the body cool during hot weather. Heat exhaustion and some respiratory problems increase. In 2010, a record-breaking heat wave in Russia resulted in 62 consecutive days of above normal temperatures, and in Moscow about 11,000 deaths were attributed to the heat. In June 2013, a heat wave in the U.S. Southwest resulted in several record temperatures and several deaths. Even though the temperatures in the U.S. Southwest were higher than those in Russia, the number of U.S. deaths was much lower because most U.S. buildings have air conditioning and most Russian buildings do not.

Heat affects air pollution Climate change will also make air pollution problems worse. Higher air temperature increases the concentration of ozone at ground level, which leads to injury of lung tissue and increases the incidence of respiratory disease, asthma, and allergies. Because children and the elderly are the most vulnerable, they are likely to suffer disproportionately from both warmer temperatures and poorer air quality.

Tropical diseases could migrate to former temperate regions Throughout the world, the prevalence of particular diseases depends largely on local climate. Several serious diseases appear only in warm areas. As the Earth becomes warmer, some of these

Focus On

Decline in Arctic Sea Ice

Since 1978, satellites have made continual observations on the area of Arctic Ocean covered by ice. There are several factors that influence the amount of ice present. During the summer months, increased air and water temperatures cause melting of the ice, and the area covered by ice shrinks to a minimum during September. The general trend since 1978 has been that the area covered by sea ice at the end of the Arctic summer has been declining. In September 2012, the Arctic sea ice reached its smallest extent since records have been kept—3.41 million square kilometers.

There are several factors that affect the rate at which the ice melts:

- Long-term trends show that the melting of ice begins earlier in the spring than it had previously. Thus, there is a longer period of time during which melting occurs, resulting in a smaller extent of sea ice at the end of the summer.

- The nature of the ice is also important. Ice that is formed during the winter is thinner than ice that has remained unmelted during the summer. This thinner ice tends to break up more easily than perennial ice. This is important because intact ice reflects sunlight, reduces warming, and slows the rate at which the ice melts. On the other hand, open water tends to absorb sunlight and is warmed, resulting in further melting of the ice.

- There is evidence that there have been changes in ocean currents that have introduced warmer water from both the Atlantic and Pacific Oceans into the Arctic Ocean, which could be contributing to the more rapid melting seen in recent years.

The extent of sea ice is important to the people and wildlife that live in the Arctic. Reduced sea ice appears to be reducing the success of polar bears in obtaining food. Similarly, native peoples typically hunt from the ice and rely on ice cover to allow them to travel from place to place. The year 2012 had the lowest extent of Arctic sea ice on record, and although 2013 rebounded to 5.35 million square kilometers, the long-term trend is toward a reduction in the extend of sea ice. (See graph.)

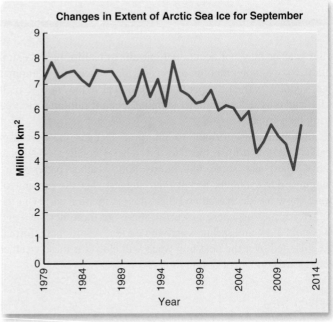

Changes in Extent of Arctic Sea Ice for September

Source: Data from NOAA Climate.gov

tropical diseases may be able to spread to parts of the world where they do not currently occur. Diseases that are spread by mosquitoes and other insects could become more prevalent if warmer temperatures enabled those insects to become established farther north. Such "vector-borne" diseases include malaria, dengue fever, yellow fever, and encephalitis. Some scientists believe that algal blooms could occur more frequently as temperatures rise, particularly in areas with polluted waters, in which case outbreaks of diseases such as cholera that tend to accompany algal blooms could become more frequent.

Changes to Ecosystems

Some of the most dramatic projections regarding the effects of climate change involve alterations to natural systems:

- **Geographic distribution of organisms** could be significantly altered by climate change. As climates warm, organisms that were formerly restricted to warmer regions will become more common toward the poles. The tundra biomes of the world will be greatly affected because of the thawing of the permafrost, which will allow the northward migration of species. The polar bear has been named as a threatened species in some areas because they hunt for seals on sea ice and reduced sea ice limits their hunting area. Similarly, mountainous areas will have less snow and earlier melting of the snow that does accumulate during the winter.

- **Coral reefs are especially challenged** because they are affected both by an increase in water temperature and by an increase in the acidity of the ocean. Many species of corals are adapted to a very narrow temperature range. When the temperature gets too high, they die. When carbon dioxide dissolves in water, it forms an acid. An increase in acidity would cause the skeletons of corals and the shells of many other organisms to tend to dissolve. This would make it more difficult for these organisms to precipitate calcium salts from the ocean to construct their skeletons and shells.

- **Low-lying islands and shorelines** will be especially impacted by rising sea level. Mangrove forests and marshes will be inundated and subjected to violent weather and storm surges.

Challenges to Agriculture and the Food Supply

Climate strongly affects crop yields. Yields will fall in regions where drought and heat stress will increase. In regions that will receive increased rainfall and warming temperatures, yields should increase. However, episodes of severe weather will cause crop damage that will affect yields. A warmer climate would reduce flexibility in crop distribution and increase irrigation demands. Expansion of the geographic ranges of insect pests could also increase vulnerability and result in greater use of pesticides. Despite these effects, total global food production is not expected to be altered substantially by climate change, but negative regional impacts are likely. Agricultural systems in developed countries are highly adaptable and can probably cope with the expected range of climate changes without dramatic reductions in yields. It is the poorest countries, where many already are subject to hunger, that are the most likely to suffer significant decreases in agricultural productivity. Table 17.2 summarizes some of the main points related to the consequences of global warming and climate change.

17.7 Addressing Climate Change

Approaches to dealing with climate change involve technological change coupled with political will and economic realities.

Energy Efficiency and Green Energy

Since the burning of fossil fuels releases carbon dioxide, which is the most important greenhouse gas, improving energy efficiency has the double impact of reducing carbon dioxide release and conserving the shrinking supplies of energy resources. It makes sense to increase energy efficiency even if global warming is not a concern. One way to stimulate a move toward greater efficiency would be to place a tax on the amount of carbon individuals and

Table 17.2 Consequences of Increases in Greenhouse Gases

Primary Change	Global Effects	Ecosystem Consequences	Cultural Consequences
Warming of the climate particularly toward poles	Permafrost melts	Tundra biome changed with melting of permafrost	Arctic native people affected
	Less sea ice	Arctic and Antarctic organisms that rely on sea ice endangered	Arctic native people who hunt marine organisms negatively affected
	Glaciers melt	Changes in volume and timing of flow of rivers	Glaciers less reliable as a source of water
	Less snowfall and accumulation	Earlier release of water from melting snow and less volume released	Impact on timing of irrigation and volume available for human use
	Changes in plant and animal populations	Shift of plant and animal distributions toward poles	Tropical disease may migrate to temperate regions
			Patterns of agriculture will need to change
Changes in weather patterns	Less rainfall in midlatitudes and some subtropical regions	Drier deserts and droughts in some regions	Patterns of agriculture will need to change
			Water shortages in arid climates
	More rainfall toward poles and parts of the tropics	Increased runoff causing erosion and increased flow in rivers	Patterns of agriculture will need to change
	More heat waves	Some species of organisms will decline because they can't deal with high temperatures	High temperatures make air pollution problems more severe
			High temperature affects health negatively
	More severe storms	Ecosystems altered by wind, flooding, and erosion	Destruction of buildings and infrastructure
			Flooding of cities and agricultural land
Warmer oceans		Coral reefs threatened by warmer water	Loss of biodiversity and marine resources
		Arctic and Antarctic marine food chains altered	Impact on fishing industry
Rise in sea level	Warm water expands leading to sea-level rise	Coastal flooding affects mangroves and salt marshes	Coastal flooding affects cities and agriculture
	Melting glaciers add to ocean volume causing sea level to rise		Cost of combating flooding
Ocean acidification	Carbon dioxide dissolved in water acidifies the ocean	Corals and organisms that make shells threatened	Loss of biodiversity and marine resources

corporations release into the atmosphere. This would increase the cost of fuels and stimulate a demand for fuel-efficient products because the cost of fuel would rise. It would also stimulate the development of alternative fuels with a lower carbon content and generate funds for research in many aspects of fuel efficiency and alternative fuel technologies.

Increases in energy efficiency and reductions in greenhouse gas emissions are likely to have important related benefits that could offset the costs. Greater energy efficiency would lead to reduced air pollution, which would result in lower health care costs and time lost from work. A study of air pollution in China determined that there were 1.2 million premature deaths in China in 2010. The cost of lost wages and increased health care costs can be converted into monetary terms so that costs of improving fuel efficiency and reducing pollution can be offset by lower health care costs and higher worker productivity. Improved energy efficiency also reduces the need for new power plants and related energy infrastructure.

Green sources of energy such as wind, solar, and hydroelectric, as well as nuclear power, do not release carbon dioxide. Thus, switching to green sources of energy is the most effective way to reduce carbon dioxide emissions.

The Role of Biomass

Since carbon is an important component of living things, what happens to biomass has a role to play in determining atmospheric carbon dioxide. Forests consist of many long-lived tree species that can tie up carbon for centuries. (See figure 17.8.) Preserving forests slows the rate of increase of atmospheric carbon dioxide. This is particularly true for tropical forests since they are the last remaining major unmodified forested areas in the world and they are very efficient at capturing carbon dioxide. The burning of tropical rainforests to provide farm or grazing land not only adds carbon dioxide to the atmosphere, it also reduces the rainforests' ability to remove carbon dioxide from the atmosphere, since the grasslands or farms created do not remove carbon dioxide as efficiently as do the rainforests. Furthermore, the grazing lands and farms in such regions of the world are often abandoned after a few years and do not return to their original forest condition. A commonly cited estimate is that 20 percent of the additional carbon dioxide entering the atmosphere is due to deforestation.

Planting trees has also been supported as a way to reduce atmospheric carbon dioxide. Certainly, long-lived plants like trees will tie up carbon for longer periods of time than grasses and other short-lived plants. Many private organizations as well as city, state, and national governments have initiated tree planting programs. At least one of the motivations for these programs is to combat global warming.

Many critics argue that this approach will provide only a short-term benefit since, eventually, the trees will mature and die, and their decay will release carbon dioxide into the atmosphere at some later time.

Technological Approaches

The U.S. Department of Energy has concluded that, relying primarily on already proven technology, the United States could reduce its

FIGURE 17.8 Forests Store Carbon Since trees use carbon dioxide to produce biomass and they live a long time, forests can store large amounts of carbon.

carbon emissions by almost 400 million metric tons per year, or enough to stabilize U.S. emissions at 1990 levels.

Many different kinds of technological approaches have been suggested as ways to limit the amount of carbon dioxide added to the atmosphere. Alternative energy sources like wind, solar, hydroelectric, geothermal, and nuclear power do not release carbon dioxide and can replace current fossil-fuel energy sources. However, since fossil fuels currently provide nearly 90 percent of the world's energy, converting to a greater reliance on non-fossil-fuel energy sources will require a great deal of new construction and technological improvements.

Another approach to the carbon dioxide problem is to prevent it from being released into the atmosphere. Carbon dioxide can be reacted with other compounds to produce solid carbonate minerals (limestone is calcium carbonate) that could be stored in landfills and prevent the release of carbon dioxide gas. It is also possible to capture and store carbon dioxide underground, particularly in saline groundwater deposits and exhausted oil and gas production wells.

However, all of these technological changes come with a cost that will be reflected in the price of energy to the consumer. Thus, as long as fossil-fuel sources of energy are less expensive than non-fossil-fuel alternatives and removing carbon dioxide adds significantly to the cost of energy, there will need to be government policies that stimulate the deployment of these technologies. Good examples from the industrialized nations include wind power

Science, Politics, & Policy

Policy Responses to Climate Change

Scientists have the basic knowledge about many environmental and technical issues that affect society, but politicians and social institutions hold the power to affect policy. By its nature, science is a self-correcting activity. Theories that once were adequate to explain activities of the natural world are replaced by different understandings when new discoveries occur. Thus, it is in the nature of scientists to state ideas in the form of probabilities—always leaving the door open to alternative explanations. This way of thinking is often misinterpreted by the public and policymakers as a lack of understanding rather than a careful statement of the reliability of the information. Furthermore, scientific knowledge is only one of the inputs that informs the thinking of policymakers. Economic, political, and social factors and even personal bias and ambition are also important.

The tension between science and policy is very evident in the debate over climate change. For example, the world authority on climate change, the Intergovernmental Panel on Climate Change (IPCC), recognizes this tension when it states in its mandate that "its scientific findings on climate change be policy neutral." IPPC provides information on the likely causes and outcomes of climate change and the reliability of the information but does not tell policymakers how to respond to the problem. For example, the IPCC clearly states that the cause of increased carbon dioxide in the atmosphere is the human activity of burning fossil fuels. It further states that science shows a link between the amount of carbon dioxide in the atmosphere and global climate change and that the Earth is getting warmer. With this information, policymakers must provide political leadership to deal with the problems resulting from a disrupted climate, whether it involves changing energy policy, capping emissions of carbon dioxide, strengthening and raising levees to prevent flooding, or moving low-lying villages.

The Kyoto Protocol of 1997, which evolved from the United Nations Framework Convention on Climate Change in 1992, has created a considerable amount of policy confusion. For example, only one-third of global emissions are subject to the Protocol. China, now the world's leading producer of heat-trapping gases, and other developing countries are not required to limit greenhouse gas emissions under the Protocol. Greenhouse gas emissions have continued to increase. On the other hand, the Protocol has provided leadership and binding international targets. In countries that signed the treaty—the United States did not sign—the Protocol has helped spur renewable-energy subsidies, tougher energy-efficiency standards, and the European Climate Exchange, a cap-and-trade market for emissions credits.

In the United States, there are a significant number of climate change policies, but they often lack coordination and coherence and there is no clear national policy. For example:

- Buildings consume 40 percent of our energy, yet building codes for energy efficiency are established locally and lack any coordination.
- Thirty-one states have renewable energy portfolio standards that require a portion of their electricity come from renewable sources by specific dates. An additional seven states have renewable energy goals that are not mandatory. However, the percentage of renewable energy required, the date for implementation, and even what is considered "renewable" varies from state to state.
- A hodgepodge of state efforts try to decouple electricity consumption from utility profits, encouraging utilities to promote energy-saving efforts rather than sell more power.

Many argue that a coordinated federal effort on energy policy that addresses climate change is needed in the United States. They believe that major federal financial incentives and loan guarantees could greatly assist emerging renewable energy industries. The lack of a clear national policy on energy and climate change makes it difficult for corporations to make important business decisions.

purchase programs in Denmark and Germany, the "10,000 rooftops" photovoltaic program in Japan, and the evolution of "green marketing" campaigns in the United States and Europe to capture consumer willingness to pay modest premiums for electricity from clean energy technologies.

International Agreements

With the recognition that chlorofluorocarbons were destroying the stratospheric ozone layer that protects us from ultraviolet radiation, international agreements were reached that led to a sharp reduction in the amount of chlorofluorocarbons (CFCs) being released. The phasing out of CFCs required significant technological changes, but the changes were very rapid once a broad international consensus was reached and a plan established. Recent data show that CFCs in the atmosphere have begun to decline. Although CFCs are a minor player in the climate change scenario, the changes made to protect the ozone layer have had the side benefit of reducing the release of a potent greenhouse gas. See the section Ozone Depletion in chapter 16.

In 1997, the development of an international agreement known as the Kyoto Protocol on greenhouse gas emissions was a

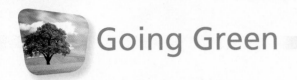

Germany's Energy Policy: Responding to Climate Change

The European Union ratified the Kyoto Treaty in 2002. Since that time, many European countries have made major commitments to reduce their production of greenhouse gases—primarily carbon dioxide. Germany, in particular, developed an energy policy that has resulted in lower carbon dioxide emissions.

The following are some of the significant policy initiatives of the German government:

- The Renewable Energy Resources Act set a goal to provide 20 percent of electricity from renewable sources by 2020.
- Low-interest loans were provided to upgrade the energy efficiency of buildings.
- Energy use certificates were required for buildings.
- A tax was imposed on the use of coal, coke, and lignite.
- A tax was imposed on motor freight that varied with the distance traveled.
- Blue Angel eco-labeling of appliances was established.
- A carbon dioxide emissions trading law was enacted.
- Low-interest loans were provided for the installation of solar energy technology.

These policies have led to significant changes in energy use and carbon dioxide emissions. Although Germany is not in the most geographically advantageous location for photovoltaics, in 2013 it had about 25 percent of the world installed capacity. Solar energy is also used for heating purposes. The growth in solar capacity is driven in large part by subsidies that are part of the government's energy policy. In 2013 it also had about 11 percent of the world wind turbine capacity. By 2013, about 29 percent of electricity was produced from renewable sources. As a result of these policies, energy efficiency has improved and carbon dioxide emissions have declined.

The success of these initiatives and the fact that goals were being reached ahead of schedule led the German government in 2010 to announce new, more aggressive targets of:

- Providing 35 percent of electricity from renewable sources by 2020 and 80 percent by 2050.
- Cutting electrical consumption 50 percent by 2050 by improving energy efficiency.
- Providing 18 percent of all energy from renewable sources by 2020 and 60 percent by 2050.

first step toward a worldwide approach to alleviating the problem. Most countries of the world ratified the treaty. Although the United States did not officially ratify the treaty, it has been an active participant in the series of climate change conferences held in recent years. However, under the Kyoto Protocol only the economically developed countries of the world were required to limit their greenhouse gas emissions to a specific percentage below 1990 levels. They were to meet their targets by 2012. Economically developing countries including nearly all of Latin America, Africa, and Asia did not have binding targets to meet. In 2006 China became the world's largest emitter of carbon dioxide and the United States was in second place. The European countries have been most successful in reducing their greenhouse gas emissions. Many of the heavily industrialized countries are actively promoting changes that reduce greenhouse gas emissions. Others are meeting their goals because of the major economic downturn in many European countries. When the economy is bad, energy use declines and carbon dioxide emissions fall. The Kyoto Protocol expired in 2012, and at a meeting in Doha, Qatar, in December 2012 an agreement was reached to extend the Protocol through 2020. While Europe in general met its goal for greenhouse gas emissions, most developing countries did not have goals to meet and greenhouse gas emissions have continued to increase.

The issue of goals became a central problem in climate meetings in Warsaw, Poland, in November, 2013. Despite the fact that about half of greenhouse gases are coming from poor countries and they have no binding greenhouse gas targets to meet, they continued to ask for funding ($100 billion/year) from rich countries to help them deal with climate change. The meeting ended with no firm targets on financing for developing countries or for reducing emissions.

One positive outcome was achieved, however. *All* countries are to publish their plans for dealing with climate change by early 2015, prior to a meeting in Paris in 2015. A new set of goals is to go into effect in 2020 and the Paris conference will be the first step in setting those goals.

Issues & Analysis

Who Should Reduce CO_2 Emissions?

The accompanying chart shows the top ten countries in terms of CO_2 emissions in 2011. These ten countries account for 67 percent of world CO_2 emissions. Thus, what these countries do will have a great impact on future CO_2 emissions and the severity of the climate change impacts that will occur as a result of an increase in the amount of CO_2 in the atmosphere. China is responsible for over 25 percent of global CO_2 emissions and the United States is responsible for about 17 percent of global emissions. However, both China and the United States have large populations so we should expect them to release more carbon dioxide than countries with small populations.

Another way to look at emissions is to compare the emissions produced per person. On a per capita basis the United States releases 2.3 times more CO_2 per person than China. Some economically developed countries, like Japan and Germany, have per capita CO_2 emissions much lower than the United States. Japan's per capita CO_2 production is 60 percent and Germany's per capita production is half that of the United States. India and Russia are both countries with large populations that currently have low per capita CO_2 emissions. If they were to undergo an economic growth similar to that of China, world CO_2 emissions would increase greatly. If we want to reduce worldwide CO_2 emissions, it is obvious that those countries that are responsible for the greatest *total* emissions must reduce emissions. It is also obvious that countries that wish to develop economically (Russia, India, and many others) must do so without increasing carbon dioxide emissions.

- What actions could the United States and China take to reduce their carbon dioxide emissions?
- What actions could the international community take to encourage economically emerging nations to develop economically without increasing carbon dioxide emissions?

- Would you support a treaty that imposed a carbon tax on all countries?

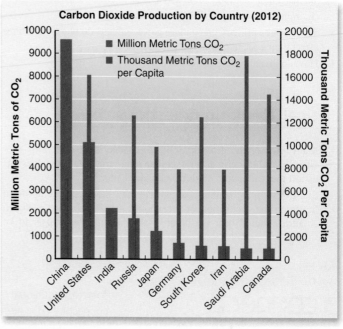

Source: Data from Global Carbon Atlas

Summary

The concept of climate change is not new. Geological studies have demonstrated that climates have changed greatly over the Earth's history. Today's climate change is different in that it is highly likely that it is being caused by human activities. The primary greenhouse gases are carbon dioxide, nitrous oxide, methane, and chlorofluorocarbons. These gases are strongly linked to an increase in the average temperature of the Earth and, consequently, are leading to major changes in the climate. These include warming of the Earth, particularly near the poles. The warming will result in melting of the permafrost, glaciers, and sea ice; changes in rain and snowfall patterns; shifts in the distribution of plants and animals; more intense heat waves and severe storms; a rise in sea level; and acidification of the oceans. Other likely effects of climate change are health effects in humans, extinction of some plants and animals, flooding of cities, and changes in agricultural productivity. The primary factor involved in climate change appears to be the carbon dioxide released from the burning of fossil fuels. Since fossil fuel use is closely tied to economic development, many developing countries are unwilling to accept limits to their use of fossil fuels.

Acting Green

1. Eat less meat—cows produce methane.
2. Purchase green energy from your electric utility.
3. Use less energy and less carbon dioxide will be released.
4. Walk or ride a bike as often as practical.

Review Questions

1. Why are geologic studies important to the understanding of climate change?
2. How does each of the following help us understand climate change?
   a. studies of the flowering times of plants
   b. measurements of the pH of the ocean
   c. satellite photos of the amount of snow in an area
   d. sea level measurements
   e. gas bubbles trapped in glaciers
   f. migration patterns of birds
3. What are the primary greenhouse gases and how do human activities affect their concentrations?
4. How do greenhouse gases cause a warming of the Earth?
5. List five changes that are likely to occur to Earth and its ecosystems as a result of global warming.
6. List three actions humans could take to reduce the release of additional greenhouse gases.
7. Describe how increased carbon dioxide in the atmosphere will alter the oceans.
8. How will climate change affect human health?
9. How effective have human efforts been at controlling carbon dioxide release?
10. List five changes likely to occur to the hydrologic cycle as a result of a warmer climate.
11. Why does a warming climate cause sea level to rise?

Critical Thinking Questions

1. Many people who deny that climate change is an issue look specifically at short-term data rather than long-term data. How does this distort the real picture of global climate change?
2. Some developing countries argue that they should be exempt from limits on the production of greenhouse gases and that developed countries should bear the brunt of the changes that appear to be necessary to curb global climate change. What values, beliefs, and perspectives underlie this argument? What do you think about this argument?
3. China and the United States are the top two countries in terms of greenhouse gas releases. Why is this true? What could be done to change this situation?

To access additional resources for this chapter, please visit ConnectPlus® at connect.mheducation.com There you will find interactive exercises, including Google Earth™, additional Case Studies, and SmartBook™, an adaptive eBook that integrates our LearnSmart® adaptive learning technology.

chapter 18

Solid Waste Management and Disposal

CHAPTER OUTLINE

As we use materials in our daily lives we generate waste. Much of that waste is sent to landfills. However, many materials that are no longer useful in their current form can be recycled for use in a different way. Most communities have recycling programs that reduce the amount of waste sent to landfills.

OBJECTIVES

After reading this chapter, you should be able to:

- Explain why solid waste is a problem throughout the world.

- Understand that the management of municipal solid waste is directly affected by economics, changes in technology, and citizen awareness and involvement.

- Describe the various methods of waste disposal and the problems associated with each method.

- Understand the difficulties in developing new municipal landfills.

- Define the problems associated with incineration as a method of waste disposal.

- Describe some methods of source reduction.

- Describe composting and how it fits into solid waste disposal.

- List some benefits and drawbacks of recycling.

- Understand the growing problem of electronic waste.

- Understand how landfill gas is turned into electricity.

- Understand the different resins used in consumer packaging.

Innovative Approaches to Solid Waste Problems

Disposal of solid waste is a problem for all societies. Often when people think "outside the box," they can come up with innovative ways to solve a problem.

1. *New Wine in Old Bottles*—Great Britain is one of the largest importers of wines in the world. Along with the wine come millions of colored glass bottles, which must be recycled. There was no market for colored glass in Great Britain. A simple solution to the problem was to import the wine in large casks and bottle it in Great Britain with bottles made in Great Britain from colored glass.

2. *Points for Recycling*—RecycleBank is a business that promotes recycling by providing home owners with "points" that can be redeemed at local merchants. Every family on a garbage route is issued a special container with a computer chip. When garbage trucks pick up the recycling, they weigh the container and record by weight how much each family is recycling and determines the number of "points" they get. RecycleBank contracts with cities to provide the service and the cities benefit by having higher recycling rates.

3. *Extended Product Responsibility*—Germany and Sweden have passed legislation that requires manufacturers to consider the waste they produce during manufacturing and to ensure safe disposal of the product at the end of its useful life. This *extended product responsibility* provides a strong economic incentive for companies to produce less waste rather than passing on the cost of waste disposal to the public. As a consequence of these actions and a general commitment to waste reduction, Germany and Sweden put less than 1 percent of their waste in landfills. Over 99 percent is recycled or burned to provide heat and electricity.

4. *Statewide Curbside Recycling*—The state of Delaware passed a uniform recycling law which requires all commercial businesses to participate in a comprehensive recycling program and all waste haulers to provide curbside recycling at least every two weeks. Recyclables include nearly all paper products, bottles, cans and other containers, and rigid plastics. Recyclables are sent to a materials recovery facility. This legislation replaces a bottle-deposit law that had been in place since 1982.

18.1 Kinds of Solid Waste

The several kinds of waste produced by a technological society can be categorized in many ways. Some kinds of wastes are released into the air and water. Some are purposely released, while others are released accidentally. Many wastes that are purposely released are treated before their release. Wastes that are released into the water and air are discussed in chapters 15 and 16, respectively. There are wastes with particularly dangerous characteristics, such as nuclear wastes, medical wastes, industrial hazardous wastes, and household hazardous wastes. Chapter 19 considers nuclear and other hazardous waste issues. The focus of this chapter is solid waste.

Solid waste is generally made up of objects or particles that accumulate on the site where they are produced, as opposed to water- and airborne wastes that are carried away from the site of production. Solid wastes are typically categorized by the sector of the economy responsible for producing them, such as mining, agriculture, manufacturing, and municipalities.

We have very good information about those waste streams that are tightly regulated (hazardous wastes, municipal solid waste, medical wastes, nuclear wastes) but only general estimates for many of the other kinds of wastes, such as mining and agricultural waste.

1. Mining waste is generated in three primary ways. First, in most mining operations, large amounts of rock and soil need to be removed to get to the valuable ore. This waste material is generally left on the surface at the mine site. Second, milling operations use various technologies to extract the valuable material from the ore. These techniques vary from relatively simple grinding and sorting to sophisticated chemical separation processes. Regardless of the technique involved, once the valuable material is recovered, the remaining waste material, commonly known as tailings, must be disposed of. Solid materials are typically dumped on the land near the milling site, and liquid wastes are typically stored in ponds. It is difficult to get vegetation to grow on these piles of waste rock and tailings, so they are unsightly and remain exposed to rain and wind. Finally, the water that drains or is pumped from mines or that flows from piles of waste rock or tailings often contains hazardous materials (such as asbestos, arsenic, lead, and radioactive materials) or high amounts of acid that must be contained or treated—but often are not. In addition, failures of the earthen dams used to form waste ponds result in the release of contaminated water into local streams. Between 2000 and 2013, about 25 such failures occurred worldwide, four of them in the United States.

It is difficult to get more than a rough estimate of the amount of mining waste produced, but the U.S. Environmental Protection Agency estimates that between 1 billion and 2 billion metric tons of mining waste are produced each year in the United States. Of the total waste produced, about 700 million to 800 million metric tons are considered hazardous.

Many types of mining operations require vast quantities of water for the extraction process. The quality of this water is degraded, so it is unsuitable for drinking, irrigation, or recreation. Since mining disturbs the natural vegetation in an area, water may carry soil particles into streams and cause erosion and siltation. Some mining operations, such as strip mining, rearrange the top layers of the soil, which lessens or eliminates its productivity for a long time. Strip mining has disturbed approximately 75,000 square kilometers (30,000 square miles) of U.S. land, an area equivalent to the state of Maine.

2. Agricultural waste is the second most common form of waste and includes waste from the raising of animals and the harvesting and processing of crops and trees. The amount of animal manure produced annually is estimated at about 1,240 million metric tons. Other wastes associated with agriculture, such as waste from processing operations (peelings, seeds, straw, stems, sludge, and similar materials), might bring the total agricultural waste to about 1.5 billion metric tons per year. Since most agricultural waste is organic, approximately 90 percent is used as fertilizer or for other soil-enhancement activities. Other materials are burned as a source of energy, so little of this waste needs to be placed in landfills. However, when too much waste is produced in one place, there may not be enough farmland available to accept the agricultural waste without causing water pollution problems associated with runoff or groundwater contamination due to infiltration.

3. Industrial solid waste from sources other than mining is variously estimated to be between 220 million and 600 million metric tons of solid waste per year. It includes a wide variety of materials such as demolition waste, foundry sand, scraps from manufacturing processes, sludge, ash from combustion, and other similar materials. These materials are tested to determine if they are hazardous. If they are classified as hazardous waste, their disposal requires that they be placed in special hazardous waste landfills. Hazardous wastes are discussed in chapter 19. In addition to solid wastes, industries produce several billion metric tons of aquatic waste. See chapter 15 for a discussion of industrial use of water.

4. Municipal solid waste (MSW) consists of all the materials that people in a region no longer want because they are broken, spoiled, or have no further use. It includes waste from households, commercial establishments, institutions, and some industrial sources and amounts to about 230 million metric tons per year. Table 18.1 summarizes estimates of the quantity of various kinds of solid waste produced in the United States. Food waste is one example of a growing municipal solid waste concern in the United States. According to the U.S. Department of Agriculture more than 25 percent of all the food produced for domestic sale and consumption ends up in landfills. This amounts to some

Table 18.1 Estimates of Solid Waste Produced per Year in the United States

Category of Waste	Amount of Waste (million metric tons)
Mining waste	1,000–2,000
Agricultural waste	1,500
Industrial waste	200–600
Municipal solid waste	230

Source: Estimates from U.S. Environmental Protection Agency, U.S. Geological Survey, and U.S. Department of Agriculture.

25 million metric tons. It is estimated that 14 percent of the food purchased by U.S. households is wasted. Some communities now pick up and centrally compost food waste from commercial and residential buildings and put the resulting nutrient rich soil to use in municipal projects or for sale to the public. The remainder of this chapter will focus on the generation and disposal of municipal solid waste.

18.2 Municipal Solid Waste

Wherever people exist, waste disposal is a problem. Some items are discarded when they are broken or worn out. Other products like magazines, catalogs, newspapers, packaging, bottles, and cans are temporarily useful and then are discarded. Those that have only temporary uses make up the majority of the solid waste stream.

The United States produces nearly 230 million metric tons of municipal solid waste each year. This equates to about 2 kilograms (4.4 pounds) of trash per person per day, or 0.73 metric tons per person per year. The amount of municipal solid waste has more than doubled since 1960, and the per capita rate has increased by nearly 70 percent in that same time, although per capita rates began to stabilize about 1990. When recycling is subtracted from total waste the net waste produced has actually fallen since 1990. (See figure 18.1.)

Nations with high standards of living and productivity tend to have more municipal solid waste per person than less-developed countries. (See figure 18.2.) Large metropolitan areas have the greatest difficulty dealing with their solid waste because of the large volume and the challenge of finding suitable landfill sites near the city.

In March 2001, New York City closed its Fresh Kills Landfill on Staten Island. (It was reopened for a time following the September 11, 2001, terrorist attack to serve as a place to take the debris from the World Trade Center so that it could be processed.) Before its closure, it was the largest landfill in the world and received about 12,600 metric tons of trash each day. Today, New York City is exporting waste to landfills in Pennsylvania, Virginia, Ohio, South Carolina, Connecticut, and New Jersey. New York City currently produces about 45,000 metric tons of residential solid waste per day. The cost of shipping all this waste to distant landfills by barge, rail, or truck is expensive. Therefore, New York City is seeking ways to reduce the amount of waste going to landfills. The ultimate goal is to divert 75 percent of the waste stream from landfills. In 2013, the city signed an agreement

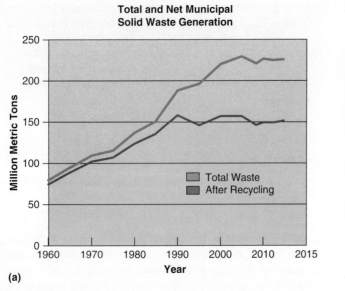

Total and Net Municipal Solid Waste Generation

(a)

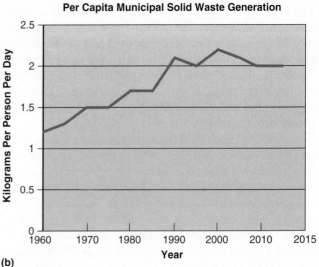

Per Capita Municipal Solid Waste Generation

(b)

FIGURE 18.1 Municipal Solid Waste Generation Rates The generation of municipal solid waste in the United States increased steadily until about 2005. However, because of increased recycling rates, the net production rates (after recyclables have been removed) have fallen slightly since 1990 (a), and the per capita rate has stabilized (b).

Source: U.S. Environmental Protection Agency, Washington, D.C.

with a waste-to-energy company to generate electricity from its waste. In addition, it has a contract with a recycling company to process glass, metal, and plastic for reuse. The city also has educational programs that encourage citizens to recycle materials and programs with restaurants to reduce the amount of food waste.

Archaeologists rely on the waste of past societies to tell them about the nature of the culture and lifestyle of ancient civilizations. In the same way today, our municipal solid waste is a reflection of

our society. Figure 18.3 shows how the composition of U.S. trash has changed since 1960. The most common items in the waste stream are paper products, food waste, yard waste, and plastics. (See figure 18.4.)

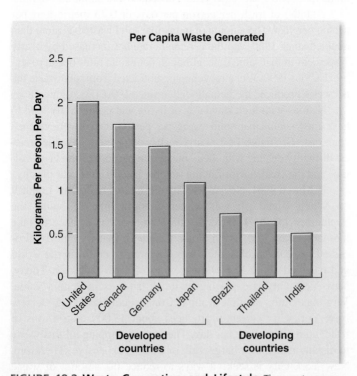

FIGURE 18.2 Waste Generation and Lifestyle The waste generation rates of people are directly related to their economic condition. People in richer countries produce more garbage than those in poorer countries.

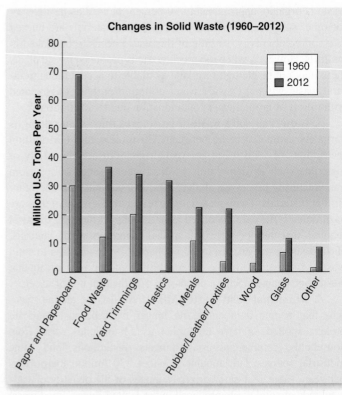

FIGURE 18.3 The Changing Nature of Trash Paper products are the largest component of the waste stream. Changes in lifestyle and packaging have led to a change in the nature of trash. Note the increase in the amount of plastics in the waste stream. Most of what is currently disposed of could be recycled.

Source: Data from the U.S. Environmental Protection Agency.

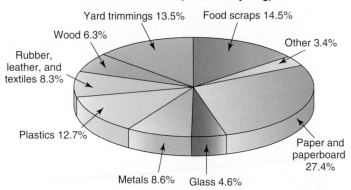

Total MSW Generation (by Material), 2012
251 Million Tons (Before Recycling)

Yard trimmings 13.5%
Food scraps 14.5%
Wood 6.3%
Other 3.4%
Rubber, leather, and textiles 8.3%
Plastics 12.7%
Paper and paperboard 27.4%
Metals 8.6%
Glass 4.6%

FIGURE 18.4 Composition of Trash in the United States (2012) Paper, food scraps, yard waste, and plastics are the most common materials disposed of, accounting for nearly 70 percent of the waste stream.
Source: Data from the U.S. Environmental Protection Agency.

18.3 Methods of Waste Disposal

From prehistory through the present day, the favored means of disposal was simply to dump solid wastes outside the dwelling or away from the city or village limits. Frequently, these dumps were in wetlands that were considered useless for other purposes. To minimize the volume of the waste, the dump was often burned. Unfortunately, this method is still being used in remote or sparsely populated areas in the world. (See figure 18.5.)

As better waste disposal technologies were developed and as values changed, more emphasis was placed on the environment and quality of life. Dumping and open burning of wastes is no longer an acceptable practice from an environmental or health perspective. While the technology of waste disposal has evolved during the past several decades, our options are still limited. Essentially, five techniques are used: (1) landfills, (2) incineration, (3) composting, (4) source reduction, and (5) recycling.

Landfills

Landfills have historically been the primary method of waste disposal because this method was cheap and convenient, and because the threat of groundwater contamination was not initially recognized. As we have recognized some of the problems associated with poorly designed landfills, efforts to reduce the amount of material placed in landfills have been substantial. Although the amount of waste has increased, composting, recycling, and combustion for energy have removed significant amounts of materials from the waste stream, and the amount of material entering landfills has declined. (See figure 18.6.)

A modern **municipal solid waste landfill** is typically constructed above an impermeable clay layer. The selection of landfill sites is based on an understanding of local geologic conditions such as the presence of a suitable clay base, groundwater geology, and soil type. In addition, it is important to address local citizens' concerns. Once the site is selected, extensive construction

FIGURE 18.5 Burning Landfills In the past, it was common practice to burn the waste in landfills to reduce the volume. Waste is still being burned in sparsely populated areas in North America and other parts of the world.

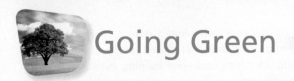

Going Green

Garbage Goes Green

When garbage is placed in a landfill, the bacteria present break down the organic matter and release a mixture of gases known as *landfill gas*. Landfill gas is about 50 percent carbon dioxide (CO_2) and 50 percent methane (CH_4). In addition, there are small amounts of other compounds, some of which cause odors. Since landfill gas is about 50 percent methane, it has about half the energy content of natural gas. The amount of landfill gas generated is influenced by the types and age of the waste buried in the landfill, the quantity and types of organic compounds in the waste, and the moisture content and temperature of the waste. Temperature and moisture levels are influenced by the local climate.

Because escaping landfill gas is an air pollutant, all landfills must control its release. In addition, both methane and carbon dioxide are greenhouse gases. Although nothing can be done with the carbon dioxide portion of the gas mixture, the methane can be burned. Some landfills simply flare (burn) the landfill gas on site to get rid of the methane and other odor-causing compounds. This simply converts one greenhouse gas into another ($CH_4 + O_2 \longrightarrow H_2O + CO_2$), although carbon dioxide is a less potent greenhouse gas than methane. However, if landfill gas is used as fuel it reduces the overall amount of greenhouse gas released because it reduces the amount of fossil fuel consumed.

In 2013, there were about 620 landfills in the United States that recovered about 100 billion cubic feet of landfill gas to generate heat or electricity or power vehicles. Although this is a tiny amount of total U.S. energy consumption, it was enough energy to provide electricity to over a million homes. The U.S. EPA lists an additional 450 landfills that have the potential to use landfill gas as a fuel source.

About 75 percent of landfill gas projects are used to produce electricity. However, the following list describes additional ways in which landfill gas is used.

Jackson County, NC—Gas from a closed landfill is burned to heat a greenhouse where the county grows plants for public landscaping.

Livermore, CA—Landfill gas is compressed to natural gas and used to fuel garbage trucks for Waste Management, Inc.

Fargo, ND—The city of Fargo initially flared the gas from its landfill to reduce odors. A local agricultural products processor, Cargill, Inc., partnered with the city to use the landfill gas in its boilers.

Greer, SC—The plant that makes the BMW Roadster powers its paint shop with landfill gas.

---

activities are necessary to prepare it for use. New landfills have an impermeable liner and complex bottom layers to trap contaminant-laden water, called **leachate,** leaking through the buried trash. The precipitation that falls into a landfill, coupled with any disposed liquid waste, dissolves water-soluble compounds and transports

particulate matter as it moves downward through the waste. The water that leaches through the site must be collected and treated. In addition, monitoring systems are necessary to detect methane gas production and groundwater contamination. In some cases, methane produced by decomposing waste is collected and used to

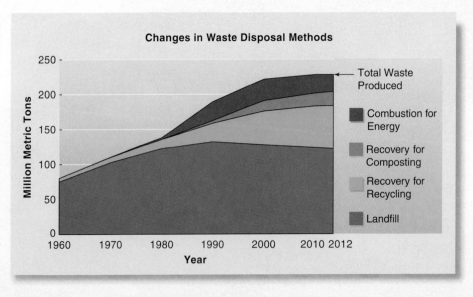

FIGURE 18.6 Changes in Waste Disposal Methods in the United States The landfill is still the primary method of waste disposal. Historically, landfills have been the cheapest means of disposal, but this is changing. Notice that recycling, composting, and energy recovery from combustion have grown substantially, while the amount of waste going to landfills has declined somewhat. A primary reason for these changes is the increasing cost of obtaining land for landfills, particularly for large cities.

Source: Data from the U.S. Environmental Protection Agency.

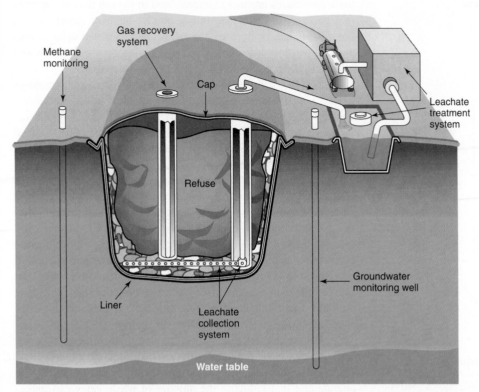

How a Modern Landfill Works

Methane monitoring

Gas recovery system

Cap

Refuse

Liner

Leachate collection system

Water table

Leachate treatment system

Groundwater monitoring well

FIGURE 18.7 A Well-Designed Modern Landfill A modern sanitary landfill is far different from a simple hole in the ground filled with trash. A modern landfill is a self-contained unit that is separated from the soil by impermeable membranes and sealed when filled. The area surrounding the landfill is monitored for methane gas and groundwater contamination to ensure that wastes are not escaping to the air or the groundwater.

Source: National Solid Waste Management Association.

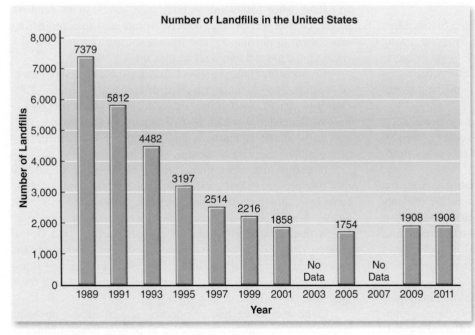

Number of Landfills in the United States

FIGURE 18.8 Reducing the Number of Landfills The number of landfills in the United States declined because they filled up or closed because their design and operation did not meet environmental standards.

Source: Data from the U.S. Environmental Protection Agency.

produce heat or to generate electricity. As a result of the technology involved, new landfills are becoming increasingly more complex and expensive. They currently cost up to $1 million per hectare ($400,000 per acre) to prepare. (See figure 18.7.)

Today, about 50 percent of the municipal solid waste from the United States goes into landfills. The number of landfills is declining. In 1988, there were about 8,000 landfills, and in 2011, there were about 1,908 active sanitary landfills. (See figure 18.8.) The number of landfills has decreased for two reasons. Many small, poorly run landfills have been closed because they were not meeting regulations. Others have closed because they reached their capacity. The overall capacity, however, has remained relatively constant because new landfills are much larger than the old ones.

Selecting sites for new landfills is extremely difficult because of (1) the difficulty in finding a geologically suitable site and (2) local opposition, which is commonly referred to as the NIMBY, or "not-in-my-backyard," syndrome. Resistance by the public comes from concern over groundwater contamination, rodents and other vectors of disease, odors, and truck traffic. Public officials look for alternatives to landfills to avoid controversy over landfill site selection.

Japan and European countries have already moved away from landfills as the primary method of waste disposal because of land scarcities and related environmental concerns. Germany and Sweden have eliminated landfills as a disposal method. European Union countries are on a schedule to remove biowaste from landfills, which is reducing the need for landfills. Instead, recycling and incineration are the primary methods. (See figure 18.9.) In addition, the energy produced by incineration can be used for electric generation or heating.

Incineration

Incineration is the process of burning refuse in a controlled manner. Today, about 12 percent of the municipal solid waste in the United States is incinerated; Canada incinerates about 5 percent. While some incinerators are used just to burn trash, most are designed to capture the heat, which is then used to make steam to produce electricity. The production of electricity partially offsets the cost of disposal. There are about

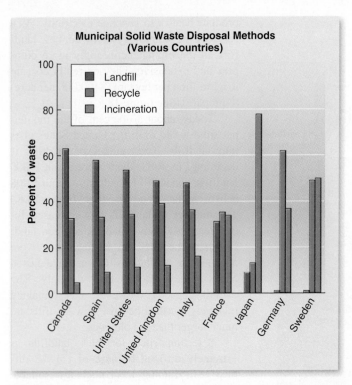

FIGURE 18.9 Disposal Methods Used in Various Countries
Although, historically, landfills have been the most commonly used method of disposing of waste, that is changing. European countries have agreed to ban biowaste (food, garden waste, etc.) from landfills and have instituted composting programs so that the amount of waste going to landfills is declining. Sweden and Germany have essentially eliminated landfills as a municipal solid waste deposal method and have high recycling and incineration rates. Generally incineration involves waste-to-energy plants. Japan has relied on incineration as its primary disposal method.

Source: Data from individual country sources.

90 combustors with energy recovery in the United States. Most incineration facilities burn unprocessed municipal solid waste. This is often referred to as **mass burn** technology. Some incinerators use refuse-derived fuel—collected refuse that has been processed into pellets prior to combustion. This is particularly useful with certain kinds of materials, such as tires.

Incinerators drastically reduce the amount of municipal solid waste—up to 90 percent by volume and 75 percent by weight. Primary risks of incineration, however, involve air quality problems and the toxicity and disposal of the ash.

Modern incinerators have many pollution control devices that trap nearly all of the pollutants produced. However, tiny amounts of pollutants are released into the atmosphere, including certain metals, acid gases, and classes of chemicals known as dioxins and furans, which have been implicated in birth defects and several kinds of cancer. The long-term risks from the emissions are still a subject of debate.

Ash from incineration is also an important issue. Small concentrations of heavy metals are present in both the fly ash captured from exhaust stacks and the bottom ash collected from these facilities. Because the ash contains lead, cadmium, mercury, and arsenic in varying concentrations from such items as batteries, lighting fixtures, and pigments, the ash is tested to determine if it should

be designated as a hazardous waste. This is a concern because the toxic substances are more concentrated in the ash than in the original garbage and can seep into groundwater from poorly sealed landfills. In nearly all cases, the ash is not designated as hazardous and can be placed in a landfill or used as aggregate for roads and other purposes.

The cost of the land and construction for new incinerators are also major concerns facing many communities. Incinerator construction is often a municipality's single largest bond issue. Incinerator construction costs in North America in 2010 ranged from $50 million to $400 million, and the costs are not likely to decline.

Incineration is also more costly than landfills in most situations. As long as landfills are available, they will have a cost advantage. When cities are unable to dispose of their trash locally in a landfill and must begin to transport the trash to distant sites, incinerators become more cost effective. The U.S. Environmental Protection Agency has not looked favorably on the construction of new waste-to-energy facilities and has encouraged recycling and source reduction as more effective ways to reduce the solid waste problem. Critics have argued that cities and towns have impeded waste reduction and recycling efforts by putting a priority on incinerators and committing resources to them. Proponents of incineration have been known to oppose source reduction. They argue that incinerators need large amounts of municipal solid waste to operate and that reducing the amount of waste generated makes incineration impractical. Many communities that have opposed incineration say that they support a vigorous waste-reduction and recycling effort.

Composting

In nature, when plants and animals shed leaves, twigs, feathers, or fur, or if organisms die, insects, worms, and microorganisms cause them to decompose and form a nutrient-rich material that retains moisture and improves soil texture.

Composting is the process of allowing these natural processes of decomposition to transform discarded organic materials—grass clippings, leaves, food waste, or soiled paper—into **compost,** a humus-like material, which is used in gardens and around plantings as a source of nutrients and to improve the porosity and water-holding capacity of the soil.

In composting operations, proper management of air and moisture provides ideal conditions for these organisms to transform large quantities of organic material into compost in a few weeks. A good small-scale example is a backyard compost pile. Green materials (grass, kitchen vegetable scraps, and flower clippings) mixed with brown materials (twigs, dry leaves, and soiled paper towels) at a ratio of 1:3 provide a balance of nitrogen and carbon that helps microbes efficiently decompose these materials.

Large-scale municipal composting operations use the same principles of organic decomposition to process large volumes of organic materials. Composting facilities of various sizes and levels of technological sophistication accept materials such as yard trimmings, food scraps, biosolids from sewage treatment plants, wood

Science, Politics, & Policy

Dealing with e-Waste

Electronic waste or "e-waste" is a growing problem worldwide, as obsolete or broken computers and other electronic equipment is discarded. About 50 million metric tons of e-waste are produced worldwide each year. The amount of e-waste is expected to grow rapidly as people in China, India, and other rapidly developing countries purchase cell phones and other electronics. In the United States, about 2.3 million metric tons of e-waste are produced each year. In the U.S. about 20 percent of electronic items are recycled. The rest are incinerated, placed in landfills, or exported.

Electronic waste contains many valuable materials that can be recycled, including copper, gold, silver, lead, cadmium, chromium, mercury, and several other metals. Many of these metals are in higher concentration in e-waste than in their ores, so recycling makes economic sense. While these metals are valuable, many of them are also toxic. When e-waste is exported to places like China, India, Pakistan, Ivory Coast, Nigeria, and other developing countries, it typically ends up in recycling centers. Laborers use a variety of crude techniques that include smashing, melting, and using acids to extract valuable materials. Another common practice is to burn the plastic coating off wiring to get the copper wire. These unregulated practices typically occur in the open where they release toxins into the environment and endanger the health of workers.

There have been several policy approaches to address the issue of e-waste and the hazards produced by disposal. The *Basel Convention on the Control of Transboundary Movements of Hazardous Wastes and their Disposal* is a United Nations agreement, which is designed to protect the health of people and the environment by preventing the dumping of hazardous materials in developing countries. It went into effect in 1992. Although it was approved by 175 countries, it is widely ignored.

The European Union (EU) in 2003 imposed a ban on the export of e-waste, along with a requirement for producers of electronic goods to take back used electronics. However, officials estimate that only about one-third of discarded e-waste is handled according to the regulations, and many unscrupulous companies have found ways around the ban and continue to ship e-waste to developing countries.

In the United States, there are few policies or laws affecting exporting e-waste. In 2010, a law was introduced in Congress to ban the export of e-waste. It never came to a vote. A similar law was introduced in 2013 and was referred to committee. Several states have banned the disposal of e-waste in landfills.

Fortunately, several consumer electronics companies have instituted take-back programs. The website *electronicstakeback.com* lists major electronics manufacturers and their take-back policies. In addition the website *Earth911.com* will help you find the nearest spot to properly dispose of used electronics in the United States.

According to the EPA, about 25 percent of electronic waste was recycled in 2011.

Focus On

Resins Used in Consumer Packaging

Thermoplastics are plastics that can be remelted and reprocessed, usually with only minor changes in their properties. About 30 percent of thermoplastic resins produced are used in consumer packaging applications. The most commonly used resins are the following:

1. Polyethylene terephthalate (PET) is used extensively in rigid containers, particularly beverage bottles for carbonated beverages and medicine containers.

2. High-density polyethylene (HDPE) is used for rigid containers, such as milk and water jugs, household-product containers, and motor oil bottles.

3. Polyvinyl chloride (PVC) is a tough plastic often used in construction and plumbing. It is also used in some food, shampoo, oil, and household-product containers.

4. Low-density polyethylene (LDPE) is often used in films and bags.

5. Polypropylene (PP) is used in a variety of areas, from yogurt containers to battery cases to disposable diaper linings. It is frequently interchanged for polyethylene or polystyrene.

6. Polystyrene (PS) is used in foam cups, trays, and food containers. In its rigid form, it is used in plastic cutlery.

7. Other. These usually contain layers of different kinds of resins and are most commonly used for squeezable bottles (for example, for ketchup).

Currently, HDPE and PET are the two most commonly recycled resins because containers made of these resins are typically recovered by municipal recycling programs. Other plastics are less frequently accepted. Most of the recycled LDPE is from commercial establishments that receive large numbers of shipments wrapped in LDPE.

PET

HDPE

PVC

LDPE

PP

PS

Other

Recycling Rates for Plastic Resins Used in Packaging (2012)

Kind of packaging	Percent recycled
Polyethylene Terephthalate (PET)	24.2%
High-density polyethylene	16.0%
Bags, sacks, and wraps (LDPE, LLDPE, and HDPE)	11.5%
Polystyrene (PS)	3.8
Polypropylene (PP)	2.1%
Polyvinyl chloride (PVC)	0

Source: Data from EPA Municipal Solid Waste (MSW) in the United States: 2012 Facts and Figures

shavings, unrecyclable paper, and other organic materials. These materials undergo processing—shredding, turning, and mixing—and, depending on the materials, can be turned into compost in a period ranging from 8 to 24 weeks. About 3,800 composting facilities are in use in the United States. Nearly 60 percent of yard trimmings is converted into mulch or composted in the United States through municipal programs. (See figure 18.10.) Most municipal programs involve one of three composting methods: windrows, aerated piles, or enclosed vessels.

- *Windrow* systems involve placing the compostable materials into long piles or rows called windrows. Tractors with front-end loaders or other kinds of specialized machinery are used to turn the piles periodically. Turning mixes the different kinds of materials and aerates the mixture.

- *Aerated piles* are large piles of material that have air pumped through them (aeration) so that no mechanical turning or agitation is necessary. They also typically are covered with a layer of mature compost or other material to insulate the pile to keep it at an optimal temperature.

- *Enclosed vessels* can also be used to compost materials very rapidly (within days). However, these systems are much more technologically complex. In such systems, compostable material is fed into a drum, silo, or other structure where the environmental conditions are closely controlled, and the material is mechanically aerated and mixed.

Source Reduction

The simplest way to reduce waste is to prevent it from ever becoming waste in the first place. **Source reduction** is the practice of designing, manufacturing, purchasing, using, and reusing materials so that the amount of waste or its toxicity is reduced.

Design changes to soft drink bottles and milk jugs are good examples of source reduction. Since 1977, the weight of a 2-liter plastic soft drink bottle has been reduced from 68 grams (2.4 ounces) to 51 grams (1.8 ounces). That translates to 114 million kilograms (250 million pounds) of plastic per year that has been kept out of the waste stream. The weight of a plastic milk jug has been reduced 30 percent.

Manufacturing processes have been changed in many industries to reduce the amount of waste produced. One of the simplest ways to reduce waste is to pay careful attention to leaks, spills, and accidents during the manufacturing process. All of these incidents generate waste, and their prevention reduces the amount of raw material needed and the amount of waste generated.

Purchasing decisions can significantly reduce the amount of waste produced. In many cases, consumers and businesses can choose to purchase things that have reduced packaging waste. You can choose to purchase products in larger sizes so that the amount of waste produced is reduced. In addition, careful planning of the quantities purchased can prevent unused surplus materials from becoming part of the waste stream.

Using materials in such a way that waste is not generated is an important means of curbing waste. Using less-hazardous alternatives for certain items (e.g., cleaning products and pesticides), sharing products that contain hazardous chemicals instead of throwing out leftovers, following label directions carefully, and using the smallest amount necessary are ways to reduce waste or its toxicity.

Reusing items is a way to reduce waste at the source because it delays or prevents the entry of reused items into the waste collection and disposal system. For example, many

(a) Individuals who compost reduce the amount of yard waste going to landfills

(b) Mechanical mixing of a windrow of compost

FIGURE 18.10 Diverting Yard Waste Through Composting Since yard waste is an important segment of the solid waste stream and can be converted to compost; many states have passed laws that prohibit the deposition of yard waste in landfills. This will extend the useful life of the landfill. Cities often encourage individuals to compost their own yard waste. To accommodate their citizens, many communities still collect yard waste but have instituted composting programs to deal with that waste.

industries participate in waste exchanges that allow a waste product from one industry to be used as a raw material in another industry. In such cases, both industries benefit because the waste producer does not need to pay to dispose of the waste and the industry using the waste has an inexpensive source of a raw material.

Most businesses and manufacturers have a strong economic incentive to make sure that they get the most from all the materials used in their operations. Any activities that reduce the amount of waste produced reduce the cost of waste disposal, the amount of raw materials needed, and the amount of pollution generated. This economic incentive also works at the consumer level. In the United States, over 4,000 communities have instituted "pay-as-you-throw" programs in which citizens pay for each can or bag of trash they set out for disposal rather than through the tax base or a flat fee. When these households reduce waste at the source, they dispose of less trash and pay lower trash bills.

Recycling

Recycling is one of the best environmental success stories of the late twentieth century. Some benefits of recycling are resource conservation, pollutant reduction, energy savings, job creation, and reduced need for landfills and incinerators. In the United States, recycling, including composting, diverted about 35 percent of the solid waste stream from landfills and incinerators in 2012, up from about 16 percent in 1990. (See figure 18.11.) Several kinds of programs have contributed to the increase in the recycling rate. However, incentives are needed to encourage people to participate in recycling programs. Several kinds of programs have been successful.

Beverage Container Deposit-Refund programs provide an economic incentive to recycle. Disposable containers are an increasing problem. In 1990 the average U.S. citizen disposed of 550 drink containers. In 2010 this had risen to 784, about a 43 percent increase. Beverage containers currently constitute 25 percent of household trash. In a deposit-refund program a cash value in the form of a deposit is placed on the glass, aluminum, or plastic beverage container. Thus, consumers have an incentive to return their containers for the redemption value. All provinces and most territories in Canada and at least 12 countries in Europe have bottle bills.

In the United States, ten states—California, Connecticut, Hawaii, Iowa, Maine, Massachusetts, Michigan, New York, Oregon, and Vermont—currently have beverage container deposit laws. They have recycling rates of 70 percent or more on containers covered by their laws. The average for the United States is about 35 percent. Benefits of recycling containers include energy savings and reduced release of greenhouse gases because it takes less energy to remanufacture containers from waste than from virgin raw materials. Less trash on roadways is another advantage. Two years after Oregon instituted its bottle law the amount of roadside trash declined 49 percent.

Many argue that a national bottle bill is long overdue. A national bottle bill would reduce litter, save energy and money, and

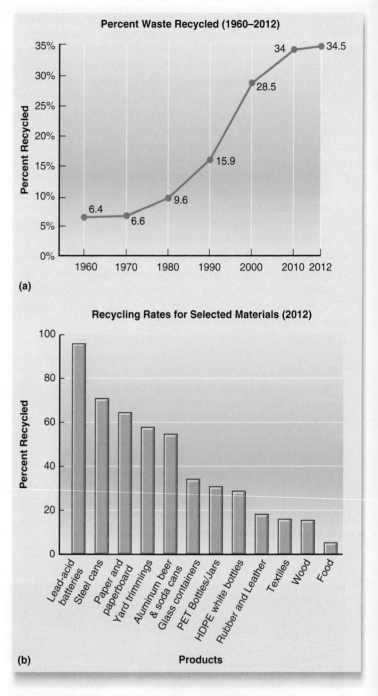

FIGURE 18.11 Recycling Rates (a) Recycling rates have increased substantially in recent years. (b) Recycling rates for materials that have high value such as automobile batteries are extremely high. Other materials are more difficult to market. But recycling rates today are much higher than in the past as technology and markets have found uses for materials that once were considered valueless.

Source: Data from the U.S. Environmental Protection Agency, *Municipal Solid Waste in the United States*, 2012 Facts and Figures.

create jobs. It would also help to conserve natural resources. But the lobbying efforts of the soft drink and brewing industries are very strong, and Congress has failed to pass a national container law.

Mandatory recycling laws provide a statutory incentive to recycle. Many states and cities have passed mandatory recycling

What's Your Take?

The growth of recycling programs has been pushed in part by municipalities' need to reduce waste and by the satisfaction people feel in taking the responsibility to recycle. These two forces have driven recycling's rise in spite of the fact that it has often not been financially profitable.

In fact, many of the increasingly popular municipal recycling programs are run at an economic loss. Formulate an argument that justifies the continuation of a community recycling program that currently operates at a loss.

laws. (See figure 18.12.) Some of these laws simply require that residents separate their recyclables from other trash. Other laws are aimed at particular products such as beverage containers and require that they be recycled. Some are aimed at businesses and require them to recycle certain kinds of materials such as cardboard or batteries. Finally, some laws forbid the disposal of certain kinds of materials in landfills. Therefore, the materials must be recycled or dealt with in some other way. For example, banning yard waste from landfills has resulted in extensive composting programs.

Those states and cities with mandatory recycling laws understandably have high recycling rates. For example, California mandated 50 percent diversion of waste through recycling and other means by 2001. Although many cities did not meet the goal,

by 2004, the statewide recycling rate was 49 percent. In 2012, California established a new goal of recycling 75 percent of its municipal solid waste by 2020.

Curbside recycling provides a convenient way for people to recycle. In 1990, a thousand U.S. cities had curbside recycling programs. By 2011, the number had grown to about 9,800 programs, which served 70 percent of the U.S. population.

Some large cities—such as Portland, Oregon; San Jose, California; Los Angeles, California; and Minneapolis, Minnesota—have achieved recycling rates of 50 percent or more. In general, these cities have curbside recycling and accept a wide variety of materials, including junk mail and cereal boxes. By contrast, cities that do not provide curbside recycling, have much lower recycling rates.

Recycling Challenges

Although recycling programs have successfully reduced the amount of material that needs to be trucked to a landfill or incinerated, there are many technical and economic problems associated with recycling. Technical questions are of particular concern when recycling plastics. (See figure 18.13.) While the plastics used in packaging are recyclable, the recycling technology differs

FIGURE 18.12 Mandatory Recycling Many states and cities have passed mandatory residential recycling laws.

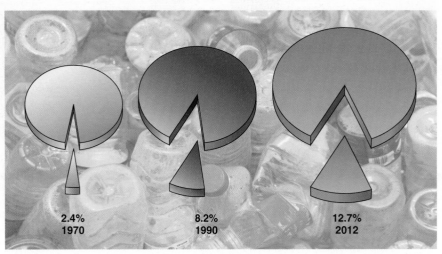

FIGURE 18.13 Increasing Amounts of Plastics in Trash Plastics are a growing component of municipal solid waste in North America.

Issues & Analysis

Paper or Plastic or Plastax?

Plastic bags are a convenience, a nuisance, and an environmental hazard. They are handy to carry items from the store. However, the lightweight bags are easily blown about and become an irritating eyesore on the landscape. They also cause environmental problems by clogging drains, which makes flooding problems worse. Many marine organisms such as turtles, whales, birds, and seals eat or become entangled in plastic bags. Sea turtles feed on jellyfish and often mistakenly eat plastic bags. Since they cannot digest the bags, they often die.

This has led to a worldwide interest in regulating the use of plastic bags. Generally the effort has taken one of two paths: outright ban on their use or a tax or fee for the use of a bag. Ireland became the first country to institute a plastax. The idea of a "plastax" or tax on plastic bags was first announced in 1999 and in 2002 the environment minister launched the program, one of the first of its kind in the world. For every bag used at the checkout counter of the supermarket a 15 euro cents (about US 25 cents) surcharge was added. This resulted in a change in the habits of shoppers. They now bring their own bags.

In 2008, China banned free plastic bags from shops and supermarkets and also banned the manufacture and use of ultra-thin bags less than 0.025mm thick. The bags are also banned from all public transportation, including buses, trains, and planes and from airports and scenic locations. The Chinese were using 3 billion plastic bags every day, which accounted for 3 to 5 percent of the total weight of landfills. It is estimated that it took 37 million barrels of crude oil a year to make all the bags needed for China.

In the United States, which has less than one-quarter of China's 1.3 billion people, almost 100 billion plastic bags are thrown out each year. Over 12 million barrels of oil are needed to manufacture the bags, of which only a fraction make it to the recycling bin. In 2008, San Francisco became the first major U.S. city to ban petroleum-based plastic bags in large grocery stores and drugstore chains. Since then, more than 100 cities, counties, and villages have either instituted bans of the bags or instituted a fee for their use.

Other countries have enacted various controls or bans on plastic bags, including the following:

Bangladesh: Banned the manufacturing and distribution of plastic bags in 2003 after it was found that they were blocking drainage systems and had been a major problem during the 1988 and 1998 floods that submerged two-thirds of the country.

Denmark: In 2004 Denmark introduced the Greentax. This tax was placed on plastic bags and paper bags and is included in the wholesale price of the bags and is not apparent to the consumer.

South Africa: South Africa passed a tax on plastic bags. Plastic bags have been dubbed the "national flower" because so many can be seen flapping from fences and caught in bushes.

India: Banned the production of plastic bags below 0.02 millimeters to prevent them from clogging drains.

Italy: Italy banned plastic bags in January 2011. Only biodegradable versions are permitted.

Mauritius: This nation has banned the import or local manufacture of non-degradable plastic bags, and has specified that only oxo-biodegradable can be considered degradable.

Barbados: Barbados charges 60 percent import surtax on non-degradable plastic bags but only 15 percent on oxo-biodegradable plastic bags.

- Do you think your community would support such a tax?
- Would you favor a tax on plastic bags?
- How would a ban on plastic bags alter your lifestyle?

from plastic to plastic. There are many different types of plastic polymers. Since each type has its own chemical makeup, different plastics cannot be recycled together. A milk container is likely to be high-density polyethylene (HDPE), while an egg container is polystyrene (PS), and a soft drink bottle is polyethylene terephthalate (PET).

The economics of recycling is a concern. Some recyclable products, like aluminum, have high value because the cost of making new aluminum from ore is much higher than making it from recycled aluminum. However, the economics of recycling glass is more difficult, particularly if the recycled material must be transported long distances. Glass is heavy and the raw material for glass (sand) is readily available. However, in recent years many businesses have developed around recycling, and there is an international business in waste materials.

Government policies are also important. When governments subsidize the mining, forestry, and fossil fuel industries, they make the use of these raw materials artificially inexpensive. For example, subsidizing the building of roads for extracting timber, taxing forest land at low rates, and selling trees from federal lands at low prices artificially lower the cost of virgin forest materials compared to recycled materials to make various kinds of paper products.

Summary

Municipal solid waste is managed by landfills, incineration, composting, waste reduction, and recycling. Landfills are the primary means of disposal; however, a contemporary landfill is significantly more complex and expensive than the simple holes in the ground of the past. The availability of suitable landfill land is also a problem in large metropolitan areas.

About 12 percent of the municipal solid waste in the United States is incinerated. While incineration does reduce the volume of municipal solid waste, the problems of ash disposal and air quality continue to be major concerns. There are several forms of composting that can keep organic wastes from entering a landfill.

The most fundamental way to reduce waste is to prevent it from ever becoming waste in the first place. Using less material in packaging, reusing items, and composting yard waste are all examples of source reduction. On an individual level, we can all attempt to reduce the amount of waste we generate.

About 35 percent of the waste generated in North America is handled through recycling. Recycling initiatives have grown rapidly in North America during the past several years. Some recyclable material such as aluminum is of high value while others like glass have lower value. Markets for low value materials are harder to develop. Plastic recycling requires separation of different kinds of plastics.

Future management of municipal solid waste will be an integrated approach involving landfills, incineration, composting, source reduction, and recycling. The degree to which any option will be used will depend on economics, changes in technology, and citizen awareness and involvement.

Acting Green

1. Buy things that last, keep them as long as possible, and have them repaired, if possible.
2. Buy things that are reusable or recyclable, and be sure to reuse and recycle them.
3. Use plastic or metal lunch boxes and metal or plastic garbage containers without throwaway plastic liners.
4. Use rechargeable batteries.
5. Skip the bag when you buy only a quart of milk, a loaf of bread, or anything you can carry with your hands.
6. Recycle all newspaper, glass, and aluminum, and any other items accepted for recycling in your community.
7. Reduce the amount of junk mail you get. This can be accomplished by writing to Mail Preference Service, Attn: Dept: 13885751, Direct Marketing Association, P.O. Box 282, Carmel, NY 10512. Ask that your name not be sold to large mailing-list companies. You may also register online by going to the website. Type Mail Preference Service into your browser. Of the junk mail you do receive, recycle as much of the paper as possible.
8. Push for mandatory trash separation and recycling programs in your community and schools.
9. Compost your yard and food wastes, and pressure local officials to set up a community composting program.

Review Questions

1. How is lifestyle related to the quantity of municipal solid waste generated?
2. What conditions favor incineration over landfills?
3. Describe some of the problems associated with modern landfills.
4. What are four concerns associated with incineration?
5. Describe examples of source reduction.
6. Describe the importance of recycling household solid wastes.
7. Name several strategies that would help to encourage the growth of recycling.
8. Describe the various types of composting and the role of composting in solid waste management.
9. Describe why electronic waste is becoming a major problem.
10. Why is food waste a growing concern? Describe how some communities are addressing the food waste issue.
11. How is landfill gas turned into electricity?

Critical Thinking Questions

1. How can you help solve the solid waste problem?
2. Given that you have only so much time, should you spend your time acting locally, as a recycling coordinator, for example, or advocating for larger political and economic changes at the national level, changes that would solve the waste problems? Why? Or should you do nothing? Why?
3. How does your school or city deal with solid waste? Can solid waste production be limited at your institution or city? How? What barriers exist that might make it difficult to limit solid waste production?
4. It is possible to have a high standard of living, as in North America and Western Europe, and not produce large amounts of solid waste. How?
5. Incineration of solid waste is controversial. Do you support solid waste incineration in general? Would you support an incineration facility in your neighborhood?

|ENVIRONMENTAL SCIENCE

To access additional resources for this chapter, please visit ConnectPlus® at connect.mheducation.com. There you will find interactive exercises, including Google Earth™, additional Case Studies, and SmartBook™, an adaptive eBook that integrates our LearnSmart® adaptive learning technology.

chapter 19

Environmental Regulations: Hazardous Substances and Wastes

The Department of Energy is responsible for the cleanup of radioactive waste at former nuclear weapons installations. These technicians are examining a drum containing depleted uranium.

CHAPTER OUTLINE

Hazardous Materials Incidents and Regulatory Response

OBJECTIVES

After reading this chapter, you should be able to:

- Distinguish between hazardous substances and hazardous wastes.
- Distinguish between hazardous and toxic substances.
- Describe the characteristics used by the U.S. Department of Transportation to identify hazardous substances.
- Describe the two methods used by the U.S. Environmental Protection Agency to identify hazardous waste.
- Name the three primary routes of entry of toxic materials into the human body.
- Describe how the threshold level of a toxic material is determined.
- Describe the difference between chronic and acute exposures to hazardous wastes.

- Understand the difference between persistent and nonpersistent pollutants.
- Describe common ways that hazardous wastes enter the environment.
- Describe why hazardous-waste dump sites developed.
- Identify the major impacts of RCRA, CERCLA, and SBLRBRA.
- Identify the major industries involved in releasing toxic chemicals to the environment.
- Describe how hazardous wastes are managed, and list five technologies used in their disposal.
- Describe the importance of source reduction with regard to hazardous wastes.

- Understand the difficulties associated with determining cleanup criteria for hazardous substances/ wastes.
- Describe the goal of the Basel Convention.
- Describe the major sources of nuclear waste.
- Describe how transuranic waste is produced and how it is being disposed of.
- Describe how uranium mining and milling waste is managed.
- Describe how high-level radiation waste is stored.
- Describe how low-level radioactive waste is stored.

Hazardous Materials Incidents and Regulatory Response

Hazardous materials are manufactured, transported, sold, and used on a daily basis. In most cases they do not cause problems, and we have become accustomed to handling them in a safe manner. For example, gasoline is a highly flammable material we use regularly. Explosions and other problems are rare because gasoline pumps are designed to reduce the probability of an accidental release of gasoline. However, hazardous materials cause accidents, injuries, and deaths every year. The following list includes some spectacular examples.

- Minamata City, Japan, 1956. The release of mercury compounds into Minamata Bay by a petrochemical company resulted in production of methylmercury by microorganisms. The methylmercury became part of the marine food chain and reached dangerous levels in fish and shellfish. People who ate the fish developed mercury poisoning. About 3,000 people developed debilitating disease including newborn children whose mothers ate contaminated seafood.

- Bhopal, India, 1984. A leak of highly toxic methyl isocyanate gas from a plant manufacturing a common insecticide (Sevin) resulted in thousands of deaths. Estimates vary, but it is probable that on the order of 20,000 people died directly as a result of the gas release or from complications of health problems caused by the release.

- Chernobyl, Ukraine, 1986. The meltdown and explosion of a nuclear power plant at Chernobyl killed over 4,000 people and spread radioactive material over a wide area. Birth defects and certain cancers (particularly thyroid cancer) have increased in those exposed to the radioactive fallout.

- Abidjan, Côte d'Ivoire, 2006. The ship *Proba Koala* offloaded 500 metric tons of toxic waste in Abidjan. The waste, which contained hydrogen sulfide, sodium hydroxide, and phenols, was dumped by local contractors at a variety of sites around the city. Seventeen people died and thousands were injured.

- Gulf of Mexico 2010. The explosion and sinking of the Deep Water Horizon drilling rig killed 11 workers and caused the release of about 4.9 million barrels of oil into the Gulf of Mexico. The oil causes several kinds of problems. Some components of the oil are toxic, the oil coats animals causing their death, and the breakdown of oil by microorganisms depletes the water of oxygen.

- West, Texas, 2013. Ammonium nitrate exploded at a fertilizer plant, killing 15 people, injuring about 200 people, and destroying many buildings. Many of the people who died were firefighters who had responded to a fire at the plant.

- Lac-Mégantic, Quebec, 2013. A runaway train hauling crude oil that was more flammable than usual crashed and exploded. The explosion and fire killed 47 people and destroyed about 30 buildings.

Typically, after disasters like these, there are investigations that lead to new rules, regulations, or legislation that makes a recurrence of the disaster less likely. A significant part of this chapter deals with laws and regulations designed to protect the public from hazardous materials.

The burning Deep Water Horizon drilling rig.

19.1 Hazardous and Toxic Materials in Our Environment

Our modern technological society makes use of a large number of substances that are hazardous or toxic. The benefits gained from using these materials must be weighed against the risks associated with their use. Other toxic and hazardous materials are produced as a by-product of industrial activities.

At sites around the world, accidental or purposeful releases of hazardous and toxic chemicals are contaminating the land, air, and water. The potential health effects of these chemicals range from minor, short-term discomforts, such as headaches and nausea, to serious health problems, such as cancers and birth defects (that may not manifest themselves for years), to major accidents that cause immediate injury or death.

Because of the health and safety issues related to the manufacture and use of hazardous and toxic materials, governments and international agencies set rules and standards to protect the health and safety of people. However, controlling the release of these substances is difficult, since there are so many places in their cycles of use where they may be released. (See figure 19.1.) Furthermore, careless, neglectful, and criminally motivated individuals cause accidental or purposeful releases.

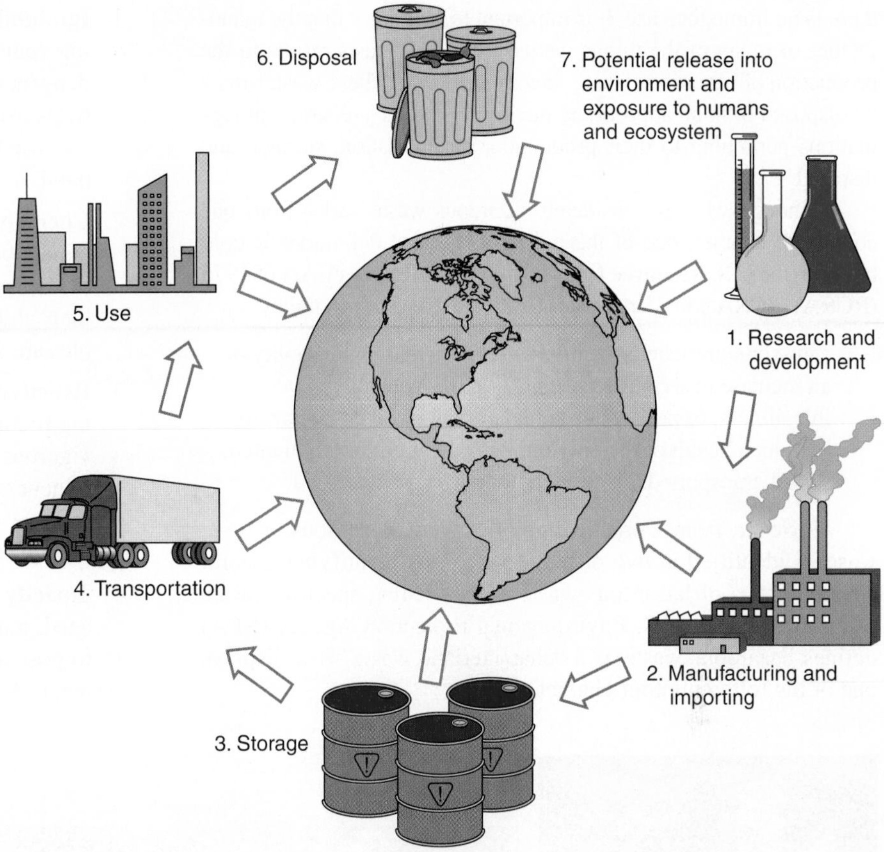

FIGURE 19.1 The Life Cycle of Toxic Substances Controlling problems related to the use of hazardous substances is complicated because of the many steps involved in a substance's life cycle.

19.2 Characterizing Hazardous and Toxic Materials

To begin, it is important to clarify various uses of the words *hazardous* and *toxic*. **Hazardous substances** or **hazardous materials** are those that can cause harm to humans or the environment. **Toxic** materials are a narrow group of hazardous substances that are poisonous and cause death or serious injury to humans and other organisms by interfering with normal body physiology. Hazardous materials, the broader term, refers to all dangerous materials, including toxic ones, that present an immediate or long-term human health risk or environmental risk. Many hazardous or toxic materials are raw materials or finished products used by industry or the public and include things such as gasoline, pesticides, industrial chemicals, and medicines.

Identifying Hazardous Materials

Different organizations and government agencies have differing definitions of what is hazardous. The U.S. Department of Transportation (DOT) lists several kinds of hazardous materials.

Explosives—materials that cause a rapid release of gas and heat
Gases—may be flammable, nonflammable, or toxic
Flammable liquids

Flammable solids

Oxidizing substances and organic peroxides—substances that release oxygen that enables or enhances the burning of materials.

Poisonous (toxic) materials and infectious substances—infectious substances may include organisms or their products that can cause disease

Radioactive material—materials that give off ionizing radiation

Corrosives—materials that cause damage to human skin

Miscellaneous dangerous goods

Some hazardous materials fall into several of these categories. Gasoline, for example, is a flammable liquid and is toxic.

Hazardous Waste—A Special Category of Hazardous Material

It is important to distinguish between hazardous substances and hazardous wastes. Although the health and safety considerations regarding hazardous substances and hazardous wastes are similar, the legal and regulatory implications are quite different. Hazardous substances are materials that are used in business and industry for the production of goods and services. Typically, hazardous substances are consumed or modified in industrial processes. **Hazardous wastes** are by-products of industrial, business, or household activities for which there is no immediate use. It is important to recognize that the manufacture of many of the things we use on a daily basis results in the production of hazardous waste. (See figure 19.2.) These wastes must be disposed of in an appropriate manner, and there are stringent regulations pertaining to their production, transportation, storage, and disposal.

Although what is considered hazardous waste varies from one country to another, one of the most widely used definitions is contained in the U.S. **Resource Conservation and Recovery Act** of 1976 (RCRA). RCRA considers wastes toxic and/or hazardous if they:

cause or significantly contribute to an increase in mortality or an increase in serious irreversible, or incapacitating reversible, illness; or pose a substantial present or potential hazard to human health or the environment when improperly treated, stored, transported, disposed of, or otherwise managed.

However, from a practical point of view, hazardous or toxic waste is identified in two different ways—by identifying general characteristics of hazardous waste and by listing specific wastes as hazardous. The U.S. Environmental Protection Agency (EPA) defines hazardous waste as a "characteristic waste" if it displays one of the following four characteristic.

1. **Ignitability**—Describes materials that pose a fire hazard during routine management. Fires not only present immediate dangers of heat and smoke but also can spread harmful particles over wide areas. Common examples of ignitable wastes are: used solvents, discarded aerosol cans of paint, oil-based paint, or solvent soaked rags.

2. **Corrosiveness**—Describes materials that require special containers because of their ability to corrode standard materials or that require segregation from other materials because of their ability to dissolve toxic contaminants. Common examples are strong acids and bases.

3. **Reactivity** (or explosiveness)—Describes materials that, during routine management, tend to react spontaneously, to react vigorously with air or water, to be unstable to shock or heat, to generate toxic gases, or to explode. Common examples of reactive wastes are: bleach solutions and cyanide-containing plating solutions.

4. **Toxicity**—Describes materials that, when improperly managed, may release toxicants (poisons) in sufficient quantities to pose a substantial hazard to human health or the environment. Almost everything that is hazardous is toxic in high

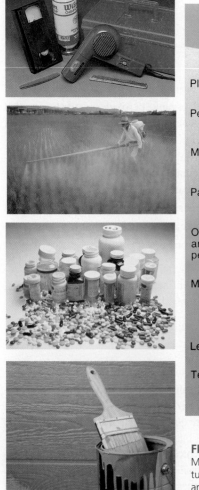

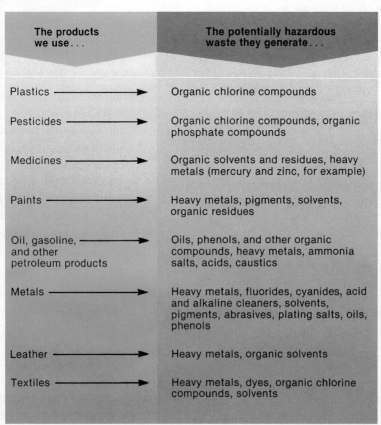

The products we use...	The potentially hazardous waste they generate...
Plastics	Organic chlorine compounds
Pesticides	Organic chlorine compounds, organic phosphate compounds
Medicines	Organic solvents and residues, heavy metals (mercury and zinc, for example)
Paints	Heavy metals, pigments, solvents, organic residues
Oil, gasoline, and other petroleum products	Oils, phenols, and other organic compounds, heavy metals, ammonia salts, acids, caustics
Metals	Heavy metals, fluorides, cyanides, acid and alkaline cleaners, solvents, pigments, abrasives, plating salts, oils, phenols
Leather	Heavy metals, organic solvents
Textiles	Heavy metals, dyes, organic chlorine compounds, solvents

FIGURE 19.2 Common Materials Can Produce Hazardous Wastes
Many commonly used materials generate hazardous wastes during their manufacture. Some consumer products such as pesticides, paints, and solvents are also hazardous and should be disposed of in an approved manner.

enough quantities. For example, tiny amounts of carbon dioxide in the air are not toxic, but high levels are. Many household waste chemicals are toxic.

Because of the difficulty in developing simple definitions of hazardous wastes, listing is the most common method for defining hazardous waste in many countries. The U.S. Environmental Protection Agency has compiled such a list of hazardous wastes (the so-called listed wastes). There are about 700 listed hazardous wastes on several lists. It should be noted that although EPA has the responsibility for developing lists of hazardous wastes, other U.S. agencies such as the Occupational Safety and Health Administration (OSHA) and the Department of Transportation (DOT) have compiled their own lists, and there is a comprehensive "list of lists" that documents many of these. If a waste appears on any other agency list, it is to be considered hazardous by the EPA.

19.3 Controlling Hazardous Materials and Waste

To regulate the use of toxic and hazardous substances and the generation of toxic and hazardous wastes, most countries draw up a list of specific substances that have been scientifically linked to adverse human health or environmental effects. However, since many potentially harmful chemical compounds have yet to be tested adequately, most lists include only the known offenders. Historically, we have often identified toxic materials only after their effects have shown up in humans or other animals. Asbestos was identified as a cause of lung cancer in humans who were exposed on the job, and DDT was identified as toxic to birds when robins began to die and eagles and other fish-eating birds failed to reproduce. Once these substances were identified as toxic, their use was regulated.

Laws and Regulations

Whether a hazardous substance is a raw material, an ingredient in a product, or a waste, there are problems in determining how to regulate it. The United States has dealt with hazardous substances and wastes by using "command-and-control" methods of governmental regulations. Major efforts to control hazardous materials began with the development of the Environmental Protection Agency and the Occupational Safety and Health Administration in 1970. Table 19.1 lists several important laws that have determined how we deal with hazardous substances and wastes.

The purpose of all environmental regulations is to control and/or stop pollution and environmental degradation. When setting regulations, governments and their regulatory agencies attempt to determine how to fairly enforce these measures to successfully control exposures to humans and the environment. Since controlling hazardous releases is expensive for industry, regulations must apply to all so that all pay the same cost for reducing their pollution. When dealing with past pollution events, governments must assess who will pay for cleanup of hazardous substances or wastes that enter the environment. The Potentially Responsible Party

Table 19.1 Key U.S. Laws Related to Hazardous and Toxic Materials

Year	Statute Name	Function/Goal/Action
1947	**FIFRA:** Federal Insecticide, Fungicide, and Rodenticide Act	Required registration of pesticides Required applicators of restricted use pesticides to be certified
1970	**CAA:** Clean Air Act	To reduce air pollutants in general Specific attention to toxic air pollutants and ozone destroying chemicals Air quality greatly improved since 1970
1970	**OSHA:** Occupational Safety and Health Act	Addresses workplace safety in general Specifically regulates worker exposure to toxic and hazardous materials
1972	**CWA:** Clean Water Act	Introduced permit system for point sources of pollutants Greatly improved water quality since its enactment
1974	**SDWA:** Safe Drinking Water Act	Set maximum contaminant levels for microbes and about 100 chemicals Greatly improved drinking water quality
1975	**HMTA:** Hazardous Materials Transportation Act	Requires that hazardous materials be identified and labeled Containers and vehicles carrying hazardous materials must have placards that identify the kinds of hazards contained in the shipment
1976	**TSCA:** Toxic Substance Control Act	Importers and manufacturers must notify EPA if they intend to import or manufacture a new substance Importers and manufacturers must also provide data on the toxicity of any new materials
1976	**RCRA:** Resource Conservation and Recovery Act	Regulates hazardous waste from production through transportation to final disposal
1980	**CERCLA:** Comprehensive Environmental Response Compensation and Liability Act (Superfund)	The goal was to clean up abandoned contaminated sites Provided funding to clean up sites Superfund sites were identified and categorized
1986	**SARA:** Superfund Amendments and Reauthorization Act	Primarily provided additional funding to clean up Superfund sites Established Innocent Landowner Defense
2002	**SBLRBRA:** Small Business Liability Relief and Brownfields Revitalization Act	Encouraged redevelopment of old contaminated industrial sites by limiting liability of those doing the redevelopment and providing funding

(PRP) will have to pay for dealing with the remediation of the site once hazardous substances/wastes are found in the environment. Regulatory agencies try to identify the PRP and force that party or parties to comply with remediation requirements. However, if an industry is put out of business due to environmental fines

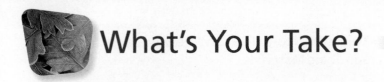

What's Your Take?

and cleanup costs, it will not be able to contribute to the control and/or remediation of the toxic material. The area will revert to an "orphaned" site, and other sources of money will have to be sought for its cleanup.

Voluntary Standards

Some voluntary industry standards have been developed and have even been incorporated into federal acts.

1. **ASTM International Phase I Environmental Site Assessment standard E-1527** is an example of a voluntary standard which involves prior assessment before beginning a project. Before financing a loan to acquire commercial real estate, or before receiving grant monies under certain federal programs, a business must first conduct an environmental site assessment to determine potential environmental liabilities associated with the real estate. This will help to determine who will pay for potential cleanup activities or how much liability the purchaser of the property will have.

2. The **International Organization for Standardization** has a standard for environmental management systems known as ISO 14000. Many organizations and corporations seek ISO 14000 certification as a way to indicate that they are controlling their environmental impact and improving their environmental performance.

19.4 Managing Health Risks Associated with Toxic Substances

Establishing the health consequences of exposure to toxic chemicals is extremely complicated. The problem of linking a particular chemical to specific injuries or diseases is further compounded by the lack of adequate toxicity data on most hazardous substances. In addition, chemicals may affect the body differently in high and low doses, may interact with one another, or differ in how long they persist in the environment.

Acute and Chronic Toxicity

The health effects of exposure to toxic materials may differ depending on whether the exposure is acute or chronic. **Acute toxicity** occurs when a person is exposed to one massive dose of a substance and becomes ill. **Chronic toxicity** occurs when a person is exposed to small doses over long periods. Acute toxicity is readily apparent because organisms respond to the toxin shortly after being exposed. Chronic toxicity is much more difficult to determine because the effects may not be seen for years. Furthermore, an acute exposure may make an organism ill but not kill it, while chronic exposure to a toxic material may cause death. A good example of this effect is alcohol toxicity. Consuming extremely high amounts of alcohol can result in death (acute toxicity and death). Consuming moderate amounts may result in illness (acute toxicity and full recovery). Consuming moderate amounts over a number of years may result in liver damage and death (chronic toxicity and death).

Another example of chronic toxicity involves lead. Lead was used in paints, gasoline, and pottery glazes for many years, but researchers discovered that it has harmful effects. The chronic effects on the nervous system are most noticeable in children, particularly when children eat paint chips.

Synergism

Another problem in regulating hazardous materials is assessing the effects of mixtures of chemicals. Most toxicological studies focus on a single compound, even though industry workers may be exposed to a variety of chemicals, and hazardous wastes often consist of mixtures of compounds. Although the materials may be relatively harmless as separate compounds, once mixed, they may become highly toxic and cause more serious problems than do individual pollutants. This is referred to as **synergism.** For example, all uranium miners are exposed to radioactive gases, but those who smoke tobacco and thus are exposed to the toxins in tobacco smoke have unusually high incidences of lung cancer. Apparently, the radioactive gases found in uranium mines interact synergistically with the carcinogens found in tobacco smoke.

Persistent and Nonpersistent Pollutants

The regulation of hazardous and toxic materials is also influenced by the degree of persistence of the pollutant. **Persistent pollutants** are those that remain in the environment for many years in an unchanged condition. Most of the persistent pollutants are human-made materials.

An example of a persistent pollutant is DDT. It was used as an effective pesticide worldwide and is still used in some countries because it is so inexpensive and very effective in killing mosquitoes that spread malaria. However, once released into the environment, it accumulates in the food chain and causes death when its concentration is high enough. (See chapter 14 for a discussion of DDT as a pesticide.)

Another widely used group of synthetic compounds of environmental concern is polychlorinated biphenyls (PCBs). PCBs are highly stable compounds that resist changes from heat, acids, bases, and oxidation. These characteristics make PCBs desirable for industrial use but also make them persistent pollutants when they are released into the environment. At one time, these materials were commonly used in electrical transformers and capacitors. Other uses included inks, plastics, tapes, paints, glues, waxes, and polishes. Although the manufacture of PCBs in the United States stopped in 1977, these persistent chemicals are still present in the soil and sediments and continue to do harm. PCBs are harmful to fish and other aquatic forms of life because they interfere with reproduction. In humans, PCBs produce liver ailments and skin lesions. In high concentration, they can damage the nervous system, and they are suspected carcinogens. Recent studies of PCB levels in fish and wildlife show that the levels are falling.

In addition to synthetic compounds, our society uses heavy metals for many purposes. Mercury, beryllium, arsenic, lead, and cadmium are examples of heavy metals that are toxic. These metals are used as alloys with other metals, in batteries, and have many other special applications. In addition, these metals may be released as by-products of the extraction and use of other metals. When released into the environment, they enter the food chain and become concentrated. In humans, these metals can produce kidney and liver disorders, weaken the bone structure, damage the central nervous system, cause blindness, and lead to death. Because these materials are persistent, they can accumulate in the environment even though only small amounts might be released each year. When industries use these materials in a concentrated form, it presents a hazard not found naturally.

A **nonpersistent pollutant** does not remain in the environment for very long. Some nonpersistent pollutants are biodegradable. Others decompose as a result of exposure to environmental factors. Still others quickly disperse to concentrations that are too low to cause harm. Biodegradable materials are chemically changed by decomposer organisms, such as bacteria and fungi. PCBs, phenol and many other kinds of toxic organic materials can be destroyed by decomposer organisms. In fact, specific kinds of microorganisms are used in remediation of contaminated sites because of their ability to degrade specific toxins.

Other toxic materials, such as many insecticides, are destroyed by sunlight or reaction with oxygen or water in the atmosphere. These include the "soft biocides." For example, organophosphates usually decompose within several weeks. As a result, organophosphates do not accumulate in food chains because they are pollutants for only a short period of time.

Other toxic and hazardous materials such as carbon monoxide, ammonia, or hydrocarbons can be dispersed harmlessly into the atmosphere (as long as their concentration is not too great), where they eventually react with oxygen.

Because persistent materials can continue to do harm for a long time (chronic toxicity), their regulation is particularly important. Nonpersistent materials need to be kept below threshold levels to protect the public from acute toxicity. They are not likely to present a danger of chronic toxicity, since they either disperse or decompose.

Setting Exposure Limits

Even after a material is identified as hazardous or toxic, there are problems in determining appropriate exposure limits. Nearly all substances are toxic in sufficiently high doses. The question is, At what dose does a chemical become a hazard? There is no easy way to establish acceptable levels. Several government agencies set limits for different purposes. The Occupational Safety and Health Administration, the Food and Drug Administration, the U.S. Public Health Service, the Environmental Protection Agency, and others publish guidelines or set exposure limits for hazardous substances in the air, water, and soil.

Organizations such as National Institute of Occupational Safety and Health (NIOSH) and American Conference of Governmental Industrial Hygienists (ACGIH), among others, also test and set exposure limits in a variety of units (PEL—Permissible Exposure Limits; STELs—Short Term Exposure Limits; TWA—Time Weighted Average; CL—Ceiling Limit; to name a few).

Furthermore, it is important to recognize that people can be exposed in three primary ways (routes of entry): by breathing the material, by consuming the material through the mouth, or by absorbing the material through the skin. Each of these routes may require different exposure classifications. Regardless, for any new compounds that are to be brought on the market, extensive toxicology studies must be done to establish their ability to do harm. Usually these involve tests on animals. (See Focus On: Determining Toxicity.) Typically, the regulatory agency will determine the level of exposure at which none of the test animals is affected (**threshold level**) and then set the human exposure level lower to allow for a safety margin. This safety margin is important because it is known that threshold levels vary significantly among species, as well as among members of the same species. Even when concentrations are set, they may vary considerably from country to country. For example, in the Netherlands, 50 milligrams of cyanide per kilogram of waste is considered hazardous; in neighboring Belgium, the toxicity standard is fixed at 250 milligrams per kilogram.

19.5 How Hazardous Wastes Enter the Environment

There are many types of hazardous waste, ranging from waste contaminated with dioxins and heavy metals (such as mercury, cadmium, and lead) to organic wastes. These wastes can also take many forms, such as barrels of liquid waste or sludge, old computer parts, used batteries, gases, and incinerator ash.

In the U.S. and most economically developed countries, the release of hazardous wastes from business and industry is tightly regulated and severe legal penalties are levied on those who willingly or accidentally violate regulations. Although industry still

Focus On

Determining Toxicity

We are all exposed to potentially harmful toxins. The question is, at what levels is such exposure harmful? One measure of toxicity of a substance is its **LD$_{50}$**, the dosage of a substance that will kill (lethal dose) 50 percent of a test population (LD$_{50}$ = lethal dose 50 percent). Toxicity is measured in units of poisonous substance per kilogram of body weight. For example, the deadly chemical that causes botulism, a form of food poisoning, has an LD$_{50}$ in adult human males of 0.0014 milligram per kilogram. This means that if each of 100 human adult males weighing 100 kilograms consumed a dose of only 0.14 milligram—about the equivalent of a few grains of table salt—approximately 50 of them would die.

Obviously LD$_{50}$ determinations are not done on humans but are typically determined in experimental animals like rats and mice. This introduces a degree of uncertainty, since test animals may react to a toxin differently from humans. For example, the group of chemicals known as dioxins is quite lethal to guinea pigs but not to humans.

Lethal doses are not the only measure of the danger from toxic substances. In recent years, efforts

have been made to determine the minimum harmful dosages of toxins. These are also known as *threshold dosages*. Although the dose may be sublethal, it may have subtle effects on the functioning of various body systems.

The length of exposure further complicates the determination of toxicity values. Acute exposure refers to a single exposure lasting from a few seconds to a few days. Chronic exposure refers to continuous or repeated exposure for several days, months, or even years. Acute exposure usually is the result of a sudden accident, such as the tragedy at Bhopal, India, mentioned in the opening of this chapter. Acute exposures often make disaster headlines in the press, but chronic exposure to sublethal quantities of toxic materials presents a much greater hazard to public health. For example, millions of urban residents are continually exposed to low levels of a wide variety of air pollutants. Many deaths attributed to heart failure or such diseases as emphysema may actually be brought on by a lifetime of chronic exposure to sublethal amounts of pollutants in the air.

---

releases substantial amounts of hazardous waste, they have greatly reduced their hazardous waste releases. Thus, individual citizens are becoming more important as sources of hazardous waste releases.

Hazardous wastes enter the environment in several ways:

Evaporation of many kinds of substances releases molecules directly to the atmosphere. Many kinds of solvents used in paints and other industrial processes fall into this category. Other materials escape from faulty piping and valves. These materials are often not even thought of as hazardous waste but are called *fugitive emissions*. Uncontrolled or improper incineration of hazardous wastes, whether on land or at sea, can contaminate the atmosphere and the surrounding environment.

Although liquid or solid wastes are more easily contained, they can still enter the environment.

Spills, leaks, or purposeful releases result in hazardous materials being released to the water. Many fluids associated with automobiles are washed from streets and parking lots into storm drains. Citizens purposely dump unwanted liquids down the toilet or into drains.

Dumping or storing on land was a common practice in the past. Wastes were placed in containers, storage lagoons, or landfills. Groundwater contamination has resulted from leaking land disposal facilities. Once groundwater is polluted with hazardous wastes, the cost of reversing the damage is prohibitive. In fact, if an aquifer is contaminated with organic chemicals, restoring the water to its original state is seldom physically or economically

feasible. Chapter 15 deals with many other aspects of groundwater contamination. Mining operations still deposit most of their waste on land. Some of it contains hazardous materials.

Improper labeling and recordkeeping procedures can result in inadvertent releases. If workers are unable to distinguish hazardous waste from other kinds of waste materials, hazardous wastes are not properly disposed of.

19.6 Hazardous-Waste Dumps— The Regulatory Response

In the United States prior to the passage of the Resource Conservation and Recovery Act (RCRA) in 1976, hazardous waste was essentially unregulated. Similar conditions existed throughout most of the industrial nations of the world. Hazardous wastes were simply buried or dumped without any concern for potential environmental or health risks. Such uncontrolled sites included open dumps, landfills, bulk storage containers, and surface impoundments. As indicated earlier, these sites were typically located convenient to the industry and were often in environmentally sensitive areas, such as floodplains or wetlands. At the time, these land areas were considered unimportant and were frequently abandoned or filled in to make room for further development. Rain and melting snow soaked through the sites, carrying chemicals that contaminated groundwater. When groundwater reached streams and lakes, they were contaminated as well. When the sites became full or were abandoned, they were sometimes left uncovered, thus increasing the likelihood of water pollution from leaching or flooding and of people having direct contact with the wastes. At some sites, specifically the uncovered ones, the air was also contaminated by toxic vapors from evaporating liquid wastes or from uncontrolled chemical reactions. (See figure 19.3.) In North America alone, the number of abandoned or uncontrolled sites was over 25,000. Recognition of the severity and magnitude of the problem led to several important pieces of legislation to address hazardous-waste sites.

Resource Conservation and Recovery Act (RCRA)

RCRA gave the EPA the responsibility for regulating hazardous waste. It required that hazardous wastes be identified. It set standards to be met by producers of waste, those who transport it, and those who treat, dispose of, or store hazardous waste. An important feature of this legislation was the "cradle-to-grave" concept of hazardous-waste management that required all participants in the hazardous-waste chain to sign documentation of their role. The participants included those who generate waste, those who transport it, and treatment, storage, and disposal facilities (TSDF). (See figure 19.4.) Later amendments to the act included regulation of underground storage tanks (USTs) and

FIGURE 19.3 Improper Toxic Chemical Storage This is a site where toxic wastes were improperly stored. Often such sites are abandoned, and it is difficult to assign responsibility for cleanup because the company that produced the waste is unknown or no longer exists.

petroleum products. Amendments also applied the regulations to smaller quantity generators and made the regulatory standards more stringent. These regulations have significantly reduced the amount of RCRA hazardous waste generated from about 435 million metric tons (480 million U.S. tons) in 2003 to about 32 million metric tons (35 million U.S. tons) in 2011—about a 93 percent reduction.

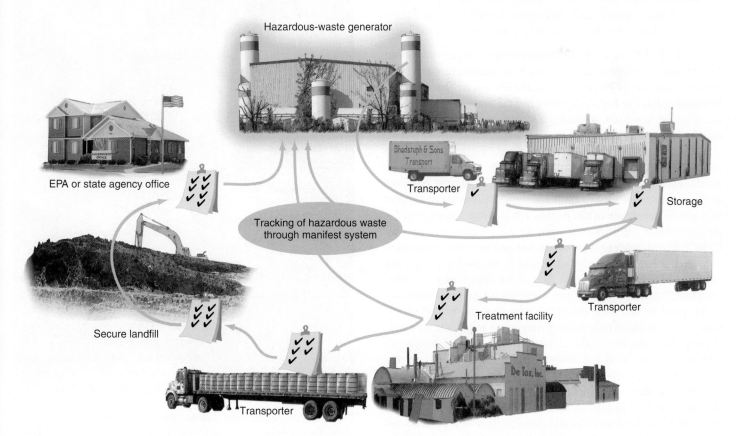

FIGURE 19.4 RCRA Requirements The Resource Conservation and Recovery Act (RCRA) made generators responsible for proper handling of hazardous waste and established a "cradle-to-grave" paper trail that assured that hazardous waste was treated and disposed of properly.

However, several kinds of waste are specifically exempt by law from RCRA regulation. These include: household hazardous waste, agricultural waste that is returned to the soil, mining overburden and processing waste, wastes associated with oil and gas exploration and production, and ash from burning coal. However, there have been efforts to change the law regarding coal ash since the failure of a dam in Kingston, Tennessee, in 2008, released coal ash sludge that destroyed several homes and polluted local rivers.

Comprehensive Environmental Response, Compensation, and Liability Act (CERCLA)

CERCLA was enacted in 1980 when Congress responded to public pressure to clean up hazardous-waste dumps and protect the public from the dangers of such wastes. An important part of the bill was a provision to finance the cleanup of large, uncontrolled hazardous-waste sites. Because of this provision the legislation became popularly known as **Superfund.**

CERCLA had several key objectives:

1. To set priorities for cleaning up the worst existing hazardous-waste sites.
2. To make those who created the hazardous waste site *(potentially responsible parties)* pay for cleanups whenever possible.
3. To set up a $1.6 billion Hazardous Waste Trust Fund—popularly known as Superfund—to support the identification and cleanup of abandoned hazardous-waste sites.
4. To advance scientific and technological capabilities in all aspects of hazardous-waste management, treatment, and disposal.

Under CERCLA, over 44,000 sites were evaluated, and about 11,000 were considered serious enough to warrant further investigation. The list of these sites became known as the **National Priorities List.** The list is still active and the number of sites listed fluctuates as new sites are added and sites are removed because they are cleaned up or deleted from the list.

Although the purpose of the Superfund program was clear, it became controversial during its early history. Millions of dollars were spent by both the federal government and industry, but most of the money went for litigation and technical studies to support or disprove the claims of the parties. Little of the money paid for cleanup. One of the primary reasons for this problem was the way CERCLA was written. It provided that anyone who contributed to a specific hazardous-waste site could be required to pay for the cleanup of the *entire* site, regardless of the degree to which they contributed to the problem. Since many industries that contributed to the problem had gone out of business or could not be identified, those that could be identified were asked to pay for the entire cleanup. Most businesses found it cost-effective to hire lawyers to fight their inclusion in a cleanup effort rather than to pay for the cleanup. Consequently, little cleanup occurred.

After a slow start, however, the Superfund program has shown significant results. By 2014, there were about 1,300 sites on the National Priorities List and about 50 were being considered for inclusion. About 1,100 sites have been cleaned up. Most of the remaining sites are in the process of being cleaned up or are under study to determine the best way to proceed. (See figure 19.5.) This has been an expensive undertaking. Total expenditures for the Superfund program have been about $27 billion. In addition, the EPA estimates that the settlements it has reached with responsible parties amount to over $20 billion.

While Superfund has been responsible for the evaluation and cleanup of thousands of sites, the fact that any party could be responsible for the entire cleanup caused problems. This resulted in two pieces of legislation that provided exemptions to two kinds of parties: Persons who bought land without knowing it was contaminated and developers who wished to use former Superfund sites. The Superfund Amendments and Reauthorization Act (SARA) included several items but one specific provision stated that if prior to the purchase of commercial real estate, a person or entity conducted "all appropriate inquiry" using "good and customary practices" they would not be held responsible for the cleanup of any contamination on the property. This was known as the "innocent landowner defense (ILD)."

Because of the provisions of CERCLA, many previously abandoned industrial sites were left vacant because developers

FIGURE 19.5 Superfund Waste Site Remediation As a result of Superfund legislation, toxic waste sites have been identified and remediation efforts have begun. This photo shows the process of remediating a site that had acid mine drainage and heavy metals contamination.

were not willing to accept the liability they would accept if they purchased the land for redevelopment. These abandoned sites became known as brownfields. The passage of the Small Business Liability Relief and Brownfields Revitalization Act (SBLRBRA) in 2002 included new language that protected purchasers from liability under CERCLA. This meant that most purchasers of commercial real estate must perform an ASTM Phase 1 Environmental Site Assessment (E-1527) prior to purchasing the land. The act also provided funding to conduct site assessment and planning for redevelopment. This act has allowed many remediated Superfund sites to be returned to a useful function rather than to sit vacant.

19.7 Toxic Chemical Releases

In 1986, Congress passed the **Emergency Planning and Community Right-to-Know Act (EPCRA),** which also became known as SARA title III. This act required that certain industries in the United States had to begin reporting the release of toxic chemicals into the environment. Any industrial plant that released 23,000 kilograms (50,000 pounds) or more of toxic pollutants per year was required to file a report. These were primarily manufacturing industries. The information collected allowed the EPA to target specific industries for enforcement action. Since the information is public, many industries were encouraged to take action to reduce their emissions. In the intervening years, changes have been made in who must report. The quantity that required reporting was reduced to 11,400 kilograms (25,000 pounds). In addition, industries that were originally exempt from reporting now must report toxic releases. These activities have been successful, and industrial emissions of toxic chemicals were reduced by about 45 percent from 2000 through 2012. About 1.6 billion kilograms (3.6 billion pounds) of toxic chemicals were reported released into the environment by industry in 2012.

Today, the primary industries involved in toxic releases to the environment are the mining, power generation, chemical, and metal manufacturing industries. (See figure 19.6.) The large amount of toxic waste produced by the mining industry is also reflected in figure 19.7. Mining waste typically contains metals and is deposited on the surface of the land. In addition, you would expect those states with large mining operations to have high amounts of waste.

19.8 Hazardous-Waste Management Choices

The regulatory legislation discussed in the previous sections (RCRA, CERCLA, SARA etc.) put pressure on business and industry to look for better and cheaper

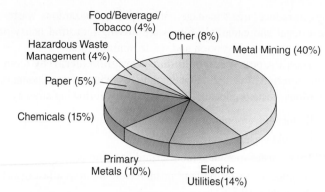

FIGURE 19.6 Sources of Toxic Releases (2012) Mining industries are responsible for about 40 percent of all toxic waste material released to the environment. These are primarily deposited on the surface of the land. Electric and chemical industries each produce about 15 percent. These are primarily releases to the atmosphere and land. Metal processing industries are also significant sources of toxic releases.

Source: Data from the U.S. Environmental Protection Agency.

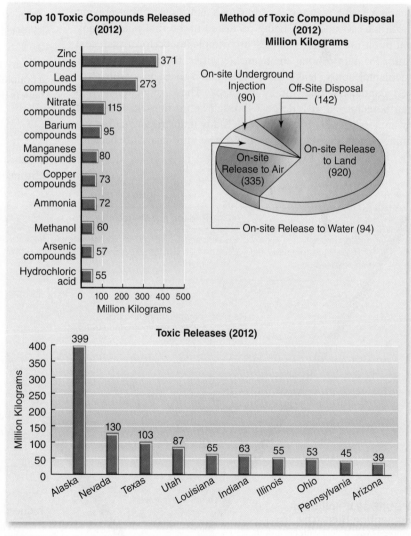

FIGURE 19.7 Toxic Releases Dominated by Mining, the Chemical Industry, and Electric Generation The large amount of waste generated by the mining industry is reflected in the kinds of waste produced (toxic metals), the environmental compartment affected (land disposal), and the states with the most toxic releases. Electric power generation and the chemical industry are responsible for large amounts of toxic materials released to the air.

Source: Data from the U.S. Environmental Protection Agency.

ways to reduce, treat, manage, and dispose of hazardous wastes. These legal and economic forces are driving industrial behavior toward pollution prevention and waste minimization.

The EPA promotes a **pollution-prevention hierarchy** (commonly referred to as "P2") that emphasizes reducing the amount of hazardous waste produced. This involves the following strategy:

1. Reduce the amount of pollution at the source.
2. Recycle wastes wherever possible.
3. Treat wastes to reduce their hazard or volume.
4. Dispose of wastes on land or incinerate them as a last resort. (See figure 19.8.)

Reducing the Amount of Waste at the Source

Pollution prevention (P2) or **waste minimization** encourages changes in the operations of business and industry that prevent hazardous wastes from being produced in the first place. Many pollution-prevention changes are simple to perform and cost little. Primary among them are activities that result in fewer accidental spills, leaks from pipes and valves, loss from broken containers, and similar mishaps. These reductions often can be achieved at little cost through better housekeeping and awareness training for employees. Many industries actually save money because they need to buy fewer raw materials because less is being lost.

Other examples include:

1. Changing a process so that a solvent that is a hazardous material is replaced with water that is not a hazardous material.
2. Using a waste produced in a process in another aspect of the process, thus reducing the amount of waste produced. For example, water used to clean equipment might be used as a part of the product rather than being discarded as a contaminated waste.
3. Using a still to clean solvents so that they can be used repeatedly results in a lower total volume of hazardous waste being produced.
4. Allowing water to evaporate from waste can reduce the total amount of waste produced. Obviously, the hazardous components of the waste are concentrated by this process.

Pollution prevention can be applied in unusual ways. The U.S. military uses about 700 million rounds of small-caliber ammunition each year at about 3,000 firing ranges. In 1998, when lead concentrated in firing berms was found to be leaching into Cape Cod's water supply, the Environmental Protection Agency ordered the Massachusetts Military Reservation to stop live-fire training. In 2000, the U.S. Army announced that it would begin issuing an environmentally friendly "green bullet" that contains no lead. The bullet contains a nonpolluting tungsten core and does not contain lead. Thus, the new bullets do not contaminate the soil and air around firing ranges.

Recycling Wastes

Often it is possible to use a waste for another purpose and thus eliminate it as a waste. Many kinds of solvents can be burned as a fuel in other kinds of operations. For example, waste oils can be used as fuels for power plants, and other kinds of solvents can be burned as fuel in cement kilns. Care needs to be taken that the contaminants in the oils or solvents are not released into the environment during the burning process, but the burning of these wastes destroys them and serves a useful purpose at the same time.

Similarly, many kinds of acids and bases are produced as a result of industrial activity. Often these can be used by other industries. Ash or other solid wastes can often be incorporated into concrete or other building materials and therefore do not require disposal. Thus, the total amount of waste is reduced.

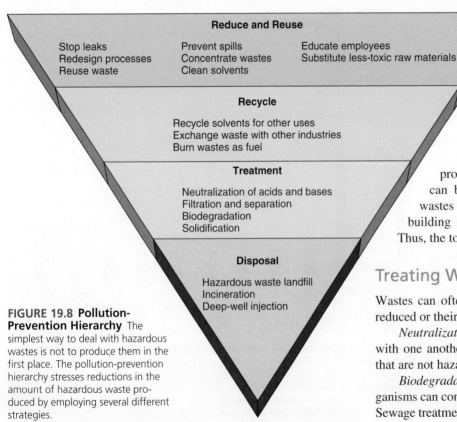

FIGURE 19.8 Pollution-Prevention Hierarchy The simplest way to deal with hazardous wastes is not to produce them in the first place. The pollution-prevention hierarchy stresses reductions in the amount of hazardous waste produced by employing several different strategies.

Reduce and Reuse
Stop leaks · Prevent spills · Educate employees
Redesign processes · Concentrate wastes · Substitute less-toxic raw materials
Reuse waste · Clean solvents

Recycle
Recycle solvents for other uses
Exchange waste with other industries
Burn wastes as fuel

Treatment
Neutralization of acids and bases
Filtration and separation
Biodegradation
Solidification

Disposal
Hazardous waste landfill
Incineration
Deep-well injection

Treating Wastes

Wastes can often be treated in such a way that their amount is reduced or their hazardous nature is modified.

Neutralization of dangerous acids and bases by reacting them with one another can convert hazardous substances to materials that are not hazardous.

Biodegradation of organic materials by the actions of microorganisms can convert hazardous chemicals to innocuous substances. Sewage treatment plants perform their function by biodegradation.

Going Green

Household Hazardous-Waste Disposal

Since the passage of the Resource Conservation and Recovery Act in 1976 and other environmental legislation, industries have been required to document the amounts of hazardous wastes they produce and to account for their appropriate disposal. As a result, the amount of industrial hazardous waste released to the environment has been significantly reduced. At the same time that citizens insist on business and industry compliance with hazardous-waste regulations, our personal behavior does not always reflect the same level of concern.

The average U.S. household generates about 10 kg (over 20 pounds) of hazardous waste each year. Frequently, these materials become waste when they become contaminated or are unwanted. Common forms of household hazardous waste are unwanted paints, varnishes, and other finishes; contaminated solvents such as alcohol and paint thinner; automobile fluids such as waste oil, brake fluid, and antifreeze; appliance and automobile batteries containing heavy metals; caustic cleaning products; unused pesticides; and fluorescent light bulbs.

We have several options to help us properly handle household hazardous waste. Most containers of such materials carry instructions for the proper disposal of the unwanted material or the empty container. In addition, thousands of communities have regular household hazardous-waste collection days staffed by people who are specially trained to assist the public in disposing of such wastes. However, these collection days are typically restricted to a few days per year, which requires that we label and store our waste, find out when and where the waste will be accepted, and deliver it to a particular place on a particular day. Often a reservation is required.

We can reduce the amount of hazardous waste we produce by:

- Substituting products that are not hazardous (water-based paints) for hazardous materials (oil-based paints)

- Carefully choosing the quantities of hazardous products we purchase, such as pesticides and solvents, so that we don't have leftover materials
- Allowing professionals to do automobile maintenance, pest control, or similar activities that involve the use of hazardous materials and the production of hazardous waste, since they are trained to handle these materials properly and have access to appropriate hazardous waste disposal facilities
- Deciding not to use pesticides and accept some inconvenience or diminished quality in lawn and garden

It is also very important to keep hazardous materials in a properly labeled container and to keep them in a secure place so that they are not accidentally spilled or dispersed. A good starting point is to identify hazardous materials in your home, secure those you will be using and dispose of unwanted, obsolete, or poorly identified materials at the next household hazardous-waste collection day.

Air stripping is sometimes used to remove volatile chemicals from water. Volatile chemicals, which have a tendency to vaporize easily, can be forced out of liquid when air passes through it. Steam stripping works on the same principle, except that it uses heated air to raise the temperature of the liquid and force out volatile chemicals that ordinary air would not. The volatile compounds can be captured and reused or disposed of, and the water is no longer considered a waste.

Carbon absorption tanks contain specifically activated particles of carbon to treat hazardous chemicals in gaseous and liquid waste. The carbon chemically combines with the waste or catches hazardous particles just as a fine wire mesh catches grains of sand. Contaminated carbon must then be disposed of or cleaned and reused.

Precipitation involves adding special materials to a liquid waste. These bind to hazardous chemicals and cause them to precipitate out of the liquid and form large particles called floc. Floc that settles can be separated as sludge; floc that remains suspended can be filtered, and the concentrated waste can be sent to a hazardous-waste landfill.

Disposal Methods

Recycling and treatment activities reduce the amount of hazardous waste that needs to be disposed of by about 20 percent. The remaining wastes are typically incinerated or disposed of on land. The primary factors that determine which method is used are economic concerns and acceptance by the public.

Incineration (thermal treatment) involves burning waste at high temperatures. A hazardous-waste incinerator can be used to burn organic wastes but is unable to destroy inorganic wastes. A well-designed and well-run incinerator can destroy 99.9999 percent of the hazardous materials that go through it. The relatively high costs of incineration (compared with landfills) and concerns about emissions affecting surrounding areas have kept incineration from becoming a major method of treatment or disposal in North America. Incineration accounts for the disposal of less than 3 percent of the hazardous wastes. In Europe and Japan, where land is in short supply and expensive, incineration is more cost-effective and is used to dispose of large amounts of hazardous wastes.

Land disposal is the primary method for the disposal of hazardous wastes in North America, because abundant land is available and it is less expensive than incineration. Land disposal can take several forms:

1. Deep-well injection into porous geological formations or salt caverns
2. Discharge of treated and untreated liquids into municipal sewers, rivers, and streams
3. Placement of liquid wastes or sludges in surface pits, ponds, or lagoons
4. Storage of solid wastes in specially designed hazardous-waste landfills

Of these methodologies, deep-well injection is the most important and is the primary method for disposing of liquid hazardous waste. About 60 percent of all hazardous waste disposed of on land is injected into deep wells. About 10 percent is released to aquatic environments, and about 4 percent is stored in landfills, ponds, and other surface sites.

19.9 International Trade in Hazardous Wastes

As hazardous waste regulations in industrialized countries have increased the cost of disposal, export of hazardous wastes to developing countries became economically advantageous. However, often the receiving countries lacked the administrative and technological resources to safely dispose of or recycle the waste, and workers used unsafe methods that resulted in their exposure to toxins. The debate over controlling hazardous-waste movements between countries culminated in 1989 with the creation of the *Basel Convention.*

The Basel Convention was negotiated under the auspices of the United Nations Environment Programme and took effect in 1992. The objectives of the convention are to minimize the generation of hazardous wastes and to control and reduce their transboundary movements to protect human health and the environment. To achieve these objectives, the convention prohibits exports of hazardous waste to Antarctica, to countries that have banned such imports as a national policy, and to nonparties to the convention (unless those transactions are subject to an agreement that is as stringent as the Basel Convention). Though not part of the original agreement, there is now a broad ban on the export of hazardous wastes from the Northern to the Southern Hemisphere. The waste transfers that are permitted under the Basel regime are subject to the mechanism of prior notification and consent, which requires parties to not export hazardous wastes unless a "competent authority" in the importing country has been properly informed and has consented to the trade.

Although most countries have approved the convention, there are still individuals or companies that are willing to ignore the rules. For example, in 1999, between 3,000 and 4,000 metric tons of mercury-contaminated concrete waste packed in plastic bags was found in an open dump in a small town in Cambodia.

The waste, labeled as "construction waste" on import documents, came from a Taiwanese petrochemical company. In this case, the waste was tracked down and returned to its point of origin. Unfortunately, most such cases are not reported or detected. (See the case of Côte d'Ivoire at the beginning of this chapter.)

19.10 Nuclear Waste Disposal

Radioactive wastes are hazardous and toxic by the definition of RCRA but are regulated in a different manner from other hazardous materials. In the United States, nuclear research was initially focused on the production of nuclear weapons. Following World War II, the Atomic Energy Commission was established to encourage the development of nuclear power. The functions of this agency were taken over by the Nuclear Regulatory Commission (NRC) and the Department of Energy (DOE), which now have the primary responsibility for dealing with nuclear waste. The NRC is primarily responsible for managing spent fuel from nuclear power plants, waste associated with the decommissioning of nuclear power plants, and low-level waste from a variety of sources. The DOE has major responsibility for the waste associated with cleaning up contaminated sites resulting from research and weapons development.

Sources of Nuclear Waste

The **nuclear fuel cycle** involves mining uranium, processing it, using it as fuel for nuclear reactors, and reprocessing or disposing of spent fuel. Each of these steps in the process results in some nuclear waste. In addition, nuclear material can be used to make weapons. The legacy of past nuclear research and weapons development resulted in a great deal of waste and contamination and requires special consideration.

Contamination from Nuclear Research and Weapons Production

In the United States, nuclear research and development was initially related to the development of weapons for use during World War II. Early in the history of nuclear research, the dangers from radioactive wastes were underappreciated or considered unimportant because of the urgent need to produce weapons. Many of the research and production facilities dealt with hazardous chemicals and *minor* radioactive wastes by burying them, pumping them into the ground, storing them in ponds, or releasing them into rivers. Following World War II, research on weapons development continued along with research on the use of nuclear reactions to produce energy. Ultimately the Department of Energy (DOE) assumed responsibility for these activities and became responsible for the stewardship of the facilities used for both nuclear energy research and weapons production. Some facilities were used for both processing nuclear fuel and providing materials for weapons. Furthermore, military uses involved the production and concentration of plutonium. The production and storage facilities invariably became contaminated, as did the surrounding land. As a result, the DOE became the steward of a large number of sites that are contaminated with both hazardous chemicals and radioactive materials.

The magnitude of the problem was huge. Over 500 sites were evaluated to determine if cleanup activities were required. Over 100 sites required some level of cleanup. The problems at the sites involved:

- Packaging and disposition of hundreds of thousands of metric tons of uranium and depleted uranium.
- Packaging and disposition of plutonium and waste associated with plutonium production.
- Disposal of hundreds of thousands of liters of liquid waste.
- Packaging and disposal of hundreds of thousands of cubic meters of transuranic waste.
- Packaging and disposal of over a million cubic meters of low-level radioactive waste.
- Remediating contaminated soils or groundwater.

Since the DOE became responsible for cleanup, about 90 major sites have been cleaned up. Each year, there are fewer sites and the remaining sites are smaller. Currently, there are 15 sites that have ongoing cleanup work. (See figure 19.9.) The waste generated at these sites is either contained on-site or shipped to disposal facilities.

Nuclear Weapons

An additional problem has arisen as a result of the reduced importance of nuclear weapons. The political disintegration of the Soviet Union and Eastern Europe made large numbers of nuclear weapons, in both the East and West, unnecessary. But how are nations of the world to dispose of their nuclear weapons? Some nuclear material can be diverted to fuel use in nuclear reactors, and some reactors can be modified to accept enriched uranium or plutonium. The security of these materials, particularly in the former Soviet Union, is a concern.

Waste from Nuclear Power Plants

The use of nuclear fuel in nuclear power plants results in the production of waste from spent fuel rods and from other materials that become contaminated because they are in contact with radioactive material. Currently, the spent fuel rods are being stored "temporarily" at the nuclear power plants.

Disposal Methods

Although there are differences in the way in which nuclear wastes are defined by different countries, there are four general categories of wastes. Each presents a different level of risk and requires particular methods of handling and disposal.

Transuranic Waste Disposal

Transuranic nuclear waste is highly radioactive waste that contains large numbers of atoms that are larger than uranium with half-lives greater than 20 years. Most of these wastes come from processes involved in the production of nuclear weapons. As the cleanup of former nuclear weapons sites in the United States takes place, transuranic waste is transported to the Waste Isolation Pilot

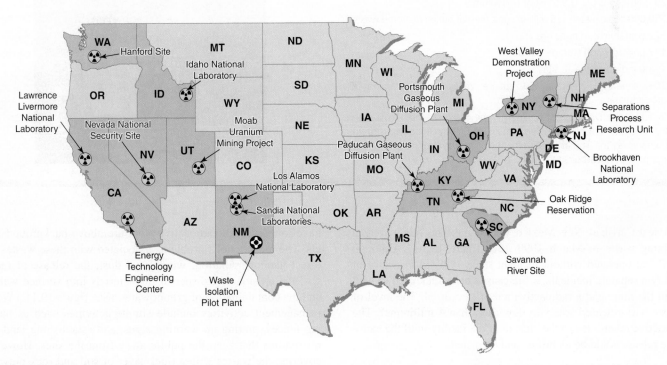

FIGURE 19.9 Contaminated Nuclear Sites Remaining The Department of Energy became the steward of over 100 contaminated nuclear weapons and research sites. As of 2014, 15 still had ongoing cleanup activities. The Waste Isolation Pilot Plant is not a cleanup site. It is a storage site for transuranic waste and will be decommissioned when all the transuranic waste from clean-up sites has been transferred to it.
Source: Data from the U.S. Department of Energy.

Focus On

The Hanford Facility: A Storehouse of Nuclear Remains

In March 1943, the U.S. government obtained 1,650 square kilometers (640 square miles) in southeastern Washington to house a large–scale plutonium production facility. The Hanford site, located on the Columbia River, offered a mild climate, isolation, and abundant water, all of which were necessary for the operation of reactors and reprocessing plants. Residents were given three weeks' notice to vacate the area, and construction began immediately.

Because it was considered crucial to the development of weapons for World War II, the facility began industrial production of plutonium without first going through a small–scale development stage. The first nuclear reactor and chemical processing plants were completed in late 1944. Over the course of the following two decades, eight additional reactors and several processing plants were constructed.

The Hanford plutonium and uranium production facility is no longer in operation. However, it has left a legacy of contamination and waste that is one of the largest and most complex environmental cleanup sites in the United States. Early in the history of operations at Hanford, low-level waste was dumped onto the ground or pumped into wells. Other liquid wastes were stored in tanks. Over the years many tanks leaked before the waste was transferred to more secure containers. Consequently, soil and groundwater have been contaminated by about 4 million liters (1 million gallons) of waste. Burial of solid waste in unlined trenches also resulted in soil and groundwater contamination.

The cleanup task was huge:

- 202,000 cubic meters (53 million gallons) of liquid radioactive waste
- 2,100 metric tons (2,300 U.S. tons) of spent nuclear fuel
- 20 metric tons (22 U.S tons) of plutonium
- 750,000 cubic meters (25 million cubic feet) of buried or stored waste
- 522 contaminated buildings
- 177 underground storage tanks
- 1,700 individual waste sites
- 208 square kilometers (80 square miles) of contaminated groundwater

In 1989, the Department of Energy, the Environmental Protection Agency, and the Washington Department of Energy signed an agreement, known as the Tri-Party Agreement, calling for cleanup and better management of hazardous materials.

The following is a partial list of accomplishments:

- Contaminated groundwater is being pumped and treated at a rate of hundreds of millions of liters per month.
- Soil contamination has been characterized.
- Liquid waste has been removed from tanks thought to be in danger of leaking.
- Seven of nine nuclear reactors have had fuel removed and have been placed in long-term storage. Decommissioning of the remaining reactors is in progress.
- A lined landfill has been constructed on site to hold low-level radioactive and mixed waste.
- Over 500 shipments of transuranic waste have been sent to the Waste Isolation Pilot Plant.
- Uranium that had been in storage has been transferred to a secure site.
- 741 buildings have been decontaminated and dismantled.
- Construction of a plant to convert low-level liquid waste to glass has begun, but it is not expected to begin operation until 2019.

However, there is still much to be done. Some contaminants are still entering the Columbia River from groundwater, although continued treatment of groundwater will further reduce the level of contamination. The more seriously contaminated sites at Hanford will require up to 40 years to clean up.

Plant near Carlsbad, New Mexico, for storage. This facility began accepting waste in March 1999. (See figure 19.10.) Although WIPP had operated without incident since 1999, in February of 2014 two separate incidents, a fire caused by a truck used to haul salt in the mine and a radioactive release, occurred. The level of radioactivity released was very low (less than 4 millirems). The radioactive release led to the closure of the facility until the cause of the release could be identified and corrected.

Uranium Mining and Milling Waste

Uranium mining and *milling waste* is produced from the preparation of uranium for both weapons and nuclear power uses. These have low levels of radioactivity but are above background levels. The most serious problems associated with these wastes are direct gamma radiation, windblown dust, the release of radon gas, the erosion and dispersal of materials into surface waters, and the contamination of groundwater. (See figure 19.11.) Waste management activities include simple activities such as building fences, putting up warning signs, and establishing land-use restrictions that keep the public away from the sites. However, covering the wastes with a thick layer of soil and rock prevents erosion, windblown particles, and groundwater contamination. The covering also greatly reduces the gamma radiation and release of radon gas.

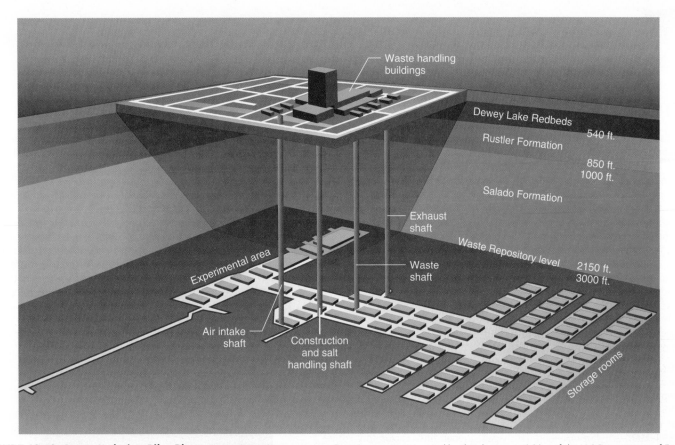

FIGURE 19.10 Waste Isolation Pilot Plant The high-level transuranic radioactive waste generated by the cleanup activities of the U.S Department of Energy is shipped to the Waste Isolation Pilot Plant near Carlsbad, New Mexico, for storage.
Source: U.S DOE.

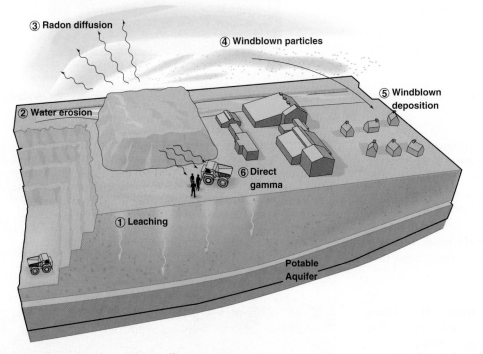

FIGURE 19.11 Uranium Mine Tailings Even though the amount of radiation in the tailings is low, the radiation still represents a threat to human health. Radioactivity may be dispersed throughout the environment and may come in contact with humans in several ways: (1) radioactive materials may leach into groundwater; (2) radioactive materials may enter surface water through erosion; (3) radon gas may diffuse from the tailings and enter the air; or (4) a person living near the tailings may receive particles in the air, (5) come in contact with objects that are coated with particulates, or (6) receive direct gamma radiation.
Source: U.S Environmental Protection Agency.

High-Level Radioactive Waste Disposal

High-level radioactive waste consists of spent fuel rods and highly radioactive materials from the reprocessing of fuel rods. The disposal of high-level nuclear waste has been a major problem for the nuclear power industry. In several countries—France, Japan, Russia, the United Kingdom, China, India, Switzerland, Belgium, and Bulgaria—accounting for just less than half of nuclear power facilities, the spent fuel rods are reprocessed to reduce the volume of the waste. Other countries, including the United States, do not reprocess spent fuel rods and must store them.

Temporary Storage In the United States there are about 70,000 metric tons (77,000 U.S. tons) of highly radioactive spent fuel rods stored at nuclear reactor sites. Most of these storage sites are water-filled containers, which were intended to provide

Science, Politics, & Policy

Disposal of Waste from Nuclear Power Plants

The disposal of spent fuel rods from nuclear power plants has been a continuing issue since the first commercial power reactor went on line in 1957. In the more than 50 years since then, it has been clear that there are only two methods of dealing with the nuclear waste from spent nuclear fuel: reprocessing the fuel to reduce the amount of waste or storing the waste at a safe site. U.S. policy has been to store rather than reprocess nuclear waste. Federal law requires the U.S. government to provide a solution to the storage of spent nuclear fuel. All nuclear power plants in the United States have been operating with the assumption that eventually their waste would be stored in a secure federal facility. The National Academy of Sciences recommended underground storage as the best way to deal with waste from nuclear power plants.

The history of U.S. efforts to establish a repository for high-level radioactive waste is long and complicated. The following provides a brief chronology of major steps in the process:

- 1982—The U.S. Congress passed legislation that gave the responsibility for finding, building, and operating a nuclear waste site to the Department of Energy with completion by 1998.
- 1987—Initially several sites were considered and Yucca Mountain was selected to receive further study.
- 1994–1997—A five-mile-long, U-shaped tunnel (the Exploratory Studies Facility) was constructed to study the suitability of the Yucca Mountain site.
- 2002—President George W. Bush signed a joint resolution of Congress designating Yucca Mountain as the site for the nuclear repository.
- 2008—The Department of Energy filed a license application with the Nuclear Regulatory Commission to construct a repository for spent nuclear fuel and high-level radioactive waste at Yucca Mountain. The citizens and political leaders of Nevada opposed the designation.

- 2010—President Obama withdrew funding for Yucca Mountain and the Department of Energy withdrew its request to the Nuclear Regulatory Commission to operate the facility. President Obama also established the Blue Ribbon Commission on America's Nuclear Future.
- 2012—The report of the Blue Ribbon Commission on America's Nuclear Future 2012 included the following statement:
  Recommendation #1: The United States should undertake an integrated nuclear waste management program that leads to the timely development of one or more permanent deep geological facilities for the safe disposal of spent fuel and high-level nuclear waste.
- 2013—A U.S. Court of Appeals ruling stated that the designation of Yucca Mountain as the nation's nuclear repository is still in effect and the Nuclear Regulatory Commission and the President cannot ignore the law and proceed with plans to close Yucca Mountain.
- The Future—The future is uncertain, but it is clear that no permanent solution for storing spent nuclear fuel is likely for decades.

Yucca Mountain

temporary storage until the United States government built a permanent storage facility. (See figure 19.12.) Since the permanent storage has not been completed, many plants are running out of storage space. Consequently, they have been authorized by the Nuclear Regulatory Commission to store older spent fuel rods, with lower radioactivity, in aboveground dry casks because the storage ponds will not accommodate additional waste. Currently about 25 percent of spent fuel rods are stored above ground. (See figure 19.13.)

Permanent Disposal of High-Level Nuclear Waste Most experts agree that a viable solution to the disposal of high-level nuclear waste is to bury it in a stable geologic formation. However, only Finland and Sweden have chosen sites. Canada is in the process of choosing a site, and the U.S. has not been able to establish a clear policy on dealing with spent fuel rods. (See Science, Politics & Policy: Disposal of Waste from Nuclear Power Plants.) It will be at least a decade before any waste will be stored in the facilities.

FIGURE 19.12 Storage of Spent Fuel Rods Spent fuel rods are initially stored in water-filled ponds.

FIGURE 19.13 Aboveground Storage of Spent Fuel Rods These casks contain older, less radioactive spent fuel rods. This kind of storage was necessitated because the promised permanent government storage site at Yucca Mountain has not been constructed and water storage pools were at capacity.

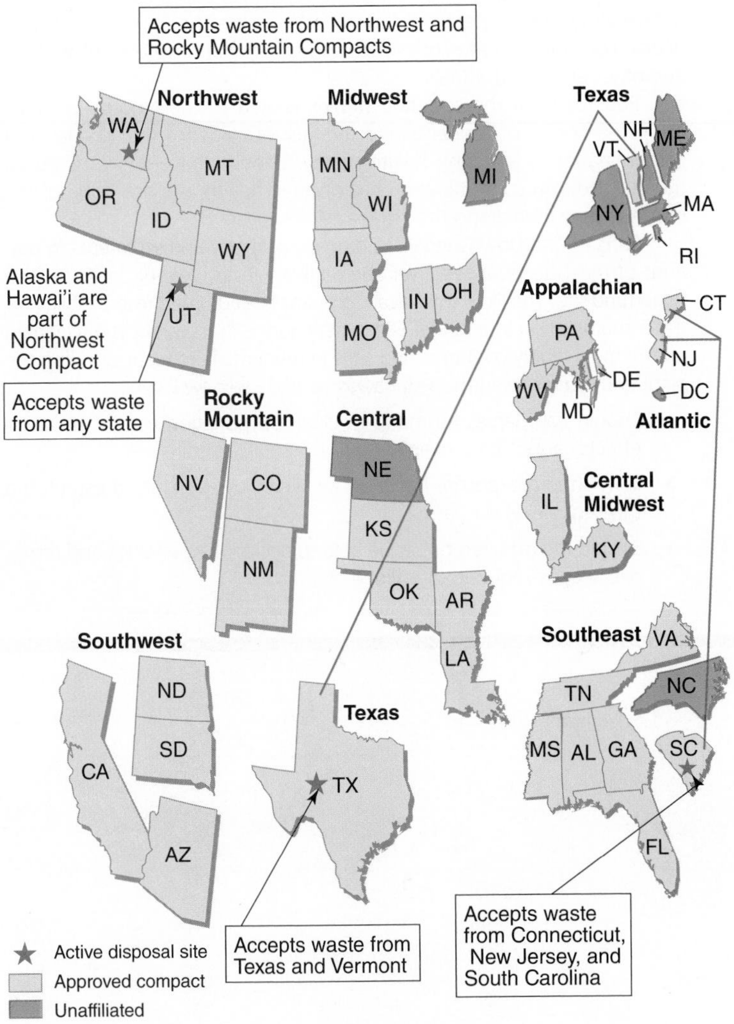

Low-Level Radioactive Waste Disposal

Low-level radioactive waste is waste containing small amounts of radioactivity that is not classified into one of the other categories. Low-level wastes come from a variety of sources, including nuclear power facilities, weapons facilities, hospitals, and research institutions, and includes such things as radioactive materials used in the medical field, protective clothing worn by persons who work with radioactive materials, contaminated cleaning materials, and materials from many other modern uses of radioactive isotopes.

The United States disposes of about 57,000 cubic meters (2 million cubic feet) of low-level radioactive waste per year. This is currently being buried in disposal sites in South Carolina, Washington, Utah, and Texas. In 1980, Congress set a deadline of 1986 for each state to provide for its own low-level radioactive waste storage site. Later, a change allowed several states to form regional compacts and select one state to provide the disposal site for all the states in the compact.

Several states are not members of any compact. Figure 19.14 shows the current compacts and the active low-level radioactive waste disposal sites. As you can see from Figure 19.14, most compacts and unaffiliated states do not have a permanently available place to deposit low-level radioactive wastes but rely on having access to the storage facilities in Utah. Figure 19.15 shows a typical storage method for low-level radioactive waste.

FIGURE 19.15 Storage of Low-Level Radioactive Waste Low-level radioactive waste is stored in specially designated land burial sites.

FIGURE 19.14 Low-Level Radioactive Waste Sites Each state is responsible for the disposal of its low-level radioactive waste. Many states have formed compacts and selected a state to host the disposal site. However, most of the designated states have not developed disposal sites. Thus most states are relying on a site in Utah to accept their waste. The map shows the current compacts and active disposal sites.

Source: Nuclear Regulatory Commission.

Map labels: Accepts waste from Northwest and Rocky Mountain Compacts. Northwest — WA, MT, OR, ID, WY, UT. Alaska and Hawai'i are part of Northwest Compact. Accepts waste from any state. Rocky Mountain — NV, CO, NM. Southwest — ND, SD, CA, AZ. Midwest — MN, WI, MI, IA, IN, OH, MO. Central — NE, KS, OK, AR, LA. Texas — TX. Accepts waste from Texas and Vermont. Appalachian — PA, WV, MD, DE. Texas — NH, ME, VT, NY, MA, RI, CT, NJ, DC. Atlantic. Central Midwest — IL, KY. Southeast — VA, TN, NC, MS, AL, GA, SC, FL. Accepts waste from Connecticut, New Jersey, and South Carolina.

★ Active disposal site
▢ Approved compact
▨ Unaffiliated

Dioxins in the Tittabawassee River Floodplain

Dioxin is the name given to a group of chemicals that are formed as unwanted by-products of industrial manufacturing and burning activities. Major sources of dioxins include chemical and pesticide manufacturing, the burning of household trash, forest fires, and the burning of industrial and medical waste products. There are approximately 210 chemicals with similar structures and properties included among the dioxins. In the environment, they usually appear in a mixture.

Trying to establish safe levels is difficult because some dioxins are very toxic and others are not. The most toxic chemical in the dioxin group is 2,3,7,8-tetrachlorodibenzo-para-dioxin (2,3,7,8-TCDD). Thus, 2,3,7,8-TCDD is the standard to which other dioxins are compared. Further complicating the picture is the fact that while dioxins are quite toxic to rats on which many toxicity studies have been done, dioxins don't appear to be very toxic to humans. Since the harmful effects of dioxins to humans are not fully known, and certain types of dioxin are more toxic than others, scientists have agreed to report dioxin levels by combining dioxin forms and converting them to an "equivalent" of that form considered most harmful—the "total toxic equivalent" (TEQ) concentration.

In the environment and the food chain the highest levels of dioxins are usually found in soil, sediment, and animal fat. Much lower levels are found in air and water.

The Dow Chemical Company has been producing chemical products in Midland, Michigan, for more than 100 years. As a result, the Tittabawassee River floodplain contains higher-than-normal levels of dioxins in soil and sediment samples. The State of Michigan Department of Environmental Quality (MDEQ), the U.S. EPA, the Centers for Disease Control (CDC), the World Health Organization (WHO), and the U.S. Agency for Toxic Substances and Disease Registry (ATSDR) have all provided guidance and toxicity profiles for dioxins. The MDEQ uses a "90 parts per trillion (ppt) Direct Contact Cleanup Criteria" standard, while the CDC action level is 1,000 ppt. However, the levels found in some sample locations along the floodplain exceeded the action levels of 1,000 ppt for dioxin. Soil samples taken from areas outside the floodplain were at normal background levels.

People living in the floodplain area could have been exposed to dioxins in the soil and sediments and in general show higher levels of dioxins than people living in other areas of Michigan. In some animal studies, low-level chronic exposure to dioxins caused cancer, liver damage, and hormone changes. In some other animal studies, dioxin exposure also decreased the ability to fight infection and caused reproductive damage, miscarriages, and birth defects. Skeleton and kidney defects, lowered immune responses, and effects on the development of the brain and nervous system were among the birth defects seen. However, other studies did not yield similar results. Human health studies of those living in the vicinity of the chemical plant and downstream of the plant do not show abnormal numbers of cancers and other health problem. Industry, federal, state, and academic institutions continue to conduct toxicity testing on animals (humans included) in an effort to determine the truth. In the meantime, state and federal agencies have issued health advisories, cautions, fish and game consumption advisories, and advice on how to reduce exposure for people living in and/or using the floodplain. In particular, consumption advisories focus on fish intake for pregnant women and other at-risk individuals.

Regardless of the scientific studies related to the amount of dioxins in the bodies of residents and the lack of evidence that the level of contamination is affecting health, many residents have concerns about the effect dioxin contamination has on their health and the value of the property they own along the Tittabawassee River.

Parts of The Dow Chemical Company property and sediments in portions of the Tittabawassee River and Saginaw River were declared to be a Superfund site. The Dow Chemical Company has admitted responsibility for the contamination and (under EPA supervision) is in the process of cleaning up the most highly contaminated sites in residential areas surrounding the plant and sediments in the Tittabawassee and Saginaw Rivers.

- Should companies be made to clean up pollution even if negative effects cannot be proven?

- How should exposure limits be set when it is difficult to establish a threshold level for harm?

- Should the affected public be able to decide the penalties and remedial actions required of polluters?

Summary

The industrialized countries of Europe and North America began major regulation of hazardous materials only during the past 40 years, and most developing countries exercise little or no control over such substances. As a result, many countries must deal with contaminated sites, which are the result of past use and uncontrolled dumping practices.

Hazardous materials are a common part of our lives. Many are used in industrial processes to create products we use. Some hazardous materials we use on a regular basis—gasoline and insecticides are examples. From a regulatory point of view there are several definitions of what is hazardous. However, those things that burn, explode, corrode, and cause oxidation are typically included along with gases, radioactive materials, and poisons. Hazardous wastes are unusable materials that must be treated or disposed of in special landfills. Hazardous wastes are identified by having one or more of the following characteristics: ignitability, corrosiveness, reactivity, or toxicity; or by being specifically identified on a list.

Laws and regulations are used by government agencies to assure that hazardous materials or wastes are handled properly. Failure to comply results in fines or other punishments.

Toxic substances kill or injure organisms because they alter normal body functioning. Establishing standards for what is toxic and the levels at which toxic materials become a hazard are determined by testing substances on animals. An important distinction is the difference between acute toxicity (immediate response to a large dose of toxin) and chronic toxicity (delayed response to the cumulative effects of repeated exposure to low doses). Toxins may also be persistent (remain unchanged for a long time) or nonpersistent (quickly disperse or degrade).

Unregulated disposal of hazardous wastes in the past resulted in a huge number of abandoned hazardous-waste sites. Recognition of the problems associated with these sites resulted in the passage of important pieces of legislation in the United States. The Resource Conservation and Recovery Act (RCRA) established a cradle-to-grave system of dealing with hazardous wastes that requires all those who generate, transport, treat, or dispose of hazardous waste to sign documentation of their role. The Comprehensive Environmental Response, Compensation, and Liability Act (CERCLA) established programs for cleaning up hazardous-waste sites. It also established the principle that those who caused the problem should pay for the cleanup. In addition, it provided money for cleanup. The Superfund Amendments and Reauthorization Act (SARA) established the "innocent landowner defense" if, before purchasing property, a purchaser made a strong effort to discover any contamination present. The Small Business Liability Relief and Brownfields Revitalization Act (SBLRBRA) provided additional funding for cleanup and made it easier for individuals to redevelop land that was previously contaminated.

Toxic chemicals are still released to the air, water, and soil. However, increasingly strict regulations have brought about a reduction in releases. The primary sources of toxic releases are mining operations and electric power plants.

The current strategy of EPA is to limit the production of hazardous waste by: reducing the waste at the source, recycling material so that it does not become waste, treating waste to reduce its hazard or volume, and, as a last resort, incinerating, injecting into a deep well, or burying the waste in a secure landfill.

Nuclear waste is primarily the result of the production of nuclear weapons and the wastes of nuclear power plants. Transuranic waste is primarily from processing uranium and plutonium for nuclear weapons. The Department of Energy has become responsible for cleaning up transuranic waste. The waste is shipped to the Waste Isolation Pilot Plant in New Mexico for storage. Uranium mining and milling waste generally has low levels of radioactivity and is usually managed by covering the waste with soil and erecting fences around the site. High-level waste from nuclear plants consists primarily of spent fuel rods that are currently stored on the site of nuclear power plants in water-filled pools or above-ground dry casks. Ultimately they are to be stored in a permanent storage site to be provided by the U.S. government. Selecting a site for storage has become a political issue and it will be many years before a permanent site is operational. Low-level wastes are materials containing small amounts of radioactive substances. They are produced by organizations such as universities, hospitals, and nuclear power plants. Although each state is responsible for managing its own waste, most have joined with other states to select storage sites. Currently there are four specially designed landfills that accept low-level wastes for storage.

Acting Green

1. Conduct a "chemical inventory" of your garage, bathroom, kitchen, or other area in your home where chemicals are stored. Properly dispose of products that are outdated or no longer of use.

2. Contact your local solid waste operator or local government solid waste office. Ask about hazardous-waste collection programs, oil recycling, and programs for e-waste collection in your community.

   If there are no programs, ask why. Become a champion for the development of programs.

3. Participate in hazardous-waste minimization programs on your campus.

4. Use rechargeable batteries and nontoxic cleaning products.

5. Call your local hospital and ask a representative how the hospital disposes of its nuclear waste.

Review Questions

1. List five different categories of hazardous materials.
2. What is a characteristic hazardous waste?
3. In addition to characteristic hazardous waste, how does the U.S. Environmental Protection Agency define a hazardous waste?
4. Distinguish between acute and chronic toxicity.
5. Give an example of synergism.
6. Give an example of a persistent pollutant and a nonpersistent pollutant.
7. What is a threshold level of exposure to a hazardous material and how is it determined?
8. List the three routes of entry of a hazardous material into the body.
9. List three ways hazardous wastes enter the environment to become a problem.
10. List three requirements of the Resource Conservation and Recovery Act (RCRA).
11. What are the goals of the Comprehensive Environmental Response, Compensation, and Liability Act (CERCLA)?
12. Why is CERCLA often known as Superfund?
13. Describe what is meant by the U.S. National Priorities List.
14. What are the goals of the Small Business Liability Relief and Brownfields Revitalization Act (SBLRBRA)?
15. List the three kinds of industries most responsible for the release of toxic materials to the environment.
16. Describe the pollution-prevention hierarchy.
17. Give three examples of how hazardous waste can be reduced at its source.
18. Give three examples of how hazardous waste can be recycled.
19. Describe five technologies for treating hazardous wastes.
20. List the two common technologies used to dispose of hazardous waste.
21. Why was the Basel Convention of the United Nations established?
22. What are the primary sources of nuclear waste?
23. What is transuranic waste and how is it disposed of?
24. What is high-level radioactive waste and how is it currently being controlled?
25. Give examples of low-level radioactive waste.

Critical Thinking Questions

1. Scientists at the EPA have to make decisions about thresholds in order to identify which materials are toxic. What thresholds would you establish for various toxic materials? What is your reasoning for establishing the limits you do? What, if any, type of testing might you conduct to arrive at these thresholds?
2. Go to the EPA's website, access the Toxic Release Inventory (TRI), and identify the major releasers of toxic materials in your area. Were there any surprises? Are there other releasers of toxic materials that might not be required to list their releases?
3. Look at this chapter's section 19.6, "Hazardous-Waste Dumps—The Regulatory Response." Do the authors present the information from a particular point of view? What other points of view might there be on this issue? What information do you think these other viewpoints would provide?
4. Many economically deprived areas, Native American reservations, and developing countries that need an influx of cash have agreed, over significant local opposition, to site hazardous-waste facilities in their areas. What do you think about this practice? Should outsiders have a say in what happens within these sovereign territories?
5. The disposal of radioactive wastes is a big problem for the nuclear energy industry. What are some of the things that need to be evaluated when considering nuclear waste disposal? What criteria would you use to judge whether a storage proposal is adequate or not?
6. Review the Issues & Analysis dealing with dioxins. How might the area be cleaned up? Who should be responsible for conducting the cleanup? To what levels would you suggest the area be remediated? Should the river water and sediments be treated as well? Should the residents in the area be consulted, and should they be compensated and given medical treatment options? Consider the plant and animal life in the floodplain—what, if anything, should be done about that?
7. Nuclear weapons testing had released nuclear radiation into the environment. These tests have always been justified as necessary for national security. Do you agree or not? What are the risks? What are the benefits?

|ENVIRONMENTAL SCIENCE

To access additional resources for this chapter, please visit ConnectPlus® at connect.mheducation.com. There you will find interactive exercises, including Google Earth™, additional Case Studies, and SmartBook™, an adaptive eBook that integrates our LearnSmart® adaptive learning technology.

chapter 20

Environmental Policy and Decision Making

Government regulation is often necessary to control the actions of uncaring individuals or corporations. The federal government has passed many environmental laws directed at forcing better behavior from individuals and corporations.

OBJECTIVES

After reading this chapter, you should be able to:

- Explain how the executive, judicial, and legislative branches of the U.S. government interact in forming policy.
- Understand how environmental laws are enforced in the United States.
- Describe the forces that led to changes in environmental policy in the United States during the past three decades.
- Understand the history of the major U.S. environmental legislation.

- Understand why some individuals in the United States are concerned about environmental regulations.
- Understand what is meant by "green" politics.
- Describe the reasons environmentalism is a growing factor in international relations.
- Understand the factors that could result in "ecoconflicts."

- Understand why it is not possible to separate politics and the environment.
- Explain how citizen pressure can influence governmental environmental policies.
- Understand the role of lobbying in the development of environmental policy.
- Describe the roles and function of the Environmental Protection Agency.

Fish Consumption Policies and Advisories

Fish are an important part of a healthy diet. However, some fish contain chemicals that could pose health risks. When contaminant levels are unsafe, consumption advisories may recommend that people limit or avoid eating certain species of fish caught in certain areas. Most of the fish in North America are safe to eat. However, all the U.S. states and Canadian provinces have fish consumption advisories in place to protect their residents from the potential health risks of eating contaminated fish caught in local waters.

A fish consumption advisory is not a regulation. It is a recommendation issued to help protect public health. These advisories may include recommendations to limit or avoid eating certain species of fish caught from specific water bodies due to chemical contamination. An advisory may be issued for the general public, including recreational and subsistence fishers, or it may be issued specifically for sensitive populations, such as pregnant women and children. An advisory for a specific water body may cover more than one affected fish species or chemical contaminant.

Most advisories involve five primary contaminants: mercury, chlordane, dioxins, DDT, and polychlorinated biphenyls (PCBs). All of these materials are chemically stable and tend to remain in the environment for years. Levels of any of the contaminants mentioned may increase as they move up the food chain, so a food chain's top predators, such as bass or walleye, may have levels several orders of magnitude higher than the water. PCBs were widely used in electronic components until banned by the Toxic Substance Control Act of 1976. Chlordane and DDT are persistent pesticides that were banned by most countries in the 1970s and 1980s. Dioxins are a by-product of the production of certain herbicides and also are produced by the burning of certain materials. Today the primary source of mercury is from coal-fired power plants. In 2004, the EPA and the U.S. Food and Drug Administration (FDA) issued advice concerning mercury in fish. The advice targeted women who might become pregnant, women who are pregnant, nursing mothers, and young children. The risks to those women and children depend on the amount of fish eaten and the amount of mercury in the fish. While nearly every state has fish consumption advisories, the following states provide examples of how advisories are used.

The Florida Fish and Wildlife Conservation Commission (FWC), the Florida Department of Environmental Protection (DEP), and the Florida Department of Health (DOH) have monitored mercury in the state's freshwater and marine environments since 1983. Fish consumption advisories for specific water bodies are issued when contaminants found in fish are at levels that may pose a risk to human health. Advisories for mercury in Florida waters have been issued since 1989.

In 2013, health officials in Oregon and Washington released a policy stating that people should protect themselves against mercury and PCB contamination by limiting consumption of certain fish species from a 200 kilometer section of the Columbia River. The advisory issued by the health departments of Oregon and Washington has a strict warning not to eat any local fish immediately upriver from the Bonneville Dam, and a more relaxed warning for the larger stretch of river that people should not eat the resident fish more than once a week.

Are there fish consumption advisories where you live? If so, what are the chemical contaminants of concern? What government agency issues the advisories?

20.1 New Challenges for a New Century

We live in remarkable times. This is an era of rapid and often bewildering alterations in the forces and conditions that shape human life. The forces and trends include the following:

Population Growth

Tomorrow's world will be shaped by the aspirations of a much larger global population. The number of people living on Earth has doubled in the last 50 years. Growing populations demand more food, goods, services, and space. Where there is scarcity, population increase aggravates it. Where there is conflict, rising demand for land and natural resources exacerbates it. Struggling to survive in places that can no longer sustain them, growing populations overfish, overharvest, and overgraze. (See figure 20.1.)

Globalization

Internationally, trade, investment, information, and even people flow across borders largely outside of governmental control.

FIGURE 20.1 Population Increase Aggravates Resource Scarcity
Struggling to survive in places that can barely sustain them with food or water, growing populations overfish, overharvest, and overgraze. The bottom photo shows people waiting for hours to collect water in India.

Domestically, deregulation and the shift of responsibilities from federal to state and local governments are changing the relationships among levels of government and between government and the private sector.

Communication Revolution

Communications technology has enhanced people's ability to receive information and influence events that affect them. This has sparked explosive growth in the number of organizations, associations, and networks formed by citizens, businesses, and communities seeking a greater voice for their interests. As a result, society outside of government—civil society—is demanding a greater role in governmental decisions, while at the same time impatiently seeking solutions outside government's power to decide.

Knowledge Economy

But technological innovation is changing much more than communication. It is changing the ways in which we live, work, produce, and consume. Knowledge has become the economy's most important and dynamic resource. It has rapidly improved efficiency as those who create and sell goods and services substitute information and innovation for raw materials. The shift to a knowledge-driven economy has emphasized the positive connection among efficiency, profits, and environmental protection and helped launch a trend in profitable pollution prevention. More and more people now understand that pollution is waste, waste is inefficient, and inefficiency is expensive.

Kinds of Policy Responses

During the 1966 fight over whether to construct dams within the Grand Canyon, the Sierra Club made their objections known through a full-page advertisement in the *New York Times*. The U.S. Bureau of Reclamation had been arguing that new reservoirs would be a boon to visitors by allowing powerboats to get closer to the canyon walls. Sierra's response: "Should We Also Flood the Sistine Chapel So Tourists Can Get Nearer the Ceiling?" This example is typical of the polarization of environmental debates since the beginning of the environmental movement in the 1960s.

Political liberals typically argue that certain places should be protected from development because of their aesthetic merit, because certain endangered species should be protected, or because some proposed development would cause pollution. Political conservatives often state that the luxury good of conservation is something taxpayers cannot afford and the wilderness must be sacrificed for the sake of economic progress.

The nature of this debate—the classic pitting of "the environment" versus "the economy"—is incomplete at the very least. Environmental problems sometimes quickly become economic or societal problems. For example, a misguided Soviet irrigation project contributed greatly to the death of the Aral Sea, turning what was once the fourth-largest lake in the world largely into useless salt flats. Or consider a recent study of air pollution in China that found it caused 1.2 million premature deaths in 2011, resulting not just in personal tragedy, but colossal expense, especially from chronic disease.

It is important that we recognize that economic, environmental, and social goals are integrally linked and that we develop policies that reflect that interrelationship. Thinking narrowly about jobs, energy, transportation, housing, or ecosystems—as if they were not connected—creates new problems even as it attempts to solve old ones.

This will require new modes of decision making, ranging from the local to the international level. While trend is not always destiny, the trend that has been evolving over the past several years has been toward more collaborative forms of decision making. Perhaps such collaborative structures will involve more people and a broader range of interests in shaping and making public policy. It is hoped that this will improve decisions, mitigate conflict, and begin to counteract the corrosive trends of cynicism and civic disengagement that seem to be growing.

More collaborative approaches to making decisions can be arduous and time-consuming, and all of the players must change their customary roles. For government, this means using its power to convene and facilitate, shifting gradually from prescribing behavior to supporting responsibility by setting goals, creating incentives, monitoring performance, and providing information.

For their part, businesses need to build the practice and skills of dialogue with communities and citizens, participating in community decision making and opening their own values, strategies, and performance to their community and the society.

Advocates, too, must accept the burdens and constraints of rational dialogue built on trust, and communities must create open and inclusive debate about their future.

Does all of this sound too idealistic? Perhaps it is; however, without a vision for the future, where would we be? As was stated previously, trend is not destiny. In other words, we are capable of change regardless of the status quo. This is perhaps nowhere more important than in the world of environmental decision making. (See figure 20.2.)

Learning from the Past

For the past quarter century, the basic pattern of environmental protection in economically developed nations has been to react to specific crises. Institutions have been established, laws passed, and regulations written in response to problems that already were posing substantial ecological and public health risks and costs or that already were causing deep-seated public concern.

The United States is no exception. The U.S. Environmental Protection Agency (EPA) has focused its attention almost exclusively on present and past problems. The political will to establish the agency grew out of a series of highly publicized, serious environmental problems, such as the fire on the Cuyahoga River in Ohio, smog in Los Angeles, and the near extinction of the bald eagle. During the 1970s and 1980s, Congress enacted a series of

FIGURE 20.2 Trend Is Not Destiny Scenes such as these were very common in North America only a short time ago. Fortunately, for the most part, such photos are today only historic in nature. Positive change is possible.

laws intended to solve these problems, and the EPA, which was created in 1970, was given the responsibility for enforcing most environmental laws.

Despite success in correcting a number of existing environmental problems, there has been a continuing pattern of not responding to environmental problems until they pose immediate and unambiguous risks. Such policies, however, will not adequately protect the environment in the future. People are recognizing that the agencies and organizations whose activities affect the environment must begin to anticipate future environmental problems and take steps to avoid them. One of the most important lessons learned during the past quarter century of environmental history is that the failure to think about the future environmental consequences of prospective social, economic, and technological changes may impose substantial and avoidable economic and environmental costs on future generations.

Thinking about the Future

Thinking about the future is more important today than ever before because the accelerating rate of change is shrinking the distance between the present and the future. Technological capabilities that seemed beyond the horizon just a few years ago are now outdated. Scientific developments and the flow of information are accelerating.

Thinking about the future is valuable also because the cost of avoiding a problem is often far less than the cost of solving it later. The U.S. experience with hazardous-waste disposal provides a compelling example. Some private companies and federal facilities undoubtedly saved money in the short term by disposing of hazardous wastes inadequately, but those savings were dwarfed by the cost of cleaning up hazardous-waste sites years later. In that case, foresight could have saved private industry, insurance companies, and the federal government (i.e., taxpayers) billions of dollars, while reducing exposure to pollutants and public anxieties in the affected communities.

Environmental foresight can preserve the environment for future generations. When one generation's behavior necessitates environmental remediation in the future, an environmental debt is bequeathed to future generations just as surely as unbalanced government budgets bequeath a burden of financial debt. (See figure 20.3.)

By anticipating environmental problems and taking steps now to prevent them, the present generation can minimize the environmental and financial debts that its children will incur.

Today, we face new classes of environmental problems that are more diffuse than those of the past and thus demand different approaches. Since the first Earth Day in 1970, the vast majority of the significant "point" sources of air and water pollution—large industrial facilities and municipal sewage systems—that once spewed untreated wastes into the air, rivers, and lakes have been controlled. The most important remaining sources of pollution are diffuse and widespread: sediment, pesticides, and fertilizers that run off farmland; oil and toxic heavy metals that wash off city streets and highways; and air pollutants from automobiles, outdoor grills, and woodstoves. Pollution from these sources cannot always be controlled with sewage treatment plants or the same regulatory techniques used to check emissions from large industries. Furthermore, we now have the global environmental problems of biodiversity loss, ozone depletion, and climate change. These problems will require cooperative international responses. We are also recognizing that controlling pollutants alone, no matter how successful, will not achieve an environmentally sustainable economy, since many global concerns are related to the size of the human population and the unequal distribution of resources.

For the most fortunate, the past 50 years have produced a quality of life unprecedented in human history. This progress notwithstanding, however, more than 20 percent of today's human population still lives in poverty, 15 percent experiences persistent hunger, and at least 10 percent is homeless. Moreover, the gap between the very rich and very poor is widening, with whole regions of the world clearly losing ground.

Through both its successes and failures, modern human development has transformed the planet. Human activities have doubled the planet's rate of nitrogen fixation, tripled

FIGURE 20.3 Environmental Debt The plight of these fishermen, resulting from the collapse of the cod fishery in the northeast United States and Canada and the salmon fishery in the northwest United States and Canada, is an environmental debt inherited from past generations of abuse and misuse.

the rate of invasion by exotic organisms, increased sediment loads in rivers fivefold, and vastly increased natural rates of species extinctions. Clearly, in the future, we will need a different vision from that which shaped our past.

Defining the Future

We are progressing from an environmental paradigm based on cleanup and control to one including assessment, anticipation, and avoidance.

Changing Technologies

Expenditures to develop technologies that prevent environmental harm are beginning to pay off. Agricultural practices are becoming less wasteful and more sustainable, manufacturing processes are becoming more efficient in the use of resources, and consumer products are being designed with the environment in mind. The infrastructures that supply energy, transportation services, and water supplies are becoming more resource efficient and environmentally benign. Remediation efforts are cleaning up a large portion of existing hazardous-waste sites. Our ability to respond to emerging problems is being aided by more advanced monitoring systems and data analysis tools that continually assess the state of the local, regional, and global environment. Finally, we are developing effective ways of restoring or re-creating severely damaged ecosystems to preserve the long-term health and productivity of our natural resource base.

Involved Public

In the long run, environmental quality is not determined solely by the actions of government, regulated industries, or nongovernmental organizations. It is largely a function of the decisions and behavior of individuals, families, businesses, and communities everywhere.

Looking ahead, consensus is growing that the next 50 years will see a world in which people are more crowded, more connected, and more consuming than at any time in human history. It will be almost certainly warmer, more polluted, and less species-rich. Many of these trends have generated a good deal of discussion recently, under headings ranging from "globalization" to "climate change." Less remarked upon is the fundamental transition under way in the growth of human populations: Rates of increase are now falling almost everywhere in the world, with the result that the number of people on the planet is expected to level off at 10 billion or 11 billion by the end of the twenty-first century, reaching around 9.5 billion—a 38 percent increase—by 2050.

Let's put all of this into another perspective that may ring closer to home. Current projections are that the U.S. population will grow by 36 percent between 2010 and 2050. Potentially, this translates into a 36 percent increase in U.S. infrastructure as well, which means 36 percent more motor vehicles and traffic; 36 percent more housing; 36 percent more capacity needed in landfills, wastewater treatment plants, hazardous-waste treatment facilities, and chemicals (pesticides and herbicides) for agriculture. (See figure 20.4.) In short, this vision is not very appealing. What happens to wilderness areas, remote and quiet places, and habitat for songbirds, waterfowl, and other wild creatures? What happens to our quality of life?

Another way of looking at the future would be along the lines of what was previously mentioned about the analogy to the agricultural and industrial revolutions. This translates into a future of profound change—a future in which virtually everthing we do will change. This future of profound change will also be one of phenomenal opportunity and excitement. It is a future in which we will farm, build, and transport in entirely new ways. This vision of the future could well become known as the environmental revolution.

FIGURE 20.4 Options and Trade-offs for the New Century Unless we become more creative in areas such as transportation and land use, it will be necessary to develop in the United States alone, in the next 50 years, an amount of farmland and scenic countryside equal to the amount developed over the past 200 years.

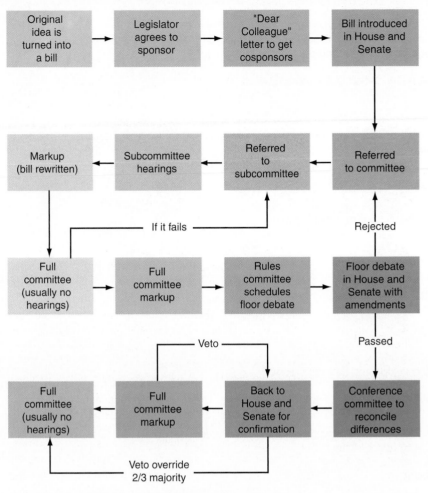

The Legislative Process

Original idea is turned into a bill → Legislator agrees to sponsor → "Dear Colleague" letter to get cosponsors → Bill introduced in House and Senate

Markup (bill rewritten) ← Subcommittee hearings ← Referred to subcommittee ← Referred to committee

If it fails / Rejected

Full committee (usually no hearings) → Full committee markup → Rules committee schedules floor debate → Floor debate in House and Senate with amendments

Veto / Passed

Full committee (usually no hearings) ← Full committee markup ← Back to House and Senate for confirmation ← Conference committee to reconcile differences

Veto override 2/3 majority

FIGURE 20.5 Passage of a Law This figure illustrates the path of a bill in the U.S. Congress from organization to becoming a law. As we can see, the process is not a quick one.

20.2 Development of Environmental Policy in the United States

Public **policy** is the general principle by which government **branches**—the **legislative, executive,** and **judicial**—are guided in their management of public affairs. The legislature (Congress) is directed to declare and shape national policy by passing legislation, which is the same as enacting law. The executive (president) is directed to enforce the law, while the judiciary (the court system) interprets the law when a dispute arises.

Legislative Action

When Congress considers certain conduct to be against public policy and against the public good, it passes legislation in the form of acts or statutes. Congress specifically regulates, controls, or prohibits activity that is in conflict with public policy and attempts to encourage desirable behavior.

When Congress passes environmental legislation, it also declares and shapes the national environmental policy, thus

fulfilling its policy-making function. (See figure 20.5.) The EPA is the concern of almost two-thirds of the House of Representatives standing committees and subcommittees and a similar percentage in the Senate. Some seventy committees and subcommittees control water quality policy, for example. Such fragmentation creates both opportunities and problems. While such a variety of committees provides enormous access for environmentalist and industry groups to lobby, the division of tasks means that no one committee or agency looks on environmental problems as a whole. (See figure 20.6.)

Role of Executive Branch

Once an environmental law is passed, it becomes the role of the executive branch to put it into force. (See figure 20.7.) The primary agency responsible for environmental legislation is the EPA. Through the years, Congress has passed and presidents have signed numerous laws to protect human health and the environment. These laws give EPA most of its authority to write regulations, and serve as the foundation for achieving the nation's environmental and public health protection goals. EPA is called a regulatory agency because Congress authorizes it to write regulations that explain the critical technical, operational, and legal details necessary to implement laws.

For example, the Resource Conservation and Recovery Act (RCRA) requires the EPA to write standards for managing hazardous waste. RCRA's central mandate requires EPA to develop standards to protect human health and the environment but does not say precisely what those standards should be. As in many other laws, Congress expects EPA to develop most of the details for regulations based on the technical and policy expertise of EPA experts.

Over 100 years ago, President Teddy Roosevelt declared that nothing short of defending your country in wartime "compares in leaving the land even better land for our descendants than it is for us." The environmental issues that Roosevelt strongly believed in, however, did not become major political issues until the early 1970s.

While the publication of Rachel Carson's *Silent Spring* in 1962 is considered to be the beginning of the modern environmental movement, the first Earth Day on April 22, 1970, was perhaps the single event that put the movement into high gear. In 1970, as a result of mounting public concern over environmental deterioration—cities clouded by smog, rivers on fire, waterways choked by raw sewage—many nations, including the United States, began to address the most obvious, most acute environmental problems.

During the 1970s, many important pieces of environmental legislation were enacted in the United States. In addition, the percent of discretionary U.S. budget spent on natural resources and the environment grew significantly. Many of the identified

House Committees and Jurisdictions

Committee on Agriculture	Pesticides
Commitee on Interior and Insular Affairs	Synthetic fuels, Conservation oversight, Energy budget, Mines, Oil shale, Outer continental shelf, Radiation (Nuclear Regulatory Commission oversight), Strip mining
Committee on Energy and Commerce	Air, Drinking water, Noise, Radiation, Solid waste, Toxics
Commitee on Natural Resources	Ocean dumping
Committee on Transporation and Infrastructure	Noise, Water pollution, Water resources
Committee on Science and Technology	Research and development
Committee on Small Business	Impact of environment regulations on small business

Senate Committees and Jurisdictions

Committee on Agriculture, Nutrition, and Forestry	Pesticides
Committee on the Budget	Budget
Committee on Commerce, Science, and Transportation	Oceans, Research and development, Radiation, Toxics
Committee on Energy and Natural Resources	Synthetic fuels, Conservation, Oversight, Energy budget, Mines, Oil shale, Outer continental shelf, Strip mining
Committee on Environment and Public Works	Air, Drinking water, Noise, Nuclear energy, Ocean dumping, Outer continental shelf, Research and development, Solid waste, Toxics, Water
Committee on Foreign Relations	International environment
Committee on Labor and Human Resources	Public health
Committee on Small Business	Impact of environmental regulations on small business

FIGURE 20.6 Committees of House of Representatives and Senate Examples of congressional committees with environmental jurisdiction.

environmental problems were so immediate and so obvious that it was relatively easy to see what had to be done and to summon the political will to do it. (See table 20.1.) Energy policy became a primary area of legislative focus during the Obama administration. Initial policy directions included:

- Renewables like wind and solar beginning to displace coal from the electricity-generating market, with solar panels mounted on the roofs of nearly all big buildings.
- Electric cars becoming more widespread, and car batteries used to store excess electricity or as a source for more electricity when it is needed.
- Energy efficiency and conservation becoming the foundation of energy policy.

Unfortunately, given the political stalemate in Washington, D.C., during the Obama administration there was little headway made on many of the earlier energy goals.

The Role of Nongovernmental Organizations

The United States is commonly recognized as having been an environmental innovator in the 1970s. Important to U.S. prominence as an environmental leader was the growth of environmental nongovernmental organizations. Environmental groups were aided by amendments to tax laws that made it relatively easy for groups to incorporate as not-for-profit organizations even while engaged in some lobbying activities. This provided them with important benefits, including the possibility of receiving contributions that are tax-deductible for the donor and favorable postage rates for reaching out to the public. Environmental groups could turn to both members and large, private financial institutions for donations; they were also eligible to receive governmental grants and contracts. U.S. environmental groups also won the important right to stand in court to sue on behalf of environmental interests.

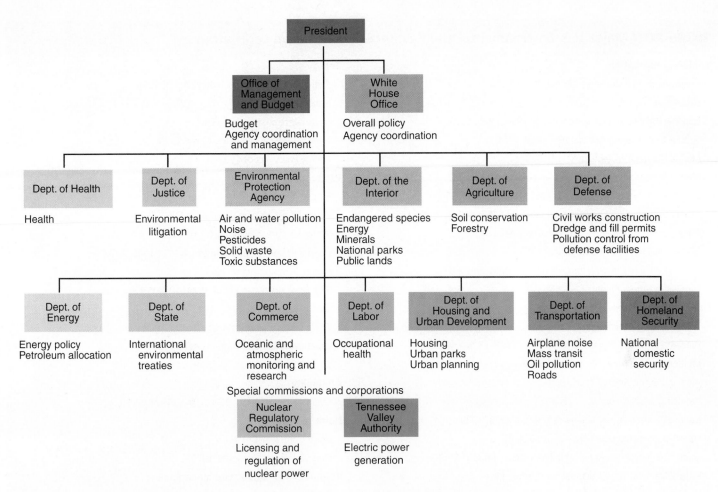

FIGURE 20.7 Environmental Responsibilites of Major Agencies of the Executive Branch Major agencies of the executive branch are shown with their environmental responsibility.

Using these new rights, environmental groups became increasingly professionalized and diversified.

The Role of Lobbying in the Development of Environmental Policy

Lobbying—the practice that seeks to alter legislation or administrative action on political issues—is big business. It is impossible to find out how many groups are lobbying specifically on environmental issues, but overall some 3,500 organizations, most of which represent businesses of some kind, have Washington, D.C., offices. Organizations that lobby in the environmental policy area fall generally into three groups: industries and their trade associations, such as Monsanto and the Chemical Manufacturing Association (CMA), which operate on a profit motive and have a vested interest in energy and environmental issues; not-for-profit public interest groups such as the Environmental Defense Fund (EDF) and Greenpeace, which depend on foundation grants or public subscriptions and are concerned about the environment's welfare; and scientific and research organizations that want to advance science with regard to environmental issues.

In addition to campaign contributions to elected officials and candidates, companies, labor unions, and other organizations spend billions of dollars each year to lobby Congress and federal agencies. Some special interests retain lobbying firms; others have lobbyists working in-house to promote their cause.

The oil and gas industry, which includes multinational and independent oil and gas producers and refiners, natural gas pipeline companies, gasoline service stations and fuel oil dealers, has long enjoyed a history of strong influence in Washington, D.C. Individuals and political action committees affiliated with oil and gas companies donated in excess of $1.3 billion to candidates and parties from 2000 to 2013, 75 percent of which went to Republican candidates. Clients in the oil and gas industry spent $141 million in 2012. Royal Dutch Shell, the industry leader in 2012 spent $14 million in lobbying expenditures that year alone.

Pro-environmental groups spent $16 million in 2012 on federal lobbying efforts. Environmental groups generally advocate against big industries such as oil and gas companies, power utilities, and chemical companies. During the 2012 election cycle, environmental groups gave $16 million to federal candidates and party committees, 94 percent of which went to Democrats. The movement's financial clout may be understated in these figures, however, as the Sierra Club, which is one of the largest contributors, has focused on spending money on direct "issue ads" rather than giving it to candidates or political parties. (See table 20.2.)

Table 20.1 Major U.S. Environmental and Resource Conservation Legislation

Wildlife conservation

Anadromous Fish Conservation Act of 1965

Fur Seal Act of 1966

National Wildlife Refuge System Act of 1966, 1976, 1978

Species Conservation Act of 1966, 1969

Marine Mammal Protection Act of 1972

Marine Protection, Research, and Sanctuaries Act of 1972

Endangered Species Act of 1973, 1982, 1985, 1988, 1995

Fishery Conservation and Management Act of 1976, 1978, 1982, 1996

Whale Conservation and Protection Study Act of 1976

Fish and Wildlife Improvement Act of 1978

Fish and Wildlife Conservation Act of 1980 (Nongame Act)

Fur Seal Act Amendments of 1983

Land use and conservation

Taylor Grazing Act of 1934

Wilderness Act of 1964

Multiple Use Sustained Yield Act of 1968

Wild and Scenic Rivers Act of 1968

National Trails System Act of 1968

National Coastal Zone Management Act of 1972, 1980

Forest Reserves Management Act of 1974, 1976

Forest and Rangeland Renewable Resources Act of 1974, 1978

Federal Land Policy and Management Act of 1976

National Forest Management Act of 1976

Soil and Water Conservation Act of 1977

Surface Mining Control and Reclamation Act of 1977

Antarctic Conservation Act of 1978

Endangered American Wilderness Act of 1978

Alaskan National Interests Lands Conservation Act of 1980

Coastal Barrier Resources Act of 1982

Food Security Act of 1985

Emergency Wetlands Resources Act of 1986

North American Wetlands Conservation Act of 1989

Coastal Development Act of 1990

California Desert Protection Act of 1994

Federal Agriculture Improvement and Reform Act of 1996

General

National Environmental Policy Act (NEPA) of 1969

International Environmental Protection Act of 1983

Energy

Energy Policy and Conservation Act of 1975

National Energy Act of 1978, 1980

Northwest Power Act of 1980

National Appliance Energy Conservation Act of 1987

Energy Policy Act of 1992

Water quality

Refuse Act of 1899

Water Quality Act of 1965

Water Resources Planning Act of 1965

Federal Water Pollution Control Acts of 1965, 1972

Ocean Dumping Act of 1972

Safe Drinking Water Act of 1974, 1984, 1996

Clean Water Act of 1977, 1987

Great Lakes Toxic Substance Control Agreement of 1986

Great Lakes Critical Programs Act of 1990

Oil Spill Prevention and Liability Act of 1990

Air quality

Clean Air Act of 1963, 1965, 1970, 1977, 1990

Noise control

Noise Control Act of 1965

Quiet Communities Act of 1978

Resources and solid waste management

Solid Waste Disposal Act of 1965

Resources Recovery Act of 1970

Resource Conservation and Recovery Act of 1976

Waste Reduction Act of 1990

Toxic substances

Toxic Substances Control Act of 1976

Resource Conservation and Recovery Act of 1976

Comprehensive Environmental Response, Compensation, and Liability (Superfund) Act of 1980, 1986, 1990

Nuclear Waste Policy Act of 1982

Pesticides

Food, Drug, and Cosmetics Act of 1938

Federal Insecticide, Fungicide, and Rodenticide Control Act of 1972, 1988

Food Quality Protection Act of 1996

Science, Politics, & Policy

The Endangered Species Act at Forty

Passed forty years ago, the Endangered Species Act (ESA) begins by stating, "The Congress finds and declares that . . . species of fish, wildlife, and plants are of esthetic, ecological, educational, historical, recreational, and scientific value to the Nation and it's people." Through the act, Congress attempted to "provide a means whereby the ecosystems upon which endangered species and threatened species depend may be conserved, to provide a program for the conservation of such endangered species and threatened species."

The first four decades of the ESA brought forth successes and conflict. Much of the discussion has been around the conflict. Major points of conflict include: Does the ESA overly burden private landowners? Does it protect species at the expense of human well-being? Is it too cumbersome and costly to implement? Does it cause increasing levels of litigation?

The ESA has played a critical role in bringing attention to species at risk of extinction. While many species remain at risk, less than 1 percent of the 2,000 listed species have actually gone extinct. The ESA has also brought about major changes in planning and land management practices. Despite its conservation achievements, the ESA faces some major challenges in the future including:

- Policy tools to engage private landowners are often very complicated and insufficiently evaluated for their actual outcomes.
- Threats to species involve complex, interconnected changes to ecosystems, such as pollution and invasive species. The ESA has, however, traditionally taken a species-by-species approach that constrains the ability to focus on the health of ecosystems.
- Funding for the ESA is increasingly becoming an issue. Public sources of conservation funding are highly constrained.

Privately owned lands play an important role in species protection: more than two-thirds of endangered or threatened species reside on private lands, and one-third are found exclusively on private lands. The ESA initially created disincentives to landowners for species protection because having a listed species on your land invoked regulations that could require costly restrictions on land use. Over the past 20 years, many changes in the ESA have lessened these disincentives and the U.S. Fish and Wildlife Service, charged with implementing the ESA, continues to develop innovative approaches to the Act that do not unjustly punish private landowners.

The conservation and recovery of species will become more complex and challenging due to climate change, water scarcity, and habitat fragmentation. The ESA does provide an effective framework to meet these challenges. The ESA must continue, however, to move away from a species-by-species approach and toward an approach that incorporates species protection within larger landscape-scale efforts that use incentives to engage private landowners. The ESA is sufficiently broad and general to accommodate most—if not all—of these improvements.

What's Your Take?

Public opinion polls in the United States continue to indicate that the public is concerned about environmental issues. Many state and local governments have been very active in adopting environmental laws and policies ranging from sustainable building practices to vehicle emission standards. The federal government, however, has not followed suit with such laws. Should the federal government play a more aggressive role in establishing environmental policy or should that role be delegated to the state and local levels of government? Develop a position paper that supports why environmental policy is best developed at either the federal or state/local level of government.

Table 20.2 Lobbying Expenditures in 2012

Environmental Organizations (Top Five)		Oil and Gas Industry (Top Five)	
BlueGreen Alliance	$2,488,000	Royal Dutch Shell	$14,480,000
Environmental Defense Fund	$1,754,000	Exxon	$12,970,000
Nature Conservancy	$1,310,000	Koch Industries	$10,550,000
Earthjustice Legal Defense Fund	$637,199	Chevron	$9,550,000
National Wildlife Federation	$620,000	BP	$8,590,000
Total	*$6,809,199*		*$56,140,000*

Lobbyists have a significant impact on the legislative process, whether they come from industry, the environmental community, or science. But many argue that the way we do lobbying needs fixing. Calls for campaign finance reform, stricter disclosure laws so the public knows where the money is coming from, and employment restrictions for government officials and employees are raised during every session of Congress.

20.3 Environmental Policy and Regulation

Environmental laws are not a recent phenomenon. As early as 1306, London adopted an ordinance limiting the burning of coal because of the degradation of local air quality. Such laws became more common as industrialization created many sources of air and water pollution throughout the world. In the United States, environmental laws often evolved from ordinances passed by local governments. Interested in protecting public health, officials of towns and cities enacted local laws to limit the activities of private citizens for the common good. For example, to have "healthy air," many communities enacted ordinances in the 1880s to regulate rubbish burning within city limits. Public health issues were the foundations on which the environmental laws of today were built.

The Significance of Administrative Law

Environmental law in the United States is governed by administrative law. Administrative law is a relatively new concept, having been developed only during the twentieth century. In 1946, Congress passed the Federal Administrative Procedure Act. This act designated general procedures to be used by federal agencies when they exercised their rule-making, adjudicatory, and enforcement powers. This is a rapidly expanding area of law and defines how governmental organizations such as agencies, boards, and commissions develop and implement the regulatory programs they are legislatively authorized to create. Some of the many U.S. federal agencies that influence environmental issues include the Environmental Protection Agency, the Council on Environmental Quality, the National Forest Service, and the Bureau of Land Management.

Administrative law applies to government agencies and to those that are affected by agency actions. In the United States, many federal environmental programs are administered by the states under the authority of federal and related state laws. States often differ from both the federal government and each other in the way they interpret, implement, and enforce federal laws. In addition, each state has its own administrative guidelines that govern and define how state agencies act. All actions of federal agencies must comply with the 1946 Federal Administrative Procedure Act.

National Environmental Policy Act— Landmark Legislation

The National Environmental Policy Act (NEPA) was enacted in 1969 and signed into law by President Richard Nixon on New Year's Day, 1970. It is a short, general statute designed to institutionalize within the federal government a concern for the "quality of the environment." NEPA helps encourage environmental awareness among all federal agencies, not just those that prior to NEPA had to consider environmental factors in their planning and decision making. Until 1970, most federal agencies acted within their delegated authority without considering the environmental impacts of their actions. However, in the 1960s, Congress seriously began to study pollution problems. Because Congress has found that the federal government is both a major cause of environmental degradation and a major source of regulatory activity, all actions of the federal government now fall under NEPA.

NEPA forces federal agencies to consider the environmental consequences of their actions before implementing a proposal or recommendation. NEPA has two purposes: first, to advise the president on the state of the nation's environment; and second, to create an advisory council called the Council on Environmental Quality (CEQ). The CEQ outlines NEPA compliance guidelines. The CEQ also provides the president with consistent expert advice on national environmental policies and problems.

NEPA has been interpreted narrowly by the federal courts. As a result, many states have passed much stronger state environmental protection acts (SEPAs) as well. Today, NEPA analysis is undertaken as part of almost every recommendation or proposal for federal action. This includes not only actions by agencies of the federal government but also actions of states, local municipalities, and private corporations. NEPA is Congress's mission statement that mandates the means by which the federal government, through the guidance of the CEQ, will achieve its national environmental policy.

Other Important Environmental Legislation

In addition to the passage of NEPA, the 1970s saw a series of new environmental laws passed, including the Resource Conservation and Recovery Act (RCRA); the Comprehensive Environmental Response, Compensation and Liability Act (CERCLA); and the Clean Air Act (CAA). All of these acts are broadly worded to identify existing problems that Congress believes can be corrected to protect human health, welfare, and the environment. Table 20.1 lists other major environmental legislation.

Protecting human health, welfare, and the environment is the national policy that Congress has chosen to encourage. For example, under NEPA, the national policy is to "promote efforts which will prevent or eliminate damage to the environment and biosphere and stimulate the health and welfare of [humans]." Similarly, under the Clean Air Act, the policy is to "protect and enhance the quality of the nation's air resources so as to promote the public health and welfare and the productive capacity of its population."

Each of the preceding statutes declares national policy on environmental issues and addresses distinct problems. These statutes also authorize the use of some or all of the administrative functions discussed previously, such as rule making, adjudication, administrative law, civil and criminal enforcement, citizen suits, and judicial review. (See figure 20.8.)

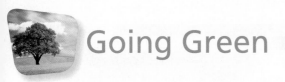

Going Green

Investing in a Green Future

In 2006, after ringing the opening bell of the New York Stock Exchange, Secretary General of the United Nations, Kofi Annan, launched the Principles for Responsible Investment (PRI). Six months later, PRI had 94 institutional investors from 17 countries representing US $5 trillion in investments.

The launch of the principles created the first-ever global network of investors who addressed many of the same environmental, social, and governance issues that the United Nations is asked to address. One of the goals of the PRI community is to work with policymakers to address issues of long-term importance to both investors and society. Investors representing more than 10 percent of global capital market value have, therefore, sent the strongest of signals to the marketplace that environment,

social, and good governance issues count in investment policy making and decision making. The PRI has evolved because investors have recognized that systemic issues of sustainability are material to long-term investment returns.

There is a range of areas where policymakers could create the necessary environment that would encourage investors to take longer-term views on environmental, social, and governance issues. Mandatory disclosure of environmental performance is one such area. Once investors are able to assess the risks involved in various activities, they are able to put pressure on companies to address those risks. But they are unable to do this if they are unaware of what the company is doing. Mandatory disclosure policies would allow investors to take action when required.

EPA Enforcement Options

1. Warning letter
- Describes alleged violation
- Outlines action that must be taken to remedy
  - Deadline for compliance or enforcement action
  - Allow for an opportunity to discuss the situation

2. Administrative order
- Requires certain activity be ceased, or
- Requires compliance by a certain date, or
- Requires certain tests be performed
- Provides rights
  - Respond by citing defenses or objections
  - Right to confer with EPA

3. Permit action
- Revoke or modify existing permit
- Seek to add conditions to permits that are being negotiated

4. Civil enforcement
- Injunctive
- Monetary relief varies by statute per violation, per day

5. Criminal proceedings
- Fines
- Incarceration

FIGURE 20.8 Enforcement Options of the U.S. Environmental Protection Agency The enforcement options of the U.S. Environmental Protection Agency range from a warning letter to a jail sentence.
Source: U.S. Environmental Protection Agency, Office of Enforcement, Washington, D.C.

Role of the Environmental Protection Agency

Administrative functions empower EPA, the states, and private citizens to take responsibility for enforcing the various authorized programs. These administrative functions not only shape environmental law but also control the daily operations of both the regulated industry and agencies authorized to protect the environment.

What Is the EPA's Future?

In the world of emotion-charged partisan politics, the Environmental Protection Agency (EPA) has been a lighting rod for zealots on both the right and the left. The conservative talk radio shows often call for the elimination of the EPA while some on the liberal end of the spectrum argue for increasing the power of the EPA. What should be the role of the agency in a changing world?

The evolving nature of the issues and challenges that the EPA will, or will likely, face in the coming years strongly suggests that the way the agency fulfills its mission will need to evolve. Historically, the EPA has demonstrated its ability to develop and implement creative solutions to new challenges, especially those that go beyond the agency's traditional pollution prevention and response role. The EPA has been successful in its efforts to move forward on what are two occasionally converging tracks, one that focuses on the agency's statutorily mandated regulatory responsibilities (the primary, or "core" track), and one that exists outside the regulatory construct and leverages the EPA's expertise in order to "pre-empt" degradation of environmental quality. The question that the EPA must now ask is whether, and if so how, those two tracks must change in order to ensure future success. In the future the EPA must also:

- Take a more holistic approach to environmental protection.
- Make environmental foresight or "future analysis" a regular component of the EPA's business-as-usual practices.

- Work with governmental and private agencies and organizations to develop strategies that link national security, environmental quality, and economic growth.

- Develop new, innovative, and integrative ways of doing business, such as collaborative problem solving (with states, tribes, and stakeholders) and regulatory innovation to increase effectiveness and efficiency of environmental programs.

- Urge the president to work with Congress to draft broad legislation outlining the evolving mission of the EPA and giving it greater flexibility to deal with broad challenges such as climate change, water resource sustainability, and ecosystem sustainability that will require the agency to use integrative approaches on a scale it has not yet undertaken.

Command and Control Philosophy

To date, much environmental law has reflected the perception that environmental problems are localized in time, space, and media (i.e., air, water, soil). For example, many hazardous-waste sites in the United States have been "cleaned" by simply shipping the contaminated dirt someplace else, which not only does not solve the problem but creates the danger of incidents during removal and transportation. Environmental regulation has focused on specific phenomena and adopted the so-called command-and-control approach, in which restrictive and highly specific legislation and regulation are implemented by centralized authorities and used to achieve narrowly defined ends.

Such regulations generally have very rigid standards, often mandate the use of specific emission-control technologies, and generally define compliance in terms of "end-of-pipe" requirements. Examples in the United States include the Clean Water Act (which applied only to surface waters), the Clean Air Act (urban air quality), and the Comprehensive Environmental Response, Compensation, and Liability Act (Superfund), which applied to specific landfill sites.

If properly implemented, command-and-control methods can be effective in addressing specific environmental problems. For example, rivers such as the Potomac and the Hudson in the United States are much cleaner as a result of the Clean Water Act. (See figure 20.9.) Moreover, where applied against particular substances, such as the ban on tetraethyl lead in gasoline in the United States, the command-and-control approach has clearly worked well.

Regardless of how environmental decisions are made, most decisions are complex and involve compromise. Large-scale and complicated ecological policy problems, such as reversing the decline of salmon, deciding on the proper role of wild fires on public lands, the consequences of declining biological diversity, and making sense about the confusing policy choices surrounding interpretations of sustainability all share several qualities:

1. **Complexity**—innumerable options and trade-offs;
2. **Polarization**—clashes between competing values;
3. **Winners and losers**—for each policy choice, some will clearly benefit, some will be harmed, and the consequences for others are uncertain;

FIGURE 20.9 Effect of the Command-and-Control Approach The environmental quality of rivers has dramatically improved over the past 40 years as a direct result of the Clean Water Act. (a) and (b) A visibly polluted Cuyahoga River in Cleveland, Ohio, actually caught fire several times. The last fire was in June 1969. (c) While challenges still remain, the water quality of the Cuyahoga has improved significantly since passage of the Clean Water Act.

Focus On

The Precautionary Principle

The precautionary principle was introduced into environmental politics in response to a perception that existing policies did not provide adequate protection to the environment. The principle, adopted at the 1992 United Nations Conference on Environment and Development, states that: "In order to protect the environment, the precautionary approach shall be widely applied by countries according to their capabilities. Where there are threats of serious or irreversible damage, the lack of full scientific certainty shall not be used as a reason for postponing cost-effective measures to prevent environmental degradation." In other words, it is better to be safe than sorry. The principle has important implications to the interpretation of science and the regulation of technology.

The principle applies to human health and the environment. And in dealing with both areas, if we wait for certainty before taking action it may be too late. Scientific standards for demonstrating cause and effect are very high. For example, smoking was strongly suspected of causing lung cancer long before the link was demonstrated conclusively. By then, many smokers had died of lung cancer. But many other people had already quit smoking because of the growing evidence that smoking was linked to lung cancer. These people were wisely exercising precaution despite some scientific uncertainty.

Serious, evident effects such as climate change, cancer, and the disappearance of species can seldom be linked decisively to a single cause. Scientific standards of certainty may be impossible to attain.

Precaution is at the basis of some U.S. environmental and food and drug legislation. For example:

- The Food and Drug Administration requires all new drugs to be tested before they are put on the market.
- The Food Quality and Protection Act mandates that pesticides be proven safe for children or removed.
- The National Environmental Policy Act is precautionary in two ways: 1) It emphasizes foresight and attention to consequences by requiring an environmental impact assessment for any federally funded project, and 2) it mandates consideration of alternative plans.

The precautionary principle is becoming the basis for reforming environmental laws and regulations. It has also been argued that preventative policies encourage the exploration of better, safer, and often ultimately cheaper alternatives—and the development of cleaner products and technologies. The following are some examples of policies specifically based on the precautionary principle:

- San Francisco adopted an environment code with the precautionary principle as article one. The city is also applying the principle to its purchasing decisions.
- The European Union has a policy based on the principle that requires all chemicals to be tested for their effects on health and the environment. It places the burden on chemical manufacturers to demonstrate their products are safe.
- The Los Angeles Unified School District adopted the principle to limit pesticide use in schools. A number of North American cities have similar ordinances.

---

4. **Delayed consequences**—no immediate "fix" and the benefits, if any, of painful concessions will often not be evident for decades;

5. **National vs. regional conflict**—national (or international) priorities often differ substantially from those at the local or regional level; and

6. **Ambiguous role for science**—science is often not pivotal in evaluating policy options, but science often ends up serving inappropriately as a surrogate for debates over values and preferences.

The EPA was not a carefully well-integrated organization in the beginning. President Nixon, by Executive Order "reorganized" the Executive Branch by transferring 15 units and their personnel from existing organizations into the now independent EPA. Four major government agencies were involved.

It was not an easy birth. Air, solid waste, radiological health, water hygiene, and pesticide tolerance functions and personnel were transferred from the Department of Health, Education, and Welfare; water quality and pesticide label review programs came from the Department of the Interior; radiation protection standards came from the Atomic Energy Commission and the Federal Radiation Council; pesticide registration was transferred from the Department of Agriculture. It took several years under the first

administrator, William Ruckelshaus, to bring relative order to the new agency.

20.4 The Greening of Geopolitics

Environmental or "green" politics have emerged from minority status and become a political movement in many nations. Issues such as transboundary water supply and pollution, acid precipitation, and global climate change have served to bolster the emergence of green politics.

International Aspects of Environmental Problems

Concern about the environment is not limited to developed nations. A 1989 treaty signed in Switzerland limits what poorer nations call toxic terrorism—use of their lands by richer countries as dumping grounds for industrial waste. In 1990, more than 100 developing nations called for a "productive dialogue with the developed world" on "protection of the environment." As was covered in chapter 2, the Earth Summit in Rio brought together nearly 180 governments to address world environmental concerns, and the 1997 conference on global warming in Kyoto brought together some 120 nations.

Environmental concern is also a growing factor in international relations. Many world leaders see the concern for environment, health, and natural resources as entering the policy mainstream. A sense of urgency and common cause about the environment is leading to cooperation in some areas. Ecological degradation in any nation is now understood almost inevitably to impinge on the quality of life in others. Drought in Africa and deforestation in Haiti have resulted in large numbers of refugees whose migrations generate tensions both within and between nations. From the Nile to the Rio Grande, conflicts flare over water rights. The growing megacities of the developing world are areas of potential civil unrest. Sheer numbers of people overwhelm social services and natural resources. The government of the Maldives has pleaded with the industrialized nations to reduce their production of greenhouse gases, fearing that the polar ice caps may melt and inundate the island nation.

Economic progress in developing nations could also bring the possibility for environmental peril and international tension. Energy consumption in China is having a major impact on world energy prices and efforts to control atmospheric carbon dioxide. China's energy consumption increased by over 180 percent between 2000 and 2012. In 2012, China consumed about 22 percent of the world's energy and became the largest energy consumer in the world. The United States became the second largest consumer of energy with about 18 percent of the world's energy. China's energy use is currently growing at over 7 percent per year while growth in energy consumption in the United States has been near zero over the 2000 to 2012 time period.

In 2012, coal accounted for about 68 percent of China's primary energy production. Because of China's extensive domestic coal resources and its wish to minimize dependence on foreign energy sources, it is expected that coal will remain the main energy source for the foreseeable future. One of the biggest concerns

regarding energy consumption in China focuses on carbon emissions and the threat of increasing global warming. China is putting a large number of new coal-fired power plants on line every year. Thus, China's carbon emissions are expected to increase by more than 10 percent per year.

More than half of China's oil imports now come from the Middle East. Consequently, China is now in the same position as most of the industrialized world—its economic stability is dependent on access to oil supplies from the world's most volatile region. China's demand for oil has been an important contributing factor in the worldwide increase in oil prices. Figure 20.10 shows the growth in total energy use and the growth in use of oil in China in recent years. Both have increased by more than 2½ times in 15 years. (See figure 20.11.)

Some experts estimate that the developing world, which today produces one-fourth of all greenhouse gas emissions, could be responsible for nearly two-thirds by the middle of the twenty-first century. Developing nations have repeatedly indicated that they are not prepared to slow down their own already weak economic growth to help compensate for decades of environmental problems caused largely by the industrialized world.

Some developing countries may resist environmental action because they see a chance to improve their bargaining leverage with foreign aid donors and international bankers. Where before the poor nations never had a strategic advantage, they now may have an ecological edge. Ecologically, there could be more parity than there ever was economically or militarily.

National Security Issues

National security may no longer be about fighting forces and weaponry alone. It also relates increasingly to watersheds, croplands, forests, climate, and other factors rarely considered by military experts and political leaders but that, when taken together, deserve

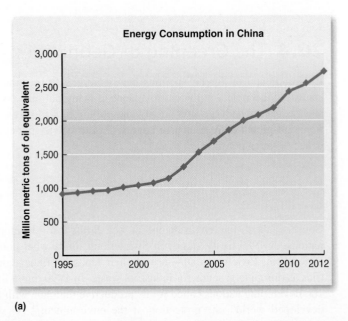

(a)

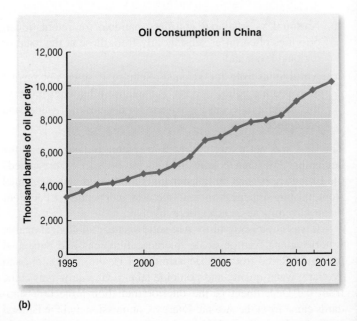

(b)

FIGURE 20.10 The Growth of Energy Consumption in China (a) China's energy consumption increased by 180 percent between 2000 and 2012. (b) China's oil consumption increased about 115 percent between 2000 and 2012.

FIGURE 20.11 Developing Concerns If China's current "modernization" continues, the boom will be fueled by coal to the possible detriment of the planet as a whole.

The Hungarian Academy of Sciences has pointed out that 90 percent of the water entering the tributaries of Hungary's rivers flows through Romania, Ukraine, Slovakia, and Austria. Numerous mines, chemical plants, oil refineries, and other sources of pollutants line those tributaries. The Hungarian government stated that the management of environmental security cannot be stopped at the borders. The government has filed lawsuits seeking monetary damage against the operators of the Baia-Mare gold extraction lagoon, and it has threatened to sue the Romanian government to help recover the cleanup costs. The accident stoked bilateral tensions: Romanian officials accused the Hungarian side of exaggerating the extent of the damage, while Hungarians asserted that the Romanians were downplaying the spill. The issue of environmental security continues to expand in the dialogue between the two nations.

The increased attention to the environment as a foreign policy and national security issue is only the beginning of what will be necessary to avert problems in the future. The most formidable obstacle may be the entrenched economic and political interests of the world's most advanced nations. If North Americans wish to stem the supply of hardwood from a fragile jungle or furs from endangered species, then they will have to stem the demand for fancy furniture and fur coats. If they wish to preserve wilderness from the intrusions of the oil industry, then they will have to find alternative sources of energy and use all fuels more efficiently.

to be viewed as equally crucial to a nation's security as are military factors. It is interesting to note that the North Atlantic Treaty Organization (NATO) has developed an Office for Scientific and Environmental Affairs, and the U.S. Department of Defense has created an Office of Environmental Security.

Recent years have shown that shifts in rainfall can bring down governments and even set off wars. The African Sahel, just south of the Sahara, provides dramatic and poignant demonstration. The deadly carnage in Darfur, Sudan, for example, which is almost always discussed in political and military terms, has roots in an ecological crisis directly arising from climate shocks. (See figure 20.12.) Darfur is an arid zone with overlapping, growing populations of impoverished pastoralists (tending goats, cattle, and camels) and sedentary farmers. Both groups depend on rainfall for their livelihoods and lives. The average rainfall has probably declined in the past few decades but is in any case highly variable, leaving Darfur prone to drought. When the rains faltered in the 1980s, violence ensued. Communities fought to survive by raiding others and attempting to seize or protect scarce water and food supplies.

Public debates tend to neglect powerful ecological effects because we focus on politics. A drought-induced famine is much more likely to trigger conflict in a place that is already impoverished and bereft of any cushion of physical or financial resources. Several studies have shown that a temporary decline in rainfall has generally been associated throughout sub-Saharan Africa with a marked rise in the likelihood of violent conflict in the following months.

20.5 International Environmental Policy

If there must be a war, let it be against environment contamination, nuclear contamination, chemical contamination; against the bankruptcy of soil and water systems; against the driving

FIGURE 20.12 Children are the most vulnerable in society to the carnage of war and environmental disasters as illustrated by this child in Sudan.

of people away from the lands as environmental refugees. If there must be war, let it be against those who assault people and other forms of life by profiteering at the expense of nature's capacity to support life. If there must be war, let the weapons be your healing hands, the hands of the world's youth in defense of the environment.

—Mustafa Tolba
Former Secretary General
United Nations Environment Programme

The Role of the United Nations

There are many institutions that address the global environment. In the United Nations system, 21 separate agencies deal with environmental issues. In addition to the United Nations, the World Bank and other institutions charged with economic development play an important role in the implementation of policies and projects that affect the global environment. Institutions that deal with trade such as the World Trade Organization (WTO) and the North American Free Trade Agreement (NAFTA) also affect the global environment.

Successes and Failures

While there have been examples of successful initiatives, global organizations have not been able to achieve significant progress in reversing global environmental degradation. There are several reasons for this. Some fail because they are controlled by a disparate membership with competing interests who are unable to reach consensus on complex and difficult issues. An example of this is the UN Food and Agriculture Organization (FAO), under whose policies only limited progress has been made in dealing with global fishing and agriculture issues. Although the FAO has pursued the collection of statistics on fish stocks and it is clear that most fisheries are being fished at or above capacity, it has not acted to develop systems that would conserve fish stocks, and individual governments continue to set catch quotas above sustainable levels.

Other institutions do not succeed because they are responsible only for specific functions or activities and are unable to address whole issues on their own. The World Bank, for example, can provide guidelines for dealing with air pollution or biodiversity preservation only for development projects that use World Bank funds. Still other institutions have failed to take even initial steps to weave the environment into their policies and programs as a permanent concern. For example, the WTO focuses on trade issues but has yet to confront the nexus of trade—sustainable economic growth and the environment.

International Coordination

International coordination and political resolve are necessary if the goals of preserving and protecting the global environment are to be realized. A step in that direction was the 1972 United Nations Conference in Stockholm, Sweden. This was the first international conference specifically dealing with global environmental concerns. The UN Environment Programme, a separate department of the United Nations that deals with environmental issues, developed out of that conference. The United Nations Conference on the Law of the Sea produced a comprehensive convention that addressed many of the issues concerning jurisdiction over ocean waters and the use of ocean resources. This treaty is viewed by many as a model for international environmental protection. Positive results are already evident from application of the agreements on pollution control, marine mammal protection, navigation safety, and other aspects of the marine environment. The issue of deep-ocean mining, primarily of manganese nodules, has to date kept many industrial nations, such as Germany and the United States, from ratifying the treaty.

Over 150 global environmental treaties have been negotiated since the start of the twentieth century. In addition, at least 500 bilateral agreements are now in effect dealing with cross-border environmental issues. These agreements generally address single issues, such as the Framework Convention on Climate Change or the Convention on International Trade in Endangered Species of Wild Fauna and Flora. History has shown that most of the existing agreements are lacking in ambition or are framed in such a way that they result in little concrete action.

However, there have been several successful international conventions and treaties. The Antarctic Treaty of 1961 reserves the Antarctic continent for peaceful scientific research and bans all military activities in the region. The 1979 Convention on Long-Range Transboundary Air Pollution was the first multilateral agreement on air pollution and the first environmental accord involving all the nations of Eastern and Western Europe and North America. The 1987 Montreal Protocol on Substances that Deplete the Stratospheric Ozone Layer addresses the ozone protective shield problem.

Agreed to in 1987, the Montreal Protocol's objective was to phase out the manufacture and use of chemicals depleting the Earth's protective ozone layer (see chapter 16). Since its adoption over a decade ago, 160 countries are now parties, representing over 95 percent of the Earth's population. Production and consumption of chlorofluorocarbons (CFCs), carbon tetrachloride, halons, and methyl chloroform have been phased out in developed countries, with reduction schedules set for their use in developing countries. In 1999, developing countries ended the production and consumption of CFCs, with phaseout scheduled for 2010.

By the late 1990s, the concentration of some CFCs in the atmosphere had started to decline, and predictions are that the ozone layer could recover by the middle of this century. Among the many reasons for these achievements, three merit attention:

- *Global agreement on the nature and seriousness of the threat.* Even the strongest skeptics could not deny the Antarctic ozone hole, which was first brought to international attention by British scientists in 1985. It was understood for the first time that emissions of ozone-depleting substances were in reality putting our lives and the lives of future generations at risk. Decisive action was required.

- *A cooperative approach, especially between developed and developing countries.* In the developed countries, it was recognized that their industries had contributed significantly to this global problem and that they had to take the lead in stopping emissions and finding alternatives. It was also recognized that solving the ozone problem required a global solution, with all countries committed to eliminating ozone-depleting substances. Innovative partnerships, including early controls for developed countries and a grace period, funding, and technology transfer for developing countries, were set in place.

- *Policy based on expert and impartial advice.* The parties to the Montreal Protocol were able to receive impartial advice from their science, technical, and economic committees. These drew together experts from around the world to evaluate the need for further action and to propose options that were technically and economically feasible.

Put these three factors together—acknowledgment of the threat, agreement to cooperate, and commitment to take effective action based on expert advice—and you have a potentially strong recipe to solve global issues, such as climate change and biodiversity loss.

Even in cases where countries have ratified treaties that have entered into force, parties to the treaties do not always comply with their provisions. An example is Russia's noncompliance with the Montreal Protocol. There are few penalties for noncompliance other than public anger, in part because countries are unwilling to give up their sovereignty.

Barriers to the Implementation of International Agreements

Why is the implementation of global environmental agreements so difficult? One answer is that for most of these accords to function effectively, participation must be truly global. Yet because of the unique nature of individual nations and their differing economies, it is difficult to reach meaningful agreement among the necessary participants. Another obstacle to implementing global environmental agreements is poorer nations' lack of capacity to comply with international treaty requirements. Some treaties have addressed this problem by offering financial assistance to countries that are unable to comply: for example, the Multilateral Fund of the Montreal Protocol. Other agreements have established more relaxed timetables for compliance for these countries. In many cases, however, the financing has not been forthcoming, and differentiating among countries based on their ability to act has become politically controversial. The Kyoto Protocol, for example, deals with emissions reductions from developed countries only, which has opened the treaty to criticism in the United States. The challenge of developing accords that are both effective and fair has proved to be a major obstacle to progress.

Despite these problems, treaties are the best traditional tools of global environmental governance. Frameworks for action that are negotiated among different nations are key to leveling the playing field and establishing the rules under which governments, businesses, nongovernmental organizations, and citizens can work together toward a common goal.

In 1997, the General Assembly of the United Nations held a special session and adopted a comprehensive document entitled Programme for the Further Implementation of Agenda 21, prepared by the Commission on Sustainable Development. It also adopted the program of work of the commission for 1999–2003. The Commission on Sustainable Development (CSD) was created in 1992 to ensure effective follow-up of the United Nations Conference on Environment and Development held in Rio de Janeiro, Brazil.

The Commission on Sustainable Development consistently generates a high level of public interest. Over 50 national leaders attend the CSD each year, and more than 1,000 nongovernmental organizations are accredited to participate in the commission's work. The commission ensures the high visibility of sustainable development issues within the UN system and helps to improve the UN's coordination of environment and development activities. The CSD also encourages governments and international organizations to host workshops and conferences on different environmental and cross-sectoral issues. The results of these expert-level meetings enhance the work of CSD and help it work better with national governments and various nongovernmental partners in promoting sustainable development worldwide.

There is no international legislature with authority to pass laws; nor are there international agencies with power to regulate resources on a global scale. An international court at The Hague in the Netherlands has no power to enforce its decisions. Nations can simply ignore the court if they wish. However, a network is growing of multilateral environmental organizations that have developed a greater sense of their roles and a greater incentive to work together. These include not only the United Nations Environment Programme but also the Environment Committee of the Organization for Economic Cooperation and Development and the Senior Advisors on Environmental Problems of the Economic Commission for Europe. Such institutions perform unique functions that cannot be carried out by governments acting alone or bilaterally.

This environmental "coming of age" is reflected in the broadening of intellectual perspective. Governments used to be preoccupied with domestic environmental affairs. Now, they are beginning to broaden their scope to confront problems that cross international borders, such as transboundary air and water pollution, and threats of a planetary nature, such as stratospheric ozone depletion and climatic warming. It is becoming increasingly evident that only decisive mutual action can secure the kind of world we seek.

Earth Summit on Environment and Development

In June 1992, representatives from 178 countries, including 115 heads of state, met in Rio de Janeiro, Brazil, at the Earth Summit. Officially, the meeting was titled the United Nations Conference on Environment and Development (UNCED), and it was the largest gathering of world leaders ever held. The first Earth Summit had been held 20 years earlier in Stockholm, Sweden. At that time, the planet was divided into rival East and West blocs and was preoccupied with the perils of the nuclear arms race. With the collapse of the eastern bloc and the thawing of the Cold War, a fundamental shift in the global base of power had occurred.

The idea behind the Earth Summit was that the relaxation of Cold War tensions, combined with the growing awareness of ecological crises, offered a rare opportunity to persuade countries to look beyond their national interests and agree to some basic changes in the way they treat the environment. The major issues are clear: The developed countries of the North have grown accustomed to lifestyles that are consuming a disproportionate share of natural resources and generating the bulk of global pollution. Many of the developing countries of the South are consuming irreplaceable global resources to provide for their growing populations.

Although the hopes of some developing nations for large commitments of new foreign assistance did not fully materialize, much was accomplished during the summit.

• The *Rio Declaration on Environment and Development* sets out 27 principles to guide the behavior of nations toward more environmentally sustainable patterns of development. The declaration, a compromise between developing and industrialized countries that was crafted at preparatory meetings, was adopted in Rio without negotiation due to fears that further debate would jeopardize any agreement.

• States at UNCED also adopted a voluntary action plan called *Agenda 21,* named because it is intended to provide an agenda for local, national, regional, and global action into the twenty-first century. UNCED Secretary General Maurice Strong called Agenda 21 "the most comprehensive, the most far-reaching and, if implemented, the most effective program of international action ever sanctioned by the international community." Agenda 21 includes hundreds of pages of recommended actions to address environmental problems and promote sustainable development. It also represents a process of building consensus on a "global work-plan" for the economic, social, and environmental tasks of the United Nations as they evolve over time.

• The third official product of UNCED was a *"non-legally binding authoritative statement of principles for a global consensus on the management, conservation, and sustainable development of all types of forests."* Negotiations on the forest statement, begun as negotiations for a legally binding convention on forests, were among the most difficult of the UNCED process. Many states and experts, dissatisfied with the end result, came away from UNCED seeking further negotiations toward agreement on a framework convention on forests.

Environmental Policy and the European Union

"The environment knows no frontiers" was the slogan of the 1970s, when the European Community—known today as the European Union—began to develop its first environmental legislation.

The early laws focused on the testing and labeling of dangerous chemicals, testing drinking water, and controlling air pollutants such as sulfur dioxide, oxides of nitrogen, and particulates from power plants and automobiles. Many of the directives from the 1970s and 1980s were linked to Europe's desire to improve the living and working conditions of its citizens.

In 1987, the Single European Act gave this growing body of environmental legislation a formal legal basis and set three objectives: protection of the environment, protection of human health, and prudent and rational use of natural resources.

The treaty reflected what many governments had already understood: that countries are part of an interconnected and interdependent world of people who are bound together by the air they breathe, the food they eat, the products they use, the wastes they throw away, and the energy they consume.

Similarly, a factory in one European nation may import supplies and raw material from several other neighboring nations; it may consume energy produced from imported gas; produce wastes that affect the air and water quality across the border or downstream; and export products whose wastes become the risk and responsibility of governments and peoples several hundred or many thousand kilometers away.

The 1992 Maastricht Treaty formally established the concept of sustainable development in European Union law. Then, in 1997, the Amsterdam Treaty made sustainable development one of the overriding objectives of the European Union. The treaty considerably strengthened the commitment to the principle that the European Union's future development must be based on the principle of sustainable development and a high level of protection of the environment. The environment must be integrated into the definition and implementation of all of the Union's other economic and social policies, including trade, industry, energy, agriculture, transport, and tourism.

New International Instruments

Over the past few decades, the global community has responded to emerging environmental problems with unprecedented international agreements—notably the Montreal Protocol, the Framework Convention on Climate Change, the Convention on Biological Diversity, and the Convention to Combat Desertification, among others. (See table 20.3.)

In the course of crafting these global agreements, many important lessons have been learned. As the experience with the Montreal Protocol shows, the scientific community can play a crucial role in two ways: first, confirming the links between human activities and global environmental problems; and second, showing what could happen to human health and the global environment if nothing is done.

When the evidence is in hand, an international consensus to act can emerge quickly. The same process is underway in the current international debate on climate change, but the process has been more difficult because the linkages between human activities and global environmental impacts are more complex and still not completely understood. Nevertheless, a global consensus for

Table 20.3 Major Environmental Treaties and U.S. Status

Treaty	Status
Convention on International Trade in Endangered Species, in force since 1975	The United States is a party.
Geneva Convention on Long-Range Transboundary Air Pollution, in force since 1983	The United States is a party.
Bonn Convention on Conservation of Migratory Species, in force since 1983	The United States is not a party.
Vienna Convention for the Protection of the Ozone Layer, in force since 1988	The United States is a party.
Montreal Protocol to the Vienna Convention, in force since 1989. This protocol set the first explicit limits to emissions of gases that erode the ozone layer.	The United States is a party.
Basel Convention on Transboundary Movements of Hazardous Wastes and Their Disposal, in force since 1992. This convention was a response to the growing practice of dumping wastes in developing countries and Eastern Europe.	The United States is a party.
Convention on Biological Diversity, in force since 1993	The United States has not ratified.
Framework Convention on Climate Change, in force since 1994	The United States is a party.
Convention on the Law of the Sea, in force since 1994	The United States is not a party.
Convention to Combat Desertification, in force since 1996	The United States is a party.
Kyoto Protocol to the Framework Convention, in force since 2005	The United States has not ratified.
Rotterdam Convention on the Prior Informed Consent Procedure for Certain Hazardous Chemicals and Pesticides in International Trade, in force since 2005	The United States has not ratified.
Stockholm Convention on Persistent Organic Pollutants. The text was adopted in 2001 but is not yet in force.	The United States has not ratified.

use flexible, market-based solutions where they are appropriate.

Finally, a fifth lesson is that agreements mark the beginning of a process, not the end. Scientists and nongovernmental organizations must continue to further global understanding of environmental problems and communicate what they have learned to the public and policymakers. Policymakers, in turn, must be flexible and respond to changing circumstances with new or modified policy solutions.

One example of new policy solutions is grounded in consumerism. *Green consumerism* is the concept of rational consumption of our scarce resources for the benefit of the environment and future generations. The old saying "the world is enough for everyone's need but not for everyone's greed" calls for a change in our behavior and lifestyle in favor of a sustainable future. Ecolabels have been introduced in a number of countries to help consumers choose products with a proven environmental edge, determined by the product's choice of raw materials, production process, product life cycle, and associated disposal problems. Ecolabels provide evidence that products have met the safety, quality, and environmental protection requirements of the authority that issues the label.

The first ecolabel was introduced in Germany in 1978. This distinctive Blue Angel label was followed by other environmental certification, including the White Swan of Northern European countries, Environmental Choice of Canada, Green Label of Singapore, and Environmental Label of China. By 2000, over a dozen countries had adopted this system of providing consumers with information enabling them to become responsible green consumers. The ecolabel of the European Union is significant because it is the world's first regional scheme to apply the same minimum standards across national markets. (See figure 20.13.)

In recent years, a new class of green certification programs has emerged. Perhaps best described as a hybrid between an environmental management system (EMS) standard and an ecolabeling program, this type of labeling is based on third-party verification of compliance with specific performance criteria for environmental management practices. Some of these programs are managed by government-approved certifiers and/or administered by state agencies, such as the Costa Rican government's Certification for Sustainable Tourism, the Energy Star program in the United States, the U.S. Department of Agriculture's Organic seal, and the U.S. Department of Commerce's Dolphin-Safe label (tuna products).

However, most of the programs have been set up by nongovernmental bodies and are inspected by certifiers accredited by those bodies. Examples include the Forest Stewardship Council, the Marine Stewardship Council, the Sustainable Tourism Stewardship Council, Green Seal (various products and services), and a wide variety of sustainable agriculture, organic agriculture, and food safety programs.

action is emerging. Another important lesson learned has to do with the structure of international agreements and the elements that can contribute to an effective structure. In the case of the Montreal Protocol, the agreement was not punitive and favored incentives and results-oriented approaches. All nations participated in the agreement but at different levels of responsibility in recognition of their differing conditions. The Kyoto Protocol to the Climate Change Convention has benefited from the experience with the Montreal Protocol and included many of the same elements.

A third vital lesson is that, to the extent possible, every interested party must have an opportunity to participate as a full partner in the process and to voice concerns. It is particularly helpful for environmental advocacy groups and the business community to be part of this process. International agreements need to provide incentives to foster public-private partnerships and to provide a role for business leaders to seek innovative technical solutions.

A fourth lesson concerns the role of governments in implementing these conventions. Government actions need to be consistent and predictable; provide sufficient lead times; favor government-led incentives over direct industry subsidies; and

1989
European Union — 'Eco-label'

1992
Singapore Green Labeling Scheme

1989
Nordic Council — 'White Swan'

1978
Germany — 'Blue Angel'

1997
'Marine Stewardship Council'

1993
'Forest Stewardship Council®'

1992
U.S. EPA Program —
'Promoting Energy Efficiency'

FIGURE 20.13 Ecolabels A sampling of ecolabels from around the world.
Sources: European Commission, Singapore Environment Council, Nordic Ecolabeling Board, Federal Environmental Agency of Germany, Marine Stewardship Council, Forest Stewardship Council, and The U.S. Environmental Protection Agency's ENERGY STAR Program. All reprinted with permission.

20.6 It All Comes Back to *You*

No set of policies, no system of incentives, no amount of information can substitute for individual responsibility when it comes to ensuring that our grandchildren will enjoy a quality of life that comes from a quality environment. Information can provide a basis for action. Vision and ideas can influence perceptions and inspire change. New ways to make decisions can empower those who seek a role in shaping the future. However, all of this will be meaningless unless individuals acting as citizens, consumers, investors, managers, workers, and professionals decide that it is important to them to make choices on the basis of a broader, longer view of their self-interest; to get involved in turning those choices into action; and, most important, to be held accountable for their actions. (See figure 20.14.)

The combination of political will, technological innovation, and a very large investment of resources and human ingenuity in pursuit of environmental goals has produced enormous benefits over the past two decades. This is an achievement to celebrate, but in a world that steadily uses more materials to make more goods for more people, we must recognize that we will have to achieve more in the future for the sake of the future. Finally, we must recognize that the pursuit of one set of goals affects others and that we must pursue policies that integrate economic, environmental, and social goals.

FIGURE 20.14 Individual Responsibility While laws protecting the environment are important, it is clear that individuals do not always respect these laws. Individuals must become accountable for their actions.

Issues & Analysis

Politics, Power, and Money

The relationship between politics, power, and money is not new. In any given year there are numerous examples of this less than noble relationship, and 2013 was no exception. In late 2013 the U.S. House of Representatives passed three climate-related bills designed to target and weaken federal environmental regulations: Together, they speed up the permitting process for cross-country natural gas pipelines and for drilling on federal lands; impose fines on those wanting to protest the government's decisions to grant drilling permits; and undercut pending federal fracking regulations. Each of the bill's Republican sponsors stated that their legislative initiative was necessary to counter the president's policies.

The largest and most sweeping of the legislative efforts, the Federal Lands Jobs and Energy Security Act, would give the Department of the Interior just sixty days to approve applications from oil and gas companies to drill on federal land. If the Department of Interior does not issue a decision in that time, the application is automatically approved. The bill also imposes a $5,000 fee on anybody who files an administrative appeal to protest the federal government's decision to issue a permit or not—a practice that citizens, consumer advocacy groups, and environmental groups often take advantage of. The Natural Resources Defense Council, an environmental organization, stated, "The notion that we have to pay for that is ridiculous. One of the few tools that citizens have on their own is this vested right to participate in the administrative process on how

their lands will be managed. Absent that you end up turning the federal lands into a quasi-oil state, where the decisions that are made are in the interests of the oil and gas industry first and everybody else second." The bill did make it through the House of Representatives but has not been taken up by the Senate. The President would likely veto the bill.

The congressmen who championed the above energy bills are also some of the biggest recipients in the House of Representatives of campaign cash from the oil and gas companies, individually accepting in excess of $200,000 in 2012. In 2013, the Citizens for Responsibility and Ethics in Washington released a study showing a 231 percent increase in the fracking industry's political contributions between 2004 and 2012. In the 2012 election cycle, companies operating hydraulically fractured wells and the trade associations representing them gave $6.9 million to congressional candidates.

- How do you feel about the relationship between politics, power, and money?

- Do some industries have too much power because of campaign contributions?

- If you were a member of Congress, what would your position be on campaign contributions?

Summary

Politics and the environment cannot be separated. In the United States, the government is structured into three separate branches, each of which impacts environmental policy.

The increase in environmental regulations in the United States over the past 30 years has caused concerns in some sectors of society.

The late 1980s and early 1990s witnessed a new international concern about the environment in both the developed and developing nations of the world. Environmentalism is also seen as a growing factor in international relations. This concern is leading to international cooperation where only tension had existed before.

While there exists no world political body that can enforce international environmental protection, the list of multilateral environmental organizations is growing.

It remains too early to tell what the ultimate outcome will be, but progress is being made in protecting our common resources for future generations. Several international conventions and treaties have been successful. In the final analysis, however, each of us has to adjust our lifestyle to clean up our own small part of the world.

Acting Green

1. Speak out. Tell your elected representatives and the businesses where you shop what you, as a voter and a consumer, want. Businesses are especially responsive to customers who have concerns. If you have compliments regarding a company's environmental performance, let the owner know.

2. Become active in an organization that works to influence an environmental or sustainability policy.

3. Volunteer on a political campaign.

4. Do not be apathetic. Do not wait for "someone else" to help bring about change. Be aware and active and be a leader. Great change always begins one step at a time.

Review Questions

1. What are the major responsibilities of each of the three branches of the U.S. government?
2. What are some of the enforcement options in U.S. environmental policy?
3. What role does administrative law play in U.S. environmental policy?
4. What are some of the criticisms of U.S. environmental policy?
5. In the past ten years, how has public opinion in the United States changed concerning the protection of the environment?
6. Why is environmentalism a growing factor in international relations?
7. Give some examples of international environmental conventions and treaties.
8. What role does lobbying play in the development of environmental policy?
9. What is the role and function of the Environmental Protection Agency?

Critical Thinking Questions

1. Does chapter 20 have an overall point of view? If you were going to present the problems of environmental policy making and enforcement to others, what framework would you use?
2. The authors of this text say that "we are progressing from an environmental paradigm based on cleanup and control to one including assessment, anticipation, and avoidance." Do you agree with this assessment? Are there environmental problems that are harder to be proactive about than others?
3. Does a command-and-control approach to environmental problems, an approach that emphasizes regulation and remediation, make sense with global environmental problems such as global climate change, habitat destruction, and ozone depletion?
4. How is it best, as a global society with many political demarcations, to preserve the resources that are held in common? What special problems does this kind of preservation entail?
5. Do you agree with William Ruckelshaus that current environmental problems require a change on the part of industrialized and developing countries that would be "a modification in society comparable in scale to the agricultural revolution . . . and Industrial Revolution"? What kinds of changes might that mean in your life? Would these be positive or negative changes?
6. New treaties regarding free trade might enable some nations to argue that other nations' environmental legislation is too restrictive, thereby imposing a barrier to trade that is subject to sanction. What special problems and possibilities might the new global economy present for environmental preservation? What do you think about that?

|ENVIRONMENTAL SCIENCE

To access additional resources for this chapter, please visit ConnectPlus® at connect.mheducation.com. There you will find interactive exercises, including Google Earth™, additional Case Studies, and SmartBook™, an adaptive eBook that integrates our LearnSmart® adaptive learning technology.

Periodic Table of the Elements

Traditionally, elements are represented in a shorthand form by letters. For example, the formula for water, H_2O, shows that a molecule of water consists of two atoms of hydrogen and one atom of oxygen. These chemical symbols for each of the atoms can be found on any periodic table of the elements. Using the periodic table, we can determine the number and position of the various parts of atoms.

Notice that atoms numbered 3, 11, 19, and so on are in column one. The atoms in this column act in a similar way, since they all have one electron in their outermost layer. In the next column, Be, Mg, Ca, and so on act alike because these metals all have two electrons in their outermost electron layer. Similarly, atoms numbered 9, 17, 35, and so on all have seven electrons in their outer layer.

Knowing how fluorine, chlorine, and bromine act, you can probably predict how iodine will act under similar conditions. At the far right in the last column, argon, neon, and so on all act alike. They all have eight electrons in their outer electron layer. Atoms with eight electrons in their outer electron layer seldom form bonds with other atoms.

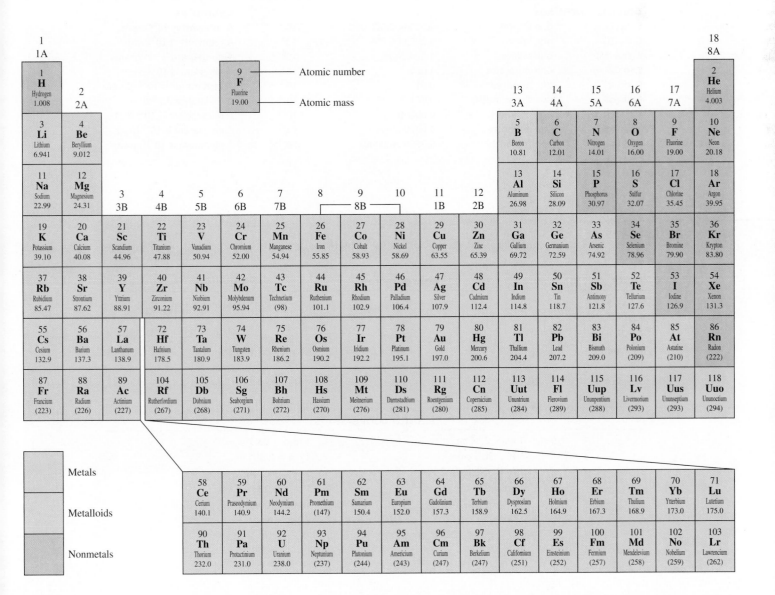

Metric Unit Conversion Tables

The Metric System

Standard metric units				Abbreviations
Standard unit of mass	Gram			g
Standard unit of length	Meter			m
Standard unit of volume	Liter			L

Common prefixes				Examples
Tera (T)	one trillion	1,000,000,000,000	10^{12}	A terawatt is 10^{12} watts.
Giga (G)	one billion	1,000,000,000	10^{9}	A gigawatt is 10^{9} watts.
Mega (M)	one million	1,000,000	10^{6}	A megagram is 10^{6} grams.
Kilo (k)	one thousand	1,000	10^{3}	A kilogram is 10^{3} grams.
Centi (c)	one-hundredth	0.01	10^{-2}	A centimeter is 10^{-2} meter.
Milli (m)	one-thousandth	0.001	10^{-3}	A milliliter is 10^{-3} liter.
Micro (µ)	one-millionth	0.000001	10^{-6}	A micrometer is 10^{-6} meter.
Nano (n)	one-billionth	0.000000001	10^{-9}	A nanogram is 10^{-9} gram.
Pico (p)	one-trillionth	0.000000000001	10^{-12}	A picogram is 10^{-12} gram.

Units of Length

Unit	Abbreviation	Equivalent
Kilometer	km	1,000 m
Meter	m	
Centimeter	cm	10^{-2} m
Millimeter	mm	10^{-3} m
Micrometer	µm	10^{-6} m
Nanometer	nm	10^{-9} m
Angstrom	Å	10^{-10} m

Length conversions

1 in. = 2.54 cm	1 mm = 0.0394 in.
1 ft = 30.5 cm	1 cm = 0.394 in.
1 ft = 0.305 m	1 m = 39.4 in.
1 yd = 0.914 m	1 m = 3.28 ft
1 mi = 1.61 km	1 m = 1.094 yd
	1 km = 0.621 mi

Units of Area

Unit	Abbreviation	Equivalent
Square meter	m^2	$1\ m^2$
Square kilometer	km^2	$1,000,000\ m^2$
Hectare	ha	$10,000\ m^2$

Area conversions

$1\ ft^2 = 0.093\ m^2$	$1\ m^2 = 10.76\ ft^2$
$1\ yd^2 = 0.84\ m^2$	$1\ m^2 = 1.19\ yd^2$
$1\ acre = 0.4\ ha$	$1\ ha = 2.47\ acre$
$1\ mi^2 = 2.6\ km^2$	$1\ km^2 = 0.386\ mi^2$
$1\ mi^2 = 259\ ha$	$1\ km^2 = 247\ acres$

Units of Volume

Unit	Abbreviation	Equivalent
Liter	L*	1
Milliliter	mL	$10^{-3}\ L$ ($1\ mL = 1\ cm^3 = 1\ cc$)
Microliter	µL	$10^{-6}\ L$

Volume conversions

$1\ tsp = 5\ mL$	$1\ mL = 0.034\ fl\ oz$
$1\ tbsp = 15\ mL$	$1\ L = 2.11\ pt$
$1\ fl\ oz = 30\ mL$	$1\ L = 1.057\ qt$
$1\ cup = 0.24\ L$	$1\ L = 0.265\ gal$
$1\ pt = 0.474\ L$	$1\ L = 33.78\ fl\ oz$
$1\ qt = 0.946\ L$	1 cubic meter (m^3) = 61,000 cubic inches
$1\ gal = 3.77\ L$	1 cubic meter (m^3) = 35.3 cubic feet
	1 cubic meter (m^3) = 0.00973 acre-inch

*Note: Many people use an uppercase "L" as the symbol for liter to avoid confusion with the number one (1). Similarly, milliliter is written mL.

Units of Weight

Unit	Abbreviation	Equivalent
Tonne (metric ton)	t	$10^3\ kg$
Kilogram	kg	$10^3\ g$
Gram	g	1
Milligram	mg	$10^3\ g$
Microgram	µg	$10^6\ g$
Nanogram	ng	$10^{-9}\ g$
Picogram	pg	$10^{-12}\ g$

Weight conversions

$1\ oz = 28.4\ g$	$1\ g = 0.0352\ oz$
$1\ lb = 454\ g$	$1\ kg = 2.205\ lb$
$1\ lb = 0.454\ kg$	
1 U.S. ton $= 0.91$ metric ton	1 metric ton $= 1.102$ U.S. tons

Temperature conversions

$$°C = \frac{(°F - 32)}{9} \times 5$$

$$°F = \frac{(°C \times 9)}{5} + 32$$

Some equivalents

$0°C = 32°F$

$37°C = 98.6°F$

$100°C = 212°F$

Glossary

A

abiotic factors Nonliving factors that influence the life and activities of an organism.

absorbed dose The amount of energy absorbed by matter, measured in grays or rads.

abyssal ecosystem The collection of organisms and the conditions that exist in the deep portions of the ocean.

acid Any substance that, when dissolved in water, releases hydrogen ions.

acid deposition The accumulation of potential acid-forming particles on a surface.

acid mine drainage A kind of pollution, associated with coal mines, in which bacteria convert the sulfur in coal into compounds that form sulfuric acid.

acid rain (acid precipitation) The deposition of wet acidic solutions or dry acidic particles from air.

activated-sludge sewage treatment Method of treating sewage in which some of the sludge is returned to aeration tanks, where it is mixed with incoming wastewater to encourage degradation of the wastes in the sewage.

activation energy The initial energy input required to start a reaction.

active solar system A system that traps sunlight energy as heat energy and uses mechanical means to move it to another location.

acute toxicity A serious effect, such as a burn, illness, or death, that occurs shortly after exposure to a hazardous substance.

age distribution The comparative percentages of different age groups within a population.

agricultural waste Waste from the raising of animals and harvesting and processing of crops and trees.

air toxics See **hazardous air pollutants.**

alpha radiation A type of radiation consisting of a particle with two neutrons and two protons.

alpine tundra The biome that exists above the tree line in mountainous regions.

alternative agriculture All nontraditional agricultural practices.

animal rights/welfare A movement that makes an ethical commitment to the well-being of nonhuman animals.

anthropocentrism A theory in ethics that views human values as primary and the environment as solely a resource for humankind.

aquiclude An impervious confining layer of an aquifer.

aquifer A porous layer of earth material that becomes saturated with water.

aquitard A partially permeable layer in an aquifer.

artesian well The result of a pressurized aquifer being penetrated by a pipe or conduit, within which water rises without being pumped.

asthenosphere Part of Earth's mantle capable of plastic flow.

ASTM International Phase I Environmental Site Assessment Standard E-1527 A voluntary standard which involves environmental assessment before beginning a project.

atom The basic subunit of elements, composed of protons, neutrons, and electrons.

auxin A plant hormone that regulates growth.

B

base Any substance that, when dissolved in water, removes hydrogen ions from solution; forms a salt when combined with an acid.

benthic Describes organisms that live on the bottom of marine and freshwater ecosystems.

benthic ecosystem A type of marine or freshwater ecosystem consisting of organisms that live on the bottom.

beta radiation A type of radiation consisting of electrons released from the nuclei of many fissionable atoms.

bioaccumulation The buildup of a material in the body of an organism.

biocentrism A theory in ethics that acknowledges the value of all living organisms.

biochemical oxygen demand (BOD) The amount of oxygen required by microbes to degrade organic molecules in aquatic ecosystems.

biocide A kind of chemical that kills many different types of living things.

biodegradable Able to be broken down by natural biological processes.

biodiversity A measure of the variety of kinds of organisms present in an ecosystem.

biogeochemical cycles The movement of matter within ecosystems, which involves living organisms, geological forces, and chemical reactions. The cycling of nitrogen, carbon, sulfur, oxygen, phosphorus, and water are examples.

biomagnification The increases in the amount of a material in the bodies of organisms at successively higher trophic levels.

biomass The weight of living material in a trophic level.

biome A kind of plant and animal community that covers large geographic areas. Climate is a major determiner of the biome found in a particular area.

biotechnology Inserting specific pieces of DNA into the genetic makeup of organisms.

biotic factors Living portions of the environment.

biotic potential The inherent reproductive capacity.

birth rate The number of individuals born per thousand individuals in the population per year.

black lung disease A respiratory condition resulting from the accumulation of large amounts of fine coal-dust particles in miners' lungs.

boiling-water reactor (BWR) A type of light-water reactor in which steam is formed directly in the reactor and is used to generate electricity.

boreal forest A broad band of mixed coniferous and deciduous trees that stretches across northern North America (and also Europe and Asia); its northernmost edge is integrated with the Arctic tundra.

brownfields Buildings and land that have been abandoned because they are contaminated and the cost of cleaning up the site is high.

brownfields development The concept that abandoned contaminated sites can be cleaned up sufficiently to allow some specified uses without totally removing all of the contaminants.

bush meat Meat from wild animals.

C

carbamate A class of soft pesticides that work by interfering with normal nerve impulses.

carbon cycle The cyclic flow of carbon from the atmosphere to living organisms and back to the atmospheric reservoir.

carbon dioxide (CO_2) A normal component of the Earth's atmosphere that is the most important greenhouse gas.

carbon monoxide (CO) A primary air pollutant produced when organic materials, such as gasoline, coal, wood, and trash, are incompletely burned.

carnivores Animals that eat other animals.

carrying capacity The maximum sustainable population for a species in an area.

catalyst A substance that alters the rate of a reaction but is not itself changed.

cause-and-effect relationship A relationship between two events or things in which a change in the first leads to a change in the second.

chemical bond The physical attraction between atoms that results from the interaction of their electrons.

chemical weathering Processes that involve the chemical alteration of rock in such a manner that it is more likely to fragment or to be dissolved.

chlorinated hydrocarbons A class of pesticides consisting of carbon, hydrogen, and chlorine; these pesticides are very stable.

chlorofluorocarbons (CFCs) Stable compounds containing carbon, hydrogen, chlorine, and fluorine. They were formerly used as refrigerants, propellants in aerosol containers, and expanders in foam products. They are responsible for destroying ozone in the upper atmosphere and are a minor contributor to climate change.

chronic toxicity A serious effect, such as an illness or death, that occurs after prolonged exposure to small doses of a toxic substance.

clear-cutting A forest harvesting method in which all the trees in a large area are cut and removed.

climax community Last stage of succession; a relatively stable, long-lasting, complex, and interrelated community of plants, animals, fungi, and bacteria.

coevolution Two or more species of organisms reciprocally influencing the evolutionary direction of the other.

combustion The process of releasing chemical bond energy from fuel.

commensalism The relationship between organisms in which one organism benefits while the other is not affected.

community All the interacting groups of different species in an area.

competition An interaction between two organisms in which both require the same limited resource, which results in harm to both.

competitive exclusion principle A theory that no two populations of different species will occupy the same niche and compete for exactly the same resources in the same habitat for very long.

compost A humus-like material produced by the decomposition of organic material.

composting The process of allowing natural processes of decomposition to transform dicarded organic materials into compost, a humus-like material.

compound A kind of matter composed of two or more different kinds of atoms bonded together.

Comprehensive Environmental Response, Compensation, and Liability Act (CERCLA) The 1980 U.S. law that addressed the cleanup of hazardous-waste sites.

confined aquifer An aquifer that is bounded on the top and bottom by impermeable confining layers.

conservation To use in the best possible way so that the greatest long-term benefit is realized by society.

conservationist approach An approach that seeks a balance between the development and preservation approaches.

conservation tillage A tillage method in which 30 percent or more of the soil surface is left covered with crop residue following planting.

consumers Organisms that use other organisms as food.

contour farming A method of tilling and planting at right angles to the slope, which reduces soil erosion by runoff.

controlled experiment An experiment in which two groups are compared. One, the control, is used as a basis of comparison and the other, the experimental, has one factor different from the control.

coral reef ecosystem A tropical, shallow-water, marine ecosystem dominated by coral organisms that produce external skeletons.

corporations Business structures that have a particular legal status.

corrosiveness Ability of a chemical to degrade standard materials.

cost-benefit analysis A formal process for calculating the costs and benefits of a project or course of action to decide if benefits are greater than the costs.

cover A term used to refer to any set of physical features that conceals or protects animals from the elements or their enemies.

criteria air pollutants Those air pollutants for which specific air quality standards have been set by the U.S. Environmental Protection Agency.

crust The thin, outer, solid surface of the Earth.

cultural relativism The view that right and wrong are to be determined from within a particular society or cultural group.

D

death phase The portion of the population growth curve of some organisms that shows the population declining.

death rate The number of deaths per thousand individuals in the population per year.

debt-for-nature exchange The purchase of a nation's debt by a third party that requires conservation on the part of the debtor nation in exchange for relief from the debt.

deceleration phase A part of the population growth curve in which the rate of population increase begins to decline.

decibel A unit used to measure the loudness of sound.

decommissioning Decontaminating and disassembling a nuclear power plant and safely disposing of the radioactive materials.

decomposers Small organisms, such as bacteria and fungi, that cause the decay of dead organic matter and recycle nutrients.

deep ecology The generally ecocentric view that a new spiritual sense of oneness with the Earth is the essential starting point for a more healthy relationship with the environment.

deferred cost Costs that are ignored, not recognized, or whose effects accumulate over time; that eventually must be paid.

deforestation Activities that destroy forests.

demand Amount of a product that consumers are willing and able to buy at various prices.

demographic transition The hypothesis that economies proceed through a series of stages, beginning with growing populations with high birth and death rates and low economic development and ending with stable populations with low birth and death rates and high economic development.

demography The study of human populations, their characteristics, and their changes.

denitrifying bacteria Bacteria that convert nitrogen compounds into nitrogen gas.

density-dependent limiting factors Those limiting factors that become more severe as the size of the population increases.

density-independent limiting factors Those limiting factors that are not affected by population size.

desert A biome that receives less than 25 centimeters (10 inches) of precipitation per year.

desertification The conversion of arid and semiarid lands into deserts by inappropriate farming practices or overgrazing.

detritus Tiny particles of organic material that result from fecal waste material or the decomposition of plants and animals.

development approach An approach that encourages humankind to transform nature as it pleases to satisfy human needs.

dioxins A general term for a group or family of chemicals containing hundreds of members (some of which are toxic) that are persistent in the environment and that are unintentionally formed by-products of industrial processes that involve chlorine and/or incineration.

dispersal Migration of organisms from a concentrated population into areas with lower population densities.

domestic water Water used for domestic activities, such as drinking, air conditioning, bathing, washing clothes, washing dishes, flushing toilets, and watering lawns and gardens.

dose equivalent The absorbed dose times a quality factor.

E

ecocentrism A theory in ethics that considers the value of ecosystems and larger wholes to be primary.

ecofeminism The view that there are important theoretical, historical, and empirical connections between how society treats women and how it treats the environment.

ecological or environmental economics An approach to economic accounting that incorporates environmental goods and harms as part of the cost of economic activity.

ecological footprint The area of the Earth's productive land and water required to supply the resources that an individual or population demands as well as to absorb the wastes that the individual or population produces.

ecology A branch of science that deals with the interrelationship between organisms and their environment.

economic growth The increase in a country's total output of goods and services.

economics The study of how people choose to use resources to produce goods and services and how these goods and services are distributed to the public.

ecosystem A group of interacting species along with their physical environment.

ecosystem diversity A measure of the number of kinds of ecosystems present in an area.

ecosystem services Beneficial effects of functioning ecosystems for people and society.

ectoparasite A parasite that is adapted to live on the outside of its host.

electron The lightweight, negatively charged particle that moves around at some distance from the nucleus of an atom.

element A form of matter consisting of a specific kind of atom.

Emergency Planning and Community Right-to-Know Act (EPCRA) An act that required certain industries in the United States to report the release of toxic chemicals into the environment.

emergent plants Aquatic vegetation that is rooted on the bottom but has leaves that float on the surface or protrude above the water.

emigration Movement out of an area that was once one's place of residence.

endangered species Those species that are present in such small numbers that they are in immediate jeopardy of becoming extinct.

endoparasite A parasite that is adapted to live within a host.

endothermic reaction Chemical reaction in which the newly formed chemical bonds contain more energy than was present in the compounds from which they were formed.

energy The ability to do work.

entropy The degree of disorder in a system. All systems tend toward a high degree of disorder or entropy.

environment Everything that affects an organism during its lifetime.

environmental aesthetics The study of how to appreciate beauty in the natural world.

environmental cost Damage done to the environment as a resource is exploited.

environmental justice The social justice expression of environmental ethics. Fair application of laws designed to protect the health of human beings and ecosystems; that no groups suffer unequal environmental harm.

environmental pragmatism   An approach to environmental ethics that maintains that a human-centered ethic with a long-range perspective will come to many of the same conclusions in environmental policy as an ecocentric ethic.

Environmental Protection Agency (EPA)   U.S. government organization responsible for the establishment and enforcement of regulations concerning the environment.

environmental resistance   The combination of all environmental influences that tend to keep populations stable.

environmental science   An interdisciplinary area of study that includes both applied and theoretical aspects of human impact on the world.

enzymes   Protein molecules that speed up the rate of specific chemical reactions.

erosion   The processes that loosen and move particles from one place to another.

estuaries   Marine ecosystems that consist of shallow, partially enclosed areas where freshwater enters the ocean.

ethics   A discipline that seeks to define what is fundamentally right and wrong.

euphotic zone   The upper layer in the ocean where the sun's rays penetrate.

eutrophication   The enrichment of water (either natural or cultural) with nutrients.

eutrophic lake   A usually shallow, warm-water lake that is nutrient rich.

evapotranspiration   The process of plants transporting water from the roots to the leaves, where it evaporates.

evolution   A change in the structure, behavior, or physiology of a population of organisms as a result of some organisms with favorable characteristics having greater reproductive success than those organisms with less favorable characteristics.

executive branch   The office of the president of the United States.

exothermic reaction   Chemical reaction in which the newly formed compounds have less chemical energy than the compounds from which they were formed.

experiment   An artificial situation designed to test the validity of a hypothesis.

exponential growth phase   The period during population growth when the population increases at an ever-increasing rate.

extended product responsibility   The concept that the producer of a product is responsible for all the negative effects involved in its production, including the ultimate disposal of the product when its useful life is over.

external costs   Expenses, monetary or otherwise, borne by someone other than the individuals or groups who use a resource.

extinction   The death of a species; the elimination of all the individuals of a particular kind.

extrinsic limiting factors   Factors that limit population size and that come from outside the population.

F

fecal coliform bacteria   Bacteria found in the intestines of humans and other animals, often used as an indicator of water pollution.

first law of thermodynamics   A statement about energy that says that under normal physical conditions, energy is neither created nor destroyed.

fissionable   The property of the nucleus of some atoms that allows them to split into smaller particles.

floodplain   Lowland area on either side of a river that is periodically covered by water.

floodplain zoning ordinances   Municipal laws that restrict future building in floodplains.

food chain   The series of organisms involved in the passage of energy from one trophic level to the next.

food web   Intersecting and overlapping food chains.

fossil fuels   The organic remains of plants, animals, and microorganisms that lived millions of years ago that are preserved as natural gas, oil, and coal.

free-living nitrogen-fixing bacteria   Bacteria that live in the soil and can convert nitrogen gas (N_2) in the atmosphere into forms that plants can use.

freshwater ecosystems   Aquatic ecosystems that have low amounts of dissolved salts.

friable   A soil characteristic that describes how well a soil crumbles.

fungicide   A pesticide designed to kill or control fungi.

G

gamma radiation   A type of electromagnetic radiation that comes from disintegrating atomic nuclei.

gas-cooled reactor (GCR)   A type of nuclear reactor that uses graphite as a moderator and carbon dioxide or helium as a coolant.

gene   A unit of heredity; a segment of DNA that contains information for the synthesis of a specific protein, such as an enzyme.

genetically modified organisms   Organisms that have had their genetic makeup modified by biotechnology.

genetic diversity   A term used to describe the number of different kinds of genes present in a population or a species.

genetic engineering   Inserting specific pieces of DNA into the genetic makeup of organisms.

geothermal energy   The heat energy from the Earth's molten core.

greenhouse effect   The property of carbon dioxide (CO_2) that allows light energy to pass through the atmosphere but prevents heat from leaving; similar to the action of glass in a greenhouse.

greenhouse gas   Gas in the atmosphere that allows sunlight to enter but retards the outward flow of heat from the Earth.

Green Revolution   The introduction of new plant varieties and farming practices that increased agricultural production worldwide during the 1950s, 1960s, and 1970s.

gross national income (GNI)   An index that measures the total goods and services generated within a country as well as income earned and sent home by citizens of the country who are living in other countries.

groundwater   Water that infiltrates the soil and is stored in the spaces between particles in the Earth.

groundwater mining   Removal of water from an aquifer faster than it is replaced.

H

habitat   The specific kind of place where a particular kind of organism lives.

habitat management   The process of changing the natural community to encourage the increase in populations of certain desirable species.

hard pesticide   A pesticide that persists for long periods of time; a persistent pesticide.

hazardous   All dangerous materials, including toxic ones, that present an immediate or long-term human health risk or environmental risk.

hazardous air pollutants   Certain airborne compounds with high toxicity.

hazardous materials or substances   Substances that can cause harm to humans or the environment.

hazardous wastes   Substances that could endanger life if released into the environment.

heavy-water reactor (HWR)   A type of nuclear reactor that uses the hydrogen isotope deuterium in the molecular structure of the coolant water.

herbicide   A pesticide designed to kill or control plants.

herbivores   Primary consumers; animals that eat plants.

horizon   A horizontal layer in the soil. The top layer (*A* horizon) has organic matter. The lower layer (*B* horizon) receives nutrients by leaching. The *C* horizon is partially weathered parent material.

host   The organism a parasite uses for its source of food.

humus   Partially decomposed organic matter typically found in the top layer of the soil.

hydrocarbons (HC)   Group of organic compounds consisting of carbon and hydrogen atoms that are evaporated from fuel supplies or are remnants of the fuel that did not burn completely and that act as a primary air pollutant.

hydrologic cycle   Constant movement of water from surface water to air and back to surface water as a result of evaporation and condensation.

hydroxide ion   A negatively charged particle consisting of a hydrogen and an oxygen atom, commonly released from materials that are bases.

hyperaccumulators   Plants that thrive when exposed to soil toxins and accumulate high levels of soil pollutants within their bodies.

hypothesis   A logical statement that explains an event or answers a question that can be tested.

I

ignitability   Characteristic of materials that results in their ability to combust.

immigration   Movement into an area where one has not previously resided.

incineration   Method of disposing of solid waste by burning.

industrial ecology   A concept that stresses cycling resources rather than extracting and eventually discarding them.

Industrial Revolution   A period of history during which machinery replaced human labor.

industrial solid waste   A wide variety of materials such as demolition waste, foundry sand, scraps from manufacturing processes, sludge, ash from combustion, and other similar materials produced by industry.

industrial water uses   Uses of water for cooling and for dissipating and transporting waste materials.

infrastructure   Permanent structural foundations of a society such as highways and bridges.

insecticide   A pesticide designed to kill or control insects.

in-stream water uses Use of a stream's water flow for such purposes as hydroelectric power, recreation, and navigation.

integrated pest management A method of pest management in which many aspects of the pest's biology are exploited to control its numbers.

International Organization for Standardization An organization that sets standards. One of their standards, known as ISO 14000, is for environmental management systems.

interspecific competition Competition between members of different species for a limited resource.

intraspecific competition Competition among members of the same species for a limited resource.

intrinsic limiting factors Factors that limit population size that come from within the population.

ion An atom or group of atoms that has an electric charge because it has either gained or lost electrons.

ionizing radiation Radiation that can dislodge electrons from atoms to form ions.

irrigation Adding water to an agricultural field to allow certain crops to grow where the lack of water would normally prevent their cultivation.

isotope Atoms of the same element that have different numbers of neutrons.

J

judicial branch That portion of the U.S. government that includes the court system.

K

keystone species One that has a critical role to play in the maintenance of specific ecosystems.

kinetic energy Energy of moving objects.

kinetic molecular theory The widely accepted theory that all matter is made of small particles that are in constant movement.

K-strategist Large organisms that have relatively long lives, produce few offspring, provide care for their offspring, and typically have populations that stabilize at the carrying capacity.

L

lag phase The initial stage of population growth during which growth occurs very slowly.

land The surface of the Earth not covered by water.

land disposal The placement of unwanted materials on the surface in landfills or impoundments or by injecting them below the surface of the land.

land-use planning The process of evaluating the needs and wants of the population, the characteristics and values of the land, and various alternative solutions before changes in land use are made.

latent heat Heat transfer that occurs when a substance is changed from one state to another—solid to liquid, gas to liquid—in which heat is transferred but the temperature does not change.

law of conservation of mass States that matter is not gained or lost during a chemical reaction.

laws Acts passed by a legislature or similar body that regulate the behavior of the public.

LD$_{50}$ A measure of toxicity; the dosage of a substance that will kill (lethal dose) 50 percent of a test population.

leachate Contaminant-laden water that flows from landfills or other contaminated sites.

leaching The movement of minerals from the top layers of the soil to the B horizon by the downward movement of soil water.

legislative branch That portion of the U.S. government that is responsible for developing laws.

less-developed countries Countries of the world that typically have a per capita income of less than US $15,000.

life cycle analysis The process of assessing the environmental effects associated with the production, use, reuse, and disposal of a product over its entire useful life.

limiting factor The primary condition of the environment that determines the population size for an organism.

limnetic zone Region that does not have rooted vegetation in a freshwater ecosystem.

liquefied natural gas Natural gas that has been converted to a liquid by cooling it to $-162°C$.

lithosphere A combination of the crust and outer layer of the mantle that forms the plates that move over the Earth's surface.

litter A layer of undecomposed or partially decomposed organic matter on the soil surface.

littoral zone Region with rooted vegetation in a freshwater ecosystem.

loam A soil type with good drainage and good texture that is ideal for growing crops.

M

macronutrient A nutrient, such as nitrogen, phosphorus, or potassium, that is required in relatively large amounts by plants.

mangrove swamp ecosystems Marine shoreline ecosystems dominated by trees that can tolerate high salt concentrations.

mantle The layer of the Earth between the crust and the core.

marine ecosystems Aquatic ecosystems that have high salt content.

marsh Area of grasses and reeds that is flooded either permanently or for a major part of the year.

mass burn A method of incineration of solid waste in which material is fed into a furnace on movable metal grates.

matter Substance with measurable mass and volume.

mechanical weathering Physical forces that reduce the size of rock particles without changing the chemical nature of the rock.

Mediterranean shrublands Coastal ecosystems characterized by winter rains and summer droughts that are dominated by low, woody vegetation with small leaves.

megacity A metropolitan area with a total population of over ten million people.

megalopolis A large, regional urban center.

methane (CH$_4$) An organic compound produced by living organisms that is the second most abundant greenhouse gas.

micronutrient A nutrient needed in extremely small amounts for proper plant growth; examples are boron, zinc, and magnesium.

migratory birds Birds that fly considerable distances between their summer breeding areas and their wintering areas.

mining waste Waste from the processing of rock from mining operations. It includes solid materials that are typically dumped on the land near the mining site and liquid wastes typically stored in ponds.

mixture A kind of matter consisting of two or more kinds of matter intermingled with no specific ratio of the kinds of matter.

molecule Two or more atoms chemically bonded to form a stable unit.

monoculture A system of agriculture in which large tracts of land are planted with the same crop.

more-developed countries Countries of the world that typically have a per capita income that exceeds US $15,000. Europe, Canada, United States, Australia, New Zealand, and Japan.

mortality The number of deaths per year.

mountaintop removal A mining method in which the top of a mountain is removed to get at a coal vein and the unwanted soil and rock is pushed into the adjacent valley.

multiple land use Land uses that do not have to be exclusionary, so that two or more uses of land may occur at the same time.

municipal solid waste landfill A waste storage site constructed above an impermeable clay layer that is lined with an impermeable membrane and includes mechanisms for dealing with liquid and gas materials generated by the contents of the landfill.

municipal solid waste (MSW) All the waste produced by the residents of a community.

mutualism The association between organisms in which both benefit.

mycorrhizae Symbiotic soil fungi, present in most soils, that attach themselves directly onto the roots of most plants. They help the host plants to absorb more water and nutrients while the host plants provide food for the fungi.

N

natality The number of individuals added to the population through reproduction over a particular time period.

National Priorities List A listing of hazardous-waste dump sites requiring urgent attention as identified by Superfund legislation.

natural resources Those structures and processes that can be used by humans for their own purposes but cannot be created by them.

natural selection A process that determines which individuals within a species will reproduce more effectively and therefore results in changes in the characteristics within a species.

nature centers Teaching institutions that provide a variety of methods for people to learn about and appreciate the natural world.

negligible risk A point at which there is no significant health or environmental risk.

neutron Neutrally charged particle located in the nucleus of an atom.

niche The total role an organism plays in its ecosystem.

nitrifying bacteria Bacteria that are able to convert ammonia to nitrite.

nitrogen cycle The series of stages in the flow of nitrogen in ecosystems.

nitrogen dioxide A compound composed of one atom of nitrogen and two atoms of oxygen; a secondary air pollutant.

nitrogen-fixing bacteria Bacteria that are able to convert the nitrogen gas (N$_2$) in the atmosphere into forms that plants can use.

nitrogen monoxide A compound composed of one atom of nitrogen and one atom of oxygen; a primary air pollutant.

nitrogen oxides (NO$_x$) A mixture of compounds that have nitrogen and oxygen in their composition.

nitrous oxide (N$_2$O), A nitrogen-containing compound that is a minor greenhouse gas.

nonpersistent pesticide A pesticide that degrades in a short period of time.

nonpersistent pollutants Those pollutants that do not remain in the environment for long periods.

nonpoint source Diffuse pollutants, such as agricultural runoff, road salt, and acid rain, that are not from a single, confined source.

nonrenewable energy sources Those energy sources that are not replaced by natural processes within a reasonable length of time.

nonrenewable resources Those resources that are not replaced by natural processes, or those whose rate of replacement is so slow as to be noneffective.

nontarget organism An organism whose elimination is not the purpose of pesticide application.

northern coniferous forest See **Boreal forest.**

nuclear breeder reactor Nuclear fission reactor designed to produce radioactive fuel from nonradioactive uranium and at the same time release energy to use in the generation of electricity.

nuclear chain reaction A continuous process in which a splitting nucleus releases neutrons that strike and split the nuclei of other atoms, releasing nuclear energy.

nuclear fission The decomposition of an atom's nucleus with the release of particles and energy.

nuclear reactor A device that permits a controlled nuclear fission chain reaction.

nucleus The central region of an atom that contains protons and neutrons.

O

observation Ability to detect events by the senses or machines that extend the senses.

oil shale A rock material that contains a high viscosity mixture of hydrocarbons that must be heated to extract the oil.

oligotrophic lakes Deep, cold, nutrient-poor lakes that are low in productivity.

omnivores Animals that eat both plants and other animals.

opportunity costs The costs associated with lost opportunities that occur when a decision precludes other potential uses for a resource.

organic agriculture Agricultural practices that avoid the use of chemical fertilizers and pesticides in the production of food, thus preventing damage to related ecosystems and consumers.

organophosphate A class of soft pesticides that work by interfering with normal nerve impulses.

overburden The layer of soil and rock that covers deposits of desirable minerals.

oxides of nitrogen (NO, N$_2$O, and NO$_2$) Primary air pollutants consisting of a variety of different compounds containing nitrogen and oxygen.

ozone (O$_3$) A molecule consisting of three atoms of oxygen that absorbs much of the sun's ultraviolet energy before it reaches the Earth's surface.

P

parasite An organism adapted to survival by using another living organism (host) for nourishment.

parasitism A relationship between organisms in which one, known as the parasite, lives in or on the host and derives benefit from the relationship while the host is harmed.

parent material Material that is weathered to become the mineral part of the soil.

particulate matter Minute solid particles and liquid droplets dispersed into the atmosphere.

passive solar system A design that allows for the entrapment and transfer of heat from the sun to a building without the use of moving parts or machinery.

patchwork clear-cutting A forest harvest method in which patches of trees are clear-cut among patches of timber that are left untouched.

pelagic Referring to those organisms that swim in open water.

pelagic ecosystem A portion of a marine or freshwater ecosystem that occurs in open water away from the shore.

periphyton Attached organisms in freshwater streams and rivers, including algae, animals, and fungi.

permafrost Permanently frozen ground.

persistent pesticide A pesticide that remains unchanged for a long period of time; a hard pesticide.

persistent pollutant A pollutant that remains in the environment for many years in an unchanged condition.

personal ethical commitment A determination of ethical right and wrong made by an individual.

pest An unwanted plant or animal that interferes with domesticated plants and animals or human activity.

pesticide A chemical used to eliminate pests; a general term used to describe a variety of different kinds of pest killers, such as insecticides, fungicides, rodenticides, and herbicides.

pH The negative logarithm of the hydrogen ion concentration; a measure of the number of hydrogen ions present.

pheromone A chemical produced by one animal that changes the behavior of another.

photochemical smog A yellowish-brown haze that is the result of the interaction of hydrocarbons, oxides of nitrogen, and sunlight.

photosynthesis The process by which plants manufacture food. Light energy is used to convert carbon dioxide and water to sugar and oxygen.

phytoplankton Free-floating, microscopic, chlorophyll-containing organisms.

phytoremediation The use of specialized plants to clean up polluted soil.

pioneer community The early stages of succession that begin the soil-building process.

plankton Tiny aquatic organisms that are moved by tides and currents.

plate tectonics The concept that the outer surface of the Earth consists of large plates that are slowly moving over the surface of a plastic layer.

plutonium-239 (Pu-239) A radioactive isotope produced in a breeder reactor and used as a nuclear fuel.

PM$_{10}$ Particulate matter that is between 10 and 2.5 microns in diameter.

PM$_{2.5}$ Particulate matter that is 2.5 microns or less in diameter.

point source Pollution that can be traced to a single source.

policy Planned course of action on a question or a topic.

pollution Any addition of matter or energy that degrades the environment for humans and other organisms.

pollution costs The private or public expenditures to correct pollution damage once pollution has occurred.

pollution prevention Action to prevent either entirely or partially the pollution that would otherwise result from some production or consumption activity.

pollution-prevention costs Costs incurred either in the private sector or by government to prevent, either entirely or partially, the pollution that would otherwise result from some production or consumption activity.

pollution-prevention hierarchy Regulatory controls that emphasize reducing the amount of hazardous waste produced.

polyculture A system of agriculture that mixes different plant species in the same plots of land.

polyploidy A condition in which the number of sets of chromosomes increases.

population A group of individuals of the same species occupying a given area.

population density A measure of how close organisms are to one another, generally expressed as the number of organisms per unit area.

population growth rate The rate at which additional individuals are added to the population; the birth rate minus the death rate.

porosity A measure of the size and number of spaces in an aquifer.

potable waters Unpolluted freshwater supplies suitable for drinking.

potential energy The energy of position.

prairies Temperate grasslands.

precipitation Removal of materials by mixing with chemicals that cause the materials to settle out of the mixture.

precision agriculture The use of computer technology and geographic information systems to automatically vary the chemicals applied to a crop at different places within a field.

predation The act of killing and feeding by a predator.

predator An animal that kills and eats another organism.

preservationist approach An approach that seeks to ensure that large areas of nature together with their ecological processes remain intact.

pressurized-water reactor (PWR) A type of light-water reactor in which the water in the reactor is kept at high pressure and steam is formed in a secondary loop.

prey An organism that is killed and eaten by a predator.

price The monetary value of a good or service.

primary air pollutants Types of unmodified materials that, when released into the environment in sufficient quantities, are considered hazardous.

primary consumer An animal that eats plants (producers) directly.

primary sewage treatment Process that removes larger particles by settling or filtering raw sewage through large screens.

primary succession Succession that begins with bare mineral surfaces or water.

probability A mathematical statement about how likely it is that something will happen.

producer An organism that can manufacture food from inorganic compounds and light energy.

profitability The extent to which economic benefits exceed the economic costs of doing business.

proton The positively charged particle located in the nucleus of an atom.

pseudoscience A deceptive practice that uses the appearance or language of science to convince, confuse, or mislead people into thinking something has scientific validity, when it does not.

R

radiation Energy that travels through space in the form of waves or particles.

radioactive Describes unstable nuclei that release particles and energy as they disintegrate.

radioactive half-life The time it takes for half of the radioactive material to spontaneously decompose.

range of tolerance The ability organisms have to succeed under a variety of environmental conditions. The breadth of this tolerance is an important ecological characteristic of a species.

reactivity The property of materials that indicates the degree to which a material is likely to react vigorously to water or air, or to become unstable or explode.

recycling The process of reclaiming a resource and reusing it for another or the same structure or purpose.

reduced tillage A tillage method that generally leaves 15 to 30 percent of the soil surface covered with crop residue following planting.

reforestation The process of replanting areas after the original trees are removed.

rem A measure of the biological damage to tissue caused by certain amounts of radiation.

renewable energy sources Those energy sources that can be regenerated by natural processes.

renewable resources Those resources that can be formed or regenerated by natural processes.

replacement fertility The number of children per woman needed just to replace the parents.

reproducibility A characteristic of the scientific method in which independent investigators must be able to reproduce the experiment to see if they get the same results.

reserves The known deposits from which materials can be extracted profitably with existing technology under present economic conditions.

Resource Conservation and Recovery Act (RCRA) The 1976 U.S. law that specifically addressed the issue of hazardous waste.

resource exploitation The use of natural resources by society.

resources Naturally occurring substances that can be utilized by people but may not be economic.

respiration The process that organisms use to release chemical bond energy from food.

ribbon sprawl Development along transportation routes that usually consists of commercial and industrial building.

risk The probability that a condition or action will lead to an injury, damage, or loss.

risk assessment The use of facts and assumptions to estimate the probability of harm to people or the environment from particular environmental factors or conditions.

risk management Decision-making process that uses the results of risk assessment, weighing possible responses to the risk, and selecting appropriate actions to minimize or eliminate the risk.

rodenticide A pesticide designed to kill rodents.

r-strategist Typically, a small organism that has a short life span, produces a large number of offspring, and does not reach a carrying capacity.

runoff The water that moves across the surface of the land and enters a river system.

S

salinization An increase in salinity caused by growing salt concentrations in soil.

saltwater intrusion The movement of saltwater into aquifers near oceans when too much water is pumped from aquifers.

savanna Tropical biome having seasonal rainfall of 50 to 150 centimeters (20–60 inches) per year. The dominant plants are grasses, with some scattered fire- and drought-resistant trees.

science A method for gathering and organizing information that involves observation, asking questions about observations, hypothesis formation, testing hypotheses, critically evaluating the results, and publishing information so that others can evaluate the process and the conclusions.

scientific law A uniform or constant fact of nature that describes *what* happens in nature.

scientific method A way of gathering and evaluating information. It involves observation, hypothesis formation, hypothesis testing, critical evaluation of results, and the publishing of findings.

secondary air pollutants Pollutants produced by the interaction of primary air pollutants in the presence of an appropriate energy source.

secondary consumers Animals that eat animals that have eaten plants.

secondary sewage treatment Process that involves holding the wastewater until the organic material has been degraded by bacteria and other microorganisms.

secondary succession Succession that begins with the destruction or disturbance of an existing ecosystem.

second law of thermodynamics A statement about energy conversion that says that whenever energy is converted from one form to another, some of the useful energy is lost.

selective harvesting A forest harvesting method in which individual high-value trees are removed from the forest, leaving the majority of the forest undisturbed.

semiconfined aquifer An aquifer in which water can pass in and out of the confining layer (aquitard).

sensible heat The heat energy stored in a substance as a result of an increase in its temperature.

seral stage A stage in the successional process.

sere The entire sequence of stages in ecological succession.

sewage sludge A mixture of organic material, organisms, and water in which the organisms consume the organic matter.

sex ratio Comparison between the number of males and females in a population.

social ecology The view that social hierarchies between groups of people are directly connected to patterns of behavior that cause environmental destruction.

soil A mixture of mineral material, organic matter, air, water, and living organisms; capable of supporting plant growth.

soil profile The series of layers (horizons) seen as one digs down into the soil.

soil structure Refers to the way that soil particles clump together. Sand has little structure because the particles do not stick to one another.

soil texture Refers to the size of the particles that make up the soil. Sandy soil has large particles, and clay soil has small particles.

solid waste Unwanted objects or particles that accumulate on the site where they are produced.

source reduction Reducing the amount of solid waste generated by using less, or converting from heavy packaging materials to lightweight ones.

speciation The process of developing a new species from a previously existing species.

species A group of organisms that can interbreed and produce offspring capable of reproduction.

species diversity A measure of the number of different species present in an area.

stable equilibrium phase The phase in a population growth curve in which the death rate and birth rate become equal.

standard of living An abstract concept that attempts to quantify the quality of life of people. Several factors included in an analysis of standard of living are: economic well-being, health conditions, and the ability to change one's status in the society.

steppe A grassland.

stormwater runoff Stormwater that runs off of streets and buildings and is often added directly to the sewer system and sent to the municipal wastewater treatment facility.

strip farming The planting of crops in strips that alternate with other crops. The primary purpose is to reduce erosion.

submerged plants Aquatic vegetation that is rooted on the bottom and has leaves that stay submerged below the surface of the water.

subsidy A gift given to private enterprise by government when the enterprise is in temporary economic difficulty and is viewed as being important to the public.

succession Regular and predictable changes in the structure of a community, ultimately leading to a climax community.

successional stage A stage in succession.

sulfur dioxide (SO₂) A compound containing sulfur and oxygen produced when sulfur-containing fossil fuels are burned. When released into the atmosphere, it is a primary air pollutant.

Superfund The common name given to the U.S. 1980 Comprehensive Environmental Response, Compensation, and Liability Act, which was designed to address hazardous-waste sites.

supply Amount of a good or service available to be purchased.

supply/demand curve The relationship between the available supply of a commodity or service and its demand. The supply and demand change as the price changes.

surface mining (strip mining) A type of mining in which the overburden is removed to procure the underlying deposit.

survivorship curve A graph that shows the proportion of individuals likely to survive to each age.

sustainability "Development that meets the needs of the present without compromising the ability of future generations to meet their own needs."

sustainable agriculture Agricultural methods used to produce adequate, safe food in an economically viable manner while enhancing the health of agricultural land and related ecosystems.

sustainable development Using renewable resources in harmony with ecological systems to produce a rise in real income per person and an improved standard of living for everyone.

swamp Area of trees that is flooded either permanently or for a major part of the year.

symbiosis A close, long-lasting physical relationship between members of two different species.

symbiotic nitrogen-fixing bacteria Bacteria that grow within a plant's root system and that can convert nitrogen gas (N_2) from the atmosphere to nitrogen compounds that the plant can use.

synergism The interaction of materials or energy that increases the potential for harm.

T

taiga Biome having short, cool summers and long winters with abundant snowfall. The trees are adapted to winter conditions.

target organism The organism a pesticide is designed to eliminate.

tar sands A combination of clay, sand, water, and a thick oil called bitumen. (Also referred to as oil sands.)

temperate deciduous forest Biome that has a winter-summer change of seasons and that typically receives 75 to 150 centimeters (30–60 inches) or more of relatively evenly distributed precipitation throughout the year.

temperate grasslands Areas receiving between 25 and 75 centimeters (10–30 inches) of precipitation per year. Grasses are the dominant vegetation, and trees are rare.

temperate rainforest Areas where the prevailing winds bring moisture-laden air to the coast. Abundant rain, fertile soil, and mild temperatures result in a lush growth of plants.

terrace A level area constructed on steep slopes to allow agriculture without extensive erosion.

tertiary sewage treatment Process that involves a variety of different techniques designed to remove dissolved pollutants left after primary and secondary treatments.

theory A unifying principle that binds together large areas of scientific knowledge.

thermal inversion The condition in which warm air in a valley is sandwiched between two layers of cold air and acts like a lid on the valley.

thermal pollution Waste heat that industries release into the environment.

threatened species Those species that could become extinct if a critical factor in their environment were changed.

threshold level The minimum amount of something required to cause measurable effects.

tight oil An oil that consists of a mixture of hydrocarbons with low viscosity which cannot flow because the rock containing it (shale or tight sandstone) is nonporous (tight).

total fertility rate The number of children born per woman per lifetime.

toxic A narrow group of substances that are poisonous and cause death or serious injury to humans and other organisms by interfering with normal body physiology.

toxicity A measure of how toxic a material is.

tract development The construction of similar residential units over large areas.

trickling filter system A secondary sewage treatment technique that allows polluted water to flow over surfaces that harbor microorganisms.

triple bottom line A method of gauging corporate success on three fronts: financial, social, and environmental.

trophic level A stage in the energy flow through ecosystems.

tropical dry forest Regions that receive low rainfall amounts, as little as 50 centimeters (20 inches) per year, and are characterized by species well adapted to drought. Trees of tropical dry forests are usually smaller than those in rainforests, and many lose their leaves during the dry season.

tropical rainforest A biome with warm, relatively constant temperatures where there is no frost. These areas receive more than 200 centimeters (80 inches) of rain per year in rains that fall nearly every day.

tundra A biome that lacks trees and has permanently frozen soil.

U

unconfined aquifer An aquifer that usually occurs near the land's surface, receives water by percolation from above, and may be called a water table aquifer.

underground mining A type of mining in which the deposited material is removed without disturbing the overburden.

uranium-235 (U-235) A naturally occurring radioactive isotope of uranium used as fuel in nuclear reactors.

urban growth limit A boundary established by municipal government that encourages development within the boundary and prohibits it outside the boundary.

urban sprawl A pattern of unplanned, low-density housing and commercial development outside of cities that usually takes place on previously undeveloped land.

V

vadose zone A zone above the water table and below the land surface that is not saturated with water.

variable Things that change from time to time.

vector An organism that carries a disease from one host to another.

volatile organic compounds (VOC) Airborne organic compounds; primary air pollutants.

W

waste minimization A process that involves changes that industries could make in the way they manufacture products that would reduce the waste produced.

water diversion The physical process of transferring water from one area to another.

water table The top of the layer of water in an aquifer.

waterways Low areas that water normally flows through.

weathering The physical and chemical breakdown of materials; involved in the breakdown of parent material in soil formation.

weed An unwanted plant.

wetlands Areas that include swamps, tidal marshes, coastal wetlands, and estuaries.

windbreak The planting of trees or strips of grasses at right angles to the prevailing wind to reduce erosion of soil by wind.

Z

zero population growth The stabilized growth stage of human population during which births equal deaths and equilibrium is reached.

zoning Type of land-use regulation in which land is designated for specific potential uses, such as agricultural, commercial, residential, recreational, and industrial.

zooplankton Weakly swimming microscopic animals.

Design Elements

Cloud chapter openers: © 123rf.com, photographer Oleg Saenko RF; Going Green photo: © 123rf.com, photographer wajan RF; Focus On photo: © Photodisc/Getty RF; Science, Politics & Policy photo: © dreamtime.com RF; What's Your Take photo: © Photodisc/Getty RF; Issues & Analysis photo: © Mark Evans/istock.com.

Chapter 1

Opener: © Comstock/PunchStock RF; p. 2a: © Tom Brakefield/Photodisc/Getty Images RF; p. 2b: U.S. National Park Services (NPS); p. 2c: © Judy Enger; p. 2d: © Getty Images RF; p. 6(top): © Jupiter RF; 1.3: © Dr. Parvinder Sethi; 1.4: © Martin Ruegner/Photodisc/Getty Images RF; 1.5: © MedioImages/Getty Images RF; 1.6: © Corbis RF; 1.7a: © Lars A. Niki; 1.7b: © Dr. Parvinder Sethi; 1.7c: © Photodisc/Getty Images RF; 1.7d: © David C. Johnson RF; 1.8a: © Digital Vision/PunchStock RF; 1.8b: Andrea Booher/FEMA News Photo; 1.8c: © ZoonarS Gibson/age fotostock RF; 1.9a: Amy Benson, U.S. Geological Survey; 1.9b: NOAA; 1.10a: Mike Powell/Getty Images RF; 1.10b: © Digital Vision/Getty Images RF; 1.10c: © Guillen Photography / Alamy RF; 1.10d: © SasPartout/age fotostock RF; p. 13: © Shib Shankar Chatterjee; p. 14(left): U.S. Fish & Wildlife Service/Claire Dobert; p. 14(right): Marci Koski/USFWS.

Chapter 2

Opener: © Digital Vision/PunchStock RF; p. 18(top): © Michael Quinton/Minden Pictures/Corbis; 2.1: © Stockbyte/Getty RF; 2.3a: © David R. Frazier Photolibrary, Inc./Alamy RF; 2.3b: © Stockbyte/Getty Images RF; 2.3c: © Image Source RF; p. 24 (Emerson): Library of Congress; 24(Thoreau): Library of Congress; p. 24 (Muir): © Bettmann/Corbis; p. 24 (Leopold): © AP Photos; p. 24 (Carson): © AP Photos; 2.5a: © Bill Kearney; 2.5b: © Bettmann/Corbis; 1.5: © MedioImages/Getty Images RF; 2.7: © Corbis RF; 2.8a: © Getty Images RF; 2.8b: © Jean-Louis Atlan/Sygma/Corbis; p. 31: © Julia Reinhart/Demotix/Corbis; 2.9 (top left): © Henk Badenhorst/Getty Images RF; 2.9 (top right): © Photodisc/PunchStock RF; 2.9 (bottom left): © McGraw-Hill Education./Barry Barker, photographer; 2.9 (bottom right): Photo by Lynn Betts, USDA Natural Resources Conservation Service; p. 35 (left): NASA/Jeff Schmaltz, MODIS Land Rapid Response Team; p. 35 (right): Courtesy of US Army/US Coast Guard/photo by Petty Officer 2nd Class Kyle Niemi; 2.10: © Micheline Pelletier/Corbis.

Chapter 3

Opener: © Corbis RF; p. 40(top): © PunchStock RF; 3.1: © McGraw-Hill Education. David Moyer, photographer; 3.4: © Brand X Pictures/PunchStock RF; 3.5(top): © Corbis RF; 3.5(left): Photo by Jeff Vanuga, USDA Natural Resources Conservation Service; 3.5 (right): © Juice Images/Alamy RF; 3.5(bottom): © Image Source/Getty Images RF; 3.6: Photo by Jeff Vanuga, USDA Natural Resources Conservation Service; 3.7: © SuperStock RF; 3.8 (water pollution): © Jeff Greenberg/PhotoEdit; 3.8 (smoke): © PunchStock RF; 3.8(cattle): © S. Meltzer/PhotoLink/Getty Images RF; 3.8(graffiti): © Ingram Publishing/AGE Fotostock RF; 3.8(smog): © Vol. 25/Photodisc/Getty Images RF; 3.8(solvents): © McGraw-Hill Education. Ken Karp, photographer; 3.8(thermal): © Steve Allen/Brand X Pictures/Alamy RF; 3.8(junkyard): © Getty Images

RF; 3.9(top left): © Corbis RF; 3.9(bottom left): © McGraw-Hill Education. Roger Loewenberg, photographer; 3.9 (top middle): U.S.Fish and Wildlife Services/Megan Durham; 3.9(bottom middle): © Glow Images RF; 3.9(top right): © Scenics of America/PhotoLink/Getty Images RF; 3.9(bottom right): © PunchStock RF; 3.10: Claire Fackler, CINMS, NOAA; p. 54: © Corbis RF.

Chapter 4

Opener 4: © STR/AFP/Getty; p. 59: EPA Burn Wise Campaign; p. 60: © Stockbyte/PunchStock RF; 4.1: © Chris Sattleberger/Getty Images RF; 4.2: © McGraw-Hill Education. John Thoeming, photographer; p. 64: © Jupiter RF; p. 69(top): © Trish Drury/DanitaDelimont.com Danita Delimont Photography/Newscom; p. 69(bottom): © AP Photo/The Penninsula Daily News, Chris Tucker; 4.10: © Corbis RF; 4.11: © Corbis/SuperStock RF; 4.13: © Hennik5000/Getty Images RF; p. 74(left): © Marcello Bortolino/Getty Images RF; p. 74(right): © Tyrone Turner/National Geographic Society/Corbis.

Chapter 5

Opener: © ComstocK Images/Alamy RF; p. 78: © Robert McGuoey/Getty Images; 5.2(left): © Creatas/PunchStock RF; 5.2(right): © Creatas/PunchStock RF; p. 80: © Digital Vision/Getty Images RF; 5.3b(top): © Getty Images RF; 5.3b(bottom): © PunchStock RF; 5.4: © Corbis RF; 5.5(beaver): © Creatas/PunchStock RF; 5.5(dam): © Getty Images RF; 5.5(gnawing): © Alan and Sandy Carey/Getty Images RF; 5.5(pond): © Brand X/PunchStock RF; 5.5(habitat): © Masterfile RF; 5.6(left): © Getty Images RF; 5.6(middle): © Don Farrall/Getty Images RF; 5.6(right): © It Stock/age fotostock RF; 5.7: © Corbis RF; 5.9: National Library of Medicine; 5.10: © image 100 Ltd. RF; 5.11: © Pixtal/age fotostock RF; 5.12(top): © Digital Vision/PunchStock RF; 5.12(bottom): © Creatas/PunchStock RF; 5.13: © Mark Wilson/Getty Images RF; 5.14: © Karen Carr Studio Inc.; 5.15: © Corbis RF; 5.16 (top): © PhotoAlto/PunchStock RF; 5.16 (bottom): © It Stock/PunchStock RF; 5.17: © Creatas/PunchStock RF; 5.18: © PunchStock RF; 5.20a: © J.H. Robinson/Science Source; 5.20b: CDC/James Gathany, William Nicholson; 5.20c: © Judy Enger; 5.21a: © Brand X/PunchStock RF; 5.21b: © ephotocorp/Alamy RF; 5.22a: © Nigel Cattlin/Science Source; 5.22b: © Photodisc/Getty Images RF; 5.23: © ER Degginger/Science Source; 5.24: © Design Pics/Richard Wear RF; 5.27a&b: © Getty Images RF; 5.27c: © Creatas/PunchStock RF; 5.27d: © Digital Vision RF; 5.27e: © Creatas Images/PictureQuest RF; 5.27f: © Getty Images RF; p. 100 (phytoplankton): © PunchStock RF; p. 100 (zebra mussels): © Eldon Enger; p. 100 (Diporeia): NOAA Great Lakes Environmental Research Laboratory; p. 100 (forage fish): © Paul Bentzen, Dalhousie University; p. 100 (sport fish): Courtesy USFWS; p. 101: © Fisheries and Ocean Canada; p. 105: © McGraw-Hill Education. Pat Watson, photographer; p. 107(top): Courtesy U.S. Fish and Wildlife Services, Charles H. Smith; p. 107(bottom): © Paul Nicklen/National Geographic/Getty Images.

Chapter 6

Opener: © Digital Vision RF; p. 110(top): © Pierre Gleizes/Greenpeace; p. 110(bottom): © MIXA Co Ltd./Getty Images RF; 6.2: © Eldon Enger; 6.5: © Stephen P. Lynch; 6.10b: © Getty Images RF; 6.10c: © Jeremy

Woodhouse/PunchStock RF; 6.10d: © Corbis RF; p. 119: Photo by Tim McCabe/USDA; 6.11b: © Creatas/PunchStock RF; 6.11c: © Getty Images RF; 6.11d: © It Stock/age fotostock RF; 6.11e: © U. S. Fish and Wildlife Service/John and Karen Hollingsworth photographers; 6.12b: © Corbis RF; 6.12c: © Brand X/PunchStock RF; 6.12d: © Digital Vision/PunchStock RF; 6.12e: © PunchStock RF; 6.13b: © Steven P. Lynch; 6.13c: © U. S. Fish and Wildlife Service/ Lee Karney photographer; 6.13d: © Judy Enger; 6.14b: © David Pedre/Getty Images RF; 6.14c: © SuperStock RF; 6.14d: © David Zurick; 6.15b: © Stephen P. Lynch; 6.15c: © PunchStock RF; 6.15d: © Getty Images RF; p. 126 (left): © David Frazier Photolibrary, Inc./Alamy RF; p. 126(middle): Lynn Betts, USDA Natural Resources Conservation Service; p. 126 (right): Terry Sohl, U.S. Geological Survey; 6.16b-d: © Getty Images RF; 6.17b: © Corbis RF; 6.17c: U.S. Fish and Wildlife Services; 6.17d: © Corel Corporation/JupiterImages RF; 6.18b: © Stephen P. Lynch; 6.18c: © Creatas/PunchStock RF; 6.18d & 6.19b: © Corbis RF; 6.19c: © Getty Images RF; 6.19d: © Radius Images/Corbis RF; 6.19e & 6.22(left): © Getty Images RF; 6.22(right): © Corbis RF; 6.23: © McGraw-Hill Education. Barry Barker, photographer; p. 137(All): © Judy Enger; p. 138: © Jim Weber/The Commercial Appeal/Corbis; p. 139: © Kike Calvo/V&W/imagequestmarine.com.

Chapter 7

Opener: © Photodisc RF; p. 143(left): © Biosphoto/SuperStock; p. 143(right): USDA, Stephen Ausmus; p. 143(bottom): © Corbis RF; 7.1a: © Digital Vision RF; 7.1b: © Purestock/PunchStock RF; 7.3a: © Getty Images RF; 7.3b: © John W. Bova/Science Source; 7.3c: © Monika Al-Mufti; 7.5(left): © Burke Triolo Productions/Getty Images RF; 7.5(middle): © Getty Images RF; 7.5(right): © Corbis RF; p. 150(left): Photo by Jeff Vanuga, courtesy of USDA Natural Resources Conservation Service; p. 150(right): © Martjan Lammertink/www.PicidPics.com; 7.10 (left): © Stephen Lackie/Corbis RF; 7.10(right): © Bengt Hedberg/Naturbild/Corbis; 7.14 (left & right): © Peter Ginter/Science Faction/Corbis; p. 158: © Bettmann/Corbis; p. 170: The Hudson Bay Project. Photo provided by Robert Jefferies.

Chapter 8

Opener: NASA/NOAA/SPL; 8.1: © Corbis RF; 8.2: © McGraw-Hill Education. Barry Barker, photographer; 8.3: Library of Congress; 8.5a: © Getty Images RF; 8.5b: © McGraw-Hill Education. John Flournoy, photographer; 8.5c: © Stockdisc/Digital Vision RF; 8.9: © Author's Image/PunchStock RF; p. 180(top) & 8.10: © Getty Images RF.

Chapter 9

Opener: © Mike Danneman/Getty Images RF; 9.2: © Comstock Images/Alamy RF; 9.9a&b: Photograph by H.E. Malde, USGS Photo Library Denver Colorado; 9.10b: Photo by C.R. Dunrud, U.S. Geological Survey; 9.11: Photo by N. Gaggiani, U.S. Geological Survey; 9.12: © Getty Images RF; 9.14a: © Corbis RF; 9.14b: © Getty Images RF; p. 200: U.S. Fish and Wildlife Service; 9.17: © Image State/age fotostock RF; 9.23: © Getty Images RF; 9.24: © Doug Sherman/Geofile RF; 9.25: © ZUFAROV/AFP/Getty Images; p. 211: U.S. Coast Guard Photo.